Agroforestry

Agroforestry

Systems and Prospects

C.B. Pandey

Presently

Head, Division of Natural Resources & Environment
Central Arid Zone Research Institute
Jodhpur – 342 003, Rajasthan

Formerly

Senior Scientist, Central Agricultural Research Institute, Port Blair
Andaman & Nicobar Islands

and

O.P. Chaturvedi

Principal Scientist & Head
Central Soil & Water Conservation Research & Training Institute
Dehradun – 248 195
Uttarakhand

NEW INDIA PUBLISHING AGENCY
New Delhi – 110 034

NEW INDIA PUBLISHING AGENCY
101, Vikas Surya Plaza, CU Block, LSC Market
Pitam Pura, New Delhi 110 034, India
Phone: + 91 (11)27 34 17 17 Fax: + 91(11) 27 34 16 16
Email: info@nipabooks.com
Web: www.nipabooks.com

Feedback at feedbacks@nipabooks.com

ISBN: 978-93-81450-97-0

Composed, Designed and Printed in India

Fax: (0542) 2368174
E-mail: singhjs1@gmail.com
jssingh@bhu.ac.in

Phone: (Off) (0542)2368399
(Res) (0542) 2369093
Mo. +919335' 178355

BANARAS HINDU UNIVERSITY
Ecosystems Analysis Laboratory

J.S SINGH PhD FNA FASc FNASc FTWAS

DEPARTMENT OF BOTANY
VARANASI – 221 005, INDIA

Foreword

Agroforestry is an emerging science in the field of natural resource management. Trees in agroforestry systems moderate climatic conditions, add organic matter in-soil through leaf and root litter, scavenge nutrients from deeper depths and deposit on the surface soil, thus placing them in an active nutrient recycling pool, protect soils from rain beating and wind erosion; all of the above help together in soil conservation. Trees also facilitate infiltration of water and reduce evaporation and thereby conserve soil moisture in all climatic conditions particularly more in arid conditions. Agroforestry systems also enhance crop diversity particularly more in homegardens, which provide different products and ensure nutritional security and systems stability. Homegardens integrate livestock and fish farming in an interactive manner, which make the system efficient in functioning.

Though suitable tree/crop combinations are worked out for all agroecological conditions under All India Coordinated Research Project (AF) of ICAR involving almost all agricultural universities of our country and proper tree spacing and management techniques are standardized and recommended, farmers still prefer traditional agroforestry systems, which are time tested, have developed over generations and fulfil their maximum day-to-day needs like fuel, food, fruit, fibre, wood *etc*. Some modern agroforestry systems, which feed industry, have also got a place of pride in farmer's fields. The present compilation takes care of putting all these information on one place in the form of a book.

The editors have done a commendable job of bringing the diverse contributions into one publication. I am sure this book will serve as a repository of information in the field of agroforestry, which can be used by planners, researchers and undergraduate and postgraduate students pursuing studies in the field of forestry/ agroforestry.

J. S. Singh
Professor Emeritus

Preface

Agroforestry is an ages old practice, being followed throughout the world particularly more in developing countries in the tropics. World over, studies are being conducted to understand structure and functioning of the ages old and time tested agroforestry practices and improve their productivity. In India, studies on agroforestry date back to 1979 when Indian Council of Agricultural Research, New Delhi organized a seminar on Agroforestry at Imphal. Keeping in view the diversity in climate and edaphic conditions in our country a network project that included scientists and institutes of different agroecological regions, were considered; this paved way to the development of All India Coordinated Research Project (AICRP) on Agroforestry in 1983. In the beginning head office of the AICRP was at Krishi Bhawan, New Delhi, which later on 1997 shifted to National Research Centre (Agroforestry) at Jhansi, U.P. There were 20 centres in the network project, which included 12 State Agricultural Universities (SAU) and 8 ICAR institutes. Later more coordinating centres were added; at present there are 27 State Agricultural Universities and several ICAR institutes, which are working systematically in different aspects of agroforestry in major agroecological zones of the country.

Today, more than 25 years after meticulous studies, agroforestry has emerged as a robust land use which advocates crop diversification, soil and soil-water conservation, cycling of organic matter and sequestration of CO_2 in plant and soil. Different area wise agroforestry models are available now, which enhances unit land productivity and fulfil, fuel, fodder, fruit and timber needs of farmers. Area specific researches undertaken by SAUs and ICAR institutes have generated volumes of information but their availability is limited; hence the information do not become available to agroforesters and students pursuing their studies in the field of agroforestry. During our tenure as a Professor we felt a need for lucid compilation of such research material for use by agroforesters and students undertaking the study of agroforestry. Thus we felt compelled towards compilation of the available information on agroforestry systems prevailing in various agro-ecological conditions across the country. We sincerely wish that this book will help planners, agroforesters and undergraduate and postgraduate students.

The book includes total 26 chapters; the first 10 chapters describe traditional agroforestry systems found in different parts of our country. Despite the fact that crop yields decline under trees in the traditional systems, farmers follow them. This is simply because trees fulfil their day-to-day basic needs and provide a handsome income after completion of their rotation cycle, which compensates loss suffered by crops during the tree growth period. Homegardens, having a multi-tire in structure, are another example of traditional agroforestry system found generally in northeast and southern part of our country. Long ago, roads were lacking and organized markets were not available. This forced homegardens to plant almost all trees species which fulfilled their fuel, fodder, fruit, spice, vegetable, and other needs. They planted these plant species according to their needs; the knowledge was acquired from their ancestors which accumulated over generations. This is reason all planted species share natural growth resources in a compatible manner and function in a complementary manner. The homegardens also serve as a storehouse of biodiversity and provide not only dietary but also nutritional security to homegardeners. Chapters 11 to 14 deal with some modern agroforestry models, and mechanism of competitive interaction found in agroforestry systems. Chapter 15 to 19 describes nutrient cycling and natural growth resources conservation by agroforestry systems. Tree has both protective as well as productive role. Tree canopies protect soils from rain beating and wind blowing whereas fine roots bind soil particles together and thereby protect soils from erosion. Trees recycle leaf and root litter and nutrients and make system at least semi-sustainable. Different models of agroforestry like shelterbelt and wind breaks are known to help protect environment from dust and sand dunes drifting; some models like medicinal plant based system that provide medicinal and other high value crops. These service oriented functions of agroforestry are described in chapter 20 to 23. Tree improvement is an important aspect of agroforestry study. It helps develop plus trees for different agroforestry models. Multiplication of propgules and planting materials are equally important for development of agroforestry. These information are provided in chapters 24 and 25. Economics is the most vital part of agriculture production system; farmers adopt only those agroforestry systems which are economically viable. Generally people study economic viability of agricultural production system using a single cost / benefit ratio parameter. But, agroforestry systems require much more parameters owing to their perennial nature. The last chapter describes economic parameters like NPV, IRR, annuity, sensitivity *etc.* in detail.

The compilation could become possible due to efforts of the contributors of the chapters who compile works nicely. I thank the Director of Central Agricultural Research Institute, Port Blair, Andaman and Nicobar Islands who always encouraged us to compile the works in the form of a book.

C.B. Pandey
O.P. Chaturvedi

Contents

Contributors

A.R. Barbhuiya
Department of Forestry
Mizoram University
Post Box: 190, Aizawl – 796009, Mizoarm

Aashutosh Sharma
Jawaharlal Nehru Krishi Vishwa Vidyalaya
Adhartal, Jabalpur – 482 004
Madhya Pradesh

A.K. Parandiyal
Central Soil & Water Conservation Research & Training Institute,
Regional Station, Kota
Rajasthan

Anup Das
Plant Sciences Division
CSWCRTI, Dehradun
Uttarakhand

Bhupendra Singh
Department of Forestry
H.N.B. Garhwal University Srinagar
Garhwal – 246 174
Utttarakhand

Baljit Singh
Department of Forestry and Natural Resources
Punjab Agricultural University
Ludhiana – 141 004, Punjab

B.N. Sathish
College of Forestry,
University of Agricultural Sciences (B)
Ponnampet - 571216, Kodag,
Karnataka

C.G. Kushalappa
College of Forestry,
University of Agricultural Sciences (B)
Ponnampet - 571216
Kodagu, Karnataka

Charan Singh
Central Soil and Water Conservation Research and Training Institute
218, Kaulagarh Road
Dehradun – 248 195, Uttarakhand

C.B. Pandey
Central Arid Zone Research Institute
Near Industrial Training Institute (ITI)
Light Industrial Area
Jodhpur - 342 003, Rajasthan

C.L. Thakur
Department of Silviculture and Agroforestry
University of Horticulture and Forestry
Solan – 173 230, Himachal Pradesh

D.K. Das
Department of Forestry
Rajendra Agricultural University
Pusa (Samastipur) – 848 125, Bihar

Eklabya Sharma
International Centre for Integrated Mountain Development
GPO Box–32 26, Khumaltar
Lalitpur, Kathmandu, Nepal

G. Singh
Division of Forest Ecology
Arid Forest Research Institute
New Pali Road, Jodhpur – 342 005, Rajasthan

G.P.S. Dhillon
Department of Forestry and Natural Resources
Punjab Agricultural University
Ludhiana – 141 004, Punjab

Ghanashyam Sharma
United Nations University
Environment & Sustainable Development Programme
Jingumae 5-chome, Shibuya-ku-150-8925
Tokyo, Japan

J.M.S.Tomar
Plant Science Division
CSWCRTI, Dehradun, Uttarakhand

K.S. Sangha
Department of Forestry and Natural Resources
Punjab Agricultural University
Ludhiana – 141 004, Punjab

Lalita Singh
Central Agricultural Research Institute
Post Box-181, Port Blair – 744 101
Andaman and Nicobar Islands

M. Din
Division of Natural Resource Management
Central Agricultural Research Institute
Port Blair – 744 101
Andaman and Nicobar Islands

M. Datta
ICAR Research Complex for NEH Region
Lembucherra – 799 210, Tripura

M. Osman
Central Research Institute for Dryland Agriculture
Santoshnagar
Hyderabad – 500 059, Andhra Pradesh

M.P. Divya
Associate Professor (Forestry)
Forest College and Research Institute
Mettupalayam – 641 301, Tamil Nadu

N. Ravisankar
Division of Natural Resource Management
Central Agricultural Research Institute
Port Blair – 744 101
Andaman and Nicobar Islands

N.P. Todaria
Department of Forestry
H.N.B. Garhwal University Srinagar
Garhwal – 246 174, Utttarakhnd

N. Chandra Sekaran
Associate Professor (Forestry)
Forest College and Research Institute
Mettupalayam – 641 301, Tamil Nadu

Navneet Kaur
Department of Forestry and Natural Resources
Punjab Agricultural University
Ludhiana – 141 004, Punjab

O.P. Chaturvedi
Head, Division of Plant Science
CSWCRTI
Dehradun – 248 195, Uttarakhand

P. K. Singh
Forestry and Fodder Development Officer
Jharkhand Tribal Development Society Welfare Department
Govt. of Jharkhand, Ranchi

P.S. Thakur
Department of Silviculture and Agroforestry
University of Horticulture and Forestry
Solan-173 230, Himachal Pradesh

Rajesh Kaushal
Plant Science Division
CSWCRTI
Dehradun, Uttarakhand

Rita Sharma
International Centre for Integrated Mountain Development
GPO Box–32 26, Khumaltar
Lalitpur, Kathmandu, Nepal

R.K. Jha
Department of Forestry, Rajendra Agricultural University
Pusa (Samastipur) – 848 125, Bihar

R.K. Singh
Central Soil & Water Conservation Research & Training Institute, Regional Station
Kota, Rajasthan

R.C. Srivastava
Division of Natural Resource Management
Central Agricultural Research Institute
Port Blair – 744 101
Andaman and Nicobar Islands

R.I.S. Gill
Department of Forestry and Natural Resources
Punjab Agricultural University
Ludhiana – 141 004, Punjab

Rajendra Prasad
National Research Centre for Agroforestry
Pahuj Dam, Gwalior Road
Jhansi-284003, Uttar Pradesh

R.S. Mertia
Central Arid Zone Research Institute
Regional Research Station
Jaisalmer-345001, Rajasthan

R. Jayaramasoundari
Associate Professor (Forestry)
Forest College and Research Institute
Mettupalayam – 641 301
Tamil Nadu

R. Kaushal
Central Soil and Water Conservation Reserach & Training Institute, Dehradun
Uttarakhand

R.L. Banik
Plant Sciences Division
CSWCRTI, Dehradun
Uttarakhand

S.M.S. Quli
Chairman,
Department of Extension & Social Forestry
Faculty of Forestry
Birsa Agricultural University
Ranchi- 834006, Jharkhand

S.D. Upadhyaya
Jawaharlal Nehru Krishi Vishwa Vidyalaya
Adhartal, Jabalpur – 482 004
Madhya Pradesh

S. Raghavendra
College of Forestry,
University of Agricultural Sciences (B)
Ponnampet - 571216
Kodagu, Karnataka

S.K. Dhyani
National Research Centre for Agroforestry
Pahuj Dam, Gwalior Road
Jhansi-284003, Uttar Pradesh

Syam Viswanatha
Institute of Wood Science and Technology
Malleswaram, Bangalore
Karnataka

S.C. Pramanik
Division of Natural Resource Management
Central Agricultural Research Institute
Port Blair – 744 101
Andaman and Nicobar Islands

S.K.Tewari
G.B. Pant University of Agriculture & Technology
Pantnagar, Uttarakhand

S. Jeyakumar
Division of Natural Resource Management
Central Agricultural Research Institute
Port Blair – 744 101
Andaman and Nicobar Islands

S.K. Chaudhari
Central Soil Salivity Resources Institute
Karnal
Haryana

S.K. Ambast
Division of Natural Resource Management
Central Agricultural Research Institute
Port Blair – 744 101
Andaman and Nicobar Islands

U.K. Sahoo
Department of Forestry
Mizoram University
Post Box: 190, Aizawl-796009
Mizoarm

1

Traditional Homegarden Agroforestry: Structural Diversity and Functional Dynamics in Aizawl District of Mizoram, Northeast India

A.R. Barbhuiya and U.K. Sahoo

Abstract: *The size of the homegardens ranged between 0.10-1.00 ha; they were located in tropical region with multi-storied structure, commonly containing three layers but sometimes with four layers. Individual households were the sampling units and for each homegarden, the area was measured and crop composition, species richness and uses of plant species was enumerated through direct observation and interviews with the farmers. The composition and richness of species varied between the homegardens but was related to site characteristics and size of the garden. A total of 231 plant species (105 tree species, 50 shrubs and 76 herbs) belonging to 88 families were recorded from the gardens among which 27% species were common to all gardens. Among the trees, as many as 24 species (23%) were common to all the homegardens. The vegetable species constituted the major functional group followed by fruits. Homegardens were found to be a good reservoir of plant genetic resources that preserve diverse landraces, cultivars, and many more ecologically and economically important plant species. Our study depicts that the homegardens can play an important role in food security and sustainable livelihood support of this hilly people in a variety of ways and can be an alternative to shifting cultivation.*

Introduction

Homegardens are well-established land use systems which may be defined as "small scale, supplementary food production systems by and for household members that replicate the natural, multifaceted ecosystem". These are the living gene banks and reservoirs of plant genetic resources that preserve landraces, cultivars, rare species and endangered species and species neglected in larger ecosystems (Eyzaguirre and Linares, 2001). Indigenous agroforestry homegardens appear to have developed independently in the Indian subcontinent, Indonesia and other parts of Southeast Asia, the tropical Pacific islands, the Caribbean, and various parts of tropical Latin America and Africa (Brownrigg, 1985; Landauer and Brazil,1990), and are found in almost all tropical and subtropical ecozones where subsistence land use systems prevail (Nair, 1993). Since these systems require virtually little economic inputs, a poor family can afford to cultivate different crops as a strategy to diversify their subsistence cash needs. Recently, concerns have developed on the long-term sustainability and environmental consequences of the intensification of agricultural systems. Increasing attention is being given to achieving stability in land utilization on long term basis while fulfilling the needs of the local population (Reijntjes *et al.*, 1992; Swift and Ingram, 1996; Matson *et al.*, 2002; Tilman *et al.*, 2002). In small holder farming systems in the tropics, the use of modern technologies might not be the first option to improve agriculture. In such areas, better use of local resources and natural processes could make farming more effective and create conditions for efficient, profitable and safe use of modern inputs (Reijntjes *et al.*, 1992; Altieri, 1995). In response to these concerns, agroforestry home gardening may play an important role in integrated land-use and sustainable agriculture in this hilly region.

The practice of indigenous agroforestry homegarden is an integral component in typical Mizo society and play crucial role in supplying vegetables, fruits, fuel wood, small timber, herbs and spices etc for their daily requirement. In the highly terrains of Northeast India particularly in Mizoram homegardens are one of the important sources of food and supply of most household requirements, and, this practice may contribute greatly to create integrated agricultural and community systems that maintain productivity, protect natural resources, minimize environmental impacts and provide economic and social needs. Despite its importance, however, detailed information on this system is almost non-existent. Nonetheless, there is a lack of indepth knowledge and information on species composition in Mizo's homegardens. The main aim of this study was, therefore, to document species diversity and richness and understand how the homegarden owners maintain and use the species at household and community level and their contributions in livelihood support.

General description

Mizoram lies in the appealing and moderate hill folds in the southern most tip of the North eastern region of India, projecting downwards between Burma and Bangladesh. It is flaked by Bangladesh on the west and Myanmar on the east and south. It has an area of 21,087 km^2 with 630 km long international boundary and *ca.* 8,91,058 populations (2001 census). Three villages *viz.*, Selesih (92°43′51.7′′-92°43′58.0′′E longitude to 23°47′41.5′′-23°48′29.4′′N latitude and altitude 1163 m msl), Sairang (92°39′06.0′′-92°39′12.6′′E longitude to 23°48′29.8′′-23°48′44.2′′N latitude and altitude 115 m msl) and Thingsulthliah (92°51′34.6′′-92°52′22.6′′E longitude to 23°41′33.5′′-23°42′42.7′′N latitude and altitude 780 m msl.) were selected in the Aizawl district of Mizoram. For the present study species composition, importance value index and some functional aspects of homegardens were studied from 45 sites positioned in three villages of Aizawl district, Mizoram Northeast India. The household size of the study area varied between 5-8 people with 2–3 earning members in the family and had average level of education. These villages are located within proximity of *ca.* 40 km from the state capital city Aizawl and inhabited by Mizo tribe, the largest and dominant tribe in the state. The study villages are characterized by moderate hill folds, remoteness, dense population with poor economic status and infrastructure. Every village has a small market but most of the gardens products are transported to be sold in a weekly (Saturday) market in the state capital. The source of cash income in the village is mostly from government services and in part from private business and subsistent agriculture. Majority of the villager (80%) normally pursue multiple livelihood option like jhum/ terrace cultivation, agroforestry, horticulture and home gardening. The home gardens feed mostly on rainwater as water harvesting technology in the villages almost non-existence due to steep slopes coupled with poor water holding capacity of the lands. Thus, access of water is an acute problem to the farms especially during winter. The climate of the area is generally moderate with mean summer and winter temperature ranged between 20°C-30°C and 8°C-18°C respectively. The average annual rainfall ranged between 1700-2500 mm, with maximum (70%) rainfall during July-August.

Sampling procedure

Individual households were the sampling units and selection of households was based on size (>0.10 ha). At the household level, data were collected using measurements, interviews and observations. Prior to the interviews, the homegardens were visited with the owner to make observations on the overall conditions of the garden. A minimum of 15 households were interviewed in each village and in total 45 homegardens were selected and sampled from the three surveyed villages. Information on crop composition was collected through direct observation and interviews with the farmers. For each homegarden, the area was measured, and the different species of crops and trees identified and enumerated.

The population of annual crops and other widely grown small plants was estimated by making sample counts on systematically selected 1m x 1m quadrants and extrapolating it to the area it covers. For tree and shrubs species the total population was counted in the garden. Information on the use, practice and management of plant/crop species was noted in the field itself by following a standard questionnaire. For quality control, the surveyed questionnaires were edited and revised in different tiers, first by the enumerator himself, then through peer review and editing among enumerators and final editing by the team of researchers on the same date. The interviews were one-on-one and respondents carefully selected to represent both male and female and it was interestingly found that women are generally the custodians of homegarden and devote much of their time in care and management of the homegarden. Men also contribute to the maintenance of homegardens, however, they often engage in introducing new crop varieties to the gardens. It is also commonly observed that men tend to introduce exotic commercial fruit trees into the homegardens whereas women prefer to maintain traditional vegetables and other plant species that are required on regular basis. Most of the plants encountered had multiple uses but classification was done based on main use that illustrated by farmers.

Data computation

In each homegarden, vegetation data like density, frequency and abundance of different plant species were measured directly (Phillips, 1959). The sum of relative frequency, relative density and abundance was calculated as importance value index (IVI) of individual species (Curtis, 1959). All species present were classified into: annual/biennial/perennial or herb, shrub and tree. Many garden species are often subspecies or cultivars; we did not attempt to classify here the plants below the species level. To determine crop species diversity, species richness and species evenness of the homegardens were calculated. Species richness is the total number of crops on a homegarden. This index does not indicate the relative proportion or abundance of a particular species in the farm. Hence, indices that incorporate both richness and the evenness of abundance were required. Shannon index (Shannon and Wiener 1949) and evenness measure, which are commonly used tools for these purposes (Pielou, 1969; Magurran, 1988; Huston, 1995) were computed. The Shannon diversity index (H') is high when the relative abundance of the different species in the sample is even, and is low when few species are more abundant than the others. It is based on the theory that when there is a large number of species with even proportions, the uncertainty that a randomly selected individual belongs to a certain species increases and thus the diversity. It is calculated using the formula $H' = \sum_{i=1}^{s} pi \ln pi$, where s is the number of species in the community and pi is the proportional abundance of species i (i= number of species i divided by total numbers in the community) (Magurran, 1988). Pielou's (1969) evenness index (e) was calculated as: e= H'/logS, where H'= Shannon's index of diversity, and S= total number of species. Also, Simpson's index [$\lambda = H' = \Sigma_{i=1}^{s}(pi)^2$] was used to describe the dominance *i.e.*, the degree that a community is dominated by one or a few very common species (Powers and McSorley, 2000).

Replicated soil samples were collected from 0-15 cm soil depth from all the homegardens and these were mixed to obtain composite samples. After removing stones, pebbles and large pieces of plant material, the samples were sieved by 2 mm mesh size and used for soil physico-chemical analysis. Soil texture was determined by Bouyouncos hydrometer method (Bouyouncos, 1962) and water holding capacity was determined according to Keen's box method given in Piper (1944), while soil moisture content was measured gravimetrically by incubating 10 g of field moist soil sample in a hot-air oven at 150°C for 24 hours. Soil organic carbon (SOC) was determined by dichromate oxidation and titration method (Walkey, 1947). Total Kjeldahl nitrogen (TKN) was estimated following semi-micro Kjeldahl procedure by acid-digestion, distillation and titration. The pH of the soil sample was determined in a soil-water suspension (1:2.5 w/v H_2O) using a digital pH meter. Available phosphorus was measured by molybdenum blue method (Jackson 1958).

Regarding income, the data obtained on annual and present currency (Rupees) basis *i.e.,* during the year 2007-2008. Annual production values per hectare for a few important individual plants were calculated by using the density values of particular species. The income from homegarden products selling were calculated separately at the farm price/local market price and multiplied with the production values to obtain the household annual farm income and also the family own consumption. Pearson correlation coefficients were worked out according to Zar (1974), wherever necessary.

Structural diversity

A wide variation in homegarden size was observed in the study area ranging between 0.10-1.00 ha; however, most of them were small in size. Studies on homegarden systems from different ecological and geographical regions showed that the worldwide average size of homegarden units is around 0.10–0.50 ha (Brierley, 1985; Fernandes and Nair, 1986; Kumar *et al.*, 1994; Das and Das, 2005). Homegarden size by and large is a function of the plant diversity and garden size. Our study has shown significant (P>0.001) positive correlation between size and total species diversity. Larger gardens had more number of species and it decreased with the decrease in size of the garden. This is because the farmers who are constrained by shortage of land concentrate on fewer species of greater utility and allocate more of their land to food crops, while large holders can afford to include different types of plants. This pattern of increasing tree species richness with increasing land holding was also reported for other homegarden systems (Kumar *et al.*, 1994; Biggelaar and Gold, 1996; Mendez *et al.*, 2001). Although few species were common to all the gardens, majority of the species differed in their composition between gardens. The variation in species composition was mostly induced by altitude and garden size.

In the present study, a total of 231 species with 105 trees, 50 shrubs and 76 herbs species were recorded from 45 indigenous agroforestry homegardens. These tree, shrub and herb species were distributed in 84 and 49; 31 and 22 and 59 and 39 genera and families, respectively (Table 1). Overall, there were 88 families, out of which 24 tree species (23%) were common to all the homegardens. Asteraceae, Caesalpinaceae, Compositae, Cucurbitaceae, Euphorbiaceae, Leguminosae, Mimosaceae, Moraceae, Musaceae, Papilionaceae, Rutaceae, Solanaceae, Verbenaceae and Zingiberaceae were the most dominant families as represented by the species form. The mean tree species in the homegardens is 40; shrubs 33 and herbs 72 individual per homegarden. The potential of these homegardens as *in-situ* reservoirs for biodiversity is even higher than the diversity reported for different homegardens of Karnataka (Shastri *et al.*, 2002), Kerala (Kumar *et al.*, 1994) and Assam, Northeast India (Das and Das, 2005). Whitmore (1971) reported 17 families of natural forest trees in a 676 hectare tropical rain forest in Malaysia with agricultural or pharmacological values. While Makaraphirom (1989) monitored over 200 species in indigenous traditional homegardens in Thailand. Ahmad and Abood (1990) documented 19 families, 30 genera, 44 species of forest species in agroforestry ecosystems in Malaysia. These figures indicates that homegardens have the potential to hold more plant diversity, possibly in response to the demand for increased production or economic return to cope with socioeconomic conditions of the society. The high number of species also indicates the significant role of these systems in the conservation of genetic diversity, as reported in homegarden studies in other regions (Michon *et al.*, 1983; Alvarez-Byulla Roces *et al.*, 1989; Soemarwoto and Conway, 1991; Kessy, 1998). The species diversity index for tree, shrub and herb in the present study was 4.295, 3.622 and 3.990 respectively. Species diversity index in our study was higher than that of the index value of 3.93 in the homegardens of Sri Lanka (Kharal, 2000), 1.9-2.7 in the homegardens of Thailand (Gajaseni and Gajaseni, 1999) and the value of 3.21 in Karnataka (Shastri *et al.*, 2002). However, our plant diversity index values are comparable with the values (4.03-4.42) reported by Sunwar *et al.* (2006) from the homegardens of western Nepal. A high species richness index in the homegardens of Aizawl indicates that the area is very high with species and the distribution of individuals is significant with low dominance. Similarly, the higher index indicates that the system is more stable and mature and therefore can be self-sustaining. The high diversity in the homegardens could be the result of selection of species by the owners with utility of the specific products as the main criterion. Many factors would influence species diversity, the notable among are: climatic and geographic location, site characteristics, plot dimension and the extent of human interaction in the past and present. Our study clearly reveals that species diversity could also be confounded by the livelihood requirements and traditional knowledge. The high diversity values in this region could also be attributed to more favourable rainfall and the temperature conditions. The evenness index for trees, shrubs and herbs also varied slightly with greater values shared by tree species followed by herbs and shrubs (Table 1). The evenness values of 0.493-0.563 indicated dominance of some tree species and the evenness values in the present study was higher than the recorded value of 0.282-0.705 in Kerala homegardens (Kumar *et al.*, 1994). Greater dominance values were recorded for shrub species, followed by herbs and trees. Nonetheless, assemblies of biological diversity in the homegarden were not random, but mostly due to the fact that the species were rather selected with their utility as the main criterion.

Table 1. Phyto-sociological and soil characteristics of the homegardens in Aizawl, Mizoram, Northeast India.

Parameters	Homegardens	
Vegetation		
No. of species	Trees	105
	Shrubs	50
	Herbs	76
No. of genera	Trees	84
	Shrubs	31
	Herbs	59
No. of families	Trees	49
	Shrubs	22
	Herbs	39
Diversity index	Trees	4.295
	Shrubs	3.622
	Herbs	3.990
Dominance index	Trees	0.264
	Shrubs	0.355
	Herbs	0.304
Evenness index	Trees	0.563
	Shrubs	0.492
	Herbs	0.493
Soil properties	Moisture (%)	29.60±0.61 (24.03±1.37)
	Water holding capacity (%)	51.81±4.03 (44.55±1.52)
	Texture	
	Sand (%)	57.55±1.89 (68.18±1.89)
	Silt (%)	27.85±1.13 (20.25±0.75)
	Clay (%)	14.60±1.22 (11.57±1.06)
	Textural class	Sandy loam
pH (1:2.5 w/v H_2O)	5.54±0.07 (5.07±0.16)	
Organic C (%)	2.97±0.06 (2.03±0.08)	
TKN (%)	0.63±0.02 (0.52±0.005)	
C/N ratio	4.71±0.16 (3.90±0.17)	
	PO^-_4-P (μg g^{-1})	12.09±0.81 (7.55±0.23)

The values in the parentheses represent the corresponding data outside the garden (in adjacent open jhum lands). ± S.E (n=45).

The tree species frequency of occurrence and importance value index show that the most dominant species in homegardens were *Albizia procera, Artocarpus heterophyllus, Azadirachta indica, Castonopsis indica, Citrus anamensis, C. maxima, C. sinensis, C. reticulata, Mangifera indica, Melia azedarachta, Persea americana, Parkia timoriana, Psidium guajava, Schima wallichi, Semecarpus anacardium* and *Tamarindus indica* (Table 2). Greater frequency and importance value index shared by shrub species like *Cajanus cajan, Carica papaya, Capsicum annum, Citrus limon, Clerodendron infortunatum, C. colebrokianum, Hibiscus sabdaiffa, Musa paradisica, M. acuminata, Ocimum sanctum, Thysanolaena maxima* and *Zea mays* (Table 3). Among the herbs, *Allium hookerii, Calamus tenuis, Centella asiatica, Colocasia affinis, C. esculenta, Cucurbita maxima, Curcuma longa, Ipomea batata, Sechium eduli, Spilenthes acmella* and *Zingiber officinalis* showed greater frequency and importance value index (Table 4).

Table 2. Tree species with principal uses in the indigenous agroforestry homegardens in Aizawl

Local Name	Botanical Name	Family	Frequency occurrence (%)	IVI	Use category
Khanghu	*Acacia nilotica* (L) Willd. Ex.Delile	Mimosaceae	24.44	2.36	Miscellaneous
Kangkhu	*Acacia pinnata* (L.) Willd.	Mimosaceae	8.89	1.43	Miscellaneous
Belthei	*Aegle mermelos* Correa ex Roxb.	Rutaceae	22.22	2.14	Fruit
Chawhmathlum	*Albizia myriophylla* Roxb.	Leguminosae	13.33	1.60	MPT
Kangtekpa	*Albizia procera* L.	Mimosaceae	64.44	4.63	Miscellaneous
Thingchawke	*Albizzia lebbeck* Benth.	Mimosaceae	22.22	2.02	Miscellaneous
Thuamriat	*Alstonia scholaris* (L.) R. Br	Apocynaceae	31.11	2.77	MPT
Athaphol	*Annona reticulata* L.	Annonaceae	44.44	3.32	Fruit
Zairum	*Anogeissus acuminate* (Roxb.) Wall.	Combretaceae	8.89	1.43	Timber
Khuva	*Areca cathechu* L.	Palmae	24.44	7.76	Fruit
Tatkawng	*Artocarpous chama* Butch-Ham.	Moraceae	11.11	1.31	MPT
Tawkte	*Artocarpous nitidus*	Moraceae	13.33	1.43	MPT
Lamkhuang	*Artocarpus heterophyllus* Roxb.	Moraceae	77.78	5.58	Fruit
Theitat	*Artocarpus lakoocha* Roxb.	Moraceae	20.00	2.04	MPT
Theiher awt	*Averrhoa carombola* L.	Oxalidaceae	48.89	4.12	Fruit
Neem	*Azadirachta indica* A. Juss.	Meliaceae	68.89	4.93	Medicinal
Pangkai	*Baccaurea ramiflora* Lour.	Euphorbiaceae	11.11	1.31	Timber

(*Contd.*)

Local Name	Botanical Name	Family	Frequency occurrence (%)	IVI	Use category
Phunchawng	*Bombax ceiba* L.	Bombacaceae	44.44	3.57	MPT
Pang	*Bombax insignae* Wall.	Bombacaceae	6.67	1.08	MPT
Siallu	*Borassus flabellifer* L.	Arecaceae	35.56	3.26	MPT
Bil	*Bursera serrata*	Burseraceae	17.78	1.80	Timber
Tuahpui	*Butea monosperma* (Lam.) Kuntze	Fabaceae	22.22	2.26	Miscellaneous
Bottlebrush	*Callistemon lanceolatus* DC	Myrtaceae	53.33	4.16	Miscellaneous
Thingpui	*Cammellia sinensis* (L.) Kuntze.	Theaceae	33.33	2.87	Timber
Berawchal	*Canarium bengalense* Roxb.	Burseraceae	4.44	0.96	MPT
Theiria	*Carallia brachiata* (Lour.) Merr.	Rhizophoraceae	8.89	1.20	Timber
Daduhlo	*Cassia alata* L.	Caesalpinaceae	31.11	2.67	Miscellaneous
Bandorlathi	*Cassia fistula* L.	Caesalpinaceae	20.00	1.91	Miscellaneous
Makpazangkang	*Cassia nodosa* L.	Leguminosae	33.33	2.77	Miscellaneous
Kel-be	*Cassia tora* L.	Caesalpinaceae	22.22	2.02	Miscellaneous
Sehawr	*Castonopsis indica* (Roxb.) Miq.	Fagaeae	55.56	4.50	MPT
Zoei	*Ceiba pentandra*	Bombacaceae	6.67	1.66	Timber
Thinghmarcha	*Celtis tomentosa* Roxb.	Ulmaceae	17.78	1.94	Miscellaneous
Zawngtei	*Chukrasia tabularis* A.Juss.	Meliaceae	31.11	2.46	Miscellaneous
Thakthing	*Cinnamomum zeylanica* Roxb.	Lauraceae	4.44	0.96	MPT
Tejpat	*Cinnamomun tamala* Nees	Lauraceae	17.78	1.94	MPT
Hatkhora	*Citrus anamensis*	Rutaceae	80.00	6.45	Fruit
Kagzinemu	*Citrus aurantifolia* (Chrisstm) Swingle.	Rutaceae	13.33	1.60	Fruit
Ser-tawk	*Citrus grandis* L	Rutaceae	75.56	6.59	Fruit
Jambura	*Citrus maxima* (Burm). Merril.	Rutaceae	73.33	6.56	Fruit
Ser	*Citrus reticulata* *Blanco*	Rutaceae	73.33	6.63	Fruit
Serthlum	*Citrus sinensis* L. Osbeck	Rutaceae	80.00	7.00	Fruit
Ba-kenfung	*Croton caudatus* Geisel.	Euphorbiaceae	6.67	1.08	Miscellaneous
Kamsahluh	*Croton jourffiah* Roxb.	Euphorbiaceae	6.67	1.95	Miscellaneous
Japanfar	*Cryptomaria japonica* (Lf) D.Don.	Abeitaceae	6.67	1.37	Timber
Thingzaizawh	*Dalbergia pinnata* (Lour) Prain	Papilionaceae	8.89	1.43	Timber

(Contd.)

Local Name	Botanical Name	Family	Frequency occurrence (%)	IVI	Use category
Aprilpar	*Delonix regia* L.	Caesalpinaceae	15.56	2.00	Timber
Kawrthindeng	*Dillenia indica* L.	Dilleniaceae	17.78	2.22	Fruit
Kaihzawl	*Dillenia pentagyna* Roxb.	Dilleniaceae	6.67	1.08	Timber
Zuang	*Duabanga sonneratioides* Buch.	Sonneratiaceae	31.11	2.77	MPT
Sarzukpui	*Elaeagnus latifolia* L.	Elaeagnaceae	22.22	2.14	Fruit
Thinglung	*Eleocarpus floribundus* Blume.	Elaeocarpaceae	40.00	3.28	Fruit
Tuahfavang	*Erythrina arborensis* Roxb.	Papilionaceae	6.67	1.08	Medicinal
Rubber	*Ficus glomerata* Roxb.	Moraceae	35.56	2.88	Miscellaneous
Hmawngsawijhr	*Ficus recemosa* L.	Moraceae	6.67	1.37	Miscellaneous
Chengkek	*Garcinia cowa*Roxb.	Guttiferae	11.11	1.70	Timber
Thlanvanwg	*Gmelina arborea* Roxb.	Verbenaceae	24.44	2.93	Timber
Silver oak	*Grevellia robusta* A.Cunn.	Protaceae	24.44	2.24	Timber
Easter	*Holarrhena antidysenterica* (L.) Wall.	Apocynaceae	15.56	2.16	Timber
Thing-dawn	*Itea macrophyla* Wall.	Saxifragaceae	4.44	0.14	Timber
Litchi	*Litchi chinensis* Sonner.	Sapindaceae	26.67	2.57	Fruit
Laisua	*Licula peltata* Roxb.	Palmae	28.89	2.88	Timber
Wild apple	*Malus pumila*	Rosaceae	44.44	3.15	Fruit
Theihai	*Mangifera indica* L.	Anacardiaceae	75.56	5.75	MPT
Haifavang	*Mangifera sylvatica* Roxb.	Anacardiaceae	17.78	1.10	MPT
Nim-suak jhr	*Melia azedaratchta* L.	Meliaceae	75.56	5.26	Medicinal
Harhse	*Mesua ferrea* L.	Clusaceae	24.44	2.24	Timber
Ngiau	*Michelia champaca* L.	Magnoliaceae	8.89	1.43	Timber
Ram kelchal thing	*Mimusops elengi* L.	Sapotaceae	62.22	4.82	Fruit
Lungli	*Morus australis* Poir.	Moraceae	20.00	2.04	MPT
Oroxylon	*Oroxylon indicum* (L.) Vent	Bignoniaceae	33.33	2.97	Medicinal
Parkia	*Parkia timoriana* (A.DC) Merr.	Leguminosae	75.56	5.47	MPT
Butter thei	*Persea americana* (Nees) Kosterm.	Lauraceae	82.22	7.24	MPT
Kawl sunhlu	*Phyllanthus acidus* (L.) Skeels.	Euphorbiaceae	40.00	3.37	Fruit
Kuam	*Premna recemosa*	Verbenaceae	11.11	1.51	Timber
Kawl thei	*Psidium guajava* L.	Myrtaceaae	82.22	6.83	Fruit
Kawrpeh	*Psychotria calocarpa* Kurz.	Rubiaceae	2.22	0.85	Fruit
Lenglap Theibufai/	*Pterygota alata*	Sterculiaceae	5.56	2.00	Medicinal

(*Contd.*)

Local Name	Botanical Name	Family	Frequency occurrence (%)	IVI	Use category
Darjelling	*Punica granatum* L.	Punicaceae	35.56	3.16	Fruit
Thlum zu	*Pyrularia edulis* A.DC.	Santalaceae	8.89	1.43	Miscellaneous
Pears	*Pyrus communis* L	Rosaceae	8.89	1.20	Fruit
Khaw thli	*Quercus griffithi* Hk.f. and Th.	Fagaeae	42.22	3.21	Timber
Raintree	*Samanea saman* (Jacq.) Merr.	Mimosaceae	42.22	3.30	MPT
Thinglawhleng	*Saprosma ternatum*	Rubiaceae	37.78	3.45	Timber
Khiang	*Schima wallichi* (DC.) Kurth.	Theaceae	75.56	5.40	MPT
Kawhtebel	*Semecarpus anacardium* Roxb.	Anacardiaceae	68.89	5.65	Fruit
Taitaw	*Spondias pinata* (L). Kurz.	Anacardiaceae	6.67	1.08	MPT
Khaupui	*Sterculia villosa* Roxb. ex Smith	Sterculiaceae	62.22	4.67	MPT
Thingvutnu	*Symplocos glomerata* King.	Styraceae	4.44	0.55	Timber
Boga-jamun	*Syzigium jambos* (L). Alston	Myrtaceae	26.67	2.57	MPT
Tengtere	*Tamarindus indica* L.	Caesalpinaceae	77.78	5.51	Fruit
Teak	*Tectona grandis* L.	Verbenaceae	62.22	4.67	Timber
Thingvandawt	*Terminalia bellerica* (Gaertn.) Roxb.	Combretaceae	35.56	3.26	MPT
Thingdawl	*Tetrameles nudiflora* R. Br.	Tetramelaceae	28.89	2.25	Timber
Teipui	*Toona ciliata* Roem.	Meliaceae	6.67	3.01	MPT
Belphuar	*Trema orientalis* Blume.	Ulmaceae	11.11	1.90	Miscellaneous
Phan	*Ulmus lancefolia* Roxb.	Ulmaceae	2.22	0.85	Miscellaneous
Sirkam	*Vaccinium acuminatum* Kurz.	Vaccinaceae	24.44	2.36	Miscellaneous
Tungtechi	*Vernicia fordii*	Euphorbiaceae	20.00	1.91	Miscellaneous
Thingkhawihlu	*Vitex peduncularis* Wall. ex. Schauer	Verbenaceae	24.44	2.24	Medicinal
Batling	*Wendlandia tinctoria* Roxb	Rubiaceae	22.22	2.14	Medicinal
Chingit	*Zanthoxylum nitidum* (Roxb.) DC.	Dip terocarpaceae	8.89	1.43	Medicinal
Borai	*Zizyphus jujubae* L.	Rhamnaceae	40.00	3.45	Fruit

MPT-Multipurpose tree species

Table 3. Shrubs species with principal uses in the indigenous agroforestry homegardens in Aizawl.

Local Name	Botanical Name	Family	Frequency occurrence (%)	IVI	Use category
Zoeng	*Amaranthus caudatus* L.	Amaranthaceae	44.44	6.61	Vegetable
Zamzo	*Amaranthus viridis* L.	Amaranthaceae	26.67	4.77	Vegetable
Sai	*Artemesia vulgaris* L.	Compositae	17.78	3.44	Vegetable
Ramser	*Atalantia monophylla* Correa.	Rutaceae	20.00	3.17	Vegetable
Zawngalehlawn	*Bauhinia purpurea* L.	Leguminosae	8.89	2.58	Miscellaneous
Kanchan	*Bauhinia variegata* L.	Leguminosae	31.11	4.14	Miscellaneous
Sarawn	*Bougainvellea spectabilis* Willd.	Nyctaginaceae	20.00	3.75	Miscellaneous
Arhar	*Cajanus cajan* (L.) Millsp.	Leguminosae	64.44	8.61	Vegetable
Zanja	*Canabis sativa* L.	Cannabinaceae	40.00	6.09	Medicinal
Hmar cha	*Capsicum annum* L.	Solanaceae	80.00	14.97	Spice
Hmar cha pui	*Capsicum frutescens* L.	Solanaceae	48.89	9.56	Spice
Papaya	*Carica papaya* L.	Caricaceae	66.67	7.64	Fruit
Kum-tlung	*Catharanthus roseus* (L.). G.Don.	Apocynaceae	15.56	3.34	Medicinal
Sertuibur	*Citrus medica* L.	Rutaceae	20.00	3.17	Fruit
Serfang	*Citrus limon* (L.) Buro	Rutaceae	77.78	8.69	Fruit
Phui hnam	*Clerodendron colebrokianum* Walp.	Verbenaceae	46.67	8.05	Medicinal
Bhati	*Clerodendron infortunatum* L.	Verbenaceae	48.89	7.25	Medicinal
Coffee	*Coffea arabica* L.	Rubiaceae	22.22	5.11	Miscellaneous
Tumthang	*Crotalaria juncea* L.	Fabaceae	22.22	3.69	Vegetable
Sunflower	*Helianthus annuus* A. Cunn. ex. R.Br.	Compositeae	37.78	5.17	Vegetable
Vaiza	*Hibiscus macrophyllus* Roxb.	Malvaceae	48.89	6.90	Vegetable
Anthur	*Hibiscus sabdaiffa* Rau Day	Malvaceae	53.33	6.98	Vegetable
Zawng anthur	*Hibiscus suranttensis* L.	Malvaceae	31.11	4.72	Vegetable
Jatropha	*Jatropha carcus* L.	Euphorbiaceae	28.89	4.58	Miscellaneous

(*Contd.*)

Local Name	Botanical Name	Family	Frequency occurrence (%)	IVI	Use category
Kawldai	*Justica adhatoda* L.	Acanthaceae	22.22	4.40	Medicinal
Buarpui	*Livingstonia chinensis* L.	Palmae	8.89	2.23	Miscellaneous
Tumbu	*Musa acuminata* Colla.	Musaceae	48.89	6.44	Fruit + Vegetable
Balah	*Musa glauca* Roxb.	Musaceae	17.78	3.02	Fruit + Vegetable
Balah	*Musa paradisiaca* L.	Musaceae	77.78	9.27	Fruit + Vegetable
Seichu	*Musa superba* Roxb.	Musaceae	20.00	2.98	Fruit + Vegetable
Changvandawt	*Musa velutina* Wendl.	Musaceae	15.56	3.34	Fruit + Vegetable
Tobacco	*Nicotiana tabacum* L.	Solanaceae	46.67	6.28	Miscellaneous
Tulsi-common	*Ocium sanctum* L.	Labiatae	68.89	10.83	Medicinal
Anbawng	*Polygonum barbata* L.	Polygonaceae	51.11	7.84	Vegetable
Bakhate	*Polygonum plebium* R.Br.	Polygonaceae	31.11	4.43	Vegetable
Mu tih	*Ricinus communis* L.	Euphorbiaceae	44.44	6.37	Vegetable
Saisiak	*Securinega virosa* Roxb. ex.Willd.	Euphorbiaceae	8.89	2.23	Vegetable
Samtawkte	*Solanum anguivi* L.	Solanaceae	11.11	3.23	Medicinal
Brinjil	*Solanum esculentum* L.	Solanaceae	17.78	6.95	Medicinal
Bekoir	*Solanum khasiana* L.	Solanaceae	75.56	10.66	Medicinal
Bawkbawn	*Solanum melongena* L.	Solanaceae	66.67	10.60	Vegetable
Anhling	*Solanum nigrum* L.	Solanaceae	57.78	7.73	Vegetable
-	*Solanum pubescens* Roxb.	Solanaceae	31.11	4.14	Medicinal
-	*Solanum ferox* L.	Solanaceae	22.22	4.22	Medicinal
Tawkpui	*Solanum torvum* Sweet	Solanaceae	22.22	3.51	Medicinal
Samtawk	*Solanum violaceum* Ort.	Solanaceae	20.00	3.36	Medicinal
Broom Grass	*Thysanolaena maxima* (Roxb.) O.Ktze	Poaceae	64.44	15.13	Miscellaneous
Bawngpupang	*Tithonia diversifolia*	Compositae	6.67	3.04	Vegetable
Kawhtebel	*Trevesia palmata* (Roxb.) Vis	Caprifoliaceae	8.89	2.23	Vegetable
Vaimin	*Zea mays* L.	Gramineae	77.78	12.55	Vegetable

Table 4. Herbs species with principal uses in the indigenous agroforestry homegardens in Aizawl.

Local name	Botanical Name	Family	Frequency occurrence (%)	IVI	Use category
Bawrhsaibe Vegetable	*Abelmoschus esculentus* (L) Moench.	Solanaceae	66.67	5.32	
Chakawkte	*Adiantum phillippense* L.	Polypodiaceae	20.00	2.20	Vegetable
Vailenhlo	*Ageratum conyzoides* L.	Asteraceae	75.56	8.46	Vegetable
Purul sein	*Allium cepa* L.	Liliaceae	68.89	8.55	Spice
Mizopuran	*Allium hookerii* Thw.	Liliaceae	80.00	8.65	Spice
Pu-runvar	*Allium sativum* L.	Liliaceae	20.00	2.75	Spice
Lakhuithei	*Annanas comosus* (L.) Merrill	Bromoliaceae	35.56	2.85	Fruit
Lelen	*Arundina graminifolia* (D.Don) Hochr.	Orchidaceae	20.00	2.56	Miscellaneous
Arkebawk	*Asparagus racemosus* L.	Asparangaceae	11.11	1.81	Medicinal
Vawkpuithal	*Bidens biternata* (Lour) Merr	Compositae	20.00	2.38	Vegetable
Buar	*Blumea alata* D.Don.	Compositae	17.78	2.83	Miscellaneous
Lenglang	*Boehmeria rugulosa* Wedd.	Urticaceae	33.33	2.56	Medicinal
Mustard	*Brasica juncea L.*	Cruciferae	68.89	7.83	Spice
Parbawr	*Brassica botrytis*	Cruciferae	44.44	5.22	Spice
Antram	*Brassica compestris* L.	Cruciferae	60.00	6.52	Vegetable
Bulbawk	*Brassica oleracea* L.	Cruciferae	66.67	6.78	Vegetable
Antam	*Brassica rapa* L.	Brassicaceae	57.78	5.42	Vegetable
Thilte	*Calamus acanthospathus* Griff.	Arecaceae	17.78	1.54	Vegetable
Hruipui	*Calamus flegellum* Griff.	Arecaceae	4.44	2.00	Vegetable
Tairua	*Calamus guruba* Buc-Ham ex. Kunth	Arecaceae	13.33	1.83	Vegetable
Thilte	*Calamus tenuis* Roxb.	Arecaceae	24.44	3.07	Vegetable
Fangra	*Canavalia ensiformis* DC.	Papilionaceae	4.44	0.77	Medicinal
Kungpuimuthi	*Canna orientalis*	Cannaceae	11.11	1.67	Medicinal

(*Contd.*)

Local Name	Botanical Name	Family	Frequency occurrence (%)	IVI	Use category
Lam-buk	*Centella asiatica* (L.) Urban.	Apiaceae	66.67	9.33	Medicinal
Ruathing	*Chimnocalamus longispiculatua*	Gramineae	8.89	1.67	Vegetable
Bai-bing	*Colocasia affinis* L.	Araceae	26.67	4.18	Vegetable
Dawl	*Colocasia esculenta* (L.) Schott	Araceae	37.78	5.91	Vegetable
Dawl	*Colocasia* sp.	Araceae	26.67	3.72	Vegetable
Coriander wild	*Coriander* sp.	Umbelliferae	24.44	3.55	Spice
Sumbul	*Costus speciosus* Smith.	Zingiberaceae	22.22	2.01	Medicinal
Binjukochu	*Cucumis sativa* L.	Cucurbitaceae	33.33	3.30	Vegetable
Maien	*Cucurbita maxima* Duchesne	Cucurbitaceae	55.56	4.18	Vegetable
Bottle guard	*Cucurbita siceraria*	Cucurbitaceae	17.78	1.54	Vegetable
Hmazil	*Cumumis melo* L.	Cucurbitaceae	33.33	2.90	Vegetable
Ailaidun	*Curcuma caesia* Roxb.	Zingiberaceae	28.89	2.73	Spice
Aieng	*Curcuma longa* Roxb.	Zingiberaceae	71.11	6.37	Spice
Aithur	*Curcumphera longiflora* Sm.	Zingiberaceae	20.00	2.10	Spice
Artelubawk	*Cyperus rotundus* L.	Cyperaceae	46.67	4.88	Medicinal
Karot	*Daucas carota* L.	Umbelliferae	51.11	5.51	Vegetable
Rambachim	*Dioscorea alata* L.	Dioscoreaceae	35.56	4.22	Medicinal
Winged bean	*Dolichos tetragonobolus*	Leguminosae	46.67	4.77	Vegetable
Leng-hmaser	*Elsholtzia communis* Coll.	Labiatae	6.67	1.53	Medicinal
Uri	*Glycine max* (L.) Merr.	Papilionaceae	15.56	2.86	Vegetable
Nuai thang	*Impatiens balsamina* L.	Balsaminaceae	22.22	2.78	Vegetable
Di	*Imperata cylindrica* Beauv.	Gramineae	24.44	3.63	Miscellaneous
Kawl Barha	*Ipomea batata* L. (Lamb)	Convolvulaceae	75.56	6.03	Fruit
Awmpawng	*Luffa cylindrica* Roem.	Cucurbitaceae	22.22	2.35	Vegetable
Pangbal	*Manihot esculanta* Krantz.	Euphorbiaceae	20.00	2.29	Vegetable
Pudina	*Mentha viridis* L.	Labiatae	20.00	3.30	Spice
Japanhlo	*Mikenia micrantha* Kunth.	Asteraceae	64.44	6.06	Medicinal
Hlonuar	*Mimosa pudica* L	Mimosaceae	60.00	5.50	Medicinal
Changkha	*Momordica charantia* L	Cucurbitaceae	26.67	2.65	Vegetable
Siakthur	*Oxalis corniculata* L.	Oxalidaceae	64.44	7.95	Medicinal
Vawihuihhrui	*Paderia foetida* L.	Rubiaceae	15.56	2.64	Medicinal
Passion fruit	*Passiflora edulis* Sims.	Passifloraceae	68.89	5.22	Fruit

(*Contd.*)

Local Name	Botanical Name	Family	Frequency occurrence (%)	IVI	Use category
French bean	*Phaseolus vulgaris* L.	Fabaceae	48.89	5.12	Vegetable
Hnahthial	*Phrynium capitatutm*	Maranthaceae	13.33	1.59	Vegetable
Pan	*Piper betle* L.	Piperaceae	20.00	2.10	Spice
Piper	*Piper boehmerifolia* (Micq.) DC.	Piperaceae	11.11	1.95	Spice
Be-pui thlanei	*Psophocarpus teragonobulus* (L.) DC.	Papillionaceae	4.44	1.08	Vegetable
Dekia	*Pteris amoena* Bl.	Pteridaceae	31.11	3.23	Vegetable
Thingba	*Pueraria peduncularis* Benth.	Papilionaceae	8.89	1.67	Vegetable
Buluih	*Raphanus sativa* L.	Cary ophyllaceae	68.89	6.33	Vegetable
Iskut	*Sechium eduli* (Jacq.) Sw	Cucurbitaeae	77.78	5.47	Vegetable
Ansate	*Spilenthes acmella* (L.) Murr.	Asteraceae	53.33	7.94	Vegetable
Ansapui	*Spilenthes oleraceae* L.	Asteraceae	75.56	9.21	Vegetable
Stevia	*Stevia rebaudiana* Roxb.	Asteraceae	4.44	1.08	Medicinal
Khangmang	*Thladiantha calcarata* Roxb.	Cucurbitaceae	6.67	1.53	Vegetable
Arawkeu	*Torenia peduncularis* Benth.	Scrop hulariaceae	2.22	0.66	Vegetable
Berul	*Trichosanthes anguina* L.	Cucurbitaceae	33.33	3.10	Vegetable
Sehnap	*Urena lobota* L.	Malvaceae	71.11	7.27	Medicinal
Blackgram	*Vinga mungo* (L) Hepper.	Fabaceae	37.78	3.89	Vegetable
Behlawi	*Vinga unguiculata* Roxb.	Fabaceae	20.00	2.75	Vegetable
Behliangthing	*Vitex negundo* L.	Vitaceae	26.67	2.35	Medicinal
Grepthei	*Vitis vinifera* L.	Vitaceae	48.89	4.45	Fruit
Sawhthing	*Zingiber officinalis* Roscoe	Zingiberaceae	80.00	5.99	Spice

Stratification

In Aizawl, Mizoram indigenous homegardens are classified from a land-use perspective as an agroforestry system with a mixture of herbs, shrubs and trees and other agricultural crops within the household boundary and under the family labour and management. Like most multistorey agroforestry systems in the tropics, the Mizo's homegardens are similar in their architecture with mixed crop composition. As these gardens are located in tropical region, they are multi-storied, commonly containing three layers but sometimes with four stratification. The uppermost canopy consisted of trees and therefore was a perennial layer. Species commonly found in this layer includes *Albizzia lebbeck, Duabanga grandiflora, Gmelina arborea, Bombax ceiba, Cassia fistula, Quercus griffithi, Schima wallichi, Areca catechu* and *Mesua ferrea*. This layer was followed by species like *Artocarpus heterophyllus, Erythrina indica, Citrus anamensis, Mangifera indica, Oroxylum indicum* and *Elaeocarpus floribundus*. Immediately below this layer occur both annual and perennial species. The commonest and most important species are *Aegle marmelos, Annona squamosa, Averhora carambola, Artocarpous lakocha, Citrus reticulata, Citrus lemon, Musa* sp., *Mimusops elengi, Phyllanthus acidus* and *Persea americana* etc. The third story consists of vegetables such as *Amaranthus* sp., *Solanum* sp., *Colocasia* sp., *Dioscorea* sp., *Manihot esculenta* and medicinal plants (*e.g. Solanum torvum, Ocimum sanctum, Costus speciosus, Clerodendron colebrokianum)* and also climbing crops like *Sechium eduli*, *Cucumis sativa, Momordica mixta*. The lowest storey consisted of species that were 20 cm or less in height like *Allium hookeri*, *Blumea alata, Colocasia esculenta, C. affinis, Coriandrum sativum, Spilenthes acmella, S. oleraceae* and creeping plants such as *Cucurbita maxima* and *Ipomoea batata* etc. From the ground layer comprising herbaceous food crops, forage, medicinal and other crops to the upper canopy of fast-growing multipurpose trees, the gradient of light and relative humidity creates different niches, enabling various species group to exploit them (Michon *et al.*, 1983). The structured layer also contributes to soil nutrient enrichment through leaf litter and prevents soil erosion in homegardens (Depommier, 2003).

Functional dynamics

The majority of the species grown in the homegardens were perennial followed by annual species and only a few biennials in all the homegardens (Figure 1). The life cycle of most of these crops is different, which makes food and other products available throughout the year. The presence of crops with different functions fulfills the nutritional and monitory needs of the farmer. The perennial nature of the systems, in which crops, fruit trees as well as other trees and shrubs shared *ca.* 58% of the community, play an important role in ecological sustainability and stability. Nonetheless, the perennial nature of these indigenous systems together with the high species diversity provides important ecological services such as nutrient recycling, soil and water conservation, and reduces environmental degradation, which in turn helps conservation of resource base necessary for future production. Hence, effective management of these homegardens with high proportion of perennial crops and trees and high species diversity could be regarded as ecologically sustainable.

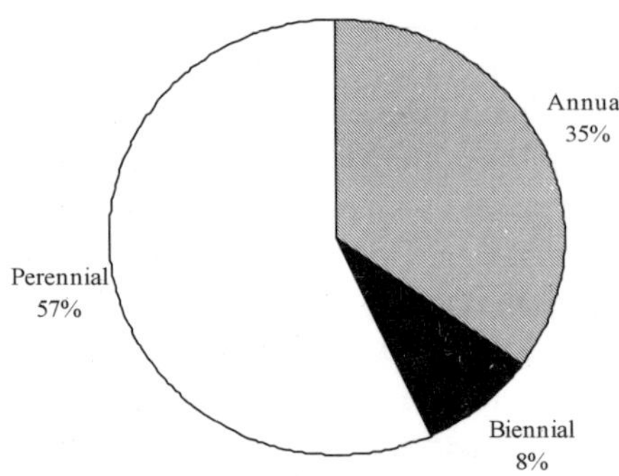

Fig. 1. Different growth habit of plant species in each home homegarden.

Seven major plant use categories were identified in these homegardens. These functional groups of crops serving different functions are fairly well represented in per homegarden. Figure 2 shows the mean number of species in each use category per homegarden, with the dominant being the vegetable followed by medicinal and spice in herbs groups, in shrubs groups it was vegetable and followed by medicinal and miscellaneous species. However, in the tree groups dominant category were fruits species followed by multipurpose and timber tree species (Figure 2). Overall, vegetables (32%) are the major constituents in the homegardens followed by fruits (15%), medicinal plants (13%), multipurpose tree species (13%), and species of miscellaneous use (11%), timber (10%) and spice (6%) (Figure 3). The number of species grown in each homegarden is an important indicator of diversity and from the utility view point it not only contributes to the richness of species but also to the species diversity and functions. Different tree species have been found to be associated with various socio-economic functions and ecological roles in the site. They provide households with timber, food, fodder, fuel, medicine and cash along with the house construction materials and different types of furniture. In this was, a tree species grown in the garden not only generates income for the farmer but also contribute to the regional economy. The ecological contribution of trees in these systems is also widely recognized. For examples, *Alstonia scholaris, Azadirachta indica, Cassia nodosa, Dalbergia pinnata, Duabanga sonneratioides, Morus australis, Schima wallichi, Sterculia villosa, Terminalia bellerica, Toona ciliata* etc grown extensively because of their roles in providing shade and mulch, control of soil erosion and in the overall improvement of microclimate. Few species like *Albizia myriophylla, A. procera, A. lebbeck, Cassia fistula, Erythrina indica* and *Parkia timoriana* were of special interest to the farmer because of their nitrogen fixing ability and easy fertility management in the systems. *Parkia timoriana* locally known as 'Jongtra' is found to be common in almost all the homegardens because of its wide economic and ecological roles in the systems, which provides good economic return every year through selling of its long tender pods most popular and delicious vegetable throughout the region especially among the tribal community. Farmers also strongly believe that these species improves soil fertility. Furthermore, final harvest of the plant provides good timber which can be used for making furniture and other house hold needs.

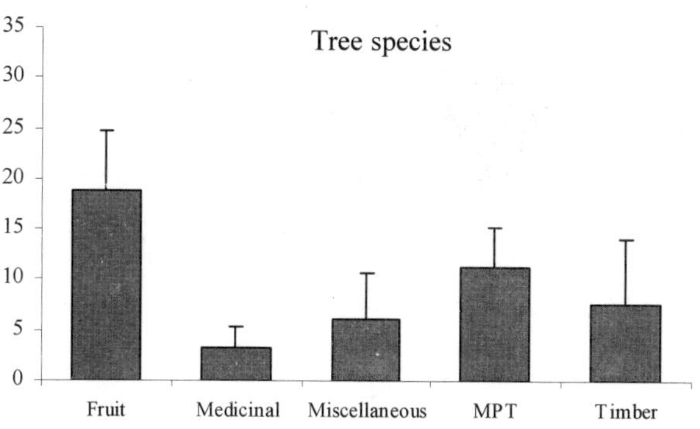

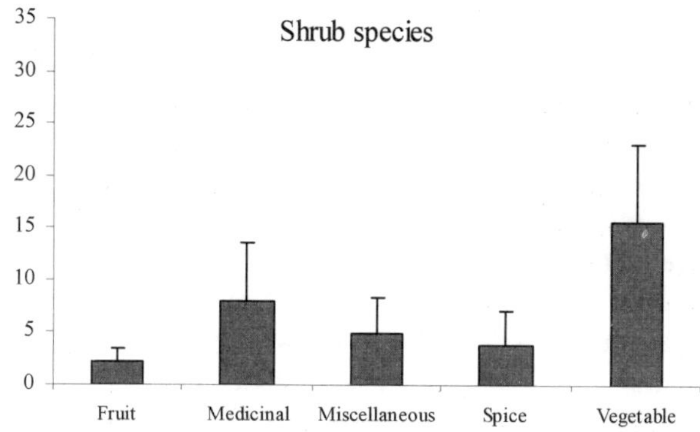

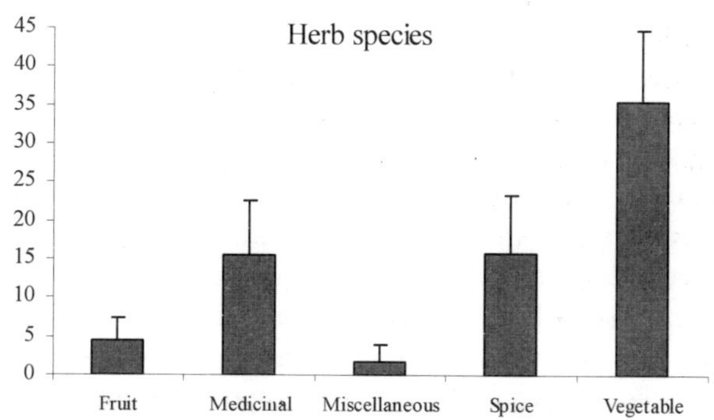

Fig. 2. Mean number of species per use category per homegarden on Y-axis. Vertical lines represents standard deviation (n=45).

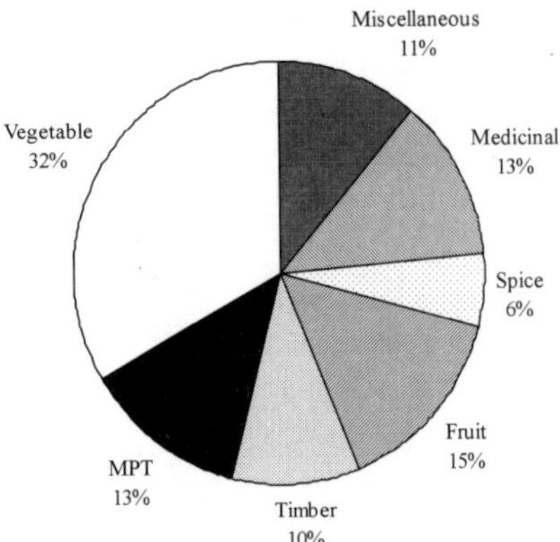

Fig. 3. Overall mean proportion of functional groups (%) of crops per homegarden.

Soil characteristic

Soil was acidic (pH=5.27-5.72) with little variation between the gardens, providing favorable fertility conditions for plant growth. In general, a relatively lower soil pH was observed in large gardens compared to the medium and small ones, which could have been due to lower rate of leaching leading to greater accumulation of reaction products in the soil. Production outputs from the homegardens were in the forms of fruits, leaves, young shoots, bark, flowers or other non-timber products. Since, there was no complete harvesting from the homegardens, there were minimal nutrient export from the system. Soil organic carbon (SOC) and total Kjeldahl nitrogen (TKN) registered lower values in the smaller gardens and increased with an increase in size of the garden. The differences in soil nutrients among the homegardens are obvious due to a combination of factors like microclimate, topography and plant species composition. A comparison of soil fertility indicators, such as the pH, TKN, SOC and available phosphorus in the surface soil inside and outside the homegardens (jhum fallows) clearly showed higher soil fertility inside the homegardens. It is generally regarded that the homegardens possess a closed nutrient cycling, much similar to the tropical forests (Soemarwoto and Conway, 1991; Nair *et al.*, 1999). This is because the systems are composed of several different tree and crop species, each differing from the other with respect to its growth, nutrients, water and other natural resource requirements, favouring a spatial pattern of nutrient cycling in tune with crop stratifications, time and space. Nevertheless, the soils of the homegardens in general are much superior compared to the adjacent jhum (shifting cultivation) fallows in the area (Table 1). Besides, the systems offer permanent soil protection, facilitate soil biological processes favourable litter decomposition and soil structure, and more efficient nutrient cycling (Schroth *et al.*, 2001). Evidences from other studies also suggest that trees as a functional group have beneficial effects on soil fertility through aboveground litter inputs, and perennial root systems that improves soil structure, reduce leaching losses, pump water and nutrients from deeper soil depths, support soil food web continually and contribute belowground detrital inputs (Fisher, 1995; Schroth and Lehmann, 1995). The positive effects of agro-ecosystem crops, their composition and richness on soil organic carbon and nitrogen management have been validated by Russell (2002) who attributes these parameters as important for designing of homegardens for soil fertility management and carbon sequestration.

Economic benefits

Home gardening in Mizoram is an integrated method of home production systems with primary function of subsistence food production and family income. Unlike jhum, homegardens provide a certainty on food supply to the farmers year round, thereby not only meeting the dietary requirement but also regular cash needs of the family through sell of various products. According to a conservative estimate, a household can earn a monthly income of Rs.3,000/ to Rs.5,000/ depending on crop composition, size of the garden and management. Table 5 presents the average annual yield and monetary values of a few important plants grown in these homegardens. The main crops sold in the local market are *Sechium edule, Parkia timoriana, Passiflora edulis, Persea americana, Carica papaya, Citrus anamensis, C. limon, C. reticulata, Allium hookerii, Areca cathechu, Ananas comosus* and *Zingiber officinalis*. The commonly sold vegetables include pumpkin, cucumber, ladyfinger etc. and seasonal fruits like mango, orange, banana, grapes, pineapple, jack fruit, amla, tamarind etc. Majority of the farmers sell more than 60% of their produce for income generation while the rest are used for family consumption. Besides many other vegetable, fruit, flowers, bulbs/corms, fonds, medicinal plants also provide them a good economic return throughout the year. The role of homegardens in generating additional cash income has been emphasized by many workers (Christanty, 1990; Torquebiau, 1992; Mendez *et al.*, 2001). Studies across different regions indicate a wide variation in the proportion of homegarden products that are used for household consumption as opposed to sale. In West Java, as much as two-thirds of the homegarden production is reported to be sold (Jensen, 1993), but only 28% of such products were sold in South African homegardens, the remainder being used for household consumption (High and Shackleton, 2000). Nevertheless, the proportion of this products sold to the market is linked to the amount of cash income available to a household. A higher proportion of homegarden products selling getting marketed reveal that the home gardening is becoming commercialized in Mizoram.

Table 5. Mean annual yield and monetary values of some of the important plants found in the homegardens in Aizawl, Mizoram.

Name of the species consumption Habit	Local unit of measurement	Average yield/ha	Average farm	Total value (Rs.)	% of consum- ption	% of sale
Abelmoschus esculentus (H)	kg	640	250	6396	39.09	60.91
Allium hookerii (H)	Bundle	1067	200	10665	18.75	81.25
Annanas comosus (H)	Unit	533	100	5330	18.76	81.24
Areca cathechu (T)	Unit	10840	1000	10840	9.23	90.77
Artocarpus heterophyllus (T)	Unit	172	75	1717	43.68	56.32
Brasica juncea (H)	Bundle	329	100	3288	30.41	69.59
Capsicum annum (S)	kg	52	20	1040	38.46	61.54
Carica papaya (S)	Unit	568	100	6816	17.61	82.39
Citrus anamensis (T)	Unit	1806	500	9030	27.69	72.31
Citrus limon (S)	Unit	2670	1000	4005	37.45	62.55
Citrus reticulata (T)	Unit	3840	1200	11520	31.25	68.75
Clerodendron colebrokianum (S)	Bundle	285	100	1423	35.15	64.85
Colocasia affinis (H)	Bundle	187	100	933	53.59	46.41
Cucurbita maxima (H)	Unit	4665	2000	13995	42.87	57.13
Dolichos tetragonobolus (H)	Bundle	400	100	2000	25.00	75.00
Glycine max (H)	Bundle	308	100	1539	32.48	67.52
Mangifera indica (T)	kg	356	150	7120	42.13	57.87
Musa paradisiaca (S)	Unit	5726	2000	11452	34.93	65.07
Parkia timoriana (T)	Bundle	618	300	6180	48.54	51.46
Passiflora edulis (H)	Unit	32200	5000	16100	15.53	84.47
Persea americana (T)	Unit	914	300	9135	32.84	67.16
Psidium guajava (T)	Unit	1580	800	1580	50.63	49.37
Sechium eduli (H)	kg	2044	1000	20440	48.92	51.08
Solanum melongena (S)	kg	240	150	3600	62.50	37.50
Solanum nigrum (S)	Bundle	391	200	1955	51.15	48.85
Spilenthes acmella (H)	Bundle	1244	650	6220	52.25	47.75
Tamarindus indica (T)	Bundle	800	200	4000	25.00	75.00
Thysanolaena maxima (S)	Bundle	107	15	1066	14.07	85.93
Zea mays (S)	Unit	160	100	962	62.34	37.66
Zingiber officinalis (H)	kg	1288	200	12880	15.53	84.47
Zizyphus jujubae (T)	kg	107	40	1065	37.56	62.44

H-herb, S-shrub and T-tree.

Ecological sustainability

The homegardens are always characterized by high species diversity, about 74% of which are indigenous to the locality. Therefore, these systems substantially contribute to genetic conservation of native species, efficient resource use and biological pest control. Besides, the perennial nature of the systems together with the high species diversity provides important ecological services such as nutrient recycling, soil and water conservation, and reduces environmental deterioration.

These facts indicate the conservation of the resource base, which is vital for future production. The maintenance of high species diversity in these systems is yet another feature which enables the farmer year round production of different crops and other products, reduces and spreads risks and expands the amount and quality of labour applied in the farm. The homegarden food and vegetable species also have multiple uses and multiple harvest times, and their year-round availability help to diversify food sources and types of micronutrients in the daily diet. For example, 'Maien' (*Cucurbita maxima*) is used for its tender shoot, flower and fruits; 'Iskut' (*Sechium edule)* is used for its tender shoot and fruits and 'Dawl' (*Colocassia sp*) is used for its leaf, petiole, corm and rhizomes. Similarly, homegardens also provide a number of green leafy vegetables, which are rich in micronutrients. Homegarden crops, vegetables and fruits are largely grown organically and therefore provide safe and healthy food for household consumption. A variety of multiple trees are found integrated with crop plants in these homegardens and these usually have multiple uses and provide food, fodder, firewood and timber for domestic uses. A number of spices, such as chili, ginger, turmeric, bay-leaf, garlic, onion, coriander and 'Mizopuron' are often found in these homegardens. Similarly, farmers maintain a number of medicinal plants such as *Aegle mermelos, Averrhoa carombola, Azadirachta indica, Alstonia scholaris, Centella asiatica, Clerodendron colebrokianum, Costus speciosus, Citrus anamensis, Mikenia micrantha, Oroxylum indicum, Solanum torvum, Terminalia bellirica* and *T. chebula* etc in their homegarden for their day-to-day ailments from common diseases.

Conclusions

The homegardens in Northeast India were found to be food-producing subsistence farming systems which improve the general well being of the society, provide ecological as well as the economic stability. These systems could be important platforms for conservation of plant diversity through use, for diversifying the nutrition of rural people that would contribute to food security at household and community level. The role of these homegardens in the *in situ* conservation, as well as the maintenance of indigenous knowledge is observed to be significant and therefore warrant priority attention. Thus, the present findings evince that the homegardens can be made to promote sustainable livelihood security and an alternative to shifting cultivation in the mountainous environment of Northeast India.

References

Ahmad, A.M., Abood, F. 1990. Selected forest tree with potential application in Malaysian agroforestry. BIOTROP Special Publication No. 39. BIOTROP, Indonesia. pp. 77-89.

Altieri, M.A. 1995. Agroecology: The science of sustainable agriculture (2nd ed). Westview Press.

Alvarez-Buylla Roces, M.E., Lazos Chavero, E., Garcia-Barrios, J.R. 1989. Homegardens of humid tropical region in Southeast Mexico: an example of an agroforestry cropping system in a recently established community. Agrofor. Syst. 8:133-156.

Biggelaar, C., Gold, M.A. 1996. Development of utility and location indices for classifying agroforestry species: the case of Rwanda. Agrofor. Syst. 34: 229-246.

Bouyouncos, G.J. 1962. Hydrometer method improved for making particle size analysis of soil. Agron. J. 54:464-465.

Brierley, J.S. 1985. West Indices kitchen gardens: A historical perspective with current insights from Grenada. Food Nutri. Bull. 7: 52-60.

Brownrigg, L. 1985. Home gardening in international development: What the literature shows: The league for international food education, Washington DC.

Christanty, L. 1990. Homegardens in tropical Asia with special reference to Indonesia. In: Landauer, K., Brazil, M. (eds), Tropical homegardens, pp. 9–20. United Nations University Press, Tokyo.

Curtis, J.T. 1959. The vegetation of Wisconsin: an ordination of plant communities. University of Wisconsin Press, Madison, Winconsin.

Das, T., Das, A.K. 2005. Inventorying plant biodiversity in homegardens: A case study in Barak Valley, Assam, North-East India. Curr. Sci. 89:155-163.

Depommier, D. 2003. The tree behind the forest: ecological and economic importance of traditional agroforestry systems and multiple uses of trees in India. Trop. Ecol. 44:63–71.

Eyzaguirre, P.B., Linares, O.F. 2001. A new approach to study and promotion of homegardens. People and Plants 7: 30–33.

Fernandes, E.C.M., Nair, P.K.R. 1986. An evaluation of the structure and function of tropical homegardens. Agric. Syst. 21:279-310.

Fisher, R.F. 1995. Amelioration of degraded rain forest soils by plantations of native trees. Soil Sci. Soc. Am. J. 59:544-549.

Gajaseni, J., Gajaseni, N. 1999. Ecological rationalities of the traditional homegarden system in the Chao Phraya Basin, Thailand. Agrofor. Syst. 46:3-23.

High, C., Shackleton, C. M. 2000. The comparative value of wild and domestic plants in homegardens of a South African rural village. Agrofor. Syst. 48: 141–156.

Huston, M.A. 1995. Biological diversity: the coexistence of species on changing landscapes. Cambridge University Press.

Jackson, M.L. 1958. Soil Chemical Analysis. Prentice Hall, Englewood Cliffs, New Jersey.

Jensen, M. 1993. Soil conditions, vegetation structure and biomass of a Javanese homegarden. Agrofor. Syst. 24: 171–186.

Kessy, J.F. 1998. Conservation and utilization of natural resources in the east Usambara forest reserves: conventional views and local perspectives. Ph.D. thesis, Wageningen University.

Kharal, D. 2000. Diversity and dynamics of tree species and its sustainability in the rural farm land: A case study in Chitawan districts in Nepal. M.Sc. Thesis, The Agricultural University of Norway, Norway.

Kumar, B.M., George, S.J., Chinnamanis, S. 1994. Diversity, structure and standing stock of wood in the homegardens of Kerala in Peninsular India. Agrofor. Syst. 25:243-262.

Landauer, K and Brazil, M. 1990. Introduction in tropical home gardens. United Nations University Press, Tokyo.

Magurran, A. E. 1988. Ecological diversity and its measurement. Croom Helm, London.

Makaraphirom, P. 1989. Checklist of species for extension in agroforestry systems. Agroforestry Research 30, Royal Forestry Department, Bangkok, Thailand.

Matson, P.A., Parton, W.J., Power, A.G., Swift, M.J. 2002. Agricultural intensification and ecosystem properties. Science 277: 504.

Mendez, V.E., Kok, L., Somarriba, E. 2001. Interdisciplinary analysis of homegardens in Nicaragua: micro-zonation, plant use and socioeconomic importance. Agrofor. Syst. 51:85-96.

Michon, G., Bompard, J., Hecketsweiler, P., Ducatillon, C. 1983. Tropical forest architectural analysis to agroforests in the humid tropics: The examples of traditional village agroforests in west Java. Agrofor. Syst. 1: 117-129.

Nair, P.K.R.1993. An introduction to agroforestry. Kluwer Academic Publishers, Dordrecht. The Netherlands.

Nair, PKR., Buresh, RJ, Mugendi, DN and Latt, CR. 1999. Nutrient cycling in tropical agroforesty systems: Myths and science. In: Buck, L.E., Lassoie, J.P., Fernandes, E.C.M. (eds), Agroforestry in Sustainable Agricultural Systems, pp. 1-31. CRC Press, Boca Raton, FL.

Phillips, E.A. 1959. Methods of vegetation study. Henry Holt & Company, New York.

Pielou, E.C. 1969. An introduction to mathematical ecology. Wiley, New York.

Piper, C.S. 1944. Soil and Plant Analysis. John Wiley and Sons, New York, USA.

Powers, E.L., McSorley, R. 2000. Ecological principles of agriculture. Library of Congress Cataloging in Publication Data. pp. 243–248.

Reijntjes, C., Haverkort, B., Waters-Bayer, A. 1992. Farming for the future: An introduction to low-external-input and sustainable agriculture. ILEIA, Leusden, The Netherlands.

Russell, A.E. 2002. Relationships between crop-species diversity and soil characteristics in southwest Indian agroecosystems. Agric. Eco. Env. 92:235-249.

Schroth, G., Lehmann, J., Rodrigues, M.R.L., Barros, E., Macedo, J.L.V. 2001. Plant-soil interactions in multistrata agroforestry in the humid tropics. Agrofor. Syst. 53: 85–102.

Schroth, G., Lehmann, J. 1995. Contrasting effects of roots and mulch from three agroforestry tree species on yield of alley cropped maize. Agric. Eco. Env. 54:89-101.

Shannon, C.E., Wiener, W. 1949. The mathematical theory of communication. The University of Illinois Press.

Shastri, C.M., Bhat, D.M., Nagaraja, B.C., Murali, K.S., Ravindranath, N.H. 2002. Tree species diversity in a village ecosystem in Uttara Kannada district in Western Ghats, Karnataka. Curr. Sci. 82:1080-1084.

Soemarwoto, O., Conway, G.R. 1991. The Javanese homegarden. J. Farm. Syst. Res. Ext. 2:95-117.

Sunwar, S., Thornstrom, C.G., Subedi, A., Bystrom, M. 2006. Homegardens in western Nepal: opportunities and challenges for on-farm management of Agrobiodiversity. Biodiv. Cons. 15:4211-4238.

Swift, M.J., Ingram, J.S.I. 1996. Effects of global change on multi-species agroecosystems. Global change and terrestrial ecosystems, Report no. 13, GCTE Activity 3.4, GCTE Focus 3 Office, Wallingford, UK.

Torquebiau, E. 1992. Are tropical agroforestry homegardens sustainable? Agri. Eco. Env. 41: 189–207.

Tilman, D., Cassman, K.G., Matson, P.A., Naylor, R., Polasky, S. 2002. Agricultural sustainability and intensive production practices. Nature 418: 671-677.

Walkey, A. 1947. A critical examination of a rapid method for determining organic carbon in soil-effects of variations in digestion conditions and inorganic soil constituents. Soil Sci. 63:251-264.

Whitmore, T.C. 1971. Wild fruit trees and some trees of pharmacological potential in the rainforest of Ulu, Kelantar. Malayan Nat. J. 24:222-224.

Zar, J.H. 1974. Biostatistical Analysis, 2nd Edition, Prentice Hall, Englewood Cliffs, New Jersey.

❑❑❑

2

Shifting Cultivation in North Eastern Region of India

J.M.S.Tomar, Anup Das, Rajesh Kaushal and O.P. Chaturvedi

Abstract: *Shifting cultivation locally called* Jhuming *is a part and parcel of tribal life of the North Eastern Hill Region of India and it has direct bearing with their socio-cultural system. It is a boon to the farmers as long as the* jhuming *cycle is sufficiently longer (10-15 years) as this would not only produce a good crop but also gives sufficient time for ecological restoration and biophysical balance. However, with the increase in population pressure on land the* jhuming *cycle is getting drastically reduced (3-6 years). As a result of this, soil health is getting degraded and overall ecosystem is in threat. Many important flora and fauna are in the verge of extinction. As we cannot run away from the problem at least thinking about posterity, we have to find a decisive solution immediately. Developing improved methods of* jhum *cultivation like proper bunding, residue recycling, use of high yielding varieties etc. are few short term options. However long term viable options would be proper land configuration like terraces, water harvesting, agroforestry, adoption of integrated farming system approach and above all education and awareness among the tribal farmers which would help in sustaining the ecosystem of the region.*

Introduction

Shifting cultivation or slash and burn agriculture locally known as "*jhuming*" is a widely practiced farming system in the hills of North Eastern India consisting of Assam, Tripura, Arunachal Pradesh, Meghalaya, Sikkim, Nagaland, Manipur and Mizoram. *Jhuming* is also practiced in southern (Andhra Pradesh, Tamil Nadu, Kerela), Eastern (Bihar, Orissa) and Central (Madhya Pradesh) part of India (Table 1). Shifting cultivation is a time-tested system of agricultural practice, most often evolved indigenously and strongly based on traditional knowledge. It is considered to be an appropriate and sustainable land use practice in diverse socio-economic setup, where the dependant human population was within the carrying capacity of a 10-15 year *jhum* cycle. Today the shifting cultivation became unsustainable due to reduced jhum cycle of 3-6 years owing to the increase in population that led to increase in food demand. This has caused decrease in productivity necessitated in bringing more virgin forest area under *jhuming.*

Jhuming is a tribe specific cultivation practice and varies widely in different parts of North East India. The system involves cultivation of crops even on steep slopes. Land is cleared by cutting of forests, bushes, etc. up to the stump level in December – January, leaving the cut materials for drying and finally burning to make the land ready for sowing of seeds of different crops before the onset of rains. The cultivation is confined to a village boundary and often after two or three years the cultivated area is abandoned and a new site is selected to repeat the process. The hutments of the village remain at the same place. Earlier whole village of some communities used to shift to the new site. After 2 – 3 years of cropping when the land losses its fertility farmers shift to another piece of virgin forest land for cultivation. After 3 – 15 years, when the vegetation in deserted land regenerates during the fallow period, the farmer again come back for farming on the same piece of land. With the rising population, the *jhum* cycle in most areas which used to be 10 – 15 years earlier has now reduced to 3-6 years only. Generally all the agricultural operations are performed manually, using only a few traditional and primitive tools. Earlier regeneration of forest and recouping of soil fertility within farming system are also achieved cost-free and effortlessly. The inherent resource degradation, low productivity, tendency to encourage large family size and little or practically no scope for adoption of modern agricultural technology are some of the drawbacks in this system (Christianty, 1986).

Extent of shifting cultivation in north east and India

According to the estimates by various agencies, shifting cultivation in North Eastern States varies between 2.80 – 7.40 million ha. According to task force on the shifting cultivation, Ministry of Agriculture 1983, the area under shifting cultivation is estimated at 3,869 Sq. km and number of families that depends on shifting cultivation for their livelihood is estimated at 60,7536 (Table 1). It is not only the source of livelihood but also accompany a high cultural importance among the people of the North east India. According to a survey of Government of Meghalaya, the average *jhum* area cultivated per family for both clear felling and *burn* cultivation had been estimated at 0.8 hectares.

Table 1. Area involved and number of families engaged in shifting cultivation in India

States	Total area (lakh ha.)	Fallow period (year)	Maximum area under shifting cultivation one time or other (sq. km)	Families involved (No.)	Districts
Arunachal Pradesh	2.61	3 - 10	2, 100	54,000	10
Assam	3.10	2 – 10	1, 392	58,000	3
Manipur	3.60	4 – 7	3, 600	70,000	5
Meghalaya	2.65	5 – 7	2, 650	52,290	5
Mizoram	0.45	5 – 4	1, 890	50,000	3
Nagaland	6.33	5 – 8	1, 913	116,046	7
Tripura	1.08	5 – 9	1, 115	43,000	3
Andhra Pradesh	1.03			23,200	4
Orissa	1.84			141,000	8
All India	22.78			60,7536	48

Source: Tomar *et al.* 2012

a) Impact of shifting cultivation: Resources, and Environment Protection

The shifting cultivation helps to conserve the rich cultural diversity as *jhum* is interwoven into the culture and tradition of more than 200 tribes inhabiting the North Eastern region. Shifting cultivation being a labour intensive and low subsidy based farming system provides an assured source of food security to the sustenance level for farmers of the region. All essential crops viz., rice (*Oryza sativa*), maize (*Zea mays*), tapioca (*Manihot esculenta*), colocasia (*Colocasia esculenta*), cucurbits, sweet potato (*Ipomoea batatas*), ginger (*Zingiber officinale*), finger millet (*Eleusine coracana*), cotton (*Gossypium* spp), tobacco (*Nicotiana* spp.) etc. are grown in the same field as mixed land use system resembling latest cafeteria system of cultivation. The multiple cropping with as many as 30 or more crop species on a plot of about 2 hectare can meet the varied needs of the family of an isolated community. A high leaf area index ensures optimal photosynthesis. Stratified root system of the crops ensures effective tapping of nutrients from the soil profile. The large non-edible crop biomass along with weed biomass recycled into the *jhum* system improves soil fertility. Since partial weeding is done, left out weeds also help in protecting the soil from losses of nutrients through sediment and water, along with the crop cover. The burning controls the weeds, reduces soil borne pathogen and corrects the soil acidity and makes soil fertile. The vast space is effectively managed through sequential harvesting all round the year. In the process, produce from a small piece of land fulfills almost all the needs of the farmers and minimizes dependency on external supplies. Under the North Eastern condition where difficult terrain with steep slopes cause hindrance in adoption of modern technologies, *jhum* provides base for low external input of agricultural technologies.

Jhum is an efficient system of agriculture from the viewpoint of energy. Unlike mechanized agricultural systems which consume five to ten units of fuel energy to produce a single unit of food energy, 17 to 20 times energy is obtained during first year cropping, and 13 to 15 times energy is obtained during second year cropping of rice. The energy efficiency may still be higher, from 41 to 48 (Toky and Ramakrishnan, 1982), or from 18 to 55 (Maikhuri and Ramakrishnan, 1991) under mixed cropping system. A typical *jhum* plot supplies the food and energy requirement of the tribal farmers through out the year is presented in (Table 2).

Table 2. Availability of food from typical *jhum* plot.

Month	Produce harvested
January February	Chillies (*Capsicum* spp.), brinjal (*Solanum melongena*), pumpkin (*Cucurbita moschata*), turmeric (*Curcuma longa*), ginger (*Zingiber officinale*), root crops, shoots, leaves and ferns from old *jhum* plot.
March	Bamboo shoots (*Dendrocalamus* spp) from new *jhum* plot that come after burning, chillies, brinjal, pumpkin, turmeric, ginger, root crops, shoots, leaves and ferns from old *jhum* plot.
April	Chillies, brinjal, pumpkin, turmeric, ginger, root crops, shoots, leaves and ferns from old *jhum* plot
May	Start harvesting watermelon (*Citrullus lanatus*), cucumber (Cucumis *sativus*), maize (*Zea mays*) from new *jhum* and continuing harvesting from old *jhum* field.
June	Celery leaves (Apium *graveolens*) and continues to harvest cucumber a n d water melon .No further harvest from old *jhum* field.
July	Harvest maize, beans, millet, pumpkin (Continues for 10 months) and early rice.
August	Main harvest of rice (*Oryza sativa*) in some areas. Continue to harvest vegetables, pumpkin etc.
September	Main harvest of rice in some areas. Continue to harvest vegetables, pumpkin etc.
October/November	Harvest vegetables, pumpkin, beans, soybean (*Glycine max*), and turmeric.

Source: Tomar *et al.* 2012

Munda *et. al.* (1996) described the short-term benefits from shifting cultivation. The *jhum* fire quickly render dense forests and foliage fit for growing crops. *Jhum* fire is a great labour saving device. Fire clears the area of extensive preponderance of fungi, insects and pests along with their larvae and eggs and acts as sterilizer. It also retards the weed growth in the *jhum* cleared by destroying the roots, tubers and seeds of weeds. The ashes correct the soil acidity and make the soil more fertile. Seeds in the fired warmed hill soils sprout early, *i.e.*, in the first fortnight of April.

(b) Socio – cultural aspects of shifting cultivation

All the operations of *jhum* are inseparably linked with the life of the tribal people. The religious rites and festivals of socio cultural importance viz, Agalmaka, Miamua, Rongchugala and Ahia Om Garo hills are celebrated in and around the *jhuming* practices The rationale behind the persistency of this system lies with its compatibility with the physico – social environment of sparse population, community land tenure system, undulating and steep topography, short crop cycle, meager resources etc. Village headman plays an important role in all round agro-economic development of the tribal farmers. Hence, training to the village headman regarding modern agricultural practices may help in adoption and diffusion of technological innovations. Tribal people also observe some major religious ceremonies with the *jhum* operations; particular mention may be made of worship and sacrifice before sowing seeds and festivals after harvesting is completed (Verma *et al.*, 2001).

(c) Soil fertility

As a result of fire nutrients locked in the organic matter may be deposited as ash on the soil surface. During burning nitrogen and sulphur, which are volatilized at relatively low temperature, are readily freed into atmosphere. Large amount of cations may be lost during burning by convection of ash Phosphorus and other cat ions viz, potassium, calcium, sodium and Magnesium are generally lost in low quantities because their volatilization temperatures are relatively higher (>500 °C). Burning of above ground vegetation showed an increase in pH and cations and a decrease in carbon, nitrogen contents in the surface soil. (Table 3). Quick release of nutrients especially cations after burning has been reported (Kellman *et al.*, 1985).

Table 3. Changes in the chemical properties of surface soil (0-5) after fire in two grassland communities of Cherrapunjee, Meghalaya.

Properties	*A.bengalensis*		*Ischaemum*	
	Before	After	Before	After
pH	5.80	6.20	5.60	5.90
Organic C (%)	2.35	2.13	1.80	1.78
N (%)	0.28	0.25	0.23	0.22
C/N ratio	8.40	8.50	8.00	8.10
NO_3-N (mg/100g)	0.28	0.29	0.23	0.24
PO_4-P (mg/100g)	0.16	0.18	0.13	0.14
K (mg/100g)	0.17	0.47	0.17	0.35
Ca (mg/100g)	1.28	2.10	1.10	1.34
Mg (mg/100g)	1.69	2.35	1.51	1.75

Source: Ram and Ramakrishnan, (1988)

Rainfall that follows fire causes heavy loss of nutrients. Major portion of nutrients is lost through run off water and relatively fewer amounts enter into the soil profile.

A study was conducted in bun cultivated areas of Mawpan watershed, Ri-Bhoi District, Meghalaya to find the effect of modified shifting cultivation (Closed burning) on soil properties and different forms of nitrogen. Results indicated a substantial increase in soil pH, effective cation exchange capacity (ECEC) and base saturation, while organic carbon and C/N ratio were reduced as a result of closed burning of organic residues irrespective of slopes (Table 4). After two years of bun cultivation, total-N and NO_3-N increased but available N, mineral N and NH_4-N were less than the initial. Different forms of N increased considerably after burning in the first year and thereafter reduced during subsequent years of cropping. Different forms of N were positively correlated with organic carbon and C/N ratio before burning but the same were negatively correlated with organic carbon and C/N ratio after burning. All the forms of N were positively and significantly correlated with pH, base saturation, EC and exchangeable cations (Ca, Mg, Na and K), but negatively correlated with exchangeable Al during the entire period of bun cultivation (Majumdar *et al*., 2004). Increase in basic cations resulted in the increase in pH in first year. In Second year, pH started decreasing and at the start of third year, it decreased again to almost its initial level. Removal of basic cations by crops and leaching, runoff and soil erosion losses from the hilly slopes may be the possible reason of such decrease in pH.

Table 4. Effect of closed burning on soil properties (At middle of the slope)

Land use Pattern	pH (1:2.5)	Organic carbon(%)	ECEC [cmol (p+)/kg]	Base saturation (%)	C/N ratio
1st year Bun (Before burning)	4.73	2.44	3.16	47.15	12.20
1st year Bun (After burning)	5.52	1.47	5.06	95.85	6.12
2nd year Bun (After 1st crop)	4.63	1.84	3.86	69.40	8.00
3rd year Bun (After 2nd crop)	4.62	1.99	4.30	66.00	9.04

Source: Majumdar *et al.*, 2004

(d) Erosion and nutrient loss

The serious adverse effect of *jhuming* is soil erosion, which is mainly of splash and washes types. As the soil in the upper reaches in a ridge are exhausted the cultivator's move to the adjoining lower elevation. The process continues till the entire ridge is exhausted. Nutrient losses from the *jhum* field through runoff and percolation are rather heavy during cropping. Soil erosion under shifting cultivation is highly erratic from year to year depending on rainfall characteristics. Studies on steep slopes (44 - 53 %) have indicated the soil loss to the tune of 40.9 tones per hectare and the corresponding nutrient losses per hectare were 702.9 kg of organic carbon, 145.5 kg of P_2O_5 and 7.1 kg of K_2O (Munna Ram and Singh, 1993). The soil loss from hill slopes (60 – 79 %) under first year, second year and abandoned *jhum* was estimated respectively to be 147,170 and 30 t/ha/year (Singh and Singh, 1981). According to an estimate annual loss of top soil, N, P and K due to shifting cultivation is respectively 88346, 10669, 0.372 and 6051 thousand tones respectively in the region (Sharma, 1998). Consequently the total production from this cultivation in paddy yield in Khasi hills (Meghalaya), Garo hills (Meghalaya), Khonsa (Arunachal Pradesh), and Siang (Arunachal Pradesh) are reported to be 1.28,5.04, 4.08 and 8.32 q/ha respectively. Singh *et al.* (1996), reported nutrient loss to the tune of 6.0 million tones of organic carbon, 9.7 tones of available phosphorus and 5690 tones of potash from the NEH region.

Although from bun cultivation good harvests of crops have been obtained, yet it leads to large amount of soil erosion. It was observed that for every one tone of potato produced by the system, the soil loss was 2 tones (Singh and Singh, 1981). It has been observed that as the time advances the horizontal spacing between the two beds goes on increasing due to loss of soil and the land is abandoned when soil is almost exhausted and even green grasses fail to grow some time exposing the bed rocks (Borthakur, 1992). Cultivation of tuber and rhizomatous crops cause soil erosion to the tune of 40 – 50 t/ha while pineapple along slopes eroded 24 to 62.6 t/ha/yr (Singh and Singh, 1981). The entire area in and around Shillong (Shillong plateau), Meghalaya is denuded by this system of *jhum* cultivation.

(e) Productivity

The total phytomass (above ground + underground) of crop and weeds at peak growth varied greatly between young (6 years fallow) and old (20 years fallow) *jhum* fields during first year cropping. The crop phytomass was 1.7 times higher in old (1473 g/m^2) than young (832 g/m^2) *jhum* field. Similarly, weed phytomass was 1.9 times higher in young (125 g/m^2) than old (66g/m^2) *jhum* field. When second year cropping was practiced, the crop and weed phytomass in young control plot was comparable to that during first year cropping (Table 5). But, in old control plot, crop phytomass decreased and weed phytomass increased significantly ($p < 0.01$) compared to that during first year cropping. The application of chemical fertilizers, FYM and their combination (CF+FYM) increased crop phytomass both in young and old *jhum* fields. The weed phytomass remained unaffected in young field but declined slightly in old field consequent to fertilizer application (Tawnenga *et al.*, 1998).

Table 5. Phytomass production in different treatment plots of young and old *jhum* fields during second year of cropping as compared to phytomass production in first year of cropping.

Ratio	Crop	Weed	Total
Young *jhum* field			
6:II: C/6:I: C	98	91	97
6:II: T/6:II: C	87	83	87
6:II: CF/6:II: C	157	95	151
6:II: FYM/6:II: C	145	102	140
6:II: CF+ FYM/6:II: C	150	97	144
Old *jhum* field			
20:II: C/20:I: C	69	264	77
20:II: T/20:II: C	99	79	96
20:II: CF/20:II: C	135	76	127
20:II: FYM/20:II: C	127	87	121
20:II:CF+FYM/20:II: C	128	93	123

I = First year, II = Second year, C =Control plot, T = Tilled plot, CF = Chemical fertilizer, FYM= Farmyard manure (*Source*: Tawnenga *et al*., 1998).

(f) Species Distribution and Biodiversity

The mixed cropping system involving 25-35 species and consisting of grain, seed, leaf and fruit vegetables, tuber and rhizome species is typical of the *jhum* system (Toky and Ramakrishnan, 1981; Mishra and Ramakrishnan, 1981).The reduction in the number of species under shorter cycles is apparently done in view of the reduced economic yield potential of the site, due to reduced fertility (Toky and Ramakrishnan, 1981). Maikhuri and Ramakrishna (1991) gave the higher important value indices (IVI) of crops in longer *jhum* cycles grown by Nishi tribe of Arunachal Pradesh (Table 6). The emphasis under longer *jhum* cycles, beyond 20 years, was more on grain and seed crops. Under shorter cycles, the emphasis was either on leaf and vegetables crops or on tuber and rhizome crops. Under short *jhum* cycle, such as 5-6 years, community is maintained more or less in permanent state of arrested succession (Kushwaha *et al*., 1981; Zinke *et al*., 1978). However, when succession progress for a longer period, such as 10 years or more, weed growth is suppressed by immigration of boreal elements (Saxena and Ramakrishnan, 1984). The vegetation of both young and old experimental *jhum* plots before slashing in Mizoram was of "Secondary successional type" with the dominance of *Melocanna bambusoides* (Tawnenga, 1990). The young field contained 45 species (16 tree, 16 shrub, 5 herb).The vegetation which grew during the intervening fallow period between first and second year of cropping was dominated by the sprouts of *Melocanna bambusoides, Ageratum conyzoides* and *Conyza auriculata* in both the fields.Herbaceous weed population was generally higher under shorter *jhum* cycles than longer ones. However, shrub and tree sprouts and saplings were more abundant under longer cycles.

Table 6. Importance Value Indices (IVI) of weeds under different *jhum* cycles in North-Eastern India.

Species	*Jhum* cycles (Year)				
	60	30	20	10	5
Herbs total	172.6	165.7	206.1	211.1	257.1
Shrubs total (Sprouts and saplings)	64.6	74.0	75.9	48.4	23.8
Trees total (Sprouts and saplings)	62.8	60.2	18.0	40.3	18.7

Source: Tawnenga, 1990

The clearing of forest areas at regular and frequent intervals results in loss of primary forests and formation of secondary forests and ultimately in loss of biodiversity. Due to shortening of *jhum* cycle, the secondary forest also do not get adequate time to regenerate. The repeated use of land with short *jhum* cycle finally converts the *jhum* fallows into degraded wastelands. After short *jhum* cycle the colonizers are mainly comprised of weedy species i.e. *Eupatorium odoratum, Imperata cylindrica, Mikania micrantha, Saccharum spontaneum* etc. Weeds particularly *E. odoratum* at lower elevations and *E adenophorum* at higher elevations often hold the ground indefinitely and form more or less permanent features of the landscape making the land unsuitable for agriculture and regrowth of other species. Under long cycles, the early colonizers within a few years are replaced by bamboo (*Dendrocalamus hamiltoni*), which comes up as sprouts from the underground rhizomes. During the process of *jhuming* a large number of edible vegetations are cut and burnt which cause great hardship to semi domestic and wild animals. A dense forest of long cycle has more tree than grasses; where as a forest of short cycle has more number of grasses (Singh *et al.*, 2000). Some of the major forest species affected in the process of shifting cultivation are *Dendrocalamus hamiltoni, Vitex penduncularis, Duabanga senersides, Goruga piñata, Terminalia belerica* etc.

Economics of shifting cultivation

Mishra and Ramakrishna (1981) studied the economic yield and energy efficiency of slash and burn agriculture, terrace cultivation, both of which involved mixed cropping and valley cultivation of a monoculture paddy, as practiced in higher elevations of Meghalaya and compared their economic yield and energetic efficiency patterns. A comparatively larger *jhum* cycle of 15 years was contrasted with two other of 10 and 5 years. From economic point of view, a 15 years cycle was found most efficient followed by a 10 years cycle; a 5 years cycle was extremely inefficient. While yield from valley cultivation was reasonable, terrace cultivation gave poor returns due to labor cost for terracing and maintenance, 15 years *jhum* cycle was found most efficient within an output/input ratio of 25.6 compared to 9.85 for 10 years cycle and only 4.63 for a 5 years cycle. Later on Tawnenga *et al.* (1997) reported that per hectare output in form of economic yield (rice grain) during first year cropping was more in old ie.20 years fallow (1586 kg or Rs.7931) than young (1462 kg or Rs.7312) field i.e. 6 year fallow. When second year cropping was done without any treatment, the output declined by 15% in young and by 20 % in old *jhum* field (Table 7). Fertilizer application (chemical fertilizer (CF)+ FYM) increased grain yield by 17% in young and 47% in old *jhum*. Thus economic yield obtained from CF+ FYM treated old field was comparable to that obtained from first year cropping in young field. The FYM treatment could enhance the economic yield up to only 5 % in young and 11 % in old field.

A study on the socio-economics of *jhum* at the lower elevations of Meghalaya (Toky and Ramakrishnan, 1981) revealed that net gain to the farmers from a 30 year *jhum* cycle is more than twice that from terrace cultivation, and that a 10 year cycle is also economically more viable than terrace cropping under comparable situation. Similar results were also obtained by Mishra and Ramakrishnan (1981) for the high elevation *jhum*. Singh *et al*. (2003) indicated that as *jhum* cycle increased the economic gains from crop production also increased. They suggested that the adoption of *jhum* cycles of 10 years is not only economically viable but also helps to sustain the hill agriculture, as increased in cycle above 10 years gave only moderate economic gains. Further, they have concluded that diversification towards allied activities particularly dairy piggery, agro-forestry and agro-based industries etc. would provide regular employment and income to the *jhumas* and also reduce human pressure on *jhum* cultivation. The examination of costs and return structure of some major crops grown by *jhumias* revealed that ginger gave the highest return per rupee investment followed by soybean. The return from paddy and maize, which dominated the cropping pattern of hill agriculture, was moderate. Benefit-cost ratio of 1.14, 1.17, 1.55 and 2.54 from paddy, maize, soybean and ginger from *jhum* cultivation in Churachandpur district of Manipur was reported by Singh *et al*. (2003).

Table 7. Output in terms of rice grain production (Kg/ha), energy (M.J/ha) and Income (Rs./ha) in different treatment plots in young and old *jhum* fields during first and second year cropping.

Output	I: C	II: C	II: T	II: CF	II: FYM	II:CF+FYM
Young *jhum* field						
Rice grain	1462	1240	1215	1375	1300	1450
Energy	21233	18003	17640	19973	18874	21052
Income	7312	6200	6075	6875	6500	7250
Old *jhum* field						
Rice grain	1586	1260	1247	1505	1402	1855
Energy	23030	18293	18112	21850	20362	26932
Income	7931	6300	6237	7525	7012	9275

I = First year, II = Second year,C=Control plot T = Tilled plot, CF = Chemical fertilizer, FYM= Farmyard manure

Strategy for sustainable development/alternatives of shifting cultivation areas

Sustainable agriculture in respect of NE region should entail development of managements systems that ensure adequate supply of food, fiber and fuel to the growing population. These systems must simultaneously ensure improving living standard of people by efficient utilization of all natural resources including land and water and external inputs in a practical and profitable manner while enhancing the environmental safety.

The concepts of development of sustainable agriculture take cognizance of the geophysical and environmental factors, which greatly govern and regulate the agricultural pattern to be adopted in situations, abound with above factors. In midst of all these above features of the North East India promise for potential development of agriculture including, horticulture, fishery, forestry, animal husbandry etc. A number of *jhum* control related development projects are being taken up. However, the present agricultural activities in this region are still not exposed by adequate scientific base in circumventing the land degradation process and scientific exploitation of water resources. There is therefore, an urgent need to further develop sustainable agricultural strategy for hill areas of North Eastern hill region to conserve soil, water and ecology while carrying out various agricultural practices. Various scientific studies and approaches suggest that mixed land use systems are better in the hilly areas, from the conservation as well as production point of view. Further, the system should be so designed as to meet the various needs effective land and water management techniques i.e., watershed management programme integrating soil conservation measures, land development, agriculture, plantation crops, horticulture, animal husbandry, fishery, forestry should be considered as vital and most important. Such actions must be carried out on a mission mode. Some of the sustainable farming strategies and alternatives are discussed as below:

Agricultural land use system

The agronomic crops can be adopted on hill slopes up to 50% gradient where soil dept is greater than 1.0 m. Contour bunding at 0.5 to 1.0 m vertical interval draining into a common grassed waterway is an essential requirement. The criteria for selection of crops should be based on the priority of crops that are already grown in the area, crops which have market potential such as spices and introduction of *rabi* crops such as mustard (*Brassica* spp), potato (*Solanum tuberosum*), pea (*Pisum sativum*), buckwheat (*Fagopyrum esculentum*) etc. in the irrigated area. Rice crop should be preferred in lower terraces. In general, ridge should be kept under fuel-fodder-timber trees, which can be planted, based on the requirement of farmers. On steep slopes about 30% of land is to be occupied under bunds and terrace risers. These areas have potential for producing of fodder crops. Amongst perennial grasses and legumes for the North East India condition,

Setaria sphacelata, Napier (*Pennisetum purpureum*), Guinea (*Panicum maximum*) and *Stylosanthes guyanensis* were found good for terrace risers.

Yield potential of rice, maize, millets, soybean (*Glycine max*), pigeon pea (*Cajanus cajan*), maize + soybean, maize + pigeon pea and maize + ginger under rainfed terraced condition at 3000 m altitude has been reported as 16.72, 21.26, 16.44, 5.25, 13.39, 83.40, 21.04 + 2.42, 17.26 + 13.51 and 17.56 + 39.10 q/ha, respectively (Awasthi, 1984). The cultivation of *kharif* crops i.e. maize, paddy, cowpea (*Vigna sinensis*), sesamum (*Sesamum indicum*), groundnut (*Arachis hypogea*), maize + cowpea, maize + soybean and during *rabi* season wheat (*Triticum aestivum*), potato, cole crops, turnip (*Brassica rapa*), tomato (*Lycopersicom esculentum*), etc. have been recommended in irrigated condition in the Nagaland situation. Paddy cum sericulture system for lowland can be practiced as this system was found more viable as the cash return were frequent with 4 cocoon crops in a year besides paddy yield.

Horticultural and plantation crops land use system

Slope of land for horticultural and plantation crops use should not be preferably more than 100%. Soil depth must be minimum 1.0 m. contour bunds at 2 meter vertical interval, half moon or crescent shape circle should be made at the location of planting, grassed waterways and making of few bench terraces at the lower slope towards foothills for growing vegetables and pineapple as essential conservation measures.

Some horticultural crops grown on hill slope gave extremely encouraging and economic return. Yield potential of newly planted Assam lemon (*Citrus* spp) orchard was found to be 11300, 12800 and 37200 fruits/ha during 3rd, 4th and 5th year after planting. It also arrests soil erosion, if field bunds/ borders are planted with economically important NTFPs bearing plants. The studies with pineapple cultivation, even when planted across the slope, resulted in soil loss to the tune of 24.0 – 62.6 t/ha/year during second year (Singh *et al.*, 1996).

The agroclimatic condition of North East India is ideal for cultivation of plantation crops. Tea (*Camellia* spp.), as a plantation crop was introduced in the region very early while other plantation crops such as coffee (*Caffea* spp), rubber (*Hevea brasiliensis*), arecanut (*Areca catechu*), black pepper (*Piper nigrum*), etc. were introduced initially with the idea of providing alternative method of livelihood for the farmers doing shifting cultivation. Various bodies such as National Committees for Agriculture and other commodity committees have recommended that in order to provide a better method of agriculture it would be useful to introduce plantation crops in the region as one of the important alternatives.

Agri–horti–silvipastoral land use system

This system comprises land use at the foothills with agricultural crops, horticulture in the mid portion of the hill slope and silvipastrol land use towards the top of the hill. Land up to 100% slope having soil depth greater than 1.0 m can be used for agri-horti-silvipastoral system. Contour bunds, bench terraces, half-moon terraces, grassed waterways and stilling basins are the conservation measures required for the treatment of land. An experiment was conducted by dividing watershed into three tier system viz., upper 1/3 area under pasture and silviculture for rearing livestock (goats and pigs), middle 1/3 area under horticulture including orange (*Citrus* spp.), guava (*Psidium guajava*) and pineapple (*Ananus comosus*) and remaining 1/3 lower area under agriculture for cultivation of cereals, pulses, vegetables, spices, fodder, *etc*. The middle portion of micro watershed of which 50% area was put under orange, 25% under guava and 25% area was put under Assam lemon. The economic evaluation of the agri-horti-silvipastrol system with guava as horticulture component has revealed that the system is economically viable (Rao, 1991). Agri-horti-silvi-pastoral system with livestock, dairy farming and Agro-pastoral system recorded a input out put ratio of 1:2.14, 1:2.08 and 1:2.05 respectively and was recommended as viable alternative to shifting cultivation in north east (Verma *et al*., 2001). Tree species in the system can also be one ones which produces NTFPs such as edible parts, gums, resins, tannins, dyes, medicine, fibers, fodders, etc. choice of species would be dependent upon local environment.

Multi-storey cropping land use system

This is highly productive, fully sustainable and very practicable system. To increase the cropping intensity multi-storey crop combination consisting of crop of varying canopy orientation and rooting have also been developed which entails differential harvesting of solar energy and utilization of soil fertility of variable depth based on the principle of canopy dimension and rooting pattern. One crop combination is coconut (*Cocos nucifera*) + black pepper + pineapple. In such high intensity cropping programmes a higher efficiency in utilization of solar energy in incident on a given area is obtained as it is intercepted at vertical intervals by the canopies of the crops (Singh and Singh, 1997).

Livestock-based land use system

For livestock-based land use system, the land up to 100% slope with minimum 0.5 m soil depth can be utilized for livestock farming. Contour bunds, trenches and grassed waterways are minimum requirement of land treatment. Crops and cropping pattern of such land uses will differ depending on the type of enterprise. The fodder production system has to ensure stability of fertility status of soil, availing the opportunity of moisture supply towards maximum fodder production for larger period during the year and conservation of fodder for lean season. Important grasses for the purpose are *Chrysopogon fulvus, Chloris gyana,*

Dichanthium annulatum, Panicum antidolate, sataria anceps, Sehima nervosum; legumes are *Atylosia scarabaeoides, Macroptilium atropurpureum, Slylosanthes gracilis, Clycine javanica* and important tree and shrub species are *Albizia chinensis, Albizia lebbek, Alizia procera, Artocarpus heterophyllus, Sesbania grandiflora, Sesbania sesban* for humid topics and for temperate and sub temperate areas, suitable trees and shrubs *are Betula alboides, Celtis australis, Morus serrata, Robinia pseudoacacia* etc., and grasses and legumes are *Bromus inermis, Dactylis glomerata, Poa pratensis, lespedeza ceraces, Lupinus augustifolis, Trifolium incernatum, Trifolium prantensis, Trifolium incernatum, Trifolium response* (Singh and Srivastava, 1990). Selection of leguminous and non leguminous annuals and perennials, shrubs and trees depend on the type of enterprises. Carrying capacity of such high land use has been estimated to be 4 to 5 livestock/unit/ha with setaria and stylo (1:1) mixture of fodder production. This system has potential for substantial income (1:1.78) from the farmyard manure and self sufficiency in fuel production through biogas plant. 90 % of annual rainfall could be retained in the watershed and soil loss was restricted to 2 t/ha/year (Verma *et al.*, 2001).

Water-harvesting for irrigation and fish production unit

The water resource potential in the North eastern Region is perhaps one of the largest in the entire country. The region, primarily because of the high rainfall, has also abundant ground water resources. The ground water resources, however, are influenced by geomorphology, culture, land use and surface hydrological features etc. There is maximum scope foe exploitation of ground water in Assam, Tripura and Arunachal Pradesh as these three states have very good scope for developing irrigation potential through ground water. Since ground water schemes are quick maturing and less costly than surface water schemes, it would be desirable to give priority on ground water development since more are could be covered under irrigation, within the limits of financial constraints, to improve agricultural production of these states which have low productivity at present. In the other states of the North East India, ground water may be developed in the small limited intermountain valleys or foothill areas, where the sediments are predominantly clayey with thin bands of fine sands. Fish production system has very high potential and can be adopted as subsidiary source of income. However, its applicability is limited in some cases only depending on the locational opportunities. Most of the water harvesting structures can be created by involvement of local manpower. Dug-out-cum embankment type of ponds can be created as seasonal or perennial water bodies for fish production. At individual and community level, this programme has tremendous potential not only in boosting the economy but also as immense value in developing water resources of the region.

Integrated approach to improvement of shifting cultivation

The following strategies for improving the system of Shifting Cultivation in north-eastern India was suggested by Ramakrishna (1992) based on a multidisciplinary study.

- The wide variations in cropping and yield patterns under *jhum* practiced by over a hundred tribes under diverse ecological situations should be continued, where transfer of technology from one tribe/area to another alone could improve the *jhum,* valley land and home garden ecosystems. Thus, for example, emphasis on potatoes at higher elevations compared to rice at lower elevations has led to a manifold increase in economic yield despite low fertility of the more acid soils at higher elevations.
- *jhum* cycle should be a minimum of ten years (this cycle length was found critical for sustainability when *jhum* was evaluated using money, energy, soil fertility biomass productivity, biodiversity and water quality as currencies) by greater emphasis on other land use system such as the traditional valley cultivation or home gardens.
- Where the *jhum* cycle length cannot be increased beyond the five-year period that is prevalent in the region, re-design and strengthen the agroforestry system incorporating ecological insights on tree architecture (e.g. the canopy form of trees should be compatible with crop species at ground level so as to permit sufficient light penetration and provide fast recycling of nutrients through fast leaf turnover rates. Local perceptions are extremely important in tree selection for introduction into the cropping and fallow phases of *jhum*, as can be seen in a major initiative in the state of Nagaland in north-east India.
- Improvement in nitrogen economy of *jhum* at the cropping and fallow phases by introducing nitrogen-fixing legumes and non-legumes. A species such as the Nepalese alder (*Alnus nepalensis*) is readily incorporated because it is based on the principal of adaptation o traditional knowledge to meet modern needs. Another such example is the lesser known food crop legume *Flemingia vestita*, traditionally used by tribes as an important species when jhum cycles decline below five years.
- Making use of some of the important bamboo species, highly valued by tribes, which can concentrate and conserve important nutrient elements such as N, P, and K. they could also be used as windbreaks to check wind-blown loss of ash and nutrient losses in water.
- Speeding up the fallow regeneration after *jhum* by introducing fast growing native shrubs and trees.

- Condensing the time-span of forest succession and acceleration restoration of degraded land based on an understanding of tree growth strategies and architecture, by adjusting the species mix in time and space.
- Improvement of animal husbandry through improved breeds of swine and poultry.

Conclusions

The shifting cultivation has become unsustainable primarily due to reduced *jhum* cycle owing to the increase in population pressure. In order to meet the growing food demand, the *Jhum* cycle is forced to get shortened which has resulted in the overall decrease of productivity. The most important negative environmental impacts of shorter *jhum* cycle realized are the degradation of biodiversity, deforestation, accelerated air pollution due to burning, soil erosion, loss of nutrients and reduction in useful flora and fauna and microbes. Considering the overall adverse impacts of the shifting cultivation, sustainable farming strategies and alternatives is the need of the day which ensures improving livelihood of people by efficient utilization of all natural resources including land, water and external input in a practical and profitable manner while enhancing the environmental safety. Integrated approach involving crop, animal husbandry, fishery, forestry with appropriate conservation measures for natural resources would be most effective in overall development of the shifting cultivation areas.

References

Anonymous 1997. State of Forest Report , Forest Survey of India, Dehradun.

Awasthi, R. P. 1984. Farming system research. Annual report. ICAR Research Complex for NEH Region, Umiam-793 103, Meghalaya.

Boerner E. J. 1982. Fire and Nutrient cycling in temperate ecosystems. Bioscience 32: 187-192.

Borthakur, D. N. 1992. Agriculture of the North Eastern Region with special Reference to Hill Agriculture. Beecee Prakashan, Guwahati.

Christanty, L. 1986. Traditional Agriculture in South East Asia. Westview Press. Boulder pp 240.

El Moursi, A.W.A. 1984. The role of higher agricultural education in the improvement of shifting cultivation systems in Africa. In: Bunting, A. H., Bunting, E.(eds.), The future of the shifting cultivation in Africa and the task of Universities, PP. 8-14, FAO, Rome.

Kushwaha, S.P.S., Ramakrishnan, P. S., Tripathi, R. S. 1981. Population dynamics of *Eupatorium odoratum* L.in successional environments following slash and burn agriculture (*Jhum*): J. Appl. Ecol.18:529-536.

Lewis, O. 1951. Life in a Mexican Village Tepoztlan restudied. University of Illinois Press, Urbana.

Maikhuri, R. K., Ramakrishnan, P. S. 1991. Comparative analysis of the village ecosystem function of different tribes living in the same area in Arunachal Pradesh in north-eastern India. Agric. Syst. 35:377-399.

Majumdar,B., Kumar,K., Saha,R., Satapathy,K.K., Patiram. 2004. Different forms of nitrogen as affected by bun system of cultivation on hill slopes of Meghalaya. Indian J. Hill Farm. (1& 2):9-14.

Moore, A. W., Jaiyebo, E. O. 1963. The influence of cover on nitrate and nitrifiable nitrogen content of the soil in a tropical rain forest environment. Emp. J. Exp. Agric. 31:189-198 (163).

Munda, G. C., Ghosh, P. K., Prasad, R. N. 1996. Adopt alternative land-use systems to shifting cultivation .Indian Farming, April: 10-14.

Munna, R., Singh, B. P. 1993. Soil fertility management in farming systems. Lectures notes, off campus training on farming system. Aizawl, Mizoram 5-7 October, pp. 46-50.

Mishra, B. K., Ramakrishnan, P. S. 1981. The economic yield and energy efficienciency of hill agro-ecosystem at higher elevations of Meghalaya in Northeastern India. Acta Oecol. 2:1369-1389.

Ramakrishna, P. S. 2000. An integrated approach to land use management for conserving agroecosystem biodiversity in the context of global change. Intern. J. Agric. Reso. Governance Ecol. 1(1): 56-67.

Ramakrishnan, P. S. 1992. Shifting Agriculture and Sustainable Development; An Interdisciplinary study from Nortyh-Estern India UNESCO-MAB Series.Paris.Parthenon Publ., Carnforth.Lanes (Republished by Oxford University Press. New Delhi.1993).

Ram, S. C., Ramakrishnan, P. S. 1988. Hydrology and soil fertility of degraded grasslands at Cherrapunji in North Eastern India. Env. Cons. 15: 29-35.

Rao, N.V. 1991. Ecopnomic aspects. Annual report. ICAR Research Complex for NEH Region, Umiam, Meghalaya, India.

Saxena, K. G., Ramakrishnan, P.S. 1984. Patterns of herbaceous vegetation development and weed potential under slash and burn agriculture (*Jhum*) in north-eastern India. Weed Res. 24: 137-144.

Singh, N.P., Singh, O.P., Jamir, N. S. 1996. Sustainable agriculture development strategy for North Eastern Hill Region of India. In: Shukla, S. P., Sharma, N. (eds.), PP 346-351., Mittal Publication, New Delhi.

Singh, N.P., Singh, A.K., Patel, D.P. 2000. Shifting cultivation and its alternate approaches towards sustainable development in North East India. J. North Eastern Council 20(2): 24-30.

Singh, A.K., Singh, S.K. 1997. Orchard management of fruit crops. Farmer and Parliament 35(9): 9-24.

Singh, Punjab, Srivastava, A. K. 1990. Forage Production Technology. IGFRI, Jhansi, Uttar Pradesh.

Singh, S.B., Datta, K.K., Ngachan, S.V. 2003. Economic analysis of shifting cultivation. IN: Bhatt, B.P., Bujarbaruah, K.M., Sharma, Y.P., Patiram (Eds.), Approaches for Increasing Agricultural Productivity in Hill and Mountain Ecosystem, pp. 323-330, ICAR Research Complex for NEH Region, Umiam, Meghalaya, India.

Singh, A., Singh, M.D. 1981. Soil erosion hazards in North Eastern Hill Region. Research Bulletin No. 10. ICAR Research Complex for NEH Region, Umiam, Meghalaya, India.

Sharma, U.C. 1998. Methods of selecting suitable land use system with reference to shifting cultivation in NEH region. Indian J. Soil Cons. 26 (3): 234-238.

Smith, W. H., Bormann, F. H., Likens, G. E. 1968. Response of chemoautrophic nitrifies to forest cutting. Soil Science 106: 471-473.

Srivastava, A.K., Gadekar, H. 1994. Shifting cultivation: Integrating indigenous knowledge for improvement, Eastern Ghats-India. CSWCTRI, Dehradun.

Srivastava, A.K. 1994. Shifting cultivation in Central Eastern India: Problem of suatainability. Agroforestry systems for degraded lands. pp.215-222. Oxford & IBH, New Delhi.

Toky,O.P., Ramakrishnan,P.S. 1982. A comparative study of the energy budget of hill agro – ecosystems with emphasis on the slash and burn system (Jhum) at lower elevations of north-eastern India. Agric. Syst. 9:143-154.

Toky,O.P., Ramakrishnan,P.S. 1981. Cropping and yields in agricultural systems of the north-eastern hill region of India. Agro-Ecosyst. 7: 11-25.

Toky, O.P., Ramakrishnan, P.S.1981a. Runoff and infiltration losses related to shifting agriculture in north-eastern India.Environ. Conserv. 8: 313-321.

Tomar, J.M.S., Das Anup, Lokho Puni, Chatuvedi O.P. and Munda G.C. 2012. Shifting cultivation in North eastern region of India-Status and Strategies for Sustainable Development. The Indian Forester 137(9):52-62.

Tripathi, R. S., Uma Shankar, Pandey, H. N. 1995. Present status and strategies for Ecorestoration of Degraded Ecosystem at Cherrapunji. In: Tiwari, B.K., Singh, S. (eds), Ecorestoration of Degraded Hills, pp. 25-30. Kaushal publications, Shillong.

Tawnenga, Uma Shankar, Tripathi, R.S. 1997. Evaluating second year cropping on *jhum* fallows in Mizoram, north-east India: Soil fertility. J. Biosci. 22:615-625.

Tawnenga. 1990. Studies on ecological implications of traditional and innovative approaches to shifting cultivation in Mizoram, Ph.D thesis, North Eastern Hill University, Shillong.

Tawnenga, Uma Shankar, Tripathi, R.S. 1998. Slash Once, Cash Twice. In: Sundriyal, R. C.,Uma Shankar, Upreti, T. C. (eds), Perspectives for planning and development in north eastern India, pp. 246-256, G. B. Pant Institute of Himalayan Environment and Development. Himavikas Ocassional Publication No.11, Almora, U.P.

Verma,N.D., Satapathy,K.K., Singh, R.K., Singh, J.L., Dutta, K.K. 2001. Shifting cultivation and alternative farming systems.. In: Verma, N.D., Bhatt B.P.(eds.), Steps towards modernization of agriculture in NEH Region, PP. 345-364, ICAR Research Complex for NEH Region, Umiam, Meghalaya.

Zinke, P.J, Sabhasri, S., Kunstadter, P. 1978. Soil fertility aspects of the 'Lua' forest fallow system of shifting cultivation. In: Kunstadter, P., Chapman, E.C., Sabhasri, S. (eds), Farmers in Forest, pp; 134-159, Hawaii: East-West Centre.

❑❑❑

3

Functional Roles and Services of Traditional Agroforestry Systems in the Sikkim Himalaya

Ghanashyam Sharma, Rita Sharma and Eklabya Sharma

Abstract: *The adoption of agroforestry systems has gained wider attention in the Multilateral Environment Agreements (MEA), international academia and policy makers for their multifunctional role and dynamics of ecosystem services. The Convention on Biological Diversity (CBD) has put forward ecosystem approach while Intergovernmental Panel on Climate Change (IPCC) emphasized agroforestry as an appropriate means to reduce emissions and enhance sinks of green house gases. Traditional agroforestry systems such as cardamom-based and farm-based practices are good examples in the Eastern Himalayas. We analyzed the management diversities and functional roles of these agroforestry systems for their productive and protective functions, biodiversity conservation, carbon sequestration, reduction of run-off rainfall, improvement of soil fertility, nutrient dynamics, nutrient release through decomposition, N_2-fixation, energetics and efficiencies. Use of N_2-fixing Himalayan alder (Alnus nepalensis) in large cardamom (Amomum subulatum) and Albizia* spp. *in mandarin agroforestry are traditional innovations. The agroforestry systems and practices are examples of how fragile and steep slopes of the mountains are converted into ecologically resilient productive zones. These adaptive traditional agroforestry provide options for sustainable management of natural resources from a unit land use types to a landscape level land use stages and offer opportunities for socioeconomic benefits. They need to be further strengthened to achieve the* *goals of sustainable development. The incremental benefits of agroforestry systems and ecosystem services bridge the linkage of upstream and downstream population and beyond to benefit for well being from goods and services to the present and future generations.*

Introduction

The adaptive management of upland cultivated and non-cultivated systems in the Eastern Himalayas by growing multipurpose trees and species and intercropping understorey crops and fruit trees in the private landholdings, livestock raising and protection of adjacent forests for variety of services is a traditional practice of the indigenous communities over many centuries. Such practices have recently been defined as agroforestry. It is now defined as a collective name for land use system and technologies involving trees combined with crops and/or animals on the same land management unit (Nair, 1993; Tamale *et al*., 1995; Ibrahim and Sinclair, 2005). The environmental services that agroforestry practices can provide, and especially their potential contribution to the conservation of biodiversity, have recently attracted wider attention among the agroforestry and conservation scientists (Mcneely and Scroth, 2006). The Intergovernmental Panel for Climate Change (IPCC, 2007) has also cited agroforestry as one of the options to reduce emissions and enhance sinks of green house gases. The IPCC Land Use Change and Forestry Report (2001) concluded that transformation of degraded agriculture land to agroforestry has far greater potential to sequester carbon than any other land use change. The ecosystems approach which is a strategy for the integrated management of land, water and living resources that presents conservation and sustainable use in the equitable way, has been emphasized by the Convention on Biological Diversity (CBD) during the Seventh Conference of the Parties (COP Decision VII/11, 2004). The Global community is now known to act on both mitigation and adaptation, and agroforestry has great potential to leverage these synergistically (World Agroforestry Center, 2006).

Most of the farming systems in Sikkim of the Eastern Himalayas are at the subsistence level and have evolved over the years from trial and error by the farmers to meet the demands for food, fodder, fuelwood and timber (Sundriyal *et al*., 1994; Sharma *et al*., 2006).

The traditional agroforestry systems in Sikkim such as the forest based and high value cash crop based practices could be the possible examples of approaches to help reduce deforestation, meet accelerating challenges of land use changes due to modern developmental policies or activities. These traditional agroforestry systems are dynamic as they are always evolving and modified. The large cardamom (*Amomum subulatum*) and mandarin orange (*Citrus reticulata*) are good commercial agroforestry products for generating high income while large number of cultivated crops, lesser known crops, fruits and tubers are for food security. The land management in catchments of mountainous regions like the Himalayas is essentially related to the ecosystem services provided by these areas to both upland and lowland population. Such ecosystem services specially involve the conservation of soil and water, the protection of land from soil quality deterioration, and the conservation of water for drinking and other farm uses (Sharma *et al.*, 2007)

In this chapter, functional roles of different agroforestry systems and practices, dynamics of management diversity, role of N_2-fixing trees, nutrient dynamics, energetics and efficiencies in agroforestry systems, and opportunities and ecosystem services will be discussed. Agroforestry systems in the Eastern Himalayas have high ecological adaptability, resilience and sustainability, provides greater economic opportunities and benefits, and supplies multiple goods and services for sustainability. The Sikkim Himalayan-traditional agroforestry systems are the result of indigenous innovations resulting from the rich agricultural biodiversity. They are the centre of *in situ* conservation of species richness and genetic diversity, of both managed crops and wild relatives, and associated trees.

Agro-ecological diversity and agroforestry system

The Sikkim Himalayan traditional agriculture system encompasses variations of agro-ecological zones that cover a range of ecosystem diversity extending between 300 m to 6000 m asl. The subtropical zones (300m above) houses rice cultivation systems in terraces and along the valleys. Above this are cardamom-based and farm-based traditional agroforestry in the subtropical to warm temperate zones (600–2500 m), extreme subsistence farming in the cool temperate and lower alpine zones (2500–4000 m), and the trans-Himalayan nomadic agro-pastoral to pastoral systems in the alpine Tibetan plateaus (4000–6000 m). The informations of the case studies presented in this chapter were carried out between the argoecological ranges between 600–2300 m. The study area is in the Indian monsoon region and has three main seasons: winter (November-February), spring (March-May) and rainy (June-October). The mean monthly maximum temperature ranged from 10 to 24 °C, the mean monthly minimum temperature from 3 to15 °C and rainfall from 2500 to 3500 mm. Relative humidity varies between 80 and 95% during the rainy season, decreasing to about 45% in spring.

The agroforestry diversity was analyzed based on management diversities, socio-ecological and socio-economical conditions, traditional knowledge systems, land use types and the land use stages associated to traditional farming and their functional roles. The analysis and classification was established based on multiple criteria such as function of trees, shrubs and the understorey multiple crops, production zones and also based on the structural dynamism, social economic benefits, ecological resilience and adaptability of the agroforestry practices (Table 1). Majority of agroforestry systems in Sikkim are practiced in private landholdings.

Table 1. Major traditional agroforestry systems under practice in the Sikkim Himalaya.

Agroforestry systems	Agroforestry practices
Agrosilvoanimal systems	***Farm-based Agroforestry***
	• Multilayered vegetation structure with fodder species, shrubs and understorey crop based garden agroforestry, multiple intercropping in terraced productive zones and woody perennials and multipurpose species at the edges of the farm landsfor soil erosion control, forming protective beltsand windbreaks·
	• Homegardens involving animal husbandry, traditional beekeeping, vegetable crops, protein banks, underutilized crops and lesser known crops
	• Trees and ground shrubs/herbs or other agroforestry species such as *Thysanolena* species for soil reclamation
Agrosilviculture systems	***Farm forest-based Agroforestry***
	• Multipurpose trees species for fodder, fuel and timber and bamboo groves, and animal feed bank and for other productive needs
	• NTFPs and minor forest products, and medical plants or other plantation crops and pasture lands, water sources conservation
	• Soil conservation and reclamation
Agrihortisilviculture	***High value cash crop based agroforestry***
Cardamom-based	1. systems agroforestry: *Alnus*-cardamom and Mix-agrofrestry tree-cardamom
	• Ecologically adaptive and socially accepted multifunctional tree crops such as *Albizia* spp., *Alnus nepalensis* and mix tree agroforestry species and understorey commercial cash crop large cardamom
	• Ground fodder species and NTFPS, medicinal plants
	• Trees, shrubs and ground grasses in soil conservation and reclamation
	2. Mandarin orange-based agroforestry
	• Multilayer arrangement of fruit orchard and fodder trees, intercropping of understorey traditional crops varieties, lesser known crops, protein banks, yams, taros *etc.* Vegetable crops, traditional beekeeping, animal husbandry and trees for fodder in the appropriately designed homegardens
	• Mandarin orange trees, multipurpose trees and N_2- fixing *Albizzia* spp.

(*Source* : Sharma G. unpublished)

Three major systems of traditional agroforestry practices in Sikkim are categorized as farm-based agroforestry practices, farm forest-based agroforestry practices, and high value crop based agroforestry practices.

Farm-based agroforestry

The farm-based agroforestry practices are primarily agrisilvicultural systems that comprise of homegardens, traditional beekeeping and livestock as the principal components of the farming system, multipurpose trees on edges of terrace crop production zones for growing a variety of landraces and local adaptive, underutilized crops and lesser known crops as protein banks. This type of agroforestry also comprises of multilayer tree gardens with fruit crops. The trees act as wind breaks and function as protective belts to the terrace risers along the edges of the crop production slopes called *sukha-bari* or terrace paddy fields. In addition, hedgerows of fodder species such as *Thysanolaena agrostis* and other ground grasses are grown for fodder and stabilizing the terrace edges.

In the farm-based agroforestry, farmers manage multipurpose tree species for fodder, fuel woodlots and timber, and for a variety of other direct and indirect uses within and around the cultivable land. Trees on the terrace risers are grown for soil stabilization, while farmers practice intercropping under tree canopies. Some such agrofrestry tree species grown at different altitudes in traditional farming systems are presented in (Table 2).

Table 2. Management of fodder trees and food production is critical in maintaining the livestock.

Agroforestry systems		Preferred Multipurpose Agroforestry Tree Species (fooder, fuelwood, timber)
Farm-based practices Elevation (m) Slope (°) Density (trees ha^{-1}) Basal area(m^2 ha^{-1})	 500–2000 25–30 250–300 3.5–6.2	*Albizia marginata, A. odoratissima, A. procera, A. lebbeck, Bauhinia purpurea, Alnus nepalensis, B. vahlii, B. variegata, B. malabarica, Brassiopsis speciosa, Gamblea ciliata, Citrus reticulata, Rhus similiata, Orozylon indicum, Grewia elastica, Melia azadirach, Garuga pinnata, Erythrina stricta, Morus alba, Terminalia myriocarpa, T. tomentosa, Garcinia stipulata, Artocarpus lacoocha,Echinocarpus aristatus, Elaeocarpus lanceae, Castanopsis hystix, C. indica, Delbergia latifolia, Erythrina sticta, Chukrassia tabularis, Litsea polyantha, Firminiana colorata, Buhinia purpurea, Sterculia villosa, Styrax serrulatum, Symplocus candata, Bridelia retusa, Callicarpa arborea, Elaeocarpus lancaefolius, Cordia* sp., *Psidium guajava, Schima wallichii, Ficus hookeri, F. roxburghii, F. elastica, F. benghalensis, F. nemoralis, F. hookeri, F. toona,*

(Contd.)

Agroforestry systems		Preferred Multipurpose Agroforestry Tree Species (fooder, fuelwood, timber)
		F. ciliata, F. semicarpa, F. hirta, F. cunia, F. benjamina, Saurauia napaulensis, Stereospermum suaveolans, Tamarindus indica etc.
Farm Forest-based practices Elevation (m) Slope (°) Density (trees ha^{-1}) Basal area (m^2 ha^{-1})	 500–2400 20–30 600–800 8.7–22.8	*Alnus nepalensis, Acer laevigatum, A. oblongum, Oestodus paniculata, Vibernum grandiflorum, Acer campbellii, A. levigatum, Juglans regia, Beilschmiedia sikkimensis, Engelhardtia spicata, Betula cylidrostachys, Chukrassia tabularis, Betula alnoides, Abies densa, Castanopsis hystrix, Cedrella toona, Brassiopsis mytis, Morus alba, Terminalia myriocarpa, Eruya acuminata, Exbucklandia populnea, Albizzia odoratissima, Symplocus ramisissima, Machilus odoratissima, M. edulis, Saurauia nepaulensis, Schima wallichi, Polynia indica, Acer papilio, Rhododendron arboreum, R. barbetum, R. falconeri, Pterygota alata, Brassiopsis speciosa, Bombax* sp., *Barberis aristata, Andromeda elliptica, Castanopsis tribuloides, Quercus pachyphylla, Q. spicata, Q. lamellosa, Leucoseptrum canum, Litsea* sp., *Machilus* spp., *Embilica officinalis, Spondias auxillaris, Magnolia campbelii, Michelia excelsa, M. champaca, Laurocerasus undulatus, Meliosma wallichii, Prunus rufa, Terminalia myriocarpa, T. tomentosa, Shorea robusta, Albizzia lucida, etc.*
Cardamom-based practices Elevation (m) Slope (°) Density (trees ha^{-1}) Basal area (m^2 ha^{-1})	 700–2200 20–30 300–550 8.2–30.2	*Alnus nepalensis, Albizzia* spp., *Acer oblongum, Schima wallichii, Toona ciliata, Ostodes paniculatus, Suymplocus theifolia, Eurya acuminata, Vibernum cordyfolium, Prunus nepalensis, Saurauia nepalensis, Juglans regia, Maesa chisia, Nyssa sessiflora, Litsea polyantha, Quercus pachyphylla, Leucoseptrum cannum, Lyonia ovalifolia, Bauhinia purpurea, Ficus* spp., *Osbeckia paniculata, Toona ciliata, Bassia butyracea, Celtis tentranda, Michelia excelsa, M. pustulata, M. indica, Quercus lamellosa, Q. ineata, Rhus semialata, Spondias auxillaris, Beilschmiedia* sp., *Cinnamomum* sp. *Alnus nepalensis, Schima wallichii etc.*
Mandarin-mix tree based agroforestry Elevation (m) Slope (°) Density (trees ha^{-1})	 600–1800 20–30 250–350	*Albizia marginata, A. odoratissima, A. procera,* practices *A. lebbeck, Bauhinia purpurea, Alnus nepalensis, Arthcarpus lakoocha, Morus indica, B. vahlii, B. variegata, B. malabarica, Brassiopsis speciosa, Gamblea ciliata, Citrus* Basal area (m^2 ha^{-1}) 6.3–20.5 *reticulata, Rhus similiata, Orozylon indicum, Grewia elastica, Melia azadirach, Garuga pinnata, Erythrina stricta,*

(*Contd.*)

Agroforestry systems	Preferred Multipurpose Agroforestry Tree Species (fooder, fuelwood, timber)
	Morus alba, Garcinia stipulata, Artocarpus lacoocha, Echinocarpus aristatus, Elaeocarpus lanceae, Castanopsis ystix, Erythrina sticta, Chukrassia tabularis, Celtis tentranda, Litsea polyantha, Firminiana colorata, Buhinia purpurea, Sterculia villosa, Styrax serrulatum, Symplocus candata, Bridelia retusa, Callicarpa arborea, Elaeocarpus lancaefolius, Cordia sp., *Psidium guajava, Schima wallichii, Ficus hookeri, F. roxburghii, F. elastica, F. benghalensis, F. nemoralis, F. hookeri, F. toona, F. ciliata, F. semicarpa, F. hirta, F. cunia, F. benjamina, Saurauia napaulensis, Stereospermum suaveolans, Tamarindus indica, Spondias auxullaris, Toona ciliata, etc.*

Source: Sharma and Liang 2006

The farm-based system is efficient land-use and resource utilization for sustaining the agricultural practices in sloping lands, recycling of organic dry matter for soil fertility maintenance and improvement of the agro-ecosystem performance and services. In addition to this, farmers also convert wastelands or abandoned lands to agroforestry that provides a variety of benefits such as green fodder during lean period.

Farm forest-based agroforestry

The farm forest-based agroforestry practices are agro-silvopastoral/agro-silvocultural systems comprising a diversity of multipurpose woodlots required for local construction and repairs, fuelwood, fodder for stall-fed livestock, bamboo groves, and pasture areas for seasonal grazing within the marginal landholdings. The forest-based practices of agroforestry are developed as support land. Some of the important agroforestry tree species cultivated at different agroecological zones along the vertical transact are given in Table 2. This system supplies nutrients and organic matter to the farm. Farmers often collect minor forest products such as fruits, fibres, medicinal plants, tubers, other wild vegetables. To overcome considerable shortage of fodder, most of the barren slope lands have been extensively converted into broom grass plantations locally called *amliso-bari* under the combination of multipurpose trees.

High value crop-based traditional agroforestry

High value crop-based agroforestry practice comprises of large cardamom (*Amomum subulatum*) based called *alainchi-bari* and mandarin orange-based (*Citrus reticulata*) called *suntola-bari*.

Cardamom-based agroforestry

About 70% of the cardamom-based agororestry practices are under N_2-fixing Himalayan alder (*Alnus nepalensis*) while 30% are under the mixed-tree agroforestry species. Some common shade trees for the mix-tree cardamom agroforestry are listed in Table 2. A total of 16,949 cardamom holdings have been recorded in Sikkim state, most of which are smaller than 1 ha. About 30% of the total area under cultivation is between 1-3 ha large (Sharma and Sharma, 1997; Sharma *et al.*, 2000). The cultivation area of large cardamom had increased by 2.3 times in the past 20 years that has now declined by almost 30% in the recent years although gap filling and replantation under the same canopy cover is a regular management practice. About 1316 ha of the reserved forest in Sikkim was used for under-canopy large cardamom cultivation on lease to farmers with no rights of cutting the trees (Singh *et al.*, 1989; Sharma *et al.*, 1994). However, most of these plantations were under protected areas and reserved forests which were withdrawn from the growers by the Department of Forest, Wildlife and Environment, Government of Sikkim during 2004-2006.

The striking constraint in the large cardamom farming has been the viral infestations locally called *Chirkey* and *Furkey* which has resulted into productivity decline in the recent years. The traditional *Lepcha* community incorporated disease resistant wild cultivars of cardamom locally called *Churumpho* in cardamom orchards infested by viruses. Two new local cultivars *Dzongu Golsai* and *Hee-Seramna* have been evolved out from the practice and are tolerant to viral diseases. Interventions on preparation of efficient, environment friendly and acceptable curing kiln, value addition to the produce at site and forward and backward linkages to the system for obtaining good outputs have been least successful and need to be strengthened.

Mandarin-based agroforestry

Another prominent and socio-economically regarded agroforestry is mandarin orange based farm agroforestry practice. Here *Albizia* is a N_2-fixing species extensively grown with other agroforestry species in the terraced cropland. Such agroforestry practices involves a number of multilayer fruit species, and mixed intercropping of maize, pulses, ginger, buckwheat, finger millet, pulses, oilseeds, taro and yam. Mandarin orange and ginger are potential cash crops of Sikkim after large cardamom. Mandarin orange is a high value, comparatively less labour intensive. Diversity of crops and other associate tree species are maintained in the system for other subsistence requirements and benefits. However, the system is considered as highly nutrient exhaustive as compared to the cardamom based system.

The interacting components and flow of services in a unit mountain farming system in Sikkim is presented in Figure 1.

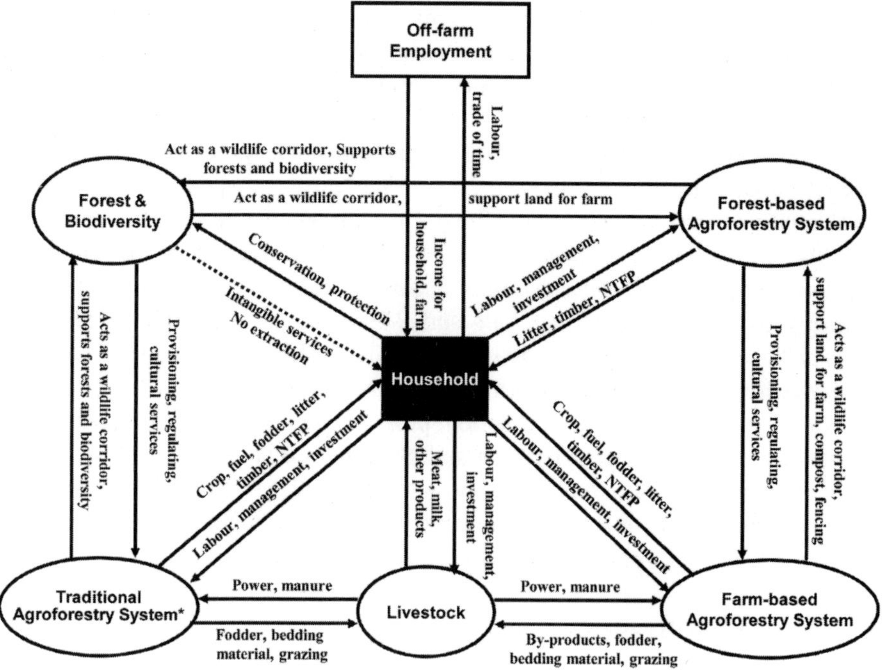

**Alnus*-cardamom agroforestry; *Albizia*-mix-tree-mandarin agroforestry

Fig. 1. Schematic model of interacting components in a traditional farming systems in the Sikkim Himalaya

Management diversity

Agrobiodiversity in traditional agroforestry

The unique mountainous terrain of the Sikkim Himalaya consist of varied physiographic factors and topography that allow location specific diverse biological microclimatic conditions giving rise to a range of ecosystem diversity rich in agrobiodiversity. The challenging mountain physical specificities have resulted into vulnerability whereas such biophysical features of diversified landscapes at a shorter distance provide conducive environment for genetic variations and landraces with sustained natural resource management. The heterogeneous landscape variability of Sikkim, location specific natural resource availability and socio-economic conditions has favoured a range of agroecosystem adaptation and production systems.

In Sikkim, the traditional ethnic communities– the *Lepchas*, *Bhutias*, *Limboos* and the *Nepalis* grow large number of landraces of rice, maize, buckwheat, beans, pulses, finger millets, yams and tubers, ginger and cardamom. The majority of the inter-cropping such as cardamom, mandarin-based intercropping, ginger-maize, paddy-soybean, pulses-turmeric, maize-potato and vegetable crops are practiced as understorey crops. The local high value cash crops in Sikkim are large cardamom, mandarin orange, ginger and potato. The conservation pattern of agrobiodiversity differs to different ethnic communities. One of the promising examples is the large cardamom which was first believed to be domesticated by *Lepchas* which was later on adopted by *Bhutias* and *Nepalese*. The cultivated species of *Amomum subulatum* is a native of the Eastern Himalayas with seven wild species such as *A. linguiforme*, *A. kingii*, *A. aromaticum*, *A. corynostachyum*, *A. dealbatum*, *A. costatum* and *A. plauciflorum* naturally occurring in the wild. Of its eight local landraces, *Ramsai*, *Sawney*, *Bharlang* and *Ramla* are cultivated above 1500 m whereas *Sawney*, *Chibey* and *Ramnang* are grown within 1000-1500 m and *Golsai*, and *Seramna* below 1000 m elevations along the vertical transect of the Sikkim Himalaya. Cultivation of large cardamom provides economic backup to maintain agrobiodiversity in the home gardens and also under other agrofrestry systems. The management of local agrobiodiversity includes all species and varieties important to the communities with emphasis on crops, plants and animal combinations, to reduce risks, increase production and enhance conservation.

Bio-geo-chemical cycling of N and P

Nutrient dynamics in Alnus-cardamom stands

Nutrient use efficiency may be expected to drop as utilization of that nutrient increases because availability of some other resource (such as water, energy, or light) limits production (Melillo and Gosz, 1983; Binkley *et al.*, 1992). The nutrient use efficiencies for both N and P in *A. nepalensis* pure monoculture plantations decreased with plantation age (Sharma, 1993). In the case of mixture of *Alnus* and cardamom plantations in the age series (5- to 40-year chronosequence), the nutrient use efficiencies were generally consistent with the above hypothesis and decreased with plantation age (Sharma *et al.* 2002a). *Alnus*-cardamom mixed stands used P less efficiently compared to pure stands of the same species of *Alnus* (Sharma, 1993). The *Alnus*-cardamom plantations also used N less efficiently compared to mixed *Alnus rubra*-conifer stands of USA. Comparison between *A. rubra* and conifers showed less efficiency in *A. rubra* than conifers (Binkley *et al.*, 1992).

The total N uptake in *Alnus*-cardamom agroforestry age was 90-239 kg ha^{-1} year^{-1} and 3.8-10.6 kg ha^{-1} year^{-1} for P, being lower for N and higher for P compared to monoculture plantations of *A. nepalensis* (Sharma, 1993). The estimated nutrient uptake in Himalayan oak forest was reported 230 kg ha^{-1} year^{-1} N and 13 kg ha^{-1} year^{-1} P (Rawat and Singh, 1988). These comparisons revealed that pure *Alnus* plantations showed higher N uptake and lower P uptake, and in the mixed stands cardamom N uptake decreased while P uptake increased. The low P uptake in pure *A. nepalensis* was attributed to a negative effect of *Alnus* on the P economy mostly by increasing soil acidity (Sharma, 1993), which causes a transition of phosphate into less soluble compounds with Fe and Al (Brozek, 1990; Sharma *et al.*, 1997). Furthermore, a heavy accumulation of organic matter in soils of pure *Alnus* plantation stands could have shifted P from a plant available pool to an organically bound pool (Sharma, 1993). The combination of *Alnus* with cardamom is a system where N and P uptakes are balanced compared to either pure stands of N_2-fixing species or non-N_2-fixing species. Therefore, plantation systems with mixture of N_2-fixing and non-N_2-fixing species like *Alnus*-cardamom are advantageous in balancing N and P cycling.

Nitrogen retranslocation of senescent *Alnus* leaves was positively related to stand age in the case of N, and negatively related in case of P. The retranslocation of N in young *Alnus* trees was minimal because this nutrient was sufficiently available through fixation; however, with advancing age the demand increased as the contribution from fixation decreased causing greater retranslocation. In the case of P, its demand for growth was high in younger stands where effective retranslocation was recorded. It decreased with advancing stand age. *Alnus* showed entirely different physiological behaviour for N and P at different ages, governed mostly by demand and availability of these nutrients (Sharma *et al.*, 2002a).

Turnover time in standing vegetation reflects the rate of nutrient cycling, and the mean turnover time of P was lower than N. The P turnover time remained between 1 to 2 years at all the plantation age, while N increased from 2 year in youngest stand to more than 8 years in oldest stand. The N turnover of the stand was mostly affected by *Alnus* component than the cardamom. Sharma (1993) also reported lower turnover time of P than N in an age series of pure *A. nepalensis* plantations; however they were greater for both N and P than the mixed *Alnus*-cardamom plantations. The turnover time of nutrients on the plantation floor however was slightly higher for P than N. This finding suggests that P cycling in vegetation was much quicker than N stands with mixture of N_2-fixing and non-N_2-fixing species.

Consistently high net primary production in the age series of *Alnus*-cardamom is in conformity with marked retention of nutrients by the plants over the annual cycles. Ratio of nutrient uptake and net energy fixation remained almost similar in all the ages of *Alnus*-cardamom plantations. However, N and P uptake per unit energy fixed was higher than results of the present study and the ratios increased with plantation age in monocultures of *A. nepalensis* stands (Sharma, 1993). Performance of cardamom under the influence of N_2-fixing *Alnus* in an age series of plantations with regards to nutrient use efficiencies, nutrient dynamics and cycling suggest the system to be sustainable up to 20 years, and adoption of replantation after 20 years for both *Alnus* and cardamom would be highly beneficial and sustainable (Sharma *et al.*, 2002a).

Decomposition studies

Decomposition studies carried out in the 40-year age chronosequence (5, 10, 15, 20, 30, 40-year) of *Alnus*-cardamom agroforestry revealed that the ash-free mass, nutrient and energy release during the first six months of decomposition were highest, which is attributable to the high rate of loss of labile fractions and most favourable environmental conditions for decomposition. The energy loss per unit area almost doubled in the 15 year-old stand compared to premature young and the oldest stands (Table 3). Each fraction of litter decomposes at a specific rate and is variably dependent on resource quality (C-to-N ratio, initial lignin-to-N ratio) and largely contributed by environmental factors. We conclude that the C-to-N ratio was better predicted to nutrient N release throughout the decomposition period, which can be compared to the report made by Seneviratne (2000). The initial lignin-to-N ratio and C-to-N ratio of the litter fractions played a more important role in determining the rate of litter decomposition than litter moisture and temperature. Osono and Takeda (2004) have also reported that the lignin-to-N ratio and lignin-to-P ratio are the indicators of N and P dynamics during decomposition. The relative loss rate of 60–90% of the ash-free mass, nutrients and energy in the present study were related to the lignin-to-N ratio, litter quality and environmental factors. In the growing age series until 20-year stand, N availability also increases through N_2-fixation by *Alnus*; during such situation high N concentrations may have an accelerating effect on lignin degradation and thus faster decomposition. The release of nutrients due to rapid decomposition from nutrient-rich litter of *Alnus* makes more nutrients available for uptake by associate cardamom, and with time, the nutrient cycle is expected to be accelerated, even in the non-N_2-fixing associate (Sharma *et al.*, 2008a).

Soil nutrient dynamics

There was a consistent decline of soil pH with agroforestry age in the age series of *Alnus*-cardamom plantations which followed the previous findings on red alder (Bormann and DeBell, 1981; Van Cleve and Viereck, 1972; DeBell *et al.*, 1983). Soil oganic-C, total-N, nitrogen availability, total-P, inorganic-P and available-P levels were highly seasonal and concentrations were higher at upper soil horizons that are readily available to the understorey cardamom crop for uptake. SOM was negatively related to pH but was positively related to moisture retention in soil. A similar trend of SOM and pH was reported by Bormann and DeBell (1981) and suggested that soil pH is strongly related to organic matter and moderately related to nitrogen weight. The increased SOM should improve soil tilth and stabilize soil N. The inclusion of N_2-fixing trees in an ecosystem often stimulates production and may lead to an increase in SOM; this increase can lower soil pH. The higher value of soil pH promoted for both N-nitrification and N-mineralization leading to increased N-availability in the younger agroforestry stands but considerably diminished in older stands by 30 to 40 years (Sharma *et al.*, 2008b). Soil organic

Table 3. Ash-free mass (AFM) remaining (g m^{-2}) and N, P (g m^{-2}) and energy (KJ m^{-2}) release after 24 months of litter decomposition in the 5-, 15- and 40-year-old *Alnus*-cardamom agroforestry stands.

Litter	5-year				15-year				40-years			
Fractions	AFM	N	P	Energy	AFM	N	P	Energy	AFM	N	P	Energy
Alnus-leaf	411	11.58	0.94	9688	664	17.49	1.76	16569	261	6.93	0.56	7499
Alnus twig	216	2.50	0.54	4868	538	6.32	1.41	12506	137	1.33	0.29	3018
Cardamom leaf	372	5.97	0.65	8286	705	8.36	1.44	14549	267	4.28	0.42	6243
Cardamom-pseudostem	384	2.10	0.48	6552	725	4.39	0.84	11401	243	1.98	0.29	4866
Total	1383	22.15	2.61	29394	2632	36.56	5.45	55025	908	14.52	1.56	21626

Source : Sharma, G. *et al.* 2008

matter (SOM) was negatively related to pH but was positively related to moisture retention in soil. A similar trend of SOM and pH was reported by Bormann and DeBell (1981) and suggested that soil pH is strongly related to organic matter and moderately related to nitrogen weight. The soil acidity increased with age, pH showed negative correlation with stand age and SOM. The nutrient dynamics across age groups vary depending on the successional stage of stand age limiting the soil nutrient availability for plant uptake after 20-years agroforestry. The performance of both alder and cardamom retarded after this age which is due to limitations of soil nutrient availability and dynamics.

The fluxes in the inorganic pools results through the difference in N-mineralization and N-nitrification of labile pools or from the differences in the rate of immobilization of released organic-N. N-nitrification rates were markedly higher in the 5-year stand and decreased in the advancement of age with significant variation while N-mineralization slightly increased with age (10 years) and decreased sharply thereafter. Further, N-nitrification was higher in winter than the rainy season when uptake by plants was lowest and chances of leaching losses were less. High rates of both N-mineralization and nitrification would be due to rapid immobilization and heterotrophic activity during winter season (Ramakrishnan and Saxena, 1984; Sharma *et al.*, 2008b). Low rates in the older stands are attributed to reduced amounts of SOM and its mineralization which in turn causes a decrease in N and P supply (Tiessen *et al.*, 1994). The potential net rate of N-transformation through N-mineralization and nitrification were more than twice in the younger agroforestry stands than the oldest and consequently more nitrogen was available for plant uptake in the younger plantations. The presence of alder increased the total-N pools and supply benefiting the understorey crop. Similar results were reported in mixed stands of *Eucalyptus saligna* and *Albizia falacataria* in Hawaii (Garcia-Montiel and Binkley, 1998). The increase of soil acidity with agroforestry age, decrease of N-transformation rates, sharp decline of soil nutrient pool sizes and limitations of organic-C, SOM, Total-N, nitrogen availability, total-P, and available-P in old stands clearly shows that the younger stands until 20 years are sustainable while older stands tend to show highly nutrient exhaustive systems (Sharma *et al.* 2008b).

N_2-fixing trees and Nitrogen fixation studies

Nitrogen fixation rates by *Alnus* root nodules were estimated in the age series of *Alnus*-cardamom agroforestry stands from three different site replicates between 1500–1800 m agroclimatic zones. Annual nitrogen fixation increased from 5-year (52 kg ha^{-1}) to reach highest at the 15-year stand (155 kg ha^{-1}) and then decreased and continued to be at a low level in 30- and 40-year stand (58 kg ha^{-1}) (Table 4). *Alnus* trees of 10 to 20 year stands added substantial amount of atmospheric nitrogen in the stands. Of the total N uptake (90–239 kg ha^{-1} $year^{-1}$) in the agroforestry age series, addition through biological fixation contributed 39–66%. Nitrogen fixation was consistently high (52–155 kg ha^{-1}) in younger agroforestry stands with corresponding high stand productivity rates (17–22 t ha^{-1}) and energy conversion efficiency (2.7–3.8 %), and significantly correlated with advancing age in the age series of *Alnus*-cardamom stands. The agronomic yield of commercial cardamom crop was proportionately high up to 20 year stand age that sharply declined in the older stands (Sharma *et al.*, 2002a). Earlier estimates of average annual fixation of nitrogen based on acetylene reduction assay was highest (130 kg ha^{-1}) in *A. rubra* stands while fixation value of 20 kg ha^{-1} was recorded in 15- to 20-year *A. sinuata* and *A. crispa* mixed stands (Binkley, 1981). Annual fixation was reported to be highest (117 kg ha^{-1}) in 7-year and lowest (29 kg ha^{-1}) in 56-year stand of pure *Alnus nepalensis* plantations (Sharma and Ambasht, 1988).

Table 4. Seasonal course of nitrogen fixation (kg ha^{-1}) by *Alnus* root nodules in the age series of *Alnus*-cardamom agroforestry based on acetylene reduction assays.

Experiments conducted	Stand age (years)					
	5	10	15	20	30	40
October-December	15 ± 2	35 ± 3	48 ± 5	41 ± 5	15 ± 3	15 ± 2
January-March	5 ± 1	14 ± 2	14 ± 1	8 ± 2	6 ± 1	5 ± 1
April-June	14 ± 1	25 ± 2	23 ± 3	19 ± 3	13 ± 2	11 ± 2
July-September	18 ± 2	54 ± 7	70 ± 8	44 ± 7	24 ± 3	28 ± 3
Total annual fixation	52	128	155	112	58	59

Values are means of three site replicates (± SE), *Source*: Sharma *et al.* 2002b

The use of N_2-fixing *Albizia* tree in large cardamom agroforestry at lower elevations (600-1200 m), mandarin based agroforestry and croplands contributed to soil fertility, and increased productivity and yield by 9–13% in rain fed land (Sharma *et al.*, 2001). The annual N_2-fixation by *Albizia* was comparatively higher in the large cardamom agroforestry (12 kg ha^{-1}) than other crop fields (3.34 kg ha^{-1}). *Albizia* grows well up to 1500 m altitude and its combination with cardamom and mandarin based agroforestry and other traditional agroforestry systems with local landraces of local crops are reported to be highly beneficial. N_2-fixing *Alnus* is extensively grown in agroforestrys systems above 1500 m. Plantation of multipurpose tree such as *Albizia* and *Alnus* in agroforestry systems are emphasized and found in the traditional practice (Sharma *et al.*, 1994a, 1994b).

Energetics and efficiencies of cardamom agroforestry stands

Energetics and efficiencies of cardamom agroforestry stands were studied in detail in the age series of *Alnus*-cardamom stands and cardamom grown under mixed forest tree species. Energy conversion efficiency (ECE) at the autotrophic level is the ratio of energy captured by vegetation to the photosynthetically active radiation reaching an area over a period of time expressed as percentage. The ECE increased from 5-year stand (2.72%) to peak at the 15-year stand (3.76%) thereafter decreased with stand age to a minimum of 1.3% in the 40-year stand. Relationships between ECE with stand age are strongly negative in all the three cases such as *Alnus* tree, cardamom and stand total. In all these three situations, curves showed slight increase in the younger stands and then sharply declined with older stands. Both production and energy conversion efficiencies were high in the younger stands and decreased with stand age to the lowest value at 40-year stand.

The ECE contribution was comparatively higher in *Alnus*-cardamom than forest-cardamom stands. The production efficiency showed a positive relationship with energy conversion efficiency. Energy fixation, storage, net allocation in agronomic yield, heat release and exit from the system were higher in *Alnus*-cardamom compared to forest-cardamom stands. Energy conversion efficiency and net ecosystem energy increment were also higher in *Alnus*-cardamom stands than forest-cardamom stands (Sharma *et al.*, 2002).

Energy utilized per kg N_2 fixed was 16′10^4 kJ in the 5-year stand which slightly decreased with stand age and remained almost similar throughout with the value of 12′10^4 kJ in the 40-year stand. Energy utilized per kg N_2 fixed dropped sharply from 5-year to 10-year stand and then more slowly thereafter. It showed negative relationship with plantation age that was converted into natural logarithmic form. Energy efficiency in N_2-fixation was lowest (64 g N_2 fixed 10^4 kJ^{-1} energy) in the 5-year stand and increased to be highest (84 g N_2 fixed 10^4 kJ^{-1} energy) in the 40-year stand. Efficiency in N_2-fixation between 10- to 30-year stands remained almost similar. Energy efficiency in N_2-fixation showed a significant negative relationship with production efficiency, indicating greater energy efficiency in N_2-fixation in the 40-year stand when production efficiency was the lowest (Table 5) (Sharma *et al.*, 2002b).

Table 5. Energy fixation, energetics and efficiencies in an age series of *Alnus*-cardamom plantations

Energy/efficiency	Stand age (year)					
	5	10	15	20	30	40
Energy storage (x 10^6 kJ ha^{-1})	1053	1435	2341	2288	2292	2635
Energy fixation (x 10^6 kJ ha^{-1} year^{-1})	322	389	444	305	261	154
Net energy allocation in agronomic yield (x 10^6 kJ ha^{-1} year^{-1})	2.19	4.47	6.02	7.05	3.40	0.78
Heat sink from the floor (x 10^6 kJ ha^{-1} year^{-1})	171	228	299	259	226	116
Energy exit (x 10^6 kJ ha^{-1} year^{-1})	2.20	56.56	56.40	18.80	22.18	28.20
Net ecosystem energy increment (x 10^6 kJ ha^{-1} year^{-1})	148.8	104.4	88.6	27.2	12.8	9.8
Energy accumulation ratio	3.27	3.69	5.27	7.5	8.78	17.11
Energy conversion efficiency (%)	2.72	3.29	3.76	2.58	2.21	1.30
Energy fixation efficiency (GJ GJ^{-1} leaf energy year^{-1})	6.25	4.59	4.39	4.28	4.17	3.62
Energy efficiency in N_2-fixation (g N_2 fixed 10^4 kJ^{-1} energy)	64	71	76	73	71	84
Energy utilized per kg N_2 fixed (x 10^4 kJ)	16	14	13	13	14	12

Source: Sharma *et al.* 2002a

Net annual energy fixation was reported highest in a young stand and declined sharply with plantation age on pure *A. nepalensis* in the region (Sharma and Ambasht, 1991). In the present study annual energy fixation and flow rates increased from 5-year stand to a peak at the 15-year stand and then sharply declined. This clearly indicated that energy flows and fixation were optimum for both *Alnus* and cardamom up to the 15-year stand age. The production efficiency and energy conversion efficiencies of the age sequence of *Alnus*-cardamom plantations showed a significant positive relationship where the younger stands performed more efficiently compared to mature stands. The production efficiency of the plantation was highest in the 5-year old stand that decreased with advancing age having similar trend in both *Alnus* as well as cardamom components. The energy fixation efficiency of cardamom decreased with advancing age which remained fairly high up to 20-year old stand supporting the system efficiency until this age. The energy efficiency of stands and separately for *Alnus* and cardamom components was highest in the youngest stand and decreased with advancing plantation age. The energy conversion efficiency of cardamom increased from the 5-year stand up to the 15-year stand and then decreased to the lowest efficiency in the 40-year stand. Energy accumulation ratio of *Alnus* increased with stand age almost reaching more than 5 times in 40-year stand compared to 5-year stand, whereas it was almost similar after 15-year age in the case of cardamom. The relationships of energy efficiency in N_2-fixation with production efficiency of the *Alnus*-cardamom plantations showed inverse function. Younger stands with higher production efficiency showed least efficiency in N_2-fixation. However, as the stands matured the production efficiency decreased and the system suddenly switched to increased energy efficiency in the N_2-fixation. The inverse relationships of production efficiency, energy conversion efficiency and energy utilized in N_2-fixation against stand age, and positive relationship between production efficiency and energy conversion efficiency suggest that younger plantations function as the most productive system, while the intermediate and mature plantations relatively less and least productive, respectively. Performance of large cardamom under the influence of N_2-fixing *Alnus* in an age sequence of plantation with regards to net primary productivity, agronomic yield, net energy fixation rates, production efficiency, energy conversion efficiency and energy efficiency in N_2-fixation suggest the optimal rotational 20 years age. Sharma *et al.* (1994a) suggested *Alnus* as an excellent associate with cardamom promoting higher performance compared to non-N_2-fixing mix tree associates. This study reveals that the *Alnus*-cardamom plantation system could be sustainable by adopting rotational cycle of 20 years.

Functional role and services from traditional agroforestry

The diverse traditional socio-cultural settings of the Khangchendzonga-Complex has been conducive to maintenance of agrobiodiversity and biodiversity of global significance; the unique ecosystems are providing services and outputs what is recently described as provisioning, regulating, and supporting to well being by the

Millennium Ecosystem Assessment (2005). A schematic representation of goods and services provided by the traditional agroforestry systems is given in Figure 2.

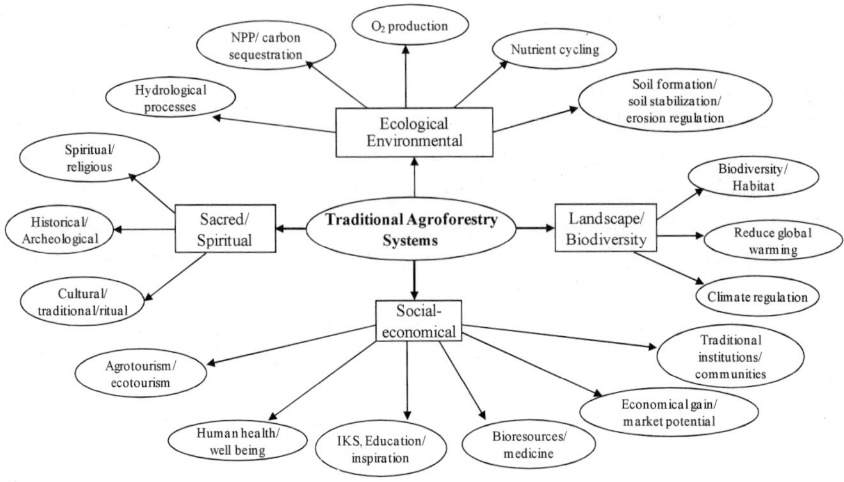

Fig. 2. Schematic representation of categorization of services and flow of resources and functions in a Traditional Agroforestry System in a mountain ecosystem in the Sikkim Himalaya. (Based on Millennium Ecosystem Assessment, 2005)

Biodiversity, wildlife habitat and PA corridor, landscape

Sikkim Himalaya is a part of the 34 globally significant biodiversity hot-spots and endemism in the Himalayas (CI, 2005) with rich agrobiodiversity as one of its principal components for livelihood security of the mountain people. It forms a part of the meeting ground of Indo-Malayan and Indo-Chinese biogeographical realms as well as Himalayan and Peninsular Indian elements that has given rise to a very rich biodiversity both wild and in cultivated landscapes. The conservation of biodiversity rich land spaces, forests, agroforestry systems, rivers and rivulets, hills and the valleys are attached to the rich culture of the communities deeply rooted and interwoven over several centuries. Landscapes form the functional unit for biological and ecological processes, since biodiversity and ecological services are normally delivered by landscape functions (Sharma, 2004).

In an account of shade trees in cardamom agroforestry and agroforests in the farms more than 92 multipurpose tree species were listed some of which are given in Table 2. The traditional communities in Sikkim practicing traditional agroforestry systems are symbolically represented by a large number of traditional practitioners, traditional knowledge systems, cultural festivities, customs and rituals, oral and traditional or customary laws. Thus culture and conservation form a basic principle to natural resources management and they consider almost all plant and animal species culturally important for a variety of reasons and benefits.

The traditional agroforestry systems such as the cardamom-based and farm-based practices have directly contributed to biodiversity through *in-situ* conservation of diversity of tree species, lower canopy shrubs, understorey vegetation, minor forest products and medicinal plants on the farms, reducing pressure on the adjacent forests. They provide suitable habitat for wild animal species that tolerate disturbance to a certain level on the farmland and large patches of agroforestry at a landscape level to support movement of these species. The agroforestry systems in the cultivated systems between the PA networks in the Sikkim Himalayan region serve as the biological corridor for the movement of wild animals designated as flagship species in the wider spaces along the Himalayas within India, and across international corridors and borders towards Bhutan in the east, Tibetan Autonomous Region of China in the north and Nepal in the west. Agricultural landscapes allow gene flow of the globally threatened and biologically restricted species and retain the biological connectivity of the discrete biological units (Sharma and Chettri, 2007). Wild biodiversity and traditional agroforestry, in the context of Sikkim Himalaya are continuous landscape elements which are characterized by the proximate interaction between natural systems and human dimensions. The high crop diversity and diversity of agroforestry systems are principal land use components for management of heterogeneous landscapes in the region that encounters the constraints of inaccessibility, fragility and marginality. The schematic model of interacting components in a unit traditional farming system is represented in Figure 2.

Protective functions

In a Himalayan watershed level land use system, overland flow, soil and nutrient losses (about 72%) were very high from the open agriculture (cropped) fields while cardamom-based and other agroforestry systems conserved more soil and nutrients under the traditional practices. Traditional interventions such as the use of multipurpose species e.g. broom grass and trees upon terrace risers, use of N_2-fixing trees *Albizia*, *Alnus* and *Erythrina* for maintenance of soil fertility on rain-fed slopes and plantation of horticulture trees have reduced the soil loss by 22% which is quite substantial in steep slopes. Soil and water conservation values in cardamom and broom grass based agroforestry were higher suggesting greater conservation potential for extensive large scale land use change from forestry to agriculture (Sharma *et al*., 2001). The adoption of sustainable agriculture and agroforestry models such as that of traditional practices of the Sikkim Himalayas has helped to reduce deforestation thereby minimizing pressure on forests and biodiversity due to human intervention. A diversity of agroforestry species, herb, shrubs and trees in the cardamom-based and forest-based agroforestry systems have been developed around the reserve forests (RFs), along the riparian habitats and adjoining to the protected areas (PAs) building up a large forest continuum thereby protecting biodiversity and forests. The agroforestry systems around the open cropped area act as windbreaks and considerably reduce soil degradation and landslide events. The local and indigenous communities have thus developed their effective watershed management techniques that involve food, fodder, fuelwood and timber production and biodiversity management in agricultural lands.

Socio-economic benefits

About 89 per cent of the population constitutes from the rural areas with their primary livelihood activity as agriculture in Sikkim. The small farmers with operational land holding of less than 2 ha constitute more than 70%, holding only 28% of the operational area, while 30% of the farmers hold 72% total operational area which shows a great disparity of land possession among the farmers in Sikkim (Subba, 2006). The per-capita land-holding in Sikkim has sharply declined over the years to 0.12 ha. All such situations cited above are causing socio-economic transition and a societal disparity. About 90 per cent of the farmers are small and marginal possessing total operational land holdings between 0.5–5 ha. Only 12.3% of the total geographical area of the state is available for agriculture while about 44% is under PA network.

Traditional agriculture system in Sikkim is still at subsistent level. The cash income earned from large cardamom in Sikkim increased from 1.9 million USD in 1975 to 13.8 million in 2005. Sikkim contributes about 40% of the world's production of large cardamom after Nepal. The total area under cardamom based agroforetry in Sikkim in 2007 was 26,734 ha of which area that gives agronomic yield were only 19,343 ha. The difference of 7,391 ha area was under replantation and gap filling. Similarly, large cardamom cultivation in the Darjeeling Hills during 2007-08 recorded to be 3,305 ha of which area that is providing agronomic yield is only 2715 ha. The total production during 2007 in Sikkim was 4,358 MT while production in Darjeeling was only 614 MT (Spices board, 2008). The market rate estimate as per local market value for Sikkim in 2007 is 13.6 million USD while for Darjeeling is 1.9 million USD. Pakistan is the single largest market importing up to 9,000 MT of large cardamom in a year (Srinivasa, 2006). It is exported to UAE, Iran, USA, Afghanistan, UK, Malaysia, South Africa, Japan, Argentina which are potential markets. The major domestic markets in India are Kolkata, Delhi and Guwahati. India is the largest market of large cardamom produced in Nepal and Bhutan. The per hectare income from *Alnus*-cardamom agroforestry was highest (2,175 USD) followed by forest-cardamom (1,275 USD) and mandarin-based agroforestry (1,165 USD) and was comparatively lowest (590 USD) farm-based (Sharma, G. unpublished data).

Watershed functions: Carbon sequestration and climate change

Agroforestry trees play a major role in sequestering carbon, the carbon stocks from different components like crop, tree, litter and soil were compared between the cardamom-based agroforestry system and the rainfed agroforestry systems with *Albizia*, *Alnus* and mix forest-tree species as shade trees. *Alnus*-cardamom system had 3.5 times more atmospheric carbon fixed compared to rainfed agriculture (Rai and Sharma, 2004).The burning of plant biomass as a management practice and decomposition of litter provide a sink for atmospheric CO_2 and also reduce emission of nitrous oxide. Promotion of such techniques that are familiar to small farmers such as crop rotation, cutting back chemical fertilizers through the use of composts can act as important sink for atmospheric CO_2 storing it below soil surface (Altieri, 2008). The total stand C stock in a Himalayan watershed measuring an area of 3,014 ha was 624 x 10^3 Mg while total carbon stock in 1 m depth soil was 456 x 10^3 Mg (Table 6) (Rai and Sharma, 2004). The total carbon stocks when compared in the cardamom based agroforestry systems showed highest stocks in the *Alnus*-cardamom stands followed by forest-cardamom and *Albizia* cardamom stands. *Alnus*-cardamom agroforestry had significantly higher floor litter carbon concentration than other stands. Contribution of litter and trees on carbon stocks was substantially higher in *Alnus* dominated stands. The contribution of soil carbon was highest in the forest cardamom stands. However, as a whole, the system under the nitrogen fixing *Alnus* showed higher carbon stocks compared to the other systems. The belowground biomass carbon in agroforestry stands also contributed more to total carbon storage (Sharma *et al.*, 1997). This greatly contributes to the carbon sequestration whereby this indigenous forest management system contributes to regulate climate change and reduce global warming (Sharma *et al.*, 2007). Lal (2004) has also clearly mentioned that agroforestry is an option of conserving soil and improving the soil organic carbon pools.

Table 6. Area-weighted total stand carbon in the mountain watershed in the Sikkim Himalaya. Values are in (x 10^3 Mg C).

Land use/cover	Vegetation	Litter	Humus	Soil	Total stand
Temperate natural forest dense	30.60	0.73	0.22	75.52	107.07
Temperate natural forest open	84.60	2.89	0.86	215.11	303.46
Subtropical natural forest open	32.70	1.10	0.23	45.60	79.63
Cardamom-based agroforestry system	4.91	0.60	0.13	29.30	34.94
Mandarin based agroforestry system	0.01	0.007	-	2.61	2.63
Open cropped area open	3.84	-	-	15.30	19.14
Open cropped area subtropical	4.15	-	-	24.30	28.45
Wasteland area temperate	-	-	-	40.00	40.00
Wasteland area subtropical	-	-	-	8.52	8.52
Total watershed	160.81	5.33	1.44	456.26	623.84

Source: Rai and Sharma 2004

Using agroforestry systems as carbon sinks, and by designing a suitable emissions carbon trading system, the Kyoto Protocol provides a new source of financial support for protection and management of biological diversity. Evolution of such mechanism would be a potential source of income to the vulnerable smallholder farmers in the Eastern Himalayas.

Sacred/cultural landscapes and aesthetic values

The Khangchendzonga landscape is traditionally described as *Beyul* (the hidden land) and *Ters* (the hidden treasure) which are linked to diversity of resources and ecological functions. The diversity of sacred landscapes extend from Sikkim on both sides along the Himalayas and has been referred to as *Beyul-Khepalung* often named for flora or fauna such as the *Dema-Dzong* (the valley of rice) for Sikkim mentioned in the Buddhist sacred text *Neysol*. The traditional conservation wisdom is based on intrinsic realization of paramount single entity consisting of the plants, animals, humans and the abiotic natural components. The Himalayan region is also described as *Ney Pemathang* or *Shangrila* (the hidden paradise on earth), and is worshipped in various religious-traditional festivals such as *Pang-Lhab-Sol* (mother deity Khengchendzonga) and *Tendon-Lho-Rum-Faat* (Tendong Hill) (Sharma and Liang, 2006). The sacred landscapes in the Ganga River System of the Central Himalaya, *Dema-Dzong* valley of the Eastern Himalayas, and the sacred mountains such as Holy Hills of the Dai tribe of Xishuangbanna in Yunnan Province of China (Ramakrishnan, 2000) are all examples of biodiversity conservation and natural resource management based on ITEK system.

Over the last two decades, Sikkim in the Eastern Himalayas has established itself as a recommended destination for contemporary tourists (Rai and Sundriyal, 1997). The scenic beauty, rich biodiversity, friendly people and rich culture are attracting two million tourists per year from all over the world. The mixed ethnic groups (*Lepcha*, *Bhutia* and *Nepali*) residing in the area with diverse traditional knowledge blended with culture and religions have made them a nature lover and to be in the wilderness (Sharma *et al.*, 2008). Conceptualised and inspired from Sikkim Biodiversity and Ecotourism project (Sharma *et al.*, 2002), the state is taking a lead role in diversifying and promoting ecotourism throughout the state and promote 'home stays' at remote areas as 'model villages'. These agroforestry systems have been the best possible land-use practices from a unit cultivated system to a landscape at spatial scale for their aesthetic, cultural, education and ecotourism values while they are the primary source of food and resource security.

The traditional agroforestry systems and practices arising from the lower altitudinal elevation of 600 m continuing up to 2500 m forming continuous forest ecosystem diversity with adjacent natural forests provide great aesthetic values for developing agro-tourism and ecotourism destinations. A large number of naturalists and scientists, educational and cultural tourists, pilgrims and nature lovers from across the world visit Sikkim every year. The region is becoming the attraction for many ecotourists mainly due to its wilderness in natural settings maintained by agroforestry and cardamom farming with rich traditional knowledge. They provide opportunities such as bird watching and trekking to the higher mountains through cardamom forests. The 'home stay' destinations such as Kewzing in south, Yuksam in west, Dzongu and Pasthang in north Sikkim have substantial area under cardamom cultivation and are promoted as a product of ecotourism package by travel agents as reported by the Ecotourism and Conservation Society of Sikkim.

Human well being

The traditional agroforestry systems of the Sikkim Himalaya provide economic and social well being to the communities (Sharma, 2006). The high value cash crops such as large cardamom, mandarin, ginger and medicinal plants are paramount important for fetching economically sound monetary benefits to the smallholder farmers for health care, education and social activities while the farm-based agroforestry provide essential products for subsistence needs such as food and nutrition. Apart from the scenic and aesthetic beauty, the recreational opportunities, mountain ecosystems with agroforestry are the main reserves of water systems for both agriculture and portable water. Agroforestry systems are the constant source of non-timber forest products, underutilized crops, clean air and water, and thus in all improving the quality of life of the communities.

The diversity of agroforestry systems are managed through organizational diversity that includes the diversity in the manner in which the farms are operated, owned and managed and the use of resource endowments from different sources in a traditional way. The management of such multifunctional agroforestry and domestication of useful species, livestock, food, air, water and material for shelter have contributed to well being of the smallholders and their society as the products of these systems are crafting the local and international market regimes.

Conclusion

In the context of several externalities of the fast changing world, policies and prospects, traditional agroforestry of Sikkim in the Eastern Himalayas show examples that can draw the global attention on sustainable management of resources, goods and services they provide and climate change related mitigation. The traditional agroforestry could be corrective option to address the challenges in cultivated systems. In socio-ecological and socio-cultural terms traditional agroforestry is a mixture of natural and man made adaptive ecosystems developed by mountain societies – an evolving dynamic landscape management system that supports human needs and ecological and economical sustainability to both upstream and downstream communities. Large cardamom and mandarin are high value cash crops that are regular source of income to small farmers while traditional crops, underutilized and lesser known crops provide food security and nutrition to mountain farmers.

Cardamom agroforestry has high potential to sequester carbon, retain biodiversity, support habitat to wildlife and above all it is considered as self sufficient system. Uses of N_2-fixing Himalayan alder and *Albizia* in agroforestry further the ecosystem services by providing appropriate shade to understorey crops, nitrogen to maintain soil fertility, and ecological role to accelerate nutrient cycling. Traditional agroforestry provide on-farm and non-farm employment opportunities such as production, and ecotourism and agro-tourism.

Traditional agroforestry of the indigenous communities is playing a vital role in securing human well being besides ecological and environmental sustainability. They support and restore adaptability, ecological resilience and enhance ecosystem services to the region and the world at large and contribute to sustainable development to the present and future generations.

Acknowledgements

Authors are thankful to the International Center for Integrated Mountain Development (ICIMOD) Kathmandu for providing facilities during the preparation of this manuscript. G. Sharma is thankful to Japan Society for the Promotion of Science, and United Nations University, Tokyo for funds and facilities provided during the postdoctoral research.

References

Altieri, M.A. 2008. Multifunctional Dimensions of Ecologically-based Agriculture in Latin America. Agroecology in Action; http://*www.cnr.berkeley.edu/%7Eagroeco3/multifunctional_dimensions.html*

Binkley, D. 1981. Nodule biomass and acetylene rates of red alder and sitka alder on Vancouver Island, B.C. Canadian J. For. Res. 11: 181-286.

Binkley, D., Sollim, P., Bell, R., Sachs, D., Myrold, D. 1992. Biogeochemistry of adjacent conifer and alder/conifer stands. Ecology 73: 2022-2033.

Bormann, B.T., De Bell, D.S. 1981. Nitrogen content and other soil properties related to age in red alder stands. Soil Sci. Soc. Am. J. 45: 428-432.

Brozek, S. 1990. Effect of soil changes caused by red alder (*Alnus rubra*) on biomass and nutrient status of Douglas-fir (*Pseudotsuga menziesii*) seedlings. Canadian J. For. Res. 20: 1320-1325.

CI. 2005. Global Hotspots Map. Washington DC: Conservation International. http:/*www.biodiversityhotspots.org/xp/Hotspots.*

COP Decision VII/11. 2004. Ecosystem Approach. COP Decision VII/11 Kaulalumpur, 9-20 February 2004. (*www.cbd.int/decision/2dec=vii/11*).

DeBell, D.S., Radwan, M.A., Kraft, J.M. 1983. Influence of red alder and chemical properties of a clay loam soil in Western Washington. Research Paper PNW- 313 Portland, OR: U.S. Department of agriculture, Forest service, Pacific North-West Forest and Range Experimental station, p 7.

Garcia-Montiel, D.C., Binkley, D. 1998. Effect of *Eucalyptus saligna* and *Albizia falcataria* on soil processes and nitrogen supply in Hawaii. Oecologia 113: 547-556.

Ibrahim, M., Sinclair, F. 2005. The History of Future of Agroforestry Research and Development: Policy Impacts and Needs. In: Forests in the Global Balance– Changing Paradigms, G.

Mery, R. Alfaro, M. Kanninen, M. Lobovikov (Eds.). IUFRO World Series Vol. 17, pp. 151-160.

IPCC. 2007. Summary for Policy Makers: Scientific-technical Analyses of Impacts, Adaptability and Mitigation of Climate Change. IPCC Working Group II.

Mcneely, J.A., Schroth, G., 2006. Agroforestry and biodiversity conservation- traditional practices, present dynamics, and lessons for the future. Bio. Conser. 15:549-554.

Millennium Ecosystem Assessment. 2005. Ecosystem and Human Well-Being. Biodiversity Synthesis. World Resources Institute, Washington DC.

Melillo, J.M., Gosz, J. 1983. Interactions of biogeochemical cycles in forest ecosystems. In: The major biogeochemical cycles and their interactions. In: Bolin, B., Cook, R. (Eds.), pp. 177-221. New York, Wiley.

Nair, P.K.R. 1993. An introduction to agroforestry. Kluwer Academic Publishers.

Osono, T., Takeda, H. 2004. Accumulation and release of nitrogen and phosphorus in relation to lignin decomposition in leaf litter of 14 tree species. Ecol. Res. 19: 593–598.

Rai, S.C., Sharma, P. 2004. Carbon flux and land use/cover change in a Himalayan Watershed. Current Science 86 (12): 1594-1596.

Rai, S.C., Sundriyal, R.C. 1997. Tourism development and biodiversity conservation: A case study from the Sikkim Himalaya. Ambio 26(4): 235-242.

Ramakrishnan, P.S. 2000. An integrated approach to land use management for conserving agroecosystem biodiversity in the context of global change. Inter. J. Agric. Res., Governance and Ecol. 1 (1): 56-67.

Ramakrishnan, P.S., Saxena, K.G. 1984. Nitrification potential in successional communities and desertification of Cherrapunji. Current Science 53: 107-109.

Rawat, Y. S., Singh, J. S. 1988. Structure and function of oak forest in central Himalaya. I. Dry matter dynamics. Ann.Bot. 62: 413-427.

Seneviratne, G. 2000. Litter quality and nitrogen release in tropical agriculture: a synthesis. Biol. Fert. Soil 31: 60-64.

Sharma, E. 1993. Nutrient dynamics in Himalayan alder plantations. Ann. Bot. 67: 329-336.

Sharma, E. 2004. Absolute advantage resources as potential neutralizers of globalization risks. International Centre for Integrated Mountain Development News Latter Mountain Risks and Hazards No. 40, Winter 2001, Kathmandu Nepal.

Sharma, E., Ambasht, R.S. 1988. Nitrogen accretion and its energetics in Himalayan alder plantations. Fun. Ecol. 2: 229-235.

Sharma, E., Ambasht, R.S. 1991. Biomass, productivity and energetics in Himalayan alder plantations. Ann. Bot. 67: 285-293.

Sharma, E., Chettri, N. 2005. ICIMOD's Transboundary Biodiversity Management Initiative in the Hindu Kush-Himalayas. Mountain Res. Dev. 25(3), 280-283.

Sharma, E., Rai, S.C., Sharma, R. 2001. Soil, water and nutrient conservation in mountain farming systems: Case-study from the Sikkim Himalaya. J. Env. Manag. 61: 123-125.

Sharma, E., Sharma, R., Singh, K. K., Sharma, G. 2000. A Boon for Mountain Populations: Large Cardamom Farming in the Sikkim Himalaya. Mountain Res. Dev. 20(2), 108-111.

Sharma, E., Jain, N., Rai, S.C., Lepcha, R. 2002. Ecotourism in Sikkim: Contributions toward conservation of biodiversity resources. pp. 531-548. In: Institutionalizing Common Pool Resources, (Eds. D. Marothia), Concept Publishing Company, New Delhi.

Sharma, E., Sharma, R., Sharma, G., Rai, S. C., Sharma, P., Chettri, N. 2008. Values and services of nitrogen-fixing alder based cardamom agroforestry systems in the eastern Himalaya. In : Smallholder Tree Growing for Rural Development and Environmental Services, (Eds. D. J. Snelder & Rodel D. Lasco), Springer Science Publications + Business Media B.V 2008 pp 391-40.

Sharma, G., Liang, L. 2006. The role of traditional ecological knowledge systems in conservation of agrobiodiversity: A case study in the Eastern Himalayas. Proceedings of International

Policy Consultation for Learning from Grassroots Initiatives and Institutional Interventions, 27-29 May 2006. Indian Institute of Management, Ahmedabad, India.

Sharma, G., Liang, L., Koji, Tanaka. 2006. On-farm agrobiodiversity management in mountain marginal farms in the Sikkim Himalaya. Proceedings of the International Workshop on Shifting Agriculture, Environment Conservation and Sustainable Livelihood of Marginal Mountain Societies, Guwahati, 23-25 September 2006.

Sharma, G., Sharma, E., Sharma, R., Singh, K. K. 2002a. Performance of an age series of *Alnus*-cardamom plantations in the Sikkim Himalaya: Nutrient dynamics. Ann. Bot. 89: 273-282.

Sharma, G., Sharma, E., Sharma, R., Singh, K. K. 2002b. Performance of an age series of *Alnus*-cardamom plantations in the Sikkim Himalaya: Productivity, energetics and efficiencies. Ann. Bot. 89: 261-272.

Sharma, G., Sharma, R., Sharma, E. 2008a. Influence of stand age in nutrient and energy release through decomposition in alder-cardamom agroforestry systems of the eastern Himlayas. Ecol. Res. 23: 99-106.

Sharma, G., Sharma, R., Sharma, E. 2008b. Impact of stand age on soil C, N and P dynamics in a 40-year chronosequence of alder-cardamom agroforestry of the Sikkim Himalaya. Pedobiol. (in press).

Sharma, H. R., Sharma, E. 1997. Mountain Agricultural Transformation Processes and Sustainability in the Sikkim Himalayas, India. Discussion Paper MFS 97/2. International Centre for Integrated Mountain Development, Kathmandu, Nepal.

Sharma, R. 2006. Traditional agroforestry and a safer mountain habitat. ICIMOD News letter-Sustainable Mountain Development in the Greater Himalayan Region. No. 50 Summer 2006, Kathmandu Nepal.

Sharma, R., Sharma, G., Sharma, E. 2002. Energy Efficiency of large cardamom grown under Himalayan alder and natural forest. Agrofor. Syst. 56:233-239.

Sharma, R., Sharma, E., Purohit, A. N. 1994a. Dry matter production and nutrient cycling in agroforestry systems of cardamom grown under *Alnus* and natural forest. Agrofor. Syst. 27(3): 293-306.

Sharma, R., Sharma, E., Purohit, A. N. 1994b. Dry matter production and nutrient cycling in agroforestry systems of mandarin grown under *Albizia* and mixed tree species. Agrofor. Syst. 29: 165-179.

Sharma, R., Sharma, E., Purohit, A. N. 1997. Cardamom, mandarin and nitrogen-fixing trees in agroforestry systems in India's Himalayan region. II. Soil nutrient dynamics. Agrofor. Syst. 35: 235-253.

Singh, K.A., Rai, R., Patiram, N., Bhutia, D. T. 1989. Large cardamom (*Amomum subulatum* Roxb.) plantations: an age old agroforestry system in Eastern Himalayas. Agrofor. Syst. 9: 241-257.

Spices Board. 2008. Preliminary crop estimate of large cardamom for 2007-2008. Official Record. Regional Office letter to the Director, Spices Board HQ, Gangtok.

Srinivasa, H.S. 2006. Large Cardamom Cultivation in India. Spices Board, Regional Office, Gangtok Sikkim, India.

Subba, J.R. 2006. SARD-M Study-Sikkim (India): Horticulture is an economically viable and environmentally sustainable driver of socio-economic development in mountainous Sikkim, Department of Agriculture, Government of Sikkim, pp 212, India.

Sundriyal, R.C., Rai, S. C., Sharma, E., Rai, Y.K. 1994. Hill agroforestry systems in south Sikkim, India. Agrofor. Syst. 26: 215-235.

Tamale, E., Jones, N., Riddihough, I. P. 1995. Participatory forestry in tropical and sub tropical countries. World Bank Forestry Series. Technical paper No. 299.

Tiessen, H., Cuevas, E., Chacon, P. 1994. The role of organic matter in sustaining soil fertility. Nature 371(6): 783-785.

Van Cleve, K., Viereck Leslie, A. 1972. Distribution of selected chemical elements in even-aged alder (*Alnus*) ecosytems near Fairbanks, Alaska. Arctic and Alpine Res. 4(3): 239-255.

World Agroforestry Center (2006). Agroforestry Science at the heart of three Environmental Conventions. Annual Report 2006. (www.worldagroforestry.org/ar2006).

❑❑❑

4

Traditional Agroforestry Systems on the Garhwal Himalaya

N.P. Todaria and Bhupendra Singh

Introduction

Agroforestry refers to practices, which deliberately or intentionally mix or retain woody perennials on the crop/animal production systems. It combines elements of agriculture, crops and/ or animals with elements of forestry in production system in a unit of land, either simultaneously or sequentially. The term woody perennials includes tree, shrubs, bushes, palm, bamboos, etc. which in agroforestry context are often referred to as multipurpose trees and shrubs (Wood, 1988). The word deliberate has significance a few trees remaining during the process of land clearance for agriculture is not agroforestry. To qualify as agroforestry, a system should actively promote the woody perennials for a particular purpose, or purposes, on the farm (Bhatt *et al.*, 2001). This land use system is particularly suitable for resource poor marginal farmers became diverse needs are fulfilled by one system with escape routes during natural calamities. In hill agro-ecosystems, agroforestry plays an important role in sustainable resource conservation and food security. It is also used to diversify and intensify farming systems through integration of indigenous tree-crops.

In broader sense, the major function of agroforestry are associated with sustainability for the farmers, stability of resources, including income, and minimization of risk. Sustainability is probably the most important function of agroforestry as it is a symbiotic relationship between the properties of the ecosystems and the management activities that result in relatively stable and/ or increasing outcomes.

Himalaya is a vast mountain system covering partly/ fully eight developing countries of south Asia: Afganistan, Bangladesh, Bhutan, China, India, Myanmar, Nepal and Pakistan. Agroforestry land use, covering 20% of the total geographical area of the Indian Himalaya is distributed as patches in the matrix of forests and agriculture covering 52% area (Nautiyal *et al.*, 1998). In Garhwal Himalaya out of 0.46 million ha of gross cropped area, agrisilvicuture, agrihorticulture and agrihortisilviculture systems occupy, respectively, 0.102, 0.03 and 0.026 million ha. It indicates that agrisilvicuture; agrihorticulture and agrihortisilviculture systems represent, respectively, 22.2, 6.5 and 5.7% area of the total cultivated land (Sachan, 2006). In traditional agrisilvicultural system, cereals, millets, pulses and oilseeds are intercropped with fodder and fuelwood yielding trees. These trees are either allowed to grow naturally or planted on the bunds of the agricultural fields. In this system, lopped foliage of trees is fed to the animals as fodder and pruned wood is used for fuel, while in interbund or interrow space annual crops are sown. The dung so obtained is applied to agriculture field as manure. It is a delicately woven organic agriculture system in mountains. On an average, 1.7 million tonnes of firewood is required per annum to meet out the energy requirements of rural folk of Garhwal Himalaya. The firewood production from different agroforestry systems has been recorded as 0.22 million tonnes/yr. It has been estimated that 17.87 million tonnes of green fodder is required to support 6.08 million populations of livestock. Agroforestry helps supplement 13% and 2% of the total firewood and fodder requirement. Rest of the firewood and fodder is collected from community/ private forest and/ or reserve forests (Sachan, 2006; Bhatt *et al.*, 1994, Negi and Todaria, 1993). Hence land development planning in the region needs to be based on integrated agroforestry and forestry systems rather than considering the two systems as independent or alternative land uses.

The Garhwal Himalaya has a long history of subsistence economy based on agriculture in which over 80% of the people are involved one way or other. Because of great variations in the altitude, topography, climate, tree resources, availability of irrigation water, and socio-economic and cultural factors, a variety of land use patterns exist in the region. If ecological conditions are superimposed on this then heterogeneity becomes more complex.

In the subsistence economy people are heavily dependent on tree resources. In addition to agriculture cattle rearing is another component of subsistence economy. But due to substantial increase in human and bovine population coupled with decreasing forests in the recent past, the very existence of the rural folk is being threatened. Recent estimates have shown that in Central Himalaya (Garhwal and Kumaon), dense and open forests constitute, respectively, 33.37 and 10.12% of the total geographical area (Anonymous, 1999). Thus, in addition to a low percentage of forest (43.49% Vs the prescribed 66.0% for the hills), the forest density is also low in this part of country (33.0%), whereas, for better soil and water conservation, the density of tree cover should not be reduced more than 50.0% (Anonymous, 1985). The available forest cover is not sufficient even to sustain the ecological balance not to speak of fulfilling the requirements of fuel, fodder and timber, etc. According to one estimate, the rate of firewood consumption for Central Himalaya has been recorded as 1.40-1.49 kg/capita/day (Bhatt *et al.*, 1994). Similarly the green fodder requirement has been estimated to be 10.34 kg/ cattle/day, irrespective of livestock reared (Sachan, 2006). Therefore, agroforestry needs to be strengthened to meet out the ever-increasing demand of fuel and fodder.

Traditional System

Mixed cropping, which is generally a characteristic feature of traditional societies throughout the world, and which is also found in the Garhwal Himalaya, has started receiving increased attention so that productivity per unit area can be increased. This technique is also significant for controlling weeds and pests and is effective for the recycling of biomass. Mixed cropping, along with the strengthening of agroforestry, effective water management, weed management, and the use of unused biomass through technological inputs, offers a considerable potential for augmenting the agricultural production of the region on a sustainable basis.

In Garhwal Himalaya, traditional agroforestry systems are well established since time immemorial. Planting and harvesting of wood products, fruits, roots, leaves and cattle fodder from trees has been done here since the earliest days of man's activities. Planting trees in land around agricultural field boundaries and homesteads aims to diversify the production and increase its magnitude. However, under present day conditions, the agroforestry techniques applied by the farmers appear to be poorly developed and do not suit present day needs. In most case, the trees are neither protected nor replanted nor managed. There is potential for improvement of traditionally managed systems to a great extent in order to realise the real production potential of agroforestry systems. Improvement is needed in both agricultural as well as tree crops. In Garhwal Himalaya, the common traditional agroforestry systems are Agrisilvicuture (Multipurpose trees and shrub + Agriculture crops), silvipastoral (Multipurpose tree + grass/ pastures), Agrihortisilviculture (Multipurpose trees and shrub+ Fruits plants + Agriculture crops), Agrihorticulture (Agricultures crops+ fruits plants). However, the components of the systems vary along altitudinal gradient. These variations are described briefly below.

Lower altitude (500 to 1000 m asl)

Along this altitudinal gradient, farmers usually cultivate ten multipurpose tree species (MPTs). On an average, 142 nos. of trees are cultivated per hectare of land in agrisilviculture system along this altitudinal gradient. Maximum density at species level has been recorded for *Grewia optiva* (35.0), followed by *Ougeinia oojeinensis* (24.0). The total tree fodder, fuelwood and fiber yield recorded was 21.1, 10.6 and 0.14 q/ha/yr, respectively. Average fodder yield was recorded highest (18.3 kg/tree/yr) in *Melia azedarach* and lowest (8.1 kg/tree/yr) in *Moringa pterygosperma*. Similarly, the fuelwood yield was recoded highest in *Celtis australis* (12.1 kg/tree/yr) and lowest in *M. pterygosperma* (3.0 kg/tree/yr). The fibre yield (extracted from *G. optiva*) was 0.40 kg/tree/yr). On an average, timber volume per tree was recorded highest (0.64 m^3) in *Toona ciliata* followed by *O. oojeinensis* (0.33 m^3). Lowest timber volume, however, was recorded in *Boehmeria rugulosa* (0.16 m^3). The total timber production was recorded highest in *O. oojeinensis* (3.7 m^3/ha) and lowest (1.7 m^3/ha) in *B. rugulosa.* (Sachan, 2006). The productivity of the grass from bunds of agricultural fields was 1.2 q/ha/yr (Table 1).

Table 1. Total productivity (q/ha/yr ±SD) in agrisilvicultural agroforestry system along an altitudinal gradient in Garhwal Himalaya

Uses	Altitudinal range (m asl)		
	500-100	1000-1500	1500-2000
Density/ha	142.1	135.5	102.0
Fodder	21.4±5.7	21.7±5.3	15.8±3.4
Fuelwood	10.6±3.0	11.8±3.0	7.9±1.8
Fibre	0.1±0.05	0.1±0.02	0.05±0.01
Timber volume (m^3/ha)	16.3±3.3	14.8±3.1	10.0±2.8
Bund grass	1.2±0.2	1.1±0.2	1.4±0.3

In this altitudinal gradient farmers grow four different crops or crop combinations in winter and nine nos. of crops during summer in Agrisilvicultural system. Wheat and barley are the major winter cereal crops, whereas, paddy, barnyard millet, finger millet, soyabean and black gram are major summer crops. Wheat is usually grown as a sole crop or intercropped with mustard and pea. Similarly, barley is grown either as a sole crop or in association with lentil. During summer, finger millet is intercropped with cowpea, horse gram, soyabean and amaranth and at some places as sole crop.

Middle altitude (1000 to 1500 m asl)

In middle attitudes also 10 multipurpose tree species (MPTs) are cultivated by the farmers in or around agricultural fields. Average tree density in this altitude was recorded 136 tree/ha. Maximum tree density at species level was noticed for *Celtis australis* (32.0), followed by *Ougeinia oojeinensis* (23.0) whereas, lowest tree density was recorded for *Ficus cunia* (5.2). The total tree fodder, fuelwood and fibre yield recorded was 21.7, 11.8 and 0.1 q/ha/yr. The average fodder yield was recorded highest (19.3 kg/tree/yr) in *C. australis* and lowest (8.5 kg/tree/yr) in *F. cunia*. Similarly, the fuelwood yield was also recorded highest in *C. australis* (10.6 kg/tree/yr) and lowest (4.6 kg/tree/yr) in *F. cunia*. The fibre yield (extracted from *G. optiva*) was 0.35 kg/tree/yr. On an average, the timber volume per tree was recorded highest (0.48 m^3) in *Toona ciliata*, followed by *M. azedarach* (0.28 m^3). Lowest timber volume, however, was recorded in *O. oojeinensis* (0.10 m^3). The total timber production was recorded highest in *T. ciliata* (5.0 m^3/ha), followed by *O. oojeinensis* (4.6 m^3/ha) and lowest (1.4 m^3/ha) in *M. azedarach* (Sachan, 2006). The productivity of the grass from bunds of the agricultural fields was 1.1 q/ha/yr (Table 1).

In this altitudinal gradient wheat and barley are the major winter cereal crops, whereas, paddy, barnyard millet and finger millet are the main summer crops. Wheat is grown either as a sole crop or intercropped with mustard. Similarly, barley is grown either as a sole crop or in association with lentil. During summer, finger millet is cultivated either as a sole crop or intercropped with cowpea, horse gram, soyabean *Cajanus cajan*, *Macrotyloma uniflorum* and amaranth.

Higher altitude (1500 to 2000 m asl & above)

The farmers of higher altitudes cultivate six multipurpose tree species (MPTs) in or around agricultural fields in agrisilvicuture system. On average, 102 nos. of trees are cultivated per hectare land in this altitudinal range. Maximum tree density at species level was noticed for *Quercus leucotrichophora* (37.0), followed by *C. australis* (19.1), whereas, lowest tree density was recorded for *O. oojeinensis* (7.3). The total tree fodder, fuelwood and fibre yield recorded was 15.8, 7.9 and 0.05 q/ha/yr, respectively. The average, fodder yield was recorded highest (21.4 kg/tree/yr) in *Q. leucotrichophora* and lowest in *Morus serreta* (9.1 kg/tree/yr). Fuelwood yield was also recorded highest in *Q. leucotrichophora* (11.5 kg/tree/yr) and lowest in *Ficus auriculata* (2.8 kg/tree/yr). The fibre yield (extracted from *G. optiva*) was 0.34 kg/tree/yr. On an average, the timber volume per tree was recorded highest (0.24 m^3) in *Quercus leucotrichophora*, followed by *O. oojeinensis* (0.15 m^3), thereby contributing 8.9 and 1.1 m^3 of timber per ha (Sachan *et al.*, 2006). The productivity of the grass from bunds of the agricultural fields was 1.4 q/ha/yr (Table 1).

In higher altitudinal range on average, three different crops or crop combinations are grown as winter crops and four nos. of crops during summer. Barley and wheat are the major winter cereal crops; whereas, paddy, barnyard millet, finger millet and foxtail millet are the major summer crops. Wheat is grown in association with lentil and mustard. During summer, finger millet is cultivated with amaranth. Growing beans as sole crop or in combination with cereals is major activity at many places. Similarly growing potato as sole crop is also characteristic feature in many areas.

Important agroforestry species of the region

Rural folk of Garhwal Himalaya, cultivate various tree/ shrub species in or around their agricultural fields, homegardens, and in fallow lands. As many as 43 tree/shrubs have been identified as multipurpose for the region (Bhatt and Verma, 2002). But 8-10 species are very common throughout the region and major studies have been carried out on them. A brief description of major agroforestry species is given below:

Bauhinia variegata L. Family: Caesalpiniaceae

Bauhinia variegata is a medium sized tree with an elongated spreading crown and bluish-green foliage. Bark is dark brown. Flowers are white or purplish and variegated. It is found distributed throughout the warmer parts of India. In Garhwal hills, it is common in mixed deciduous forests, and crop fields up to an elevation of 1200 m (Gaur, 1999).

Silviculture Methods

B. variegata is moderate light demanding species. It is drought and frost hardy and withstands dry and rocky sites. It is fair coppicer and can withstand heavy lopping. Pollarding is practiced for sustained harvesting of fodder. Seeds fallen on the ground germinate during rainy season and survival of seedlings is satisfactory. Seedlings establish quickly, owing to their resistance to drought and frost. Seed germinates 90 percent in laboratory (Nautiyal and Thapliyal, 1987). The leaf and bark extracts of *B. variegata* was found toxic to barnyard millet, finger millet, maize, cowpea and soyabean. Among these crops cowpea and soyabean were most susceptible and barnyard millet, finger millet and maize were most resistant to leaf and barks extracts of *B. variegata*. The leaf extract of *B. variegata* was found more toxic as compared to bark extracts (Kaletha *et al.*, 1996).

Uses

Fuelwood contains 18.3 kJ/g calorific value, 0.74 density (g/cc), 2.8% ash, 51.7% moisture, 0.61% nitrogen and 935 Fuelwood Value Index (Bhatt and Todaria, 1990c). Leaves provide good quantity fodder. Foliage contain, 13.3% crude protein, 29.1% crude fibre, 46.3% nitrogen free extract, 9.13% ash, 0.3% phosphorus and 2.9% calcium(Ghosh and Bohra, 1984). Crude fibre and crude protein contents increase in the leaves with maturity. Leaves also contain 2.02% total tannins with 1.04% condensed tannins (Lohan *et al.*, 1983).

Boehmeria rugulosa Wedd. Family: Urticaceae

Boehmeria rugulosa is small or medium sized evergreen tree, 8-10 m in height with short bole. Leaves elliptic or ovate-lanceolate (Gaur, 1999). Wood red and moderately hard, very smooth and even grained. *B. rugulosa* is common on open slopes, stream banks as well as crop fields between 700 –1200m asl. Once abundant in forests but now seldom found in nature because of heavy exploitation earlier. Natural regeneration is very poor as tree is pruned repeatedly year after year for fodder.

Silviculture methods

B. rugulosa is moderate light demander. It is hardy to drought and frost. It coppices and pollards profusely. It requires shade for regeneration and initial development (Luna, 1996). *B. rugulosa* can be propagated through direct sowing, nursery-raised seedling, coppice shoot and through branch cuttings. The branch cuttings of *B. rugulosa* treated with 500ppm IBA produced maximum (70 %) rooting (Bhatt and Todaria, 1990b). Nautiyal and Thapliyal (1987) observed 80 % germination in *B. rugulosa* seeds under laboratory conditions. Leaf extracts of *B. rugulosa* caused maximum reduction in *Hordeum vulgare* as compared to the *Triticum aestivum* and *Brassica compestris*. While bark extract was inhibitory to germination of *T. aestivum* as compared to the *B. compestris* and *H. vulgare*. Leaf extracts significantly reduced the radicle length in *B. compestris* and plumule length in *H. vulgare* as compared to the control (Todaria *et al*., 2005).

Uses

B. rugulosa is a excellent wood curving species and once utensils of this species were used to store milk products throughout these hills. Leaves provide palatable fodder, free from tannins (Bhatt and Badoni, 1995). On average, foliage contain 14.24% crude protein, 12.9% crude fibre, 43.3% nitrogen free extract, 4.89% calcium, 0.08% phosphorus, 0.82% magnesium and 1.23% potassium. It is also used as fuelwood, which contains 16.56 kj/g calorific value, 1.17g/cc density, 2.6% ash, 36.66% moisture, 0.53% nitrogen and Fuelwood Value Index of 2033.96 (Bhatt and Badoni, 1990).

***Celtis australis* L. Family: Ulmaceae**

Celtis australis is a moderate sized, deciduous tree, attaining 25-30 m height and 60-80 cm diameter. Flowers greenish, polygamons (Gaur, 1999). It is found distributed extensively in Uttarakhand at altitudes between 800-2500 m asl. *Celtis* is the only agroforestry species, which has probably widest range of ecological amplitude (Gaur, 1999; Luna, 1996).

Silviculture methods

It is moderate light demanding but seedling can withstand moderate shade. However, seeds require full light for optimum growth. It is a frost hardy species. It is fair coppicing and pollarding species. Seeds are shed in November-December and germinate in March or April. It is also raised by planting out nursery-raised seedlings. Vegetative propagation by branch cutting is one of the best methods for propagation of *C. australis* (Singh and Uniyal, 2005). Seed germination in *C. australis* is strong temperature dependent and 25 °C constant temperature induced optimum (46.00 %) germination (Singh *et al.*, 2004). This species has been found allelopathic to some crops. Bhatt and Todaria (1990a) reported that leaf extracts of *C. australis* did not affect the germination of *Eleusene coracana* and *Hordium vulgare*, while, germination of *Glycine max* was suppressed by *C. australis*. Bhatt *et al.* (1993) reported that leaf and bark extracts of *C. australis* inhibited the germination and radicle growth of *Dolochos biflorus*, *Glycine max*, *Phaseolus lunatus* and *Phaseolus mungo*. The leaf and bark extracts of *C. australis* also inhibited the germination of *Zea mays*, *Vigna unguiculta* and *Glycine max*. Leaf and bark extract of *C. australis* adversely affected the radicle growth of *Vigna unguiculata*, *Glycine max*, *Echnochloa frumentaceae, Eleusene coracana* and *Zea mays* (Kaletha *et al.*, 1996).

Uses

Its foliage is lopped for fodder and is rated as an excellent fodder based on its palatability and free from tannin contents (Lohan *et al.*, 1980; Sharma and Gill, 1969). On average, digestibility of protein of Celtis foliage has been noticed 63% (Sharma and Gill, 1969). The foliage contains 18.21% crude protein; 21.1% crude fibre; 14.3% ash; 4.24% calcium; 0.19% phosphorus; 0.57% magnesium and 1.11% potassium (Negi and Todaria, 1994). Its wood is also used as fuelwood, which contains 16.81 kJ/g calorific value, 0.54 g/cc density, 3.4% ash, moisture, 0.40% nitrogen and 464 Fuelwood Value Index (Purohit and Nautiyal, 1987).

Ficus cunia Buch-ham ex. Roxb. Family: Moraceae

Ficus cunia is evergreen tree, attaining a height up to 15 m. Bark is dark-grey and leaves alternate. It is common along the margins of forest, rocky slopes, bank of streams, edges of agricultural field up to an elevation of 1400m (Gaur, 1999).

Silviculture methods

It grows on variety of soils, but thrives well on loam soil. It is mainly propagated by seeds. It can also be multiplied through branch cuttings (Gupta, 1986). The leaf and bark extracts of *F. cunia* inhibited the germination and radicle growth of *Echnochloa frumentaceae, Eleusene coracana*, *Zea mays*, *Vigna unguiculta* and *Glycine max*. Leaf and bark extract of *F. cunia* adversely affected the germination of *Vigna unguiculta* and *Glycine max* more severely as compared to *Echnochloa frumentaceae, Eleusene coracana, Zea mays*, which showed little resistance against the leaf and barks extracts of *F. cunia* (Kaletha *et al.*, 1996). The higher concentration of leaf and bark extracts of *F. cunia* is more toxic to germination, radicle and plumule growth of *Triticum aestivum* and *Hordium vulgare*. Rhizospheric soil of *F. cunia* also reduced germination of *T. aestivum*. Reduction in root dry weight was also recorded maximum in *T. aestivum* in the rhizospheric soil of *F. cunia*, while rhizospheric soil of same species stimulated root dry matter production in *Lens culinaris*. On the other hand, maximum reduction in shoot dry weight was recorded in *L. culinaris* under rhizosphere soil of *F. cunia* (Singh *et al*., 2008).

Uses

Fruits are edible and also converted into Jam. Foliage provide excellent feed to livestock. Bark yield fibre, which is used for making ropes. Wood provides excellent fuel, which contain 17.7 Kj/g calorific value, 0.80g/cc wood density, 5.2 % ash, 51.0% moisture and 0.30% nitrogen with fuelwood value Index of 1470 (Bhatt and Todaria, 1992).

Ficus auriculata Lour. (Syn. *F. roxburghii* Wallich ex. Miq.) Family: Moraceae

Ficus auriculata is sub-deciduous tree attaining a height up to 10 m. Bark grey-warty. Young shoots hollow, leaves alternate, broadly ovate or sub orbicular, green, globose above and pubescent beneath. It is common around ravines, rivers bank, crop fields up to an elevation of 1500m asl. (Gaur, 1999).

Silviculture methods

It grows on variety of soils but thrives well on loam soil. It is propagated through seed and branch cuttings. It is also fair coppicer. This tree crop was found least toxic to germination and growth of barnyard millet, finger millet, maize, cowpea and soyabean (Kaletha *et al*., 1996). Leaf and bark extracts of *F. auriculata* were found least toxic to germination of *Triticum aestivum*, *Hordium vulgare* and *Brassica compestris* and more toxic to radicle growth of the same crops. The leaf and bark extracts of trees from lower altitude were more toxic to radicle growth of same crops as compared to leaf and bark extracts from higher altitude (Todaria *et al*., 2005). Rhizospheric soil under *F. auriculata* was toxic to germination of *L. culminaris* and *H. vulgare* in pot culture and root dry matter of *Triticum aestivum* and *Hordium vulgare* were also influenced by rhizospheric soil of *F. auriculata*.

Uses

Its foliage provide palatable fodder, which is lopped during winter throughout the hills. On average, its foliage contain 13.24% crude protein, 10.3% phosphorous and 0.415% potassium (Negi, 1995). Wood is also used as firewood, which contains 21.87kJ/g calorific value, 0.88 g/cc wood density, 3.5% ash, 57.72% moisture, 0.34% nitrogen with Fuelwood Value Index of 757 (Purohit and Nautiyal, 1987).

Grewia optiva **J.R. Drummond ex Burret in Notizbl (Syn. G.** ***oppositifolia*** **Buch.-Ham ex D. Don) Family: Tiliaceae**

Grewia optiva is deciduous tree, attaining a height up to 12m, leaves ovate-lanceolate. It flowers are in axillary or leaf opposed cymes (Gaur, 1999). *G. optiva* is found distributed in northwest Himalaya and adjacent plains of India up to an elevation of 1600m asl. It is frequently associated with agricultural fields and rarely found along margins of forests (Gaur, 1999).

Silviculture methods

G. optiva is strong light demander species but tolerates frost. It has a strong pollarding efficiency and also coppices well. It is mainly propagated by seed, transplanting of nursery-raised seedlings, by cutting or stumps (Chandra and Sharma, 1977). Due to hard seed coat, germination is poor. Seeds require pretreatments to improve germination (Uniyal *et al.*, 1999). Seeds soaked either in hot water for 48 hrs, or incubated in rumen fistula improved germination. Bark and leaf extract of *G. oppositifolia*, were found toxic to food crops. *G. oppositifolia* extracts greatly suppressed the radicle elongation of *Dolochos biflorus*, *Glycine max*, *Phaseolus lunatus* and *Phaseolus mungo* (Bhatt *et al.*, 1993). The leaf and barks extracts of *G. oppositifolia* significantly inhibited the germination and radicle growth of barnyard millet, finger millet, maize, cowpea and soybean (Kaletha *et al.*, 1996). The leaf and bark extracts of trees from lower altitude were more toxic to germination, radicle and plumule growth of *Triticum aestivum*, *Hordium vulgare* and *Brassica compestris* as compared to leaf and bark extracts from higher altitudinal trees (Todaria *et al.*, 2005).

Uses

It is an important agroforestry tree crop of the region. Its bark yields fibre used for making ropes. Firewood represents 16.87 kJ/g calorific value, 0.52g/cc wood density, 0.90% ash, 67.22% moisture, 0.62% nitrogen and Fuelwood value index of 1450 (Nautiyal and Purohit, 1987). Foliage is excellent fodder. Young leaves (March-May) contain up to 32% crude protein but it decreases remarkably with leaf maturation (Khosla *et al.*, 1992). On average crude protein digestibility has been reported 72% (ICAR, 1962).

Haldinia cordifolia **(Roxb.) Ridsdale in Blumea (Syn.** ***Adina cordifolia*** **Roxb.) Family: Rubiaceae**

Haldinia cordifolia is a deciduous tree, attaining height up to 25m. Flowers are cream-yellow, in globose pedunculate heads (Gaur, 1999; Luna, 1996). In Garhwal hills, It is common in open miscellaneous forests, along margins of crop fields and in sub Himalayan tracts up to an elevation of 1000m (Gaur, 1999; Luna, 1996).

Silvicultural methods

It is a strong light demanding and resistant to drought. Young plantations are tender to fire and seedlings are very much sensitive to drought and frost. Its natural regeneration from seeds is fairly well under favourable conditions and seeds get dispersed from wind from one place to considerable distance. Germination takes place early in the rainy season and on average 20 to 30% seedlings survive. Planting out of nursery raised seedling is most reliable method for cultivation of this species (Beniwal *et al.*, 1990)

Field soil mulched with dry leaf of *Adina cordifolia* suppressed the germination and growth of *Glycine max*. Leaf and bark extracts were not found toxic to growth attributes of cereals, millets and other food crops in bioassay study (Bhatt and Todaria, 1990a). However, bark extracts were found harmful to the germination and radicle extension of *Dolochos biflorus*, *Glycine max*, *Phaseolus lunatus* and *Phaseolus mungo*. Similarly, the leaf extracts of *A. cordifolia* significantly inhibited the germination and radicle extension of *D. biflorus*, *Glycine max*, and *Phaseolus lunatus* except *Phaseolus mungo* (Bhatt *et al.*, 1993) (Table 2).

Table 2. Chemical constituent and nutritive value index (NVI) of some agroforestry tree species

Agroforestry tree species	Availability	Lopping time	Crude protein	P	K	Ca	NVI
Bauhinia variegata	*High*	Oct-Dec.	16.92	0.41	0.31	5.03	0.02
Boehmeria rugulosa	Medium	Mar-May	14.24	0.08	1.23	4.89	0.15
Celtis australis	High	Apr. June	18.21	0.19	1.11	4.24	0.06
Ficus roxburghii	High	June-Sept.	13.24	0.19	0.41	2.06	0.07
Grewia optiva	High	Nov.-Mar.	20.21	0.24	0.45	3.26	0.14
Kydia calycina	Low	Mar.-Apr.	13.61	0.19	1.24	4.97	0.09
Ougeinia oojenensis	Low	June-Aug.	16.07	0.15	0.92	2.45	0.10
Quercus leucotrichophora	Medium	Apr.-May	9.39	0.07	0.42	1.60	0.04
Terminalia tomentosa	Low	Dec.-Feb.	14.2	0.08	0.67	4.56	0.10

Uses

Its foliage are not considered good fodder. Leaves contain 15.35 crude proteins, 12.7% fibre, 60.2% nitrogen free extract, 3.9% fat, 7.9% total ash 2.4% calcium and 0.26% phosphorus (P.I.D., 1988). Its wood is used as firewood, which exhibits 18.46 kJ/g calorific value, 0.68g/cc wood density, 0.70% ash, 63.25% moisture,0o.195 nitrogen with Fuelwood Value Index of 2835 (Purohit and Nautiyal, 1987). Wood is also used in house construction.

***Holoptelea integrifolia (Roxb). Planchon.* Family: Ulmaceae**

Holoptelea integrifolia is a large spreading tree, attaining a height up to 30m. Bark grey, leaves obliquely ovate. Flowers green, in numerous clusters on the leafless branches (Gaur, 1999; Luna, 1996). It is abundant in dry miscellaneous forests of sub Himalayan tract, and on the edge of crop fields up to an elevation of 1000m (Gaur, 1999; Luna, 1996).

Silviculture methods

It is a moderate light demanding species and grows on comparatively dry soils. *Holoptelea* is sensitive to frost. It coppices and pollards well. Seed is dispersed by wind in the summer and germination takes place during the rainy season. It is a slow growing species. Artificially it can be raised either by direct sowing or by planting out the seedlings raised in the nursery. The rate of germination in *H. integrifolia* as in most tree species seems strongly temperature dependents and 25°C constant temperature is most suitable for its uniform and faster germination (Singh *et al*., 2007). The leaf and barks extracts of *H. integrifolia* significantly inhibit the germination and radicle extension of leguminous crops *Dolichos biflorus*, *Pheseolus lunatus* and *P. mungo*. The leaf extracts of *H. integrifolia* completely inhibited the germination and radicle growth of *Dolichos biflorus*. Leaf extract is more toxic as compared to bark extracts (Bhatt *et al*., 1993).

Uses

It is not considered a good fodder tree and leaves contain 13.7% crude protein, 14.52% ash, 4.5% calcium, and 0.15% phosphorus. It is used as fuelwood, which contains 15.56 kJ/g calorific value, 0.54% g/cc wood density, 2.0% ash, 67.41% moisture, 0.57% nitrogen with fuelwood value index of 522 (Purohit and Nautiyal, 1987).

***Kydia calycina* Roxb. Family: Malvaceae**

Kydia calycina is deciduous tree of 15-20m height, leaves are suborbicular or ovate rounded. It flowers in axillary or terminal panicles. (Gaur, 1999). In Garhwal hills, it is common in miscellaneous forests of tarai-bhabhar belts, mixed forests and on the edges of crop fields up to an elevation of 1200m (Gaur, 1999).

Silviculture methods

Kydia is a strong light demanding species. It is fairly frost hardy and drought resistant. It coppices and pollards very well. Its natural regeneration is quite satisfactory. Seeds germinate after 1 month of sowing and the germination is more under moist places than those of dry places. Changing colour is the best index for field collection of fruits. Dark brown coloured fruit are best for collection and produce maximum germination (Negi and Todaria, 1995). It may be propagated by direct sowing or planting out methods (Chauhan *et al.*, 1993). The Kydia seeds give 46.67% germination at 25°C constant temperature (Negi *et al.*, 1995). Aqueous leaf and bark extracts of *K. calycina* are phytotoxic to germination of *V. unguiculata* and *Glycine max* as compared to *Zea mays* and *E. coracana* while *E. frumentaceae* was found more resistance to toxic response of this tree crops. Radicle extension of crops (*E. frumantaceae*, *E. coracana*, *Z. mays*, *Vigna unguiculata*, and *G. max*) was significantly reduced by aqueous leaf and bark extracts of this tree crops (Keletha *et al.*, 1996).

Uses

It is a good fodder species and foliage contain 13.6% crude protein, 19.6% crude fibre, 14.6% ash, 4.97% calcium, 0.19% phosphorus, 0.925 magnesium and 1.24% potassium (Negi and Todaria, 1994). Firewood exhibits 17.2kJ/g calorific value, 1.5% ash, 48.55 moisture percent, 0.72% nitrogen with fuelwood value index of 768.26 (Negi and Todaria, 1993).

Moringa oleifera Lam. (Syn. Pterygosperma gaertner) Famiy

Moringaceae, *Moringa oleifera* is deciduous tree. It attains a height of 10m. Flowers pedicelled, about 2.5cm across, white or pinkish, honey scented (Gaur, 1999). In Garhwal hills, it is commonly found along the agriculture field and in miscellaneous forests of sub-Himalayan tracts up to an elevation of 1200m asl. (Gaur, 1999).

Silviculture methods

It is a strong light demander. It coppices and pollards very well and is hardy against mechanical injuries like cutting/brushing and also against frost/fire. *Moringa* can be propagated by direct sowing, stump planting or by branch cuttings. The leaf extracts of *M. oleifera* completely inhibited the germination and radicle extension of *Dolichos biflorus* and *Phaseolus lunatus*. Similarly bark extracts of *M. oleifera* completely inhibited the germination and radicle extension of *Dolichos biflorus* (Bhatt *et al.*, 1993).

Uses

It is lopped for fodder. Foliage contains 15.6% crude protein, 17.9% crude fibre, 48.7% nitrogen free extract and 13.43% total ash. Its foliage is rated as good fodder having high rate of palatability. Its wood is also used as firewood, which contains 16.86 kJ/g calorific value, 0.39 g/cc wood density, 1.8% ash, 71.77 % moisture, 0.34% nitrogen with fuelwood value index of 508 (Purohit and Nautiyal, 1987).

Ougeinia oojeinensis (Roxb.) Hochreutiner (Syn. _O.dalbergioides_ Benth.)
Family: Papilionaceae

Ougeinia oojeinensis is a medium sized deciduous tree. It attains a height up to 12m. Leaf is redish. Flowers pale pink or pinkish, (Gaur, 1999; Luna, 1996). In Garhwal hills, it is common along exposed slopes and crop fields up to elevation of 1500m asl (Gaur, 1999). Once it formed dense forest stands in these hills but because of its continuous use as fodder and for making utensils especially to store milk products, it is now restricted to small patch at some places only. Because of heavy lopping for fodder, natural regeneration is negligible.

Silviculture methods

Ougeinia is moderately light demanding species and the tree is resistant to drought and frost. It coppices well and produce root suckers in large numbers. Under natural condition, germination in this species is profuse and seeds germination early during rainy season. Artificially, the species may be propagated by root cuttings, stump-planting or by young nursery raised entire seedlings with ball of earth. Soaking of seeds in water before sowing may hasten or improve germination. In laboratory its seed germinate 100 percent but under field conditions only 87 percent germination was recorded (Nautiyal and Thapliyal, 1987). The bioassay study of *Ougeinia* leaf and bark reveals no significant reduction in the germination percentage of food crops. However, leaf and bark extracts of *Ougeinia* were found toxic to plumule and radicle growth of many crops (Negi *et al*., 2007). In pot culture, germination of *Triticum aestivum* and *Hordium vulgare* was significant inhibited by rhizospheric soil under *Ougeinia*, however, no reduction was recorded in *Brassica compestris*. Rhizospheric soil reduced the shoot-root growth and dry matter yield of many crops (Negi *et al*., 2007).

Uses

Its foliage is rated as an excellent fodder, having 11.62-15.1% crude protein, 21.9-26.5% crude fibre, 10.0-14.9% total ash, 2.4-3.4% calcium and 0.2-0.42% phosphorus (Luna, 1996). Firewood contains 16.79kj/g calorific value, 0.68g/cc wood density, 0.60% ash, 67.78% moisture, 0.38% nitrogen with fuelwood value index of 2807 (Purohit and Nautiyal, 1987).

Quercus leucotrichophora A. Camus (Syn. Q. incana Roxb.)

Family: Fabaceae

Quercus leucotrichophora is a evergreen tree attaining a height up to 40 m. Leaves ovate lanceolate, glossy dark green above, densely white pubescent beneath. (Gaur, 1999). *Q. leucotrichophora* is found distributed from sub montane to montane Himalaya in Garhwal. It is abundant on North East slopes or otherwise, usually associated with *Rhododendron arboretum* and *Myrica esculanta*, and commonly on agricultural fields. Its range of distribution extends from 800-2000m asl (Gaur, 1999). In the middle and higher hills this species is the major fodder as dense forest stands of this species still exist but degradation of these forests is common scence everywhere. Regeneration is poor due to heavy lopping for fuel-fodder.

Silviculture methods

It is a moderate light demanding species. The seedlings and sapling can withstand shade but trees prefer to grow by overhead light. Seedlings are sensitive to drought and frost particularly during first 2 year of growth. Acorns germinate during June to July in moist shady places. It can be raised either by direct sowing or by planting nursery-raised seedlings. Its germination was best at 30°C as compared to the 20 and 25°C (Todaria *et al.*, 2003).

The leaf and barks extracts of *Q. leucotrichophora* significantly inhibited the germination of *Z. mays*, *V. unguiculata* and *G. max*, while, the germination of *E. frumentaceae* and *E. coracana* was least affected. Radicle extension of *Z. mays*, *V. unguiculata, G. max*, *E. frumentaceae* and *E. coracana* were suppressed by the leaf and bark extracts of *Q. leucotrichophora*. Most published work has showed that foliage leachates are most potent source of toxic metabolites, however, in *Q. leucotrichophora* leaf and barks leachates were found equally toxic (Kaletha *et al.*, 1996). Although, aqueous leaf extract and soil mulched with leaf litter suppressed the germination of *Triticum aestivum* and *Lens culinaris* but the production was comparable with control plots, hence the tree is considered least phytotoxic to growth attributes of these food crops (Bhatt and Chauhan, 2000).

Uses

Oak leaves provide valuable fodder for livestock in hills and leaves can be harvested throughout the year, when there is scarcity of other foliage. On average, leaves contain 10.2-11.4% crude protein, 31.3-32.1% crude fibre, 5.1-6.4% total ash, 0.9-1.65% calcium and 0.11-0.15% phosphorus. Foliage contains 4.2% total tannins (dry matter basis) and 2.5% condensed tannins with crude protein digestibility of 57% (ICAR, 1962). In this species tannins increase with leaf maturation but protein- precipitation phenolic contents are higher in young leaves and decrease as leaf mature (Makkar *et al.*, 1988). Thus young leaves would be more toxic than mature leaves.

Fuelwood contains 15.74% kJ/g calorific value, 1.0g/cc wood density, 1.6%ash, 47.47% moisture, 0.24% nitrogen and fuelwood value index of 2072 (Purohit and Nautiyal, 1987).

Terminalia bellirica **Roxb. (Vern. Behra –family Combretaceae)**

Description

It is a large, deciduous tree attaining 40-50m height, 3-4m girth. It occurs throughout the sub-Himalayan tract upto 1200-1300m and is usually common in sal (*Shorea robusta*) and miscellaneous forests of sub –tropical zone (Gaur, 1999)

Silviculture methods

The natural regeneration in *T. bellerica* is through seeds. Bright brown is the best colour for field fruit collection (Negi and Todaria, 1995). In laboratory 100 percent germination is recorded in *T. bellerica* at 25°C constant temperature (Negi *et al.*, 1995). Bhatt *et al.* (1997) reported that percent germination and radicle extension of *Hordium vulgare*, *Eleusine coracana*, *Glycine max* and *Brassica campestris* were significantly reduced by aqueous extracts of dried leaf of *Terminalia bellerica* in bioassay studies. The leaf and fruit pulp of *T. bellirica* significantly suppressed the radicle and plumule growth of *E. coracana*, *G. max* and *B. campestris*. The fruit pulp of *T. bellerica* completely inhibited the germination and radicle growth of *Hordeum vulgare* (Tables 3,4 and 5).

Table 3. Allelopathic effect of some multipurpose trees on germination and radicle growth (cm) of leguminous field crops.

Agroforestry tree species	Leguminous crops							
	Dolichos biflorus		Glycine max		Phaseolus lunatus		Phaseous mungo	
	% G	*RL*	*% G*	*R L*	*% G*	*RL*	*% G*	*RL*
			Tree bark					
Adina cordifolia	12.0	2.13	48.0	2.63	16.0	9.3	0.0	0.0
Celtis australis	32.0	2.92	76.0	7.40	40.0	7.62	72.0	6.29
Grewia oppositifolia	0.0	0.0	20.0	3.75	24.0	4.53	36.0	3.87
Holoptelea Integrifolia	56.0	2.92	68.0	3.64	48.0	1.54	24.0	1.57
Moringa oleifera	0.0	0.0	60,0	1.62	76.0	4.43	68.0	2.48
Ougeinia oojenensis	40.0	2.57	60.7	5.42	68.0	7.17	72.0	7.42
			Tree Leaves					
Adina cordifolia	48.0	2.34	64.0	4.64	20.0	1.92	92.0	2.83
Celtis australis	32.0	3.33	76.0	6.43	40.0	5.2	72.0	4.94
Grewia opp7ositifolia	0.0	0.0	44.0	3.33	0.0	0	24.0	1.49
Holoptelea Integrifolia	0.0	0.0	52.0	2.58	36.0	3.32	48.0	3.83
Moringa oleifera	0.0	0.0	44.0	1.82	0.0	0.0	16.0	1.23
Ougeinia oojenensis	48.0	3.54	72.0	61.3	88.0	6.63	88.0	6.48

G %= germination percent, R L =Radicle length

Table 4. Allelopathic effect of some multipurpose trees on germination and radicle growth (cm) of summer field crops

Agroforestry tree species	*Echinochoa frumentacea*		*Eleusine coracana*		*Zea mays*		*Vigna unguiculata*		*Glycine max*	
	G%	R L	G%	R L	G%	RL	G%	R L	G%	R L
				Tree Bark						
Bauhinia variegata	96.0	5.2	40.0	5.6	71.0	12.2	68.0	4.9	73.0	4.4
Celtis australis	100.0	4.4	78.0	3.5	87.0	12.3	54.0	4.9	30.0	3.9
Ficus cunia	98.0	4.9	94.0	3.0	78.0	9.5	42.0	5.1	59.0	5.2
Ficus roxburghii	97.0	5.2	65.0	2.9	97.0	11.8	88.0	3.3	34.0	4.3
Grewia oppositifolia	94.0	3.6	74.0	2.8	71.0	13.8	80.0	4.0	37.0	5.2
Kydia calycina	97.0	4.2	91.0	4.2	42.0	14.6	62.0	6.0	39.0	5.0
Q. leucotrichophora	93.0	5.0	88.0	5.7	63.0	14.8	59.0	4.2	44.0	3.8
				Tree Leaves						
Bauhinia variegata	96.0	4.7	56.0	4.8	87.0	11.8	75.0	5.1	36.0	5.4
Celtis australis	100.0	6.3	88.0	6.3	99.0	10.6	40.0	4.0	91.0	4.2
Ficus cunia	95.0	5.2	72.0	5.2	81.0	16.8	25.0	4.5	51.0	6.5
Ficus roxburghii	95.0	4.1	90.0	4.0	68.0	11.0	83.0	4.5	45.0	3.3
Grewia oppositifolia	98.0	3.7	98.0	2.7	36.0	8.3	80.0	2.9	17.0	3.4
Kydia calycina	94.0	6.5	69.0	3.9	90.0	17.0	78.0	5.1	28.0	3.8
Q. leucotrichophora	99.0	5.8	78.0	4.7	45.0	8.0	44.0	4.4	38.0	7.3
Terminalia bellerica	-	-	57.0	0.63	-	-	-	-	47.0	6.77

G %= germination percent, R L =Radicle length

Table 5. Allelopathic effect of some multipurpose trees on germination and radicle growth (cm) of winter field crops

Agroforestry tree crops	*Triticum aestivum*		*Hordeum vulgare*		*Brassica campestris*	
	G%	R L	G%	R L	G%	RL
		Tree Leaves				
Boehmeria rugulosa	76.0	4.51	68.0	4.01	87.0	3.19
Celtis australis	82.0	5.92	69.0	4.45	78.0	4.1
Ficus roxburghii	80.0	4.63	53.0	3.56	88.0	4.0
Grewia oppositifolia	83.0	3.48	77.0	3.14	87.0	2.95
Ougeinia oojenensis	85.0	3.8	95.0	4.0	99.0	2.5
Terminalia bellerica	70.0	1.98	19.0	2.64	26.0	0.68
		Tree Bark				
Boehmeria rugulosa	72.0	5.78	73.0	5.2	91.0	5.27
Celtis australis	75.0	5.93	37.0	3	93.0	4.96
Ficus roxburghii	79.0	6.02	43.0	5.65	89.0	6.33
Grewia oppositifolia	73.0	5.15	46.0	3.4	88.0	4.44
Ougeinia oojenensis	89.0	3.2	94.0	3.7	100.0	3.1
Terminalia bellerica	65.0	6.28	-	-	-	-

G %= germination percent, R L =Radicle length

Fig. 1. A traditional agroforestry system in Garhwal Himalaya

Fig. 2. *Celtis australis* in combination with Agriculture crop

Fig. 3. Multipurpose trees *Grewia oppositifolia*, *Ficus auriculata* growing along the agriculture field

Fig. 4. Mix cropping practices along with *Ficus auriculata* and *Celtis australis*

Fig. 5

Fig. 6

Fig. 5&6. *Boehmeria rugulosa* and *Ougeinia oojenensis* successful combination with agriculture crops

Uses

Seed is edible; exocarp is one of the important components of the medicine commonly known as "Triphala churan". Leaves are lopped for fodder (Luna, 1996). The leaves are highly valued as fodder for milk cattle.

Conclusion

It is pertinent to mention here that all the above-mentioned tree crops are used extensively as multipurpose farm trees in Garhwal Himalaya. Moreover, *O. oojenensis*, as a leguminous species, has a most promising future in the improvement of agroforestry systems in the area. On the bases of their utility and allelopathic interactions the importance of these trees in agroforestry may be considered to decrease in the order *O.oojenensis*>*C.australis*>*B. rugulosa*<*F. auriculata*<*K.calycina*<*H.integrifolia*>*M.Oleifera*>*A.cordifolia*>*G. oppositifoliai*< *Q. leucotrichophora*<*B. variegata*<*F. cunia*.

Our data demonstrated that legume crops viz. *Dichohos biflorus*, *G. max*, *P. lunatus* and *P. mungo* could be cultivated in association with *O. oojenensis* and *C.australis* in traditional agroforestry systems to enhance the sustainable productivity of rural poor farmer of Garhwal Himalayan region. Among the summer crops *E. frumentaceae* was found most resistance and *E. coracana* moderately resistance crop and could be grown under tree crops. However, winter millets *T. aestivum*, *H. vulgare* and *B. compestris* can be cultivated along with *O. oojenensis* in the same field. Among winter crops *B. compestris* is more resistance and can be grown successfully under tree canopy with least allelopathic effects.

References

Anonymous, 1985. Advisory Board on Energy. Towards a Perspective on Energy Demand and Supply in India. Govt. of India, New Delhi.

Anonymous, 1999. State of Forest Report. Published by Ministry of Environment and Forests, D. Dun., Uttaranchal.

Beniwal, B.S., Dhawan, V.K., Joshi, S.R. 1990. Effects of shade and mulch on germination in Arunachal Pradesh. Indian For. 144: 650-655.

Bhatt, B.P., Chauhan, D.S. 2000. Allelopathic effects of *Quercus* spp. On crops of Garhwal Himalaya. Allelo. J. 7: 265-272.

Bhatt, B.P., Badoni, A.K. 1990. Characteristic of some mountain firewood shrubs and tree. Energy 15: 1069-1070.

Bhatt, B.P., Badoni, A.K. 1995. Remarks on the fodder plants of Garhwal Himalaya. In: Chadha, S.K.(ed.) Echoes of Environment, Himalayan Publishing House, New Delhi, pp. 52-75.

Bhatt, B.P., Todaria, N.P. 1990a. Studies on the allelopathic effects of some agroforestry tree crops of Garhwal Himalaya. Agrofor. Syst. 12:251-255.

Bhatt, B.P., Todaria, N.P. 1990b. Vegetative propagation of tree species of social forestry value in Garhwal Himalaya. J. Trop. For. Sci. 2:195-210.

Bhatt, B.P., Todaria, N.P. 1990c. Fuelwood characteristics of some mountain trees and shrubs. Biomass 21: 233-238.

Bhatt, B.P., Todaria, N.P. 1992. Fuelwood characteristics of some mountain trees and shrubs. Commonwealth For. Rev. 71:183-185.

Bhatt, B.P., Verma, N.D. 2002. Some Multipurpose Tree Species for Agroforestry Systems. Published by ICAR Research Complex for NEH Region, Umiam, Meghalaya, pp.148.

Bhatt, B.P., Chauhan, D.S., Todaria, N.P. 1993. Phytotoxic effects of tree crops on germination and radicle extension of some food crops. Trop. Sci. 33: 69-73.

Bhatt, B.P., Kaletha, M.S., Todaria, N.P. 1997. Allelopathic exclusion of understory by some agroforestry tree crops of Garhwal Himalaya. Allelo. J. 4: 321-328.

Bhatt, B.P., Misra, L.K., Tomar, J.M.S., Singh, M., Chauhan, D.S., Dhyani, S.K., Singh, K.A., Dhiman, K.R., Datta, M. 2001. Agroforestry research and practices in NEH Region. An overview. *In:* Verma, N.D. and Bhatt, B.P. (eds.) Steps Towards Modernization of Agriculture in NEH Region. Published by ICAR Research Complex for NEH Region, Umiam, Meghalaya, pp. 365-392.

Bhatt, B.P.; Negi, A.K., Todaria, N.P. 1994. Fuelwood consumption pattern at different altitudes in Garhwal Himalaya. Energy 19: 465-468.

Chandra, J.P., Sharma R.K. 1977. Note on nursery technology of biul (*Grewia oppositifolia*). Indian For. 103: 684-685.

Chauhan, D.S., Bhatt, B.P., Todaria, N.P. 1993. Vegetative propagation studies in some tree and shrub species of Garhwal Himalaya. Ind. J. Plant Physio. 36:112-114.

Gaur, R.D. 1999. Flora of the District Garhwal Northwest Himalaya with Ethnobotanical Notes. Transmedia Publication Center, Srinagar Garhwal, Uttaranchal, India, pp. 811.

Ghosh, P.K., Bohra, H.C. 1984. Platability, digestibility and nutritive value of some important feeds of arid and semi arid regions of india. In Shankarnarayan, K.A. (ed) Agroforestry in Arid and Semi Arid Zones, C.A.Z.R.I., Jodhpur, India, Publication No. 24.

Gupta, R.K. 1986. Establishment and management of fodder trees in temperate and sub-temperate regions for agro/social forestry. In: Khosla *et al.* (eds.) Agroforestry systems- A New Challenge, Indian Society of Tree Scientists, Solan, H.P., India. pp. 161-180.

I.C.A.R. 1962. Animal nutritive: Appendex 1, table C, pp.188-194. In: Research in Animal Husbandry: A Reviw 1929-1954. Indian Council of Agricultural Research, New Delhi.

Kaletha, M.S., Bhatt, B.P., Todaria, N.P. 1996. Allelopathic effects of some agroforestry tree crops of Garhwal Himalaya. Range Mgmt. Agrofor. 17: 193-196.

Khosla, P.K., Toky, O.P., Bisht, R.P., Hamidullah, S. 1992. Leaf dynamics and protein contents of six important fodder trees of the Western Himalaya. Agrofor. Syst. 19:109-118.

Lohan, O.P., Lall, D., Pal, R.N., Negi, S.S. 1980. Note on tannins in tree fodders. Indian J. Ani. Sci. 50: 881-883.

Lohan, O.P., Lall, D., Negi, S.S. 1983. Partitioning of total tannins in some fodders in condensed and hydrolysable forms. Ind. J. Ani. Sci. 53: 1333-1335.

Luna, R.K. 1996. Plantation Trees. International Book Distributors, Dehradun, pp. 975.

Makker, H.P.S., Dawa, R.K., Singh, B. 1988. Change in tannin content, polymerization and protein precepitation capacity in Oak (*Quercus incana*) leaves with maturity. J.Sci. Food Agric. 44: 301-307.

Nautiyal, A.R., Thapliyal, P. 1987. A note on seed germination in Indian mountain tree species. Himalayan Res. Dev. 6:41-43.

Nautiyal, S., Maikhuri, R.K., Semwal, R.L., Rao, K.S., Saxena, K.G. 1998. Agroforestry systems in the rural landscape- a case study in Garhwal Himalaya, India. Agrofor. Syst. 41:151-165.

Negi, A.K., Todraia, N.P. 1993. Fuel wood evaluation of some Himalayan trees and shrubs. Energy 18: 799-801.

Negi, A.K., Todaria, N.P. 1994. Nutritive value of some fodder species of Garhwal Himalaya. In: Higher Plants of Indian Subcontinent (Additional Series of Indian Journal of Forestry, No

VI Vol. IIIrd), Published by Bishen Singh Mahendra Pal Singh, Dehradun, Uttrakhand, India, pp.117- 123.

Negi, A.K., Todaria, N. P. 1995. Effects of seed maturity and development on germination of five species from Garhwal Himalaya, India. J. Trop. For. Sci. 8: 255-258.

Negi, A.K., Chauhan, S., Todaria, N.P. 1995. Effect of temperature on germination pattern of some trees of Garhwal Himalaya. J. Tree Science 14: 11-13.

Negi, B.S., Chauhan, D.S., Todaria, N.P. 2007. Allelopathy effects of *Ougeinia oojeinensis* Rox.b (Fabaceae) on the germination and growth of wheat, barley and mustard. Allelo. J. 20: 403-410.

Negi, G.C.S. 1995. Phenology, leaf and twig growth, pattern and leaf nitrogen dynamics of some multipurpose tree species of Himalaya: Implementations towards agroforestry practices. J Sus. Agric. 6: 43-60.

P.I.D. 1998. The Wealth of India: A Dictionary of Indian Raw Material and Industrial Products, Vol. 2B, pp.173, Publication and Information Directory, C.S.I.R., New Delhi.

Purohit, A.N., Nautiyal, A.R. 1987. Fuelwood Value Index of Indian mountain tree species. Internat. Tree Crops J., 4: 177-182.

Sachan, M.S. 2006. Structure and functioning of traditional agroforestry systems along an altitudinal gradient in Garhwal Himalaya. D.Phil thesis. H.N.B. Garhwal University, Srinagar Garhwal, Uttaranchal.

Sharma, D.D., Gill, R.S. 1969. Chemical composition and nutritive value of Khirk (*Celtis australis*) tree leaves. J. Research PAU. 6: 388-393.

Singh, Bhupendra, Uniyal, A.K., Todaria, N.P. 2007. Effect of water, salinity stress with varying temperature regimes on seed germination and early growth of *Holoptelia integrifolia* (Roxb.) Planchon. Journal of Plant Biology 34: 87-93.

Singh, Bhupendra, Uniyal, A.K. 2005. Rooting response of branch cutting of *Celtis australis* L. to hormonal application. Forests Trees and Live. 15: 305-308.

Singh, Bhupendra, Uniyal, A.K., Todaria, N.P. 2008. Phytotoxic effects of *Focus species* on traditional food crops. Range Mgmt. Agrofor. 29(2):xx-xxx

Singh, Bhupendra, Bhatt, B.P., Prasad, P. 2004. Effect of seed source and temperature on seed germination of *Celtis australis* L. A promising Agroforestry tree-crop of Central Himalaya. For. Trees Livel. 14: 53-60.

Todaria, N.P., Singh, B., Dhanai, C.S. 2005. Allelopathic effects of tree leachate on germination and seedling growth of field crops. Allelo. J. 15: 285-294.

Todaria, N.P., Chauhan, S., Singh, M., Singh B. 2003. Seed source variation in relation to morphophysiology of *Quercus leucotrichophora* in Garhwal Himalaya. Indian J. Trop. Biodiv. 11: 22-27.

Uniyal, A.K., Bhatt, B.P., Todaria, N.P., 1999. Provenance characteristics and pretreatment effects on seed germination of *Grewia oppositifolia* Roxb.: A promising agroforestry tree-crop of Central Himalaya, India. Intern. Tree Crop. J. 10: 203-213.

Wood, P.J. 1988. Agroforestry and decision making in rural development. Forest Ecol. & Mgmt. 24: 191-201.

❑❑❑

5

Acacia nilotica Based Traditional Agroforestry System in Central India

C.B. Pandey

Abstract: *Acacia nilotica (L.) Willd. Ex Del is an important multipurpose tree of traditional agroforestry system in the central India. The present chapter reports influence of tree canopy positions, i.e. mid canopy, canopy edge and canopy gap, of Acacia nilotica (12-yr) on texture, organic C, total and mineral N and P and soil pH, in different depths of soils, on productivity of rice during rainy and wheat during winter season. It also reports how residual nitrogen under the tree influence yield of rice following its felling after completion of rotation cycle. Sand particles are found to decline by 10% and 9% whereas clay particles increase by 14% and 10% under mid canopy and canopy edge, respectively, than in open. Clay particles are not influenced by the canopy positions. Soil organic C, total N, total P, mineral N (NO_3^- -N and NH_4^+- N) and P are greater under mid canopy and canopy edge positions than in open. Soil organic C and N pool sizes are maximum in 0-10 cm and decline with depths. Total and mineral P contents are nearly uniform across the depths. C/N ratio increases with soil depth. Despite greater nutrients pools under the tree, density, aboveground biomass, belowground biomass, aboveground biomass / belowground biomass ratio and yield of the rice decline by 27.5%, 27.5%, 19.1%, 14.8% and 24.9%, respectively. Like the rice, yield of wheat also declines under the tree. But, the decline is maximum (52%) under the tree canopy in irrigated condition and minimum (33%) in non-irrigated condition. Moreover, following removal of the tree after completion of rotation cycle, maximum (53%) of the residual nitrogen is released quickly for the first rice cropping season and remaining part (37%) gradually until the fifth* *cropping season. The release of the residual N increases yield of the rice crop by 73% for the first cropping season, 52% for the second, 45% for the third, 41% for the fourth and 26% for the fifth cropping season. The crop yield, soil organic C and total N increase up to 5m from the stump in the first cropping season. By the fifth cropping season, the increase due to tree removal is confined to 2m from the stump. Total increase in the crop yield is 12.5 t ha^{-1} over five years, which is nearly equal to the reduction in the crop yield suffered during the tree growth period.*

Introduction

Acacia nilotica (L.) willd. Ex. Del, grows naturally and also has been widely planted since generations by farmers in crop fields in Asian countries like Pakistan, Bangladesh, and India (Singh *et al.*, 1990). It is a most preferred tree species in a traditional agroforestry in central India, because being a hardy tree species it survives in the water logging conditions in rice (*Oryza sativa*) fields for a pretty long time and provides fuel, fodder, timber, gums etc.

One of the tenets of agroforestry is that trees maintain soil fertility (Palm, 1995). Accumulation of nutrients under trees in agroforestry systems are well documented (Young, 1997; Nair, 1993; Palm, 1995; Pandey *et al.*, 2000). The nutrients generated by trees can be exploited within production system, either simultaneously, as in intercropping, or sequentially, as in rotational fallow systems (Rhoades, 1997). However, nutrients generated by trees generally do not support growth of intercrops in simultaneous system (Kesser, 1992; Kater *et al.*, 1992). *Acacia nilotica* tree builds up nutrients under its canopy and forms islands of nutrients in the fields (Pandey *et al.*, 2000). However, the under storey crop particularly rice are unable to utilize these nutrients due to limited light (Pandey *et al.*, 1999). Farmers in Chhattisgarh generally do not follow canopy management of the tree, rather they cut and sell >12-yr-old trees after completion of the rotation cycle. Crop yields are reported to increase when trees are cut or coppiced in agroforestry (Nye and Greenland, 1960; Tilander *et al.*, 1995; Szott *et al.*, 1999). This is one of the options for nutrients build under the tree.

This chapters describes effect of trees on nutrients build up and influence of the tree on growth and yield of rice and wheat growing under it's canopy and examines how residual nitrogen increase rice yield after removal of *Acacia nilotica* trees, after completion of rotation cycle, in central (Chhattisgarh), India.

Changes in physical properties of soils under the tree

Sand particles decline under mid canopy (10.34%) and canopy edge (9.29%) position of *Acacia nilotica* than in open (Table 1). Proportion of sand particles is maximum in 0-10 cm and declines with depths. Unlike sand particles, the proportion of clay particles increases (13.80 %) under mid canopy and canopy edge (10.43%) positions than in open. Clay particles under canopy edge does not differ from that under mid canopy. Silt particles are not influenced by the tree canopy positions. Reduced proportion of sand with simultaneous increases in the proportion of clay particles under the tree canopy occur due to protection of soils from the impact of rain drops, which otherwise increase defloculation and erosion of clay particles.

Table 1. Sand, silt and clay soil particles in different soil depths under mid canopy, canopy edge and open field in a traditional agroforestry in central India

Soil depth	Sand particles (%)			Silt particles (%)			Clay particles (%)		
	Mid canopy	Canopy edge	Open field	Mid canopy	Canopy edge	Open field	Mid canopy	Canopy edge	Open field
0-10	^{x}37.33^{a}	^{x}37.89^{a}	^{y}42.15	^{x}20.78^{a}	^{x}21.46^{a}	^{x}21.04^{a}	^{x}41.89^{a}	^{x}40.65^{a}	^{y}36.81a
10-20	^{x}36.63^{a}	^{x}36.89^{a}	^{y}41.54	^{x}22.03^{a}	^{x}21.49^{a}	^{x}20.93^{a}	^{x}41.42^{a}	^{x}41.62^{a}	^{y}37.53^{a}
20-30	^{x}36.39^{a}	^{x}36.84^{a}	^{y}39.37	^{x}20.80^{a}	^{x}21.25^{a}	^{x}23.06^{a}	^{x}42.81^{a}	^{x}42.91^{a}	^{y}37.57^{a}
Mean	^{x}36.78	^{x}37.21	^{y}41.02	^{x}21.20	^{x}21.40	^{x}21.68	^{x}42.04	^{x}41.73	^{y}37.30

Values in a column for a parameter suffixed with different superscripts are significantly different at P<0.05

Values in a row for a parameter suffixed with different superscripts are significantly different at P<0.05

Source: Pandey *et al*. 2000

Soil pH declines (7.71%) under mid canopy than in open (2). Soil pH does not vary with depths under mid canopy, but, it varies (P<0.05) among the depths under the other two canopy positions. Organic C and total N vary among the canopy positions (P<0.0001) in each soil depth (Table 2). Organic C is greater (60.66%) under mid canopy and canopy edge positions (36.1%) than in open (P<0.05). Organic C is maximum in 0-10 cm depth under all canopy positions and declines with increasing depths. Total N is greater (124.7%) under mid canopy and canopy edge (71.13%) positions than in open (P<0.05). Like organic C, total N is also maximum in 0-10 cm soil depth under all canopy positions and declines with the increasing depth of soil (P<0.0001). Total P does not differ significantly among canopy positions and soil depths (2). C/N ratio varies significantly among canopy positions (P<0.0001) and soil depths (P<0.0001). C/P ratio is maximum in 0-10cm soil depth under all canopy positions and declines with depth (P<0.0001). Higher soil organic carbon and total N pool under mid canopy and canopy edge positions compared to open field reflect primarily the accumulation of above- and belowground organic matter for 12 years (tree age), through leaf litter and dead roots. The decrease in organic C and total N towards canopy edge compared to mid canopy position is due likely to relatively low inputs of leaf litter as the canopy of *Acacia nilotica* is thin towards canopy edge (Pandey *et al*., 1999). Higher concentration of total P under mid canopy and canopy edge compared to that

Table 2: Soil pH, organic C, total N and total P concentration under in different soil depths under mid canopy, canopy edge and open field in a traditional agroforestry in central India

Soil	pH			Organic C (%)			Total N (%)			Total P (%)		
depth	Mid canopy	Canopy edge	Open field	Mid canopy	Canopy edge	Open field	Mid canopy	Canopy edge	Open field	Mid canopy	Canopy edge	Open field
0-10	$^{x}6.64^{a}$	$^{x}6.75^{a}$	$^{y}6.98^{a}$	$^{x}1.18^{a}$	$^{y}1.04^{a}$	$^{z}0.82^{a}$	$^{x}0.118^{a}$	$^{y}0.091^{a}$	$^{z}0.059^{a}$	$^{x}0.066^{a}$	$^{x}0.066^{a}$	$^{y}0.062^{a}$
10-20	$^{x}6.52^{a}$	$^{y}6.97^{b}$	$^{y}7.15^{b}$	$^{x}1.09^{b}$	$^{y}0.91^{b}$	$^{z}0.58^{b}$	$^{x}0.102^{b}$	$^{y}0.079^{b}$	$^{z}0.040^{b}$	$^{x}0.065^{a}$	$^{x}0.064^{a}$	$^{x}0.064^{b}$
20-30	$^{x}6.59^{a}$	$^{y}7.17^{c}$	$^{z}7.25^{b}$	$^{x}0.66^{c}$	$^{y}0.55^{c}$	$^{z}0.44^{c}$	$^{x}0.035^{c}$	$^{y}0.026^{c}$	$^{z}0.016^{c}$	$^{x}0.064^{a}$	$^{x}0.061^{a}$	$^{x}0.061^{c}$
Mean	$^{x}6.58$	$^{x}6.96$	$^{y}7.13$	$^{x}0.98$	$^{y}0.83$	$^{z}0.61$	$^{x}0.085$	$^{y}0.065$	$^{z}0.038$	$^{x}0.065$	$^{x}0.064$	$^{x}0.062$

Values in a column for a parameter suffixed with different superscripts are significantly different at $P<0.05$

Values in a row for a parameter suffixed with different superscripts are significantly different at $P<0.05$

Source: Pandey *et al.*, 2000

under canopy gap, observed in this study, may be due to greater amount of clay particles which are known to fix phosphorus (Lee *et al.*, 1980). Greater organic C content under tree canopy bind soil particles together and protect them from erosion probably thereby increases fine soil particles (Pandey *et al.* 1995). Mann and Saxena (1980) report an increase in clay content beneath the canopy of *Prosopis cineraria* under aridisols of Rajasthan, India.

NH_4^+-N is higher (39.92%) under mid canopy and canopy edge (33.90%) than in open (Table 3). However, the difference in NH_4^+-N between mid canopy and canopy edge in 0-10cm depth of the soil is not significant. Nitrate N is greater (65.10%) under mid canopy and canopy edge (55.5%) than in open. Both, NH_4^+-N and NO_3^- -N are maximum in 0-10 cm under all canopy positions and decline with depth. Higher NH_4^+-N and NO_3^- -N under mid canopy and canopy edge is mainly due to lower C/N ratio at these positions as compared to open. Ntayombya (1993) report that available NO_3^- -N pool in a tree based intercropped treatment is higher (43%) than in sole crop. NO_3^- -N / NH_4^+-N ratio varying significantly among the canopy positions is maximum in 0-10 cm depth and declines with depths (Table 4). High NO_3^- -N / NH_4^+-N ratio under mid canopy and canopy edge positions indicates greater nitrification and tree induced mineralization. Higher nitrification potential of soil directly beneath the canopy may also be partly attributed to the higher clay and carbon contents as the clay mineral surfaces with adsorbed NH_4^+ ions are preferentially colonized by nitrifiers (Kunc and Stotzky, 1980). Available P, averaged across the soil depths, is greater (70.27%) under mid canopy and canopy edge (44.32%) positions than in open.

Table 3. Ammonium N, nitrate N and available P concentration in different depths under mid canopy, canopy edge and open field in a traditional agroforestry in central India

Soil depth	NH_4^+-N (μg g^{-1})			NO_3^- -N (μg g^{-1})			Available P		
	Mid canopy	Canopy edge	Open field	Mid canopy	Canopy edge	Open field	Mid canopy	Canopy edge	Open field
0-10	$^{x}15.55^{a}$	$^{x}14.96^{a}$	$^{y}10.57^{a}$	$^{x}4.44^{a}$	$^{y}4.05^{a}$	$^{z}2.17^{a}$	$^{x}3.30^{a}$	$^{x}2.84^{a}$	$^{y}2.12^{a}$
10-20	$^{x}13.04^{b}$	$^{x}12.73^{b}$	$^{y}8.90^{b}$	$^{x}3.39^{b}$	$^{x}3.27^{b}$	$^{y}2.15^{a}$	$^{x}3.16^{a}$	$^{x}2.66^{a}$	$^{y}1.91^{a}$
20-30	$^{x}11.23^{c}$	$^{x}10.36^{c}$	$^{y}8.95^{b}$	$^{x}2.19^{c}$	$^{y}2.12^{c}$	$^{z}1.75^{b}$	$^{x}3.00^{a}$	$^{x}2.51^{a}$	$^{y}1.52$a
Mean	$^{x}13.27$	$^{y}12.68$	$^{z}9.47$	$^{x}3.34$	$^{y}3.15$	$^{z}2.02$	$^{x}3.15$	$^{y}2.67$	$^{z}1.85$

Values in a column for a parameter suffixed with different superscripts are significantly different at P<0.05

Values in a row for a parameter suffixed with different superscripts are significantly different at P<0.05

Source: Pandey *et al.* 2000

Table 4. C/N ratio, NO_3^- -N/NH_4+-N ratio and C/P ratio under mid canopy, canopy edge and open field in a traditional agroforestry in central India

Soil depth	C/N ratio			NO_3^- -N/NH_4^+-N ratio			C/P ratio		
	Mid canopy	Canopy edge	Open field	Mid canopy	Canopy edge	Open field	Mid canopy	Canopy edge	Open field
0-10	$^{x}10.02^{a}$	$^{y}11.40^{a}$	$^{z}13.79^{a}$	$^{x}0.29^{a}$	$^{x}0.27^{a}$	$^{y}0.21^{a}$	$^{x}17.80^{a}$	$^{y}15.77^{a}$	$^{z}13.03^{a}$
10-20	$^{x}10.60^{a}$	$^{x}11.48^{a}$	$^{y}14.81^{a}$	$^{x}0.26^{b}$	$^{x}0.26^{a}$	$^{x}0.24^{b}$	$^{x}16.64^{b}$	$^{y}14.19^{b}$	$^{z}9.08^{b}$
20-30	$^{x}18.93^{b}$	$^{y}20.92^{b}$	$^{z}27.26^{b}$	$^{x}0.20^{c}$	$^{x}0.20^{b}$	$^{x}0.20^{c}$	$^{x}10.19^{c}$	$^{y}8.84^{c}$	$^{z}7.05^{b}$
Mean	$^{x}13.18$	$^{y}14.60$	$^{z}18.62$	$^{x}0.25$	$^{y}0.24$	$^{z}0.22$	$^{x}14.88$	$^{y}12.93$	$^{z}9.72$

Values in a column for a parameter suffixed with different superscripts are significantly different at P<0.05

Values in a row for a parameter suffixed with different superscripts are significantly different at P<0.05

Source: Pandey *et al.* 2000

Change in light under *Acacia nilotica* trees

Generally height and CBH of *Acacia nilo*tica trees found in the traditional agroforestry of Chhattisgarh are 7.75 ± 0.37 m and 77.83 ± 4.12 cm, respectively. The crown cover, ranging from 38.3 to 162.8 m^2 $tree^{-1}$, averaged for 65 m^2 $tree^{-1}$. The intensity of light is reduced maximum (8.5 times) at 2 m and minimum (1.6 %) at 6 m distance than in open (25 m from tree) (Fig. 1). The intensity of light (Y) increases with increasing distance (X) up to 6 m and stabilizes thereafter (Y = 82.53 + 740.90 ln (X), r^2 = 0.795, P<0.001). The distance traveled by tree shade depends upon the sun angle to tree crown and the height of the tree itself.

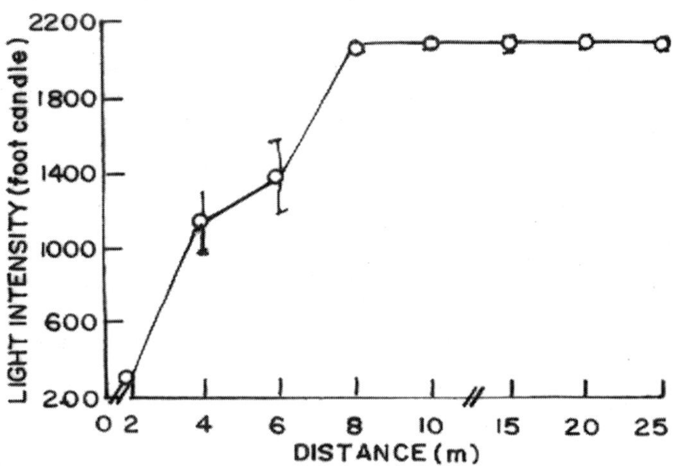

Fig 1: Light intensity at different distance from *Acacia nilotica* tree in the traditional agroforestry system in central India

Source: Pandey *et al.* 1999

Lower is the sun angle greater is the distance traveled by the shade. At 90^0 sun angle, the shade is confined below the tree crown. Increase in the light intensity at greater distance is, perhaps, due to side light.

Effects of *Acacia nilotica* trees on rice crop

Density (tiller m^{-2}) of the rice varies (269-472 m^{-2}) across the distances from the tree trunk (Fig. 2 a); being minimum at 2m distance, it increases up to 8 m. There is no significant increase in the density beyond 8 m. An average density is reduced (27.5 %) due to tree crown with maximum (43%) occurring at 2m and minimum (16.3 %) at 8m distance. The density is positively correlated to the distance. Aboveground biomass and belowground biomass, varying from 316-586 g m^{-2}, and 83-122 g m^{-2}, respectively across the distance are the lowest at 2 m distance (Figs. 2 b, c). Aboveground biomass increases significantly from 2 m to 10 m, however, no significant increase is found beyond 10m. Like the density aboveground biomass is reduced across the distances, with maximum (46.1%) at 2 m and minimum (13.5 %) at 15 m distance with over all 19.1% reduction across the distances. Compared the aboveground biomass, belowground biomass suffers reduction for greater distance (up to 15 m). The yield of the paddy is also the lowest (339 g m^{-2}) at 2 m and the highest (606 g m^{-2}) at 25 m distance from the tree trunk (Fig. 2 d). The yield is reduced maximum (44%) at 2 m and minimum (14%) at 8 m distance than in open (25 m away from the tree). The growth and yield parameters were positively correlated with intensity of light (Fig.3 a,b,c,d).

In spite of greater organic C and total N and available N in the soil directly beneath the tree canopy, reduction in growth and yield of rice crop occurs under the tree canopy results from decreased availability of light (44 to 62%) under the canopy. In paddy fields gravimetric soil-water is above the water holding capacity and C/N ratio is low (20%) beneath the tree crown than in open (C/N ratio = 20.4) indicating greater availability of mineral N (Pandey *et al.*, 1999). Studies examining the resource sharing ability of multipurpose trees in an intercropping system, crop growth and yield are found to be depressed under tree crown, where soil nutrient pool and their availability, and soil water are adequate, due to competition for light (Farrell, 1990; Kesser, 1992). In sub-humid condition where soil water is not critical to plant growth and its productivity, light becomes limiting under tree canopy. Contrary to that of sub-humid condition, productivity is reported to be encouraged under tree canopy in arid condition (Tiedemann and Klemmedson, 1977; Ovalle and Avendano, 1987). These findings support the concept of multiple limiting factors (Blackman, 1905), that plants productivity responds to increased inputs of only the most limiting factors until another becomes limiting. Crop may exploit the greater amounts of nutrients to increase productivity in simultaneous agroforestry, if tree canopy is broken open to facilitate greater light availability during the cropping season. In a nursery experiment it is observed that shoot cutting caused root nodule shedding and fine root mortality in *Acacia nilotica*.

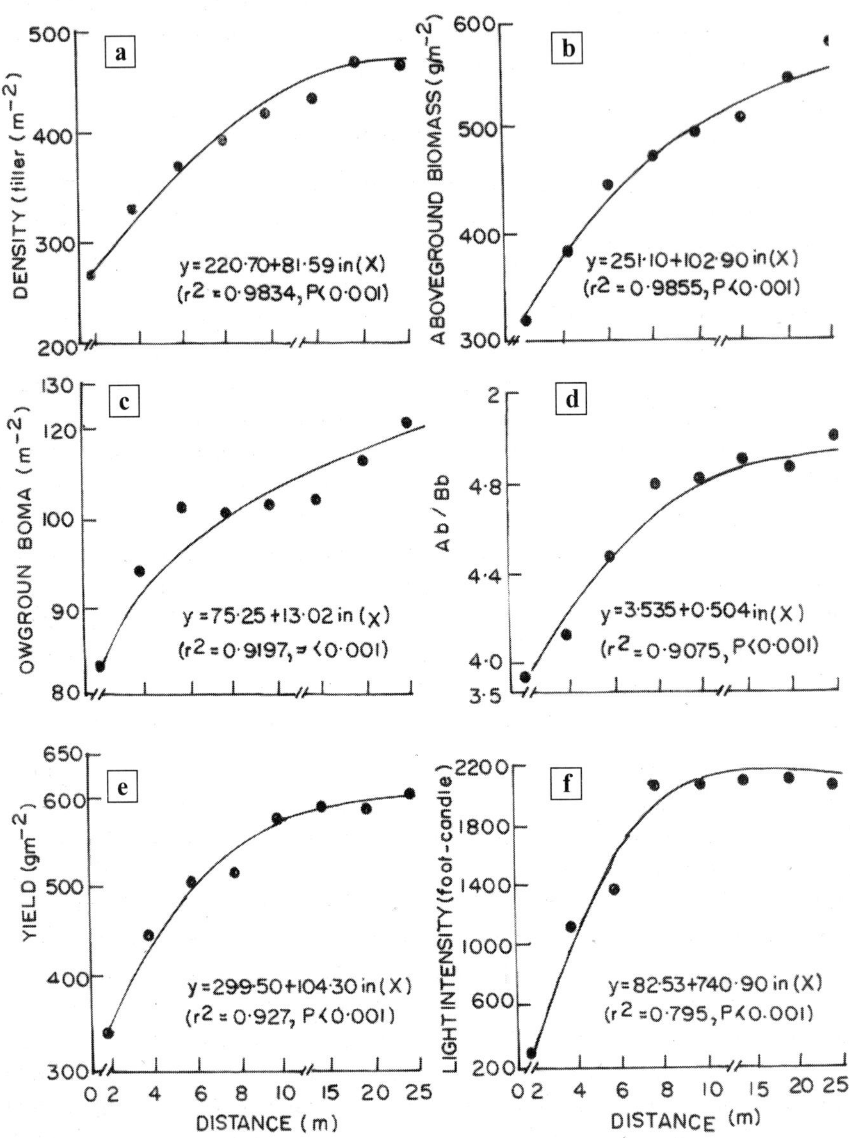

Fig. 2. (a) Density, **(b)** aboveground biomass, **(c)** belowground biomass, **(d)** Ab/Bb ratio, **(e)** yield of rice and **(f)** light intensity in relation to distance from *Acacia nilotica* tree in the traditional agroforestry system in central India

Source: Pandey *et al.* 1991

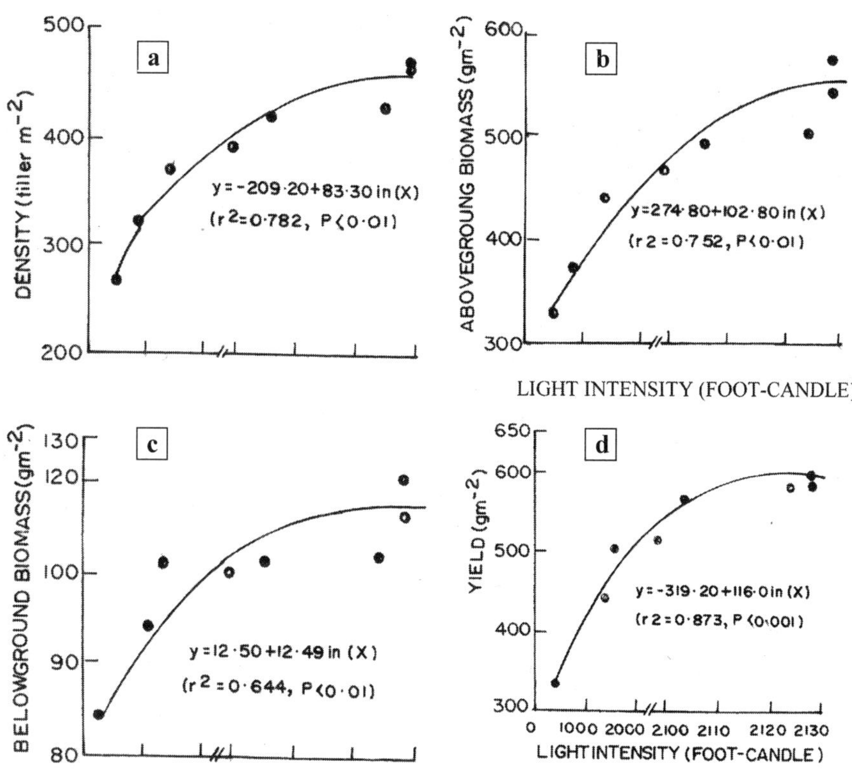

Fig. 3. Relationships between light intensity and **(a)** density, **(b)** above ground biomass, **(c)** below ground biomass, and **(d)** yield of rice in the traditional agroforestry system in central India.

Source: Pandey *et al*. 1999

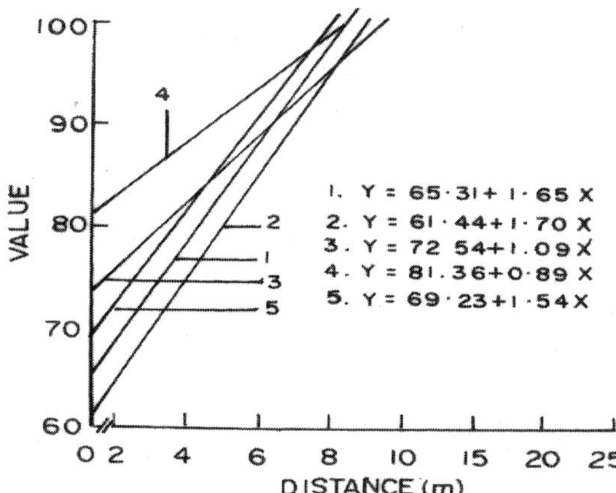

Fig. 4. Regression lines and equations of normalized mean values of different parameters: 1. density; 2. aboveground biomass; 3. belowground biomass; 4. aboveground/belowground biomass ratio; 5. yield.

Source: Pandey *et al*. 1999

Therefore, removal of branches may increase not only the availability of light to the understorey crop but also elevate the substrate level for mineralization. In arid environment soil water, being critical, set an upper limit for productivity, which is made available under tree crown as it is protected from excessive evaporation. Mordelet and Menaut (1995) report that soil-water dynamics is lower beneath the tree clumps.

To know which parameter is affected the most, sensitivity of different parameters to the distances is compared by regression analysis of normalized values (Fig. 4). The slope of the regression line shows the sensitivity of the parameters (i.e. the steeper the slope, the greater the effect). It is apparent from the figure that the inclination of the slope for aboveground parameters (density, aboveground biomass and yield) is greater than that for belowground parameter (root biomass). This implies that effect of the tree is greater on the aboveground biomass than root biomass. Other studies have also reported that the effect of tree is greater on the aboveground attributes than on belowground attributes (Puri *et al.*, 1994). From each respective regression equation the degree of response of the parameter can be calculated at any given distance from the tree. It appears from the equation that each parameter is drastically affected nearer to the tree trunk. The tree crown of the tree is spread up to 4.55 m, however, the effect of the tree was up to 8m indicating that the impact of tree is not confined within its crown cover, rather than beyond to it. Greater reduction of the intensity of light towards the tree trunk is perhaps due to thick cover of the crown. However, the positive correlation between distance and the intensity of light suggests that penetration of light to ground through crown (sunflecks) increases towards the crown edge.

Biomass allocation in rice in relation the tree shade

Aboveground biomass/belowground biomass ratio (Ab /Bb) is minimum at 2 m from the tree trunk, and increases with the distance (Fig. 5 a). Ab/Bb ratio is positively correlated to the distance. Pattern of dry matter allocation to aboveground versus belowground structure influence the competitive ability of a plant in a given habitat, and also the relative competitive ability of different component species in an agroforestry system (Anderson and Sinclair, 1993). Under *A. nilotica* trees aboveground biomass / belowground biomass ratio is low under tree crown due to proportionately greater accumulation of biomass underground plant part in the stress condition of light (Fig. 5 b). Puri *et al.* (1994) report greater root biomass compared to that of shoot biomass in the shelterbelt plantation in arid condition in India.

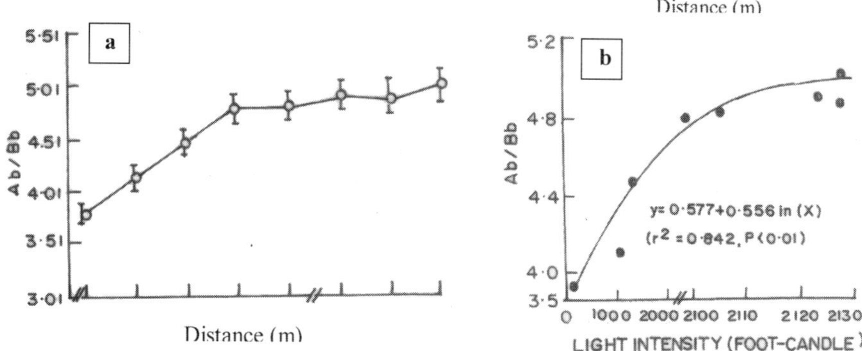

Fig. 5. Variation in aboveground (Ab) / belowground biomass (Bb) ratio, and relationship between Ab/Bb ratio and light intensity under *Acacia nilotica* tree in the traditional agroforestry system in central India.
Source: Pandey *et al.* 1999

Acacia nilotica/wheat interaction under irrigated and no-irrigated conditions

Effect of *Acacia nilotica* tree on growth and productivity of wheat crop under irrigated condition is similar to that of the rice crop, but in no-irrigated condition it is relatively lower (Fig.6 a, b and c). Yield decline is maximum (52%) under the tree canopy in irrigated condition and minimum (33%) in non-irrigated condition. Beyond the tree canopy, reduction in the yield was 8% greater in irrigated than in non-irrigated condition. Soil organic C (SOC) and total N are 35% and 13%, respectively greater in irrigated and 43% and 46%, respectively greater in no-irrigated condition than in open. SOC under the tree canopy under irrigated condition did not differ from that in no-irrigated conditions. However, total N in soil under the tree is 2 times greater in irrigated than in no-irrigated condition because Nitrogen fertilizer is done in irrigated condition. SOC and total N being maximum at 2 m declines with distance from the tree. On the contrary, C/N ratio at all distance is lower in irrigated than no-irrigated condition (Fig.7).

Three zones under the tree in the agroforestry can be demarcated. First zone is found beneath the tree where both aboveground competition for light and belowground competition for soil-water and nutrients occur. The second zone occur beyond the tree canopy where belowground competition is found, and third zone is open field where neither aboveground nor belowground competition occur (Rao *et al.*, 1998). Decline in yield, aboveground biomass and density of wheat crop under the canopy compared to open in irrigated condition may be due to aboveground competition for light. Torquebiau and Akyeampong (1994) argues that when soil-water is enough plants compete for the next factor that is most likely to be limiting light. The results described above are in keeping with the observations made in Burkina Faso, where Sorghum yield under Karite and Nere trees are reduced an average 50% and 7%, respectively (Kessler, 1992), and in India wheat yields are reduced up to 60% (Puri and Bangarwa, 1992) and mustard

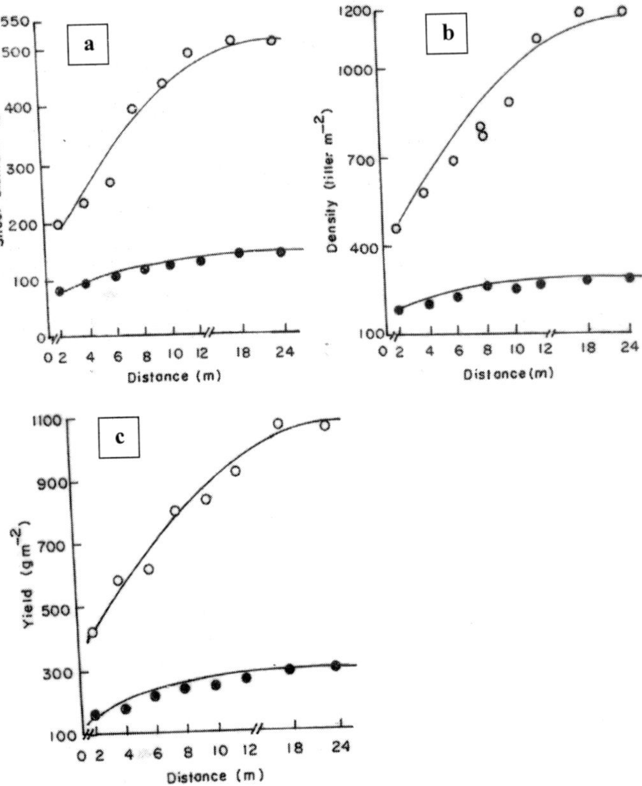

Fig. 6. (a) Shoot biomass, **(b)** density, and **(c)** yield of wheat under *Acacia nilotica* tree in irrigated and non-irrigated conditions in the *Acacia nilotica* based traditional agroforestry systems in central India.

Source: Pandey *et al*. 1999

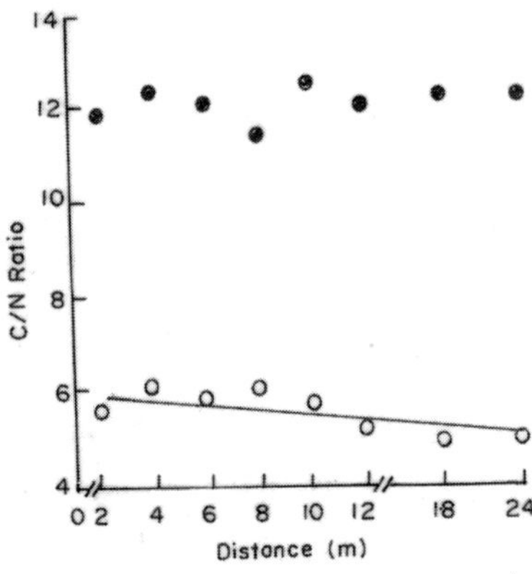

Fig. 7. Carbon/nitrogen (C /N) ratio in soil under *Acacia nilotica* tree in irrigated and non-irrigated conditions in the *Acacia nilotica* based traditional agroforestry system in central India.

Source: Pandey *et al*. 1999

(Brassica sp.) yields up to 65% (Yadav *et al.*, 1993) under *Acacia nilotica* tree. However, lower reduction in yield and growth parameters of the crop under the tree in non-irrigated condition indicated that soil-water might have become a primary limiting factor. The tree seems to have reduced evaporation and thereby increased soil-water under the tree, which became available to the crop (Fig. 8). This indicates that soil-water, if lowest, off set the effect of tree shade.

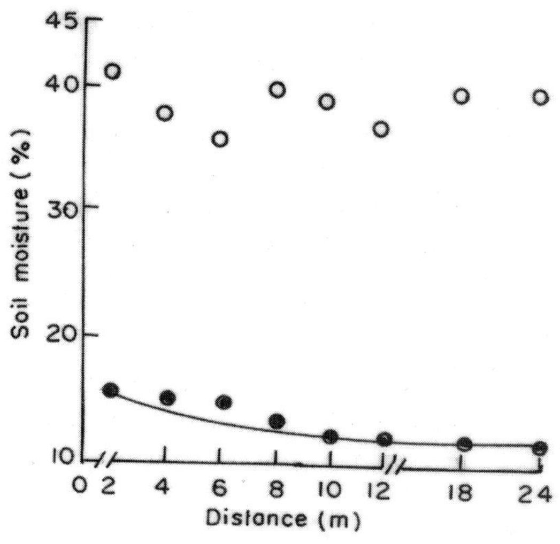

Fig. 8. Soil moisture under *Acacia nilotica* tree in irrigated and non-irrigated conditions in the *Acacia nilotica* based traditional agroforestry system in central India.
Source: Pandey *et al.* 1999

Wheat yield of crop with increases with increasing intensity of light up to 12 m from the tree in irrigated condition indicating that effect of the tree is not limited to its canopy rather beyond to it. The yield under the tree canopy was 3 times greater in irrigated condition than in non-irrigated condition. Greater yield of the crop at the each distance in irrigated condition than in non-irrigated condition suggests that the crop utilizes reduced light under the tree relatively more efficiently due to fertilization in irrigated condition. Though, harvest index increases with increasing intensity of light in both irrigated and non-irrigated conditions, the increase is exponential in irrigated condition (Fig. 9). This implies that aboveground competition for light is greater in irrigated condition. Reduction in harvest index due to shade is well known (Ong, 1993).

Reduction in yield of the crop beyond the tree canopy indicates that belowground competition occurs between the tree and crop in both irrigated as well as no-irrigated conditions. One can expect that greater belowground competition in no-irrigated compared to irrigated condition as soil water and nutrients are lower in no-irrigated condition. In contrast, greater reduction (19%)

in the yield under the canopy and beyond the tree canopy (8%) is found in irrigated condition, perhaps due to resource preemption by the tree. Resource preemption theory of competition states that competition between two inter-specific species occur because one of the species preempt resources asymmetrically greater and make them limiting to other (Grime, 1973, Keddy, 1989). The growth resource preemption is reported to be greater in productive habitat because it supports higher growth rates (Grime, 1973). Higher soil water and nutrients in irrigated conditions probably stimulates greater growth in *Acacia nilotica* tree. The tree is evergreen, hence growth most likely occurs during the crop growing season. Greater amount of total soil N and lower C/N ratio under the tree in irrigated condition indicates greater availability of mineral N. Since fertilization is done after irrigation leaching does not seem to occur. The tree have extensive lateral root spread and therefore can mine soil water and nutrients from a large surrounding area. Nere tree with crown of 7 m radius extends lateral roots up to 20 m from the tree base. Lateral roots of *Acacia seyal* extended up to 26 m and those of *Sterocarya birrea* extends up to 50m (Groot and Moumare, 1995). Stabilization of yield of the crop from 18 m onwards from the tree indicates that lateral spread of the roots may be at least up to 12 m. Greater reduction in yield beyond the tree canopy in irrigated condition than in no-irrigated condition suggests greater utilization of growth resources by tree component in nutrient rich habitat.

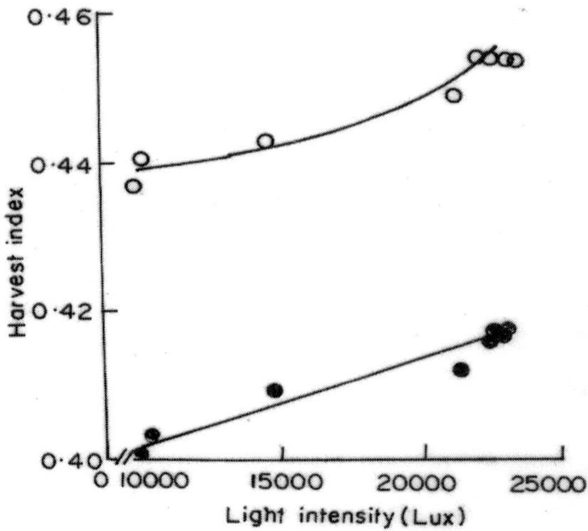

Fig. 9. Harvest index of wheat crop under *Acacia nilotica* tree in irrigated and non-irrigated conditions in the *Acacia nilotica* based traditional agroforestry system in central India.

Source: Pandey *et al.* 1999

Residual effect of nutrients on rice yield under *Acacia nilotica* trees

Following removal of the tree, maximum (53%) of the residual nitrogen under the *Acacia nilotica* trees is released quickly for the first rice cropping season and remaining part (37%) gradually until the fifth cropping season (Fig. 10 a); soil organic carbon follows the pattern of total N (Fig. 10 b). The gradual release of residual N after first year of the tree felling is due to low concentration of nitrogen and high amount of tannins in leaf and roots of *A. nilotica*. Tannins is known to bind nitrogen of plants, making it insoluble (Mafongoya, 1998). Pandey and Sharma (2003) find low concentration of nitrogen (1.51%) in the leaf of *A. nilotica* compared to other leguminous trees. Mafongoya *et al.* (1998) report slow release of N from low quality materials. This suggest that slow decomposition of litter or other organic materials (suberized tree roots) provide humus which bind soil particles and increases soil-moisture and thereby help increase in crop yields. Relatively higher amount of soil- moisture under the felled tree is evident from high ratio of bold and chaffy grains. Pandey *et al.* (2000) find 14% greater clay soil particles under 12 yr or more old trees of *A. nilotica.* Pandey and Singh (1991) find positive relation between fine soil particles and soil-moisture in a dry tropical savanna. Denitrification, leaching and oxidation of organic C probably are another phenomenon, besides export of mineralized N in the crop harvest, for the loss of organic C and total N in the soil. Rice field remained water-logged most of the time during the cropping season, whereas, temperature rises maximum 46°C during summer season. Temporal pattern of SOM and N are correlated because most of the soil N exists in organic form. Loss of soil organic C due to increased soil temperature is well documented (Parton, 1987). Following felling of the tree, grain yield of rice-crop increases under the area covered by the trees during its growing period, but the increase is differential with years and the distance from tree stump (Fig. 11). Following the tree felling, grain yield increases maximum (126%) for the first year and minimum (30%) for the fifth year at 1 m distance from the stump and declines with the distance. Grain yield increases maximum up to 5m from the stump for the first year which recedes to 2m in the fifth year. Grain yield, across the distance, increases 73% for the first yr, 52% for second yr, 45% for third yr, 41% for forth yr and 26 % for fifth yr and stabilizes at seventh year after the tree removal. Grain yield is inversely related to the years ($r = -0.9692$, $p<0.01$). Pattern of decline in growth parameters like crop density under the felled tree follows the pattern similar to that of yield. It increases dramatically (109%) for the first year following the tree removal. This increase is up to 5m from the tree stump. Density is inversely related to the distance for all the years ($p<0.01$). Density, across the distance, declines with the year ($r = -0.89$, $p< 0.01$).

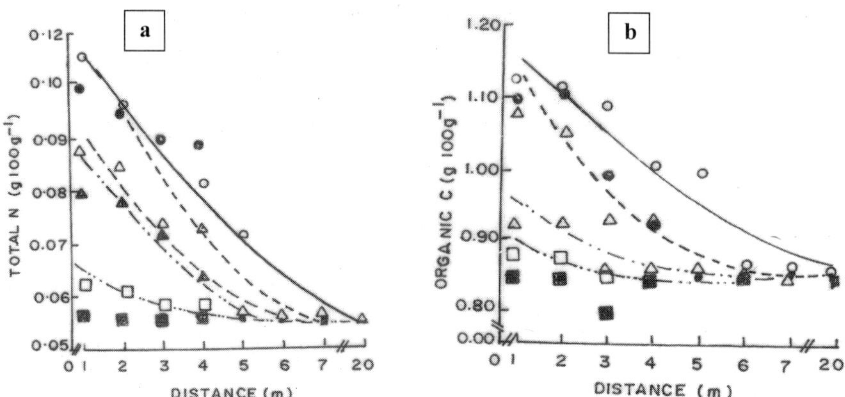

Fig. 10. Relationships between **(a)** Relationships between (i) total N (N_1, g/100g) in soil and distance (X, m) under one year old felled trees of *A. nilotica* according to : $N_1 = 0.15493\ X^{-0.27872}$, ($r^2 = 0.8289$, $p<0.001$), (ii) total N (N_2, g/100g) in soil and distance (X, m) under two years old felled trees of *A. nilotica* according to: $N_2 = 0.10544\ X^{-0.26098}$, ($r^2 = 0.7046$, $p<0.001$),(iii) total N (N_3, g/100g) in soil and distance (X, m) under three years old felled trees of *A. nilotica* according to: $N_3 = 0.0883\ X^{-0.19049}$, ($r^2 = 0.7648$, $p<0.001$), (iv) total N (N_4, g/100g) in soil and distance (X, m) under four years old felled trees of *A. nilotica* according to: $N_4 = 0.08565\ X^{-0.16145}$, $p<0.001$), (v) total N (N_5, g/100g) in soil and distance (X, m) under five years old felled trees of *A. nilotica* according to: $N_5 = 0.067354\ X^{-0.08664}$, ($r^2 = 0.9457$, $p< 0.001$), (vi) total N (N_7, g/100g) in soil and distance (X, m) under seven years old felled trees of *A. nilotica* according to : $N_7 = 0.05672\ X^{-0.0042}$, ($r^2 = 0.16$). **(b)** (i) organic C (C_1, g / 100g) and distance (X, m) under one year old felled trees of *A. nilotica* according to : $C_1 = 1.16626\ X^{-0.11084}$, ($r^2 = 0.8010$, $p<0.001$), (ii) organic C (C_2, g / 100g) and distance (X, m) under two years old felled trees of *A. nilotica* according to : $C_2 = 1.10797\ X^{0.108794}$, ($r^2 = 0.8784$, $p<0.001$), (iii) organic C (C_3, g/100g) and distance (X, m) under three years old felled trees of *A. nilotica* according to : $C_3 = 1.07986\ X^{-0.09156}$, ($r^2 = 0.7614$, $p<0.001$), (iv) organic C (C_4, g/100g) and distance (X, m) under four years old felled trees of *A. nilotica* according to : $C_4 = 0.96969\ X^{-0.05772}$, ($r^2 = 0.6916$, $p<0.001$), (v) organic C (C_5, g/100g) and distance (X, m) under five years old felled trees of *A. nilotica* according to : $C_5 = 0.892396\ X^{-0.019403}$, ($r^2 = 0.5918$, $p<0.001$), (vi) organic C (C_7, g/100g) and distance (X, m) under seven years old felled trees of *A. nilotica* according to : $C_7 = 0.85228\ X^{-0.00637}$, ($r^2 = 0.0638$). Symbols denote: open circle for first cropping season, solid circle for second cropping season, open triangle for third cropping season, solid circle for fourth cropping season, open squace for fifth cropping season and solid square for seventh cropping season.

Source: Pandey and Sharma 2003

The highest increase in the density and yield of the crop for the first year under the felled tree is consistent with the lowest C/N ratio. C/N ratio is lower under the felled tree compared to open for the first year and declines to minimum for the fifth year ($r = 0.962$, $p<0.01$). The highest amount of organic C and total N in the soil, and the lowest C/N ratio indicates the highest soil N mineralization for the first year following the tree removal. Rice crop is grown in the rainy season and trees are felled in June before the onset of rains.

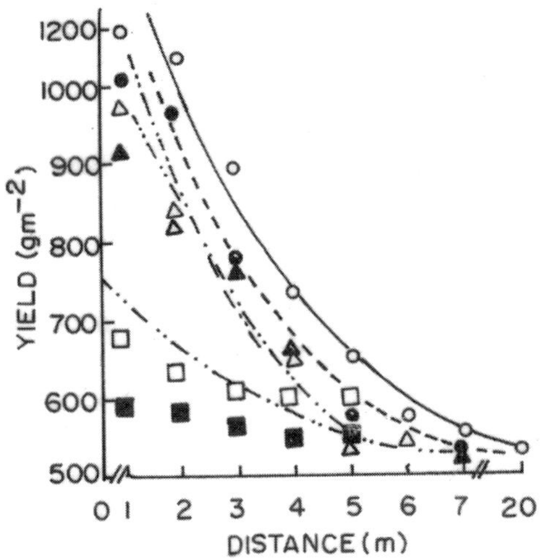

Fig. 11. Relationships between (a) yield (Y_1, gm^{-2}) and distance (X, m) under one year old felled trees of *A. nilotica* according to: $Y_1 = 1202\ X^{-0.32813}$, ($r^2 = 0.8579$, $p < 0.001$), (b) yield (Y_2, gm^{-2}) and distance (X, m) under two years old felled trees of *A. nilotica* according to: $Y_2 = 859.29564\ X^{-0.161177}$, ($r^2 = 0.8087$, $P<0.001$), (c) yield (Y_3,gm^{-2}) and distance (X,m) under three years old felled trees of *A. nilotica* (according to :$Y_3 = 934.39549\ X^{-0.24324}$ ($r^2 = 0.8092$, $p<0.001$) (d) yield (Y_4, gm^{-2}) and distance (X,m) under four years old felled trees of *A. nilotica* according to: $Y_4 = 897.8868\ X^{-0.2280677}$, ($r^2 = 0.7904$, $p<0.001$), (e) yield (Y_5, gm^{-2}) and distance (X,m) under five years old felled trees of *A. nilotica* according to: $Y_5 = 679.44017\ X^{-0.09828}$, ($r^2 = 0.84395$, $p<0.001$) (f) yield (Y_7, gm^{-2}) and distance (X,m) under seven years old felled trees of *A. nilotica* (solid rectangle Ò) according to: $Y_7 = 592.47096\ X^{-0.0435559}$, ($r^2=0.8647$,$p<0.001$).
Symbols are same as in figure 10.

Source: Pandey and Sharma 2003

Rainfall following the tree removal probably causes mineralization which support tillering and yield production. Dead roots and leaf litter seem major source of organic matter for nutrient build up under the tree (Pandey *et al.*, 2000). Comparing the pool sizes of total nitrogen at different times (years) following the tree removal, it is estimated that 90% of the residual nitrogen is released in the five years of which maximum 53% during the first year. Concentration of total nitrogen under the felled tree for the seventh year reaches equal to that of open field indicating that 10% of the residual nitrogen is released during sixth year. Highest release of nitrogen for the first year immediately after the tree removal is attributed to dead fine roots of the tree and other nitrogen present in active fraction of soil organic matter. It is most likely that tree felling causes death to fine roots which elevate substrate level to the mineralization. Tilander *et al.* (1995) report that residual nutrients enhances crop yields under coppiced trees of *Parkia biglobosa* in a parkland on the central plateau of Burkino Faso in West Africa. Ladd *et al.* (1981) report that 22% of the N from labeled alfalfa is recovered by a first crop of wheat and additional 4% by second crop. Sisworo *et al.* (1990) follow the N added from a cowpea crop and more than 70% is recovered in six consecutive crops. In our study residual nitrogen is released gradually until five years. Crop yield coinciding with the release of residual nitrogen attains plateau at the seventh year indicating that residual nitrogen supports the crop productivity for 5 years after removal of 12 yr or more old tree of *A. nilotica*. Bishop (1982) describes an agrosilvopastoral system from Ecuador, in which eight years old fallow of *Inga edulis* tree support productivity of food crops for two years. Padwick (1983) and Drechsel *et al.* (1996) report that residual benefit of both natural fallow and short-term improved fallows on crop yields occasionally lasts for three post fallow crops, but is usually negligible after two crops.

Inverse relation between the crop yield and distance under the felled tree is attributed to greater concentration of organic C and total N in the soil nearer to the tree stump. Mazzarino *et al.* (1991); Kater *et al.* (1992) and Pandey *et al.*(2000) report that soil organic matter, nitrogen and other nutrients decline with radial distance from tree bole or centre of crown. Similar pattern of decline in grain yield and soil organic C and total N suggests the importance of residual nutrients on the crop productivity. Influence of residual nitrogen on the crop productivity is not limited to the tree canopy. Canopy radius of the tree is 4.25m, but effect of residual nitrogen on the crop yield is much beyond the limit of the canopy in 12 year old trees. Van Noordwijk *et al.* (1996) report lateral root expansion seven times of crown radius in *Acacia seyal* and 10 times of crown radius in *Sclerocarya birrea* trees. In the traditional agroforestry system influence of residual nitrogen is not much away of the canopy mainly because of the presence of greater suberized primary tree roots (anchoring roots), and their location in deeper layers. Intensive ploughing is done by tractor in the field, which probably causes greater distribution of the tree roots vertical. However, gradual receding pattern of the nitrogen with the distance may be attributed to greater biomass of suberized tree roots nearer to the stump. Van Noordwijk *et al.*(1996) reported greater root biomass nearer to the tree trunk.

Recovery of crop yield loss after felling of *Acacia nilotica* trees in the system

Comparing the reduction in the crop yield during 15 years of the tree growth and gain in the yield after the tree removal, it is apparent that residual nitrogen almost compensates for the loss in the rice yield during the tree growing period (Table 5). No reduction in the crop yield in the beginning for 3 years seems mainly owing to smaller and thinner canopy of the tree. Tree produces timber and fuel biomass as additional benefits to the farmers. Farmers get fuel wood from the pruned branches about 0.6 t ha^{-1} per year from trees below the 10 years. From the age of 10 years and onwards farmers cut the thicker branches about 0.7 t ha^{-1} per year for local agricultural implements. At the time of felling, total wood production is estimated 167 t ha^{-1}, which includes 45 m^3 merchantable timber. Survey of local timber market reveals that farmers fetch US S 30.74 from one tree (>15 years) and US S 922 from 1ha land. This makes the system economically viable ultimately. Kass *et al.* (1993) and Kass and Somarriba (1999) report that *Senna guatemalensis* and *Mimosa scabrella* tree fallows in Latin America are more profitable than more intensive alternatives (e.g. fertilization of annual crops) because they support crop yield and provide fire wood and forage biomass as well.

Table 5: Reduction in rice yield under *Acacia nilotica* tree canopy during 20 years of rotation cycle and gain in yield following the 15-yr-old tree removal in a tradition agroforestry in central India

Age of tree (yr)	Rice yield reduction during the tree growing period			After the tree felling following completion of rotation cycle (15-yr)			
	Under tree canopy (t ha^{-1})	Open field (t ha^{-1})	Reduction in yield (t ha^{-1})*	Cropping season	Felled tree (t ha^{-1})	Open field (t ha^{-1})	Gain in yield (t ha^{-1})*
1.	5.01	5.02	0.01[a]	1.	9.20	5.32	3.88
2.	5.28	5.31	0.03[a]	2.	8.04	5.29	2.75
3.	4.91	4.93	0.02[a]	3.	7.69	5.30	2.39
4.	4.80	5.40	0.60	4.	7.40	5.25	2.15
5.	4.68	5.44	0.76	5.	6.60	5.24	1.36
6.	4.69	5.48	0.79	7.	5.61	5.30	0.31
7.	4.69	5.49	0.80				
8.	4.48	5.33	0.85				
9.	4.50	5.50	1.00				
10.	4.53	5.53	1.00				
11.	3.57	5.07	1.50				
12.	3.50	5.0	1.50				
13.	3.57	5.07	1.50				
14.	3.57	5.07	1.50				
15.	3.57	5.07	1.50				
16.	3.60	5.12	1.52				
17.	3.48	5.02	1.54				
18.	3.54	5.07	1.53				
19.	3.43	4.97	1.54				
20.	3.56	5.10	1.54				
Total	82.96	103.99	20.97	7	44.54	31.70	12.53

*Significant at P<0.05

[a]Not significant

Source: Pandey and Sharma, 2003

Conclusions

Acacia nilotica increases SOC under its canopy through organic matter inputs via leaf and dead roots, which bind soil particles and reduce runoff. As a result an apparent increase in the proportion of soil clay occurs under the canopy. Increased soil C and clay contribute interactively to increased total and available forms of N and P. This increase occur gradually over a long period (growing period of the tree). Despite greater amount of nutrients under the tree, yield of rice during rainy season declines during the growing time of the tree because light becomes limiting. Wheat yield during winter season also decline under the tree in both in irrigated and no-irrigated condition but, the decline is lower in the no-irrigated than irrigated condition. Following felling of the tree (after completion of rotation cycle), maximum proportion of residual nitrogen is released quickly for the first cropping season, and remaining part is released gradually until the fifth cropping season. This release increases crop yield. The increase is maximum for the first cropping season and minimum for the fifth cropping season. Total increase in the crop yield, across the cropping season, following the tree removal compensates the loss in the yield during the tree growing period.

References

Anderson, L.S., Sinclair, F.L. 1993. Ecological interactions in agroforestry system. Agrofor. Abstr. 6:57-91.

Bishop, J. P. 1982. Agroforestry systems for humid tropics east of the Andes. In: Hecht, S.B. (Ed.), Amazonia, Agriculture and Land Use Research. CIAT, Cali, Colombia, pp. 403-416.

Blackman, F.F. 1905. Optima and limiting factors. Ann. Bot. 19:281-295.

Drechsel, P., Steiner, K. G., Hagedorn, F. 1996. A review on the potential of improved fallows and green manure in Rwanda. Agrofor. Syst. 33: 109-136.

Farrell, J. 1990. The influence of trees in selected agroecosystems in Mexico. pp. 167-183. In: Gliessman, S.R. (Ed.) Agroecology: Researching the Ecological Basis for Sustainable Agriculture. Spriger-Verlag, New York.

Grime, J.P. 1973. Competitive exclusion in herbaceous vegetation. Nature 242:344-347.

Groot, J.J.R., Soumare, A. 1995. The roots of the matter: soil quality and tree roots in the Sahel. Agrofor. Today 7(1): 9-11.

Kass, D.C.L, Foletti, C., Szott, L. T., Landaverde, R., Nolasco, R. 1993. Traditional fallow systems of the Americas. Agrofore. Syst. 23:207-218.

Kass, D.C.L, Somarriba, E. 1999. Traditional fallows in Latin America. Agrofor. Syst. 47:13-36.

Kater, L. J. M., Kante, S., Budelman, A. 1992. Karite (*Vitellaria paradoxa*) and nere (*Parkia biglobosa*) associated with crops in South Mali. Agrofor. Syst. 18:89-105.

Keddy, P.A. 1983. Shoreline vegetation in Axe Lake, Ontario: effects of exposure on zonation patterns. Ecology 64: 331-344.

Kessler, J. J. 1992. The influence of karite (*Vitellaria paradoxa*) and nere (*Perkia biglobosa*) trees on sorghum production in Burkina Faso. Agrofor. Syst. 17: 97-118.

Kunc, F., Stotzky, G. 1980. Acceleration by montmorillonite of nitrification in soil. Folm Microb. 25: 106-125.

Ladd, J. N., Oades, J.M., Amato, M. 1981. Distribution and recovery from legume residue decomposing in soils sown to wheat in the field. Soil Biol. Biochem. 15:231-238.

Lee, D., Han X.G., Jordan, C. F. 1980. Soil phosphorus fraction, aluminium, and water retention as affected by microbial activity in an ultisol. Plant Soil 121: 125-136.

Mafongoya, P. L., Giller, K. E., Palm, C. A. 1998. Decomposition and nitrogen release pattern of tree prunings and litter. Agrofor. Syst. 38: 77-97.

Mann, H. S., Saxena, S. K. (eds.). 1980. Khejri (*Prosopis cineraria*) in the Indian Desert. CAZRI Monograph No. 11, Jodhpur. Central Arid Zone Research Institute, Mimeo 77 pp.

Mazzarino, M. J., Oliva, L., Nunez, A., Nunez, G., Buffa, E. 1991. Nitrogen mineralization and soil fertility in the dry Chaco ecosystem (Argentina). Soil Sci. Soc. Am. J. 55:515-522.

Mordelet, P., Menaut, J. C. 1995. Influence of trees on aboveground production dynamics of grasses in a humid savanna. J. Veg. Sci. 6:223-228.

Nair, P.K.R. 1993. An Introduction to Agroforestry. Kluwer Academic Publishers, Dordrecht, The Netherlands, 499 pp.

Nye, P.H, Greenland, D.L. 1960. The Soil Under Shifting Cultivation. Technical Communication 51. Commonwealth Agricultural Bureaux, Harpendon, UK, pp. 156.

Ong, C.K. 1993. Cropping systems approaches to agroforestry research. pp. 51-69. In: Bentley, W.R., Khosla, P. K., Seckler, K. (eds.) Agroforestry in South Asia: Problems and Applied Research Perspectives. International USA and Oxford and IBH Publishing Co. Pvt. Ltd., New Delhi.

Ovalle, C., Avendano, J. 1987. Interactions de la strate ligneuse avecla strate herbaceev dans less formations d' Acacia caven (Mol.) Hook. et Arn. Chili. I. Influence de l arbre surla composition floristique, la production et la phenologiede la state herbacee. Act. Oecol. 8:385-404.

Padwick, G.W. 1983. Fifty years of experimental agriculture in the maintenance of soil fertility in tropical Africa: a review. Exp. Agric. 19:293-310.

Palm, C.A. 1995. Contribution of agroforestry trees to nutrient requirements of intercropped plants. Agrofor. Syst. 30: 105-124.

Pandey, A. N., Purohit, R. C., Rokad, M. V. 1995. Soil aggregates and stabilization of sand dunes in the Thar desert of India. Environ Conser 22: 69-71.

Pandey, C. B., Pandya, K. S., Pandey, D., Sharma, R. B. 1999. Growth and productivity of rice (*Oryza sativa*) as affected by *Acacia nilotica* in a traditional agroforestry system. Trop. Ecol. 40(1): 109-117.

Pandey, C.B., Singh, A.K, Sharma, D.K. 2000. Soil properties under Acacia nilotica trees in a traditional agroforestry system in central India. Agroforestry Systems 49: 53-61.

Pandey, C.B., Singh, J.S. 1991. Influence of grazing and soil condition on secondary savanna vegetation in India. J. Veg. Sci. 2: 95-102.

Pandey, C.B., Sharma, D.K. 2003. Residual effect of nitrogen on rice productivity following tree removal of Acacia nilotica in a traditional agroforestry system in central India. Agri. Eco. Env. 96:133-139.

Parton, W.J., Schimel, D.S., Cole, C.V., Ojima, D.S. 1987. Analysis of factors controlling soil organic matter levels in Great Plains grasslands. Soil Sci. Soc. Am. J. 51:1173-1179.

Puri, S., Bangarwa, K.S.1992. Effects of trees on the yield of irrigated wheat crop in semi arid regions. Agrofor. Syst. 20:229-241.

Puri, S., Singh, S., Kumar, A. 1994. Growth and productivity of crops in association with an Acacia nilotica tree belt. J. Arid Environ. 27:37-48.

Rao, M.R., Nair, P.K.R., Ong, C.K. 1998. Biophysical interactions in tropical agroforestry systems. Agrofor. Syst. 38:3-50.

Rhoades, C.C. 1997. Single-tree influence on soil properties in agroforestry: lessons from natural forest and savanna ecosystems. Agrofor. Syst. 35:71-94.

Singh, G., Arora, Y.K., Narain, P., Grewal, S.S. 1990. Agroforestry Research in India and other Countries. Surya Publications, Dehradun, India.

Sisworo, W.H., Mitrosuhardjo, M. M., Rasjid, H., Myers, R.J.K. 1990. The relative role of nitrogen fixation, fertilizer, crop residues and soil in supplying N in multiple cropping systems in humid, tropical upland cropping system. Plant Soil 121: 73-82.

Szott, L.T., Palm, C.A., Buresh, R.J. 1999. Ecosystem fertility and fallow function in the humid and sub-humid tropics. Agrofor. Syst. 47: 163-196.

Tiedemann, A.R., Klemmedson, J.O. 1977. Effect of mesquite tree on vegetation and soils in the desert grasslands. J. Range Manage. 26:27-29.

Tilander, Y., Ouedraogo, G., Yougma, F. 1995. Impact of tree cropping on tree-crop competition in park-land and alley farming systems in semiarid Burkina Faso. Agrofor. Syst. 30: 363-378.

Torquebiau, E., Akyeampong, E. 1994. Shedding some light on shade, its effects on beans, maize and banana. Agrofor. Today 6(4):14-15.

Van Noordwijk, M., Lawson, G., Soumare, A., Groot, J. J., Hairiah, H. 1996. Root distribution of trees and crops: competition and or complementarity . pp. 319-364. In: Ong, C. K., Huxley, P. (eds.) Tree-crop Interaction: A Physiological Approach. CAB International Wallingford, U.K.

Yadav, J. P., Sharma, K. K., Khanna, P. 1993. Effect of Acacia nilotica on mustard crop. Agrofor. Syst. 21:91-98.

Young, A. 1997. Agroforestry for Soil Management. CAB International, Wallingford, UK.

6

Homegarden Agroforestry: Structure and Function in the Andaman and Nicobar Islands

C.B. Pandey and Lalita Singh

Abstract: *The study reports the influence of biophysical and socio-cultural factors on composition, diversity and distribution of plant species in the homegardens of the Andaman and Nicobar archipelago of India. In Nicobar mostly an aboriginal tribe the Nicobari predominates. The Nicobaries are Christians. However, in the Andamans people from different parts of Indian subcontinent have been rehabilitated in the mid twentieth century. Mixed culture prevails in South Andaman, but in the other Andaman islands either Bengali or Tamil culture predominates. Biophysically, South Andaman is hilly whereas Nicobar is flat. Rests of the islands lie in-between the two extremes. But, all are similar climatically. Twelve to thirty-four plant species are found in the homegardens that are planted, cared and harvested. These plants are: palm, fruit, spice and forest trees, formed three-storey structure in Andaman and two-storey structure in Nicobar's homegardens. Floristic similarity among the homegardens of the Andamans is 82 to 92%. However, it is only 12-18% between the homegardens of Andaman and Nicobar. Compared to Nicobar, species richness is greater in the homegardens of the Andamans. But, diversity is higher and evenness lower in the homegardens of Nicobar. All plant species in the Nicobar's homegardens are spontaneous in regeneration. Among the Andamans, proportion of spice trees is higher in the homegardens of South Andaman, mango and citrus in North Andaman, and pineapple and vegetables in Little Andaman. High plant diversity in these* *homegardens serves the subsistence needs and provides income to most of the households as well. Net income in the homegardens of Andaman was 6.9 times higher compared to that in the Nicobar's homegardens.*

Introduction

Andaman and Nicobar Island were totally covered with forest until it was rehabilitated with refugees of East Pakistan, Srilanka repatriates, Burma repatriates, land less people from different parts of country and ex-serviceman families (Bandhopadhyay, 1998). The settlers were provided with 2ha of paddy land, 2ha of hilly land and 0.4ha land for house construction. Settlers raised coconut, arecanut and fruit trees in the premises of their house, which developed gradually into homegarden and became a permanent system of rural economy. During third five-year plan, State Department of Agriculture directed development of coconut, arecanut, pineapple and cashew nut cultivation and rejuvenation of old coconut plantation in the homegarden of south Andaman (50,000 ha) in the islands (Ghosh, 1994). Homegarden in additions to plantation and fruit trees include poultry, livestock and fishpond. These components being an in integral part of the homegarden function in an interactive manner.

Homegardens are an integral part of farmer's farming system, an adjunct to the house, where selected trees, shrubs and herbs are grown for edible products and cash income (Kumar and Nair, 2004). The homegardeners are perpetual experimenters, they are constantly trying and testing new species and varieties and their management (Ninez, 1984; Kumar and Nair, 2004). Over centuries, they have selected specific species and manipulated their physical and ecological locations in a homegarden planting for maximizing space and production (Gillespie *et al.*, 1993), based on their accrued wisdom and insights acquired through interactions with environment without access to exogenous inputs, capital or scientific skills (Kumar and Nair, 2004). The crop combinations found in the homegardens of a region, however, are strongly influenced by the biophysical and socio- cultural factors besides the specific needs and preference of the household and nutritional complementarity with other major food sources (Vogl *et al.*, 2002; Kumar and Nair, 2004). As a consequence, homegardens vary greatly in species, species richness, structural complexity and size, but general principles are broadly similar (Gillespie *et al.*, 1993). From the ground layer to the upper canopy, the gradient of light and humidity determine different niches that species exploit according to their own requirements (Fernandes and Nair, 1986). The position, height and shade tolerance of plants are important traits that are acquired with time (evolutionary). Therefore, study of these parameters gives an idea of the temporal and spatial positioning of plants, species interaction and mixed species silviculture that are pertinent for designing a multistrata agroforestry and management of its productivity (Gillespie *et al.*, 1993; De Clerck and Negreros-Castillo, 2000).

A great deal of study has been conducted to know the structure and functioning of homegardens in tropics (Fernandes and Nair, 1986; Gillespie *et al.*, 1993; Alvarez-Buylla Roces *et al.*, 1989). However, there is a general lack of information regarding how biophysical and socio-cultural factors determine floristic composition, species diversity and their distribution in the homegardens. The present study provides such information with the aim of testing the hypotheses that plant species and their proportions in the homegardens of a region are influenced by the biophysical and sociocultural factors as well as preferences of the homegardeners. However, distribution of species and crop management depend upon the accumulated knowledge of the homegardeners. The present study examines component interactions and productivity also in the homegardens of the Andaman and Nicobar islands, India. Our homegardens are unique where bunded rice fields, lying adjacent, encircle the former. Therefore, for the purposes of this study, homegardening is defined as a more or less specialized agricultural production or farming system close to and usually surrounding the homes of the homegardeners, composed of one or a set of fields managed by a farming family (Ceccolini, 2002). Family members work all or part of the time to produce food and sometimes other production for subsistence and/ or sale (Okigbo, 1990; Ceccolini, 2002).

General description of the Andaman and Nicobar Islands

The Andaman and Nicobar islands are located in the Bay of Bengal (6° to 14° N lat., and 92° to 94° E long.); terrain is mostly undulating with altitude ranging from 0 to 365 m above sea level. The rocks occurring in Andaman and Nicobar group of islands are marine deposits dating back from late *Mesozoic* to *Pleistoceine* and evolution of Andaman appears to be linked with the Himalayan ranges (Srinivasan, 1986). The archipelago is divided, geographically, into five groups of islands namely, North Andaman, Middle Andaman, South Andaman, Car Nicobar and Nancowry. Administratively the former three form Andaman district and the latter two Nicobar district. Total geographical area of the islands is 8249 sq km of which the Andamans share 78%. Predominant soils in the islands are alfisols (ferric luvisols, FAO), entisols (dystric fluvisols), inceptisols (dystric cambisols) and mollisols (humic acrisols). The soils are loose, well drained, generally gravelly loamy to sandy loamy in texture, mostly slightly acidic in reaction and low to moderate in nutrients (NBSSLUP, 1991).

The climate is equatorial humid tropical. Temperature varies from 23.1°C to 30.1°C, being maximum in May and minimum in December. Average annual rainfall is 3000 mm, distributed over 8 to 9 months. High rainfall (73%) prevails from May to October, whereas four months from December to March are relatively dry. Humidity ranges from 71 % to 85 %, and is maximum in September and minimum in February.

Forest (87 %), homegarden (4.6 %) and rice fields (1.3 %) covering together around 93% of the total geographical areas are three major land uses in the islands (Basic Statistics, 2001). In these islands homegardens developed in the mid twentieth century when government of India rehabilitated the convicts of British occupied India and Burma, refugees of East Pakistan, ex-servicemen and poor people of India (Majumdar, 1975). The homegardens are located generally on hillocks whereas the rice fields in flat-bottomed valleys (Fig. 1). Size of the homegardens ranges from 0.05 to 2 ha in Andaman and 0.5 to 5 ha in Nicobar district. Rice fields are lacking in Car Nicobar. Potential natural vegetation in the islands is a tropical rainforest. *Pterocarpus dalbergiodes* DC., *Dipterocapus grandiflora* (Blanco) Blanco, *D. alatus* Roxb. ex G. Don, *D. gracilis* Blume, *Artocarpus chaplasha* Roxb., and *Calophyllum innophylum* L. are common forest species found abundantly in the study region. On interface of the sea and landmass, mangrove vegetation is found which protects the islands from the sea invasion (Forest Working Plan, 1980).

Two distinct types of villages are found in the Andamans, first type in which homegardens congregate along either sides of roads and the rice fields lie quite apart, and another type where rice fields lying around the homegardens separate one homegarden from the other. However, in Nicobar a typical village comprises of 12-15 Nicobaries huts arranged almost in a circle and their doors face one another. The space enclosed by the huts is used commonly for social works. Population in the villages ranges from 365 to 525 person village^{-1} in Andaman and 113 to 1208 person village^{-1} in Nicobar and corresponding population density 24 to 49 person km^{-2} in Andaman and 12 to 150 person km^{-2} in Nicobar. The population has increased 17 times in Andaman and 7 times in Nicobar mainly due to its high growth rate and partly due to immigration from mainland of India. Two types of

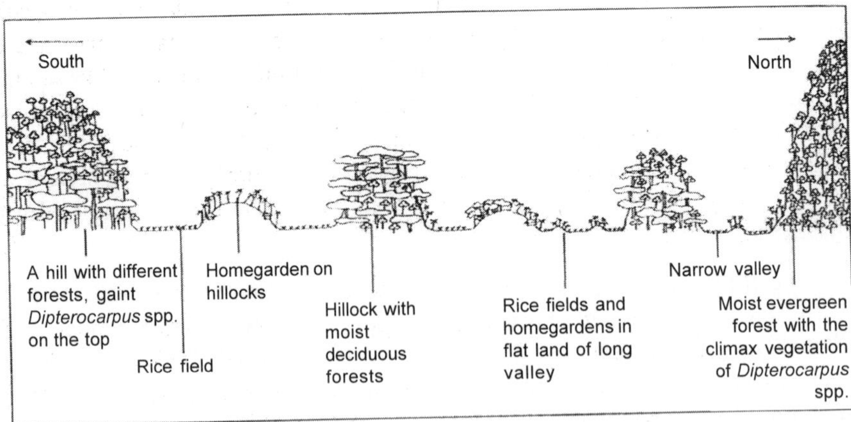

Fig. 1: Schematic presentation of distribution of homegardens, and land-use pattern in the Andaman and Nicobar Islands. *Source*: Pandey *et al.* 2007

societies are found in the Andamans. First type is characterized by mixed culture in which followers of different religions live together, regard religions and celebrate festivals of one another. However, the second type strictly follows a particular religion, rites and rituals. But, social unit is individual family in both the cases, which includes parents and their siblings. Both the societies are patrilineal and ownership is individual. In Car Nicobar, the aboriginal Nicobaries are a tribe. Social unit is a community and land ownership belongs to the community. The villagers select a head for each village called 'captain' who is empowered for all socio-economic decisions. He allots land to a person who gets married. After death of the person the allotted land goes back to the community.

At the time when homegardens developed, transport and market facilities in these islands were either poor or completely lacking. Now, a well-developed intra- and inter- island transport facilities are available both by sea and by roads in the islands. Villages are well connected with metalled roads. Among the Andamans, inter-island transport of goods is made maximum by ships and to some extent by lorries. But, the transportation from Andaman to Nicobar is performed only by sea. Small vegetable markets are organized temporarily along the roads during evening. Farmers sell non-perishable homegarden's products like coconut kernel, arecanut and spices to local vendors but perishable items like vegetables and fruits to either local vendors or directly in the vegetable markets. In Nicobar, markets are similar to that found in the Andamans, but there mostly fishes are sold. However, non-perishable items like coconut kernel is sold at community level by the village captain. Permanent markets, having market yards, are found only at Port Blair, the capital city of these islands, where import and export of goods are materialized.

Biophysical and sociocultural characters of the islands

Biophysical and sociocultural characters of the studied islands are given in Table 1. South Andaman is characterized by hillocks having steep to moderate slopes and flat bottomed vallies. Rests of the Andaman islands were either flat or moderately sloppy in topography. Mixed culture predominated in South Andaman. But, either Tamil or Bengali culture prevailed in the other Andaman islands. In Nicobar, the Nicobaries are Christians. People from all the cultures practice the homegardening traditionally.

Components of the homegardens

Orchard, poultry, dairy, fish pond and rice fields are components of homegardens in Andaman and Nicobar islands.

Table 1. Biophysical and socio-cultural characteristics of the Andaman and Nicobar islands.

Site	Soil depth (cm)	Soil type	Topography	Dominant homegardener
South Andaman	0-30	Gravely-fine sandy-loamy	Undulating (steep to moderate slopes)	Hindu (Tamilian, and Bengali), Muslim
Middle Andaman	38-75	Gravely-fine	Undulating (moderate sandy-loamy	Hindu (Bengali) slopes to flat)
North Andaman	> 100	Loamy	Relatively flat	Hindu (Bengali)
Little Andaman	15-30	Fine sandy-loamy	Relatively flat	Hindu (Bengali)
Car Nicobar	>100	Fine sandy	Flat	Christian
Nancowry	50 -75	Loamy to fine sandy- loamy	Flat	Christian, Hindu (Punjabi)

Source: Pandey *et al.* 2007

Species composition and structure of the orchard

Six homegarden types namely arecanut-coconut-banana-pineapple, arecanut-coconut-banana-mango, areanut-coconut-mango-banana, arecanut-coconut-pineapple-banana, coconut-banana-arecanut and coconut-arecanut-banana are found that correspond the homegardens of South Andaman, North Andaman, Middle Andaman, Little Andaman, Car Nicobar and Nancowry, respectively (Table 2).

Total 33 to 34 plant species were found in the homegardens of Andaman and 9 to 12 in Nicobar that are planted, cared and harvested. They are classified as palms, fruit trees, spice trees and agroforestry / forestry trees (Table 2). Though the species are common, their density varied within and among the sites (Table 3). Arecanut is dominant in the homegardens invariably in all the islands of Andaman. However, coconut dominates in the homegardens of Nicobar (Table 2). Dominance of most of the fruit and spice trees is higher in Andaman compared to that in Nicobar's homegardens. Among the Andamans, almost all the fruit trees are encountered in the homegardens, but their proportions differed. Proportion of mango, banana, cashewnut, papaya and sapota is highest in South Andaman's homegardens and lowest in that of the Little Andaman. However, proportion of custard apple and pineapple is highest in Little Andaman and citrus in North Andaman. Proportion of mango in North Andaman is nearly equal to that found in South Andaman. Like fruit trees, proportion of spice trees is also highest in South Andaman. Forest trees are found in almost all the homegardens of the studied islands, but their numbers are higher in the homegardens of Andaman compared to that found in Nicobar district.

Plant species found in the studied homegardens are similar to that found in Kerala, India (Nair and Sreedharan, 1986), Bengladesh (Millate-E-Mustafa *et al.*, 1996) and Thailand (Boonkird *et al.*, 1984). Common occurrence of the palm,

Table 2. Dominanace (IVI) of important plant species in the homegardens, of the different islands of Andaman and Nicobar archipelago.

Common name	Scientific name	Dominance (IVI, %)					
		South Andaman	North Andaman	Middle Andaman	Little Andaman	Car Nicobar	Nancowry
Palm trees							
Coconut	*Cocos nucifera* L.	9.7 (3.8)	8.0 (2.4)	9.0 (3.2)	6.4 (2.1)	109.5 (12.6)	96.9 (11.1)
Arecanut	*Areca catechu* L.	92.1 (18.4)	97.2 (17.7)	100.9 (21.3)	98.4 (15.4)	14.4 (4.8)	18.4 (5.3)
Fruits trees							
Cashewnut	*Anacardium occidentale* L.	4.7 (0.7)	3.7 (0.3)	3.9 (0.4)	4.3 (0.7)	0	0
Mango	*Mangifera indica* Linn.	6.2 (0.9)	6.0 (0.8)	5.6 (0.3)	4.8 (0.2)	1.2 (0.2)	2.1 (0.3)
Lemon	*Citrus lemon* Burm f.	3.6 (0.3)	4.9 (0.4)	2.4 (0.2)	3.5 (0.2)	0	0
Banana	*Musa paradisiaca* Linn.	8.2 (1.0)	7.1 (1.0)	5.4 (0.3)	5.1 (0.4)	21.0 (3.7)	14.4 (3.2)
Papaya	*Carica papaya* Linn.	5.7 (0.4)	4.6 (0.4)	3.4 (0.2)	3.3 (0.2)	8.0 (1.1)	6.0 (0.4)
Sapota	*Achrus sapota* Linn.	3.5 (0.2)	2.6 (0.4)	2.9 (0.5)	0	0	0
Custard apple	*Annona squamosa* L.	1.5 (0.1)	1.9 (0.2)	2.3 (0.3)	4.8 (0.4)	0	0
Sweet orange	*Citrus aurantiifolia* (Christm.) Swingle	1.0 (0.1)	4.4 (0.5)	0	0	0	0
Passion fruit	*Passiflora edulis* Sims f.	1.2 (0.2)	0.5 (0.19)	1.9 (0.2)	0	0	0
Tamarind	*Tamarindus indica* Linn.	1.3 (0.1)	3.0 (0.3)	3.1 (0.3)	2.6 (0.2)	0	0
Jack fruit	*Artocarpus hetrophyllus* Lamk.	3.7 (0.4)	3.3 (0.3)	3.6 (0.4)	4.6 (0.5)	9.6 (0.7)	9.1 (0.9)
Pineapple	*Ananas comosus* (L.) Merr.	7.7(2.8)	5.9 (1.7)	4.6 (1.6)	13.6 (3.1)	0	0
Other species		6.6 (2.2)	9.1 (3.5)	10.4 (4.0)	6.5 (2.1)	4.0 (1.0)	5.1 (2.0)
Spice trees							
Clove	*Eugenia cariophyllata* Thunb	5.8 (1.3)	3.7 (1.0)	4.1 (1.1)	2.0 (1.0)	0	0
Cinnamon	*Cinnamomum zylanicum* Nees.	4.2 (1.2)	3.5 (1.0)	3.3 (1.0)	2.0 (0.4)	0	0.8 (0.2)
Nutmeg	*Myristica fragrans* Houtt.Nees.	3.6 (1.5)	2.7 (1.0)	3.2 (1.0)	1.1 (0.5)	0	0
Black pepper	*Piper niger* L.	4.9 (1.2)	2.7 (0.8)	3.7 (1.1)	1.0 (0.9)	0	0.2 (0.1)
Bay leaf	*Cinnamomum tamala* (Buch-Ham.) Nees & Eberm.	3.2 (1.0)	2.0 (0.1)	2.2 (0.2)	0.8 (0.1)	0	0
Agroforestry/forest trees							
Silk cotton	*Ceiba pentandra* (L.) Gaertn.	4.2 (1.8)	3.7 (1.4)	3.7 (1.0)	4.6 (1.5)	9.3 (2.5)	8.1 (2.2)
Kari leaf	*Murraya koenigii* (L.) Spreng.	3.6 (1.1)	2.7 (1.0)	2.9 (1.0)	3.6 (1.2)	7.8 (2.2)	4.3 (1.2)

(Contd.)

Common name	Scientific name	Dominance (IVI, %)					
		South Andaman	North Andaman	Middle Andaman	Little Andaman	Car Nicobar	Nancowry
Jindaballi	*Gliricidia sepium* (Jacq.)Kunthex Walp.	5.2 (1.4)	5.7 (1.6)	5.2 (1.2)	4.1 (1.1)	11.2 (3.1)	13.0 (3.5)
Padauk	*Pterocarpus dalbergioides* DC.	3.2 (1.2)	2.5 (1.0)	3.1 (1.0)	3.1 (1.2)	0	5.5 (1.4)
Garjan	*Dipterocarpus grandiflora* (Blanco) Blanco	2.4 (0.9)	2.3 (0.8)	2.3 (0.8)	2.7 (0.7)	0	4.8 (1.2)
	Other species	3.0 (0.9)	6.3 (2.2)	6.9 (2.8)	17.1 (3.5)	4.0 (1.1)	11.3 (2.8)
@Vegetable#							
Bottle gourd	*Lajenaria siceraria* (Mol) Standl.	5.2 (1.3)	7.1 (2.8)	6.3 (2.1)	0	0	0
Ridge gourd	*Luffa acutangula*(Roxb.)L.	9.8 (2.2)	9.0 (2.3)	9.5 (2.4)	9.6 (2.5)	0	9.5 (2.8)
Cucumber	*Cucumis stivas* L.	8.1 (2.0)	7.8 (1.8)	8.1 (2.1)	8.1 (1.8)	0	0
Pumpkin	*Cucurbita moschata* Duch ex Poir.	8.6 (2.4)	8.1 (2.2)	8.5 (2.4)	9.1 (2.7)	0	0
Cowpea	*Vigna unguiculata* L.	11.4 (3.2)	10.9 (3.1)	11.3 (3.3)	12.2 (3.2)	0	0
Okra	*Abelmoschus esculentus* L.	12.1 (3.2)	10.4 (2.9)	11.0 (2.5)	13.0 (2.4)	0	25.8 (4.1)
Bringal	*Solanum melongina* L.	1.4 (0.4)	2.0 (0.4)	2.1 (0.8)	6.3 (2.2)	0	5.5 (2.1)
Tomato	*Lycopersicon esculentum* Mill.,nom.cons.	1.9 (0.5)	3.7 (1.8)	2.9 (1.0)	8.1 (2.8)	0	0
Chilli	*Capsicum anuum* vr. *Capitata* L.	2.4 (0.7)	6.0 (1.9)	4.4 (1.4)	7.0 (2.2)	0	5.8 (2.2)
Amaranthus	*Amaranthus tricilor* L.	13.1 (3.4)	11.7 (3.1)	13.8 (3.4)	14.0 (3.7)	0	34.7 (6.8)
Nicobari tuber	*Tecca lentipetiloides* (Linn.) Kuntze.	0	0	0	0	100 (20.3)	0
Others		26.0 (7.5)	23.3 (6.8)	22.1 (4.8)	12.6 (3.1)	0	18.7 (3.8)
No. of crops		27	27	27	23	1	12

@ Relative cover is used to know the relative dominance of vegetable crops.
# Vegetables were grown in the rice fields.
Note: Data in parentheses are standard errors.

Source: Pandey *et al.* 2007

Table 3: Density, diversity, species richness, evenness and concentration of dominance of homegarden trees and vegetable crops in the different islands of Andaman and Nicobar archipelago.

Islands	Density (No ha^{-1})	No. of species sampled	Diversity	Species richness	Evenness	Concentration of dominance (CD)
South Andaman (n=100)						
Palm trees	1210 (33.4)	2	0.01	0.3	1.0	0.9
Fruit trees	80 (5.8)	15	2.1	7.0	3.4	0.3
Spice trees	13 (0.7)	5	2.2	3.6	2.3	0.2
Forest trees	40 (2.7)	11	2.0	6.3	3.5	0.4
*All species	1343(35.3)	34	1.2	10.4	5.1	0.7
Vegetables	-	-	3.6	10.0	4.8	0.1
North Andaman (n=100)						
Palm trees	1119 (51.8)	2	0.3	0.3	1.0	0.9
Fruit trees	71 (6.9)	14	2.3	7.0	3.9	0.3
Spice trees	5 (0.3)	5	2.1	5.6	2.3	0.4
Forest trees	38 (2.4)	11	2.1	6.3	3.5	0.4
*All species	1233 (52.5)	33	1.1	10.3	5.0	0.7
Vegetables	-	-	4.0	9.9	4.8	0.1
Middle Andaman (n=100)						
Palm trees	1104 (49.8)	2	0.3	0.3	1.0	0.9
Fruit trees	53 (3.1)	14	2.6	7.7	3.9	0.2
Spice trees	8 (0.4)	5	2.1	4.4	3.3	0.2
Forest trees	38 (2.2)	11	2.2	6.3	3.5	0.4
*All species	1203 (50.7)	33	1.1	10.4	5.0	0.7
Vegetables	-	-	4.0	9.9	4.8	0.1
Little Andaman (n=30)						
Palm trees	1408 (167.3)	2	0.2	0.3	1.0	1.0
Fruit trees	42 (4.7)	11	1.3	4.8	3.4	0.6
Spice trees	3 (0.2)	5	2.3	9.9	2.3	0.2
Forest trees	34 (3.5)	11	2.8	6.6	3.5	0.2
*All species	1487 (169.8)	30	1.1	9.0	4.9	0.7
Vegetables	-	-	3.8	8.4	4.5	0.1
Car Nicobar (n=50)						
Palm trees	177 (4.8)	2	0.4	0.5	1.0	0.8
Fruit trees	25 (3.1)	7	1.6	4.3	2.8	0.5
Spice trees	0(0)	0	0	0	0	0
Forest trees	14 (1.7)	4	1.1	2.6	1.0	0.6
*All species	216 (12.0)	13	1.4	5.2	3.7	0.6
Vegetables	-	-	-	-	-	-
Nancowry (n=50)						
Palm trees	144 (8.9)	2	0.5	0.5	1.0	0.8
Fruit trees	29 (3.3)	6	1.5	3.4	2.6	0.5
Spice trees	1 (0.8)	2	1.0	-3.7	1.0	0.5
Forest trees	20 (1.8)	8	1.8	5.4	3.0	0.4
*All species	194 (9.1)	18	1.9	7.4	4.2	0.5
Vegetables	-	-	2.7	5.2	3.6	0.2

*include pineapple (not tree) also-indicates not applicable

Notes: Diversity (Shannon index), species richness (Margalef number) and evenness or equitability (Whittaker number) are indices of species diversity whereas concentration of dominance (Simpson index) is an index of species dominance. Formulae of these indices are given in methodology section. Data in parentheses are standard errors.

Source: Pandey *et al.* 2007

fruit and forest species in the homegardens across the islands is probably due to the similar climatic condition (Kaya *et al.*, 2002), but the differences in their proportions is due to the variations in the biophysical and socio-cultural characters of the islands. Greater proportion of the spice trees like clove, nutmeg and cinnamon in South Andaman primarily is due to its hilly terrain. Kaya *et al.* (2002) observe that nutmeg performs well on steep slope, however, clove on gentle slope. Partly it may also be attributed to the preferences of the homegardeners. Tamilians, who predominate in the island, are very much intended to make money from the homegardens. Since, this island is nearer to Port Blair, the capital city of these islands, tourists visit frequently and therefore the spices have a high market demand. Andaman is known for its best quality spices. In the present study, though it is difficult to separate, the influence of the biophysical characters appears greater than that of the sociocultural factors. For instance, both North and Little Andaman are dominated by Bengali population, but fruit trees like mango and citrus predominate in the former whereas pineapple and vegetables in the latter. Fruit fly and fruit borer are severe pests to the mango crop in almost all the islands, but North Andaman is almost free from them (Gangwar and Bandyopadhyay, 1993). Similarly, bacterial wilt is a major constraint to vegetable cultivation, particularly to solanaceous crops, in all the islands, but Little Andaman rarely registers the disease owing to neutral soil pH. Soils of the island are rich in calcium (NBSS LUP, 1991). The bacterial wilt is found more pronounced in acidic soils (Gangwar and Bandyopadhyay, 1993).

Species diversity in the orchard across islands

Diversity in different categories of trees differs among the homegardens across the islands (Table 3). Palm trees registered lowest diversity but highest concentration of dominanace in the homegardens of all the islands. Among the Andamans, fruit trees records the highest diversity in Middle Andaman and lowest in Little Andaman. Similarly, spice trees registers the highest diversity in Little Andaman and lowest in North Andaman. All species diversity is the highest in the homegardens of Nicobar and lowest in that found in the Andamans. In general, species richness increased whereas evenness declines with increasing diversity in the homegardens.

Plant diversity within an orchard

In depth analysis of homegardens reveals that arecanut contributes maximum (54 to 76%) followed by coconut to the density in the homegardens. Arecanut: coconut ratio is higher in the homesteads indicating that farmers prefer arecanut most as it provides economic security to household, whereas coconut serves the subsistence. Arecanut was a highly remunerative crop when free trade was not effected. Arecanuts are distributed in closer spacing (= 60 cm) because natural growth from fallen seeds (nuts) always supplement the population of the species which result into high density. Generally arecanut are distributed under coconut, but in bigger size homegardens a separate block of arecanut are grown mainly for maximization of household income. Banana particularly var. champa, locally known as "Cheena kela" is most common contributing 85% to the total banana population. Pineapple is another fruit crop found relatively greater in the homegarden. It's density was high but frequency low indicating uneven distribution across the homegardens. Only few peasants (6%) are found to grow the fruit species for commercial purpose but maximum (94 %) for household consumption. In maximum homegardens labour intensive cumbersome cultivation of pineapple is avoided because Govt. services are major source of household income in south Andaman. An average 1.5 person (range 1-2) per family is engaged in Govt. offices. But, perennial ligneous species like tamarind and *Ceiba pentendra* are found generally in each homegarden. The former serves as food whereas later provides flosses which are used for making beds and pillows. Tree spices like clove, nutmeg and cinnamon are found in the homegarden, but relative frequency of cinnamon is the lowest perhaps due to higher labour input, in harvesting, debarking and drying and comparatively low return (Rs. 200 per kg) at two years interval. Rotation cycle of cinnamon is generally 2 years, but few farmers make delayed harvesting for higher yields. However, the former species once starts flowering produce yields continuously for around 50 years at minimum input and give reasonably good return (Rs. 300 per kg; Rs. 290 per kg or Rs. 1 per nut, respectively). Mohan Kumar *et al.*(1994) find 127 woody species across the homestead of 14 districts and 3 to 25 species per homestead in Kerela. However, Nair and Sreedharan (1986) reported 30 arboreal taxa from the selected home gardens of Kerala. High number of tree and shrub (301 species) have been reported also in Mayan homegardens of Yucatan, Mexico (Rico-Gray and Wienen, 1963), 168 species in Santa Rosa in the Peruvian Amazon (Padoch and de Jong, 1991) and 179 species in homegardens of Java (Soemarwoto, 1987). In present homegarden natural forest tree species are almost absent perhaps due to high forest cover (86%) in near by area, which fulfill, though illegal, the fuel wood requirement of the farmers. Cooking gas is easily available to farmers in the islands. However, each family brings an average 5 kg fuel wood daily from near by forests (Pandey *et al.*, 2002).

Species diversity in present homegardens corresponds with the diversity value (0.739) reported for the homegardens of Kerala (Mohan Kumar *et al.*, 1994), but is lower from that of evergreen forests (0.90) of Western ghats (Pascal, 1988). Thus, the homegardens ensure crop diversification, provide diversified products though low in amount but nutritious in nature (Pandey and Venkatesh, 2007), conserve plant genetic resources and evolutionary processes (Esquivel and Hammer, 1992) in the Andaman islands.

Despite greater species richness, lower diversity of fruit trees is found in the homegardens of the Andaman islands because of their more homogeneous distribution. It is evident from their higher evenness. The islands were isolated for a long period of time from the mainland due to poor transport facilities. Commercial airport still does not exist in the islands. This seems to have prompted the homegardeners to accommodate greater number of fruit trees in a relatively more systematic manner to fulfill their dietary and nutritional requirements as evident from their maximum consumption by the family members. Lower species richness of fruit trees in the homegardens of Nicobar is probably due to the liking of Nicobaries only for banana. Higher plant diversity despite lower number of species and species richness in Nicobar's homegardens has occurred probably because of their relatively more haphazard distribution owing to spontaneous regeneration. There, the coconut trees too are natural in origin. The homegardens are developed in the islands, long back before the rehabilitation was done, from the seeds flown in from Burma. The island lies at the sea level (Majumdar, 1975). Nicobaries still do not know crop husbandry and therefore never make planting to supplement the regeneration in the homegardens.

Spatial arrangement of plant species in the orchard

The palm, fruit, spice and forest trees form three-storey structure in the homegardens of the Andamans (Fig. 2a & b). Top storey (12 to 16 m) is always occupied by arecanut, coconut and forest trees. The coconuts are planted generally at wider spacing in the beginning and some arecanuts in their interspaces. Trees like mango, jackfruit, neem, tamarind and *Ceiba pentendra* form the second storey (4.50 to 9.50 m). Fruit trees are planted mostly closer to house. Spice trees like nutmeg and cinnamon, and fruit trees like papaya and lemon occupy the first storey. They are grown mostly under the coconut and occasionally under the arecanut. Clove is found in both the first and the second storey as well. Cinnamon is grown commonly under the arecanut. Banana is grown always relatively in open where water from the house drains. Distribution of the house have no specific pattern but is preferred by up lands to avoid stagnation of rainwater. Ground cover, generally, is not cultivated for annual crops, but occasionally it is cultivated for pineapple. Few smallholders (2 to 3%) grow *Curcuma longa* (L.) Roxb, *Zingiber officinalis* Adams. (Boehm.), *Manihot esculenta* Crantz and *Ammorphophallus campanulatus* Blume in the homegardens in South Andaman and only 8 to 10% in North and Little Andaman mainly for household consumption. Flowers like crossandra (*Crossandra infundibuliformis* (L.) Nees), tuberose (*Polyanthes tuberosa* L.), marigold (*Tagetes erecta* L.) and jasmin (*Jasminum grandiflorum* L.) are grown a little in front of the houses and *Ixora parviflora* L. on the homegarden's boundary. Unlike Andaman, coconut forms the second storey whereas banana the first storey in Nicobar's homegardens. A wild tuber (*Tacca lentipetiloides* (Linn.) Kuntze.) and grasses constitut the ground flora in these homegardens.

Fig. 2a: Profile diagram of a typical homegarden in the Andaman and Nicobar islands
Source: Pandey *et al.* 2007

Fig. 2b. View of Homegardens in the Andaman and Nicobar islands.

Most distinguishing feature of the homegardens is a mixed distribution of species in three distinct tiers like other tropical homegardens (Nair and Sreedharan, 1986; Millate-E-Mustafa *et al.*, 1996). The homegardeners seem to have years of accumulated knowledge of ecological requirements of the trees (Kumar and Nair, 2004); therefore, they plant cinnamon, clove and nutmeg under the trees as these require shade (Wezel and Bender, 2003). On the contrary, like Indonesia they grow banana in open near the water channel because it requires light as well as plenty of water (Michon *et al.*, 1986). Fruit trees, they plant mostly closer to the house where relatively greater intensity of light is available. But, Andaman's homegardens like old traditional Javanese homegarden (Michon *et al.*, 1986) are different from most of the tropical homegardens because their ground cover is not cultivated for annual crops mainly due to giant African Snails, which are nocturnal and voracious and cause severe loss to the crops and tender seedlings of the homegarden trees. Nonetheless, they are never a threat to the homegardens. They, in fact, control the stocking rate of the trees in the homegardens, as the smallholders never remove the natural growths from fallen seeds. Major threat to the homegardens is, however, from soil erosion, which is estimated 12 t ha^{-1} yr^{-1} under the palm trees. Moreover, tilth preparation on sloppy land (>45 ^{0}) provokes the soil erosion to 122 t $ha^{-1}yr^{-1}$ (Pandey and Venkatesh, 2007).

Temporal organization of the plant species in the orchard

Temporal organization of crops is an important feature of homestead agroforestry. Temporal organization ensures round the year production in the system because of the difference in the biological cycle in the crops (Table 4). Plantation crops like coconut and arecanut flower and bear nut round the year. Harvesting in arecanut generally occurs from September to December but maximum yields are found from October to November. Harvesting in coconut occurs from January to February for copra, but tender coconuts are harvested round the year.

Floristic similarity in the homegardens of different islands

Floristically the homegardens of Car Nicobar are 92% similar with that found in Nancowry. However, they are only 10 to 18% similar with that found in the Andamans. Among the Andamans, the highest (97%) similarity is observed between the homegardens of North and Middle Andamans and the lowest (82 %) between the homegardens of Little and Middle Andamans (Table 5).

Table 4. Calendar of events of economically important plant species in the homegardens of Andaman and Nicobar archipelago

Sl No. Species	Month											
	Jan.	Feb.	March	April	May	June	July	Aug.	Sept.	Oct.	Nov.	Dec.
1. Plantation crops												
Coconut	FH	FH	FH	FH	FH	FH	FH	FH	FH	FH	FH	FH
Arecanut	FH	F	FH	F	F	F	F	F	F	FH	FH	F
2. Fruits crops												
Cashewnut	F	-	-	H	H	H	-	-	-	-	-	F
Mango	F	F	-	H	H	H	-	-	-	-	-	F
Lemon	-	-	F	F	F	-	H	H	H	-	-	
Banana	FH	FH	FH	FH	FH	FH	FH	FH	FH	FH	FH	FH
Papaya	FH	FH	FH	FH	FH	FH	FH	FH	FH	FH	FH	FH
Pineapple	F	F	-	-	H	H	H		-	-	-	-
Jack fruit	-	-	H	H	H	H	-	-	-	-	F	F
3. Spices												
Clove	H	-	-	-	-	-	F	F	-	-	-	-
Cinnamon	-	-	-	-	-	-	-	-	-	H	H	-
Nutmeg	FH	FH	FH	FH	FH	H	FH	FH	FH	FH	FH	FH
Black pepper	H	-	-	-	-	F	F	-	-	-	-	H
Bay leaf	H	H	H	H	H	H	H	H	H	H	H	H
Rice	H	-	-	-	-	-	T	T	-	-	-	H
Vanilla	-	F	F	-	-	-	-	-	-	-	H	H

F = Flowering, H = Harvesting, T = Transplanting
Source: Pandey *et al.* 2002

Table 5. Similarity (%) in the floristic composition of the homegardens of different islands of Andaman and Nicobar archipelago.

Islands	South Andaman	North Andaman	Middle Andaman	Little Andaman	Car Nicobar	Nancowry
South Andaman	-	92.5	89.1	89.7	14.9	16.2
North Andaman	92.5	-	96.8	84.4	14.3	15.8
Middle Andaman	89.1	96.8	-	82.1	15.9	17.6
Little Andaman	89.7	84.4	82.1	-	10.2	11.4
Car Nicobar	14.9	14.3	15.9	10.2	-	91.9
Nancowry	16.2	15.8	17.6	11.4	91.9	-

Source: Pandey *et al.* 2007

Livestock

Cows being major animal of livestock averaged for 2 cows per homegarden ranged from 1 to 6 per homestead (Table 6). All farmers are not having livestock. Forty percent of the homegardens are not having livestock. Some farmers used he buffalo as a draft animal. Hired tractor is also used for tilth preparation particularly for paddy and vegetable cultivation. An average one goat is found per homegarden. Goats are used mostly for meat. Livestock feeding totally depend or grazing. They are trained in such a way that they go for grazing in morning in nearby forest after milking and comeback in evening at a fixed time. Milking cows are provided concentrate during evening.

Poultry

An average one poultry shade is found in the 80 % homegardens. Poultry birds are comprised of hen, cock and duck. Hen ranging from 2 to 55, average 13 birds per homegarden (Table 6). Sixty seven percent homegardens are having poultry birds like hen and cock. Duck are observed in 38% homegardens. An average 2340 eggs are produced by hen and 633 eggs by ducks in a year for each homegarden. Nearly 25% of the total eggs are consumed by households whereas 75% are sold. Though duck produces relatively greater number of eggs and cost of eggs are also high but numbers of duck are lower because all farmers do not own pond. All the poultry birds feed among the trees in homegarden. Household refuge and a little amount of grains, around 20g per individual, are provided to them daily. In Malaysia, poultry rising in rubber plant stand is a common and remunerative practice (Ismail, 1986). Chickens freely feed among the plants in homegarden at Java (Michon, 1983).

Table 6. Input and output ratio of pond, livestock and poultry birds in homegarden of the South Andaman islands.

Sl No	Component	Nos.	Input (Rs)	Output (Rs)	Input/Output Ratio
1.	Fish Pond *	1	1984	19475	9.82
Livestock					
2.	Cow	2	9836	26638	2.70
3.	Goat	1	100	710	7.1
Poultry birds					
4.	Hen & Cock	13	1138	8046	7.07
5.	Duck	2	198	1901	9.60

*Pond size: 25 feet x 25 feet
-Not estimated
Source: Pandey *et al.* 2002

Fish pond

Fish ponds are found maximally in the homegarden, owned by Bengali family (Table 6). Maximum Bengali family has small fish pond locally known as diggy. Digges are generally 0.5 to 1 m in radius and 1.5 to 2m in depth. This small diggy is used particularly for household use. However, 8% homegardens are having big size ponds. Size of big ponds ranges from 10 x 10m or more than that. Big ponds serve both domestic and commercials purposes. Small diggies are located within the homegardens whereas big ponds are found adjacent to homegarden. Rohu, katla.etc are the fish, which are found in the ponds. Owner of big ponds adds fingerling every year during rainy season. Number of fingerlings depends upon the size of the ponds. Fishes are fed mostly on micro flora. Farmers add cow dung and rice husk to support growth of micro flora in the pond. Cow dung was available from livestock population. Those who owns big size pond purchase cow dung from the neighbour farmers. Farmers apply diammonium phosphate at 3 year interval. This inorganic fertilizer seems to have supported growth of micro flora which is fed upon by fish. In west Java, fish production in homegarden pond is common, with an income of 2 to 2.5 times that of rice fields in the same area (Soemarwoto and Convey, 1991). A fish pond is a common feature of garden of Java and various kinds of fish are fed with vegetable and human waste (Michon, 1983).

Rice fields

Rice fields are located quite close to coconut-arecanut orchards in the Andamans. Smallholders grow mostly long duration rice (6 months) cultivar C14-8, in July-Aug and harvest it in December. The rice is mostly transplanted with one-month-old seedlings raised in the nursery. Generally between the two rice crops they grow vegetables in the rice fields. Though, the vegetable crops are common among the islands, their proportions as well as acreages vary. Proportion of tomato, bringal and chilli is always higher in Little Andaman than that in any other islands. However, proportion of cucumbers, gourds, cowpea, okra and amaranthus is nearly equal among the islands (Table 2). Maximum 0.5 to 2 ha household^{-1} are cultivated for vegetables in Middle Andaman followed by 0.25 to 1 ha household^{-1} in North Andaman. Avreages of vegetable cultivation are lowest 0.02 to 0.25 ha household^{-1} in South Andaman. Cole crops are grown from November to January, whereas, the remaining crops round the year. Nicobaries in Nicobar do not know vegetable and rice crop husbandry. Paddy cultivation in Nancowry is similar to that in the Andamans, but the vegetable cultivation is done in meager.

Management of the homegarden

Soil working, weeding and an occasional little fertilization (150 g tree^{-1}) are performed in all the newly planted palms, spice and fruit trees in the homegardens, until flowering starts (4-5 years). Once the bearing starts only weed cutting with local made swords is done. During the weed cutting fallen leaves of respective trees are made into pieces and put around the tree trunk for decomposition. Two years old cinnamon is harvested and debarked and to avoid invasion in the gaps formed due to the harvesting, weed cutting is done particularly during the rainy season. Planting is always supplemented by natural regeneration particularly of arecanut, coconut and cinnamon. Generally diseases are not reported in the homegarden trees except occasional appearance of stem borer of coconut caused by rhinoceros beetle. Though not very common, smallholders resort to a bio-control measure by putting pheromone traps. The smallholders generally do not cut the old trees; rather let them die naturally in the homegardens. The gap formed due to the death of the trees is filled with natural growths. But, applications of inorganic fertilizers are avoided, as the smallholders scare that these may spoil fertility of the soil. Most of the smallholders (97%) neither perform weeding nor apply fertilizers to the rice crop. However, they perform weeding and soil working and apply insecticides and fertilizers frequently to the vegetable crops. In Nicobar, Nicobaries never perform the management operations in the homegardens.

The homegardeners seem to have a high skill of the soil and crop management. They always perform weeding by mowing by the local made swords but never uproot them to protect the top fertile soils from erosion. Soils of the islands are loose hence prone to erosion (Pandey and Venkatesh, 2007). They harvest two-year old cinnamon because when it grows older it becomes infested with diseases (Michon *et al.*, 1986). Similarly, they cultivate a long duration rice to avoid unfavorable long rainy months for post harvest operations. The homegardeners generally do not apply pesticides to control diseases in the coconut, which occur occasionally, rather resort to a bio-control measure by putting pheromones traps indicating that they are aware of environmental hazards of pesticides. But, unlike Northwestern Tanzania tending operations in trees they never perform which help change in microclimate suitable to the spices (Rugalema *et al.*,1994a).

Functioning of the homegardens

Resources input (labour and capital)

An average family size in the Andamans is 5, of which 1 to 2 are young men of the age group of 20 to 35 year. The family members work in the homegardens and consume maximum (60 %) of the fruits produced (Table 7). A clear-cut division of labour was observed in the islands. Transplanting, harvesting and winnowing are done maximum by women. Men prepare land, collect seedlings from the nursery and transport these to the rice fields. Men and women both make fire wood collection from the nearby forests. In the vegetable cultivation, men do cumbersome works like plowing, fertilizers and pesticides applications and marketing. Women perform weeding and picking. Nevertheless, men and women do irrigation jointly during dry spells. The men perform cleaning of ground cover in the homegardens and harvesting of nuts, fruits, and spices. Mostly women carry out collection of nuts, dehusking of arecanut and post harvest operations of

Table 7: Percent work that men, women and children perform in the homegardens, and rice and vegetable cultivation in the Andaman and Nicobar islands.

Work	Andaman			Nicobar		
	Men (%)	Women (%)	Children (%)	Men (%)	Women (%)	Children (%)
Homegarden						
Weeding	5 (1.5)	85 (3.0)	10 (1.7)	26 (1.9)	53 (1.7)	21 (0.6)
Hoeing	80 (3.2)	0 (0)	20 (2.9)	0 (0)	0 (0)	0 (0)
Nut dehusking	36 (1.5)	62 (3.0)	2 (1.0)	62 (1.7)	11(1.4)	27 (1.5)
Fruit picking	19 (1.4)	0 (0)	81 (1.9)	48 (1.5)	0 (0)	52 (1.7)
Spice picking	65 (2.0)	25 (2.0)	10 (1.2)			
Spice post harvest	13 (2.0)	71 (3.0)	16 (1.8)			
Rice cultivation						
Soil working	95 (3.2)	0 (0)	5 (1.9)	96 (5.6)	0 (0)	4 (1.0)
Nursery raising	86 (3.0)	5 (1.2)	9 (1.1)	90 (4.8)	2 (0.8)	8 (2.8)
Transplanting	0 (0)	90 (3.8)	10 (1.5)	0 (0)	88 (7.5)	12 (3.5)
Harvesting	0 (0)	87 (2.3)	13 (1.7)	10 (2.1)	70 (8.2)	20 (3.6)
Transportation	80 (3.9)	0 (0)	20 (1.7)	82 (3.2)	0(0)	18 (2.8)
Winnowing	10 (1.5)	85 (2.1)	5 (1.3)	8 (2.1)	86(6.5)	6 (1.3)
Vegetable cultivation						
Soil working	90 (4.0)	0 (0)	10 (1.9)			
Sowing	5 (1.1)	85 (2.0)	10 (1.9)			
Cultural operations	3 (1.0)	91 (3.7)	6 (1.4)			
Picking	0 (0)	89 (2.7)	11 (1.8)			
Selling	100 (1.2)	0 (0)	0 (0)			

Note: Data in parentheses are standard errors
Source: Pandey *et al.* 2007

clove, black pepper and nutmeg. But, coconut dehusking is performed by men exclusively.

Mechanization is lacking in the homegardens. The tools used commonly include spade, pickaxe, axe, local made sword and Dav. Tilth preparation in the rice fields is done maximum 80% by draft animals (he buffalos and bullocks) and 20 % by hired power tillers and tractors.

Production

Total production among the homegardens of the Andaman islands did not differ significantly; therefore, their production data are pooled. Arecanut and coconut are major commercial crops in the homegardens, which production figures have been obtained from the smallholders, whereas for the consumed items they have been estimated (Table 8). Maximum 70 % coconut and 98 % arecanut of the total production are sold to local vendors in the Andamans. The family members consume the remaining quantity. Arecanut is used for chewing whereas coconut as an oil seed and for several other food items like sweets and sauce. An average production of the coconut is 30 nuts palm^{-1} (10-50 nuts palm^{-1}) in Andaman and 18 nuts palm^{-1} (6 to 30 nuts palm^{-1}) in Nicobar. Arecanut production is, an average, 0.75 kg palm^{-1} (0.5 to 2.8 kg palm^{-1}). Rice production is quite low 2.69 t ha^{-1}yr^{-1}, but sustainable. Total income from all the components, including rice and vegetables, is estimated US $ 2357.00 in Andaman. Coconut and fish are major commercial commodities, which earn livelihood to Nicobaries in Nicobar. They prepare and sell coconut kernel. Nicobaries either do not eat or eat a little amount of the coconut kernel, but feed (25%) it always to their pigs. They never sell the pigs rather distribute them to their neighboring villagers. Total production of the system in Nicobar is estimated 6.9 times lower than that found in Andaman.

Black pepper cultivation on *Gliricidia* standard in homegardens increases productivity

Organic black pepper, being a constituent of many traditional / ethnic medicines, fetches a premium price in international market (Anonymous, 1999). Farmers in the Andaman Islands are organic black pepper growers by default in their homegardens. Farmers grow black pepper traditionally on a few arecanut palms only for household consumption. Some progressive farmers grow it on *Gliricidia* standards in their coconut plantations at commercial level. The standard provides 3 to 4 times greater canopy cover than arecanut and adds nitrogen rich pruning biomass to the soil. Generally sufficient organic matter is a constraint for commercial organic cultivation of agricultural crops (Magat, 1993). But, homegardens are known to produce a reasonably good amount of leaf litter which recycles nutrients required to sustain productivity at low level (Pandey and Singh, 2009). In natural systems fine root production and its turnover provide an ample amount of nutrients which contribute to the sustainability of the system (Nadelhoffer *et al.*, 1985).

Table 8: Economic (Rs.) characteristics of homegardens of the Andaman and Nicobar islands.

Components of @Homegarden	Andaman			Nicobar		
	Gross income† (Rs.)	#Input(Rs.)	Net income (Rs.)	Gross income (Rs.)	Input(Rs.)	Net income (Rs.)
Arecanut	61495 (280)	10000 (260)	51495 (278)	0	0	0
Coconut	14000 (247)	5116 (75)	8884 (220)	12000(197)	4000 (146)	8000 (176)
Fruit & spices	10492 (330)	1559 (51)	8933 (308)	0	0	0
Vegetables	48370 (480)	30721 (281)	17649 (389)	0	0	0
Rice	13350 (198)	5875 (182)	7475 (473)	12500 (198)	6750 (158)	5750 (147)
Total	1,47,707(9470)	53,271 (4645)	94,456 (7468)	24,500(2685)	10,750 (1240)	13,750 (1539)

@Land management unit is ha^{-1}

Family labour is included in "Input." Inputs in rice include transplanting, harvesting, threshing, tilth preparation.

† 1$ (US) = Rs.45.40, August 04

Note: Data in parentheses are standard errors

Source: Pandey *et al.* 2007

Component interaction in homegardens

Components of the homegardens are inter- related and inter- dependant. Dead roots, leaves, inflorescence, nut husks and household refuge (vegetables and fruits peels) provide substratum to bio- chemical cycle similar to that of natural ecosystem which is further supplemented with organic matter inputs from poultry and livestock (data not shown) and humus carried from the adjoining forests that lie on the hilltop. It is observed that pool sizes of nutrients in the homegardens are either equal to or greater than that of the nearby forests depending on altitudinal location (C, N, P, K, Na = 0.91%, 0.32%, 25.46 $\mu g\ g^{-1}$, 0.034 % and 0.027 %, respectively in homegardens and 0.96%, 0.41%, 21.18 $\mu g\ g^{-1}$, 0.01 %, 0.013 %, respectively in the evergreen forest). Rugalema *et al.* (1994a) also find cyclic interaction between crop residues, cattle manure and tree litter in the homegardens of North – Western Tanzania. Complementary interaction is evident from changed microclimatic conditions under the coconut and arecanut trees which help growth in the underneath crops. Temperature is found 2 °C lower (30 °C vs. 28 °C) in the homegardens compared to that in open. Intra- specific competition is common in arecanut patch owing to closer spacing (= 60 cm), which causes reduction in nut size and yields. Inter-specific competition occurs between the coconut and fruit trees for soil-water, nutrients and light, but the greatest for the last one, that changes the geometry of the trees (lower canopy and greater height). The light is reduced 4.8 times under trees in the homegardens compared to that in the open (2439 Lux in the homegardens vs. 11636 Lux in the open). Thus, components of the homegardens i.e. trees, livestock and household members interact together in space and time that result in a steady flow of energy, which leads to the cycling of nutrients. Such type of nutrient cycling distinguishes homegardens from other agroecosystems.

Economics of the homegardens

Economic relations of different components of the homegarden are given in Fig. 3. It is apparent from the Table 8, that economy of the homegarden depends maximum on arecanut. It contributed maximum 66% to the total economy of the homestead. But, out put/input ratio is greater for spices and fruits because they require relatively low input. Total annual output of the homegarden, including its all components, is worth rupees 1,37,206 of which 83% is contributed by orchard, 5% by livestock, 3 % by poultry birds and 9% by fish pond. Total expenditure in the garden is Rs. 62,021. Of the total labour, 56% is done by the family members and rest 44% by hired laboures in the orchard. Coconut, arecanut, fruits, spices and silk cotton are consumed worth rupees 15,599 by the household members. Net return for Rs. 1, 14,126 goes to household from orchard via market. Livestock contributes 9.11% to the net income of the household. Livestock requires total annual expenditure for rupees 9936 as feed. Total income from livestock is rupees 27,238 of which 95% is contributed by milk, 2% by meat and 3% by cow dung.

Household utilizes 40% of the total milk production. Maximum 75% cow dung is used in pond and 25% in orchard. But those who do not own pond use cow dung for orchard. The net income to the household from poultry birds is estimated Rs. 4032. Poultry birds like cock and hen are fed in orchard whereas ducks are fed in ponds or paddy fields. They are provided additional household refuge and grains equivalent to Rs. 1336. Limited farmers particularly more the Bengali family, owns pond. Govt. is encouraging the farmers for aquaculture and providing subsidies for 25x25 m size ponds. Ponds require annual expenditure for Rs. 1774 in the form of cow dung, rice husk and fingerlings. Livestock do not meet out the requirement of cow dung for the pond. Cow dong worth Rs 900 is purchased from the market. Total return from the pond is estimated Rs. 19,475 of which 28% is consumed by family members. Poultry birds and ponds require less management expenditure compared to livestock. For each rupee expenditure there

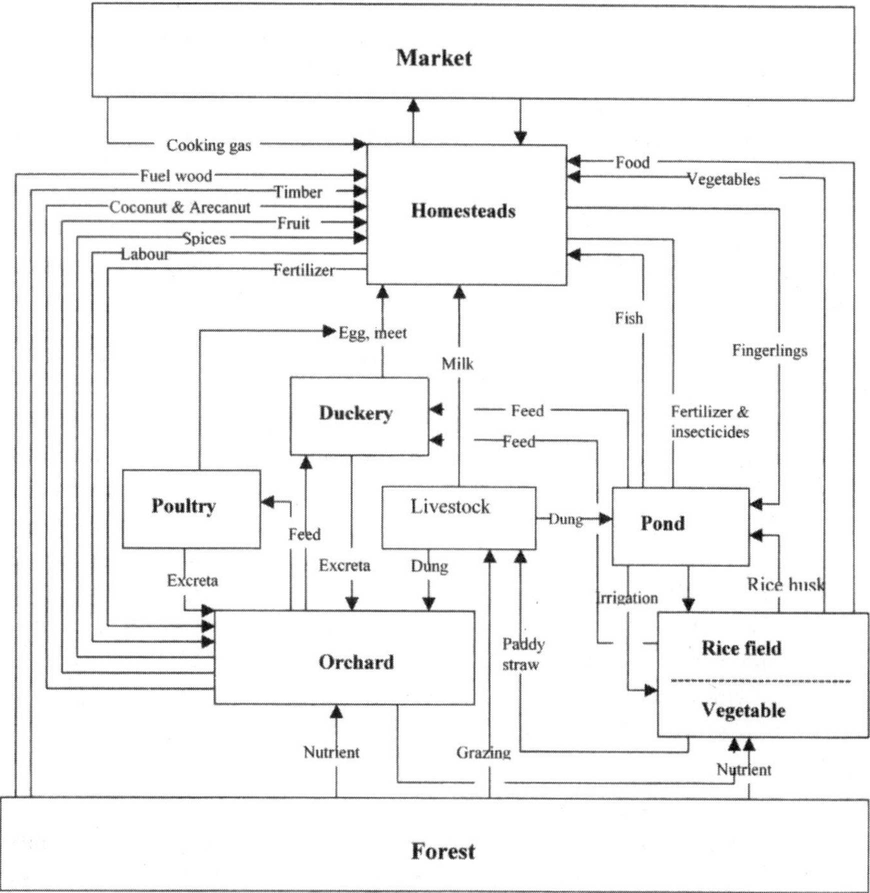

Fig. 3: Flow diagram showing components and their inter-relations in the homegardens of the Andaman and Nicobar islands

is a output for 8 rupee from poultry birds and 9 rupees from the pond. However, this amount is nearly half for the livestock compared to pond and poultry birds. Since homegarden is arecanut based system, maximum food grains and other household items are purchased from the market. Food grains and other house hold items are estimated worth rupees 41619 each year. Thus total net income from all components of the homegarden are estimated Rs. 1,37,206. Farmers utilize an average Rs. 41675 for purchasing food grains, cloths and other household items from the market.

Threat to the homegarden

The homegardens in the islands are a main source of economy to many households. There is a limited option for out-side job because these islands lack industries and factories. Therefore, the homegardening is characterized by subsistence as well as commercial farming like other tropical homegardens (Dury *et al*., 1996; Alvarez-Buylla Roces *et al*.,1989; Michon *et al*., 1986). Arecanut and coconut are two major cash crops in the homegardens but former the greater which is evident from its highest IVI. Total income from the homegardens is well within the range reported in Mexican (0 to 6525 $ US) (Alvarez-Buylla Roces, 1989) and Javanese homegardens, Indonesia (365 to 5000$ US) (Dury *et al*., 1996). However, the extremely low income from the homegardens of Nicobar is mainly because the Nicobaries are an aboriginal tribe and they are still inclined towards hunting and gathering and rely more on wild crops. This is also one of the reasons why the system survives.

Population increase and rural-urban migration seem two major factors that may upset these systems and the livelihoods that depend on them. The high population growth and patrilineal inheritance system, where land property is divided equally among sons, are expected to cause fragmentation of the homegardens like Pekarangans of Indonesia (Arifin *et al*., 1997) that may lead losses in species similar to Chhagga homegardens of Tanzania (Rugalema *et al*., 1994b) and ultimately may convert the systems to either into cropping systems consisting of less number of crops or small-scale plantations of the palms and fruit trees depending upon frequency of division like the homegardens of Kerala, India (Ashokan and Kumar, 1997). The tribal society (community land ownership) that provides safeguards from excessive land fragmentation, however, may prove itself a better option for protection of the traditional homegarden systems in future.

Migration of work force (youth) to urban areas for employment, though low at moment, is expected to increase with time due to the population increase and decline in arable land due to developmental activities like roads, dams, buildings etc., resulting into decline in per capita income from the system. It will affect agricultural tasks adversely that ultimately impoverish the system badly like the traditional homegardens of Catalonia (Spain) (Agelet *et al*., 2000). However, such changes may be averted through employment generation at village level. Diversified products from homegardens are known to provide opportunities for the development of small-scale rural industries, which create off-farm employment that may reverse the negative effects of the migration (Torquebiau, 1992). Therefore, financial institutions should provide soft loans for setting up the industries and government should provide marketing facilities for the poverty alleviation in the islands.

Conclusions

The study concludes that homegardens of all the islands include palms, fruit and spice trees distributed in a haphazard manner. However, their proportions depend on the biophysical and sociocultural factors and preferences of the homegardeners. High stocking rate particularly in the palms resulted from natural regeneration, low or absence of high yielding varieties of the palms, and poor or no crop management cause low productivity in the system. Lack of knowledge of crop husbandry, however, exacerbates the effects that reduce the productivity many folds in the tribal homegardens of Nicobar.

The islands are endowed with a high amount of rainfall (3000 mm) distributed over 8-9 months, but the homegardeners do not utilize it properly due to lack of knowledge. For achieving increased and stable crop production of the system, techniques of integrated farming system have to be applied. Some of the techniques are: (1) double cropping of rice with, a medium duration (120 days) (*Oryza sativa* var. QL No.1 or 148) followed by a short duration (105 days) variety (*O. sativa* var. *Taichung-Sen-yu*), improved cultivation methods like SRI (system of rice intensification), and mechanized harvesting and post harvest operations. The SRI method of rice cultivation requires low inputs, prevents lodging and ensures high yields in the crop; (2) integration of a pond that optimizes the use of rainwater through an aquaculture (composite fish culture) and helps in conservation of rain-water which falls short during the dry months (March and April) in these islands; (3) replacement of low yielding palms with high yielding cultivars, like king coconut (80 nut palm^{-1}) for coconut and samarudhi (3 kg palm^{-1}) for arecanut, and intercropping of the spice and fruit trees, including highly remunerative black pepper on live standards of *Gliricidia sepium* that provides 6 times greater yields compared to that on the arecanut palm. However, the intercropping should ensure more interspecific than intraspecific competition. Though precise information is not available on homegarden system, personal observation suggests that intraspecific competition (arecanut) reduces yields greater than that of the interspecific competition (coconut-spice trees). However, appropriate stocking rate of the trees for the multistorey cropping system should be worked out. Similarly, wilt resistant varieties of vegetable (solanaceous) should be developed, and fruit fly and fruit borer control measures of mango should be evolved to increase the productivity of the former and marketability of the latter; and (4) integration of animals (livestock and poultry) that will ensure milk and egg production for domestic consumption and sale, provide draft power for tilth preparation and maintain fertility of soils. Since these components (homegarden, livestock, poultry and pond) are inter-related and inter-dependant, their appropriate size for unit land (1 ha) should be worked out for optimum production from the system on sustainable basis.

References

Agelet, A., Angles, B.M., Valles, J. 2000.Homegardens and their role as a main source of medicinal plants in mountain regions of Catalonia(Iberian peninsula). Eco. Bot. 54:295-309.

Alvarez-Buylla Roces, M. E., Chavero, E. L., Garcia Barrios, J. R. 1989. Homegardens of a humid tropical region in southeast Mexico: an example of an agroforestry cropping system in a recently established community. Agrofor. Syst. 8: 133-156.

Anonymous 1999. Product and market development – organic food and beverages: world supply and major europian markets, Geneva, Switzerland, International Trade Centre.

Arifin,H.S., Sakamoto K., Chiba, K. 1997. Effects of the fragmentation and change of the social and economical aspects on the vegetation structure in the rural homegardens of West Java. J. Japanese Inst. Landsca Archite. 60:489-494.

Ashokan,P. K., Kumar, B. M. 1997.Cropping systems and their water use. In: Thampi K.B., Nayer N. M., Nair, C. S., (eds.), pp.200-206, he Natural Resources of Kerala. World Wide Fund for Nature- India, Kerala State Office,Thiruvananthapuram 14, Kerala, India.

Bandtopadhyay, A.K 1998. Agro forestry Systems in Andaman: Home garden. Agro-climate zonal planning Unit for the Islands Zone_XV. CARI, Port Blair, A&N Islands.

Basic Statistics, 2001. Andaman and Nicobar Islands: Basic Statistics. Directorate of Economics and Statistics, Andaman and Nicobar Administration, Port Blair, India, 461 pp.

Boonkird, S., Fernandes, E., Nair, P. 1984. Forest villages: an agroforestry approach to rehabilitating forest land degraded by shifting cultivation in Thailand. Agrofor. Syst. 2: 87-102.

Ceccolini, L. 2002. The homegardens of Soqotra island, Yemen: an example of agroforestry approach to multiple land-use in an isolated location. Agrofor. Syst. 56: 107-115.

De Clerck, F. A. J., Negreros-Castillo, P. 2000. Plant species of traditional Mayan homegardens of Mexico as analogs for multistrata agroforests. Agrofor. Syst. 48: 303-317.

Dury, S., Vilcosqui, L., Mary, F. 1996. Durian trees (*Durio zibethinus* Murr.) in Javanese home gardens: Their importance in informal financial system. Agrofor. Syst. 33: 215-230.

Esquivel, M., Hammer, K. 1992. The Cuban homegarden 'conuco': A perspective environment for evolution and in-situ conservation of plant genetic resources. Genetic Res. Crop Evo. 39:9-22.

Fernandes, E.C.M., Nair, P.K.R. 1986. An evaluation of the structure and function of tropical home gardens. Agrofor. Syst. 21:279-310.

Forest Working Plan. 1980. Working Plan for the Middle Andaman Forest Division. Port Blair, India, 238 pp.

Gangwar, B., Bandyopadhyay, A.K., (eds.), 1993. Annual Report. Central Agricultural Research Institute, Port Blair, India, pp.132.

Ghosh, A.1994. A Study on the Development Strategy for the Andaman and Nicobar Islands. Classical Publishing Company, New –Delhi.

Gillespie, A. R., Knudson, D. M., Geilfus, F. 1993. The structure of four homegardens in the Peten, Guatemal. Agrofor. Syst. 24: 157-170.

Ismail, T. 19986. Integration of animals in rubber plantations. Agrofor. Syst. 4: 55- 66.2

Kaya, M. L., Kammesheidt, Weidelt, H. J. 2002. The forest garden system of Saparua island, Central Muluku, Indonesia, and its role in maintaining tree species diversity. Agrofor. Syst. 54: 225-234.

Kumar, B. M., Nair, P. K. R. 2004. The enigma of tropical homegarden. Agrofor. Syst. 61: 135-152.

Magat, S. S. 1993. Knowing soil fertility management and fertilizers. Quick Notes Review. Quezoncity, M. Manila, DOST, PCA, pp.50.

Majumdar, R. C. 1975. Penal Settlement in Andamans. Ministry of Education and Social Welfare, Govt. of India Press, New Delhi, India, pp. 339.

Michon, G.1983. Village- Forest – Garden in West Java.Pp.13-24. In : P.A. Haxley (ed.) Plant Research and Agroforestry, ICARF, Nairobi, Kenya.

Michon, G., Mary, F., Bompard, J. 1986. Multistoried agroforestry garden system in West Sumatra, Indonesia. Agrofor. Syst. 4: 315-338

Millate -E- Mustafa, M. D., Hall, J. B., Teklehaimanot, Z. 1996. Structure and floristics of Bangladesh homegardens Agrfor. Syst. 33: 263-280.

Mohan Kumar, B., George, S. J., Chinnamani, S. 1994. Diversity, structure and standing stock of wood in the home gardens of Kerala in peninsular India. Agrofor. Syst. 25:243-262.

Nadelhoffer, K. J., Aber, J. D., Melillo, J. M. 1985. Fine root, net primary production and soil nitrogen availability: a new hypothesis. Ecology 66: 1377-1390.

Nair, M. A., Sreedharan, C. 1986. Agroforestry farming systems in the homegardens of Kerala, Southern India. Agrofor. Syst. 4: 339-363.

NBSS LUP. 1991. Soil Resource Atlas: Andaman and Nicobar Islands. National Bureau of Soil Survey and Land Use Planning (ICAR) and Directorate of Agriculture, Andaman and Nicobar Administration, Port Blair, India, pp. 64.

Ninez, V. K. 1984. Household gardens: theoretical considerations on an old survival strategy. International Potato Center. Potatoes in Food Systems Research Series Report No. 1. Lima. Peru.

Okigbo, B. N. 1990. Homegardens in tropical Africa. In: Landauer, K., Brazil, M. (eds.), pp. 21-40, Tropical Homegardens. United Nation University Press, Tokyo.

Padoch, C., de Jong, W. 1991. The house gardens of Santa Rosa: Diversity and variability in an Amazonian agricultural systems. Eco. Bot. 45:166-175.

Pandey, C. B., Singh, L. 2009. Soil fertility under homegarden trees and native moist evergreen forest in South Andaman, India. J. Sust. Agri. 33:303-318.

Pandey C B, Venkatesh A. 2007. Agroforestry for Sustainable Biomass Production in the Andaman and Nicobar Islands. Technical Report, Central Agricultural Research Institute, Port Blair, India, pp 147.

Pandey, C. B., Lata, K., Venkatesh, A., Medhi, R.P. 2002. Home garden: its structure and economic viability in South Andaman, India. Indian J. Agrofor. 4:17-23.

Pansey, C.B., Rai, R.B. Singh, L. Singh, A.K. 2007. Homegardens of Andaman and Nicobar, India. Agric. Syst. 92:1-22.

Pascal, J.P.1988. Wet Evergreen Forests of the Western Ghats of India. French Institute, Pondichery, India.

Rico-Gray, C. E., Wienen, W. 1963. The Mathematical Theory of Communication. Urban University Press, IL, USA.

Rugalema, G. H., Johnsen, F. H., Oktingati, A., Minjas, A. 1994a. The homegarden agroforestry system of Bukoba district, North – Western Tanzania. 3. An economic appraisal of possible solutions to falling productivity. Agrofor. Syst. 28: 227-236.

Rugalema,G.H., Johnson F.H., Rugambisa J. 1994b.The homegarden agroforestry system of Bukoba district, North-Western Tanzania.2.Constraints to farm productivity. Agrofor. Syst. 26: 205-214.

Soemarwoto, O. 1987. Home gardens: a traditional agroforestry system with a promising future. In: Steppler, H. A., Nair, P.K.R. (eds.), Agorofrestry: A Decade of Development. ICRAF, Nairobi, Kenya. pp. 157-172.

Soemarwoto, O., Conway G. R. 1991. The Javanese home garden. J. Farm. Syst. Res. Ext. 2:95-118.

Torquebiau, E. 1992. Are tropical agroforestry homegardens sustainable? Agric. Ecosyst. Environ. 41:189-207.

Vogl, C. R., Vogl-Lukraser, B., Caballero, J. 2002. Homegardens of Maya Migrants in the district of Palenque, Chiapas, Mexico: Implications for Sustainable Rural Development. *In:* Stepp, J. R., Wyndhum, F. S., Zarger, R. K., (eds.), Ethnobiology and Biocultural Diversity, University of Georgia Press, Athens, GA. pp. 1-12

Wezel, A., Bender, S. 2003. Plant species diversity of homegardens of Cuba and its significance for household food supply. Agrofor. Syst. 57: 39-49.

7

Integrated Farming System for Island Conditions: A Case Study from The Andaman and Nicobar Islands

N. Ravisankar, S.C. Pramanik, S. Jeyakumar, M. Din, S.K. Ambast and R.C.Srivastava

Abstract: *Agro-ecosystem analysis of the island farming system revealed four distinct micro farming situations (MFS) viz., Hilly (MFS I), Slopping hilly upland (MFS II), Medium upland valley (MFS III) and Low lying valley (MFS IV). The enterprise combination study indicated the suitability of plantation based cropping sequences + backyard poultry and livestock in hilly areas, Short and medium duration paddy, vegetables, floriculture, plantations, fodder + backyard poultry + goat, fish cum poultry cum duck and cattle in slopping hilly uplands, vegetables, plantations + backyard poultry, fish cum poultry cum duck in medium upland valley and long and short duration paddy, vegetables + backyard poultry, fish cum poultry and cattle in low lying valley micro farming situations. Productivity and profitability analysis of different farming systems under various micro farming situations revealed that net return obtained from cropping was Rs 3,26,408, livestock and poultry was Rs 64,300 in MFS I where as under MFS II, it is Rs. 1,53,989 from cropping, Rs 77,176 from livestock and backyard poultry and Rs 12,465 from poultry + duck + fish. In MFS III, net return obtained from cropping was Rs 9,297 livestock was Rs 72,306 and poultry was Rs. 2,630 and poultry + duck + fish was Rs 24,957. Similarly, in MFS IV, net return obtained from cropping was Rs 57,760 where as Rs 90,893 was obtained from livestock, Rs 11,000 from backyard poultry and Rs 14,872 from poultry*

+ duck + fish. Energy analysis of various systems infers that effective conversion of input energy in to out put energy. For example in MFS II, the total input and output energy from field crops, vegetables, plantation and spices was 40,149 and 1,35,092 MJ where as in cattle component, it is 200190 and 22465 MJ respectively. Component analysis of Integrated Farming system under various farming situations indicated that cropping contributed more to net returns (69 to 83 %) in hilly and slopping hilly uplands where as livestock components (Cattle, poultry and fish) contributed more to net returns (49 to 66 %) in medium upland valley and low lying valley areas. Study on bacterial load in fish cum poultry cum duck system revealed, bacterial load especially Salmonella sp *in the pond gets increased during monsoon and further increased during the summer inferring unsuitability of pond water for house hold purposes.*

Introduction

Emerging production scenario, higher economic growth, population explosion and shifts in dietary pattern has changed the supply and demand profiles of food from agriculture and its allied sectors in the new millennium. Integrated Farming System (IFS) seems to be the possible solution to meet the continuous increase in demand for food, stability of income and diverse requirements of food grains, vegetables, milk, egg, meat etc., thereby improving the nutrition of the small-scale farmers with limited resources (Fig. 1&2). Integration of different agriculturally related enterprises with crops provides ways to recycle products and by-products of one component as input through another linked component and reduce the cost of production and thus raises the total income of the farm. Multiple land use through integration of crops, livestock and aquaculture can give the best and optimum production from unit land area as the scope for horizontal expansion of crop area is bleak. In other words, farming system is a resource management strategy to achieve economic and sustained production to meet diverse requirement of farm household while preserving natural resource base and maintaining a high-level environment quality.

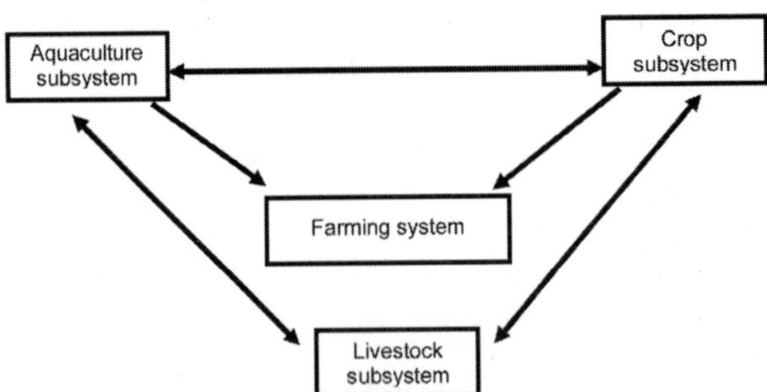

Fig. 1. Linkages of different components in farming system

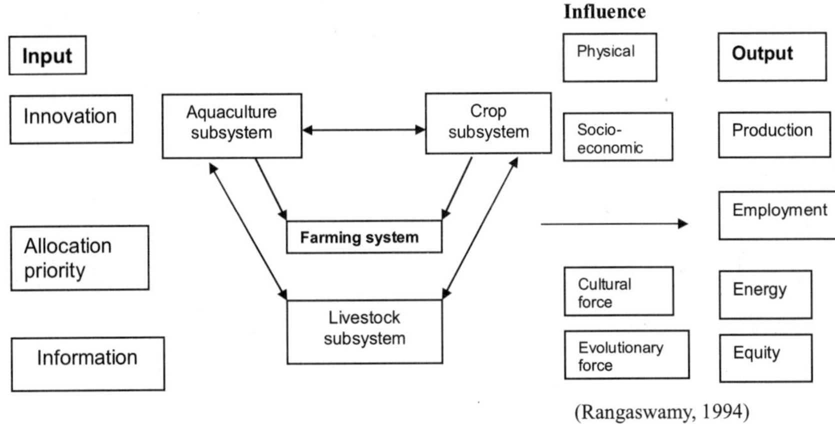

Fig. 2. Integration of subsystems into a sustainable farming system

Thus considering the economic stability of production and sustainability of the fragile ecosystem of Bay islands, study on development of Integrated Farming System model under different resource conditions in humid tropics of Bay Islands was carried out at Central Agricultural Research Institute, Port Blair during 2004 to 2007 under which farmers from four different villages, namely Calicut, Mannarghat, Tushnabad and Port Mout which are distinct in terms of ecology, cropping system and enterprises, were selected for study and development of IFS model (Table 2).

Weather

On an average, the islands receive 3074 mm (Table 1) of rainfall distributed over 8 months. From June to September, the rainfall is intense and may have even up to 30 rainy days per month. From January to April, the rainfall is scanty or nil, particularly in March (Fig. 3). The undulating terrain results in severe water crisis during the period which is aggravated by higher evapo transpiration (Fig 3).

Table 1. Weather parameters of Andaman Nicobar Islands

Average Rainfall (mean of 48 years)	3074.2 mm
Number of rainy days (mean of 26 years)	143
Mean minimum temp (mean of 48 years)	23.0^0 C
Mean maximum temp(mean of 48 years)	30.1^0 C
Mean relative humidity (mean of 31 years)	80 %
Mean wind speed (Mean of 22 years)	9.7 KMPH
Evapotranspiration	1588.2 mm
Mean daily sunshine hours (mean of 1 year)	6.16 h

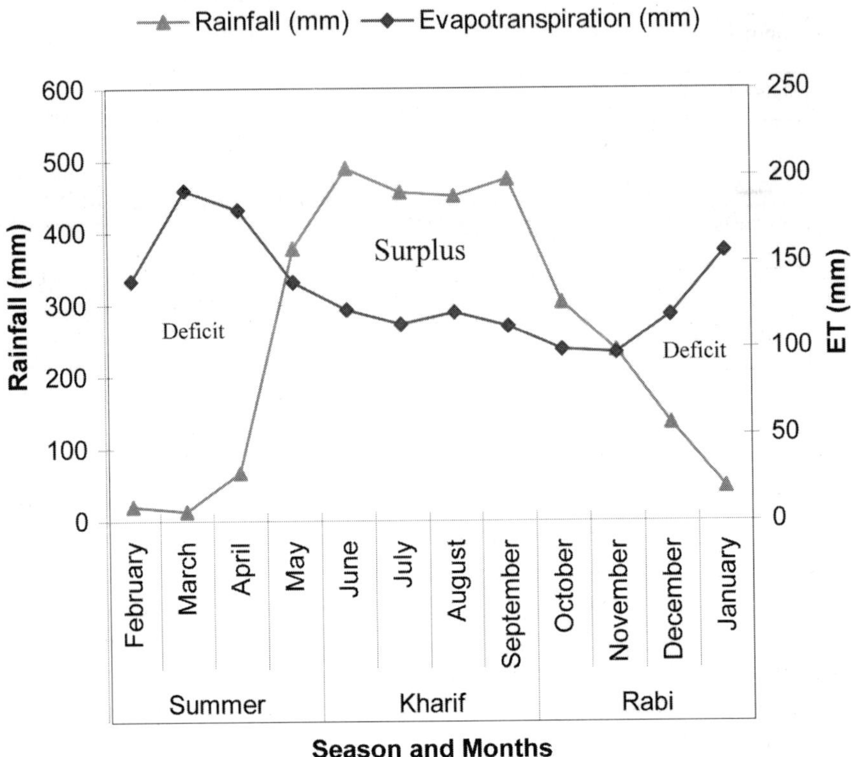

Fig. 3. Water and deficit periods

Table 2. Number, area and average size of operational holdings in A&N Islands

Category	Numbers	Area (ha)	Av. size (ha)
Marginal	2428	916	0.38
Small	2424	3435	1.42
Semi-medium	3343	8210	2.46
Medium	2119	9128	4.31
Large	68	4835	71.10
Total	**10382**	**26524**	**2.55**

Development of water resources

After massive earthquake followed by tsunami that occurred on 26th December 2004, the island has shifted towards south westerly direction and has tilted which has resulted in lowering of ground water table. Thus after tsunami there was an acute water shortage in some parts of the island including Calicut village. To over come this problem, water resources was developed through ring well. Shallow

wells of 2 m and 2.5 m diameter with a depth of 3.2 and 4 m respectively was constructed in the valley area in slopping hilly upland situation with a capacity to harvest 10.05 m^3 and 19.64 m^3 of water per day. Due to creation of water resources (ring well) through participatory mode, area under cropping increased apart from growing crops such as sugarcane, chillies, Bhendi, ginger, marigold even during the water scarcity period.

Characterization of resource

Agro-ecosystem analysis of the farming systems revealed four distinct micro-farming situations.

Farming situations

Hills (MFS-1): Coconut and arecanut is the main crop with fruit trees like jackfruit, guava, sapota as intercrops in the backyard of homesteads: Topography is sloppy, soils were gravel and coarse sand, soil depth was 30-45 cm only with underlying stones and boulders. Major problems were soil erosion. Small ponds are present in the lower reaches of the slope.

Slopping Hilly upland (MFS-II): Arecanut is the main plantation crop, banana, pineapple grown as fruit crop. Majority of the cash crops like vegetables, flowers, sugarcane, are cultivated on this land, topography is slopping 15-25%, soils are coarse sand, gravel, soil depth is 50-70 cm., soil erosion is the major problem on these field. Unmanaged fish ponds are present.

Medium upland valley (MFS-III) : Paddy, vegetables and pulses are the main crop, topography is less slopping. This land mainly used for vegetable cultivation as cash crops, soil are sandy loam and silty due to colluvial deposition. Soil depth more than 1.5 m.

Low lying valley (MFS – IV): Long duration paddy is the main crop on these land. Fields are adjacent to seacoast. Soil depth is more; soil is clay to clay loam. Water logging during monsoon is the main problem on these lands.

Characterization of livestock and poultry

The farmers had jungli fowl as backyard poultry. The birds were fed with wheat, broken rice, rice bran, kitchen waste and rest of the time they scavenge around the household. The average egg production of jungli fowl was 70 eggs per annum. Each farmer had 35-40 numbers of jungli fowl. The farmers also had cow and work bullocks. The cattle were fed with paddy straw and were let for grazing in near by lands. The production of milch cow was low (average 4 litre/day) due to non-availability of green fodder.

Characterisation of fish ponds

In Andaman ponds are relatively small and shallow bodies of impounded water with limited wind action. Both drainable and undrainable pond are found in the farmers field. Selection of the farm has been done based on the potential availability or development of almost all component like fish cum poultry integration, fish cum duck integration, Fodder cum plantation integration etc. Bay Island receives eight months rainfall in a year and remaining months are in the form of summer characterized by high humidity and temperature ranging from 30-37 degree Celsius. With the onset of monsoon, torrential downpours sweep across the land and the amount and frequency of rain decrease towards the end of the monsoon. Though these ponds are basically constructed for minor irrigation purpose and storing water in such areas for multiple uses, ranging from supplying drinking water for human population, livestock, etc., recent trend is to utilize them for fish culture.

General characteristic of pond

Ponds in Andaman are relatively small, perennial or seasonal water bodies constructed or excavated for multiple uses. Some of these ponds have proper embankments. They are mostly in the dimensions ranging from 0.03 ha to over 0.1 ha in water surface area and 0.5 cm to 5 m depth. Larger ponds are relatively deeper while smaller and seasonal ponds are shallower. Unlike shallow seasonal ponds, the bottom of the small perennial ponds is never exposed to sunlight and therefore the whole ecosystem of such ponds is quite different from those of shallow and seasonal ponds. Use of these ponds for Integrated Farming System will provide the source of organic enrichment. As soon as the monsoon ceases, water level starts decreasing gradually and shortage of water is quite common during the pre-monsoon season. Water is lost from the pond through evaporation, seepage and transpiration by aquatic macrophytes and the trees and shrubs planted along the pond sides.

While designing the fish sub-system, both fish production conditions as well as the type of wastes/by-products expected from other sub-systems that are to be recycled in fishpond has been taken into account. Except for modifications in the design to accommodate the poultry/crop sub-system, the rest remains more or less similar to normal polyculture system.

a. Size of pond

Considering rural conditions and farmers' capacity to invest capital, no modification was made in pond size (0.036 ha), which was almost ideal for fish cum poultry cum duck culture. Smaller ponds can be used for integrated fish culture. The size of the pond identified for Andaman and Nicobar island condition is 0.036 ha for Integrated Farming System which is the optimum size of ponds present in the farmers field.

b. Depth of pond

Any pond that retain 2 to 3 m water can be considered as suitable. However, the determining factor is the water depth in dry season. Minimum of 1.5m of water depth is essential during the summer season. Risk of organic overloading is high in the shallow water ponds when poultry-duck sub system is integrated which may restrict growth of fingerlings and even death of fishes. In the ponds under IFS, water level of 3.0 m was found to be optimum during rainy season and 1.0-1.5 m in summer season for poultry unit of 25 birds and 5 ducks.

c. Fishpond management

The basic management practices in integrated fish pond are more or less similar to that of simple polyculture system. Pond preparation, daily routines, sampling, harvesting, and health care are same. However, fish species combination has to be adjusted according to the type of the livestock sub-system to be integrated. There should be no supplementary feeding and fertilization of the pond water. After earthquake and Tsunami in 2004 in A and N islands, major and minor cracks developed in the ponds which caused the low water level during summer months hence, early harvesting was advised to the farmers to avoid any disease outbreak.

d. Feeding

As advised in fish cum poultry integration that supplementary feeding should be avoided in order to have healthy fish, as there is always possibility of high organic production due to poultry dropping.

Table 3. Status of the pond before integration of IFS in different resource condition

Micro farming situation	Village Pond (ha)	Area of	Nature of Feed	Size of Stocking (mm)
MFS II, model 1	Calicut	0.036	Kitchen waste	120-150
MFS II, model 2	Calicut	0.036	Kitchen waste	120-150
MFS III	Tushnabad	0.06	Rice bran, Kitchen waste,	300
MFS IV	Port Mourt	0.037	Rice bran, Kitchen waste,	120-150

In general, the yield of the pond was very poor due to the following factors

i. Farmer's knowledge about fish farming was not scientific.
ii. In some cases, they had no idea of fish farming and its economical impacts in the overall farming status
iii. Feeding was not given any priority, because they had the impression that fish will grow marketable size naturally.
iv. Management practices like regular netting, liming was not given due attention

Integration of systems and evaluation

In an effort for a holistic integration of different farming enterprises with cropping with the objectives of increasing income and recycling of farm wastes and by-products to sustain the soil productivity under various agro ecological situations *viz.,* Low land valley, medium valley, sloping hills and hills, studies on integrated farming system involving various components were carried out. The identified components and their integration are briefly given below. Integration and size of component are presented on 1 ha area basis except MFS III.

Micro Farming Situation – I : Hilly

Cropping sequence

Coconut (0.01)+ sapota (0.05)+ banana(0.05)	:	0.20 ha
Arecanut + black pepper+ fodder	:	0.50 ha
Sapota + banana (0.03 ha)	:	0.10 ha
Banana	:	0.10 ha

The coconut is the unique plantation crop in these islands, which occupies nearly 50% of the total cultivable area. The land holdings of the farmers are very small (1ha and below) and the return per unit area is very less due to lack of technical know-how. To increase the return per unit area, the concept of coconut based cropping system was introduced. Thus coconut based cropping/farming systems involving cultivation of compatible crops in the interspaces of coconut and its integration with other enterprises like dairy, poultry etc. led to considerable increase in production and productivity per unit area, by more efficient utilization of sunlight, soil, water and labour.

Integrated farming systems involving Poultry + Cattle + Crop was taken in Mannarghat village. The normal cropping programme followed was Arecanut, coconut + sapota, banana as sole crop. But in the integrated approach, a modification was made in the existing cropping pattern by integrating black pepper with arecanut. Cropping was undertaken in 0.90 ha and 0.10 ha was allotted for backyard poultry and livestock. Vanraja (20 no.) and ducklings (5 no.) were supplied to the farmer for rearing as backyard poultry. Birds were fed with grains, broken rice, wheat, coconut waste, and vegetable waste etc. that are available within the system. Dung served as source of FYM for fruits and other plantation crops thereby reducing the cost of production. The milk production on an average from two buffaloes is 5.5 l/day. The meat & egg from poultry, milk from the buffaloes served as an enriched source of protein supplement for the farmer's family.

Economics and employment generation of crop component

Maxi mum net return of Rs. 1,50,350/- was obtained from Arecanut + black pepper (though black pepper has not yet come to fruiting stage) followed by sapota + banana (Rs 99,863/-) with B:C ratio of 4.32 and 18.25.Though the net return was more in case of arecanut +black pepper integration, but the B:C ratio was less as compared to sapota + banana integration, due to additional mandays required in dehusking of arecanut. The employment generated was 365 mandays in case of cropping from 0.90 ha area. Farmwoman was involved in livestock and poultry management. The results are presented in Table 4.

Energy flow from crop component in MFS I

Recycling of resources in terms of energy is one of the main beneficial effects of integrated farming system. Seed, manure, fertiliser, pesticide, insecticides and labour used are considered as input energy and output energy was calculated in terms of yield (fruits, nuts, etc.) obtained from the crop. The energy flow in the system is depicted in the Table 5. The total output energy was 1,33,018 MJ. This indicates an efficient conversion of input energy into by products. Energy efficiency and Specific energy for different cropping sequence also presented in Table 5.

Table 4. Yield and economics of different cropping sequences of MFS – I

Cropping sequence	Yield (q)			Cost of cultivation (Rs)	Gross return (Rs)	Net return (Rs)	B:C ratio
	1st crop	2nd crop	3rd crop				
Coconut + sapota nuts + banana	770	37.5	1.12	5,280	80,950	75,670	14.33
Arecanut + black pepper (vegetative stage)	37.03	-	-	34,800	1,85,150	1,50,350	4.32
Sapota + banana	52.50	0.67	-	5,472	1,05,335	99,863	18.25
Banana	2.25	-	-	600	1,125	525	0.87
Total				**46,152**	**3,72,560**	**3,26,408**	**7.07**

Table 5. Energetic of MFS - I

Cropping sequences	Energy input (MJ)	Energy output (MJ)	Energy efficiency	Specific energy (MJ/kg)
Coconut + sapota + banana	689	38,125	55.26	0.16
Arecanut + black pepper (vegetative stage)	4,547	67,764	14.90	1.23
Sapota + banana	715	25,601	35.80	0.13
Banana	78	1,525	19.45	0.35
Total	**6,030**	**1,33,018**	**22.05**	

Livestock component

Two adult Murrah cross buffalo and three female calves. The milk production on an average from two buffaloes is 5.5 l/day. The animals are fed with green fodder, kitchen waste (rice-ganji *etc.*), vegetable waste, and wheat flour with groundnut oil cake apart from grazing. He sold the milk directly to the other villagers by himself. The selling price of milk is Rs. 28 /l. Dry fodder like straw is scarce so the animals are rarely fed with straws. The farmer has been motivated to grow fodder in his field to reduce the fodder cost and for improving milk production. The manure obtained from the animal is used in the field as FYM. Economics and energetics has been evaluated.

Economics of cattle and poultry component (MFS I)

The net return obtained from the cattle and poultry component was Rs. 50,780/- and Rs. 13,520/- respectively. The employment generation in cattle component was 140 mandays/year whereas in poultry component, it was 23 mandays. The benefit cost ratio was estimated to be 1.96 and 1.16 in cattle and poultry respectively (Fig. 4,5 & 6).

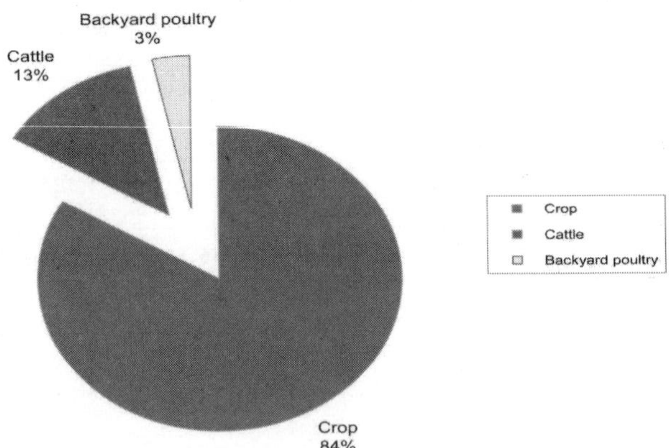

Fig. 4. Percent contribution of different components under IFS in MFS I

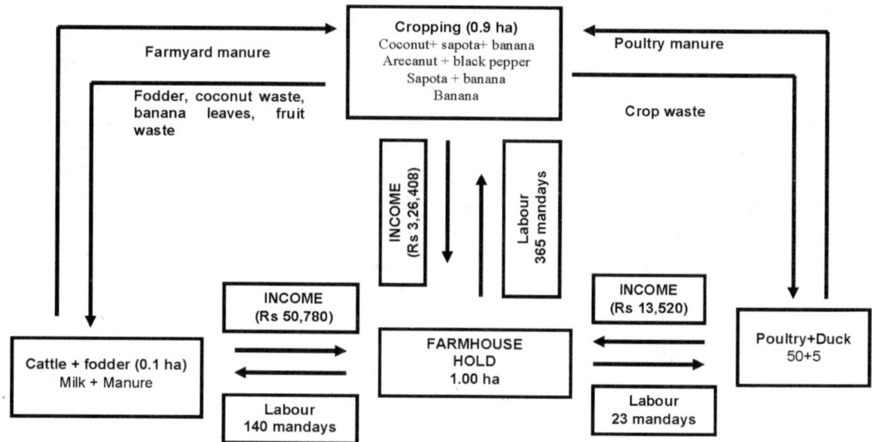

Fig. 5. Resource Recycling in Crop+ Poultry + Cattle in IFS in MFS I

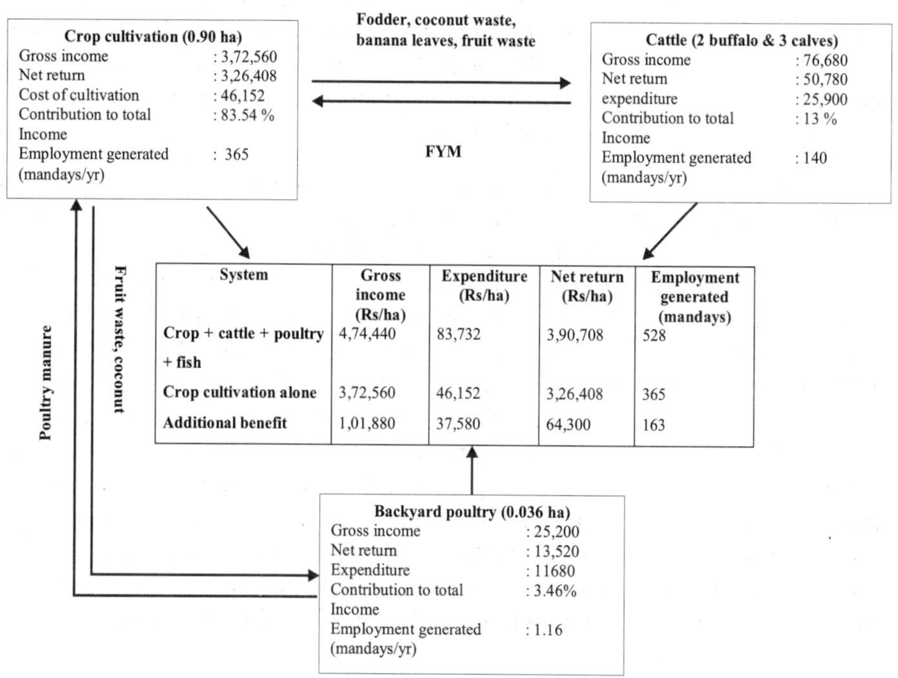

Fig 6. Productivity and Profitability linkages in MFS I

Backyard poultry

About 50 local birds are reared as backyard out of which ten laying hens, growers and chicken. The economics for the backyard poultry has been estimated and presented in the Table 6.

Table 6. Economics of cattle and poultry component (MFS I)

Particulars	Cattle	Poultry
Gross income (Rs)	76680	25200
Expenditure (Rs)	25900	11680
Net income (Rs)	50780	13520
Benefit cost ratio	1.96	1.16
Contribution to net return (%)		
Employment generated (mandays)	140	23

Micro farming situation II

Crop + Fish + Poultry + Cattle + Goat Integrated model

Integrated Farming System comprising of crop and livestock has been sustainable over centuries. The system could be able to meet food needs of the ever-increasing population. In this system animals are raised on agricultural waste and the animal power is used for agricultural operation and voids are used as manure and fuel. Summarily, the waste of one enterprise becomes the input of another leaving no waste to pollute the environment. This model is suitable for slopping hilly upland areas of Bay Islands.

The normal cropping programme followed was Rice – Rice – vegetable, Sugarcane, Ginger- leafy vegetable – ginger. But in the integrated approach, a modification was made in the existing cropping pattern by including floriculture in arecanut garden. Cropping was undertaken in 0.90 ha and 0.036 ha was allotted for fish pond and the poultry shed was placed above the pond. Birds were fed with grains, broken rice, coconut waste, and vegetable waste etc. that are available within the system. Poultry litter and cow dung served as fish feed reducing the feed cost of fish. No other supplement was provided for fish. The cattle were fed with crop residue, paddy straw, vegetable waste and fodder grass. The farmer prepared concentrate mixture utilizing the locally available ingredients for feeding cattle. Bullocks were used for ploughing the field. The manure obtained by the cow and bullocks was used in field after converting it into compost manure. Milch cow produced 4-6 lit of milk per day and was milked morning and evening by the farmer. The meat & egg from poultry, milk from the cow served as an enriched source of protein supplement for the farmer's family.

Cropping sequence of model 1

Floriculture – leafy vegetable – okra	0.10 ha
Paddy – sugarcane (Chewing canes)	0.22 ha
Ginger – leafy vegetable – leafy vegetable	0.16 ha
Paddy – vegetable – vegetable	0.19 ha
Coconut + fodder	0.08 ha
Arecanut + black pepper	0.05 ha
Coconut + crossandra (.025)+ marigold (.025)	0.05 ha
Coconut + ginger	0.05 ha
Fish cum poultry cum duck	0.036 ha
Livestock and fodder	0.064 ha

Economics and employment generation of crop component under MFS – II

Maximum net return of Rs. 76,208/- was obtained from ginger – amarantus – okra cropping sequence followed by coconut - crossandra Rs 24,256/-with B:C ratio of 4.12 and 7.51 which was less compared to that of arecanut + black pepper 14.55 cropping sequence, because arecanut involves less cost of cultivation as compared to others. Vegetable cultivation was less because of water scarcity which resulted in less employment generation i.e. 180 mandays in case of cropping from 0.90 ha area as compared to last year. Farmwoman was involved in all the agriculture activity like sowing, manure application, harvesting and selling of produce. The results are presented in Table 7.

Table 7. Yield and economics of different cropping sequences of model 1

Cropping sequence	Yield (q)			Cost of cultivation (Rs)	Gross return (Rs)	Net return (Rs)	B.C. ratio
	1st crop	2nd crop	3rd crop				
Paddy – Sugarcane	9.44	3239 nos.	-	25480	39,942	14,462	0.57
Paddy – leafy vegetable (amaranthus) – leafy vegetable (amaranthus)	8.15	15.25	15.25	10,099	15,670	5,571	0.55
Ginger – Amaranthus – okra	19.24	12.84	7.34	18,484	94,692	76,208	4.12
Floriculture (marigold) – leafy vegetable (Amaranthus) – leafy vegetable (Amaranthus)	2.14	8.02	8.02	7,893	11,232	3,339	0.42
Coconut +fodder	787	16.0	-	1694	4,749	2,394	1.41
Arecanut + black pepper	1.28	0.32	-	639	9,938	9,299	14.55
Coconut + crossandra + marigold	492.	0.63	0.53	2,985	20,784	17,799	5.96
Coconut + ginger	492	6.01	-	3,228	27,484	24,256	7.51
Total				**70,502**	**2,24,491**	**1,53,989**	

Energy flow from crop component in MFS II

Recycling of resources in terms of energy is one of the main beneficial effects of integrated farming system. Seed, manure, fertiliser, pesticide, insecticides and labour used are considered as input energy and output energy was calculated in terms of yield (grain, straw etc.) obtained from the crop. In MFS II, the total input energy for the two models was 40149.27 and 32,760 MJ respectively. The total output energy was 1,35,092.20 and 1,80,504 MJ respectively. This indicates an efficient conversion of input energy into by products.

Livestock and Poultry Subsystem

Energy flow from livestock component

Recycling of resources in terms of energy is one of the main beneficial effects of integrated farming system. Feed and labour used are considered as input energy and output energy was calculated in terms of egg, meat and manure production in case of poultry and in cattle milk, work power and dung. In MFS II, the total input energy for the two models from cattle sub system were 2,00,190 MJ and 2,01,784 MJ respectively. The total output energy was 22,465 MJ and 25,405 MJ respectively. Energy efficiency ratio in both cases was almost equal i.e. 0.11 band 0.13. The total input energy for poultry component for both the models were 6,380 MJ and 6,690 MJ. The total output energy was 1,612 MJ and 1,739 MJ. The energy ratio was 0.25 and 0.26. This indicates an efficient conversion of input energy into by products. In cattle the energy ratio was lesser than poultry.

Economics and employment generation by livestock component

Integration of various enterprises has great potentialities in the agricultural economy. Inclusion of livestock enterprise in farming system allows for more effective use of natural resources like climate, land, soil and vegetation. Livestock converts weeds, crop residues and by-products into food stuffs for human consumption, produces meat and milk and provides power in the farm besides its manure, which improves the soil productivity. The economics worked out for the livestock component in 0.064 ha of IFS is presented in Table 8. The net return from cattle for the first and second model was Rs.29,343/- and Rs.62,331/- respectively In the first model milk production was less compared to the second model.

In cattle 34 mandays was generated for bullocks alone and the milching cow added an extra 57 mandays (Table 8). The above model has been developed for two work bullocks and three milch cows in 0.064 ha.

Table 8. Economics of cattle component MFS II

Particulars	
Gross income (Rs)	62,307
Expenditure (Rs)	32,964
Net income (Rs)	29,343
Benefit cost ratio	0.89
Contribution to net return (%)	14.7
Employment generated (mandays)	91

Poultry cum Fish cum duck integration

In Integrated Fish cum Poultry System, farmers were supplied with Nicobari fowl growers and ducklings. The poultry shed was constructed above the pond and the birds were reared under semi intensive system of rearing. The production and economics of the integration as a single unit has been calculated for both the models.

Model 1

In this there are 17 chicken and 8 ducks integrated with 200 fingerlings in 0.036 ha of pond area. The egg production from poultry was 150 eggs / bird / annum. Duck egg production was 91 eggs/bird/annum. The birds attained maturity at about 22nd week of age and weighed 1070 g (average). Net income from poultry and duck was Rs.10,153/- and fish yielded Rs 1,840/-. The net income from this integration was Rs 11,993/- and generated an employment of 38 mandays for the farmwoman. Eggs are sold @ Rs 5/egg. The fish production was less than previous year this is due to unavailability of fingerlings in time and late release. The pond dried during the month of January so the fish was harvested earlier. The total production was 48 kg, which was sold @ Rs. 80/kg. Total manure added to the pond was 948 kg / year (Table 9 & 10).

Table 9. Economics of poultry + duck component in IFS

Particulars	
Gross income (Rs)	15,485
Expenditure (Rs)	5,332
Net income (Rs)	10,153
Benefit cost ratio	1.99
Contribution to net return (%)	5.3
Employment generated (mandays)	23

Table 10. Economics of poultry + duck + fish component in IFS - MFS II

Particulars	
Gross income (Rs)	19,245
Expenditure (Rs)	7,172
Net income (Rs)	12,073
Benefit cost ratio	1.68
Contribution to net return (%)	5.46
Employment generated (mandays)	46

Production cost for egg and milk

Production in terms of egg is low in IFS model when compared to commercial poultry, but the cost of production is much lesser. Labour cost appears to be more but in actual terms, it is nil, since the family member is involved in the management and it generates employment for family labour. Farmwoman was involved in the poultry management. This integrated approach provides an opportunity for the women to involve in income generating activities. Selling price of egg is more (Rs. 4.00/egg) when compared to the commercial poultry (Rs. 2.50), hence the net return is more in IFS. Cost of feed was very less in IFS due to the feed ingredient used was mainly crop waste like broken rice, rice bran etc. Apart from this birds were let free for grazing around the household to feed on kitchen waste, insects, pests, coconut gratings and fruit waste. Cost of production per litre of milk is Rs.12.67 and the selling price varied from Rs.18 to 25 / litres. Low cost of production makes the enterprise profitable even with less production (Table 11).

Table 11. Production cost of poultry egg in integrated farming system (MFS II) and commercial poultry farm

Parameters	Model 1	Commercial poultry
No. of eggs / bird / year	150	285
Cost of labour / bird / year (Rs)	92	5.00
Cost of feed / bird / year	134.40	380
Total cost of production / bird / year (Rs)	226.4	385
Cost of production / egg (Rs)	1.50	1.80

Backyard poultry

10 growers, 10 chicks, 2 laying hens and 1 cock were integrated. Egg production is 80 eggs/bird/annum. The hatchability of egg is 89.74 % under farm condition. Due to early chick mortality the survival is less. Feed supplement like Vimeral, Groviplex etc. has been substituted to meet out any deficiency. Apart from this regular deworming of birds has been given and the farmer has been informed about regular deworming of birds. Birds are sold @ Rs 110/kg. The net return obtained is Rs. 3275/. (Table 12).

Table 12. Economics of backyard poultry component in IFS

Particulars	Model 1
Gross income (Rs)	4175
Expenditure (Rs)	900
Net income (Rs)	3275
Benefit cost ratio	3.63
Contribution to net return (%)	1.48
Employment generated (mandays)	15

Duck farming

Duck egg production is 91 eggs /bird/ annum, initially many eggs were lost in the pond as the ducks laid directly in the pond. The farmer was advised to open the birds after 9 a.m. so that all the egg laying is completed. Six ducklings hatched and 5 survived. The farmer has culled 7 birds from old stock and kept 1 old and 5 new ducks for next batch. The rest of the eggs were sold for both hatching and table purpose. Hatchability of duck egg is 90% under farm condition.

Guinea fowl production

Guinea fowl served as a snake repellant as well as pest and insect control. They scavenge around the homestead and feed on worms and insect. But sometimes they damaged vegetable plants in early stages. This can be prevented by keeping the birds in shed for few days till the vegetable grows to a stage when the vegetable grows to a stage when they cant damage or by keeping a watch on them. The egg production was seasonal it started around June and lasted till September. Initially many eggs were lost as the bird laid egg near bushes in the open and not n the shed. Guinea fowl were successfully hatched in farm condition (by setting eggs under broody hen). The incubation period was 28 days. The hatchability of egg was 50 % in one model and 83.3 % in the other. Early chick mortality was 50%. In model 1 there are 5 adult birds in the second model there are 3 adults and 3 growers.

Goat production

Total stock of goat is 9. The goat supplied attained maturity at 7 months of age. Artificial insemination was performed twice but the goat did not conceive. So it was let for natural service. One of the female was crossed with Boer cross male (F1 generation) (Table 13).

Production traits

Kid birth weight- 1.2 kg

Wight at one month – 5 kg

Weight at sexual maturity- 17.5 kg

Age at sexual maturity- 7 months

Total manure production 1.9 tonnes/year

Table 13. Economics of goat component in IFS

Particulars	Model 1
Gross income (Rs)	60,169
Expenditure (Rs)	22,279
Net income (Rs)	37,890
Benefit cost ratio	1.70
Contribution to net return (%)	10.08
Employment generated (mandays)	92

Aquaculture sub system

The basic principle of fresh water aquaculture is to improve the natural productivity of pond, soil and water for increasing the quality and quantity of fish food organism (Plankton, benthos etc.) for selected fast growing fish species and its subsequent conversion to animal protein through fish production. In this context water quality was analysed from model I & model II of micro farming situation II for successful venture of fish cum poultry cum duck integration (Table 14).

Table 14. Recommended fish species combinations and stocking in integrated fish-cum-poultry farming system in the MFS II (Model 1& 2).

Trophic niche	Fish species	Stocking ratio (%)	No. of poultry / ducks recommended
Surface feeder	Catla	50 (25)	20 birds + 5 ducks
Mid-water feeders	Rui	100 (50)	
Bottom feeders	Mrigal	50 (25)	
Total stocking		200	

* Recommendation was worked out based on the performance of fish species, poultry droppings, water quality parameters of pond and area of pond

In IFS model, profitable and cost effective fish culture depends upon judicious distribution of Indian Major Carps according to their respective niches. For example Rohu fish has the habit of staying in the column of the pond utilising all the food matter in that part of niche like zooplankton and other organism. In the same way Catla is a surface feeder and will consume all the phytoplankton for their grazing, Since Mrigal is bottom feeder so it will consume all the benthos and other organism for their growth. Another important factor that has contributed to this combination was the availability of fingerlings of Indian Major Carps. Rohu was added in maximum number primarily because of its availability in large number. Best possible combination was made based on the availability and pond fertility. Recommendation is given for both models 1 and 2 under MFSII, because of the similar pond size and most importantly physicochemical parameter of both the pond are almost equal and the data shown here is an average of all data recorded in every month (Table 15).

Table 15. Economics of Fishery component in IFS in (MFS II)

Particulars	Model 1	Model 2
Gross Income (Rs)	3760	4400
Expenditure (Rs)	1920	1920
Net income (Rs)	1840	2480
Benefit cost ratio on tribution to net return (%)	0.96	1.29

Gross income generated in model 1 was slightly higher than Model 2. labour requirement in both the model was found to be equal in terms of money due to similar nature of the pond like size, no. of fingerling, feed *etc*. Contribution to net return although is mediocre in terms of economy to the overall farming system but cost of input in terms of feed was completely avoided. In this case, poultry manure and farm waste fruit was used as prime source of nutrient requirement (Fig. 7,8 & 9).

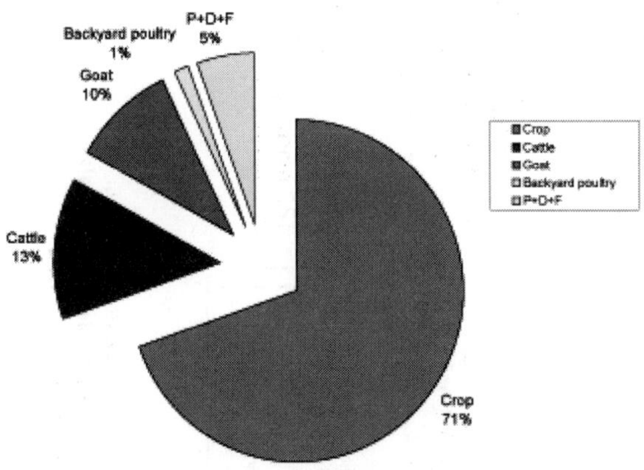

Fig. 7. Percent Contribution of different components under IFS, MFS - II, (Model 1)

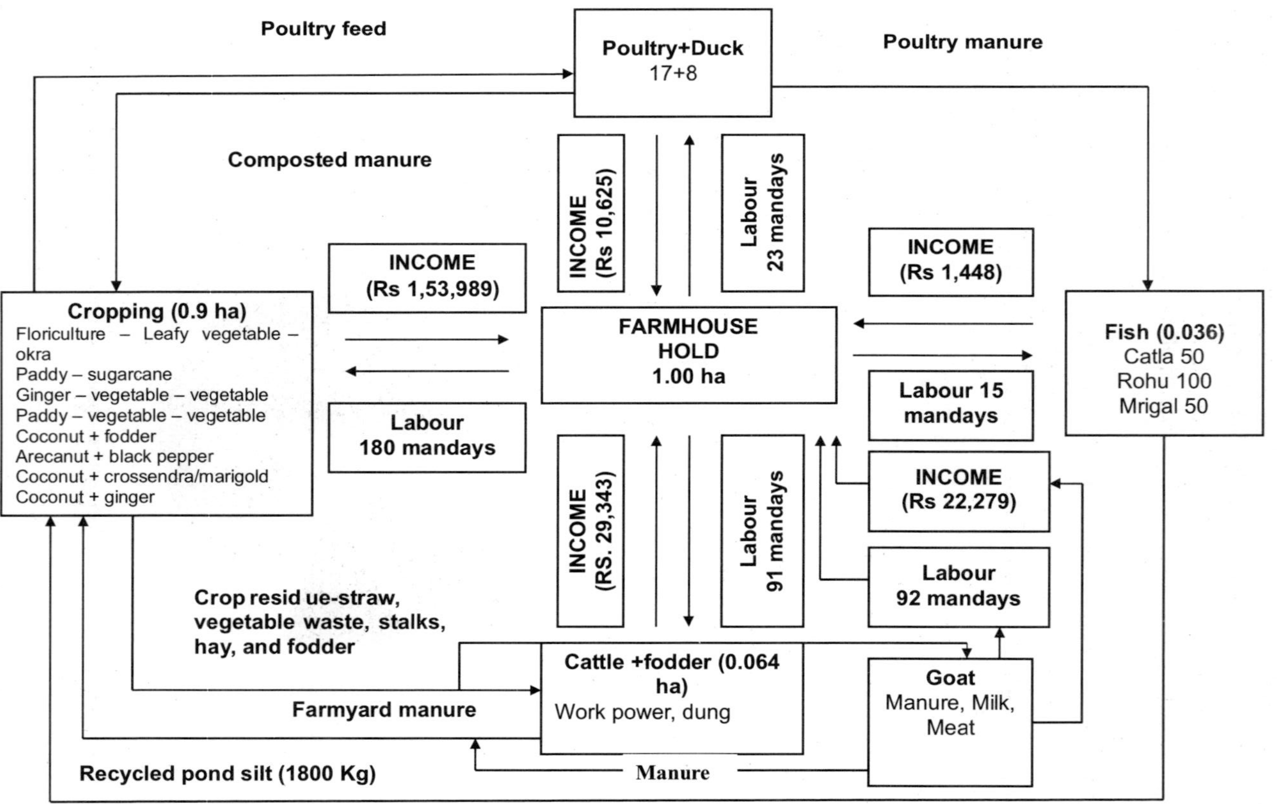

Fig. 8. Resource Recycling in Crop + Poultry + Fish + Cattle in IFS in MFS II (Model 1)

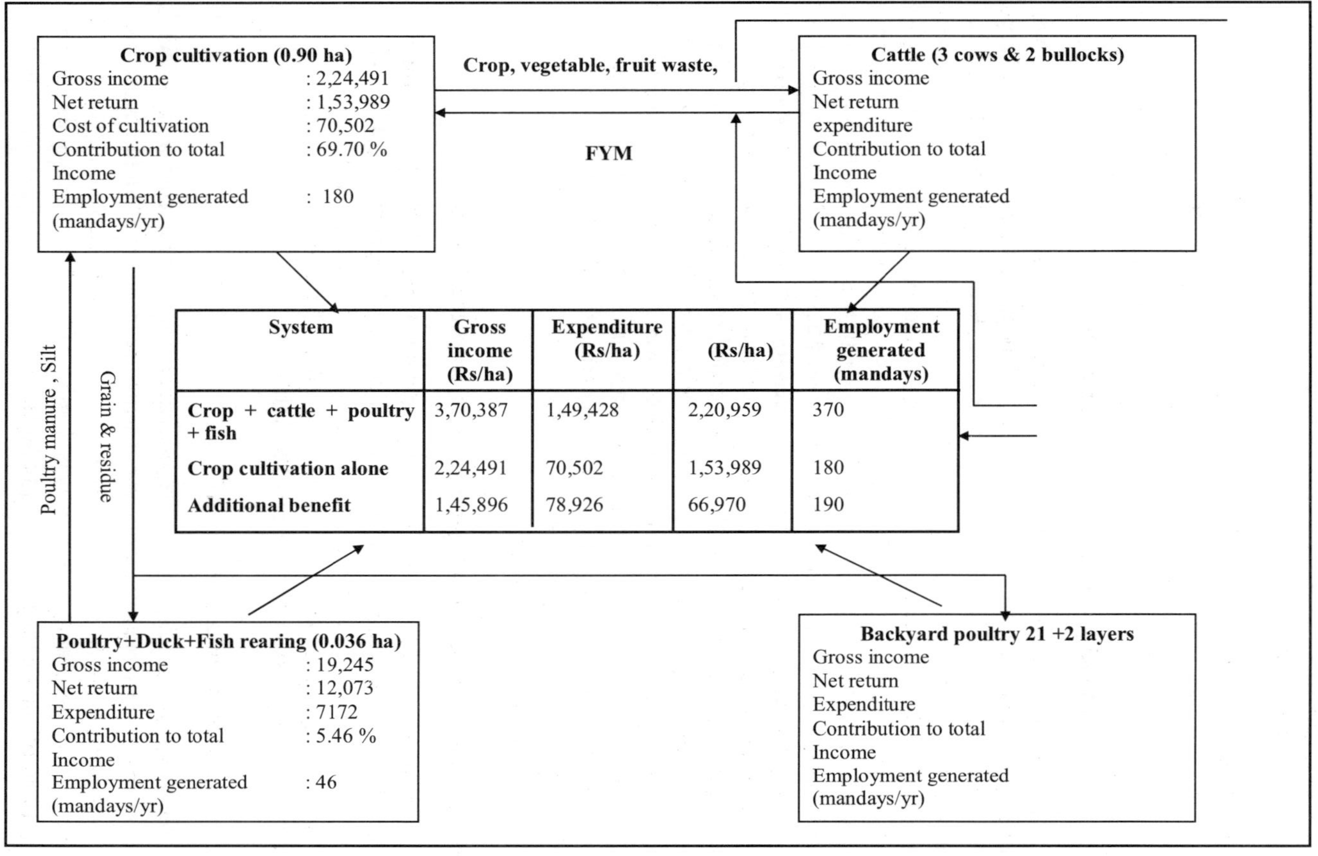

System	Gross income (Rs/ha)	Expenditure (Rs/ha)	(Rs/ha)	Employment generated (mandays)
Crop + cattle + poultry + fish	3,70,387	1,49,428	2,20,959	370
Crop cultivation alone	2,24,491	70,502	1,53,989	180
Additional benefit	1,45,896	78,926	66,970	190

Fig. 9. Productivity and Profitability linkages in MFS II (Model 1)

Micro Farming Situation III

Cropping sequence of MFS – III in 0.04 ha	
Vegetable (okra)	0.02 ha
Vegetable (cowpea)	0.01 ha
Vegetable (bottle gourd)	0.01 ha
Livestock and fodder	0.05 ha
Fish cum poultry cum duck in	0.06 ha

Crop + Fish + Poultry + Cattle Integrated model

An integrated farming system involving Poultry+ Fish + Cattle + Crop was taken in Tushnabad village. Farmer had only fish pond, cattle and plantation crops in small area which was in an unmanaged condition. Farmer had one well managed cattle shed which showed, his main interest was on cattle farming. After intervention of IFS, the pond dykes was cleared, vegetable and fodder cultivation was taken on that. Low cost poultry shed was constructed above the pond with locally available trees within the farm. Cropping was undertaken in 0.04 ha and 0.06 ha was allotted for fish pond. Poultry litter and cow dung served as fish feed reducing the feed cost of fish. No other supplement was provided for fish. Milch cow produced 6.5 lit of milk per day and was milked morning and evening by the farmer. The fingerlings were released in the pond as per the recommended stocking density. Accordingly 300 fingerlings were released in the pond of 0.06 ha. No separate feed was purchased from the market apart from poultry dropping. Twenty-five growers were kept in the shed to meet the feed requirement of fingerlings released in the pond. The farmers had only backyard poultry, before intervention and after intervention ducks have been introduced. Duck has been introduced to check the unwanted fish called tilapia and to aerate the pond.

Economics and employment generation of crop component

Maximum net return of Rs.7173/- was obtained from bottle gourd followed by okra Rs 1,437/-and cowpea Rs 620/-. Production of vegetables was less due to more insect pest attack and extreme summer. The employment generated was 20 mandays in case of cropping from 0.04 ha area. There was no involvement of farmwoman. The results are presented in Table 16.

Energy flow from crop component

Seed, manure, fertiliser, pesticide, insecticides and labour used are considered as input energy and output energy was calculated in terms of yield (fruits, nuts, etc.) obtained from the crop. The energy flow in the system is depicted in the (Table 5). The total output energy was 3760.1 MJ . This indicates an efficient conversion of input energy into by products.

Table 16. Yield and economics of vegetables grown on pond dykes of MFS – III

Crop	Yield (q)	Cost of cultivation (Rs)	Gross return (Rs)	Net return (Rs)	B:C ratio
Okra	1.02	603	2,040	1,437	2.38
Cowpea	0.45	280	900	620	2.21
Bottle gourd	18.32	1987	9160	7173	3.61
Total		**2,870**	**12,100**	**9,297**	**3.31**

Livestock component

Before intervention 2 cows, 2 calves, few backyard poultry and unmanaged pond with plantation crops were present. The farmer was motivated to adopt integrated fish cum poultry cum duck system. After cleaning and preparing of pond fingerlings were released. The pond area was 0.06 ha. The required number of birds was worked out to be 25 birds and 5 ducks. Poultry shed of 7.5 sq.m was erected above the pond for poultry. The birds (25 nos of Vanaraja chicks) were released. The birds are fed with rice bran, broken wheat/rice, groundnut oil cake *etc.* Fodder cultivation on the dyke has been taken up in the rainy season (Table 17).

Table 17. Economics for cattle and poultry component in MFS III

Parameters	Cattle	Backyard poultry	Poultry +duck
Gross income	1,22,496	5930	25,765
Expenditure	50,190	3300	6,607
Net income	72,306	2630	19,157
% Contribution to net income	66.22	2.41	17.54
B : C	1.44	0.80	2.90
Employment generation (mandays)	205	15	30

Cattle component

The farmer had two lactating Jersey crossbred cows and two male calves. The average milk production is 6.5 L / day. The milk is sold @ Rs. 28/l. The farmer fed the cows with concentrate mix prepared using wheat, groundnut oil cake and rice bran and sometimes cows are let loose for grazing. Under this MFS, a model for 3 cows and 2 calves has been developed. The productivity, profitability and employment generation has been estimated. The net income from cattle component is Rs. 72306/- with a benefit cost ratio of 1.44. Energetics for the above subsystem has been worked out taking feed and labour as input energy and milk and manure as ouput energy (Table 17).

Poultry cum fish cum duck integration

25 Vanaraja growers and 5 ducklings were integrated with 300 IMC fingerlings. The net income from this component as a single unit was Rs.24,957/- and has generated an employment of 65 mandays/year . The total manure added from poultry component was 1022 kg /year (Table 18).

Table 18. Economics of poultry cum fish duck component in IFS (MFS III)

Particulars	Poultry cum fish cum duck
Gross income (Rs)	33,765
Expenditure (Rs)	8,807
Net income (Rs)	24,957
Benefit cost ratio	2.83
Contribution to netReturn (%)	22.86
Employment generated (mandays)	50

Aquaculture sub system

Poultry cum fish cum duck has been integrated. Like in MFS II, here also pond water quality was assessed before introduction of Indian Major Carp fingerlings. Water parameter in this pond was found satisfactory (Table 19, Fig. 10,11 & 12).

Table 19. Economics of Fishery component (MFS III)

Particulars	MFS III
Gross Income (Rs)	8000
Expenditure (Rs)	2200
Net income (Rs)	5800
Benefit cost ratio	2.63
Contribution to total net return (%)	

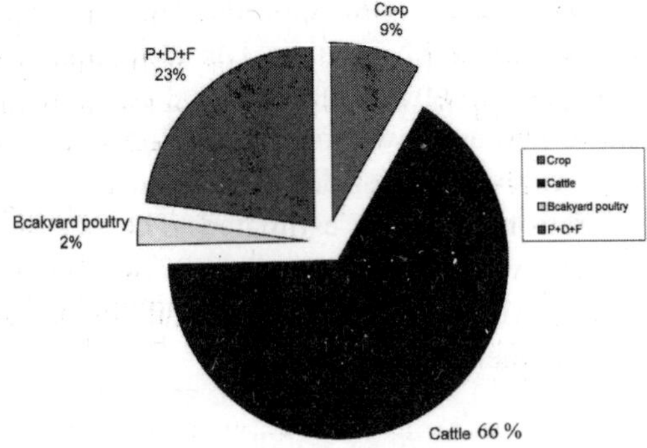

Fig. 10. Percent contribution of different components under IFS, MFS-III

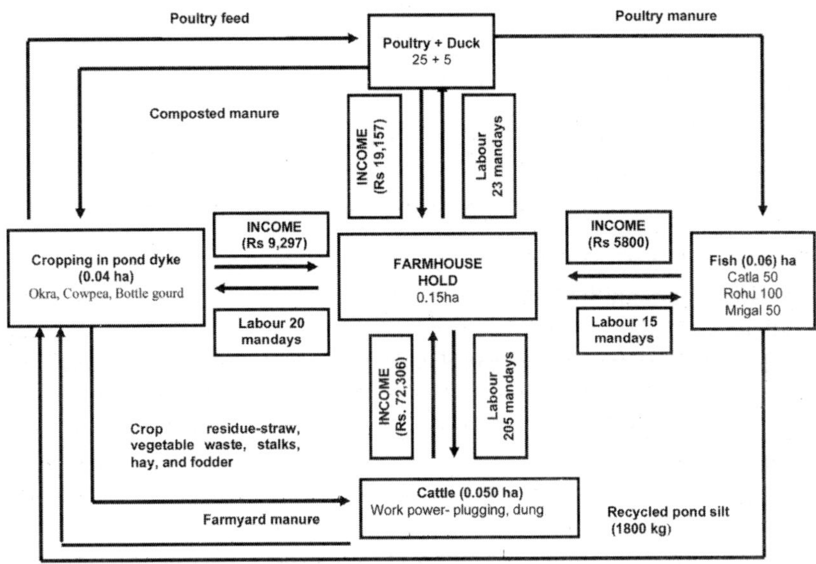

Fig. 11. Resource Recycling in Crop+ Poultry + Fish + Cattle in IFS in MFS III

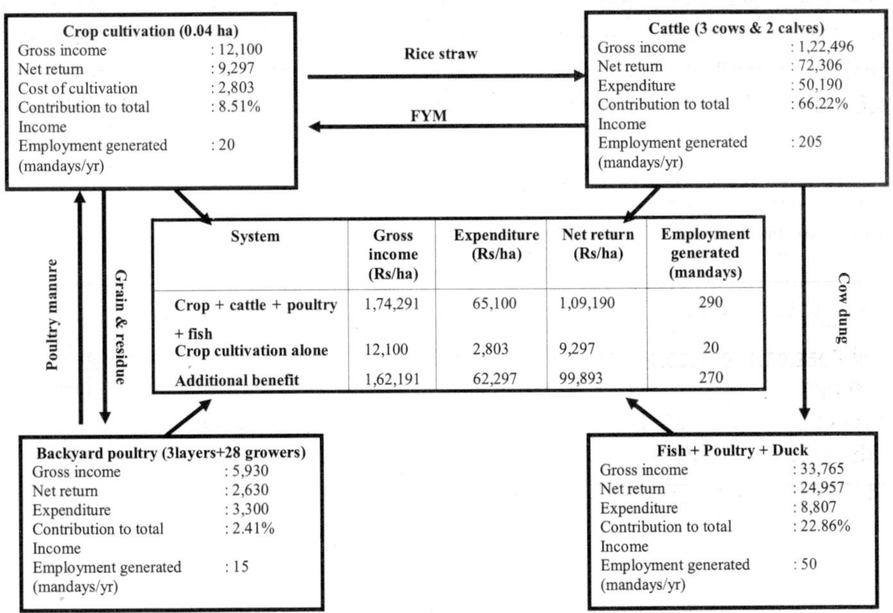

System	Gross income (Rs/ha)	Expenditure (Rs/ha)	Net return (Rs/ha)	Employment generated (mandays)
Crop + cattle + poultry + fish	1,74,291	65,100	1,09,190	290
Crop cultivation alone	12,100	2,803	9,297	20
Additional benefit	1,62,191	62,297	99,893	270

Fig. 12. Productivity and Profitability linkages in MFS III

Micro farming situation IV

Cropping sequence of MFS IV in 0.9 ha

Paddy (long duration)	0.20 ha
Paddy (Short duration) – vegetables*	0.61 ha

**Vegetables (okra - 0.12 ha, bottle gourd - 0.14 ha, pumpkin – 0.10 ha, ash gourd – 0.10 ha, cowpea – 0.15 ha)*

Coconut + orange + fodder	0.02 ha
Arecanut + black pepper	0.05 ha
Vegetables on pond dykes (bottle gourd and ridge gourd)	0.02 ha

Livestock and fodder 0.064 ha

3 cows + 2 bullocks	0.004 ha
Fodder	0.06 ha

Fish cum poultry cum duck 0.036 ha

Existing monocropping of paddy (C14-8) were replaced with cropping systems involving high yielding varieties of paddy and vegetables along with dairy, fish and poultry. The normal cropping programme followed was Long duration paddy and in small area short duration paddy followed by vegetables. But in the integrated approach, a modification was made in the existing cropping pattern by including few more vegetables like okra, cowpea, ash gourd. Farmer was also motivated to grow vegetables on the pond dykes. Cropping was undertaken in 0.90 ha and 0.036 ha was allocated for fishpond and poultry shed. Duck, poultry and fish were integrated.

Economics and employment generation of crop component

Maximum net return of Rs. 41,394/- was obtained from paddy (medium duration) – vegetable cropping sequence followed by Arecanut + black pepper. The employment generated was 58 mandays in case of cropping which was less compared to that of MFS II because in MFS IV less importance is given for vegetable cultivation due to its low lying nature. Farmwoman was involved in all the agriculture activity like sowing, manure application, harvesting, cleaning of produce and poultry feeding (Table 20).

Energy flow from crop component

The total input and output energy was 6942 MJ and 68837MJ respectively. This indicates an efficient conversion of input energy into by products.

Table 20. Yield and economics of different cropping sequences of MFS IV

Cropping sequence	Yield (q)			Cost of cultivation (Rs)	Gross return (Rs)	Net return (Rs)	B:C ratio
	1st crop	2nd crop	3rd crop				
Paddy (long duration)	7.16	-	-	2298	5728	3,430	1.49
Paddy (short duration)	25.37		-	38,242	79,636	41,394	1.08
– vegetable (okra		6.00					
+bottle gourd +		11.90					
pumpkin + ash		9.25					
gourd +		20.00					
cowpea)		8.02					
Coconut + orange +fodder	160	1.06	4.20	628	6360	5,732	9.13
Arecanut + black pepper	1.15	0.13	-	621	7,825	7,204	11.60
Total				41,789	99,549	57,760	1.38

Economics and employment generation by livestock component (MFS IV)

The net return from poultry+duck and cattle was Rs.14525/- and Rs. 87800/-. The poultry manure was used as fish feed and dung was used as farmyard manure. In cattle an employment of 148 mandays/year was generated whereas in poultry+duck it was 23 mandays.

Backyard poultry

There are 5 hens, 1 cock and 15 chicks in the backyard stock. The egg production is low. On an average the farmer gets 2 eggs per day. They sell the birds for meat purpose. The farmwoman takes care of backyard poultry, collects egg and feeds the birds. Rice bran and broken rice makes the major feed constituent. The economics has been worked out for this present stock. The gross income from this component is Rs. 7075/- and the expenditure incurred is Rs. 2600/-. Net income is Rs. 4475/- with B:C ratio of 1.72 farm backyard poultry (Table 21).

Table 21. Economics of livestock and poultry component in IFS - MFS IV

Particulars	Backyard poultry (21 No.s)	Poultry+ duck (13+7 No.s)	Cattle (3cows +2buffaloes)
Gross income (Rs)	7,075	14,525	1,38,755
Expenditure (Rs)	2,600	2,800	50,955
Net income (Rs)	4,475	11,725	87,800
Benefit cost ratio	1.72	4.19	1.72
Contribution to net Return (%)	2.71	7.10	53.23
Employment generated (mandays)	15	23	148

Conclusion

Andaman and Nicobar Islands are blessed with rich natural resources of land, water and weather. Proper utilization of these resources lies in the hands of proper planning, execution, monitoring and implementation. In Agriculture, year after year, returns changes as mainly due to the price fluctuations. Pest and disease incidence, reduced returns, higher cost of cultivation are the inherent characters of the mono cropping / culture where as in Integrated Farming System (IFS), the risk is buffered through integrating many components such as different crops, livestock, poultry, duckery and goat *etc.* Apart from the reduction in risk, the natural resources can be managed effectively through optimum integration of components under various resource conditions or micro farming situations. The wastes are considered as gold for other components in integrated system; hence, less investment, more profit, and less environmental hazards are associated in it. Successful implementation of integrated models requires the attention of farmers, agricultural extension agencies and policy makers. The main friction in adoption of Integrated Farming System (IFS) is advising the farmers on the fixed components rather flexibility in selection of components should be given as per the end users wish. The size of components may be fixed on the basis of scientific information by taking in to consideration of input and out put relationships evaluated.

Involvement of women in many activities like garland making of flowers, harvesting of fruits, and management of poultry birds including duck could be promoted for successful implementation of the models. The successful implementation of the integrated farming system in island ecosystem will certainly bring the financial & nutritional security among the farmers of Island ecosystem.

References

Andaman and Nicobar Adminstration. 2004. Statistical Hand book, A & N Administration, Port Blair.

Ganeshamurthy, A.N., Dinesh, R., Ravisankar, N., Nair, A.K., Ahlawat, S.P.S. 2002. Land Resources of Andaman and Nicobar Islands, CARI, Port Blair.

Rangasamy, A., Jyanthi, C. 1994. Recycling of organic wastes in integrated farming systems. P.114-118. In: Proc. Natl. Training on organic farming. GOI & Tamil Nadu Agric. Univ., Coimbatore. 1-8, Sep., 1994.

Ravisankar, N., Nawaz, Shakila, Bibi, Nabisat, Biswas, T. K., Sajibala, B. 2006. Role of spices in IFS. Paper presented in National Workshop on Integrated Development of Spices from 12th – 14th October 2006 organized by Horticulture and Forestry Division, Central Agricultural Research Institute, Port Blair, Sponsored by Directorate of Arecanut and Spices, Ministry of Agriculture, Calicut, Kerala.

Ravisankar, N., Zameer Ahmed, S.K., Medhi, R.P., Srivastava, R.C. 2006. Training manual on "Integrated farming system for rural prosperity" published by CARI in collaboration with KVK, Port Blair.

Ravisankar, N., Biswas, T.K., NabisatBibi, Shakila Nawaz, Pramanik, S.C. 2007. Integrated Fish – Poultry-Duck farming in BayIslands, A & N. Fishing Chimes 26 (10): 137-141.

8

Agroforestry Practices in Tamil Nadu

M.P. Divya

Introduction

Agroforestry has vast potential not only to meet the demands of fuel, fodder, timber, medicine and other Non-wood Forest Products but also to enhance green cover, soil and water conservation and reduce pressure on forests. The objective of agroforestry is to take advantage of the complimentary relationships between trees, crops and livestock in such a way that the productivity, stability and sustainability of the total system are maintained. It has great potential in improving degraded and affected sites and has been recognized as a restoration agent, rehabilitation process, and bioremediation mechanism, in addition to increasing farm income.

Agroforestry is not a new practice for the farmers of Tamil Nadu. It has been in practice in various forms e.g. growing *Acacia leucophloea* in the grazing lands with *Cenchrus ciliaris* in the Kangayam tract of Tamil Nadu exists from time immemorial to meet the mixed farming in that tract. In late 60's farmers in coastal areas of Chengalpattu and Cuddalore districts started growing Casuarina to be sold as firewood. This effort was entirely of the farmers without any support from the Government. In the last two and a half decades, large financial outlays have been made by the ICAR to develop suitable agroforestry models for various regions and promote it among the farmers for adoption.

Suitable agroforestry practices for Tamil Nadu

The various agroforestry practices followed in different agroclimatic zones of TamilNadu are described below.

1. North Eastern Zone

This agroclimatic zone comprises of the following districts viz., Thiruvallur, Vellore, Kancheepuram, Thiruvannamalai, Villupuram, Cuddalore (excluding Chidambaram and Kattumannarkovil taluks),Perambalur and Ariyalur. The climate is semi arid and tropical. The major crops grown in this zone are paddy, sorghum, cumbu, ragi, greengram, blackgram, redgram, cowpea, groundnut, castor, sugarcane, cotton, sesame and cashew. North east monsoon provides maximum amount of rainfall. The major soil types in this zone are red loam and clay loam. Black soils are also present in limited extent and coastal alluvial soils occur along the sea coast. Saline and alkaline soils are also found in patches.

The major agroforestry practices being followed in this zone are detailed below.

a) Monoculture of *Casuarina equisetifolia*

Casuarina was introduced into the Indian subcontinent in the sixties of the last century and it was found most successful. Since then it has spread along the east and west coast of the Indian peninsula .Among tree crops, the first choice of the farmer in this zone is Casuarina. Under irrigated condition, it gives about 100 ton ha^{-1} at the end of third year. It is a highly remunerative crop and planted at a spacing of 1.2mx1.2m.

b) Intercropping with Casuarina

Many farmers take up intercropping in the interspaces of the tree during the first year. Groundnut in sandy soil, sesame in red soil and pulses in heavy soil are recommended. Intercropping in the second year is seldom done.

c) Bund planting

In some pockets of this zone, bund planting of *Thespesia populnea* and *Lannea coromandelica* is resorted to.

d) Intercropping with cashew

This tree is mostly grown in red loam soil. The trees are raised either from nuts or from epicotyl grafts and planted at an espacement of 7mx7m. The row interspaces are used for growing intercrops like groundnut, pulses and minor millets up to 2 to 3 years until the canopy closes.

2. North Western zone

This zone consists of the following districts *viz*., Salem, Dharmapuri (excluding hilly area, Krishnagiri, Namakkal (excluding Tiruchengode taluk) and Perambalur taluk of Perambalur district. The climate ranges from semi arid to sub humid. About 40% of rainfall is received during Southwest monsoon and the remaining rainfall during Northeast monsoon. The major soil types are red loam and black soils. The major crops grown in this zone are rice, sorghum, cumbu, ragi, samai, maize, greengram, blackgram, redgram, cowpea, groundnut, sunflower, castor, cotton, gingelly, horsegram, sugarcane, tapioca, tomato, beans, cabbage, potato, turmeric, water melon, onion, mango, banana, betelvine, grapes,horsegram and coconut.

The important agroforestry practices being followed in this zone are described below.

a) Bund planting of *Pongamia pinnata*

It is a native tree of western ghats of the Indian subcontinent and grown wild on sandy rocky soils. It grows in variety of soils and can tolerate salinity. Its pod contains 30-40% oil content. Oil is used for many purposes like soap making and paint industries.

b) Monoculture of *Eucalyptus tereticornis*

The monoculture of *Eucalyptus tereticornis* under rainfed and irrigated conditions is a common practice in this zone. Because it is major tree for pulpwood, many farmers grow this tree on contract farming with paper industries.

c) Intercropping with *Eucalyptus tereticornis*

Tapioca, the dominant crop in Salem district is intercropped with *E.tereticornis* during first year. This tree is generally raised at a spacing of 2mx2m and felled at the end of 5-7 years depending on the site quality. The agricultural crops like sorghum and finger millet are grown as intercrops.

d) Block plantations of Tamarind, Mahua and Pungam

The tree species viz., *Tamarindus indica, Bassia latifolia and Pongamia pinnata* are grown as block plantations or as scattered trees in the farmlands. The agricultural crops *viz*.,sorghum, groundnut are grown during the initial years.

3. Western Zone

This zone consists of Coimbatore and Erode districts (excluding hilly areas), Nilakottai and Palani taluk of Dindigul district, Thiruchengode taluk of Namakkal district, Usilampatti, Uthamapalayam and Periyakulam taluks of Theni districts,

Karur Taluk of Karur district and northern parts of Madurai district. The climate of this zone is from semiarid to sub humid. The major crops in this zone are rice, sugarcane, cotton, sorghum, ragi, turmeric, banana, groundnut, bengal gram and tobacco.The major soil types are shallow red soils and deep black soils. Loamy red soils occur in some areas of southern parts of Coimbatore district.

The major agroforestry practices being adopted in this zone are given below.

a) Woodlots of Kapok

In Coimbatore and Erode districts, the Singapore kapok, *Ceiba pentandra* var.*caribaea* whose pods yield valuable floss mainly used in upholstery is raised as commercial crop. It is usually raised as woodlot at an espacement of 6m x 6m to 8m x 8m.It is grown under irrigated condition and comes to bearing in 4 years. However the optimum yield is obtained from seventh year. Intercropping is done during first two years.

b) Silvipasture

This system is characteristics of Erode district where the primary vocation is livestock production. *Acacia leucophloea,* a legume tree is selectively managed in farmlands at a density ranging from 40 to 60 trees ha^{-1} dispersed in a diffuse manner. Fodder crops like fodder sorghum, *Pennisetum glaucum, Dolichos uniflorus*, pulses, cereals, *Cenchrus setigerus* and *Cenchrus ciliaris* are grown in the interspaces. *C. ciliaris* is a hardy grass which dry up in summer but regenerate naturally with the onset of monsoon. During the off- season, cattle are let in to the field and fed with the pods of the tree. Feeding of the pods is believed to give extra boom to the coat of the animal which enhances its sale value. The leaves are also lopped and fed to goat and sheep.

c) Bund planting

Ailanhtus excelsa is grown in bunds of the field. It is an excellent splint timber which contributes to bulk of the raw materials used by the match industries in and around the zone. *Hardwickia binata* is a fodder cum timber species that pollards well. These species are much favoured for bund planting.

d) Intercropping with *Ailanthus excelsa*

The intercropping is also being done with *A.excelsa.* For which, ailanthus is planted at 4mx4m spacing. The agricultural crops viz., cowpea, horsegram and sesame are grown as intercrops. For established ailanthus plantations, fodder cowpea and guinea grass are found suitable.

e) Intercropping with Simaruba

For intercropping, simaruba is planted at 4mx4m. The following agricultural crops *viz.*,blackgram, cowpea and greengram are found to be suitable. Simaruba is oil bearing tree and in addition, its wood is used for making match splints. It is very hardy species and suitable for variety of soils.

4. Cauvery delta zone

This zone includes cauvery delta area in Thanjavur, Nagapattinam, Musiri, Lalgudi, Thuraiyur and Kulithalai taluks of Tiruchirapalli , Aranthangi taluk of Pudukottai district and Thiruvarur. The irrigation source for this zone is Cauvery river through canal system.The major crops grown in this zone are rice , sorghum setaria , pearl millet, sesame, maize, greengram, blackgram, redgram ,soybean, groundnut , sunflower castor, cotton , sugarcane and chillies. The soils of this zone consist of river alluvium which is characterized by layers of fine and coarse sand which lie alternately. Saline coastal alluvium varying in texture from sandy to pure sand occurs along the coastal regions of this zone. Nearly 50 per cent of the annual rainfall is recorded during the north east monsoon. This zone which is known for its rich soil and adequate water supply is the rice bowl of TamilNadu.

The agroforestry practices prevailing in this zone are given below.

a) Bund planting of *Acacia nilotica* and *Bambusa bambos*

The tree species viz.,*Acacia nilotica* is grown in field bunds. This tree is planted mainly for fuel and fodder purpose. Bund planting of bamboos is age old practice in Tanjore district. It also serve as a live fence.

Many farmers are cultivating *Tectona grandis* along the flanks of water courses. As it is a very good timber, it fetches high price. Hence, this is the very common practice in cauvery delta zone.

b) Live fence of *Lannea coromandalica*

Growing *Lannea coromandelica* is common in the coastal areas of Nagapattinam. This tree is raised around the houses and used as live fence post to which barbed wire is hinged.

c) Woodlots of *Terminalia arjuna*

Terminalia arjuna is a lofty timber species capable of tolerating excess soil moisture and water logging. It is mostly planted in low lands.

5. Southern Zone

This zone comprises the following districts *viz.*, Virudhunagar, Ramanathapuram, Sivagangai, Thoothukudi, Tirunelveli, Madurai, Dindigul and

Natham taluks of Dindigul district, Melur, Tirumangalam, Madurai South and Madurai North taluks of Madurai district and Pudukottai district (excluding Aranthangi taluk).The climate is generally semi arid and majority of the rainfall is duing northwest monsoon.The major crops in this zone are rice, sorghum, cumbu, ragi, redgram, groundnut, blackgram,greengram, cowpea,gingelly, sunflower and cotton.This zone consists of saline coastal alluvium and river alluvium along the east coast , black soil in the middle of the zone ,red sandy soil in the north eastern and south western regions with deep red soil on the western side.

a) Wood lots of Neem and Tamarind

Although tamarind occurs naturally throughout the state, deliberate planting of the tree on agricultural lands is popular in Dindigul.The variety from vembarpatti inDindigul possesses high acid content and an appealing golden yellow pulp which fetches high price in the market. One to 2 year old containerized seedlings are planted in large pits of 45cm x 45cm x 45cm at an espacement of 8m x 8m to 12m x 12m and irrigated in the early stages. The tree comes to bearing in 8th year after planting and economic yield commences from 10th year and continue up to 60th year. Pod yield per tree varies from 50kg in young trees to 200kg in an adult tree. Pulp makes about 35% to 50% of the pod weight.

Neem offers the widest assortment of uses but its primary utility lies in the oil extracted from the kernels which is used in soap manufacture. Nine to 10 months old seedlings are planted at an espacement of 10mx10m. The tree comes to bearing in 5th year and each tree yield about 50kg fresh fruits every year.

b) Bund planting with Kapok

The kapok, *Ceiba pentandra* is preferred by several farmers for floss yield. It is planted in bunds.

c) Intercropping with Kapok

Ceiba pentandra var. indica is very popular in Madurai district which accounts for the largest area in this region. The trees are planted at a spacing of 7mx7m and the intercrops viz., pulses or cereals are grown up to 3 years.

d) Monoculture of *Eucalyptus tereticornis* and *Casuarina equisetifolia*

E. tereticornis is extensively grown in this zone as a monoculture and is raised at a spacing of 2mx2m.*C. equisetifolia* is raised at an espacement of 1mx1m.In alkaline pockets with pH of more than 9.0, growing casuarinas for three years reduced pH to 8.0 and made cultivation of paddy feasible thereafter.

6) High Rainfall Zone

It comprises entire Kanyakumari district. This zone is characterized by three major soil types. A narrow belt of sandy alluvial soil is found along the coast while the plains have a deep red loam and the hilly regions have lateritic soil. Acidic soils are found in Thovalai, Thuckalay and Thiruvattar. Due to proximity of Western Ghats and the sea, the climate is influenced by southwest and northeast monsoons. The average annual rainfall in this zone is above 1400mm and the distribution of rainfall is equal during the two monsoon seasons. The major crops grown in this zone are rice, tapioca, coconut, pepper, clove, nutmeg, cardamom and coffee.

The important agroforestry practices prevailing in this zone are described below.

a) Homegardens

The traditional home gardens are followed in Kanyakumari district. In these gardens, many plant species of differing statures and canopy architecture planted in a disorderly manner form multistoried and close cover vegetation. Three to five stratifications of vegetation are usually recognized. The ground layer up to 1.5m is dominated by vegetables and ginger. The second layer is occupied by banana and papaya. The third layer is formed by small trees like guava and citrus. The fourth layer is dominated by fruit trees like jack and mango. The fifth layer consists of tall emergent trees like coconut. The layered structure of the home garden vegetation imposes a gradient of light and humidity that the plants have to exploit according to their ecological and phytosociological needs. These gardens help in year round supply of some produce or the other. The maintenance of the garden involves only a small fraction of the working time.

b) Bund Planting of *Albizia falcataria*

Albizia falcataria is one of the very fast growing trees in the world. It was introduced by the State Forest Department as a wind break around rubber plantations. Its soft timber is ideally suited for making boats, catamarans and match splints. In view of these, the species has gained wide popularity among farmers and is planted on field bunds.

c) Scattered trees in farmlands

Oil bearing trees like *Calophyllum inophyllum, Bassia latifolia* and *Pongamia pinnata* are grown as scattered trees on farmlands. Presently, these trees are being exploited for biodiesel production apart from its other major uses.

7. Hilly zone

This zone covers the hilly regions like the Nilgiris, Shevroys, Elagiri-Javadhi, Kollimalai, Pachaimalai, Anamalais, Palanis and Podhigai malai. The rainfall varies from 1000m at the foot hills to 5000mm at the hill tops. The soils are mainly lateritic. Plantaion crops like tea, coffea and horticultural crops like cabbage, cauliflower are predominately grown. *Grevillea robusta, Eucalyptus globulus*, temperate acacias, terminalias are the dominant tree species

The common agroforestry practices followed in this zone are given below.

a) Intercropping with *Eucalyptus globulus*

This tree species was first introduced by the British in 1860 in the Nilgiri hills for cooking and heating purpose for the soldiers. Farmers grow this tree in systematic rows and raise potato in the interspaces. Though the yield of potato is depressed up to 20%, this yield loss is more than compensated by the revenue from the tree.

b) Shade trees in tea and coffea gardens

Grevillea robusta and *Erythrina indica* are extensively used as shade trees in tea and coffee garden. The litter fall helps maintain soil fertility of these gardens.

9

Production and Functional Aspects of Agroforestry for Enhancing Livelihood Support in Indian Dry Regions

G. Singh

Abstract: *Dry land ecosystems cover about 41% of the earth's land surface including hyper arid region. There is a considerable difference between them, in which arid region covers 7% of global lands. These are barely productive and provide a meager existence to small populations, usually nomadic with little settlements as in Thar desert. In addition to hyper arid, other types are arid, semi and dry-subhumid, which are in danger of becoming desertified. Environmental conditions are harsh affecting agriculture production. People of Indian arid zone are well acquainted about agroforestry- an age old practice in India, but maintaining only a few trees on farmlands does not serve the growing demand of fodder, fueldwood and food. Adoption of alternate land use systems like alley cropping, lay farming, tree farming, dryland horticulture etc. may intensify crop production and sustain the rural livelihoods. The selection of tree/shrub species may be based on the tolerance towards drought, salitnity/alkalinity prevailing in the region. Based on arrangements and combinations of trees/shrubs with associated crops, agroforestry are categorized into: agri-silviculture, agri-horticulture, agripasture, silvipasture, hortipasture and silviherbal. Common trees are P. cinreraria, T. undulata, Salvadora oleoides, Acacia nilotica, A. tortilis, Ailanthus excelsa, Zizyphus nummularia, Z, mauritiana etc. In general, being a diverse system agroforestry utilize full available space and above and below resources and help maintain soil microbial and faunal activity and thus have positive* *effects on soil fertility. In absence of net positive effects of tree on the companion crop, many of the interacting factors can be manipulated in favour of increased production. Agroforestry minimize the risks of drought and famines by way of providing diversified products and help in ameliorating the land under varying types of degradation like salinity, alkalinity, water-logging etc. Many studies demonstrate the usefulness of tree in increasing yield of agriculture crops/ and thus total productivity of land, but there are reports of decreased crop production under tree integration in farmlands, and are the main constraints in its adoption on large scale particularly among the developing countries. In facts trees are removed from the pasture/farmlands for uninterrupted cultivation, without thinking that soil improvement under tree integration in agriculture land is in great part related to increases in soil nutrients and more importantly the organic matter. Besides their role in above-ground carbon sequestration, tree integrated system also have a great potential to increase carbon stocks in the soil and certainly merit consideration in mechanisms that propose payments for mitigation of greenhouse gas emissions to reduce land degradation and minimize the effect of climate change. Convincing people regarding adoption and promotion of agroforestry is a great challenge and can be overcome by capacity building, providing suitable incentives and utilization of public-private partnership. But major roles are expected from the governments focusing this system toward protective and supportive role of agroforestry in desert/arid regions.*

Introduction

Natural resource base for agricultural production, namely water and soil resources are increasingly in short supply on a per capita basis and are under varying degree of degradation in dry areas (Kolarkar *et al.*, 1983). Agriculture already takes 70% of the available freshwater resources and by 2015 the growing demand for water may exceed its availability. Most agricultural soils are in a degraded state which means that they do not anymore reach their full productive capacity. Clearing of native vegetation and replacing it with arable crops, cropping without replenishing soil nutrients, intensive tillage leading to erosion and soil organic matter depletion, and destroying the vegetation cover by overgrazing have led to widespread soil degradation endangering livelihoods of rural populations. As a consequence of tillage-based agricultural land uses, biodiversity at all trophic levels is being increasingly reduced, leading to a deceased functionality and capacity of ecosystems to adapt to changing environmental conditions. Subdivision and fragmentation of land holdings caused by the sharing among sons based on the succession laws and increase in the village population also results a shortfall of food on small farms; 12% in cereals and 42% in pulses, promoted continuous cultivation and the increase of monoculture, and deteriorated the land productivity through its effect on the soil fertility and land management (Ram *et al.*, 1999).

Dry land ecosystems cover about 41% of the earth's land surface including hyper arid region. There is, however, a considerable amount of difference between them. Some of them, covering 7% of global lands, are hyper-arid. These are barely productive and provide a meager existence to small populations, usually nomadic, who depend upon water sources in oases and wells. Hyper-arid regions include the Sahara-Arabian deserts, the Gobi desert, the Thar etc. Other types are arid, semi and dry-subhumid, which are in danger of becoming desertified. The arid regions of India cover an area of 317,090 km^2 and lies between 24^0 to 29^0 latitude and 70^0 to 76^0 longitude. The arid region is spread over seven states, viz., Rajasthan, Gujarat, Punjab, Harayana, Maharashtra, Karanataka and Andhra Pradesh. The north-western part of the country constitutes almost 90% of the total arid zone area in India, where Rajasthan alone accounts for about 61% of the arid zone. The mean annual rainfall varies from 100 mm in the northwest to 450 mm in the eastern part of Rajasthan. Over 90% of the total annual rainfall occurs between June and September. The mean maximum temperature during the summer goes upto 40-42^{0}C. The mean winter season temperature varies from 14^{0}C to 16^{0}C. The potential evapotranspiration during summer is 7 to 9 mm/day, during the monsoon 5.3 to 6.4 mm/day and in winter ranges from 1.8 to 2.9 mm/day. The evapotranspiration far exceeds precipitation throughout most of the year. High wind velocity is characteristics during summer and monsoon season throughout the northwestern region, whereas it is minimum during the post monsoon period. Because of such variability in climatic conditions, the traditional cropping leads to a high degree of uncertainty in yield, income and employment (Radhamani *et al.*, 2003). A survey conducted in 110 villages covering an area of about 1600 km^2 and enquiry made of 390 villagers in Barmer district of Rajasthan indicated almost two third population of small ruminant i.e. sheep and goat as compared to the total livestock population with average land holding and family size of 9.23 ha and 9.1 individuals per family, respectively (Table 1).

Table 1: Land holding, family size and livestock population in Barmer district of Rajasthan surveyed during 2006-07.

Variable	Barmer area	Gudamalani area	Sindhari area	Average
Land holding (ha)	10.41	8.42	8.86	9.23
Family size (nos.)	8.04	9.49	9.80	9.11
Goat (nos.)	15.00	13.00	10.00	12.67
Sheep (nos.)	11.00	12.00	7.00	10.00
Cow (nos.)	4.00	3.00	2.00	3.00
Buffalo (nos.)	3.00	4.00	2.00	3.00
Camel (nos.)	2.00	2.00	1.00	1.67

To minimize the effects of land degradation and the risk of drought and diversify farm produce, the people of arid western Rajasthan adopted integrated farming systems, which includes a change in the farming techniques for maximum productivity by way of optimal utilization of resources, judicious mix of agricultural crops and other enterprises suited to the region and enhance socio-economic status by improving prosperity in the farming system (Kapoor, 2001). Thus a more sustainable approach to agricultural production and agro-ecosystem management are mandatory to reverse the degradation processes in dry areas (FAO, 2011). Integrating trees in the farmlands has potential for simultaneously satisfying the important needs of protecting and stabilizing the ecosystems; producing a high level of output of economic goods; and improving income and basic materials to rural population (Dhyani *et al.*, 2009). Besides, this system is capable to conserve natural resources, livelihood security by fulfilling basic human needs *viz.*, food, fuel, fodder, and employment generation. Agroforestry has also potential in meeting out the deficit of demand and supply in timber, fodder supply, bioenergy sector through tree biomass and meeting the food/fruit security.

Agroforestry definitions and its presence in Indian context

The International Council for Research in Agroforestry (ICRAF, now World Agroforestry Center) defined agroforestry as: "a dynamic, ecologically based, natural resource management system that, through the integration of trees on farms and in the agricultural landscape, diversifies and sustains production, enhancing social, economic and environmental benefits for land users at all levels". Popularization of multipurpose trees and the domestication of trees for products other than wood have made this system easier to measure and promote the potential use of trees in non-forest situations. Even granting that agroforestry is an ancient art, the current interest in trees and their development is unquestionably responsible for some of the newly enhanced awareness about growing trees outside of the forest areas and help control land degradation (Bellefontaine *et al.*, 2002). The people of western Rajasthan, since ages has developed a variety of site specific agroforestry system (Gupta *et al.*, 2000). Here farmers allow growing scattered trees and shrubs on their farmlands/grazing fields to sustain their life. The farmers consider these trees as boon in the region particularly during drought when rainfed crops fail. The trees provide fodder, fruit, vegetable, fuelwood, timber and fiber for sustaining rural livelihood. It is aclaimed that agroforestry provides 62% of the fodder, fuelwood and timber requirement of the rural people (Tewari *et al.*, 1999). However, the striking feature of traditional agroforestry systems is their ability to interchange in time and space. Traditionally, farmers use their farmlands as silvi-pastoral system during the years of deficient rainfall and changing to agri-silvicultural system during normal or good rainfall years. Density of trees/ shrubs also varied from one agroforestry system to other depending upon availability of the resources (Singh *et al.*, 2007).

Traditional agroforestry in western India

There are ample examples of permanent, slow and fast traditional adaptations i.e., intercropping, sequential cropping and agroforestry to seasonal variability for coping strategies and food security throughout the world. Agroforestry is an age old practice in India- in the form of the shifting agriculture, a variety of cereal cropping, home garden systems, traditional plantation systems, etc. The role of many common trees such as khejri (*Prosopis cineraria*), aswattha (*Ficus religiosa*), palasa (*Butea monosperma*), Varana (*Crataeva roxburghii*) and Afoye or Rohida (*Tecomella undulata*) in Indian folk-life has been mentioned in ancient literature of Rig Veda, Atharva Veda and other ancient scriptures. The efficient community strategies in this region have exemplified the intelligent and sustainable use of natural resources without causing damage to the resilience and functioning of the surrounding ecosystem (Mukhopadhyay, 2008). For example, the Bishnoi community of Rajasthan has over the centuries combined a unique blend of ecological sense and religious sensibility in the Thar Desert in India. Living amidst the barren wastelands interspersed with the isolated trees of Khejri (*Prosopis cineraria*) and Babool *(Acacia nilotica)*, these people are proud of it. Amrita Devi a Bishnoi woman, who along with more than 366 others, died saving *Prosopis cineraria* trees. The Bishnois are an example of people living in harmony with nature where they maintain groves, locally known as orans, where animals graze and birds feed. Orans, which account for about 9% of the desert area, serve as important rechargers of rainwater in the desert aquifers, where every single drop of water is precious. The tree species, *Prosopis cineraria* or Khejari, is worshipped for its immense ecological and economic value (Khatri *et al*., 2010). The peoples of this regions are so associated with nature/ trees that they use to protect and promotes whatsoever trees or shrubs comes up naturally on their farmlands. In a survey conducted in Barmer district of Rajasthan indicated that peoples had more affinity with *P. cineraria* tree indicated by 67 individual of *P. cineraria* per family (Table 2) followed by *Capparis decidua* (38 individual per family) and *Zizyphus nummularia* (28 individual per family).

Table 2: Number of trees/shrubs on farmlands of an individual family in Barmer district of Rajasthan.

Species	Barmer area	Gudamalani area	Sindhari area	Average
Acacia tortilis	3	4	-	2.3
Prosopis cineraria	46	75	49	57.3
Tecomella undulata	18	16	5	13
Salvadora oleoides	16	5	5	8.7
Acacia senegal	26	4	4	11.3
Capparis decidua	61	29	24	38.0
Zizyphus nummularia	38	43	3	28.0
Cordia myxa	2	1.5	-	1.2
Azadirachta indica	3	6	-	3
Acacia nilotica	8	30	-	12.7
Eucalyptus camaldulensis	2	5	-	2.3
Dalbergia sissoo	1	-	-	0.3
Prosopis juliflora	5	51	-	18.7
Acacia jacquemontii	-	25	-	8.3

Farmers in western Rajasthan traditionally grow arable crops in association with perennial trees, shrubs and grasses growing naturally on their farmlands as a strategy to avoid the risk of frequent drought. Thus tree integration on farmlands manifests rural people's knowledge and methods to benefit from complimentary uses of such annuals and woody perennials on the sustained basis. Besides ensuring some production benefits even during the worst drought, the traditional agroforestry components fulfill the most of the requirement from the land, including food, fodder, fuel, timber, etc. Traditional agroforestry systems generally have scattered trees and shrubs of *Prosopis cneraria, Tecomella undulata, Ailanthus excelsa, Acacia senegal, Capparis decidua* and *Leptadenia pyrotechnica* etc (Harsh *et al.*, 1992; Harsh and Tewari, 2007; Tewari and Singh, 2006). In these traditional agroforestry systems (Table 3), *Prosopis cineraria* (Khejri) and *Ziziphus nummularia* (Bordi) are the two most important multipurpose woody components in traditional agro-forestry system of the region. Unfortunately, many of these traditional systems have started breaking down for a variety of reasons and in many cases the high diversity based agroforestry systems have been replaced by low diversity simplified cash crop systems, which is a question in terms of sustainability (Depommier, 2003).

Table 3: Traditional agroforestry systems of Thar Desert.

Rainfall zone (mm)	Agroforestry system	Trees/shrubs (nos ha^{-1})	% D[illegible] prom[illegible]
>400	*P. cineraria – A. nilotica* based	31.4	80.5
300-400	*P. cineraria* based	14.2	80.0
200-300	*Zizyphus* spp. – *P. cineraria* based	91.7	100.0 (91.7% *Zizyphus* spp.)
<200	*Zizyphus* spp. – *P. cineraria* –*Salvadora* spp. based	17.2	87.2 (65% *Zizyphus* spp.)

Source: Narain and Tewari, 2004

Goyal and Sharma (2009) put emphasis on traditional wisdom and value addition prospects of arid foods of desert region, which is endowed with valuable natural resources. Arid foods are largely used by the native people of north-western India as a prime source of food with their own traditional wisdom with less effort for value addition. However, processing of traditionally important arid foods into more useful and convenient product improves livelihood security of the people residing in the desert regions.

Types of agroforestry

About 70 per cent of the Indian arid zone is covered by sand dunes, desert, and grey brown desert soils, and is vulnerable to wind action and erosion. General control measures include shelterbelt plantations, tree plantations, and grasses on marginal and sub-marginal lands (Singh, 199[illegible]). Thus various forms of agroforestry includes tree on farmlands, live-hedge, farm boundary plantation, roadside/canal side plantation, shelterbelts, silvipasture etc. Site specific measures include grassland development and afforestation using multipurpose trees and shrubs in zones with less than 200 mm rainfall (i.e., Jaisalmer and Barmer); afforestation by using multipurpose trees, grassland development, silvipasture, and wind strip cropping of legumes with grasses in 200-300 mm rainfall zones (i.e., Barmer and Bikaner); agroforestry, pasture development, wind strip cropping, minimum and proper tillage-set row ridge/furrow cultivation in 300-400 mm rainfall zones (i.e., Jodhpur, Nagaur, Jalor and Pali); and agroforestry, silvipasture, strip cropping, optimum tillage-set row cultivation, and stubble mulching in >400 mm rainfall areas (i.e., Churu, Jhunijhunu, Sikar, Pali, Nagaur). In these agroforestry systems, there are a lot of variations in species exists in the farmlands mainly due to varying amount of rainfall and soil moisture availability (Table 4).

Table 4: Rainfall range and species distribution in farmlands in western Rajasthan

Rainfall	Species
>200 mm	*Acacia tortilis, A. senegal, Prosopis cineraria, Zizyphus nummularia, Calligonum polygonoides, P. juliflora, Tecomella undulata, Salvadora oleoides,*
250-400	*P. cineraria, Hardwickia binata, Colophospermum mopane, Dichrostachys nutans, Ailanthus excelsa, Acacia leucophloea, Grewia tenax, Acacia nilotica, Zizyphus mauritiana*
[illegible]	*Acacia nilotica var. cupressiformis, Albizzia amara, A. lebbek, Bahuhinia racemosa, Cassia siamea, Emblica officinialis, Hardwickia binata, Ailanthus excelsa, Moringa oleifera, Holoptelia integrifolia, Anogeisus rotundifolia*

However, trees/shrubs in a farmland can also be presented in an infinite number of arrangements and species combinations depending mostly on farmers' objectives as well as the environmental characteristics of the region providing a number of improved agro-forestry packages. Trees by virtue of their perennial nature impart stability in production along with improvement in microclimate and soil fertility. *Prosopis cineraria*, *Holoptelia integrifolia* and *Hardwickia binata* are some of the tree species suitable for agrisilviculture system in dry land area of this region (Muthana and Arora, 1977; Shankarnarayan *et al.*, 1987). Some of them are:

- Agri-silviculture
- Agri-horticulrure
- Agripasture
- Silvipasture
- Hortipasture
- Silviheral

Agri-silviculture

Growing agriculture crops in association of tree crops is generally known as agri-silviculture (Fig. 1). The interrelationship between an agricultural crop and a tree crop on a single piece of land may be supplementary, complementary or competitive. In the former two cases this system is always an attractive option (Nair, 1993; Boffa, 1999). However, there are some examples where agroforestry increase risk that includes: (i) tree products are mainly sold in an uncertain future, where prices are not known with certainty during the planting period, and the long time taken by trees to realize output as compared to annual crops; (ii) a combination of trees and crops may promote diseases and pests; and (iii) tree product markets in developing countries are still underdeveloped thus acting as a potential source of risk in agroforestry income (Senkondo, 2000). However, if we can show or convince that an agroforestry system does reduce risk and enhance land productivity, it will have great economic value to agriculture (Senkondo, 2000).

Tecomella undulata + Wheat

Ailanthus excelsa + Musturd

Fig. 1. Improved agroforestry involving *Tecomella undulata* and *Ailanthus excelsa* with associated wheat and mustard crop, repectively under irrigation in western Rajasthan

Scientific evidences demonstrate that integrating trees in agriculture lands increases the total productivity per unit of land as compared to sole arable cropping. Agri-silvi system consisting *Tecomella undulata* as woody component with various dryland crops was found promising due to improved soil fertility, microclimate and moisture availability (Harsh and Tewari, 1993). A study conducted in arid region of Haryana on *Prosopis cineraria* based agro forestry system revealed that the tree influenced grain and fodder yield of associated crops. Fodder yield and net return was more in association with tree as compared to sole cropping. Maximum net returns of Rs. 15197 ha^{-1} was obtained with pearlmillet-*Brassica tournefoila* cropping system in association of *P. cineraria* (Kaushik and Kumar, 2003). In arid region of Gujarat, *Azadirachta indica* and *Ailanthus excelsa* based system involving *Vigna aconitifolia*, *V. radiata*, *Cymopsis tetragonoloba* and *Sesamum indicum*, on an average fetched 59.3% and 25.7% more income, respectively than sole cropping. Besides higher economic return, the system also improved soil organic carbon along with catering diverse need of farmer viz. fodder, fuel wood and timber (Patel *et al.*, 2008). Preliminary results of an experiment conducted at Bikaner, Rajasthan indicated that integration of perennial vegetation like *Calligonum polygonoides* and *Acacia jacquemontii* imparted stability in production. For example, during four years of experimentation (2003- 2007), the arable crops failed to attained harvest maturity due to paucity of rainfall, whereas the perennial woody components provided fodder and fuel wood. Further the growth of *Calligonum polygnoides* was influenced by associated component and it attained maximum growth in association with legume (*C. tetragnoloba* and *V. aconitifolia*) and least with association of grasses (*Lasiurus sindicus* and *Cenchrus ciliaris*). Studies conducted at Pali revealed that strip cropping of *Lawsonia inermis* consisting 4 rows of Henna alternated with 4 rows of *C. tetragnoloba* at 60 cm spacing provided higher return than their sole planting (Singh *et al.*, 2005; Vyas, 2005). About 20% increases in yield of *V. radiata* has been reported when intercropped with lopped *Holoptelia integrifolia* (12 years old plantations) tree as compared to the un-lopped ones (Harsh, 1995). A study conducted in rainfed stands of *Pennisetum glaucum* sown as intercrop with *Zizyphus mauritiana, Prosopis cineraria* and *Acacia nilotica* indicated an average crop height at harvest of 107, 105 and 67 cm, respectively (Singh, 2003). Here the grain yield of pearl millet was 190, 370 and 522 kg ha^{-1} sown with the respective tree species.

In another study where silvi component namely *Eucalyptus*, *Leucaena* and *Casuarina* were intercropped with black gram, kalmegh, tobacco and guinea grass, the yield of later did not vary significantly with tree species in the first year, but in second and third years, the yields of intercrops except guinea grass and kalmegh declined drastically. The yield of tobacco declined to 29, 33 and 54 percent of sole crop with *Eucalyptus, Leucaena* and *Casuarina*, respectively in the third year, indicating higher impact of eucalyptus. The yield of blackgram declined to 33, 33 and 52 percent of sole crop with *Eucalyptus, Leucaena* and *Casuarina*, respectively in the third year. Kalmegh and guinea grass yielded up to 85 percent of sole crop in association with trees even in third year. The results suggested that annual crops can be grown up to two years with perennials without any adverse impact on yield (Osman *et al.*, 2008). A four years study on tree-crop interactions (in Jodhpur) as affected by varying tree density (1666, 833 and 417 stems ha"^{1}) and intercrops, i.e. *Vigna radiata* (mungbean), *Cyamopsis tetragonoloba* (clusterbean), *Vigna acontifolia* (mothbean) and *Pennisetum glaucum* (pearlmillet) indicated that during the tree establishment phase, *P. cineraria*/ *T. undulata* did not compete with the associated agricultural crops nor was its growth affected by the intercrops (Gupta *et al.*, 1998; Singh *et al.*, 2005). At four years of age, 417 stems ha"^{1} (4 × 6 m^2 spacing) was found optimum tree density. Higher stem density adversely affected tree growth. Pulses (mung bean, clusterbean and mothbean) were better suited than pearlmillet as intercrops. When *Emblica officinalis*, *Hardwickia binata* and *Colophospermum mopane* were intercropped with *Vigna radiata* (mungbeans) in 1994, *Cyamopsis tetragonoloba* (cluster beans) in 1995 and *Pennisetum glaucum* (pearl millet) in 1996, Singh *et al.* (1999) did not observe variation in crop yield due tree components. Rather the production of *Vigna radiata* was significantly higher in 1996 a better rainfall year than in 1994 (Table 5).

Table 5: Influence of trees on crop production in agri-silviculture/agri-horticulture system in Indian arid zone of Rajasthan.

Tree component	Agriculture crop	Yield change than to sole crop	Source
Prosopis cineraria	*V. radiata/ Pearlmillet*	Increase/decrease	Singh *et al.* (2007); Singh (2010)
Tecomella undulata	*V. radiata/ Pearlmillet*	Increase/decrease	Singh *et al.* (2005); Singh (2010)
Holoptelia integrifolia	*V. radiata*	Increase	Harsh (1995)
C. mopane	*V. radiata*	Decrease	Singh and Rathod (2007)
Hardwickia binata	*V. radiata*/ Sorghum	Decrease/Increase & chick pea	Singh (2010)/ Devaranavadgi *et al.* (2003); Gill (2000)
Faidherbia albida	Caster/sorghum	Increase	Osman and Rao (1988)
Acacia cupressiformis	Chick pea	Increase	Gill (2000)
Z. mauritiana	Guara/Mung	Increase	Singh *et al.*, (2003); Gupta, 1992; Sharma and Gupta (2001)
D. sissoo/A. nilotica	Wheat	decrease	Puri *et al.* (1995)
Aonla/Terminatia	Ashwagandha /Andrographis	decrease	Lata *et al.* (2012)

A tree density of 100-200 plants ha^{-1} was found optimum for minimum interference with yield of dry land crop like *Cyamopsis tetragonoloba* under *P. cineraria* canopy shade. *Vigna sinensis* and *Pennisetum typhoides* showed better performance with tree species than *Vigna radiata* and *Vigna aconitifolia*. Besides good yield of dry land crops, bonus yield of dry leaves and twigs (650-1050 kg ha^{-1}) and fuel wood (1.8-2.6 t ha^{-1}) could be obtained from *Prosopis* tree through annual lopping (Bhati *et al.*, 2008). However, studies conducted during June 1995 and June 2002 to find out optimum tree density of *Prosopis cineraria* and *T. undulata* by maintaining tree density at 417, 278 and 208 stem ha^{-1} at five year age and intercropped with *V. radiata* and *P. glaucum* Singh *et al.* (2004, 2007) recorded lower grain and straw yield under intercropping than with a sole

Table 6. Crop yield (kg ha^{-1}) under the influence of varying density of *P. cineraria* and *T. undulata* trees. D*: density in D_1, D_2 and D_3 were 416, 278 and 208 during June 1995 to 2002 and 208, 138 and 104 during June 2002-2004, respectively.

D*	2000 (*V. radiata*)			2001 (*P. glaucum*)			2003 (*V. radiata*)		
	Grain	Straw	Total	Grain	Straw	Total	Grain	Straw	Total
P. cineraria									
D_1	153	1666	1819	133	4330	4463	454	2211	2665
D_2	162	1699	1861	141	4430	4571	674	2747	3421
D_3	193	1803	1996	158	5420	5578	1295	2713	4008
T. undulata									
D_1	123	1457	1580	122	3827	3945	422	2023	2445
D_2	140	1423	1563	131	4100	4231	644	2427	3071
D_3	154	1510	1664	136	4560	4696	1216	2493	3709
Control	180	1720	1900	165	5470	6635	1616	4539	6145

agricultural crop (Table 6). Production was significantly low with tree density of 416 tree ha^{-1}. The yield was highest with tree density of 278 tree ha^{-1} at 6-7 years of tree age and with tree density of 208 tree ha^{-1} at 10-11 year of tree age. In addition the utilizable trees biomass from *P. cineraria* was 19.1, 15.8 and 10.3 tones/ha and dry leaf mass of 0.85, 0.67 and 0.50 tones/ha, respectively with tree density of 208, 278 and 417 tree ha^{-1} at the age of 12 years. The trees of *T. undulata* produced an utilizable biomass 3.63 to 4.08 t ha^{-1} at the age of 9 years. Tree density plays an important role not only on the yield of agricultural crops but also on the tree growth. For example, reduction in tree density by 75% (416 from 1666 tree ha^{-1}) resulted in a significant increase in height and collar diameter of *P. cineraia* and *T. undulata* between June 1995 and June 2002 as compared to the tree density of 278 tree ha^{-1} (from 833 tree ha^{-1}, i.e., 66% reduction) and 208 tree ha^{-1} (from 416 tree ha^{-1} i.e., 50% reduction) (Singh *et al.*, 2004, 2007). In a khejri (*Prosospis cineraria* trees planted at 3×3 m spacing) based system tree attained a height and DBH (diameter at breast height) of 528 cm and 14.8 cm, respectively at the 12 years of age, whereas leaf fodder and fuelwood production from lopping up to 50% of height was 6.6 kg and 14.0 kg/tree, respectively (Jaimini *et al.*, 1998). Singh *et al.* (2005) recorded taller and thicker plants of *T. undulata* at the age of 5 years that produced greater dry biomass per *Tecomella* plant at tree density of 416 tree ha^{-1} than those at 1666 tree ha^{-1} and 833 tree ha^{-1}. However, higher plant density at 1666 tree ha^{-1} resulted in biomasses of 1.39 and 1.46 Mg ha^{-1} compared to 0.68 and 0.83 Mg ha^{1} at 416 tree ha^{-1} in no-crop and cropped plots, respectively.

Potential competition for resource between the trees and associated crop was analyzed by measuring soil water contents, soil organic matters and NH_4-N at different depths of soil layers i.e., 0-25 cm, 25-50 cm and 50-75 cm in the subsequent experimentation by maintaining tree densities of 208, 138 and 104 trees ha^{-1} after June 2002 for both *P. cineraria* and *T. undulata* (Singh, 2009). Results showed that tree height increased by 3% to 7% during June 2002 to June 2004. Collar diameter increased by 30% and 11% at 208 trees ha^{-1}, 23% and 19% at 138 trees ha^{-1} and 18% and 36% at 104 trees ha^{-1}, respectively, in *P. cineraria* and *T. undulata* in two years period. The increase in crown diameter was 9% to 18% in *P. cineraria* and 11% to 16% in *T. undulata.* Tree growth was relatively greater in 2002 than in 2003. Yield of *V. radiata* increased linearly from tree density of 208 tree ha^{-1} to 104 tree ha^{-1} (Table 6) *P. cineraria* was more beneficial than *T. undulata* in improving soil conditions and increasing crop yield by 11.1% and thus more suitable for its integration in agricultural land. The yield of agricultural crop increased when density of tree species was optimum.

Facilitative effects have been observed under *Hardwickia binata*, where Devaranavadgi *et al.* (2003) observed better compatibility of *Hardwickia binata* with rabi sorghum due to minimum crown diameter in southern India. Bheemaiah and Subrahmanyam (2004) studied intercropping of sunflower with *Psidium guajawa* (2004), where growth characters like plant height, leaf number, leaf area and dry matter production and yield were significantly higher under intercropped sunflower over sole crop. Gill (2000) also recorded highest seed yield of chickpea when grown as intercrop in *Madhuca latifolia, Acacia cupressiformis* and *Hardwickia binata* and the yield was increased with increasing distance from the tree component. Growing of castor and sorghum for two consecutive years under two years old nitrogen fixing trees (NFTs) showed 23% and 13% increase in yield of castor and sorghum respectively under *Faidherbia albida* as compared to respective sole crops (Osman and Rao, 1988). However, a shift from facilitation to competition with tree age has also been recorded (Singh and Rathod, 2007), when *Colophospermum mopane* and *Hardwickia binata* trees were intercropped with *P. glaucum, V. radiata, Sesamum indicum* and *Cymopsis tetragonoloba.* Mean annual increment in tree biomass of *C. mopane* was 1.2 tonnes ha^{-1} $year^{-1}$. The highest grain and dry matter yield in tree-integrated plots from 1996 – 2000, and in the control plots in 2001 and 2003 indicated a shift from facilitation to competition with tree age. Tree-integrated plots had lesser soil water than in the control indicating tree water use. Added litters and nutrient uptake by tree increased soil organic matter and decreased available PO_4-P, NH_4-N and NO_3-N in the tree-integrated plots. These tree species enhanced land productivity through increased system production and agriculture yield in the initial 4 – 5 years. However, these both species found highly competitive at 9 year of age i.e., 2003, when rainfall was sufficient only for the agriculture crop) because of competitive use of soil water (Singh, 2010). Lowest SWC in 0-25 cm soil layer and at 1 m from tree suggested competitive use of soil water by the trees reducing crop yield by 97%.

In absence of such positive effects, various management practices are adopted to favorably alter the interactive relationship in agro-forestry system (Patil and Channabasappa, 2008; Singh and Sukla, 2012). Some of these are manipulation of tree crowns and roots through management operations, mainly by pruning and thinning. For fodder tree species, lopping done for harvesting forage periodically can at the same time increase light availability to the under canopy crops and improve their performance. For example, reduction of 15% to 90% has been observed in the yield of greengram and clusterbean under unlopped *Colophospermum mopane, Hardwickia binata* and *Tecomela undulata* trees in agroforestry (Singh, 2004). Resource management in agroforestry systems may potentially provide options for improvement in livelihoods through simultaneous production of food, fodder and firewood with additional benefits of carbon sequestration. Trenching around tree to reduce root overlapping between tree and agriculture crop and therefore competition for resources are also beneficial in enhancing crop/forage yield (Singh and Shukla, 2012).

Agri-horticulture

Shifting towards horticulture based production system is considered effective strategy for improving productivity, employment opportunities, economic condition and nutritional security (Pareek, 1999; Chadha, 2002) because of risks involved in crop production in this hot arid region. *Cappirs decidua*, *Salvadora oleoides*, *Cordia dichotoma*, *Cordia gharaf*, *Zizyphus nummularia*, *Z. mauritiana* etc are suitable horticultural species for the area receiving rainfall <300 mm. Besides providing fruit these plant produce moisture laded nutritious leaves for animal. Other fruit crops are *Emblica officinialis*, *Punica granatum*, *Aegle marmelos*, *Phoenix dactylifera* and *Tamarindus indica,* but these can be grown with supplemental irrigation (Pareek and Awasthi, 2008). Results of different agri-horticultural research indicated an increase in total productions. For example, a trial at Pali showed highest yield of mungbean (2.70 quintal ha^{-1}) with *Citrus* spp. and of clusterbean (3.8 q ha^{-1}) with *Carissa carandus* plants. Maximum increment in height was found in pomegranate and collar diameter was in *Z. mauritiana*. Intercropping of bottle guard during *kharif* season and pea and kasuri methi during *rabi* with *Z. mauritaiana* did not show adverse effect on three year old *Z. mauritaiana* and produced 40900 kg ha^{-1}, 5200 kg ha^{-1} and 8100 kg ha^{-1} green leaves and 880 kg seeds ha^{-1} of bottle guard, peas and methi respectively (Singh and Kumar, 1993). An agri-horti system involving *Z. rotundifolia* + *V. radiata*/*V. aconitifolia*/*C. tetragonoloba* and *Z. mauritiana* + *V. radiata*/*C. tetragonoloba* have been found environmental friendly and economically viable even during drought years. In *Ziziphus* + *V. radiata,* the yield of mungbean was reduced by 44% as compared to 51% reduction in sole crop during subnormal year when rainfall was 51% less than long term average of 360 mm. This system provided round the year supply of fodder for 5 goat/sheep ha^{-1} and fuel wood for family off 4 members, besides efficient nutrient cycling and increase in economic stability (Faroda, 1998). Gupta *et al.* (2000) reported that 3-years old plantation of *Z. maruitiana* @ 400 plants ha^{-1} in association with green gram performed well with seasonal rainfall of 210 mm and fruit yield from intercropped increased net profit to Rs. 288.6 ha^{-1}, minimizing risk and helps impart economic stability. Saroj *et al.* (2003) recommended *C. tetragonoloba* – *B. juncea* and Indian aloe as ground storey component in *Z. mauritiana* to optimize productivity and profitability in arid areas. Integration of *C. tetragonoloba*, *V. radiata* and *Sesamum* with *Z. mauritiana* increased fruit yield by three fold (14.8 kg $tree^{-1}$) as compared to pure orchard i.e., 5.2 kg $tree^{-1}$ (Singh, 1997). Intercropping in newly planted *Z. mauritiana* orchard had no adverse effect on plant growth up to 5 years. The intercrop also exhibited higher yield when planted with ber compared to monoculture under rain fed conditions. Agri-horti system comprising *Zizyphus mauritiana* and mungbean or legume crops provided fruit, fuel wood and round year employment even in below average rainfall year and enhanced agricultural production (Gupta, 1992; Sharma and Gupta, 2001; Singh *et al.*, 2003). An experiment conducted in Bikaner indicated that intercropping of annual crops with fruit trees provides the extra income to farmers when fruit trees are in their juvenile phase. The highest total income and net profit was realized with *Agle marmelos* + groundnut intercropping followed by *Z. mauritiana* +groundnut and kinnow + groundnut (Yadava *et al*., 2006). The highest B: C ratio was recorded with *Agle marmelose* + clusterbean followed by *Agle marmelose* bael + Mothbean.

A kinnow based agri-horti system under irrigated arid condition at Sriganganagar showed a non-significant effect on fruit yield. The yield was highest under mung bean and lowest under cotton intercropping, the fruit yield was at par with intercropping of mungbean, cotton-barley and cotton- chickpea (Bhatnagar *et al*., 2007). Lal (2005) reported that integration of arable crops (clusterbean, horse gram, mungbean and henna) with pomegranate improve the profitability over sole pomegranate on medium soil of Pali, whereas pomegranate was found compatible with pearl millet, mung bean, isabgol, sorghum and cumin in Jalore district of Rajasthan (Gupta, 2000). In an aonla based multi storey production system comprising aonla-ber-brinjal-mothbean- fenugreek and aonla-bael-karonda-mothbean-gram showed that ground storey crops did not affect growth of over storey crop and vice versa and these systems was found promising under arid conditions (Awasthi *et al*., 2005; 2008). The net return obtained from above cropping through ground storey crop during first year was to the tune of Rs. 23614 and 25662 ha^{-1}, respectively, which increased up 20 and 15%, respectively, in 2nd and 3rd years. This indicates that during juvenile phase of fruit tree, there are ample opportunities for raising annual, biennial and perennial crops which can meet diversified need of farmers. *Emblica officinalis* spaced at 5 x 5 m intercropped with *V. radiata, P. glaucum, C. tetragonoloba* and *S. indicum* at Jodhpur indicated less soil water content in the tree-integrated plots than in the control plots and similar trend was for soil organic matter and available NH_4-N (Singh *et al*., 2008). Crop yield was higher in tree-integrated plots between 1996 and 2000 but thereafter a reverse trend was observed suggesting that the beneficial effect of trees on crops depended upon tree age and available soil water.

Among the management practices, use of suitable rootstocks for improved cultivar, bio-fertilizer (Meghwal *et al*., 2006), seed treatment (Meghwal, 2007; Meghwal and Vashishtha, 1998), pruning management (Pareek and Vishalnath, 1996; Pareek, 2001) and use of water harvesting (Sharma *et al*., 1982) are important in enhancing the productivity of this system. Circular catchments around each tree in 1.5 m radius with 5% slope towards the tree trunk and covering the catchment with black polyethylene sheet found effective to conserve soil moisture, whereas providing supplemental irrigation at 1.0 IW: CPE ratio at 10 cm depth with 11 to 12 irrigation from June to February was found best for pomegranate enhancing fruit yield (Anonymous, 1990-91). Drip irrigation resulted in early commercial fruit production in *Ziziphus mauritiana, Punica granatum* and *Emblica officinalis* and application of water equal to 60% of pan evaporation showed significantly higher fruit yield as compared to 40% pan evaporation (Anon., 2007). In an irrigated agri-horti-silvi system Singh *et al*. (2012) recorded higher growth and survival of *C. mopane* and *Ailanthus excelsa* as compared to sole tree crop, where the yield of wheat grown with *C. mopane* and *Cordia mixa* indicated significant reduction in crop yield.

Agri-pasture

Studies conducted in Rajasthan on cropping between grass strips laid out against prevalent wind direction revealed that average yield of *C. tetragonoloba* at Jodhpur was 418 kg ha^{-1} under unprotected plot, which increased up to 503 kg ha^{-1} in protected plot. Strip cropping produced at least some biomass even in low rainfall years besides reducing soil erosion. In good rainfall year, production of arable crops increased along with increased forage yield of grasses. At Bikaner, average yield of cluster bean during normal and low rainfall were 589 and 181 kg ha^{-1}, respectively. Dry forage yield of *Lasirus sindicus* with strip cropping system obtained during normal and low rainfall year were 6400 and 2650 kg ha-1, respectively. The practice of strip cropping of legumes and grasses seems beneficial in the dry region (Singh, 1989; 1995). Ley farming also increased grain yield of crop significantly over control (Singh and Gupta, 1997). Dauley (1994) reported that intercropping of arid legumes with *Cenchrus ciliris* gave higher yield, return and moisture use efficiency than sole pasture. The study of Bhati (1997) on net economic returns of agri-pasture indicated highest benefits from agri-pastoral system (B: C ratio 1.87) as compared to the other forms i.e., agroforestry (1.69), silvi-pasture (1.66), agri-horticulture (1.46) and sole production of crops (1.24). The higher returns from agripastoral system are especially because of the importance of livestock component in the farming systems prevailing in the region. It also safeguards the vulnerable terrain from drought and desertification. Growing of pearl millet and clusterbean with grasses *L. sindicus* and *C. ciliaris* between rows of *A. tortilis* and *Z. rotundifolia* have been found highly compatible and remunerative in terms of grain and fodder yield. The most highly compatible trees with grasses are: *Acacia senegal, A. tortilis, Albizzia lebbek, Tecomella undulata, Colophospermum mopane, Dychrostasis nutans, Hardwickia binnata, Z. nummularia and Z. rotundifolia*. Among the pasture, legumes *Clitoria ternatea* and *Lablab purpureus* showed good compatibility with *L. sindicus* and *C. ciliaris*.

Silvipastoral system

Traditionally every village or cluster of villages used to set aside certain areas for livestock grazing, called goucher, and protected other areas from grazing, where the grass was allowed to grow long, to be subsequently harvested and stored for later use. These grasslands were crucial to rural economy as the fodder produced was used both during the lean summer months and to tide over drought periods. However, with the promulgation of Land Ceiling Act, and increasing population pressures, the land use practices have been altered. The land set aside for grazing purpose has been encroached upon for agriculture, industrial development and urbanization, as a result protected fodder producing grasslands came under increasing pressure of livestock grazing. Silvipasture refers to a combination of pasture grasses/legumes with trees help optimizing land productivity, conserving plants, soil and nutrient to produce forage, fuel wood, timber etc on sustainable basis (Shanker *et al.*, 1976; Dhurvanarayan, 1993; Gajja *et al.*, 1999; Gajja and Harsh, 2002).

Considering the degradation pattern in dry areas, it is essential to renovate the degraded land through perennial vegetation to check further degradation and to meet requirement of livestock and human. Silvipasture system has been found as an ideal alternative for development of such degraded land in our country (Rai, 2008). Keeping 10-15% of total land holding as fallow for 2-3 years is normal practice among farmers of arid areas. For silvi-pasture development, *A. lebbeck, T. undulata, C. mopane, A. senegal, Z. numularia* and *Z. rotundifolia* are some woody species compatible with grass component, whereas *Clitoria ternata* and *L. purpureus* are legume crops and *L. sindicus* and *C. ciliaris* are the important grass species (Bhati *et al.*, 1986). Shankar (1980) opined that silvii-pasture systems are more remunerative as compared to rainfed farming in arid and semi-arid region. However, for getting optimum production from this system, knowledge of species, planting techniques, fertilizer application and harvesting schedule are of prime importance (Shankar, 1995). For example, production from a silvipastoral system with *A. tortilis* and *C. ciliaris* was higher than that from a pure pasture (Muthana *et al.*, 1985). While, the carrying capacity of a pure pasture recorded to be 3.9 sheep ha^{-1} after 9 years of establishment, a silvi-pastoral system showed carrying capacity of 8.5 sheep after 7 years (Tewari and Harsh, 1998). Based on a study, Gupta and Mohan (1982) concluded that silvipasture system provided more net annual return than arable crop, and recommended that multiple use of silvipasture system to be economically attractive in addition to many ecological benefits.

Shankarnarayan *et al.* (1987) worked out the productivity and economics of *Acacia tortilis* based silvi-pastoral system and reported higher revenue generation (Rs. 3895 ha^{-1}) from silvipasture system compared to sole *Acacia tortilis* (Rs. 3000 ha^{-1}) and sole grass (Rs. 1150 ha^{-1}).

Integration of *Z. nummularia* with *Cenchrus ciliaris* strips in 1:2 ratio gave higher live weight (33 kg ha^{-1} yr^{-1}) and wood production (5.65 kg ha^{-1} yr^{-1}) over sole pasture there by high return Rs. 1326 ha^{-1} year^{-1} from grazing of mixed flock of sheep and goat (Bhati *et al.*, 1987). Silvipasture of *Z. rotundifolia* and *Cenchrus ciliiris* could sustain 554 tharparkar cattle days ha^{-1} with 60% pasture utilization (Naraian and Bhati, 2004). Economics of different silvipasture system at CSWRI, Avikanagar revealed that three tier systems provided an income of Rs. 2056.3 ha^{-1} followed by two tier system (Rs. 1913.47 ha^{-1}), single tier system (Rs. 1616.29 ha^{-1}) and natural pasture (Rs.922.42 ha^{-1}) (Anon, 1998). In a combination of trees like neem, Acacia and subabul and grasses like *C. ciliaris* and *C. setegerus*, Neem + grasses combination was most productive system for the Kachchh region in term of both grass yield and tree growth. Among the grasses, *Cenchrus ciliaris* was found to be superior to *C. setigerus* in terms of fodder production (Dayal *et al*., 2008) and the soil of silvipastoral systems enhanced the soil organic carbon and K content of the degraded lands (Shamsudheen *et al*., 2009). In a study at Jodhpur, Singh and Shukla (2011) recorded observations on the effects of three year old *Azadirachta indica* plants and nitrogen fertilizer on herbaceous diversity, productivity and soil improvement in desert pasturelands. Intact plants of *A. indica* reduced photosynthetically active radiations by 53-55%, whereas nitrogen fertilization increased soil nitrogen, species richness, diversity and biomass but negatively affected soil water. Plots with intact plants showed highest soil available nitrogen, species richness, diversity and evenness indices. However, soil water and species dominance were highest in the control plots. Herbage biomass increased by 1.4- to 2.0-fold over control. Both tree association and nitrogen fertilization enhanced the productivity. In another study the effects of *Prosopis juliflora* on vegetation composition and productivity were investigated adopting management strategies like canopy removal and trenching around *P. juliflora* trees (Singh and Shukla, 2012). Out of 20 herbage species, 17 were under *P. juliflora*. Species diversity, population size, herbage dry biomass, soil organic matter, and nutrients concentrations were highest in treatments with intact tree and trench. Soil water content was 17% less under the intact tree than in the lopped tree. *P. juliflora* utilized deeper soil water during drought in 2002. *Peristrophe paniculata* contributed 64% of the total biomass. Canopy retention of *P. juliflora* trees and trenching around the trees improved soil resources, diversity, and herbage production.

Horti-pastoral system

This involves growing of fruit crops and grasses together to meet the growing demand of fruits and fodder, and therefore identified as a potential alternate land use option in shallow to medium deep soils (Singh and Osman, 1995). In view of the assured income from fruit trees and greater demand of fodder, several attempts were made to integrate fruit trees and pasture so as to make the system more productive and lucrative to the farming communities. Sharma and Diwakar (1989) recorded1.2 t ha^{-1} forage yields in *Cenchrus ciliris-Z. mauritiana* based horti-pastoral system that did not affect fruit yield of *Z. mauritiana*, whereas a study conducted at Avikanagar indicated higher yield of grass, when cultivated with Ber (Singh and Jain, 1990). A long term study conducted at Samadari areas in Pali district of Rajasthan with plantation of *Z. rotundifolia* @ 280, 140 and 170 tree ha^{-1} with *C. ciliris* produced 624 to 824 kg ha^{-1} forage yield, where a density of 280 plants ha^{-1} of *Z. rotundifolia* found best (Sharma and Vashishta, 1985). In a semi arid region of Jhansi, introduction of pasture did not show adverse effect on growth parameters of orchard during establishment phase up to 5 years and produce 3-7 tones ha^{-1} dry forage in initial 3 years (Kumar *et al*., 2002).

Gajja *et al*. (1999, 2004) believed that hortipasture system are more profitable than arable cropping under arid condition and they opined that horti-pasture comprising ber tree and *C. ciliaris* grass are highly economic on the basis of B:C ratio, net present worth and annuity value at 10% rate of interest under arid ecosystem. Whereas performance of jujube (*Ziziphus mauritiana*)-based hortipasture system under rainfed conditions in Jhansi, Uttar Pradesh, India, showed significantly higher dry matter and crude protein yields of pasture. Intercropping of Guinea grass (*Panicum maximum*) with Caribbean stylo (*Stylosanthes hamata*) gave significantly higher dry forage and crude protein yields compared to other grass-legume combinations. Dry leaf forage yield and fuel wood production were not significantly influenced with different pasture combinations. However, significantly higher fruit yield was recorded under natural pasture than other grass-legume intercroppings (Ram *et al.*, 2005).

Silviherbal

Diversifying land uses are important strategies in present scenario of climatic abrasions, and growing medicinal crops either alone or in associations of tree may help diversification, combat desertification and better use of resources is part of it (Kiran *et al.*, 2009). *Acacia nilotica, Prosopis cineraria* and *Azadirachta indica* when intercropped with herbal plants viz. *Cassia angustifolia* (Saina) and *Myrtus communis* (Heena), there were no significant difference in the production of *C. angustifolia* crop in the first year. *Myrtus communis* yielded highest leaves when grown with *P. cineraria*. However, *M. communis* competed strongly with the tree species affecting the growth of all the three species but percent decrease was maximum in case of collar diameter compared to height of the plants. The decrease was greater in *A. nilotica* plants and lesser in *P. cineraria* (Anon. 2000). Arya (2003) recorded higher yield of *C. angustifolia* when grown in association with *Ailanthus excelsa* as compared to that when grown in association of *Acacia nilotica*. In another experiment, yield of *Cassia angustifolia* was recorded along with the other crops grown under sequence and in association of *P. cineraria* and *T. undulata*. The highest grain yield was for mungbean (398 kg ha^{-1} with *P. cineraria* and 438 kg ha^{-1} with *T. undulate*) in 2000, whereas the production was highest when mungbean was grown in rotation with sesame in *P. cineraria* plots. In *T. undulata* plot, the grain yield was highest for *C. angustifolia* followed by the clusterbean indicating more compatibility of *C. angustifolia* with *T. undulata* (Singh, 2003). *Cassia angustifolia* when grown in association with *A. tortilis, P. juliflora* and *Calligonum polygonoides* produced dry leaves of 0.76, 0.96 and 1.39 tones ha^{-1} $year^{-1}$, respectively (Singh *et al.*, 2003). Considering the economic return from *C. angustifolia* leaves, Rs 16720 ha^{-1} $year^{-1}$ could be obtained from the plots in which *C. polygonoides* are integrated with *C. angustifolia* as compared to 9120 ha^{-1} $year^{-1}$ from the plots in which *A. tortilis* was integrated (Table 7).

Table 7. Dry mass (tones ha^{-1}) of leaf of *Cassia angustifolia* and the economic benefits (Rs ha^{-1} @ Rs 12 kg^{-1}) under different tree species in a shifting dune

Tree species	Nov 1998	May 1999	Nov 2000	Mean	Income (Rs ha^{-1})
A. tortilis	0.76	0.80	0.71	0.76	9120
P. juliflora	0.97	1.04	0.87	0.96	11520
C. polygonoides	1.39	1.56	1.23	1.39	16720

Source: Singh *et al.*, 2003

Growing aswagandha and andrographis with four year old amla and terminalia plantations (Lata *et al.*, 2012) showed that plant height, dry matter production and leaf area per plant of aswagandha/andrographis were markedly higher under sole cropping situation when compared to intercropping situation both in amla and terminalia. Days to physiological maturity of aswagandha was delayed by 9-10 days, whereas that in andrographis delayed by 3-4 days in intercropping situation in terminalia when compared to intercropping in amla. Root and seed yields (kg ha^{-1}) and withanolide content of aswagandha were the highest with sole cropping situation compared to either of the intercropping situations. Aswagandha performed better to some extent as an inter-crop in amla as compared in terminalia. Herbage yield (kg ha^{-1}) of andrographis and andrographolide content were the highest in sole cropping of andrographis compared to intercropping situations. The total gross and net monetary returns from the system (tree + crop) were the highest with andrographis intercropped in terminalia when compared to sole cropping of andrographis.

Functional aspects of agroforestry

Contribution of agroforestry is not limited to crop and tree yield and thus economic returns only rather it contribute variously to ecological, social and economic functions. In his review Pandey (2007) analyzed the role of agroforestry in: (i) biodiversity conservation; (ii) yield of goods and services to society; (iii) augmentation of the carbon storage in agroecosystems; (iv) enhancing the fertility of the soils, and (v) providing social and economic well-being to people. The suggested points were to eliminate many of the uncertainties that remain, and also carefully test the main functions attributed to agrforestry against alternative land-use options. Various factors like competition for land due to intense population pressure, rapidly depleting fuel wood resources, time- scale for orchard operations, dispersed distribution of benefits from forestry and seasonal shortage of labour are putting enormous pressure on non- renewable land resource leading to shift in emphasis towards agroforestry based land use system (Rekha *et al.*, 2007). Despite of these growing of trees did not find a suitable place in rural economy and rangeland development programmes. Therefore, making efforts convincible for adoption of agroforestry particularly in dry areas is a great challenge to scientific and administrative communities and converting it to a successful storey may be a great achievement. Some of the functional aspects helpful in propagating agroforestry adoption are knowledge sharing and extension about the extent up to which agroforestry system is beneficial in terms of minimizing risks, interactive effects of different component and their management for a positive effects, climate amelioration and help control land degradation-desertification.

Risk minimization effects

The concept of optimization of the production of various products and services, with minimum risk, from a unit of land as compared to "maximization" of a single commodity in monocultural production systems, is considered to be new in agroforestry (Lundgren, 1897). However, tree planting may among others reduce salinity, improve soil fertility, control and prevent erosion, control water logging, reduce the greenhouse effect, reduce catchment eutrophication, possibly check acidification and probably increase local biodiversity (Prinsley, 1993). Woody plants can play a significant role in the transition phase of agrosilvopastoral systems in semi-arid regions from extensive systems to intensified systems. Woody plants provide buffering functions, stabilizing ecosystem dynamics, and allowing effective use of additional nutrient and water inputs, or allowing effective use of these resources, where they occur naturally. So far woody plants have been predominantly used for productive purposes. Substantial changes are required to change the focus to protective and supportive functions of these agroforestry systems (Breman and Kessler, 1995; Chambers and Leach, 1989). For agrosilvicultural systems, three situations have been described where the integration of woody plants with cropping has a synergistic insurance effect (Breman and Kessler, 1995). These are: (i) In dry sandy soils where nutrients have become the limiting factor for plant production, woody plants oriented as windbreaks improve germination and establishment of crops by reducing the impact of wind; (ii in the more humid parts of sub-humid zones, where run-off and leaching are significant processes, evenly distributed woody plants can lead to improved nutrient availability for crops i.e., selection of woody species and management of the woody plants can be oriented at the combination of minimum light reduction and maximum woody plant production; and (iii) woody plants can lead to a prolonged growing season for crops. Most suitable are relatively humid conditions, and the presence of fertile and deep soils.

Microclimate amelioration

A reduction in the light level under trees or shrubs canopies lowers air, leaf, canopy, and soil surface temperature (Table 8). It is the extreme temperatures, which are modified i.e., decrease and increase in mean maximum and mean minimum temperatures, respectively. But removal of upper storey of shrub increases in the maximum soil temperature by 8°C in the summer (Ovalle and

Avendano, 1988). Belsky *et al.* (1989, 1993) reported a reduction in maximum soil temperature by 5 to 12°C under isolated trees of *Adansonia digitata* and *Acacia tortilis* in African Savanna. A reduction up to 10°C has been recorded by Wilson and Wild (1995) on litter and in surface soil (top 2 cm). The temperature reduction play an important role in conserving surface moisture and might influenced soil fauna/earthworms, which are effective in litter breakdown and nutrient cycling (Tripathi *et al.*, 2012). Reductions in the temperature of the foliage or the grassland canopy are smaller, and some 1 to 2 °C less in the shade (Wong and Stur, 1996). Singh (2003) recorded a lower air temperature inside the crop canopy *Zizyphus mauritiana, Prosopis cineraria* and *Acacia nilotica* by 0.2 to 1.2°C and high relative humidity by 2 to 7%, respectively, in comparison to uncropped (open) field. Radiation interceptions by pearl millet at peak day hour during vegetative stage varied between 32 and 64% in different treatments, the highest interception was 77% when intercropped with *Z. mauiritiana* followed by *P. cineraria* (74%) and *A. nilotica* (67%) during the crop reproductive stage.

Table 8: Increase/decrease in air/soil temperature under tree/shrub in different region of the world.

Tree/Shrub	Increase/Decrease	Source
Shrub	Increase by 8°C	Ovalle and Avendano (1988)
Acacia tortilis & *Adansonia digitata*	Decrease by 5 to 12°C	Belsky *et al.* (1989, 1993)
Artificial shade	Decrease by 2 to 6°C	Wilson *et al.* (1986) and Wong and Stur (1996)
Tree	Decrease up to 10°C	Wilson and Wild (1995)
Faidherbia albida	Decrease up to 10°C	Van den Beldt and Williams (1992)
Z. mauritiana, P. cineraria and *A. nilotica*	Decrease by 0.2-1.2°C	Singh (2003)

Though light is not a limiting factor particularly in desert environment but it limits forage yield, which is directly proportional to the photosynthetic active radiation (PAR) falling through tree canopy in an agroforestry system (Acciaresi *et al.*, 1994; Shukla and Hazara, 1994, Singh, 2003; Pandey *et al.*, 2010; Singh and Shukla, 2012). Tree/shrub modify the micro-climate through various processes i.e., radiation interception by tree stands, effect of canopy structure, effect of tree orientation and spacing, effect of latitude and time of year on solar paths, shade from a single crown and spectral quality of sunlight under partial shade (Reifsnyder and Darnhofer, 1989). The reduction in PAR ranges from 20-88% at different places depending upon the species and canopy density (Table 9). The evaluation of morphological, physiological, and yield responses of *Tricachna californica, Setaria macrostachya, Bouteloua eriopoda* and a bush *Muhlenbergia porteri* under *Prosopis juliflora* showed that all plants made their best growth in full sunlight; but *T. californica*, *M. porteri*, and *S. macrostachya* displayed greater ability than *B. eriopoda* to adapt to shade (Tiedmann *et al*., 1971). In arid region of India the radiation penetration under *Acacia tortolis* was 14% to 30% of that in the open during the daytime (Ramakrishna and Shastri, 1977). Below crown species are more responsive to reduction in light intensity than open-grassland species. Below crown species reduced their water loss whereas shaded species conserved soil water for later growth. Open grassland species appeared to be more efficient than below- crown species in extracting water from dry soil under full sun condition (Amundson *et al*., 1995). However, the positive effect of shade are visible when resources like nitrogen and water are limiting in open areas (*i.e*., *D. aristatum* growth), whereas some species reduced leaf area for increased understorey production and facilitated growth of *Prosopis glandulosa* seedlings by enhancing soil water (Anderson *et al*., 2001). Increased biomass of *Enchylaena tomentosa* was facilitated under shade of *Atriplex bunburyana*. *Maireana brevifolia* increased above-ground biomass allocation, whereas *M. georgei* reduced root biomass in response to shade (Jefferson and Pennacchio, 2005). While Singh (2003) recorded 67% to 77% reduction in light under *Zizyphus mauritiana, Prosopis cineraria* and *Acacia nilotica* during the crop reproductive stage of pearlmillet indicating positive relation in crop height with intensity of light interception, Pandey *et al*. (2010) suggested management of *A. indica* tree to increase intensity of PAR and thus the yield of *Phaseolus mungo*.

Table 9. Reduction in Photosyntheically active radiations (PAR) (μ mol m^{-2} s^{-1}) under isolated tree species as compared to control.

Species	PAR (μmol m^{-2} s^{-1}) (%)	Climatic Zone	Source
P. cineraria, P. juliflora, A. indica, A. nilotica	84 to 88%	Indian Desert	Singh *et al*. (2008)
Azadirachta indica	53-79%	Indian Desert	Pandey *et al*. (2010); Singh and Shukla (2011)
Prosopis juliflora	>85%	Indian Desert	Singh and Shukla (2012)
Acacia tortilis	45	African savanna	Belsky *et al*. (1993)
Adansonia digitata	65	African savanna	Belsky *et al*. (1993)
Vitellaria trees	45	Sahelian savanna	Kessler (1992)
Balanites aegyptiaca	20	Sahelian savanna	Akpo and Grouzis (1996)
Z. mauritiana, P. cineraria and *A. nilotica*	67-77	Indian Desert	Singh (2003)

Soil resource utilization

Biotic and abiotic components of ecosystems affect each other through complex interactions and processes. These dynamic interactions give ecosystems their distinct identities and provide ecosystem services critical to human survival. However, human activities have placed increasing demands on specific ecosystem services. Some of these have determined man's impact on plant-water use, biomass production (energy) and water use efficiency *i.e.*, biomass produced per unit of water transpired, termed productive green-water use (Gupta and Saxena, 1978). For example, measurements of evapotranspiration from different vegetation types showed that annual water use is strongly related to the proportion of the year in which a dense canopy of transpiring leaves is maintained. Likewise, evergreen vegetation such as riparian fynbos and plantations of introduced tree species exhibit a relatively high annual ET_a, when compared to seasonal grasslands and deciduous trees that only maintain their transpiring canopy during summer (Everson, 2011). The environmental modifications caused by trees or shrubs may be positive or negative on the companion crops. Whenever the net outcome is positive, the overall interaction is said to be facilitation; where it is negative, the interaction is competitive. Since the strength of modifications decreases with distance from the plant, a mixed tree-vegetation community consists of a spatial patchwork of different degrees of competition and facilitation (Vetaas, 1992; Dye *et al.*, 1995). The mechanisms of facilitation include improved soil fertility (Moro *et al.*, 1997), microclimatic amelioration (Franco and Nobel, 1989), and increased water availability (Raffaele and Veblen, 1998), whereas competition occurs when more than one species has to share the resources from a limited pool of the growth resources. Lehmann *et al.* (1998) studied the effect of intercropping and tree pruning on root distribution and soil water depletion in an alley cropping system with *Acacia saligna* and *Sorghum bicolor* in northern Kenya. The agroforestry combination used the soil water between the hedgerows more efficiently than the sole cropped trees or crops, as water uptake of the trees reached deeper and started earlier after the food irrigation than of the Sorghum, whereas the crop could better utilize topsoil water. Under the experimental conditions, the root system of the alley cropped Acacia and Sorghum exploited a larger soil volume utilizing soil resources more efficiently than the respective monocultures. Competition from an established tree is a major factor limiting the successful selection of a model in agroforestry system. Competition is mainly for light, with little competition for nutrients in fertile agro-ecosystems. In a case study, *Jatropha curcas* evaporated during December to February on clear hot days 3 mm d^{-1} to 4 mm d^{-1}, however, due to the deciduous nature of the species, water use was negligible (< 1 mm d^{-1}) during winter months of May to August (Everson, 2011). At a catchment scale, studies in a montane grassland ecosystem showed that the partitioning of the main hydrological fluxes into streamflow and evaporation was dependent on the wetness of the hydrological year. In average to wet years (>1 200 mm precipitation) the hydrological flux was equally split between evaporation (650 mm) and runoff

(550 mm), while in drier years evaporation became the dominating component of the water balance (752 mm vs. 356 mm, respectively).

Results of a variety of studies on the growth and water use of indigenous trees growing in natural forest and plantation systems suggest that, compared to introduced tree species, indigenous species use substantially less water, show lower water use efficiency, and grow more slowly. Advantages to such indigenous systems potentially include lower management costs, higher product values, a wider range of non-wood products and a lower hydrological impact. Their usefulness may be greatest on sensitive sites, where land use systems with a reduced environmental impact are required. While growing in association, root competition usually affect the balance between competing species more than shoot competition through resource depletion (Toky and Bisht, 1992; Singh, 2010). For example *Dalbergia sissoo* and *Acacia nilotica* based agroforestry systems reduced the yield of wheat (Puri *et al.*, 1995) and the reduction of crop yield in agroforestry system was due to competitive use of resources, but it may also be compensated in the long run by microclimate modification (Kohli and Saini, 2003) and residual nitrogen after removal of old trees as a result of enhanced N fixation under the Acacia. Muthuri *et al.* (2005) tested the hypothesis that deciduous (*Paulownia fortunei*) and semi-deciduous (*Alnus acuminata*) trees are less competitive with crops than evergreen species (*Grevillea robusta*) due to their differing leafing phenology. The presence of trees affected maize growth and yield 2.5 years after planting to an extent which depended on tree species and location. A positive interaction between *A. acuminata* and maize was apparent at first site,, but growth was suppressed in the first two crop rows at other. *G. robusta* reduced maize yield by 36% close to the tree rows at first, whereas yield reductions were negligible adjacent to *P. fortunei*. These findings suggested complementarity of resource use between *A. acuminata* and maize at the first, and neutral or competitive interactions between trees and crops in all other treatments. However, there are a lot variation in soil water and nutrient uses, which varied due to species and availability of soil resources (Singh *et al.*, 2007). Soil water content recorded in September 2003 (0-75 cm soil layer), indicated 67.3 mm of soil water in sole crop plot whereas it was 56.1 mm in the plot having grasses as the outgrowth. The soil water in *H. binata* and *C. mopane* plot was 48.0 mm and 41.8 mm, respectively. It was low in fixed crop (FC) plot than in the rotational crop (RC) plot indicating a decrease in soil water in *H. binata* and *C. mopane* plots by 19.9 mm and 25.5 mm per 75 cm of soil depth, respectively (Table 10). The soil water use of *H. binata* was lesser than *C. mopane* trees.

Table 10. Decrease in soil water status (mm) in tree integrated plots (0-75 cm soil layer) as compared to that in the control plots of crop monoculture. RC: rotational cropping plot with *C. tetragonoloba*, *S. indicum*, *P. glaucum* and *V. radiata* intercrops respectively in 1999, 2000, 2001 and 2003 ; FC: fixed cropping plot (*Vigna radiata*).

Species	1999		2000		2001		2002		2003	
	RC	FC	RC	FC	RC	FC	RC	FC	RC	FC
E. officinalis	18.1	16.9	8.6	5.8	5.5	2.6	2.2	2.6	1.5	2.4
H. binata	12.0	9.1	3.6	5.0	4.7	3.5	2.1	2.4	17.8	20.9
C. mopane	17.8	16.0	6.3	7.7	5.4	4.5	2.0	2.0	24.9	26.1

Amelioration of soil fertility

Woody perennials are recognized to improve soil conditions by improving concentration of organic matter and nutrient (Singh and Lal, 1969; Tiedmann and Klemmedson, 1973; Garcia-Miragaya *et al.*, 1994). Nutrient enrichment occurs across a broad array of tree and shrub growth-forms and species inhabiting diverse climatic conditions (Virginia, 1986). Trees may act as nutrients pumps, drawing nutrients from deep horizon and laterally from areas beyond the canopy and depositing them mainly beneath tree canopy via litter fall and canopy leaching (Scholes, 1990). The second mechanism is that the tall, aerodynamically rough tree canopy acts as an effective trap for atmospheric dust (Szott *et al.*, 1991). The dust contains nutrients, which wash of the leaves during rainstorms and drip into the subcanopy area (Singh *et al.*, 2000a). Though not well quantified, a third mechanism may be of importance where trees are sparse and may serve as focal points attracting roosting birds and mammals seeking shade or cover. Herbivores, microbial and faunal biomass that take refuse in the shade of trees also enhance the local nutrient cycle and soil organic carbon (Mordelet *et al.*, 1993; Menezes *et al.*, 2002; Tripathi *et al.*, 2005). However, nutrient level under tree cover decreased with increase in soil depth while under grass cover nutrient decreased marginally up to 30 cm depth without further change with depth (Sharma *et al.*, 1990). Gender differences also results in variation in soil nutrients i.e., an increased P cycling in fruit or litter or difference in P transformation beneath female *Simarouba amara* trees (Rhoades *et al.*, 1994).

Inclusion of trees and woody perennials on farm lands can, in the long run, result in marked improvements in the physical conditions of the soil. The chemical characteristics of soil under different land use systems exhibited higher organic carbon and available NPK under *P. cineraria* (Aggarwal, 1980; Aggarwal and Kumar, 1990, Singh *et al.*, 2007) and *T. undulata* (Singh *et al.*, 2004) and Ardu (Patel *et al.*, 2005). Singh *et al.*, (1982) observed a decrease in soil pH and bulk density and increase in organic carbon, EC and available N, P and K under tree canopy of *Hardwickia binata* over sole agriculture on rainfed Alfisols. Kumar *et al.* (1998) also reported higher organic carbon content, moisture availability and nutrient status (N, P and K) under the canopies of *Prosopis cineraria, Tecomella undullata, Faidherbia albida* and *Azadirachta indica* as compared to outside the tree canopies. Roy (1999) reported that nitrogen, phosphorus, potassium and calcium status of soil increased with increase in tree density of *Hardwickia binata* at Jhansi. Higher build up of amino acids, amino sugars, hydrolysable NH_4-N and total hydrolysable N has also been reported under *Prosopis cineraria, Colophospermum mopane* and *Hardwickia binata* over rest of the species (Burman *et al.* (2002). Crop suitability study showed that the advantage of growing crop with *Ailanthus excelsa* was maximum during initial 5-6 years which later on supplementing a part of nitrogen requirement of the associated crops (Patel *et al.*, 2009). Meena *et al.* (2005) observed the similar trend of *A. excelsa* under canopy silvipasture system in semiarid areas of Rajasthan. Effect of *P. cineraria* and *T. undulata* tree density on availability of nutrient to crop reveals that availability of P and NO_3-N were significantly low in plot with 274 stem per ha than in 208 and 416 stem per ha plots (Singh , 2004). Available P and NO_3-N increased with distance from the tree base, whereas NH_4-N displayed the reverse trend. Comparing the soil depth, there was a decrease in the availability of P and NH_4-N with soil depth. But NO_3-N was high in deeper soil layer than in the top 0-25 cm soil layer. Soil organic matter did not indicate difference due to *P. cineraria* and *T. undulata*, though it was higher under *P. cineraria* than under *T. undulata* based agroforestry (Singh *et al.*, 2005; 2007). Soil organic matter was higher in 0-25 cm soil layer and it decreased with soil depth being lowest in 50-75 cm soil layer. SOC was comparatively higher in mungbean-mungbean plot (legume rotated by legume) as compared to mixed cropping of mungbean and sesame. SOM was started increasing after 6-8 year of *E. officinalis*, *H. binata* and *C. mopane* agroforestry plots.

During a study on organic carbon and nutrient availability in arid zone agroforestry systems comprising *Emblica officinalis Hardwickia binata* and *Colophospermum mopane* tree species with *Vigna radiata* as inter-crop, Singh *et al.* (2000) recorded significant temporal variations in soil organic carbon (SOC), extractable P and soil available N. Increased SOC coincide with the pe-riods of litter production and the harvest of agricultural crop. The levels of PO_4-P, NO_3-N and NH_4-N increased during summer. However, the increase in the concentration of some nutrients was also observed in winter months which may be attributed to either plant senescence or change in the soil biological processes. Higher availability of NH_4-N in soil under *E. officinalis* indicated relatively lower rate of nitrogen mineralization than under the legume species *H. binata* and *C. mopane* soil. Approximately two to three fold higher concentration of NH_4-N was observed during monsoon (cropping period) and which probably was due to utilisation of nitrate by the agricultural crops and/or due to nitrogen fixation by *V. radiata* and moisture controlled net-nitrifications (Table 11). An eight-year study on soil organic carbon dynamics related to three-years litter dynamics study in *E. officinalis*, *H. binata* and *C. mopane* based agroforestry systems indicated a decreasing trend in soil carbon content by 56% during the study period (Singh, 2005). However,

Table 11. Soil nutrients under the influence of different tree species and agricultural crops in Indian Desert. RC: rotational cropping plot with *C. tetragonoloba*, *S. indicum*, *P. glaucum* and *V. radiata* intercrops respectively in 1999, 2000, 2001 and 2003; FC: fixed cropping plot (*Vigna radiata*). (Source: Annon. 2003)

Treespecies	Soildepth	NH_4-N (mg kg^{-1})				NO_3-N (mg kg^{-1})				PO_4-P (mg kg^{-1})			
		1999		2003		1999		2003		1999		2003	
		RC	FC	RC	FC	RC	FC	RC	FC	RC	FC	RC	FC
E. officinalis	0-25	6.62	6.39	4.17	4.20	4.55	4.67	3.73	3.66	13.12	12.81	16.39	15.99
	25-50	5.35	5.35	3.09	3.25	3.71	3.53	2.52	2.52	11.54	11.50	12.30	12.02
	50-75	3.53	3.39	2.58	2.55	2.82	2.70	2.29	2.19	10.12	10.18	10.56	10.97
H. binata	0-25	6.33	6.36	4.55	4.64	4.26	4.37	3.23	3.25	12.91	13.22	14.45	14.18
	25-50	5.17	5.18	3.02	3.16	3.33	3.32	2.40	2.25	11.27	11.47	11.26	11.33
	50-75	3.33	3.24	2.57	2.35	2.72	2.67	2.16	2.11	10.44	10.37	10.04	10.48
C. mopane	0-25	6.16	6.05	4.83	4.79	4.12	4.35	3.01	3.14	13.00	12.63	12.12	12.05
	25-50	4.23	5.22	3.39	3.24	3.24	3.20	2.22	2.19	11.71	11.40	10.69	10.57
	50-75	3.26	3.44	2.36	2.59	2.48	2.23	2.04	2.10	10.86	10.41	9.85	9.82
Control	0-25	6.41		4.44		4.10		3.99		13.55		13.21	
	25-50	4.85		3.50		3.28		2.28		11.29		11.09	
	50-75	4.12		2.13		2.58		1.96		10.70		9.86	

integration of trees allowed a loss of SOC only by 3.2% in *E. officinalis*, 22% in *H. binata* and 35.5% in *C. mopane* plots, indicating greater sequestration of carbon in *E. officinalis* plot and least in *C. mopane* plot. The study suggested that integration of tree in agricultural land will be an important strategy to sequester carbon not only in the form of biomass but also in soil and may therefore maintain soil productivity and help control land degradation.

During study of soil organic matter, a decreasing trend was observed with soil depth both in the control and agroforestry plots being lowest in control plots. In the tree-associated plots, the difference in SOM due to species was significant. SOM was highest in *E. officinalis* plots and lowest in *C. mopane* plot. The rotation-cropping plot had high SOM to the year 1999 whereas fixed cropping plot had high SOM after 1999. Temporal changes in SOM (0-75 cm soil layer) indicated a decreasing trend from 1994 to 2002 in 0-25 and 25-50 cm soil layers in the control plot, whereas there was an increasing trend in SOM from 1998 to 2000 in 0-25 cm and 25-50 cm soil layer, respectively in *E. officinalis* plot. In *H. binata* and *C. mopane* plots, the increase in SOM was started from the year 2000 (Table 12). However, in 2003 there was a decrease in SOM in all the plots as compared to that in the year 2002, except in the control plot and was due to tillage activity and drought in 2002 influencing mineralization and loss of SOM.

Table 12. Soil organic matter (%, w/w) under the influence of different tree species and agricultural crops recorded after crop harvest in October. RC: rotational cropping plot with *C. tetragonoloba*, *S. indicum*, *P. glaucum* and *V. radiata* intercrops respectively in 1999, 2000, 2001 and 2003 ; FC: fixed cropping plot (*Vigna radiata*). (Source: Anon. 2003)

Species	Soil depth	1999		2000		2001		2002		2003	
		RC	FC	RC	FC	RC	FC	RC	FC	RC	FC
E. officinalis	0-25	0.48	0.451	0.520	0.525	0.527	0.534	0.529	0.536	0.312	0.309
	25-50	0.295	0.311	0.287	0.312	0.292	0.315	0.292	0.316	0.254	0.241
	50-75	0.122	0.145	0.133	0.137	0.132	0.153	0.134	0.132	0.121	0.122
H. binata	0-25	0.475	0.399	0.434	0.436	0.467	0.443	0.443	0.446	0.352	0.354
	25-50	0.267	0.288	0.214	0.233	0.260	0.236	0.233	0.243	0.194	0.236
	50-75	0.121	0.132	0.105	0.110	0.161	0.115	0.118	0.127	0.115	0.134
C. mopane	0-25	0.449	0.429	0.340	0.361	0.345	0.376	0.349	0.387	0.328	0.333
	25-50	0.244	0.255	0.190	0.173	0.194	0.173	0.198	0.180	0.193	0.175
	50-75	0.120	0.119	0.105	0.097	0.108	0.102	0.121	0.110	0.119	0.112
Control	0-25	0.351		0.226		0.212		0.231		0.369	
	25-50	0.194		0.136		0.172		0.162		0.229	
	50-75	0.108		0.095		0.080		0.089		0.124	

Improvement in soil biological activities

Trees in a variety of agroforestry systems enhance nutrient cycling, improve soil conservation and soil faunal and microbial activities. It serves as models for the protective-productive rehabilitation strategies of agrarians. The nature of the organic materials present and their decomposition governs nutrient availability in soil systems. The higher percentage loss of non-fodder plant litter, and the comparatively high phosphorus mineralisation offer the possibility of improving the biological characteristics of arid soils (Tripathi *et al.*, 2005; Sundaramoorthy *et al.*, 2010). Changes in microflora due to addition of *Acacia leucophloea* and *Prosopis juliflora* litter in wetting-drying cycles in microcosm experiments revealed that a higher proportion of litter enhanced population levels of cellulolytic fungi, with a further rise following rewetting. However, populations of lipolytic fungi, proteolytic bacteria and actinomycetes were higher during the initial periods of decomposition. The results indicate a strong relationship between lipolytic fungi and sugar fungi, lignolytic fungi, actinomycetes, and proteolytic bacteria. Bacteria and actinomycetes play a synergistic role during decomposition of desert soils (Purohit *et al.*, 2002). Whereas Verma *et al.* (2009 reported temporal changes in AM population (correlated with soil organic matter) under canopy of *P. cineraria*); Tarafdar (2008) reported improvement in microbial population of different microorganism in different farming system viz. *P. cineraria*, *T. undulata* and *Z. mauritiana*, of arid ecosystem, where the highest improvement was recorded in *Z. mauritiana* based system (Table 13).

Table 13. Percent increase in microbial population and biomass in different agri-slivisystem in Indian desert as compared to the sole crop.

Tree component	Microorganism			
	Fungi	Bacteria	Actinomycetes	Microbial biomass
P. cineraria	13.-24	21-37	9-18	18-23
T. undulata	18-27	27-39	10-16	19.25
Z. mauritiana	19.39	23-53	23-53	20-77

In a work on dehydrogenase activity and VAM fungi in tree-rhizosphere of six agroforestry trees (*Azadirachta indica*, *Acacia tortilis*, *Eucalyptus camaldulensis*, *Prosopis cineraria* and *Tecomella undulata*) Prasad and Mertia (2005) recorded higher dehydrogenase activity (9.5 to 16.8 p kat g soil^{-1}), root colonization (58.3 to 68.5%) and spore density (132.5 to 234.7 spores 100 g soil^{-1}) as compared to that in non-rhizosphere (7.4, 37.7 and 44.4). Irrigation had increased dehydrogenase activity by 22.1% while it reduced root colonization and spore density by 14.2% and 16.2%, respectively. *A. indica* registered maximum growth while *E. camaldulensis* the least. Noureen *et al.* (2008) studied improvement in fertility of nutritionally poor sandy soil of Cholistan desert, Pakistan by C*alligonum polygonoides.* The results of the study clearly showed the existence of differences in the soil nutrient status between the various subhabitats, which occurred at various places from stem base area of the plant towards the open, uncanopied area. The results confirmed that electrical conductivity, nitrogen, organic matter, potassium, calcium, magnesium, sulphur, phosphorus, carbonates and bicarbonates were significantly high in the soils under the canopy cover of the plant as compared to those in uncaopied area (open area) but pH value was comparatively low in the soils under the canopy cover of the plant. It is evident from the results that canopy cover of *C. polygonoides* is playing a significant role in enrichment of nutrient poor soils of Cholistan desert. Rathore *et al* (2009) emphasized the role of indigenous microbes associated with economically or medicinally important native pants *Mucuna pruriens* and *Pueraria tuberosa*. Application of non-conventional approaches may further provide information regarding nodulation and plant microbe interactions which will be useful for upgrading the available information on this aspect. Yadav *et al.* (2010) recorded significant and substantial improvement in soil biological activity in terms of microbial biomass C, N and P, dehydrogenase and alkaline phosphatase activity under different tree based agroforestry systems as compared to a no tree control (cropping alone), where soil microbial biomass C, N and P under agroforestry varied between 262–320, 32.1–42.4 and 11.6–15.6 mg kg^{-1} soil, respectively, with corresponding microbial biomass C, N and P of 186, 23.2 and 8.4 mg kg^{-1} soil under a no tree control. Fluxes of C, N and P through microbial biomass were also significantly higher in *P. cineraria* based land use system followed by *D. sissoo*, *A. leucophloea* and *Acacia nilotica* in comparison to a no tree control. *P. cineraria* based system found best which brought maximum and significant improvement in soil biological activity.

Conclusions

Climatic conditions are one of the most important determining factors affecting agriculture production in the region with irregular rainfall and high temperatures. People of Indian arid zone are well acquainted about agroforestry- an age old practice in India- in the form of the shifting agriculture, variety of cereal cropping, home garden systems, traditional plantation systems, etc. Many common trees like *Prosopis cineraria*, *Ficus religiosa*, *Butea monosperma*, *Crataeva roxburghii* and *Tecomella undulata* are very much associated with in Indian folk-life. However, maintaining only a few tree on farmlands is not going to solve the need of growing populations and their requirements. Forests in India are already under pressure to fulfill the ever increasing demands. Alternate land use systems like alley cropping, lay farming, tree farming, dryland horticulture and agroforestry etc. may intensify crop production, sustain the rural livelihoods and help control land degradation in dry areas. It not only help in generating much needed off-season employment in monocropped dry areas but also minimize risk of high off-season rains which may otherwise go waste as runoff prevent degradation of soils and restore balance in the ecosystem. Integration of tree species, field crops, forage grasses and shrubs, aromatic and medicinal species, bio-fuel crops, and fruit tree are profitable and suitable in terms of local utilization. Crops diversification based on the tolerant species to the degraded lands of arid region are likely to be the key to future agricultural growth in regions. However, dry lands requires special attention particularly the rainfed agriculture, which can be taken care by improving degraded, marginal and sub-marginal lands by introduction of suitable alternate land use systems, which not only help in generating much needed off season employment in monocropped dryland but also minimize risk tallies off - season rains preventing land degradation and restoring balance in the ecosystem.

The selection of tree/shrub species may be based on the tolerance towards drought, salinity/alkalinity prevailing in the region. Some of the characterisris are:

- deep root system to draw water from deeper soil profile,
- leaf shading in summer to conserve moisture.
- water binding mechanism, and
- xerophytic characteristics to minimize the loss of water through transpiration.

Because of arrangements and combinations of trees/shrubs and the associated crops, agroforestry are categorized into: agri-silviculture, agri-horticulture, agripasture, silvipasture, hortipasture and silviherbal. Most common trees are *P. cinreraria*, *T. undulata*, *Salvadora oleoides* under rainfed and *Acaia nilotica* and *Ailanthus excelsa* under irrigation. The common agriculture crops are *Vigna radiata*, *V. aconitifolia*, *Cymopsis tetragonoloba*, *Seasamum indicum*, *Penesetum glaucum* under rainfed conditions. *Cuminum cymium*, *Plumbago ovata* are grown under irrigation. Now growing *Arachis hyposea* in desert region and *Gossypium* spp, and rice in Ganganagar, and Hanumangarh area is common. In general, it is safe to say that a greater diversity of species will be more favourable, as it results in a more complete occupation of space above and below the soil, and the variation in the characteristics of the litter produced can maintain a greater level of soil biodiversity, with positive effects on soil fertility. However, there is always a complex gradient of various resources and disturbances while moving from the open to the canopy of a tree or shrub. Over this gradient, some factors (*e.g.*, nutrients, water) changes for the better, whereas others (*e.g.*, light, allelopathic exudates) changes for the worse. The net effect of these correlated changes depends on the combined response of plants to all factors involved. The general pattern is probably that the negative effects of one limiting factor often can be compensated for, to a certain extent, by improvement in other environmental conditions stressing the importance of trees in desert ecosystem. Further, in absence of net positive effects of tree on the companion crop, many of the interacting factors can be manipulated in favour of increased production i.e., by adopting suitable management strategies facilitative effects of trees and shrubs may be utilized in enhancing the productivity of desert ecosystem. Trees/shrubs in a farming systems help improves soil faunal and microbial activities, which seems to facilitates litter decomposition and nutrient cycling.

Agroforestry minimize the risks of drought and famines by way of providing diversified products and help in ameliorating the land under varying types of degradation like salinity, alkalinity, water-logging *etc*. Many studies demonstrate the usefulness of tree in increasing yield of agriculture crops/ and thus total productivity of land, but there are reports of decreased crop production under tree integration in farmlands, and are the main constraints in its adoption on large scale particularly among the developing countries. Despite of traditionally well aware about the multifaceted benefits of trees/shrubs, growing of trees did not find a suitable place in rural economy and rangeland development programmes and peoples are least enthusiastic about tree planting on their farmlands in Indian arid zone. In facts trees are removed from the pasture/farmlands for uninterrupted cultivation, without thinking that soil improvement under tree integration in agriculture land is in great part related to increases in soil nutrients and more importantly the organic matter (as compared to sole crop), whether in the form of surface litter or soil carbon. Besides their role in above-ground carbon sequestration, tree integrated system also have a great potential to increase carbon stocks in the soil and certainly merit consideration in mechanisms that propose payments for mitigation of greenhouse gas emissions to reduce land degradation and minimize the effect of climate change.

Some of the recommendations are

- Making convincible efforts for adoption of agroforestry particularly in dry areas to make this endeavour a great success.
- Propagating agroforestry adoption through knowledge sharing and extension about the extent up to which agroforestry system is beneficial in terms of minimizing risks, positive effects on companion crops, climate amelioration and help control land degradation.
- Identification of community lands like oran, gauchar through proper surveys to find out village-wise extent and development of plans for their conservation, management and improvement on high priority.
- Minimization of uncontrolled grazing with the active cooperation of the people and promotion of silvi-pastoral system by integrating indigenous species with supply of seeds of improved variety of grasses, legume, fodder crops, *etc*.
- Curbing grazing on the stabilized sand-dunes during the first two years and grasses have to be harvested manually to stall feed the animals.
- Encouraging forest, pastures or water catchment reserves on the marginal and sub-marginal lands.
- Erection of fences of resistant shrubs and trees as barriers against the coming sand.

- Promotions of indigenous species, the products of which is of high commercial value.
- Promotion under public-private partnerships, with substantial changes focusing protective and supportive role of agroforestry particularly in arid region where the economic returns are relatively less.

References

Agarwal, R.K. and P. Kumar, 1990. Nitrogen response to pearl millet grpwn on soil underneath *P. Cineraria* and adjacent open site an arid environment. Annals of Arid Zone, 29:289-293.

Aggarwal, R.K. 1980. Physico-chemical status of soils under 'Khejri (*Prosopis cineraria* Linn.). In: Mann H.S and Saxena S.K (Eds), Khejri (*Prosopis cineraria*) in the Indian desert & Its Role in Agroforestry, 32-37. CAZRI Monograph, 11:78.

Annonymous 2000. Research and development support for desert development programme in the hot arid regions of Rajasthan and Gujarat. Project Report. Arid Forest Research Institute, Jodhpur. 200p.

Anonymous 1998. 35 Years of Research. Central Sheep and Wool Research Institute, Avikanagar, p.128.

Arya, R. 2003. Yield of *Cassia angustifolia* in combination with different tree species in a silvi-herbal trail under hot arid condition in India. Bioresource Technology, 86:165-169.

Awasthi, O.P., Saroj, P.L., Singh, I.S. and More, T.A. 2008. Fruit based Diversified Cropping System for Arid Region. CIAH Technical Publication No. 25 p.18.

Awasthi, O.P., Singh, I.S. and Dhandhar, D.G. 2005. Aonla based mulitier cropping s stem for sustainable production in Western Rajasthan. In: Proceedings of National Seminar on Globalization of Aonla, organized by Aonla Grower Association of India, Madurai (TN) October 7-9, 2005.

Bellefontaine, R., Petit, S., Pain-Orcet, M., Deleporte, P. and Bertault, J.C. 2002. Trees outside forests. Towards better awareness. FAO Conservation Guides 35, Rome, 216 pp.

Bhatanagar, P., Kaul, M.K. and Singh, J. 2007. Effect of intercropping in kinnow based production system. Indian Journal of Arid Horticulture. 2(1):15-17.

Bhati, T.K. 1997. Management of dryland crops in Indian arid ecosystem. *In:* Desertification Control in the Arid Ecosystem of India for Sustainable Development (Eds. Surendra Singh and Amal Kar), pp. 298-307. Agro-Botanic Publishers, Bikaner.

Bhati, T.K., Bhati, G.N. and Shankarnarayan, K.A. 1986. Legume-shrub grasslands in western Rajasthan. Annual Progress Report, CAZRI, Jodhpur.

Bhati, T.K., Tewari, J.C. and Rathore S.S. 2008. Productivity nad dynamics of integrated farming systems in western Rajasthan. *In:* Diversification of Arid Farming System (Eds. Pratap Narain, M.P. Singh, A. Kar, S. Kathju and Praveen-Kumar), pp. 23-30. Arid Zone Research Association of India and Scientific Publisher (India) Jodhpur, India,

Bheemaiah, G. and Subrahmanyam, M.V.R. 2004. Effect of green-leaf manuring on productivity of sunflower (*Helianthus annuus*) under guava (*Psidium guajava*)-based agrihorticultural system. Indian J. Agric. Sci., 74:5154-520.

Boffa, J.-M. 1999. Agroforestry parklands in Sub-Saharan Africa. FAO Conservation Guide 34, Rome, pp 230.

Breman, H., and Kessler, J.J. 1995. Woody plants in agro-ecosytems of semi-arid regions (with an emphasis on the Sahelian countries). Springer, Berlin *etc*, 340 pp

Burman, U., Kumar, P. and Harsh, L.N. 2002. Single tree influence on organic forms and transformation of Nitrogen in arid soils. Indian Soc. Soil Sci., 50:151-158.

Chadha, K.L. 2002. Diversification to horticulture for food, nutrition and economic security. Indian Journal of Horticulture 52: 137-140.

Chambers, R. and Leach, M. 1989. Trees to Meet Contingencies: savings and security for the rural poor. World Development,17: 329-432.

Dauley, H.S. 1994. Intercropping systems for the Indian Arid zone. *In:* Sustainable Development of the Indian Arid Zone (Eds. R.P. Singh and S. Singh,), pp. 251-259. Scientific Publisher, Jodhpur.

Dayal, D., Swami, M.L., Bhagirath Ram, Meena, S.L. and Shamsudheen, M. 2008, Production potential of grasses under silvipastoral system in Kachchh region of arid Gujarat. *Abstracts.* National symposium on Agroforestry knowledge for sustainability, climate moderation and challenges ahead" (15-17, December, 2008) NRC-Agroforestry, Jhansi (UP). pp. 195.

Depommier, D. 2003.The tree behind the forest: ecological and economic importance of traditional agroforestry systems and multiple uses of trees in India. Tropical Ecology, 44: 63-71.

Dhurvanarayan,V.V. 1993. Land capability classification. *In*: Soil and water Conservation Research in India, pp. 57-61. Indian Council of Agricultural Research, New Delhi.

Dhyani, S.K., Kareemulla, K., Ajit and Handa, A.K. 2009. Agroforestry potential and scope for development across agro-climatic zones in India. Indian Journal of Forestry, 32: 181-190.

Dye, K.L., Ueckert, D.N. and Whisenant, S.G. 1995. Red berry Juniper- herbaceous understory interactions. Journal of Range Management, 48: 100-107.

Everson, C.S. 2009. The effect of the introduction of agroforestry species on the soil moisture regime of traditional cropping systems in rural areas. Phase II: on farm trials of alternative agroforestry systems. Annual Progress Report for WRC Project No. K5/1351, 5 March 2009. Water Research Commission, Pretoria, South Africa.

Everson, C.S., Dye, P.J., Gush, M.B. and Everson, T.M. 2011. Water use of grasslands, agroforestry systems and indigenous forests. Water SA, Water Research Commission. Accessed on 13th August 2012. http://www.scielo.org.za/scielo.php?script=sci _arttext&pid=S1816-79502011000500019&lng=es&nrm=iso.

FAO 2011. Save and Grow; a policymaker's guide to the sustainable intensification of smallholder crop production; FAO, Rome, 102 pp.

Faroda, A.S. 1998. Arid Zone Research: An overview. *In:* Fifty Years of Arid Zone Research in India (Eds. A.S. Faroda and M. Singh), pp. 1-16. CAZRI, Jodhpur.

Franco, A.C. and Nobel, P.S. 1989. Effect of nurse plants on the microhabitat and growth of cacti. Journal of Ecology, 77: 870-886.

Gajja, B.L. and Harsh, L.N. 2002. Economic performance of pasture management under alternative farming system in arid zone of Rajasthan. Current Agriculture, 26: 35-40.

Gajja, B.L., Bhati, T.K., Harsh, L.N. and Khan, M.S. 1999. Comparative economics of silvipasturue, horticulture and annual crops on marginal lands of arid zone of Rajasthan. Annals of Arid Zone, 38: 173-180.

Gajja, B.L., Meghwal, P.R. and Harsh, L.N. 2004. Economic performance of dryland horticultural crops in arid zone of Rajasthan-ex-ante analysis. Current Agriculture, 28: 59-62.

Garcia-Miragaya, J., Flores, S. and Chacon, N. 1994. Soil chemical properties under individual evergreen and deciduous trees in a protected Venezuelan savanna. Acta Oecologica, 15: 477-484.

Gill, A S. 2000. Effect of fruit trees and their spacing on the yields of intersown wheat and silvi component. Range Mgmt and Agroforestry, 21:153-159.

GoI 2006. Report of the Task Force on Grasslands and Deserts. Government of India, Planning Commission, New Delhi. Accessed on 9th August 2012. http://planningcommission.nic.in/aboutus/committee/wrkgrp11/tf11_grass.pdf.

Goyal, M. and Sharma, S.K. 2009. Traditional wisdom and value addition prospects of arid foods of desert region of North West India. Indian Journal of Traditional Knowledge, 8:581-585.

Gupta, G.N., Singh, G. and Kachwaha, G.R. 1998. Performance of *Prosopis cineraria* and associated crops under varying spacing regimes. Agroforestry Systems, 40:149-157.

Gupta, J.P. 1992. Role of agroforestry for sustainable production in arid areas. 2nd Annual Group Metting cum Symposium on Agroforestery, College ofAgriculture, Nagpur.

Gupta, J.P. 1997. Some alternative production systems and their management for sustainability. *In*: GuptaJ.P. and Sharma B.M. (eds.) Agroforestry for sustained productivity in arid regions. Scientific Publishers, Jodhpur, India. 31-39 pp.

Gupta, J.P. 2000. Agroforestery for sustained production. *In:* Technonolgy Approach for Greening Arid Lands, CAZRI, Jodhpur.

Gupta, J.P. and Saxena, S.K. 1978. Studies on the monitoring of the dynamics of moisture in the soil and the performance of ground flora under desertic communities of trees. Indian Journal of Ecology, 5: 30-36.

Gupta, J.P., Joshi, D.C. and Singh, G.B. 2000. Management of Arid Agro-ecosystem. In Natural Resource Management for Agricultural Production in India (Eds. J.S.P. Yadav and G.B. Singh), pp.551-668. International Conference on Managing Natural Resources for Sustainable Agricultural Production in the 21st Century hold on Feb. 14-18, 2000, New Delhi, India.

Gupta, T. and Mohan, D. 1982. Economics of afforest ration in hot arid zone of India. Proceedings of Second Forestry Conference, FRI, Dehradun.

Harsh, L.N. 1995. Alternate land use system for sustainable development: Silviculture. In: Land Resources and their Management for Sustainability in Arid Region (Eds. A.S. Kolarkar, D.C. Joshi and A. Kar), pp 187-196. Scientific Publisher, Jodhpur.

Harsh, L.N. and Tewari, J.C. 1993. Sand dune stabilization, shelterbelts and silvi-pastoral plantation in dry zones. *In:* Desertification and Its Control in the Thar, Sahara and Sahel Regions (Eds., A.K. Sen and Amal Kar). Scientific Publishers, Jodhpur, pp. 269-279.

Harsh, L.N. and Tewari, J.C. 2007. Agroforestry systems in arid regions of India. In: Puri, S. and Panwar, P. (eds). Agroforestry: systems and practices. Pp. 175-189

Harsh, L.N., Tewari, J.C., Burman, U. and Sharma, S. 1992. Agro-forestry in arid regions. Indian Farming, 45: 32-7.

Jaimini, S. N. and Tikka, S.B.S. 1998. Khejri (*Prosopis cineraria*) based agroforestry system for dryland areas of north and north-west Gujarat. Indian Journal of Forestry, 21:331-332.

Kapoor, B.B.S. 2001. An assessment of natural resources of the Indian Desert. Madhu Publications, 292p.

Kaushik, N. and Kumar, S. 2003. Khejri (*Prosopis cineraria*)- based agroforestry system for arid Haryana, India. Journal of Arid Environments, 55:433-440.

Khatri, A., Rathore, A. and Patil, U.K. 2010. *Prosopis cineraria* (L.) Druce: a boon plant of Desert- an overview. International Journal of Biomedical and Advance Research, 1: 141-149.

Kiran, R.K., Rani, M. and Pal, A. 2009. Reclaiming Degraded Land in India Through the Cultivation of Medicinal Plants. Botany Research International 2(3):174-181, 2009

Kohli, A. and Saini, B.C. 2003. Microclimate modification and response of wheat planted under trees in a fan design in northern India. Agrofor. Syst., 58:109-117.

Kolarkar, A.S., Murthy, K.N.K. and Singh, N. 1983. Khadin - A method of harvesting water for agriculture in the Thar Desert. Journal of Arid Environments, 6: 59-66.

Lata A.M, Raju, M.S., Joseph, B., Shankar, A.S. and Rao, P.C. 2012. Performance of medicinal herbs in tree based cropping system under dryland conditions. Journal of Pharmacognosy, 3: 92-95.

Lehmann, J., Peter, I., Steglich, C., Gebauer, G., Huwe, B. and Zech, W. 1998. Below-ground interactions in dryland agroforestry. Forest Ecology and Management, 111:157-169.

Lundgren, B.O. 1987. Institutional aspects of agroforestry research and development. *In:* Steppler HA, Nair PKR (eds), Agroforestry: A decade of development. ICRAF, Nairobi, 150 pp.

Meena, L.R., Mann, J.S., Sharma, S.C. and Mehta, R.S. 2005. Performance of Ardu (*A. excels* Roxb) under Silvipasture system in semiarid areas of Rajasthan. Indian J. Agroforestry, 7: 32-3.

Meghwal, P.R. 2007. Propagation studies in lehsua(*Cordia myxa).* Indian Journal of Agricultural Science*,* 77: 765-767.

Meghwal, P.R. and Vashishtha, B.B. 1998. Effect of time of planting and auxin on rooting of cuttings of ker (*Capparis decidua* L. Forsk). Annals of Arid Zone*,* 37: 401-404.

Meghwal, P.R., Aseri, G.K. and Rao, A.V. (2006). Raising bioinoculant aided rootstocks for higher yield in ber. Indian Horticulture, 51: 16-17.

Menezes, R.S.C., Salcedo, I.H. and Elliott, E.T. 2002. Microclimatic and nutrient dynamics in a silvipastoral system of semiarid northeastern Brazil. Agroforestry Systems, 56: 27-38.

Mordelet, P., Abbadie, L. and Menaut, J. C. 1993. Effects of tree clumps on soil characteristics in a humid Savanna of West Africa. Plant and Soil, 153:103-111.

Moro, M.J., Pugnaire, F.I. and Puigdefábregas. J. 1997b. Mechanism of interaction between *Retama sphaerocarpa* and its understorey layer in a semi-arid environment. Ecography, 20:175–184

Mukhopadhyay, D. 2008. Indigenous knowledge and sustainable natural resource management in the Indian desert. In: C. Lee and T. Schaaf (eds.). The Future of Drylands. Netherlands, Springer: pp. 161-170.

Muthana, K.D. and Arora, G.D. 1977. Studies on the establishment and growth of the economic tree species viz. *Holoptellia intrifolia* in integrated land use pattern in arid environment. Annual Progress Report, CAZRI, Jodhpur.

Muthuri, C.W., Ong, C.K., Black C.R., Ngumi, V.W. and Mati, B.M. 2005. Tree and crop productivity in Grevillea, Alnus and Paulownia-based agroforestry systems in semi-arid Kenya. Forest Ecology and Management, 212:23-39.

Nair, P.K.R. 1993. An introduction to agroforestry. Kluwer Academic Publ, Dordrecht etc, 499 pp

Narain, P. and Tewari, J.C. 2004. Trees on agricultural fields: a unique basis of life support in Thar Desert. *In:* Tewari VP and Srivastava R.L. (eds), Proceeding: Multipurpose Tees in the Tropics: Management and Improvement Strategies, IUFRO Conference, 22-25 Nov 2004 at AFRI, Jodhpur, pp 516-523.

Noureen, S., Arshad, M., Mahmood, K. and Ashraf, Y.M. 2008. Improvement in fertility of nutritionally poor sandy soil of Cholistan desert, Pakistan by C*alligonum polygonoides* Linn. Pak. J. Bot., 40(1): 265-274, 2008.

Osman, M., Sreenivasulu, R. and Narsimlu, B. 2008. Agroforestry systems as alternate land use options to tobacco in rainfed areas. Indian Journal of Dryland Agricultural Research and Development, 23:61-66.

Pandey, A. K., Gupta, V. K. and Solanki, K. R. 2010. Productivity of neem-based agroforestry system in semi-arid region of India. Range Management and Agroforestry, 31: 144-149.

Pandey, D. N. 2007. Multifunctional agroforestry systems in India. Current Science, 92(4): 455-463.

Pareek, O.P. 1999. Dryland horticulture. *In:* Fifty years of Agricultural Research in India, pp. 475-484. CRIDA, Hyderabad.

Pareek, O.P. 2001. Ber. International Center for Underutilized Crops. Southampton U.K.

Patel, J.M., Jaimini, S.N. and Patel, S.B. 2009. Evaluation of Neem and Ardu based agri-silvi systems. *Eds*: Pratap Narain, M.P. Singh, Amal Kar, S. Kathju and Parveen Kumar. *In*: Diversification of Arid farming systems. 114-118 pp.

Patel, J.M., Jamini, S.N. and Patel, S.B. 2008. Evaluation of neem and ardu based agri-silvi system. *In:* Diversification of Arid Farming Systems (Eds. Pratap Narain, M.P. Singh, A. Kar, S. Kathju and Praveen-Kumar), pp 114-118. Arid Zone Research Association of India and Scientific Publisher (India) Jodhpur, India.

Patil, M.B. and Channabasappa, K.S. 2008. Effect of tree management practice on *Acacia auriculiformis* based agroforestry system on growth and yield of associated black gram. Karnatak J. Agric. Sci., 21:538-540.

Prasad, R. and Mertia, R.S. 2005. Dehydrogenase activity and VAM fungi in tree-rhizosphere of agroforestry systems in Indian arid zone. Agroforest Systems, 63:219-223.

Prinsley, R.T. 1993. The role of trees in sustainable agriculture. Kluwer, Dordrecht *etc*, 186 pp.

Puri, S., Bangarwa, K.S. and Singh, S. 1995. Influence of multipurpose trees on agricultural crops in arid regions of Haryana, India. Journal of Arid Environment, 30:441-451.

Purohit, U., Mehar, S.K. and Sundaramoorthy, S. 2002. Role of *Prosopis cineraria* on the ecology of soil fungi in Indian desert. Journal of Arid Environments, 52:17-27.

Radhamani, S., Balasubramanian, A., Ramamoorthy, K. and Geetjalaxmi, V. 2003. Sustainable integrated farming systems for drylands – a review. Agric. Rev., 24:204-210.

Raffaele, E. and Veblen. T.T. 1998. Facilitation by nurse shrubs on resprouting behavior in a post-fire shrubland in northern Patagonia, Argentina. J. Vegetation Science, 9:693–698.

Rai, P. 2008. Management of arid rangelands. *In:* Diversification of Arid farming systems (Eds. Pratap Narain, M.P. Singh, A. Kar, S. Kathju, and Praveen- Kumar), pp. 79-102. Arid Zone Research Association of India and Scientific Publisher (India) Jodhpur, India.

Ram, K.A., Tsunekawa, A., Saha, D.K. and Miyazaki, T. 1999. "Subdivision and fragmentation of land holdings and their implication in desertification in the Thar desert, India. Journal Arid Environments, 41:463-477.

Ram, S.N., Kumar, S. and Roy, M.M. 2005. Performance of jujube (*Ziziphus mauritiana*)-based hortipasture system in relation to pruning intensities and grass-legume associations under rainfed conditions. Indian Journal of Agronomy, 50: 181-183.

Rathore, M.S., Shekhawat, N.S. and Gehlot, H.S. 2009. Need of Assessing Rhizobia for Their Plant Growth Promoting Activities Associated with Native Wild Legumes Inhabiting Aravalli Ranges of Rajasthan, India. Botany Research International 2 (2): 115-122, 2009

Rhoades, C.C., Sanford Jr. R.L. and Clark, D.B. 1994. Gender dependant influences on soil phosphorous by dioecious lowland tropical tree. Simarouba amara. Biotropica, 26: 362-368.

Roy, M.M. 1999. Litter fall and nutrient store in a silvopastoral system based on *Hardwickia binata* Roxb. in semi-arid India. Annals of Forestry, 7:70-76.

Saroj, P.L., Dhandhar, G., Sharma, B.D., Bhargava, R. and Purohit, C.K. 2003. Ber (*Ziziphus Mauritiana* L.) based agri-horti system: a Sustainable land use for arid ecosystem. Indian Journal of Agroforestry, 5:30-35.

Scholes, R.J. 1990. The influences of soil fertility on the Ecology of African Savannas. Journal of Biogeography, 17:417- 419.

Senkondo, EMM 2000. Risk attitude and risk perception in agroforestry decisions: the case of Babati, Tanzania. Mansholt Studies 17. Mansholt Institue, Wageningen University, 213 pp

Shamsudheen, M., Devi Dayal, Meena, S.L. and Ram, B. 2009. Improvement of soil properties under silvipastoral systems in the Kachchh region of arid Gujarat. 4th World Congress on Conservation Agriculture: Innovations for Improving Efficiency, Equity and Environment, (February 4-7, 2009), National Academy of Agricultural Sciences, New Delhi, India. 253-254 pp.

Shankar, V. 1980. Silvipasture research: a review. Forage Research, 6:107-122.

Shankar, V. 1995. Ecological consideration of rangeland development. *In:* Land Resources and Their Management for Sustainability in Arid Region (Eds. A.S. Kolkar, D.C. Joshi and A. Kar), pp 87-104. Scientific Publisher, Jodhpur (India).

Shankar, V., Dadhich, N.K. and Saxena, S.K. 1976. Effect of khejri tree (Prosopis cineraria) on the productivity of range grasses growing in its vicinity. Forage Research, 2:91-96.

Shankarnarayan, K.A., Harsh, L.N. and Kathju, S. 1987. Agroforestry in the arid zones of India. Agroforestry Systems, 5: 69-88.

Sharma, A.K. and Gupta, J.P. 2001. Agroforestery for sustainable production under increasing biotic and abiotic stresses in arid Zone. Abstract In: Impact of Human Activities on Thar Environment, pp 95-96, Jodhpur.

Sharma, L. and Pant D.C. 2008. Diversification of cropping pattern through horticulture crops in Rajasthan. Indian Journal of Agriculture Economics, 62:43-48.

Singh, A.K., Singh, J., Reddy, K.K. and Singh, R. 2003. Drought proofing mechanism in Mahi command area through linear programming model: A case study. Indian Journal of Agricultural Research, 37:187-192.

Singh, G. 2003. Agroforestry research for sustainable production in arid and semi arid regions of Rajasthan. Project Report, Arid Forest Research Institute, Jodhpur. 43 p.

Singh, G. 2004. Influence of soil moisture and nutrient gradient on growth and biomass production of *Calligonum polygonoides* in Indian desert affected by surface vegetation, J. Arid Environment, 56:541-558.

Singh, G. 2004. Project concluding Report of the project entitled "Agroforestry research for sustainable production in arid and semi arid region of Rajasthan. AFRI, Jodhpur, India. pp– 43.

Singh, G. 2005. Carbon sequestration under an agri-silvicultural system in the arid region. Indian Forester, 131:543-552.

Singh, G. 2009. Comparative productivity of *Prosopis cineraria* and *Tecomella undulata* based agroforestry systems in degraded lands of Indian Desert. Journal of Forestry Research, 20:144-150.

Singh, G. 2010. Rainfall dependent competition affected productivity of *V. radiata* in *Hardwickia binata* agroforestry in Indian desert. Indian Forester, 136:301-315.

Singh, G. and Gupta, G.N. 2000. Living barriers in Indian arid zone and their ecological Benefits. Indian Forester, 126: 257-268.

Singh, G. and Rathod, T.R. 2007a. Growth, production and resource use in *Colophospermum mopane* based agroforestry system in northwestern India. Archive of Agronomy & Soil Science, 53: 75-88.

Singh, G. and Rathod, T.R. 2007b. Productivity and soil resource availability in *Hardwickia binata* based agroforestry system in Indian desert. Arid Land Research and Management, 21: 193-210.

Singh, G. and Shukla, S. 2011. Effects of *Azadirachta indica* canopy manipulation and nitrogen fertilization on diversity and productivity of herbaceous vegetation in an arid environment of India. Arid Land Research and Management, 25: 129-150.

Singh, G. and Shukla, S. 2012. Effects of *Prosopis juliflora* (DC.) tree on under canopy resources, diversity and productivity of herbaceous vegetation in Indian desert. Arid Land Research Land Management, 26:151-165.

Singh, G. and Shukla, S. 2012. Effects of *Prosopis juliflora* (DC.) tree on under canopy resources, diversity and productivity of herbaceous vegetation in Indian desert. Arid Land Research Land Management, 26:151-165.

Singh, G., Bala, N., Kuppusamy, V. and Rathod, T.R. 2003. Adaptability and productivity of *Cassia angustifolia* in sandy soil of Indian Desert. Indian Forester, 129:213-223.

Singh, G., Bala, N., Mutha, S., Rathod, T.R. and Limba, N.K. 2004. Biomass production of *Tecomella undulata* agroforestry in arid India. Biological Agriculture & Horticulture 22: 205-216.

Singh, G., Gupta, G.N. and Kuppusamy, V. 2000a. Seasonal variations in organic carbon and nutrient availability in arid zone agroforestry systems. Tropical Ecology, 41:17-23.

Singh, G., Gupta, G.N. and Kuppusamy, V. 2000b. Changes in soil properties and nutrient accumulation in Neem under the canopy of *A. tortilis*. International Tree Crop Journal, 10:237-246.

Singh, G., Kuppusamy, V. and Rathod, T.R. 1999. Effect of intercropping on the multipurpose trees and the associated crop in Indian desert. Range Management & Agroforestry, 20: 26-33.

Singh, G., Mutha, S. and Bala, N. 2007b. Growth and productivity of *Prosopis cineraria* based agroforestry system at varying spacing regimes in the arid zone of India. J. Arid Environment, 70:152-163.

Singh, G., Mutha, S., Bala, N., Rathod, T.R., Bohra, N.K. and Kuchhawaha, G.R. 2005. Growth and productivity of *Tecomella undulata* based agroforestry system in Indian desert. Forests, Trees and Livelihood, 15:89-101.

Singh, G., Rathod, T.R., Mutha, S., Upadhyaya, S. and Bala, N. 2008. Impact of different tree species canopy on diversity and productivity of understorey vegetation in Indian desert. Tropical Ecology, 49:13-23.

Singh, G., Rathod, T.R., Singh, B. and Chouhan, M. 2008. Component interactions and productivity in *Emblica officinalis* based agri-horticulture system in Indian Desert. Biological Agriculture & Horticulture, 25:253-268.

Singh, H.P. 1994. Soil management in the Indian arid zone. In: H.P. Singh and S. Singh (eds) Sustainable development of the Indian arid zone: a research perspective. Accessed on 2nd August 2012. http://www.cabdirect.org/abstracts/19941905358.html

Singh, K.C. 1989. Importance of grasses in management of sandy soils-strip cropping of grasses. In: International Symposium on Managing Sandy Soils. CAZRI, Jodhpur, pp 561-564.

Singh, K.C. 1995. Management of pasturelands in arid regions. In: Sustainable Development of Dryland Agriculture in India (Eds. R.P. Singh), pp. 415-423. Scientific Publishers, Jodhpur.

Singh, K.C. and Gupta, J.P. 1997. Ley farming for sustainable production in arid region of Rajasthan. In: Proceedings of the Symposium on Recent Advances in Management of Arid Ecosystem, held at CAZRI, March 3-5, 1997, pp. 95.

Singh, K.S. and Lal, P. 1969. Effect of Khejri (*Prosopis cineraria*) and Babool (*Acacia nilotica*) trees on soil fertility and profile characteristics. Annals of Arid Zone, 8: 33-36.

Singh, R.S. 2003. Influence of water stress and micro-meteorological changes on pearl millet productivity under arid zone agroforestry systems. Jour. Agric. Physics, 3:130-135.

Singh, R.S. and Kumar, A. 1993. Agri-horticulture systems under semi-arid conditions. Agroforestry News, 5:3-5.

Singh, S. 1982. Types and formation of sand dunes in the Rajasthan desert. In: Perspectives in Geomorphology, (Ed., H.S. Sharma). Concept Publishing Company, New Delhi, 4:165-183.

Sundaramoorthy, S., Mehar, S.K. and Suthar, M.S. 2010. Soil biology in traditional agroforestry systems of the Indian Desert. In: Desert Plants-Biology and Biotechnology (Ed. K.G. Ramawat). Springer. pp. 91-110.

Szott, L.T., Fernandes, E.C.M. and Sachez, P.A. 1991. Soil plant interactions in agroforestry systems. Forest Ecology and Management, 45:127-152.

Tarfdar, J.C. 2008. Biological manipulation through farming syatems for the arid regions of India. In: Diversification of Arid Farming System (Eds. Pratap Narain, M.P. Singh, A. Kar, S. Kathju and Praveen-Kumar), Arid Zone Research Association of India and Scientific Publisher (India) Jodhpur, India, pp 200-209.

Tewari, J.C. and Harsh, L.N. 1998. Forestry research in arid tract of India. *In:* Fifty Years of Arid Zone Research in India (Eds. A.S. Faroda and Manjit Singh), CAZRI, Jodhpur, pp. 307-322.

Tewari, J.C., Sharma, A.K., Pratap, N. and Singh, R. 2007. Restorative forestry and agroforestry in hot arid region of India: A Review. J. Tropical Forestry, 23:1-16.

Tewari, J.C., Bohra, M.D. and Harsh, L.N. 1999. Structure and production function of traditional extensive agroforestry systems and scope of intensive agroforestry in Thar Desert. Indian J. Agroforestry, 1:81-94.

Tewari, V.P. and Singh, M. 2006. Tree-crop interaction in the Thar desert of Rajasthan, India. Secheresse, 17:326-332.

Tiedmann, A.R. and Klenmedson, J.O. 1973. Nutrient availability in desert grassland soils under mesquite (*Prosopis juliflora*) tree and adjacent open areas. Soil. Sci. Soc. Am. Proc., 37:107-111.

Toky, O. P. and Bisht, R.P. 1992. Observations on the rooting patterns of some agroforestry trees in an arid region of north-western India. Agroforestrv Systems, 18: 245-263.

Toky, O.P. and Arya, S. 2005. Intraspecific Biodiversity in Important Trees of Arid India and its Conservation. In: Gupta *et al.* (Editors): Ecology and Environmental Management: Issues and Research Needs. Bulletin of the National Institute of Ecology 15: 19-24.

Tripathi, G., Ram, B., Sharma, B.M and Singh, G. 2005. Soil faunal biodiversity and nutrient status in silvipastoral systems of Indian deserts. Environmental Conservation, 32: 178-188.

Verma, N. Tarafdar, J.C. and Srivastava, K.K. 2009. Periodic changes in *Prosopis cineraria* associated AM population at different soil depth and its relationship with organic carbon and soil moisture. African Journal of Microbiology Research, 4:115-121.

Vetaas, O.R. 1992. Micro-site effects of trees and shrubs in dry Savannas. J. Vegetation Science, 3: 337-344.

Virginia, R.A. 1986. Soil development under legume tree canopies. Forest Ecology and Management, 16: 69-79.

Vyas, S.P. 2005. Performance of *Lawsonia inermis* L. in arid parts of Gujarat. Indian Journal of Soil Conservation, 33:73-75.

WAC 2009. Agroforestry options for Tanzania. World Agroforestry Centre Policy Brief 03. http://www.worldagroforestry.org/af1/downloads/publications/PDFs/BR09007.

Yadav, R.S., Yadav, B.L., Chhipa, B.R., Dhyani, S.K. and Ram, M. 2010. Soil biological properties under different tree based traditional agroforestry systems in a semi-arid region of Rajasthan, India. *Agroforestry Systems*, DOI: 10.1007/s10457-010-9277-z.status in silvipastoral systems of Indian deserts. Environmental Conservation, 32: 178-188.

Verma, N. Tarafdar, J.C. and Srivastava, K.K. 2009. Periodic changes in *Prosopis cineraria* associated AM population at different soil depth and its relationship with organic carbon and soil moisture. African Journal of Microbiology Research, 4:115-121.

Vetaas, O.R. (1992). Micro-site effects of trees and shrubs in dry Savannas. J. Vegetation Science, 3: 337-344.

Virginia, R.A. 1986. Soil development under legume tree canopies. Forest Ecology and Management, 16: 69-79.

Vyas, S.P. 2005. Performance of *Lawsonia inermis* L. in arid parts of Gujarat. Indian Journal of soil Conservation, 33:73-75.

WAC (2009). Agroforestry options for Tanzania. World Agroforestry Centre Policy Brief 03. http://www.worldagroforestry.org/af1/downloads/publications/PDFs/BR09007.

Yadav, R.S., Yadav, B.L., Chhipa, B.R., Dhyani, S.K. and Ram, M. 2010. Soil biological properties under different tree based traditional agroforestry systems in a semi-arid region of Rajasthan, India. Agroforestry Systems, DOI: 10.1007/s10457-010-9277-z.

10

Status and Scope of Agroforestry Practices in Bihar

D.K. Das, R.K. Jha, and O.P. Chaturvedi

Abstract: *Economy of Bihar is solely dependent on agriculture. Only 7% area is under forest as against a minimum 33% required for ecological balance. To counter the environmental degradation and to provide fuel, fodder and timber, different tree-crop combinations were tried for accessing their compatibility. As a result some agroforestry systems have been recommended and are discussed in this paper. Economics of some of the successful system has also been discussed. A detailed account of prevalent agroforestry practices in Bihar is also given. Agri-horticulture system is most popular system adopted by large and medium farmers in the region. Finally, constraints in adopting and carrying out research in agroforestry along with future strategies to extend agroforestry in Bihar are also given in the paper.*

Introduction

Agroforestry land use system is one of the most important means for improving productivity of rural resources. People's perception about agroforestry varies with their use perspectives. Only 7 % area is under forest as against a minimum 33 % required for ecological balance. Per capita consumption of wood fuel in Bihar is 0.27 m^3 per year and industrial wood is 0.054 m^3. One of the studies, on demand and supply of wood in Bihar conducted by Division of Forestry, Krishi Vigyan

Kendra, Munger indicated a negative wood balance of 76.07 million cubic meters by 2050. Analysis of data revealed that at present the total requirement of fuel wood in the state is around 16.61 million tones. Out of these, roughly 1.98 million tones come from the farm forestry sector including common lands. As per estimates, to fulfill the demand of fuel need, 4.6 lakh hectare area should be covered by high calorific valued fuel producing species. The situation is more-more complex than it appears because of more demand for small timber, poles for housing and agricultural implements in rural Bihar. The utilization of tree leaves for feed to livestock is very common during drought and flood in Bihar also. Population of livestock in Bihar is 47.93 million. State livestock population constitutes 10.2% of the country's livestock population. A careful review of the supply and demand scenario and its projections upto 2050 shows a huge deficit which is likely to increase in view of the increasing demands of the products (Table 1). The available alternative point is to be the area under range management, degraded lands and improved forage production technologies. While the fodder trees serve major needs in the Bihar, their introduction in the agroforestry system will be required to augment green fodder supply with high nutrition.

Table 1. Per capita supply and demand of top feed of a cattle (Projected estimates in kg) in Bihar

Year	Supply	Demand	Deficit as % of demand
2005	807.88	2123.78	61.96
2010	818.97	2199.20	62.76
2020	841.12	2350.04	64.20
2050	895.67	2842.41	68.48

Source: Jha, R.K. (2006)

Per capita availability of forest in the state is 0.01 ha which is much lower than the world and Indian average of 0.80 ha and 0.08 ha, respectively. The pressure on existing forest is too high to meet the demand of people for timber, fuelwood, fodder etc. in adequate quantity. Area of waste land is about 6.59 lakh ha categorized under barren and uncultivable land (4.37 lakh ha), cultivable waste land (0.51 lakh ha), fallow lands other than current fallows (1.52 lakh ha), grassland/permanent pastures *etc*. (0.19 lakh ha) (Table 2).

Table 2. Land use utilization pattern in Bihar (in lakh ha)

Sl.No.	Items	Area
1.	Total geographical area	93.60
2.	Forests	6.16
3.	Barren and unculturable land	4.37
4.	Land put to non-agricultural uses	
	(i) Land area	12.71
	(ii) Water area	
	(a) Permanent	2.07
	(b) Temporary	1.38
	Total (i + ii)	16.16
5.	Culturable Wastes land	0.51
6.	Permanent pastures and other grazing lands	0.91
7.	Land under miscellaneous tree crop and groves not included in area sown	2.33
8.	Follow lands other than current follows	1.52
9.	Current fallows	6.32
10.	Total uncultivable land	37.56
11.	Net sown area	56.03
12.	Total cropped area	79.46
13.	Area sown more than once	23.43

The flood besides causing water stagnation in certain areas washes out field crops causing its shortage. The water stagnation delays the sowing of succeeding crops. Not only crops are damaged but also in flood prone area, stored food grains and seeds are also spoiled. The flood damage in Bihar accounts 20-40% of total damage in the country. Like flood, drought also affects several districts of the state regularly. The drought causes the failure of standing kharif crops and affects the possibilities of taking good rabi crops in rainfed areas. About 2.24 lakh hectare of land in Bihar is salt affected (Table 3) (Prasad, 1995). They are either saline-alkali or alkali soils. The high water expenditure, faults in soil and water management, excessive seepage from canals, high monsoon rainfall, undulating topography and recurring floods have resulted in a large area of Bihar under water logging. Such condition has also rendered quite a large area unproductive on account of secondary salinization. These salt affected soils are often put to arable farming by our farmers despite the fact that these are not suitable for arable cropping. This has resulted in agriculture being not sustainable because of the fact that crops grown in most of these areas are under rainfed situation and have a poor organic matter and water holding capacity resulting poor productivity.

Table 3. Extent of salt affected area in the various districts of Bihar

Districts	Salt affected area (ha)
Muzaffarpur	31,600
Vaishali	30,270
Samastipur	42,840
East Champaran	24,310
West Champraran	4,330
Saran	26,570
Siwan	33,750
Gopalganj	30,600
Total	2,24,270

Diara and tal lands are flood prone areas. When perennial rivers during monsoon are accompanied by high and continuous rainfall, it causes deluge. Diara and tal land in Bihar occupy nearly 10.25 lakh ha (Table 4) (Yadav, 2000). These areas remain submerged from mid June to November. However, chaur lands may remain submerged from June to mid January. Agriculture in these areas is risky and so people remain poor.

Table 4. River-wise flood prone areas

Sl.No.	River	Area (lakh ha)	Affected zones
1.	Ganga	2.40	I, II, III
2.	Gandak	1.40	I
3.	Burhi Gandak	2.25	I
4.	Sone	1.15	III
5.	Kosi	1.50	I, II
6.	Bagmati	0.80	I
7.	Adhwara group	0.75	I
	Total :	10.25	

Source: Yadav (2000)

Such environmental degradations have brought human population under socio-economic crisis and ecological imbalance. Under such situations, it is imperative to minimize the farmers' risk through better utilization of natural resources. There is need to improve land productivity through development of sustainable land use models and farming systems. Agroforestry is viewed as practice to overcome these crises by bringing the wasteland and other unutilized land into productive use.

Climate

The climate of Bihar is subtropical humid to sub-humid type having three distinct seasons, *viz.*; winter (November-February), summer (March-June) and rainy season (July-October) is sub-tropical. Minimum and maximum temperature varies between 7.1°C in January to 41.8°C in June. The state is divided into three agroclimatic plain zones.

1. ***North-West Alluvial***: It comprises of Saran, Tirhut and Darbhanga divisions. The average annual rainfall is about 1211 mm.
2. ***North-East Alluvial***: This comprises of Saharsa, Purnia and part of Begusarai divisions. The average annual rainfall is about 1405 mm.
3. ***South Bihar Alluvial***: This zone is located south of river Ganga and comprises of Patna, Gaya and Bhagalpur divisions. The average annual rainfall is about 1110 mm.

Soils

The soils in Bihar are an alluvial deposit with very good soil depth, high fertility and are of recent origin. There are occasional patches of problematic soils with high amount of calcium carbonate and salinity problems. The soil is light to medium textured except in those districts away from influence of rivers where the soil is medium to heavy. The soils under influence of rivers are mostly calcareous. Average soil pH, organic carbon, N, P_2O_5 and K_2O varied in the state between 6.8-8.3, 0.38-0.55%, 235-290 kg ha^{-1}, 14-25 kg ha^{-1}and 115-165 kg ha^{-1}, respectively.

Traditional agroforestry practices

Different types of agroforestry practices are in vogue in different agroclimatic zones of Bihar. Some commonly seen agroforestry systems practiced in Bihar are listed bellow (Table 5):

(i) Agri-horticultural/Agri-horti-silvicultural system

Bihar is a major supplier of fruits in India. A total of 2.86 million hectares of land in Bihar is under fruit orchards, of which mango, litchi, guava and lime occupy areas of 1.55 m ha. 0.26 m ha, 0.31 m ha and 0.29 m ha, respectively. The fruit trees take a long time to bear fruit and are generally spaced widely. Till they grow and cover the entire land farmers raise cereals, pulses, vegetables, tubers, medicinal plants and fodder crops in the interspaces. Papaya and dwarf varieties of banana are also intercropped depending upon the edaphic conditions. With increase in canopy cover of trees, tuber crops are preferred. Improvement in soil conditions through various agronomical managements required for intercropping helps in improvement of fruit bearing capacity of trees.

Table 5. Trees and crops under different agroclimatic zones of Bihar.

Important Tree Species	Cropping Systems		
	Kharif	Rabi	Summer
1. North-West Gangetic Plain			
Irrigated	Early paddy	Maize	-
Dalbergia sissoo	Maize	Potato	Moong
*Tectona grandis*Maize/early	Paddy	Rai	Moong
Swietenia mahogany	Paddy	Wheat	-
Bombax ceiba	Maize Potato	Vegetables	
Wendlandia exserta	Groundnut	Wheat	-
Populus deltoides	Paddy	Wheat	Moong
Syzygium cumini	Ragi	Maize	-
Anthocephalus cadmba	Paddy	Sugarcane	Sugarcane
Artocarpus heterophyllus	Maize	Potato	Onion
Gmelina arboea	Paddy	Berseem	-
Cocos nucifera	Maize	Wheat	Moong
Rainfed			
Mangnifera indica	Paddy	Gram	-
Litchi chinensis	Paddy	Wheat	-
Borassus flabellifer	Paddy	Paira lentil	-
Phoenix sylvestris	Paddy	Sugarcane	Sugarcane
	Maize	Sweet Potato	Moong
	Ragi	Sweet Potato	-
	Mazie + turmeric /Ginger	-	-
	-	Tobacco + Garlic	-
	-	Tobacco	-
Irrigated			
2. North - East Gangetic Plain			
Dalbergia sissoo		Paddy	Wheat
Bombax ceiba	Paddy	Paddy	-
Acacia lenticularis	Early Paddy	Potato	-
Accacia nilotica	Jute	Maize	-
Wendlandia exerta	Jute	Rai	Maize
Tectona grandis	Paddy	Potato	Onion
Azadirachta indica	Maize	Tori	Groundnut
Artocarpus heterophyllus	Jute	Wheat	-
Cocos nucifera	Jute	Pulses	-
Borassus flabellifer	Jute	Wheat	-
	Paddy	Gram	-
3. South Gangetic Plain	Paddy	Wheat	-
Rainfed			
Dalbergia sissoo		Paddy	Wheat Urd
Bombax ceiba	Paddy	Maize	-
Acacia lenticularis	Paddy	Potato	Moong
Albizia lebbeck	Paddy	Onion	Moong
Gmelina arboea		Paddy	Toria

(*Contd.*)

Important Tree Species		Cropping Systems		
		Kharif	Rabi	Summer
Moong				
	Terminalia arjuna	Paddy	Oilseed	-
	Mangifera indica	Maize	Vegetable	-
	Zizyphus mauritana	Maize	Berseem	
Rainfed				
	Spondias pinnata	Maize	G. Nut	
	Emblica officinalis	Paddy	Paira Lentil	-
	Artocarpus heterophyllus	Maize	Oilseed	-
	Borassus flabellifer	Fallow	Gram	-
		Fallow	Lentil	-
		Urd	Wheat	-

Farmers often put a trench on the boundary of their orchards and on the ridge of the trench multipurpose trees like shisam (*Dalbergia sissoo*), bankat (*Wendalandia exserta*), khajur (*Phoenix sylvestris*), tar(*Borassus flabellifer*), sagwan (*Tectona grandis*) etc are planted. These trees protect the orchard from the high velocity of wind and desiccating heat and provide fuel, fodder and timber to the farmers. The interspaces in the orchard are utilized with cultivation of crops.

(ii) Agri-silvicultural system

Under this agroforestry practices, the trees are planted at wide spacing and cultivations of crops are done in the inter-spaces till the canopy closes and restrict light reaching the ground. Under this system trees are also planted along the boundaries of agricultural fields. In Bihar babool (*Acacia nilotica*), *Dalbergia sissoo*, semal (*Bambax ceiba*), chah (*Acacia lenticularis*), *Borassus flabellifer, Phoenix sylvestris* and kathal (*Artocarpus heterophyllus*) are the important species grown with cereals and pulses. Recently, *Bombax ceiba*, kadamb (*Anthocephalus cadmba*) and poplar (*Populus deltoides*) have received greater attention due to their fast growth and many benefits.

(iii) Silvipastoral/hortipastoral system

Under this system, naturally occurring grasses are available in the interspaces of the horticultural or silvicultural component. Most of the livestock graze freely. Scattered tree species like *Acacia nilotica*, khair (*Acacia catechu*), Jamun (*Zizyphus mariationa*) *Dalbergia sissoo* and mango (*Mangifera indica*) are often found growing on these grazing lands covering a large portion of flood prone and salt affected areas. The grasses associated with these grazing lands are *Dichanthium annulatum, Chrysopogon fulvus, Heterpopogon contortus* and *Cenchrus ciliaris.* Lopped pruning from trees are often used for fuel and fodder. The tree also provides shelter to the grazing animals during summer and rainy seasons. Controlled grazing can help in improvement of the productivity of these grazing lands.

(iv) Block planting

Dalbergia sissoo, Tectona grandis, Bombax ceiba, Acacia lenticularis, gamhar (*Gmelina arborea*), *Populus deltiodes,* bamboo (*Dendrocalamus strictus, Bambusa balcoa, B. bambos* etc) and *Acacia nilotica* are commonly grown species in the woodlots. These species provide fuel, fodder and timber, apart from helping in conservation and amelioration of soils.

(v) Homegardens

Homegardens contain mixed vegetation with various perennial trees and annual vegetable crops. Selection of the species is determined by the environmental factors and needs of the family. *Dendrocalamus strictus, Dalbergia sissoo, Mangifera indica,* Litchi (*Litchi chinensis*), banana (*Musa paradisica*) and papaya (*Carica papaya*) are very common species associated with homegardens. A good number of vegetables and species are found growing in homegardens.

(vi) Apiculture with trees

The main purpose of this system is the production of honey where various honey (nectar and pollen) producing tree species frequently visited by the honeybee are planted on the boundary or mixed with agricultural crops. In Bihar, the potential rewards of bee keeping are great but apiculture is not achieving their full potential mainly due to shortage of flora. Agroforestry may offer some solutions based on symbiotic relationship between bees, field crops and trees. The increased presence of bees could lead to better pollination of both tree and field crops and thus to greater yields of timber, fruit, crop and honey. Consequently, it would bring farmers increased cash income. *Bombax ceiba, Litchi chinensis*, palas (*Butea monosperma*), *Dalbergia sissoo*, karanj (*Pongamia pinnata*), chakundi (*Cassia siamia*), imli (*Tamarindus indica*), siris (*Albizia lebbeck*), subabul (*Leucaena leucocephala*), guava (*Psidium guajava*), *Zizyphus* mariatiana, arjun (*Terminalia arjuna*), *Anthocephalus* cadmba, safeda (*Eucalyptus tereticornis*) etc. are suitable for apiculture in agroforestry system.

(vii) Aquaforestry

Fishponds are an important and intensive form of production in many parts of Bihar, particularly in Chaurs (low-lying waterlogged areas). Planting trees such as *Dalbergia sissoo, Pongamia pinnata, Tectona grandis, Acacia lenticularis, Bombax ceiba, Wendlandia exserta, Pongamia pinnata*, *Zizyphus* mariatiana, *Terminalia arjuna* etc. around fishponds augment the fish diet. The trees are managed by the farmers for production of fuel, fodder, timber and soil conservation.

Research finding

Some of the research findings obtained under different projects have been listed below:

Litchi (*Litchi Chinensis* Sonn.) based agri horticultural system

Litchi is a major commercial crop in Bihar. Growing a number of other crops in association with litchi plantation is a widespread practice in all litchi-growing areas. Litchi trees, which produce a much valued tropical fruit, are grown in an area of 18543 ha in north Bihar, India. Litchi is useful as wind break and bee forage. Plantations receive benefits by manure and fertilizers given to crops and weeds are eliminated.

Chaturvedi and Jha (1998) conducted an experiment on crop production following the two cropping patterns: pattern-I Paddy-wheat-green gram-ginger and pattern-II Maize-toria-green gram-pointed gourd under litchi orchard. There was a marked reduction in yield of all intercrops with increase in age of the plantation. In first cropping pattern, paddy, wheat and green gram were cultivated till 6 years of planting followed by cultivation of shade loving ginger crop upto 8 years of planting. In second cropping pattern maize, toria and greengram were cultivated upto 5 years of planting followed by pointed gourd upto 8 years of planting.

At the end of 9th year, the cumulative benefit : cost ratio under rotation-I (Paddy-wheat-green gram-ginger) and fruit system was 2.17. For the Rotation-II (Maize-toria-green gram-pointed gourd) and fruit it was 2.73, indicating that both the intercropping systems are profitable. On a nine-year cycle of a litchi orchard, establishment, management and harvest of fruits and intercultural operations of intercrops generate employment opportunities of 130-140 person days ha^{-1} year^{-1}.

Poplar based agri-silvicultural systems

In recent years poplars (*Populus deltoides*, G-3) received popularity among farmers in Bihar due to their fast growth and ability to attain substantial biomass production in short rotations of 9 to 10 years. Poplar mostly occurs in low-lying, moist alluvial ground, and tolerates flooding for a short time during the rainy season, and can be propagated vegetatively. In the inter space of poplar plantation, several seasonal crops are cultivated for enhancement of overall productivity of the land and generation of supplementary income. Poplar's attributes such as high propagation potential through cuttings, easy establishment, and rapid growth, straight cylindrical bole and high volume returns make this species suitable for cultivation under agroforestry systems.

Chaturvedi and Pandey (2001) studied on two cropping patterns under poplar plantation. In cropping pattern I, maize (*Zea mays*) *cv* "Suwan" followed by wheat (*Triticum aestivum*) *cv* "HUW 234" were cultivated until the fourth year and after that, the shade loving crop turmeric (*Curcuma domestica*) *cv* "Rajendra Sonia" was cultivated from the fifth to the ninth year. In cropping pattern II, pigeonpea (*Cajanus cajan*) *cv* "Bahar" was cultivated upto fifth year and turmeric *cv* "Rajendra Sonia" from the sixth to the ninth years. Yield decreased with the age of the plantation. Turmeric was cultivated when the yield reduction was 65% in maize, 41% in wheat and 67% in pigeonpea in relation to the yield in the open field. The yield of turmeric which was cultivated from the 5th to the 9th year declined drastically to just over 60% of the control in the 9th year of the plantation. The crops under different ages of plantations with varying light intensity (21.4 to 89.4% in June and 61.6 to 90.2% in December) received normal recommended doses of fertilizers, irrigations and other cultural practices. The average girth and height of poplar trees in the 10-year-old plantation were 1.42 $\pm$ 0.14m and 24.87 $\pm$ 1.30m, respectively. The cumulative benefit: cost ratio at the end of 10th year after tree planting with cropping pattern I was 5.01 and under cropping pattern II and poplar was 6.68.

In the same agroforestry systems, Das and Chaturvedi (2005a) and Das and Chaturvedi (2005b) found that annual transfer of litter nutrient to the soil by vegetation was 37.3 to 146.2 N, 5.6 to 17.9 P and 25.0 to 66.3 K kg ha^{-1} $year^{-1}$ in young (three-year-old) and mature (nine-year-old) plantations. Turnover rate and time for different nutrients ranged between 0.86 to 0.99 $year^{-1}$ and 1.01 to 1.16 year, respectively. This study shows that *Populus deltoides* has a potential of cropping under an agroforestry land use system and can provide standing biomass of 90.6 Mg ha^{-1} at a short rotation of 9 years. The disappearance of litter to the tune of 3.71 to 4.13 (young) and 10.55 to 10.74 Mg ha^{-1} (mature) protects the soil health from further deterioration and thus maintains sustainability.

Agri-silvicultural systems with bund trees

In another experiment it was found that soil fertility was higher near to tree base of shisham (*Dalbergia sissoo*), bankat (*Wendlandia exserta*), tar (*Borassus flabellifer*), khajur (*Phoenix sylvestris*) growing on field bund along with two crop rotations "Paddy-wheat" and "Maize-tori-mung", while grain yield increased with increase in distance from tree base. All the fields associated with different bund trees had sown greater availability of N, P and K in the soil as compared to the field without any association of trees. Grain yields were adversely affected in association of trees in the pattern tar< khajur < bankat < shisham. The average reduction value of grain yield across all crops and associated trees varied between 16.3 to 37.3%. Besides, these trees provide fuelwood and timber. The juice extracted from tar and khajur is the primary economic product, which is used for preparation of gur (molasses). Thus, these trees strengthen economic security of poor people in case of disaster and emergencies (Chaturvedi and Das, 2002a).

Study on the effect of bamboo row on the yield of some agricultural crops in Bihar showed that turmeric, ginger and dinanath grass produced 32.4 to 40.5% relative yield under full light. Rice and finger millet gave the lowest yield near bamboo row and their yield increased with increasing distance from bamboo row. Thus ginger, turmeric and dinanath grass can be grown successfully in bamboo shade depending upon grower's choice and economic return (Ali *et al.*, 2006)

Root biomass recovery and rooting characteristics of some agroforestry tree species

Knowledge of the quantitative assessment and structural development of root system is essential to improve and optimize the productivity under agroforestry systems. A comparative study of the rooting depth of 4-year-old agroforestry tree species namely, poplar, bankat, shisham and chah inter planted on boundary of wheat crop field was conducted. The results confirm the general view that bankat has a deep root system, which interferes less with shallow-rooted wheat crops. Shisham and chah with more fine and bigger roots density and biomass in top soil (0-20 cm), have potential for efficient soil and water conservations, whereas not much preferred in agroforestry due to more competition with crops for nutrients and moisture from top soil. If competition for these species is to be minimized, tree planting must be combined with more appropriate management practices (Chaturvedi *et al.*, 2005).

Root distribution for five, four-year-old agroforestry tree species, namely, Akashmuni (*Acacia auriculiformis*), Neem (*Azadirachta indica*), Kachnar (*Bauhinia variegata*), Semal (*Bombax ceiba*) and Bankat (*Wendalendia exserta*) were studied (Das and Chaturvedi 2008). Maximum rooting depth was recorded in *W. exserta* (2.10 m) and minimum, in *B. variegata* (1.00 m). Variation in horizontal root spread was 2.05 m in *B. ceiba* and 8.05 m in *A. auriculiformis*. Root spread exceeded the crown cover for all tree species. The length and diameter of the main root were highest in *A. indica* (108.3 cm) and *B. ceiba* (23.2 cm) respectively. Maximum length of lateral roots was recorded in *B. variegata* (201.6 cm) and maximum diameter in *A. indica* (1.8 cm). Total root phytomass among different species accounted for 18.38% of the total tree biomass. This study infers that deep rooted species like *W. exserta* and *A. auriculiformis* are suitable for dry areas and may be mixed with shallow rooted species in plantations or in agroforestry systems. Since all species in the present study had most of their root biomass in the upper soil layer, the deep rooted species like *W. exserta* and *A. auriculiformis* will be preferred for agrisilviculture and will have potential to pump nutrients from deeper layer of the soil. However, to reduce the strong competition for water and nutrients with the intercrops, the lateral roots of the tree have to be pruned. Although, all the tree species showed great potential for improving soil fertility status and soil water level, *A. auriculiformis* as a N-fixing species was most effective for promoting soil fertility.

Studies on rooting patterns of 5-year-old important agroforestry tree species indicated that deep rooted species like babool, shisham and jangali jalebi are suited to dry areas and may be mixed with shallow rooted species in plantations or in agroforestry systems (Chaturvedi and Das, 2002b). Since all the species had most of their root biomass in the upper soil layer, the deep rooted species like babool and shisham will be preferred for intercropping as they will provide greater anchorage and will have potential to pump nutrients from deeper zone of the soil. However, to reduce root competition with companion crops the lateral roots of the tree have to be pruned. Jangali jalebi, safed siris, babool and shisham have high values of the soil binding factor and thus have potential for soil conservation.

Chah (*Acacia lenticularis*) Based Agri-Silvicultural System: *Biomass Production, Root Distribution and Nutrient Status*

A study of an agrisilviculture system comprising *A. lenticularis* and turmeric (*Curcuma domestica*) local cv. Rajendra Sonia was conducted (Chaturvedi *et al.*,2008). Above and below ground biomass production and distribution of coarse and fine roots were studied in 4-year-old *A. lenticularis* planted at a spacing of 2 x 2 m, 2 x 3 m, 2 x 4 m and 2 x 5 m. The shoot biomass varied from 22.0 t ha^{-1} to 30.0 t ha^{-1} depending on the tree density. Among the different above ground tree components, stem wood contributed maximum biomass (62.5-70.4 %), followed by branches and leaves. Root distribution pattern showed that most of the coarse roots were distributed in the top 40 cm of soil, whereas fine roots were concentrated in the top 20 cm. Coarse root biomass decreased with an increase in spacing. The spread of roots was asymmetrical in trees planted at 2 x 2 m and 2 x 3 m spacing, while it was symmetrical in trees planted at wide spacing. No significant difference was observed in the fine root biomass in different stands. The root : shoot ratio increased with an increase in spacing. Yield of turmeric varied significantly from 160 q ha^{-1} to 220 q ha^{-1} and it increased with a decrease in tree density.

Chemical properties of the top soil (0-15 cm) showed the slight improvement in soil fertility over initial soil properties in terms of pH, EC, organic carbon, available N, available K_2O. Among the plantations of different density, pH, EC and available P_2O_5 did not show any specific trend while the rest parameters showed increasing trend with the increase in the density of the plantations.

Results in the present study indicated that improved soil fertility under different density plantations did not translate into higher yield of turmeric than open due to another limiting overriding factor which in this case was greater shading. This study suggests that a strong competition for nutrients and moisture is expected if the trees are inter planted with field crops and it was due to allocation of large proportion of fine roots near the soil surface. It is therefore necessary to alleviate the expected root competition between the trees and crop by pruning of their lateral roots lying in the surface soil horizon.

Silvipastoral/hortipastoral systems

In a study on seasonal variation in ground vegetation under 3-year-old shisham and chah plantations spaced at 5 x 5 m, it was noted that the above ground live shoot ranged from 312-673 g m^{-2} under semal, 185-551 g m^{-2} under chah and 118-550 g m^{-2} in open field. Peak value was recorded in September. Standing dead biomass increased in February, thereafter it declined in subsequent months due to transfer into litter. The ground vegetation were represented by *Cynodon dactylon, Panicum repens, Setaria glauca, Eleucene indica, Desmostachya cynosuroides* and *Dicanthium annualatum* (Chaturvedi and Das, 2002c).

Seasonal variation in biomass, net primary productivity and turnover of dry matter of herbaceous layer under *Albizia lebbeck* (L.) Benth and *Populus deltoides* G-3 Marsh plantations and in an adjacent open field was studied at Pusa in Bihar, eastern India (Das *et al.*, 2008). Live shoot biomass attained peak values of 682, 665 and 512 g m^{-2}, respectively under *A. lebbeck, P. deltoides* and in open field in September. Corresponding values of net abovegroundproduction were 749, 692 and 678 g m^{-2} $year^{-1}$. The belowground net primary production of herbaceous layer was also highest under *A. lebbeck* (89 g m^{-2} $year^{-1}$) followed by *P. deltoides* (110 g m^{-2} $year^{-1}$) and open field (67 g m^{-2} $year^{-1}$). Shoots shared 86 to 91% of annual total net primary production. This study concludes that the herbaceous vegetation under monsoon subtropical condition shows seasonality in biomass production. Under the plantations, herbaceous biomass attains higher net primary production, which can be utilized to sustain secondary production of livestock. Thus the herbaceous vegetation under the monsoon subtropical condition exhibited a strong seasonality in biomass production. Under the plantations, herbaceous biomass achieved higher net primary production compared to the open field.

Studies were conducted to evaluate the seasonal variation in dry matter and nutrient concentration in fodder of two commonly associated tree species in the agroforestry systems viz., kachnar and bamboo (Jha and Chaturvedi, 1990). The growing season between March-June was assessed to be suitable period for harvesting the bamboo leaves to feed the animals. The suitable period for kachnar appeared to be July/August, when the crude protein was moderate and tannin content was lowest. Among the two tree fodders, bamboo was better than kachnar with greater value of nutrients and crude protein.

In another study, effects of cutting intervals on the forage yield from guinea, napier, dinanath and orchard grasses under a 6-year-old plantation of chah showed that guinea, napier and dinanath produced maximum dry matter and crude protein yield when they were cut at 60 days interval, whereas orchard grass showed maximum dry matter and crude protein yield at 40 days of cutting intervals. Crude protein, nitrogen free extract, ether extract, total ash and insoluble ash showed the increasing trend with increasing cutting intervals, while crude fibre showed reverse trend (Das and Chaturvedi, 2004).

Bamboo [*Bambusa bambos* (L.) Voss] Plantations (Kanta Bans): *Culm Recruitment, Dry Matter Dynamics and Carbon flux*

Culm recruitment, standing bamboo biomass, net production, litterfall, floor littermass and carbon flux were estimated at the ages 3, 4 and 5 years of bamboo plantation (Das and Chaturvedi, 2006a). The number of culms produced per clump varied from 22-31. The recruitment to culm population varied between 16 and 21% and shoot mortality from 3-5% per year. Net accumulation of green culms between 3rd and 4th year was 675 and between 4th and 5th year 975 ha^{-1}. The total biomass was 170.8 t ha^{-1} in the 3-year-old to 257.3 t ha^{-1} in the 5-year-old plantation with an average per cent contribution of culms (65%), branches (21%) and leaves (4%). Total net primary production (NPP) ranged between 44.6 t ha^{-1} (4-year-old) and 60.5 t ha^{-1} (5-year-old), of which aboveground net production was 41.2 to 55.6 t ha^{-1} (between 3-4, and 4-5 years, respectively). Short-lived components (leaves and roots) contributed about 20% of net production of bamboo. Annual litterfall increased with the age of the plantation from 5.06 t ha^{-1} (3-year-old) to 6.94 t ha^{-1} (5-year-old). Percentage contribution of leaf and woody litter varied between 89-91% and 9-11%, respectively. The bulk of litterfall (80-89%) occurred during winter season (November to February). The mean litter mass on the bamboo plantation floor ranged from 3.2 to 4.0 t ha^{-1}. Turnover rate and time for litter ranged from 0.77-0.80 year^{-1} and 1.25-1.30 years, respectively.

The total carbon storage in standing biomass of 4- and 5-year-old bamboo plantation varied from 83.3 to 103.8 t ha^{-1}, of which 68% was distributed in stem, 21% in branch and 9% in belowground biomass. Moreover, carbon dioxide, the most important greenhouse gas is converted in to structural carbon in the plantation. Since most of the products of bamboo stems are long lasting, the accumulated carbon would remain sequestered for a long time. A substantial amount of carbon varying from 2.3 (4-year-old) to 2.8 t ha^{-1} year^{-1} (5-year-old) was transferred to soil through litterfall and root. The nutrient concentration in the various biomass components of the bamboo was generally in the order of leaf > rhizome > root > branch>stem, with the nutrient elements in the order of N>K> P (Das and Chaturvedi, 2006b).The maximum amount of all nutrients was accumulated in the stems, followed by branches, rhizomes, leaves and roots. Considerable reductions (55-62%) in concentration of nutrients (N, P and K) in leaves occurred during senescence. The uptake of nutrients by bamboo with and without adjustment for internal recycling has been calculated separately. Annual transfer of nutrients through litters and roots to the soil was 49.2-58.7 N, 2.7-3.1 P and 40.4-48.9 K kg ha^{-1} year^{-1}. Annual turnover rate of nutrients on the floor of different aged bamboo plantation ranged from 69 to 93%.

Litter decomposition

Study on decomposition rates of leaf biomass of six multipurpose tree species viz. *Populus deltoides, Dalbergia sissoo, Bombax ceiba, Acacia auriculiformis,*

Acacia lenticularis and *Tectona grandis* showed that leaf of all the tree species showed higher rate of decomposition during early stages (Das *et al.*, 2004). However, after initial one month of leaf biomass disappearance in case of *Acacia auriculiformis* was the slowest (4.4%). The highest rate of decomposition was observed in the leaf of *A. lenticularis* followed by *T. grandis, P. deltoides, B. ceiba, D. sissoo* and *A. auriculiformis*. Complete disappearance of leaf biomass in *A. lenticularis, T. grandis, P. deltoides, B. ceiba* and *D. sissoo* was observed in 5th, 7th, 8th, 9th and 10th month of placement in soil, respectively. Only 91.8% weight loss had occurred in *A. auriculiformis* by the end of 10th month. Nitrogen and phosphorous passed through the net immobilization phase in all the species except *A. lenticularis.* However, no immobilization was observed in K in all the species.

From this study, it may be concluded that incorporation of nutrient-rich *A. lenticularis* leaves, which result in quick loss of dry matter, has important implications for synchronizing nutrient release with crop uptake in low input cropping systems.

Collection and evaluation trials

On the basis of the growth performances of twelve tree species in the compact blocks and twenty two in the arboretum, Safed siris (*Albizia procera*), Eucalyptus (*Eucalyptus tereticornis*), Arjun (*Terminalia arjuna*), Chah (*Acacia lenticularis*), Mahogony (*Sweatenia mahagony*), Sagwan (*Tectona grandis*), Deshi Semal (*Bombax ceiba*), Kadamb (*Anthocephalus cadamba*), Poplar (*Populus deltoides*) and Bankat (*Wendlandia exserta*) have been found promising MPTs for cultivation in north Bihar.

Reclamation forestry

The twelve tree species studied in general were effective in bringing about improvement in the soil properties as reflected by the changes in pH, EC, organic carbon, available nitrogen, phosphorus and potassium (Das *et al.*, 2007). Higher available N, P_2O_5 and K_2O as well as higher organic carbon percentage were noted under canopy of *Albizia procera* followed by *Leucaena leucocephala*. The pH and EC were lowest under *Albizia procera* and changes were observed from 8.7 to 7.7 and 0.76 to 0.40 dS m^{-1}, respectively in the span of 12 years. *Albizia procera* produced maximum litter fall (13.95 t ha^{-1}year^{-1}) followed by *Leucaena leucocephala* (13.25 t ha^{-1}year^{-1}) plantation. Percentage contribution of leaf and non-leaf litter to the total litter of different tree species varied between 90 to 97% and 3 to10%, respectively. The nutrients returned through litter fall followed the order N (60-319 kg/ha/yr)>K(45-120 kg/ha/yr)>P(7-36 kg/ha/yr) in all the tree species and was helpful in reclamation of high pH soils.

In another study conducted under 18-year-old forest tree plantations of six multipurpose tree species, the highest DTPA extractable Cu (2.87 ppm) was observed under *E. tereticornis* and the value was at par with *A. procera* and *P. pinnata* plantations (Laik *et al.*, 2008). DTPA-Fe was also maximum (11.81 ppm) under *E. tereticornis*. *T. arjuna*, *P. pinnata* and *E. tereticronis* plantations showed higher and similar improvement of DTPA-Zn in soil. On the other hand *P. pinnata*, *A. procera* and *T. arjuna* showed higher available Mn. *E. tereticornis* produced maximum litter fall (8.46 Mg ha^{-1}) which was at par with *A. procera* (8.32 Mg ha^{-1}) and *T. arjuna* (8.20 Mg ha^{-1}). The data in respect of timber volume indicated that the *A. procera* proved to be the best followed by *T. arjuna*. By and large, considering timber production, carbon sequestration and improvement in soil fertility, *A. procera* was found to be the best among the tested tree species. This study, thus, suggested that inclusion of *A. procera* in agroforestry system may be the viable option for natural resource management among the selected tree species and could sustain long-term soil productivity in calcareous recent alluvial soil of Bihar.

Residual effect of some tree species on soil productivity and fertility

An attempt was made to study the productivity and fertility of cropped marginal lands as influenced by residual effect of some tree species (Das *et al.*, 2008). A field experiment was carried out comprising four tree species viz., *Leucaena leucocephala, Sesbania grandiflora, Wendlandia exserta, Eucalyptus tereticornis* and three crops viz., wheat, mustard and maize. The tree species were felled after 8 years of their plantation and then three crops were grown for judging the residual effect of tree species. Considerable variation in biomass of components for different tree species was noticed. A positive residual effect of all the tree species on crop yield was recorded and the increase in yield was in between 10-31% more as compared to control. Electrical conductivity and pH were lower under *L. leucocephala* and decreased from 0.81 to 0.43 dSm^{-1} and 8.6 to 7.8, respectively. Available N, P_2O_5, K_2O and organic carbon were also noted relatively high with *L. leucocephala* followed by *S. grandiflora*. The total aboveground biomass was maximum in *L. leucocephala* (245.4 t ha^{-1}) followed by *S. grandiflora* (182.9 t ha^{-1}), *W. exserta* (143.7 t ha^{-1}) and *E. tereticornis* (94.5 t ha^{-1}). The accumulation of wood biomass (bole+branch) per tree after 8 years of growth was found in the order of *S. grandiflora* (97.0%) > *L. leucocephala* (96.5%)>*W. exserta* (95.5%) > *E. tereticornis* (93.1%).

From this study, it is concluded that sodic soils could be reclaimed/ameliorated considerably through afforestation during a reasonable periods of growth of the legumes. *L. leucocephala* and *S. grandiflora* have been found to be superior in soil reclamation and subsequent crop production due to relatively good build-up soil fertility.

Nitrogen supplementation by leguminous tree leaves for rice cultivation

In an attempt to reduce the dose of inorganic N for rice cultivation, the leaves of leguminous tree species were incorporated into soil (Kumar *et al.*, 2008). Leaves of six leguminous tree species namely, Safed Siris [*Albizia procera* (Roxb.) Benth], Subabul [*Leucaena leucocephala* (Lam.) de Wit.], Chah (*Acacia lenticularis* L.), Shisham (*Dalbergia sissoo* Roxb.), Karanj [*Pongamia pinnata* (L.) Pierre] and Chakundi (*Cassia seamia* Lamk.) were applied @ 5 ton ha^{-1} along with half dose of nitrogen and full dose of P and K. The treatments were compared with the plots receiving full dose of N, P and K and with control. *L. leucocephala* treated plots showed maximum grain yield (44.28 q ha^{-1}) which was at par with plots receiving *A. procera* leaves. These two treatments along with *P. pinnata* leaf applied plots had significantly more grain yield than that of the plots which was applied with recommended doses of N, P and K. *D. sissoo* treated plots had the lowest grain and straw yield and N uptake, among the leaf treated plots and the plots with fertilizer 100% NPK. Yield attributing characters like height of plant, panicle length, number of grains per panicle and thousand grain weights were also maximum in *L. leucocephala* incorporated plots. Incorporation of leaves of *A. procera* and *P. pinnata* could increase thousand grain weights over the absolute chemical fertilizer treated plot. Hence green manuring with the leaves of *L. leucocephala, A. procera* and *P. pinnata* at 5 ton ha^{-1} can save half of the dose of inorganic N for rice cultivation.

Electricity generation through poplar and bamboo

The energy content of stems, branches, roots and litter was determined and these data were used to estimate energy storage, net energy fixation and energy transfer within poplar (*Populus deltoides* G-3 Marsh) plantations of two ages (Das and Chaturvedi, 2008). The energy stored in the above-ground tree components from 2131.87 ha (5-year-old) and 1002.88 ha (7-year-old) or in the above-ground net annual production from 3924.15 ha (5-year-old) and 2386.37 ha (7-year-old) of poplar plantations is sufficient to operate a 5 MW generating station for 1 year. Above ground biomass and net production from1 ha of 5-year-old and 7-year-old poplar plantations is sufficient to meet the energy need of an average household in eastern India for 8.5 and 18.0 years and 4.6 and 7.6 years, respectively.

Biomass accumulation and stored energy content of *Bambusa bambos* (L.) Voss were assessed for different age classes at Pusa on the northwest alluvial plain of Bihar (Jha and Das, 2008). Assuming an electricity consumption of 210 kwh per month per household, then 1 hectare of this bamboo species stores enough energy for approximately 330, 400, or 500 households after 3, 4 or 5 years. To operate a 50 MW power station continuously for a year would require 9,011,412 GJ year^{-1}. This would require the energy stored in 3-year-old culms from some 6,000 ha, or from 4-year-old culms 5,000 ha, and from 5-year-old culm from over 4,000 ha a year, for a continuous supply employing a culm-selection silvicultural system.

Agricultural studies on raising of til with Shisham

Dalbergia sissoo Roxb. (Shisham) is important nitrogen fixing multipurpose tree species. It is widely grown in Bihar particularly in Gangetic plain (Jha, 1995). While raising of *Dalbergia sissoo*, the interspaces between the rows of plants are not utilized. The light canopy of *D. sissoo* favours to grow Til (*Sesamum indicum*) upto 5 years. Krishna-258 gave maximum yield 4.79 q/ha followed by AVT-18 (2.98 q/ha) and AVT-7 (2.76 q/ha). Krishna-258 was significantly superior in grain yield.

Studies on fuel production with Subabul (K8)

Leucaena locally known as subabul is one of the fast growing tropical tree species and capable of producing high biomass (Jha, 1992). As regards the yield of *Leucaena*, 21.6t/ha/year fuel wood were estimated, when row to row and plant to plant distance was 2m. The pit size was 45cm x 45cm x 45cm. Benefit: cost ratio of subabul (*Leucaena*) is 1 : 4.45. It indicated that a farmer will get benefit of Rs. 4.45 on the investment of Rs. 1/- only.

Growth and green biomass Production of five fast growing tree species in munger

High human and animal population has exerted immense pressure on existing vegetative cover for fuel, fodder, wood, lumber and timber. Rural energy needs are mostly met by conventional sources such as fire wood, cow dung and agricultural residues (Jha, 1999a). Shortage of fuel wood, fodder and small timber magnifies the need to identify and select promising tree species for the region. The results indicated that the maximum biomass production and net growth i.e. height, collar diameter, and diameter at general breast height among five tree species, the most promising one to provide fuel wood and fodder in this region was *Leucaena leucocephala* followed by *Cassia siamea, Gliricidia sepium, Albizia lebbeck* and *Pongamia pinnata*.

Evaluation of young nitrogen fixing tree species with and without *Rhizobium* in South Central Bihar

Degraded or marginal uplands with grass covers are nitrogen deficient. This suggests that low status of soil nitrogen is a common problem that would limit plant growth (Jha, 1999b). To minimize or eliminate the need for nitrogen fertilization of marginal grassland areas, the planting of tree legumes seems ideal. This may attributed to the fact that most agroforestry legumes are capable of fixing atmospheric nitrogen through their specific *Rhizobium*. To evaluate the growth performance of young nitrogen fixing tree species with and without *Rhizobia* in South Central Bihar, an experiment was conducted at K.V.K. Munger, Bihar, India. The results clearly indicate that growth attributes are very much influenced by *Rhizobium* species.

Silvi-horticultural studies on rising of papaya with Shisham

The intercropping experiment of five varieties of *Carica papaya* namely Pusa giant, Pusa majesty, Pusadelicious, Pusa dwarf and Pusa nanha with *Dalbergia sissoo* was carried out at K.V.K. Munger (Jha, 1998). The pit size for *Carica papaya* was 60 cm^3 and for *Dalbergia sissoo* was 45 cm^3 and row to row and plant to plant distance was 4m for both species. The maximum average yield was recorded in Pusa giant 120 kg/plant followed by Pusa majesty110 kg/plant, Pusa delicious 95 kg/plant, Pusa dwarf 60 kg/plant and Pusa Nanha 40 kg/plant upto initial two years. The initial five years of yield of juvenile wood of *Dalbergia sissoo* was estimated as10 m^3/ha/year.

Silvi-horticultural studies on rising of Bhindi with Chakundi

Bund planting of *Cassia siamea* was done with a pit size of 45 m^3 and plant to plant distance was 4m (Jha, 1997). The variety of *Abelmoschus esculentus* i.e. *Pusa Pravani Kranti* was seeded at the rate of 8 kg/ha. The recommended dose of FYM (200 q), N (120 kg), Phosphorus (60 kg) and Potash (60 kg) were applied. The row to row spacing was 50 cm and plant to plant spacing was 45 cm. The yield of Bhindi (*Abelmoschus esculentus*) i.e. *Pusa Pravani Kranti* was recorded 95 q/ha and yield of juvenile wood of *Cassia siamea* was15 m^3/ha/year on the same unit of land.

Agri-silvicultural studies on raising of Maize with Chakundi

Cassia siamea trees attain required basal girth and height at the end of the third year and can be harvested. Each tree yields a pole of size 3 – 3.5 m length, 28 cm basal girth and 14 cm top girth (over bark) (Jha,1996). No perceptible reduction in maize yield was noticed up to third year. The average grain yield of maize amounted to 1.1 t /ha/year and stover yield 4.5 t/ha/year. The net income from the tree component (Poles and fuel wood) amounted to Rs. 23,935/-. This is in addition to the net income from agricultural crop which amounted to Rs. 10,150/ha. The average annual net income from the model amounts to Rs. 23,935/ha/year and NPV of model works out to Rs. 27,360/-.

Screening of multipurpose tree species for dry land farming of Munger

The mean annual increment of wood of thirteen-year-old tree revealed that *Albizia procera* produces 34.98 m^3 wood per year on hectare basis followed by *Cassia siamea* (27.99), *Dalbergia sissoo* (20.64), *Terminalia arjuna* (18.02), *Tectona grandis* (16.18), *Gmelina arborea* (16.06), *Pongamia pinnata* (12.41) and *Dalbergia latifolia* (5.77). The mean annual increment of wood of *Dalbergia sissoo, Terminalia arjuna, Tectona grandis* and *Gmelina arborea* were at par to each other; however *Albizia procera* and *Cassia siamea* were significantly superior to all other species (Jha, 2006a).

To select suitable clones of poplar of agro-climatic zone Iii-A

The mean annual increment of four-year-old *Populus deltoides* clones showed that S13 C11 clone produces maximum wood 9.02 m^3/year on hectare basis followed by RD-01 (6.34), 23-N (6.12), UD-88 (5.86), A-194 (5.62), 41-N (5.55), 43-N (5.50) and 40-N (5.40). Wood produced by S13 C11 clone was significantly superior to all other clones (Jha 2006b).

Carbon sequestration by different tree species after thirteen years of plantation

The data on carbon sequestration potential for thirteen-year-old tree species revealed that *Albizia procera* store maximum bole carbon 18.67 t/hectare/year followed by *Cassia siamea* (16.69), *Dalbergia sissoo* (12.45), *Tectona grandis* (9.76), *Terminalia arjuna* (9.46), *Pongamia pinnata* (6.78), *Gmelina arborea* (5.68) and *Dalbergia latifolia* (4.070). Bole carbon stored by *Albizia procera, Cassia siamea* and *Dalbergia sissoo* were at par to each other but significantly superior to all other species (Jha, 2006c).

Culm dynamics, carbon sequestration and potential for electrical

Energy from lathi bans plantation in south-eastern Bihar, India

Out of many forms of renewable energy, bio-energy could solve the village energy needs through distributed generation to a great extent. An attempt is made to analyse carbon capture and carbon credit generation from bamboo plantations (Jha and Das, 2007). The total energy stored as biomass by a 5-year-old plantation of *Dendrocalamus strictus* was 2.375 TJ ha^{-1} and that by 7-year-old plantation was 3.579 TJ ha^{-1}. Energy capture efficiency for the above ground and total clump vegetation after 5 and 7 years was 1.32 percent, 1.41 per cent, and 1.74 per cent and 1.87 per cent, respectively. The total carbon storage in standing biomass of 5-year-old and 7-year-old bamboo plantations varied from 56.75 to 84.51 Mg ha^{-1}, of which 48 per cent was distributed in stem, 14 per cent in branches and 25 per cent in below ground biomass. The total biomass was 118.23 Mg ha^{-1} in the 5-year-old and 176.31 Mg ha^{-1} in the 7th-year-old plantation. The number of culms produced per clump varied from 58 to 88. The recruitment to culms population varied between 20 and 21 per cent and shoot mortality between 4 and 5 per cent per year. Net accumulation of green culms between 4th and 5th year was 2,160, and between 6th and 7th year 3477 ha-1. Energy stored in biomass of 5-year-old culms from 9,283 ha, or biomass of 7-year-old culms from 6,253 ha is sufficient to operate a 50 MW generating station for one year. Total energy production in 5- and 7-year-old culms from 1 ha of *D. strictus* is sufficient to meet energy needs (in terms of electrical energy) of an average house hold of Bihar, India for 21.5 and 31.3 years, respectively.

Biomass production, carbon sequestration and electrical energy from safed siris plantations of Bihar

In this study biomass production, energy stored and net energy fixed at two sites of plantations of *Albizia procera* (Roxb.) Benth in Bihar were estimated at 9-year-old trees (Jha and Das, 2008). Total biomass and net production averaged 210.62 Mg ha^{-1} and 28.5 Mg $ha^{-1}year^{-1}$, respectively. The total energy stored as biomass (up to 9 years) and fixed during last one year were 4.575 TJ ha^{-1} and 0.624 TJ ha^{-1} $year^{-1}$, respectively. Energy capture efficiency was 1.95 and 2.29% of photo-synthetically active radiation, respectively, for the above ground trees and total tree stand. Energy stored in the biomass of the above ground trees from 2392 ha or in the net annual above ground production from 16,971 ha is sufficient to operate 50 MW generating station for one year. Total biomass and net production from 1 ha of *Albizia procera* (Roxb.) benth plantation are sufficient to meet the energy needs in terms of electrical energy of an average household of Bihar for 74.46 years and 10.6 years, respectively. Bole carbon was about 52.04% of total carbon storage of plantation.

Genotype X environment interaction of selected clones of poplar in Bihar

On the basis of mean performance, the high yielding clones for each agro climatic zones of Bihar were superior (Jha, 2007). However, the consistency in their overall stability was identified for different locations. Clones 25-N, 26-N and RD-01 were superior for agro climatic region of Pusa site. It showed consistency in performance i.e. higher in juvenile wood production and overall stability. RD-01 was also better for Basudeopur site for juvenile wood volume as it had significantly higher mean values than the grand mean at both the site. It stood first in pool mean, showed consistency in performance and overall stability for juvenile wood production. Other clones for this site were A-194, S4C2 and 41-N; whereas the clones S7C15, S4C2, A-194, RD-01, 40-N and UD-88 were better for Sabour site as it showed consistency in performance in yield (Juvenile wood) which were superior on pooled basis with overall stability for this trait. Based on principal component analysis to conclude as to which clones are the stable with respect to the traits this analysis revealed that the clones studies, 40-N, 43-N, A-194, G-48, 38-N, UD-88, 26-N, 22-N, L-34/82, 42-N and L-49/82 are stable for all three ecological regimes of Bihar. Thus, these clones may be recommended for cultivation under these different site conditions. Therefore, these results conclusively demonstrated that these clones should be introgressed in the on going silvicultural programme of the Bihar state for their evaluation in achieving high wood production in the state.

Future Projections

1. Nutrient dynamics and soil fertility management for sustainable yield.
2. Tree crown and root management studies to optimise growth and production of trees and crops under agro-forestry.
3. Effect of environmental factor, viz., light, temperature, relative humidity, soil moisture on productivity of trees and crops.
4. Bio-economic modelling of system interactions for investment and decision-making.
5. Tree nursery propagation and multiplication techniques for mass production of superior planting stock.
6. To standardize micro proliferation technique for mass propagation of local bamboo species.
7. Training extension activities and transfer of technology to end-users.

Constraints

1. Farmers, in general, are reluctant to grow trees on their farm-lands owing to long gestation period in getting returns from trees.
2. Rigid legal laws restricting harvesting, transporting and sale of trees.
3. Lack of assured financial support for popularizing agroforestry.
4. Non-availability of assured market and support price for agroforestry produce.
5. Lack of mechanized processing unit.
6. Small and marginal farmers rarely adopt agroforestry practices except for few trees on the field bunds. However, some large and medium farmers prefer to cultivate trees due to shortages of labourers and surplus nature of landholdings.

References

Ali, M.S., Das, D.K., Chakraborty, A. K., Sattar, A. 2006. Performance of intercrops under bamboo based agroforestry system in Bihar. J. Agromet. 8 (2): 266-268.

Chaturvedi, O.P., Das, D.K., Kumar, R. 2005. A comparative study of the rooting depth of four agroforestry tree species interplanted on boundary of wheat crop field in Bihar. Inter. J. Ecol. Env. Sci. 31: 49-52.

Chaturvedi, O.P., Das, D.K. 2002a. Effect of bund trees on soil fertility and yield of crops. Range Mgmt. Agrofor. 23 (2): 90-94.

Chaturvedi, O.P., Das, D.K. 2002b. Studies on rooting patterns of 5-year-old important agroforestry tree species in north Bihar, India. For. Trees and Live. 12: 329-339.

Chaturvedi, O.P., Das, D.K. 2002c. Influence of seasons and tree plantations on biomass and primary productivity of herbaceous vegetation in eastern India. J. Trop. For. 18 (II & III): 11-22.

Chaturvedi, O. P., Das, D. K., Chakraborty, A. K. 2008. Biomass Production and Root Distribution of *Acacia lenticularis* (L.) Willd under an Agri-silvicultural System in North-West Alluvial Plain of Bihar. Indian J. Agrofor.10(1): 24-29.

Chaturvedi, O.P., Jha, M.K. 1998. Crop production and economics under *Litchi chinensis* Sonn. plantation across 1 to 9 year age series in north Bihar, India. Inter. Tree Crop J. 9: 159-168.

Chaturvedi, O.P., Pandey, I.B. 2001. Yield and economics of *Populus deltoides* G_3 Marsh based intercropping system in Eastern India. For. Trees and Livel. 11: 207-216.

Das, D.K., Ali, M. S., Mandal, M. P., Kumar, R. 2008. Residual effect of some tree species on soil productivity and fertility in north-west alluvial plain in Bihar. Indian J. For. 31(2): 187-192.

Das, D.K., Chakraborty, A. K., Chaturvedi, O. P., Kumar, R. 2004. Growth, biomass production and leaf decomposition pattern of some agroforestry trees. Range Mgmt. Agrofor. 25 (1): 30-36.

Das, D.K., Chaturvedi, O. P. 2004. Effect of cutting intervals on the grass production and quality under *Acacia lenticularis* based silvipastoral system. J. Trop. For. 22 (I & II): 36-39.

Das, D.K., Chaturvedi, O. P. 2005a. Structure and function of *Populus deltoides* agroforestry systems in eastern India: 1. Dry matter dynamics. Agrofor. Syst. 65: 215-221.

Das, D.K., Chaturvedi, O. P. 2005b. Structure and function of *Populus deltoides* agroforestry systems in eastern India: 2. Nutrient dynamics. Agrofor. Syst.65: 223-230.

Das, D.K., Chaturvedi, O. P. 2006a. *Bambusa bambos* (L.) Voss plantation in eastern India: I. Culm recruitment, dry matter dynamics and carbon flux. J. Bamboo and Ratt. 5 (1 & 2): 47-59.

Das, D.K., Chaturvedi, O. P. 2006b. *Bambusa bambos* (L.) Voss plantation in eastern India: II. Nutrient dynamics. J. Bamboo and Ratt. 5 (3 & 4): 105-116.

Das, D.K., Chaturvedi, O. P. 2008. Energy dynamics and bioenergy production of *Populus deltoides* G-3 Marsh plantation in eastern India. Biom. Bioen. (In press)

Das, D.K., Chaturvedi, O. P. 2008. Root biomass and distribution of five agroforestry tree species. Agrofor. Syst. 65: 223-230.

Das, D.K., Chaturvedi, O. P., Mandal, M. P., Kumar, R. 2008. Effect of tree plantations on biomass and primary productivity of herbaceous vegetation in Eastern India. Trop. Ecol. 49 (2): 95-101.

Jha, A.N., Chaturvedi, O.P. 1990. Seasonal variation in dry matter and nutrients in the leaves of *Bauhinia variegata*. L. and *Dendrocalamus strictus* Roxb. Range Mgmt. Agrofor. 11: 207-212.

Jha, R.K. 1992. Studies on fuel production with Leucaena leucocephala (K-8) in south eastern Bihar. Annual Report of Dry Land and Rainfed Research Sub Station, Munger.pp. 20-22.

Jha, R.K. 1995. Agricultural studies on raising of Sesamum indicum with Dalbergia sissoo. Range Mgmt. Agrofor. 16 (1): 109-111.

Jha, R.K. 1996. Agri-silvicultural studies on raising of maize with chakundi Annual Report of Dry Land and Rain fed Research Sub Station, Munger. pp.18-23.

Jha, R.K. 1997. Silvi-horticultural studies on raising of bhindi with chakundi. Annual Report of Dry Land and Rain fed Research Sub Station, Munger.pp.16-20.

Jha, R.K. 1998. Silvi-horticultural studies on raising of *Carica papaya* with *Dalbergia sissoo*. Annual Report of Dry Land and Rain fed Research Sub Station (Rabi), Munger.pp.14-17.

Jha, R.K. 1999a. Growth and green biomass production of five fast growing tree species with and without rhizobia in south eastern Bihar. Annual Report of Dry Land and Rain fed Research Sub Station (Rabi), Munger. pp. 8-9.

Jha, R.K. 1999b. Evaluation of young nitrogen fixing tree species with and without *rhizobium* in south eastern Bihar. J. Appl. Biol. 9 (2): 185-188.

Jha, R.K. 2006a. Screening of multipurpose tree species for dry land farming of munger. Annual Report of Dry Land and Rain fed Research Sub Station (Kharif), Munger. 10-11 pp.

Jha, R.K. 2006b. Carbon sequestration by different Clones of Poplar after four years of plantation in south eastern Bihar. Annual Report of Dry Land and Rain fed Research Sub Station (Rabi), Munger. 8-9 pp.

Jha, R.K. 2006c. Carbon sequestration by different tree species after thirteen years of plantation in South Eastern Bihar. Annual Report of Dry Land and Rain fed Research Sub Station (Rabi), Munger. 10-11 pp.

Jha, R.K. 2006d. Perspective landuse planning of forestry in Bihar, Directorate of Research, R.A.U., Pusa, Samastipur (Bihar)-848 125. 9-10 pp.

Jha, R. K. 2007. Genotype x Site interaction of selected clones of Poplar (Populus deltoids Bartr. Ex Marsh) in Bihar. Range Mgmt. Agrofor. Vol.28. Number-1, 57-62 pp.

Jha, R.K., Das, D. K. 2007. Culm dynamics, carbon sequestration and potential for electrical energy from *Dendrocalamus strictus* Nees plantation in South-eastern Bihar, India. J. Bamboo and Ratt. 6 (3 & 4): 183-191.

Jha, R.K., Das, D. K. 2008. Biomass production, carbon sequestration and electrical energy from *Albizia procera* (Roxb.) Benth plantations of Bihar. Indian J. Power River Vall. Dev. 58 (1&2): 13-15.

Jha, R.K., Das, D.K. 2008. Biomass and bioenergy from *Bambusa bambos (L) Voss, Besch.* Plantation of Bihar, India. For. Trees Liveli. (In press)

Kumar, P., Das, D.K., Laik R., Chaturvedi, O. P. 2008. Nitrogen Supplementation by Leguminous Tree Leaves for Rice Cultivation. Indian J. Agrofor. (Submitted).

Laik, R., Koushlendra, K., Das, D.K. 2008. Organic carbon and nutrient build- up in a calciorthent soil under six forest tree species. For. Trees Liveli. (In press).

Prasad, R. 1995. Distribution and characterisation of salt affected soils in Bihar and their remedial measures. In Refresher Course on *Soil and Crop Management for Sustained Productivity.* B. Roy (ed), RAU, Pusa, Samastipur. 130-148 pp.

Yadav, C.P. 2000. Crop System Planning for Flood Prone Areas. In Refresher Course on *Agro-technologies in relation to ecological changes. A-Agricultural Development and Planning.* B.N. Verma (ed.), RAU, Pusa, Samastipur. 22-26 pp.

11

Poplar Based Agroforestry Models on Trans-Gangetic Plains of India

R.I.S. Gill, Baljit Singh, K.S. Sangha, G.P.S. Dhillon and Navneet Kaur

Abstract: *The Trans Gangetic Plains of India is a large fertile tract of north-west India where intensive rice-wheat based farming has led to soil and water problems. Agroforestry is among the viable options to overcome these problems. It has already shown positive effects such as increasing tree cover, augmenting wood needs of industries and ameliorating socio-economic status of farmers. Poplar based agroforestry models have been widely accepted by farmers. Site specific superior poplar clones have been developed for superiority, growth and resistance to pest and diseases. Farmers successfully grow the annual crops (sugarcane, turmeric), seasonal crops in winter (wheat, potato, mustard, mentha etc.) and summer season (cotton, moong, pearl millet, sorghum etc). The reduction in wheat grain yield varied from 10.2 to 46.6 per cent from first to sixth year as compared to open field. Selection of shade tolerant crop variety, appropriate sowing time, irrigation, seed and fertilizer dozes help to mitigate these losses. Wheat sown during first fortnight of November produced significantly higher grain yield than other sowing times. Wheat variety PBW 502 performed better amongst the six wheat varieties tested under poplar. Tree crop interactions influence the damage caused by insects and diseases. Cultural practices reduce the number of hibernating pupae of poplar leaf defoliators (PLD) which subsequently leads to lower incidence of these defoliators. Percent damaged leaves were more in fallow (5.44 - 60.53) and less in intercropped plantation (2.49 - 49.20). PLD adult emergence percentage was significantly less in intercropped and ploughed (49.5 and 67.5) than fallow (83.0) plantations. The litter production was* *highest (5.94 t ha^{-1}) at six year and lowest at one year (0.39 t ha^{-1}) age of poplar plantation. The concentration of nutrients (N, P, K and Ca) in the litter decreased whereas nutrient return increased with increase in plantation age. The total quantity of 20.1 t ha^{-1} leaf litter and 176, 21.7, 133 and 368 kg ha^{-1} N, P, K and Ca, respectively were returned through litterfall in six years. Zinc deficiency in poplar plantations appeared in light textured soils and can be corrected by soil application or spray of zinc sulphate. The comparative economics of poplar based agroforestry model in block and boundary plantation revealed that these were 2.8 and 1.6 times more profitable respectively than rice-wheat rotation.*

Introduction

Trans Gangetic Plain comprises of irrigated areas of Punjab, Haryana and Rajasthan. Annual rain fall in this region varies from 360-890 mm. Rice-wheat cropping system predominantly followed in the fertile plains of the country has ensured food security of the nation and enabled it to achieve the goal of food self sufficiency. But both rice and wheat are exhaustive crops, and this double cropping system can quickly deplete soil of its nutrient content. The cropping sequence that yields 7 t/ha of rice and 4 t/ha of wheat takes up more than 300 kg nitrogen (N), 30 kg phosphorus (P), and 300 kg potassium (K) per ha from the soil (Narang and Virmani, 2001). The productivity of this system has stagnated or even declined in certain areas due to depletion of native nutrient reserves, under-ground water table, emergence of multinutrient deficiencies and resurgence of insects and diseases. To compensate the loss in productivity, farmers have started applying greater doses of fertilizers, insecticides, fungicides, weedicide, etc. Such indiscriminate use of chemicals has further worsened the nutrient balance in soil-plant systems, besides increasing the pest incidence, cost of production and environmental problems.

Such over-exploitation has accentuated the need for diversification from this over exhaustive cropping system. One of the viable options is the adoption of agroforestry. Likewise, Chinese government has introduced agroforestry as one of different ecological agriculture methods recognizing its potential (Hildreth, 2008). The National Forest Policy (1988) of India envisages that a minimum of 20 per cent of plains should be under tree cover to maintain the ecological balance. However, the forest and tree cover in the state of Punjab, Haryana and Rajasthan is only 6.3, 6.6 and 7.1 per cent respectively, of the total geographical area. There is a little scope to increase the forest cover under existing cropping system, as it is not possible to divert the fertile arable lands to forest as such. The adoption of agroforestry in these regions could play an important role to fulfill the needs of food and wood from the same piece of land and to increase the area under tree cover.

Trees have always remained an integral part of farming system and the society of this region. Traditionally, farmers used to retain naturally growing trees like *Dalbergia sissoo, Acacia nilotica, A. catechu, Azadirachta indica* and *Prosopis juliflora* in their agricultural fields. Most of these trees are slow growing and incompatible with agriculture being having spreading crown. Thus with commercialization of agriculture, farmers shifted to fast growing tree species. Economic returns from fast growing tree species such as *Populus deltoides, Eucalyptus tereticornis* and *Melia composita* have encouraged farmers to adopt systematic agroforestry systems.

Existing agroforestry systems in the region

Scattered trees on farm

This practice is common in rainfed regions where no assured source of irrigation is available. Farmers usually retain the local trees like *Acacia catechu, A. nilotica var. indica, A. modesta, Butea monosperma, Dalbergia sissoo, Prosopis cineraria, Mangifera indica, Azadirachita indica*, etc., on their farms wherever they happen to regenerate naturally. These plants provide fuel and fodder during lean period and additional income. They are also helpful in soil and moisture conservation.

Boundary plantation

This is growing trees along the farm boundaries in single or double rows of trees on field bunds and irrigation channels. This practice is suitable to all categories of farmers especially the small farmers who can grow trees and supplement their agricultural income without diverting any land from agricultural operations. Moreover, the overall yield reduction of the system is small in boundary plantation (Rao *et al.*, 1998). Boundary row of trees should preferably be planted in north-south direction to minimize the adverse effect of shade on adjoining crops. Farmers usually prefer *P. deltoides, Eucalyptus tereticornis, Leucaena leucocephala, D. sissoo, Melia azedarach* and *M. composita* for boundary planting. Trees are grown alongside the usual seasonal crops cultivated by the farmers.

Block plantation

Another option for tree planting on farmlands is block planting of trees. Farmers having large land holding mostly adopt this arrangement of tree planting. Absentee land-lords and industrialists owning land near cities have taken up block planting of *P. deltoides, E. tereticornis* and *M. composita*. Block planting of trees is distributed over fairly productive agricultural lands to marginal lands and riverine belt.

Owing to its compatibility with prevailing cropping practices and potential economic benefits, poplar has been widely adopted by farmers for cultivation on farmlands along with crops. Poplar can be successfully planted on farm boundaries as well as in block plantation. Under well managed conditions, the block plantation of poplar recorded 49.6 m^3/ha/yr above ground biomass production in a short rotation of 6 years (Dhanda, 1999). In this chapter, various aspects of poplar such as breeding, agroforestry systems, nutrient dynamics, insect pest management and economics have been discussed.

Genetic improvement of poplar

Populus deltoides Bartr. Ex Marsh, a native of USA and one of the fast growing species of this genus, was introduced in India during 1950 in the tarai regions of Uttar Pradesh (U.P.) through U.P. State Forest Department. *P. deltoides*, also known as eastern cottonwood or poplar, has shown good performance in the areas lying north of 28^0 N latitude in the Indo Gangetic plains and to a lesser extent in the hills of north India. It thrives well on well-drained, nutrient rich and deep sandy loam soils of neutral pH in subtropical to temperate habitat. It is a light demanding, drought susceptible species and avoids water logging.

Introduction and breeding

Realizing the importance of poplars in modern plantation forestry, India became member of the International Poplar Commission in 1965 and constituted a National Poplar Commission with the objective of cultivating poplars extensively. Indigenous clones were also developed by many institutes mainly from the superior seedlings raised from open-pollinated seeds of available clones (mainly G-3, G-48 and D-121). However, some clones have also been developed by controlled pollination among various clones of *P. deltoides* and also with other *Populus* spp. (interspecific hybrid of *P. deltoides* x *P. ciliata*) by Forest Research Institute, Dehra Dun, Uttar Pradesh State Forest Department, WIMCO seedlings Limited, Rudarpur and Univeristy of Horticulture and Forestry (UHF), Solan. The clones were named “L” series of UP State Forest Department, Lal Kuan, “WSL” series of WIMCO Seedlings Limited and “U” series by UHF, Nauni.

Punjab Agricultural University (PAU) Ludhiana started genetic improvement work in late eighties of foregone century by introducing clones from various institutes of the country viz. U.P. State Forest Department, WIMCO Seedling Ltd., University of Horticulture and Forestry, Solan and Forest Research Institute, Dehra Dun. At present PAU Ludhiana is maintaining 250 clones in the clonal bank.

Clonal testing for growth and yield

Various nursery and field trials of *Populus* spp. have been conducted in many parts of the country. *P. deltoides* performed better than any other *Populus* species. There is high variability in the growth performance of many clones. Kumar *et al.* (1999) ranked 108 clones of *P. deltoides* on the basis of growth performance. Clones S_7C_8, 82-35-4, 113324 and G-48 were the four best clones at both the test sites. Clone G-3 was at 54th position. Singh *et al.* (2001) found significant variation in all growth parameters among 50 clones of *P. deltoides*. The expected genetic gain was 30.28 per cent for volume by selecting 5 best clones (40-N, UD 9116, 25-N, 63-N and UDH 1002). However, to keep broad genetic base they recommended clonal mixture by using five additional promising clones based on volume production.

A field trial conducted at two sites (Ludhiana and Bathinda) of Punjab concluded that clones G-48, L-313/85, L-39/84, L-154/84, and L-71/84 produced 35.3, 35.3, 33.9, 33.9, and 32.5 per cent higher wood volume, respectively than that of control (G-3) in central plain region (Ludhiana). Clones L-154/84, 113324 and L-188/84 performed better in semi-arid zone (Bathinda) with respective higher volume production of 223.7, 216.2 and 206.2 per cent over that of control (Table 1). On the basis of their growth and stem quality traits, the clones L-39/84, L-71/84, L-154/84, L-313/85, G-48 and L-154/84, L-188/84 and 113324 were released by Punjab Agricultural University, Ludhiana during 1996 for commercial cultivation for central plain region and semi-arid region, respectively.

Table 1. Mean volume production of different clones of *Populus deltoides* on two sites in Punjab (Age 4 years).

Clone	Volume production (m^3/tree)	
	Ludhiana (central plain region)	Bathinda (semi-arid region)
G-3 (Control)	0.221^{d}	0.0308^{d}
G-48	0.299^{a}	-
1467	0.223^{d}	0.0478^{cd}
L-188/84	0.280^{ab}	0.0943^{a}
L-71/84	0.293^{a}	0.0789^{a}
L-39/84	0.296^{a}	-
L-62/84	0.268^{bc}	-
L-313/85	0.299^{a}	0.0798^{ab}
L-200/84	0.284^{ab}	0.0618^{bc}
S_7C_8	0.295^{a}	0.0964^{a}
L-51/84	0.255^{c}	0.0801^{ab}
64/243-1	0.228^{d}	0.0701^{b}
113324	0.279^{ab}	0.0974^{a}
L-154/84	0.296^{a}	0.0997^{a}
110702	0.267^{bc}	0.0676^{bc}
$CD_{5\%}$	0.020	0.0213

On the basis of ten multi-location clonal trials established in each of the agro-climatic regions of Punjab, several clones have been identified owing to their superiority for wood volume (Table 2). Clone L-48/89 recorded 41 to 65 per cent volume superiority over control, whereas L-17/92 registered 28 to 40 per cent higher wood production than control. These clones are in pipe line for recommendation to tree growers.

Table 2. Superior poplar clones on basis of multi-location testing

Clone	Location	Age (yrs)	Volume (m^3)/ tree	Relative to control* (%)
L-48/89	Hambran (Ludhiana)	6	0.3534	164.9
L-48/89	PAU, Ludhiana	6	0.4811	141.0
72/58	Dhindsa (Jalandhar)	8	0.4270	118.3
L-17/92	Hambran (Ludhiana)	6	0.3011	140.5
L-17/92	PAU, Ludhiana	6	0.4400	128.9

Poplar based agroforestry system

Poplar is a preferred tree in agro forestry due to its straight bole, small crown and winter deciduous nature that allows winter cropping with only marginal negative effect on crop yields. Moreover, the ability of poplar root system to adapt to the wheat competition by distorting their root architecture is an important feature that improves complementarity between trees and crops (Cannel *et al.*, 1996). Poplar based agroforestry system is a big success in riverine tract of trans gangetic plain. Farmers can harvest and market it to the industry within 5-6 years in well-irrigated conditions. Increasing demand for poplar for the production of pulp, veneer and composite structural wood products has generated interest in their cultivation on farmland (Balatinecz *et al.*, 2001). The carbon sequestration potential of poplar based intercropping system is four times more than that reported for conventional agricultural fields (Thevathasan and Gordon, 2004). Adoption of poplar based agroforestry system is driven by economic incentives (as wood production is often more valuable or less labour intensive) and by national policies (to save natural forests from further deforestation) (Zomer *et al.*, 2007).

In the first two years of plantation any crop can be grown without much reduction (10-15 percent) in the yield. However during summer season, paddy is not recommended for planting with poplar as its shallow root system results in uprooting of trees (Dhanda *et al.*, 2008). Third year onwards, during the winter season, different crops like wheat, turmeric, oats, berseem etc. can be grown while during summer season, only fodder crops such as pearl millet, cowpea etc are recommended (Table 3).

Table 3. Crops that could be successfully grown under different aged poplar

Tree age	Summer	Winter
1.	Sugarcane, Cotton, Turmeric, Mentha, Mung, Ginger, Colocasia	Sugarcane, Wheat, Mustard, Potato, Marigold, Fenugreek, Fennel
2.	Sugarcane, Cotton, Turmeric, Mentha	Sugarcane, Wheat, Mustard, Mung, Ginger, Colocasia Barseem, Oats, Potato, Fenugreek, Fennel Marigold
3.	Turmeric, Sorghum, Pearl millet, Ginger	Wheat, Mustard, Barseem, Colocasia, Mentha, Sesbania Potato, Oats, Fenugreek, Fennel, Marigold
4.	Sorghum, Pearl millet, Sesbania	Wheat, Mustard, Barseem, Oats
5.	Not economical to grow crops	Wheat, Mustard, Barseem, Oats
6.	Not economical to grow crops	Wheat, Mustard, Barseem, Oats

(Dhanda *et al.* 2008)

Tree crop interface

There will be above- and below-ground interactions at the crop-tree interface when crops and trees are co-cultivated. Growing a crop between rows of trees can improve total productivity per unit land area. But trees reduce photosynthetically active radiation (PAR) and air temperature, and increase relative humidity (RH) in their understory. PAR reductions below the optimum for photosynthesis are detrimental to understory crop production. Conversely, decreased temperature and wind, and increased RH can reduce leaf-to-atmosphere vapour pressure differences and thereby promote photosynthesis (Hogg *et al.*, 2000). Trees also compete with crops for soil nutrients (Campbell *et al.*, 1994) and soil water (Burner and Mac Kown, 2005).

In an experiment to study relative effects of irradiance and soil water on productivity of alley cropping, it was observed that PAR had a greater effect on herbage specific leaf weight, leaf elongation rate, tillers plant^{-1}, mass tiller^{-1}, and total nonstructural carbohydrate concentration than soil water (Burner and Belesky, 2008). Thus PAR was the scarcer of the two resources according to the Sprengel-Liebig Law of the Minimum (van der Ploeg *et al.*, 1999) because irrigation generally failed to improve herbage productivity under intense shade. It was observed that economical yield of herbage should be possible if PAR is maintained at 40–50% of full sun, and this may be achieved by appropriate planting design, and management to increase solar penetration and rainfall reception to the alley. While irrigation probably would not be economical either for herbage or tree production, growing the trees in wider alleys would simultaneously increase PAR and soil water to the alley crop (Burner and Belesky, 2008). Many studies conducted in this region have reported reduction in yield of companion crops.

In an agroforestry system, the competition among different components often increase with increase in age of tree components and it leads to reduction of crop yield as compared to crop grown in open. In poplar wheat based agroforestry system, the percent yield reduction of crops increased with the age of trees and it varied from 10 to 46 per cent from first to six-year-old plantation (Table 4), (Gill *et al.* 2007).

Table 4. Wheat yield (PBW 343) under different aged poplar plantations

Age of poplar (years)	Height(m) yield (q/ha)	DBH(cm)	Grain reduction (%)	Yield
1.	8.3	7.4	44.0	10.2
2.	12.6	14.6	40.5	17.5
3.	15.7	17.3	36.7	25.3
4.	18.2	20.1	32.0	24.8
5.	18.9	21.2	30.7	37.5
6.	21.1	22.8	26.2	46.6

Control (under open conditions): 49.0 q/acre

Thus, overall the challenge for managers is to balance complementary and constraining factors to maintain an economically sustainable system to meet production goals.

Choice of poplar clone

Many studies have reported clonal variation in poplar for branching pattern, crown shape, date of leaf flush and leaf fall. These can be exploited to identify promising clones for agroforestry systems. Karnatak *et al.* (1994) observed significant differences among 30 poplar clones for leaf emergence. They identified five clones which flushed late and also shed leaves early for introduction into agroforestry systems.

Dhillon *et al.* (2003) categorized 40 poplar clones planted at PAU, Ludhiana on basis of leaf shedding and leaf flush pattern the clones into five groups as given in (Table 5).

Table 5. Different categories of clones on basis of leaf fall and leaf initiation

Clone/ Category	Date of		Number of
	Leaf fall	Leaf emergence	leafless days
Early leaf fall and early leaf initiation			
3167	December 15	March 22	97
S_7C_{20}	November 26	April 4	120
113520	December 20	April 9	110
Early leaf fall and late leaf initiation			
ST-124	December 6	May 1	146
2502	December 14	April 23	130
Late leaf fall and late leaf initiation			
PIP-201	January 8	April 16	98
PL-5	January 9	April 16	97
PIP-212	January 1	May 1	120
Late leaf fall and early leaf initiation			
S_7C_4	January 11	March 22	70
PIP-215	January 8	March 22	73
Intermediate leaf fall and leaf initiation			
Remaining 30 clones			

The spacing and direction of row influence the tree crop interaction greatly. In general, wider spacing under agroforestry systems reduces both below and above ground competition between trees and crops, allows the mechanized harvesting of crops with tractors or crop harvesters and improves the PAR that reaches under storey crops. Study conducted at PAU Ludhiana with three spacings and two row directions of poplar block plantation revealed that percent yield reduction of all the intercrops viz sorghum (*Sorghum vulgare*), cowpea (*Vigna unguiculata*), turmeric (*Curcuma longa*), wheat (*Triticum aestivum*), berseem (*Trifolium alexandrinum*), pearl millet (*Pennsetum typhoides)* and flowers declined significantly with the increase in spacing of trees (Gill *et al*., 2007). This may be attributed to significant differences in light intensity available to the under canopy crops during both summer and winter season. Comparing the two tree row directions, wheat grain yield was recorded to be more in N-S direction (28.7 q/ha) than E-W direction (26.8 q/ha) in all the spacing of poplar, especially wider inter-row spacing. The results are in conformity with Dhillon *et al*. (1984) where higher crop yield was observed in N-S orientation of tree rows.

In a comparative study of different densities of poplar in wheat based agroforestry system in Pakistan, it was observed that intercropping has no significant effect on wheat grain quality. However, it has negative effect on grain weight and size that can be minimized with the increase in tree spacing from 6 to 9 m between tree rows (Chaudhry *et al*., 2003). Likewise, eleven crops were screened for seven consecutive years under poplar planted at three spacing (5 x 4, 10 x 2.5, 15 x 2.5 m) at HAU, Hisar. Yield of all crops decreased with increasing age of poplar and increased with increasing poplar spacing. Nevertheless, a spacing of 10 m x 2.5 m seems to be the ideal spacing for getting optimum growth and yield of agricultural crops (Nandal and Hooda, 2005). In another study on poplar wheat based agroforestry system established at 8 x 3, 6 x 4, 5 x 5 and 4 x 6 m, the growth parameters of trees were significantly higher in the 8 x 3 m spacing compared with the other spacing treatments (Chauhan and Dhiman, 2007). Overall the wider inter row spacing (7-9 m) and N-S orientation of tree rows is recommended for poplar based agroforestry system.

Pruning and management

Pruning is essential for improving quality and quantity of useful timber. This helps to maintain the shape of tree and minimize both below and above ground competition for limited resources. Earlier pruning was recommended based on total height of tree. Such as no pruning during first year, pruning lower one third of tree height during 2nd and 3rd year, up to one half during 4th and 5th year and up to two third of tree height after 5th year (Sidhu *et al.*, 1990). However, recent studies have recommended only selective pruning of thick branches irrespective of tree height that compete with main tree stem (Hara, 2004). The below ground competition can be minimized by root pruning. Tree root systems can be partially managed via root pruning (Sudmeyer *et al.*, 2002; Woodall and Ward 2002) or barrier systems with varied success. The use of these strategies depends on the relative economic value of tree and crop products and the value put on secondary tree benefits. Such practices often increase the yield of crops and pastures in the competition zone but at the same time reduce the tree growth by 14-43 % as reported in various *Eucalyptus* species of W. Australia (Sudmeyer and Fluggee, 2005). Such strategies should cautiously be applied so that main component of system is not adversely affected. Thus in a poplar based agroforestry system shoot pruning should be restricted to thick branches only and root pruning should not be performed.

In a comparative study of neglected and managed poplar plantations, it was observed that with timely execution of various cultural and management practices (such as pruning, irrigation and fertilization) about six times higher productivity could be achieved in managed plantations compared to neglected plantations (Table 6), (Dhanda, 1999). Irrigation to poplar plantations is necessary to achieve high productivity from the trees. It should be applied at 7-10 days interval during high temperature of summers. Due to high biomass production of poplar, deficiencies of Zn in poplar plantations have been observed in Punjab on the coarse textured soils. Out of the different levels of Zn applied (through $ZnSO_4.7H_2O$) to soil, the application of 25 g plant^{-1} Zn to 1-year-old plants and 40 g plant^{-1} Zn to 2-year-old plants was found more effective than the lower doses for amelioration of Zn deficiency. In 4-year-old plants, the application of 30 kg ha^{-1} Zn in 3 m strip along tree rows completely corrected the deficiency. Deficiency can also be cured by

5-6 weekly sprays of 0.5 per cent $ZnSO_4.7H_2O$. In the ongoing experiments on nutrient requirement of poplar it has been observed that tree growth can be enhanced by applying the nutrients (N and P) to trees during different years at the appearance of new leaves and growing period of trees in addition to nutrients applied to intercrops.

Table 6:

Productivity	Ownership: Management / Care			
	Absentee Neglected	Resident (R) Neglected	Resident Good	R + NPKGood
Timber vol. (over bark) m^3	47.4	146.5	202.9	295.4
Timber wt. (tonnes)	37.8	119.3	165.5	240.1

Source: (Dhanda 1999)

Crop management

For moderation of competition between poplar and under crops, there is a need to adopt various cultural and management practices that can minimize competition among the different components. This includes the selection of an appropriate crop, its variety, time of sowing, seed rate, fertilizer rate etc. Such management practices vary with tree age, spacing, site conditions and marketing scenario of products.

Under an agri-silvicultural trial at PAU Ludhiana Gill *et al.* (2008) evaluated six wheat varieties with four sowing dates for three years and found that wheat cultivar PBW 502 out yielded the rest of wheat varieties over different time of sowings (Table 7). In PBW 502, highest grain yield (30.5 q/ha) was recorded when crop was sown on 10-11 Nov over the three years. All the varieties performed better when sown early compared to one-month delay. The various growth parameters like tiller height and number, spike length and 100 grain weight contributed significantly towards higher grain yield in early sown conditions (10-11 November) under poplar irrespective of the age of the plantation.

The emergence of poplar leaves during late February or early March cast shade to the late sown crop due to its delayed maturity and as a consequence it hampered the translocation of photosynthates from source to sink (Pannu and Dhillon, 1999). In addition, the incidence of bird damage was more in late sown variety as compared to early sown crop. Randhawa *et al.* (1992) reported that in the late sown crop the flowering period is shortened as the atmospheric temperature starts rising by the time the crop comes to flowering. During the leafless period, the impact of the canopy on direct and diffuse radiation is small (Douglas *et al.*, 2006). Therefore, the late sown crop is forced to flower and mature early resulting in for reduction of yield under late sown conditions.

Table 7. Effect of time of wheat sowing on the yield (q/ha) of 6 wheat varieties (Pooled mean of three years)

Variety	Time of sowing				Mean
	T1 (25-26 Oct)	T2 (10-11 Nov)	T3 (25-26 Nov)	T4 (10- 11 Dec)	
PBW 502	23.2	28.4	22.7	15.4	22.43
PBW 509	22.8	22.8	20.8	17.7	21.03
PDW 274	24.1	26.8	21.0	16.2	22.03
PBW 343	22.6	25.8	18.4	14.8	20.40
PBW 373	21.2	24.4	20.2	13.6	19.85
WH 542	21.7	24.2	19.6	14.5	20.00
Mean	22.60	25.40	20.45	15.37	

CD(p=0.05): T= 1.91, V = 1.24 T x V= 2.49
(Gill *et al.*, 2008)

Likewise, partially shade tolerant cultivars of sugarcane (Chauhan and Dhiman, 2003), mung bean (Pandey and Tewari, 2004), *Mentha arvensis* (Chauhan *et al.* 2000), *Curcuma longa* (Nandal and Hooda, 2005), rice (LuZhiYing *et al.*, 2006) *etc.* were screened under poplar. In addition, various aromatic crops viz. lemon grass (*Cymbopogon flexuosus*), citronella java (*C. wintrianus*), palmarosa (*C. martinii*) and Japanese mint (*Mentha arvensis*) were screened for five consecutive years under poplar. In general, herb and oil yield from a sole crop was higher compared to the intercropped yields. Significant reduction in herb and oil yield started after the third year in citronella java, palmarosa and Japanese mint, while that of lemon grass yield decreased slightly after the fourth year of the crop cycle. Lemon grass showed the best performance with respect to sustained herb and oil yield during entire growth period (Chauhan, 2000).

To make the poplar based agroforesty system more remunerative, the performance of various high value crops such as flowers were evaluated under three year old poplar plantation having three spacing [5 x 4 m, 8 x 2.5 m and 11 x 3 m, (diagonally paired at 2 m)]. Among the nine flowering plants screened, dimorphotheca, dianthus and calendula performed significantly better in closer row spacing (5 x 4 m) as compared to wider row spacing (8 x 2.5 and 11 x 3 m). On the other hand, growth of marigold plants was stunted under closer row spacing and significantly higher number of buds were observed in wider spacing of poplar (Dhanda *et al.*, 2007). In a preliminary study at PAU, Ludhiana, to screen various high value crops (such as fenugreek, celery, fennel, dillseed and coriander) under two and three years of poplar plantation, it was observed that the yield of fenugreek and fennel was comparatively higher than the other high value crops tested.

In order to find out optimum seed rate of wheat under poplar block plantation, three seed rates of wheat i.e. recommended (100 kg/ha), 25 per cent more (125 kg /ha) than recommended and 50 per cent more (150 kg/ha) than recommended were compared at PAU, Ludhiana. Fifteen per cent higher grain yield was recorded with 125 kg/ha seed rate. Six fertilizer doses for wheat crop and sorghum fodder sown under poplar were compared at PAU, Ludhiana. During the winter season, significant increase in grain yield of wheat was recorded when 50 per cent additional N (185 kg/ha) with recommended P (62 kg/ha) (grain yield increased from 3.57 to 3.99 t /ha). During summer season, the yield of sorghum fodder increased significantly with application of 50 per cent additional N along with 50 per cent more P (20.3 t/ha) over the recommended dose of these fertilizers (17.8 t/ha).

Nutrient dynamics

Nutrient cycling through tree species consists of uptake of nutrients from soil, their utilization in the tree biomass, retention in the biomass, return through litterfall/roots and changes in nutrient status of soil under poplar plantations. Nutrients tied up in tree biomass are removed through harvesting and this removal depends upon the intensity of harvesting. Unlike pure agriculture, a considerable amount of nutrients taken by trees is returned to soil through aboveground litterfall and belowground root turnover. Litterfall in poplar based agroforestry systems contains a large proportion of nutrients that maintains and even enhances the organic matter and nutrient status of underneath soil. The concentrations of nutrients in the added residues and their release on decomposition determine the transfer of nutrients to the soil and nutrient status of soil in agroforestry systems. The major parts in a system include uptake, retention, internal cycles, return of nutrients and their influence on nutrient status of soil.

Nutrient uptake and retention by trees

Nutrient uptake by trees is influenced by age and growth of plantation, soil type and climatic conditions of the site. Annual uptake of nutrients although commensurate with the annual crops but because a major portion of taken nutrients is returned to the soil or translocated within the tree, so the amount of nutrients retained is relatively small. The uptake of N, P and K was observed by Durai (1996) [through pruning of trees branches from 2 to 6 years old plantations and through whole tree harvesting at the age of 6 years] in poplar based agroforestry system having wheat-fodder rotation in Punjab. The uptake of nutrients was observed The total removal of K (539.1 kg/ha) was highest followed by N (499.1 kg/ha) and P (82.7 kg/ha). Nutrient uptake in poplar increases as the productivity of plantation increases.

Net annual retention of nutrients is the difference between total nutrient uptake and that returned to the soil in the form of litter and dead roots. The accumulation of nutrients usually increases linearly or exponentially during rapid growth and at a diminishing rate as the stand reaches maturity. Singh (1998) observed uptake, retention, retranslocation and return of N, P and K by poplar (G-3 and G-48 clones) in 8th year of their growth. The uptake of these nutrients was almost similar in both clones but retention and retranslocation of N was significantly higher in case of G-3 as compared to G-48. The total quantity of N in the wood was 40% higher in G-3 than G-48, indicating higher requirement of N for G-3 to produce the same quantity of wood than for G-48 (Figure 1). Tandon *et al.* 1991 observed that out of the total removal of 51.0 kg/ha N, 11.5 kg/ha P, 32.1 kg/ha K, 69.3 kg/ha Ca and 14.2 kg/ha Mg after 7 year growth of poplar stand, about 54.9% N, 56.5% P, 34.9% K, 67.2% Ca and 23.6% Mg was returned through litterfall.

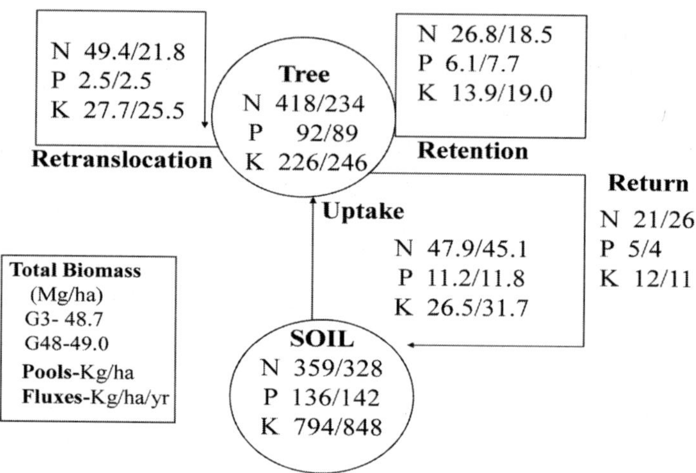

Fig. 1: Cycling of nutrients in two poplar clones(G3/G48)at 8 years of age
Source: Singh (1998)

Nutrient return through litterfall and roots

Litterfall addition leads to recycling of nutrients taken by the tree species from different soil layers, thus, enriching the nutrient status of the surface soil layer. In an agroforestry system, the litterfall production varies from the pure tree stands because of the response of trees to different cultural and agronomical practices given to the crops like irrigation, fertilizers, etc.

Poplar is a winter deciduous tree and sheds major portion of its leaves during November and December. Litter production and nutrient return through poplar litterfall varies with age of trees, intercrops, spacing and other management practices (Singh *et al.*, 1989; Mohsin *et al.*, 1996; Singh *et al.*, 2007). Litter production and nutrient return through litterfall were studied in a poplar (*P. deltoides*) plantation having pearlmillet (*Pennisetum americanum*)-wheat (*Triticum aestivum*) rotation as intercrops at Forestry research area at PAU Ludhiana. Litter production increased significantly with increase in the age of plantation; it was highest at six year age and lowest at one year. Nutrient return through litterfall was enhanced as the plantation aged. The total quantity of 20.1 t ha^{-1} leaf litter, and 176, 22, 133 and 368 kg ha^{-1} N, P, K and Ca, respectively were returned through litterfall in six years (Table 8). Increase in the growth of trees as well as its leafy biomass leads to increase in leaffall from the trees. The concentration of nutrients was highest in one year old plantation and lowest in six year old. Amongst the various nutrients, the concentration of Ca at various ages was highest (1.77-2.12%) and that of P the lowest (0.092-0.163%). Singh *et al.* (2007) also reported the highest concentration of Ca and the lowest of P amongst the macronutrients in poplar plantation having three spacings and two row directions. The total litterfall addition was significantly higher in 5 x 4m and 8 x 2.5 m spacings than 11 x 3m, paired (Table 9). Concentration of Ca and Fe was the highest among the different macro and micronutrients, respectively during different months. The return of different nutrients was significantly higher during November and December than other months. The total return of nutrients through litterfall was significantly higher in 5 x 4 m and 8 x 2.5 m spacings than 11 x 3 m (paired). The total return of Ca in different spacings was maximum followed by N amongst different macronutrients, whereas the return of P was minimum. Among the different micronutrients, the total return of Fe was maximum and that of Cu, minimum in the different spacing.

Table 8. Litterfall and nutrient return through litterfall at different ages of poplar plantation

Plantation age (years)	Litterfall (Mg ha^{-1}) and Nutrient (kg ha^{-1}) return				
	Litterfall	N	P	K	Ca
1.	0.39	4.88	0.62	3.55	8.25
2.	1.34	15.32	1.86	11.47	27.15
3.	2.77	27.13	3.69	21.01	53.68
4.	4.19	38.25	4.91	29.21	77.48
5.	5.37	43.81	5.29	32.72	96.74
6.	5.94	46.98	5.37	34.52	104.33
Total	20.1	176.4	21.7	132.5	367.6
CD (P=0.05)	0.44	3.93	0.57	1.96	6.08

Quantity of litterfall, its chemical composition, nutrient addition and changes in chemical constituents of soil were studied under agroforestry systems involving *Populus deltoides* and *Eucalyptus* hybrid trees with intercrops of *Cymbopogon martinii* and *C. flexuosus* in the tarai tract of Kumaon hills of Uttar Pradesh (Singh *et al.*, 1989). On an average, dry litter production of poplar was 5.0 kg/tree/year whereas of eucalypt was 1.5 kg/tree/year. Litter production increased with increase in age of trees. The addition of N, P and K through *P. deltoides* litter was 36.6, 91.6 and 70.0 per cent more than *Eucalyptus* hybrid litter, respectively which added 88.8 kg N, 43.6 kg P and 78.0 kg K as against 65 kg N, 24 kg P and 46 kg K by *Eucalyptus* hybrid per hectare annually. Mohsin *et al.* (1996) observed litterfall addition by poplar (*P. deltoides*) of different ages intercropped with *Mentha* and *Cymbopogan* species. The litterfall addition increased significantly as the age of poplar increased from two to seven years. The addition was higher in intercropped plantation as compared to pure poplar, which was attributed to the response of poplar to the various cultural operations, like weeding, hoeing, fertilization and irrigation given to the intercrops. Similarly, the addition of N, P and K increased with increase in age of poplar and addition was higher in intercropped than pure poplar. Sharma *et al.* (2000) assessed the leaf litter production and quantity of nutrients returned to the soil through litterfall under *Populus deltoides* based agroforestry system under irrigated condition in a boundary plantation. Litterfall was restricted upto 9 m from the tree line and total amount of 913.8 and 1291.8 kg/ha was estimated in 3 and 4 year old plantation, respectively. The return of nutrients through leaf litter into soil decreased in the order of Ca>N>Mg>K>P. It was estimated that highest Ca of 18.3-24.3 kg/ha, followed by N of 11.2-12.5 kg/ha, Mg of 7.8-10.0 kg/ha, K of 6.9-8.9 kg/ha and P of 2.45-2.59 kg/ha was returned to soil during winter leaf fall of poplar.

Table 9. Litterfall and nutrient addition through litterfall as influenced by spacings and months in five year old poplar plantation.

Treatments	Litterall	N	P	K	Ca	S	Zn	Fe	Mn	Cu
			kg ha^{-1}					g ha^{-1}		
a) Spacings										
5 x 4 m (S_1)	5621	53.1	7.30	41.7	103.5	15.68	473	7767	1053	115.6
8 x 2.5 m (S_2)	5479	51.0	6.67	41.3	98.5	15.46	462	7559	1017	115.3
11 x 3 m (S_3) (paired at 2 m)	4912	44.4	5.97	36.0	86.7	13.82	411	6854	919	103.1
C.D. (P=0.05)	228	1.22	0.43	3.36	5.33	1.11	14.4	339	68.7	5.94
b) Months										
September	569	5.80	0.75	4.51	11.40	1.85	47.6	833	116	12.37
October	686	7.39	0.90	5.54	13.14	2.49	58.4	1000	136	15.46
November	1721	14.92	2.45	12.72	29.81	4.10	152.5	2420	307	37.34
December	1768	16.36	1.89	12.66	31.62	4.92	143.4	2341	337	34.16
January	593	5.04	0.65	4.26	10.28	1.62	46.7	798	101	11.99
C.D. (P=0.05)	62.4	0.67	0.128	0.76	1.38	0.26	5.34	86.7	14.9	1.59

(Singh *et al.* 2007)

Roots, mainly fine roots have a significant contribution towards biomass addition. Intercropping, soil depth and different seasons influenced fine root addition in poplar (Singh, 1999). Intercropping of wheat increased the addition of fine roots and effect of intercultivation was more on 0-30 cm layer of soil as compared to lower layers. Fine root biomass addition was highest on surface layer of soil. Amongst the various seasons, fine root biomass addition was more in the month of October than July, which was attributed to retranslocation of photosynthates from leaves to roots (producing more biomass) whereas in July leaves development take place (leading to less investment of photosynthates in roots). Singh (1994) observed that fine root biomass of poplar increased with increase in spacing from

2x2 m to 6x6 m which was attributed to higher carbohydrates production and its translocation to roots with increase in spacing.

Influence on nutrient status of soil

Soil properties undergo a considerable change due to addition and subsequent decomposition of litterfall in poplar plantations. Soil organic carbon (OC), and concentration of available macronutrients (N, P and K) and micronutrients (Zn, Fe, Mn and Cu) in surface soil (0-15 cm depth) were determined in plantations having fodder {sorghum (*Sorghum bicolor*)/pearlmillet (*Pennisetum americanum*) in summer}-wheat (*Triticum aestivum*) (in winter) rotation throughout the poplar age and those having sugarcane (*Sachharum officinarum*) initially during two years and fodder-wheat rotation thereafter (Singh and Sharma 2007). Soil OC was significantly greater in older (6.83 g kg^{-1}) than the younger (5.35 g kg^{-1}) plantations. Available macronutrients in soil increased at successive sampling times. The average Zn concentration at final sampling was 17 per cent lower compared to initial sampling, whereas the other micronutrients tended to increase. The increase was higher in four year old plantations than one year due to higher inputs of organic matter. Poplar sheds its leaves in winter season from September to January and its decomposition increases under favourable moisture conditions and high temperature, thus, releasing nutrients into soil. Nutrient contents were high in 0-15 cm horizon and decreased with increasing depth of the soil positively due to nutrient cycling and surface enrichment by the biomass (Singh *et al.*, 1989). In 0-15 cm layer, there was 33.3 to 83.3 per cent enrichment in organic carbon, 38 to 69 per cent in available N, 3.4 to 32.8 per cent in available P and 5.9 to 24.3 per cent in available K over control. There was higher fertility build-up with respect to these elements under sole and intercropped poplar than *Eucalyptus* hybrid but both were significantly superior over control. Mohsin *et al.* (1996) observed that the available N, P and K contents of the soil under pure stands of poplar was higher than the soil of intercropped stands. Maximum amount of N (32.4-35.0%) and K (25.0-28.5%) was found in the superficial layer of the soil (0-15 cm) which decreased with increasing depth. Large part of P was accumulated in the soil at a depth of 15-30 cm. Bhardwaj *et al.* (2001) observed the chemical characterization of soils in high density poplar plantation. Organic carbon content at both the depths (0-30 and 30-60 cm) increased significantly with reduction in spacing (mean value of 1.43% at 60 x 60 cm and 0.59% at 120 x 120 cm spacing). Available N, P and K decreased with increasing depth and increased with increasing spacing. These were 248.3, 53.94 and 217.60 kg/ha in 60 x 60 cm spacing which increased to 309.3, 84.4, 275 kg/ha at 120 x 120 spacing respectively.

On the other hand, Lodhiyal and Lodhiyal (1997) observed that in a poplar plantation spaced at 5 x 3 m, the soil nutrient concentration decreased with plantation age, possibly as a consequence of nutrient depletion caused by ever increasing uptake by the vegetation (shrubs, herbs and trees) and leaching of soil nutrient from the plantations. Plantation raised during different years and having age of 1, 2, 3 and 4 years indicated that N, P and K concentration were 0.169, 0.015 and 0.068 per cent respectively in one year old plantation and these decreased to 0.149, 0.012 and 0.060 per cent, respectively in four years old plantation.

Insect pest management

Trees are susceptible to insect damage just like agricultural crops. The extent of damage however, depends on the stage at which they are infested and the nature of the insect. Trees are usually attacked by leaf eaters in nursery, termites and root grubs during the early establishment stage in the field, defoliators, sap-suckers, gall farmers, stem borers, pod or fruit borers and seed eaters, perpetually during the later stages. Potential benefits of agroforestry cannot be realized unless the pest problems are tackled effectively. The extent of insect damage in an agroforestry system is determined first by the primary interaction between the plant species and second by the interactions among components (tree, crop, soil and environment) of the system. The interactions among the components of a agroforestry system may have positive, negative or neutral effect on pests of trees and crops depending upon whether the pest's activity is enhanced, reduced or remains unaffected, respectively. They are regarded as negative when pest problems are increased in an agroforestry system as compared to a monoculture block plantation. A decrease in pest activity under agroforestry indicates a positive interaction, while no change in pest intensities between monoculture and agroforestry denotes neutral interaction from insect pest management point of view. Many climatic changes and management practices influence insect activity individually or collectively and directly or indirectly. The pest situation in an agroforestry system will be influenced by the degree of interaction between the components, the type of system and the composition of the plant communities in each component (Rao *et al.*, 2000).

An important area of research in agroforestry entomology is to study the effect of interactions among the components on insect dynamics. A comparison of insect abundance on a plant species in monoculture and in agroforestry will provide first hand information about the effect of co cultivation on insect dynamics. The mechanisms that govern pest populations in agroforestry are central to understanding the insect activities in these systems.

Insect pest problems in exotic poplars

Poplar is cultivated as boundary plantings, blocks and rows of trees in cropped areas and also intercropped with annual crops throughout this region. Large scale defoliation of poplar plantation by its major defoliators [*C. fulgurita* (Walker), *C. cupreata, C. restitura*] has been reported by various workers (Singh *et al.*, 1983; Sohi *et al.*, 1987; Singh *et al.*, 2005; Sangha *et al.*, 2007). These defoliators have now spread throughout the distributional range of *P. deltoides* in north-western India. In some cases, complete loss of leaves and defoliation render the plant susceptible to attack by other pests and diseases.

Eight insect species were identified as important pests of poplar in agroforestry system in north-western India (Singh *et al.*, 2004). Three species *C fulgurita, C. cupreata* and *Apriona cinerea* were ranked as major pests and five species were identified as potential pests (*Ascostis selenaria, Eucosma glaciata, Phalanta phalanta, Nodostoma water housie* and white grubs). In Punjab, *C. fulgurita* (Walker), *C. restitura* (Walker), *Asphadastis cryphomycha* (Meyrich), *Indarbela quadrinotala* are major pests, whereas, stem borer *Apriona cinerea*, leaf hopper, *Kusala salicis* (Ahmad) and *Phlanta phalanta* (defoliator) are pests of minor importance (Sangha *et al.*, 2007). The incidence of major and minor pests varied from 85 – 100 and 60-70 per cent, respectively.

Pest management: Monoculture vs agroforestry

The practices for the management of an agroforestry system need to be developed or modified with a view to minimizing insect pest incidence and injury. Depending on the behaviour of major insect pests in the system, agronomic practices (tillage, irrigation and intercropping) may be tailored to discourage the pests and favour the natural enemies.

Adult emergence of poplar leaf defoliator (PLD) (*Clostera* spp.) was significantly affected by intercropping, tillage and irrigation (Table 10). Minimum adult emergence (49.5 per cent) was recorded in poplar +wheat plot and maximum (83.0 per cent) in poplar+ fallow (unploughed). Poplar + wheat (49.5 per cent) and poplar + barseem (54.0 per cent) which were statistically at par with each other but showed significant less adult emergence compared to poplar + fallow (ploughed) (67.5 per cent) and poplar + fallow (unploughed)(83.0 per cent). Further, adult emergence in poplar +fallow (ploughed) (67.5 per cent) was significantly less than poplar + fallow (unploughed) (83.0 per cent) (Sangha 2004). To confirm the above trend, pupae of PLD were buried in soil at three depths (5 cm, 10 cm and 15 cm). No adult emergence occurred when the pupae were buried, irrespective of the depth at which they were buried.

Table 10. Effect of tillage, irrigation and intercropping of poplar on the adult emergence of *C fulgurita.*

Crop	Adult emergence (Per cent)
Poplar + Wheat	49.50 (44.69)
Poplar + Barseem	54.00 (47.29)
Poplar+ Fallow (ploughed)	67.50 (55.25)
Poplar+ Fallow (unploughed)	83.00 (66.10)
CD (p= 0.05)	(7.45)

Figures in parentheses are Arc sine values and Mean of four replications

Comparison of intercropped and fallow poplar plantations w.r.t. per cent damaged leaves, larvae and pupae per meter branch length revealed that intercropping significantly reduced the PLD incidence throughout the year. Mean per cent leaves damaged in fallow plantations (29.37) was more in comparison to intercropped plantations (21.49). Per cent leaves damaged during different months also varied significantly with maximum damage during October, 2003 (54.87 per cent) and minimum damage during March (3.90 per cent). Mean number of larvae per m branch length in intercropped plantations (12.36) were significantly lower than fallow plantation (16.5). The number of larvae also varied significantly during different months with maximum population during March (0.3). Similarly, mean number of pupae in fallow plantation (1.75) were significantly higher than intercropped plantation (0.86). Maximum average number of pupae per m branch length in fallow and intercropped plantation(3.13) was recorded in October which was significantly higher than all other months (Table 11).

Table 11. Per cent leaves damaged, no. of larvae and pupae per meter branch length of Clostera fulgurita in fallow and intercropped agroforestry plantations of *Populus deltoides* during 2003

Month	Per cent damaged leaves			No. of larvae / mt branch length			No. of pupae / mt branch length		
	Fallow	Inter cropped	Mean	Fallow	Inter cropped	Mean	Fallow	Inter cropped	Mean
March	5.44	2.49	3.96	0.57	0.02	0.30	0.07	0.05	0.62
April	8.33	2.89	5.61	8.12	4.92	6.52	0.52	0.27	0.40
May	11.49	3.70	7.59	11.12	8.42	9.77	0.55	0.30	0.42
June	24.12	17.52	20.82	17.57	13.60	15.58	1.62	0.95	1.28
July	46.67	25.52	36.10	23.87	18.77	21.32	2.50	1.20	1.85
August	38.43	21.60	30.02	18.63	13.63	16.13	1.50	0.86	1.18
September	30.07	29.45	29.76	18.55	15.77	17.16	2.25	0.92	1.58
October	60.53	49.20	54.87	32.85	22.47	27.66	4.12	2.15	3.13
November	39.24	41.03	40.13	17.26	13.63	15.45	2.68	1.08	1.88
Mean	29.37	21.49		16.50	12.36		1.75	0.86	

Factor	Per cent Leaves damaged	No. of larvae / mt branch length	No. of pupae / mt branch length
Cultural Practice	0.43	1.18	0.25
Month	0.91	2.51	0.53

Burying of pupae/leaf folds with pupae results in negligible adult emergence. This is further corroborated by low incidence of the PLD in intercropped plantations (also fallow plantations which were ploughed and kept clean of weeds and other undergrowth). Recommended cultural practices followed in raising intercrops in an poplar agroforestry system leads to damage/ burying of pupae/leaf folds particularly in winter crops (prior to hibernation as poplar is deciduous) and hence, the population build up in the subsequent year is less. Also, poplar raised along with crops gets proper care resulting in healthy and robust stand which is inherently less prone to defoliators attack. In case of fallow plantations, deep ploughing especially during winter (December) reduces the defoliator population as in the case of intercropped plantations.

Insects as influenced by Microclimatic changes

The microclimatic changes caused by trees in agroforestry systems include shading of under storey crops, increased humidity, reduced air and soil temperature, and decreased wind speed. The activity of shade loving insects is usually increased under trees. Lower temperatures and higher humidity under trees in warmer regions reduce desiccation of the soft bodied larvae and sap sucking insects like aphids and increase their activity.

Aphid incidence was more on wheat sown under poplar plantation compared to open sowing. Mean number of aphids/tiller were minimum on wheat variety PBW 502 and maximum on PBW 509 during both the years. The population did not vary significantly among the varieties and sowing times (Table 12, Gill *et al.*, 2008)

In nutshell, integrated pest management in agroforestry should rely on the following strategies: (1) selection off tree species based on pest problems and management opportunities (2) Selecting tolerant/resistant tree species (3) modification of agronomic practices such that they reduce populations of insects but favour natural enemies (4) Identification of potential biocontrol agents (5) use of environmentally safe chemicals and botanicals. All said, the field of agroforestry entomology is still virtually unexplored. The effect of co-cultivation of crops and trees on insect dynamics needs to be studied in depth in different agroforestry systems.

Table 12. Mean number of aphids/tiller on different wheat varieties under poplar block plantation during 2005 & 2006

Wheatvariety	Number of Aphids / tiller*											
	2005 (4 yr poplar plantation)						2006 (5 year poplar plantation)					
	Open	T1	T2	T3	T4	Mean	Open	T1	T2	T3	T4	Mean
PBW 502	0.50	0.66	1.13	0.87	1.30	1.17	0.40	1.88	1.29	2.12	1.41	1.16
PBW 509	0.60	1.38	1.68	1.13	2.40	1.54	0.50	1.80	1.70	0.75	1.32	1.41
PDW 274	1.10	1.51	1.08	1.01	0.79	1.33	0.70	0.55	1.43	1.01	0.87	1.29
PBW 343	0.50	1.01	1.47	0.82	1.17	1.25	0.50	0.94	2.13	1.20	0.87	1.23
PBW 373	0.70	1.15	1.06	1.86	1.16	1.28	1.10	1.02	1.10	1.05	0.87	1.14
WH 542	0.90	2.11	1.45	1.56	1.48	1.26	0.80	1.28	1.22	0.71	1.07	1.19
Mean	0.71	1.17	1.32	1.11	1.60		0.50	1.70	0.99	1.21	1.03	

LSD (p=0.05) Sowing Time – NS Varieties – NS Sowing Time x varieties - NS
Open vs intercropped- 0.35
Sowing times: T1 – Second fortnight of October; T2 – First fortnight of November; T3 – Second fortnight of November; T4 – First fortnight of December
No termite, armyworm and gram pod borer attack was observed during both years under poplar plantation
* Mean of five plants per plot and 10 tillers/ plant

Economics

Intensive surveys and research trials were conducted to validate the assumption that poplar based agroforestry system is economically viable as compared to wheat – rice cropping system,. The comparative economics has been worked out taking three representative situations viz. agriculture [rice-wheat rotation], agroforestry [poplar in block (500 trees/ha) + pearl millet fodder + wheat], agroforestry [poplar on boundary (140 trees/ha) + pearl millet fodder + wheat] (Table 13). The annuity value and BC ratio of agroforestry block plantation was highest among cropping systems screened (Gill *et al*., 2006). Although the agroforestry system on boundary is not economically viable compared to agroforestry in block, but for small farmers it can be a viable option. This land-use system is also capable of providing employment opportunities on farms. Sensitivity analyses indicate that the poplar based agroforestry system is not highly risky (Jain and Singh, 2000; Kareemulla *et al*., 2005; Gill *et al*., 2006). Poplar based system with sugarcane (*Saccharum officinarium)* for the first two or three years and then wheat + summer fodder (*Sorghum vulgare)* for the successive years was found more remunerative than wheat + summer fodder system (Kumar *et al*., 2004; Singh and Dhaliwal, 2005).

Table 13. Comparative economics of agriculture, forestry and agroforestry models along with sensitivity analysis

	Agriculture	Agroforestry	Agroforestry	Sensitivity Analysis	
	(Rice/Wheat)	(Block)	(Boundary)	(Block)	(Boundary)
Gross Returns*	473955	1383581	750198	944870	634998
Total (Variable) Costs*	194373	297192	209519	297192	209519
Net Returns*	309582	1086389	540679	647678	425479
PNW[#]	224718	630530	341498	382889	276471
AV[#]	51605	144796	78422	87927	63489
BCR[F]	2.88	4.14	3.32	2.91	2.88

1. *Figures are in $Rs.ha^{-1}$ at 2006-07 prices,
   [#] Figures in $Rs.ha^{-1}$ at 10 percent discount rate and
   Figures are (unit-less) at 10 percent discount rate.
2. *Poplar-wood price ($Rs.q^{-1}$) for Sensitivity Analysis – 366, 408 (>24"girth), 183, 204 (10"- 24" girth) and 61, 68 (<10" girth) i.e. 39% and 32% decrease respectively in case of Agroforestry (Block) and Agroforestry (Boundary) plantations.*

Future directions

The traditionally followed cropping system in the Trans Gangetic plain needs improvement to protect over-exploitation of our natural resources. Agroforestry seems to be an economically viable and ecologically sound option for crop diversification.

The following initiatives would be helpful for the large scale adoption of poplar based agroforestry system in this region:

- **Technical** – Sufficient information is available regarding tree crop cultural and management practices. Doubts have been raised regarding the benefits accrued from poplar based agroforestry. Contrarian reports exist in literature suggesting both beneficial and detrimental effects of wheat intercropping in poplar (Burgess *et al.* 1996). Further, even the sustainability of these systems has been questioned by some workers (Hsiung *et al.*, 1995; Hildreth, 2008; Xu *et al.*, 2006). The myth that fast growing tree species deplete water more rapidly has gained ground (especially eucalypts). Zomer *et al.* (2007) have tried to bust this myth by studying poplar based agroforestry system on irrigated lands in northern India (where about 10 per cent of the irrigated area is under poplar). They infer that the impact on annual irrigation requirement for the region is estimated to be minimal (1.6% increase), whereas the contribution of poplar trees to the local economy and farmer livelihoods is quite significant and well established. In view of above, comprehensive study involving all components of the system (tree, crop, soil, environment, insect, disease, management) need to be carried out to alleviate the doubts.
- **Market** - The overall economics of any system is governed by the marketing scenario of its components. Large fluctuations in poplar wood prices have been observed during the last 5-6 years. The market price (Rs./t) of poplar wood dropped from 4250 to 1530 from 1999 to 2004 and presently it is showing an upward trend (Rs. 5500/t in 2008). It has been observed that poplar + pearl millet + wheat based agroforestry system is economically more remunerative that wheat + rice system provided the price of poplar wood remains more than Rs. 2800/t. More tree crop combinations need to be tested to enhance the sustainability and economics of the agroforestry model. To popularize this system government needs to device some policy so that price of poplar wood remains stable.
- **Socio – political** – There is lot of scope for poplar based agroforestry system in this region. Foremost objective should be to establish commercial scale or working farm demonstration units in prospective areas and to encourage the farmers to adopt this system. Credit availability (low interest loans from NABARD or cooperative societies) and other incentivies are needed to provide impetus to adoption of agroforestry as it will solve twin purposes of enhanced tree cover and ecological stability of this region.

References

Balatinecz, J.J., Kretschmann, D.E., Leclercq, A. 2001. Achievements in the utilization of poplar wood guideposts for the future. For. Chron. 77:217–392.

Bhardwaj, S.D., Panwar, P., Gautam, S. 2001. Biomass production potential and nutrient dynamics of *Populus deltoides* under high density plantations. Indian For.127: 144-153.

Burgess, P.J., Stephens, W., Anderson, G., Durston, J.1996. Water use by a poplar-wheat agroforestry system. Aspects Appl. Biol. (44): 129-136.

Burner, D. M., Belesky, D. P. 2008. Relative effects of irrigation and intense shade on productivity of alley-cropped tall fescue herbage. Agrofor. Syst. 73: 127-139

Burner, D.M., MacKown, C.T. 2005. Herbage nitrogen recovery in a meadow and loblolly pine alley. Crop Sci. 45:1817–1825.

Campbell, C.D., Atkinson, D., Jarvis, P.G., Newbould, P.1994. Effects of nitrogen fertiliser on tree/pasture competition during the establishment phase of a silvopastoral system. Ann. Appl. Biol. 124:83–96.

Cannel, M.G.R., van Noordwijk, M., Ong, C.K. 1996. The central agroforestry hypothesis: The trees must acquire resources that the crop would not otherwise acquire. Agrofor. Syst. 34:27-31.

Chaudhry, A.K., Khan, G.S., Siddiqui, M.T., Akhtar, M., Aslam, Z. 2003. Physico-chemical characteristics of wheat variety Inqalab-91 under poplar (*Populus deltoides*) based agroforestry system. Pakistan J Agri. Sci. 40(1-2): 77-81.

Chauhan, H.S. 2000. Performance of poplar (*Populus deltoides*) based agroforestry system using aromatic crops. Indian J. Agrofor. 2:17-21.

Chauhan, H.S., Kalra, A., Mengi, N., Rajput, D.K., Patra, N.K., Singh, K. 2000. Performance of menthol mint (*Mentha arvensis*) genotypes to varying levels of nitrogen application under poplar based agroforestry system in Uttar Pradesh foot hills. J. Med. Arom. Plant Sci. 22(1B): 447-449.

Chauhan, V.K., Dhiman, R.C. 2003. Yield and quality of sugarcane under poplar (*Populus deltoides*)-based rainfed agroforestry. Indian J. Agric. Sci. 73(6): 343-344.

Chauhan, V.K., Dhiman, R.C. 2007. Atmospheric humidity and air temperature studies in wheat-poplar based agroforestry system. Indian For. 133(1): 73-78.

Dhanda, R.S.1999. Performance of farm forestry plantations in Punjab. Poplar case studies. In: Proc. National Symposium on role of Agri-Business Enterprises in Agroforestry and Wasteland Development. Association of Agri-plantation companies in India, New Delhi. pp. 48-60.

Dhanda, R.S., Gill, R.I.S., Singh, B., Kaur, N. 2008. Agroforestry models for crop diversification in Punjab plains. Dept of Forests and Wildlife Preservation, Punjab p. 24.

Dhanda, R.S., Kaur, N., Gill, R.I.S., Singh, B. 2007. Performance of flowers under block plantation of poplar (*Populus deltoides* Bartr.) in Punjab. Indian J Agrofor. 9(2): 77-80.

Dhillon, M.S., Singh, S., Atwal, A.S., Dhillon, G.S. 1984. Developing agri-silvicultural practices : studies on the shading effect of *Dalbergia sissoo* and *Acacia nilotica* on the yield of adjoining crops. Indian J. Ecol. 11: 249-253.

Dhillon, G.P.S., Singh, B., Sidhu, D.S. 2003. Genetic variation in leaf emergence and leaf shedding trend in *Populus deltoides.* Published in Absracts "National Symposium on Agroforestry in 21st century", Feb. 11-14, 2003. p. 84.

Douglas, G.B., Walcroft, A.S., Hurst, S.E., Potter, J.F., Foote, A.G., Fung, L.E., Edwards, W.R.N., Dijseel van den, C. 2006. Interactions between widely spaced young poplars (*Populus* spp.) and the understorey environment. Agrofor. Syst. 67:177-186.

Durai, V. 1996. Nutrient uptake through prunings and tree harvesting in poplar based agroforestry system in Punjab. M.Sc Thesis, PAU Ludhiana, p. 95.

Gill, R.I.S., Kaur, N., Singh, B., Khullar, V. 2006. Economic evaluation of poplar based agroforestry system in comparison to rice – wheat cropping system. National seminar on "Trees Outside Forests" at CII, Chandigarh. 25-26 April. p. 7.

Gill, R.I.S., Singh, B., Kaur, N., Dhillon, G.P.S., Sangha, K.S., Rattan, G.S. 2008. Wheat varieties and their time of sowing under poplar plantations for inclusion in the package of practices of Punjab Agricultural University, Ludhiana. Proceeding of Research Evaluation Committee of PAU.

Gill, R.I.S., Singh, B., Kaur, N., Luna, R.K.2007. Evaluation of crops in poplar plantation with three spacing in two row directions. Indian For. 133(2A): 45-57.

Hara, S.S. 2004. Agroforestry in Northern India: A unique success story. Paper presented at first world congress of agroforestry held at Orlando Florida, USA.

Hildreth, L.A. 2008. The economic impacts of agroforestry in the Northern Plains of China. Agrofor. Syst. 72: 119-126.

Hogg, E.H., Saugier, B., Pontailler, J.Y., Black, T.A., Chen, W., Hurdle, P.A., Wu, A. 2000. Responses of trembling aspen and hazelnut to vapour pressure deficit in boreal deciduous forest. Tree Physiol. 20:725–734.

Jain, S.K., Singh, P. 2000. Economic analysis of industrial agroforestry: poplar (*Populus deltoides*) in Uttar Pradesh (India). Agrofor. Syst. 49(3): 255-273.

Kareemulla, K., Rizvi, R.H., Kumar, K., Dwivedi, R.P., Singh, R. 2005. Poplar agroforestry systems of Western Uttar Pradesh in Northern India: a socio-economic analysis. For. Trees Liveli. 15(4): 375-381.

Karnatak, D.C., Khanna, P., Chandra, A.1994. Growth performance of some poplar clones. Indian J. For. 47: 314-318.

Kumar, D., Singh, N.B., Rawat, G.S., Srivastava, S.K., Mohan, D. 1999. Improvement of *Populus deltoides* in India I. - Present Status. Indian For. 125 (3): 245-263.

Kumar, R., Gupta, P.K., Gulati, A. 2004. Viable agroforestry models and their economics in Yamunanagar district of Haryana and Haridwar district of Uttaranchal. Indian For. 131-148.

Lodhiyal, L.S., Lodhiyal, N. 1997. Nutrient cycling and nutrient use efficiency in short rotation, high density central Himalayan tarai Poplar plantation. Ann. Bot. 79: 517-527.

Lu ZhiYing, Luo-Cheng Bin, Fang-Sheng Zou, Liang-Yong Hu. 2006. Effects of poplar-crop agroforestry patterns on rice yield and quality. J. Nanjing For., University Nat. Sci. 30(6): 83-86.

Mohsin, F., Singh, R.P., Singh, K. 1996. Nutrients cycling of poplar plantation in relation to stand age in agroforestry system. Indian J. For. 19: 302-310.

Nandal, D.P.S., Hooda, M.S. 2005. Production potential of some agricultural crops under different spacings of poplar. Indian J. Agrofor. 7(1): 16-20.

Narang, R.S., Virmani, S.M. 2001. Rice-Wheat Cropping Systems of the Indo-Gangetic Plain of India. Rice-Wheat Consortium Paper Series 11. p. 34.

Pandey, A., Tewari, S. K. 2004. Yield dynamics of mungbean (*Vigna radiata* L. Wilczek) varieties in poplar based agroforestry system. Indian J Agrofor. 6(2): 89-91.

Pannu, N. S., Dhillon, M. S.1999. Production potential of poplar-wheat based agroforestry system in relation to wheat varieties and their dates of sowing. Indian J. For. 22(3): 257-262.

Ramankutty, N., Foley, J.A. 1999. Estimating historical changes in land cover: North American croplands 1850 to 1992. Glob. Ecol. Biogeo. 8:381-396

Randhawa, A.S., Kahlon, P.S., Dhaliwal, H.W. 1992. Rate and duration of grain filling in wheat. Indian J. Gen. Plant Breed. 52: 161-163

Rao, M.R., Nair, P.K.R., Ong, C.K. 1998. Biophysical interactions in tropical agroforestry systems. Agrofor. Syst. 38: 3-50.

Rao, M.R., Singh, M.P., Day, R. 2000. Insect pest problems in tropical agroforestry systems: Contributory factors and strategies for management. Agrofor. Syst. 50: 243-277.

Sangha, K.S., Rattan, G.S., Luna, R.K., Singh, P., Makkar, G.S. 2007. Status and distribution of insects and diseases of poplar (*Populus deltoides*) Indian For. 133(2A): 166-177.

Sharma, N.K., Singh, H.P., Dadhwal, K.S. 2000. Nutrient returns through litterfall in *Populus deltoides* based agroforestry system. Indian For. 126: 295-299.

Sidhu, D.S., Hans, A.S., Dhanda, R.S.1990. Poplar Cultivation, PAU Publication. p.38.

Singh, A.P., Bhandari, R.S., Verma, T.D. 2004. Important insect pests of poplars in agroforestry and strategies for their management in northwestern India. Agrofor. Syst. 63: 15-26.

Singh, B. 1998. Biomass production and nutrient dynamics in three clones of *Populus deltoides* planted on Indo Gangetic plains. Plant Soil 203: 15-26.

Singh, B., Sharma, K.N.2007. Tree growth and nutrient status of soil in a poplar (*Populus deltoides* Bartr.)-based agroforestry system in Punjab, India. Agrofor. Syst. 70: 125-134.

Singh, B., Gill, R.I.S., Kaur, N. 2007. Litterfall and nutrient return in poplar plantation varying in row directions and spacings. Indian J. Agrofor. 9: 33-37.

Singh, J. 1999. Fine root dynamics in poplar plantation having different regimes of irrigation, N and intercultivation. M.Sc Thesis, PAU Ludhiana, pp. 145.

Singh, K., Chauhan, H.S., Rajput, D.K., Singh, D.V. 1989. Report of a 60 month study on litter production, changes in soil chemical properties and productivity under *Populus deltoides* and *Eucalyptus hybrid* interplanted with aromatic grasses. Agrofor. Syst. 9: 37-45.

Singh, N.B., Kumar, D., Rawat, G.S., Gupta, R.K., Singh, K., Negi, S.S. 2001. Clonal evaluation of poplar (*Populus deltoides* Bartr.) in eastern Uttar Pradesh. II- Estimates of genetic parameters infield testing. Indian For. 127 (2): 163-172.

Singh, P., Rawat, D.S., Misra, R.M., Massarat, F., Prasad, G., Tyagi, B.D.S.1983. Epidemic defoliation of poplars in central Tarai Forest division and its control. Indian For. 109 (9): 675-95.

Singh, S., Dhaliwal, H.S. 2005. An economic analysis of poplar based agroforestry systems in Punjab, India. Indian J. For. 28(4): 381-388.

Singh, V. 1994. Root distribution in poplar plantations in an arid regions in north-western India. Trop. Ecol. 35:105-113.

Sohi, A.S., Singh, H., Sandhu, G.S. 1987. *Clostera restitura* (Walker) (Lepidoptera: Notodontidae) a defoliator of Poplar (*Populus* sp.) in the Punjab and its chemical control. J. Tree Sci. 6 (1): 30-33.

Sudmeyer, R., Flugge. 2005. The economics of managing tree-crop competition in windbreak and alley systems. Aust. J. Expt. Agric. 45: 1403-1414.

Sudmeyer, R.A., Hall, D.J.M., Eastham, J., Adam, M. 2002. The tree crop interface: The effects of root pruning in south-western Australia. Aust. J. of Expt. Agric. 42: 763-772.

Tandon, V.N., Pandey, M.C., Sharma, D.C., Rawat, H.S. 1991. Biomass production and mineral cycling in plantations of *Populus deltoides* in UP. Indian For. 117: 596-608.

Thevathasan, N.V., Gordon, A.M. 2004. Ecology of tree intercropping systems in the North temperate region: Experiences from southern Ontario, Canada. Agrofor. Syst. 61: 257-268.

van der Ploeg, R.R., Bohm, W., Kirkham, M.B. 1999. On the origin of the theory of mineral nutrition of plants and the Law of the Minimum. Soil Sci. Soc. Am. Proc. 63:1055–1062.

Woodall, G.S., Ward, B.H. 2002. Soil water relations, crop production and root pruning of a belt of trees. Agric. Water Manage. 53: 153-169.

Xu Cui, Sui Peng, Xie Guang Hui, Gao WanSheng. 2006. Soil water effect and productivity in poplar and wheat-corn agroforestry system. Scientia Agricult. Sinica. 39(4): 758-763.

Zomer, R.J., Bossio, D.A., Trabucco, A., Yuanjie, L., Gupta, D. C., Singh, V.P. 2007. *Trees and water: Smallholder agroforestry on irrigated lands in Northern India.* Colombo, Sri Lanka: International Water Management Institute. 47p. (IWMI Research Report 122).

12

Silvopastoral System in the Andaman Islands

C.B. Pandey

Abstract: *Stall feeding to cattle is not practiced in the Andaman Islands since generations. In the morning, after milking, cattle are allowed to go for free range grazing in the nearby forest or coconut arecanut orchards. But, like mainland territory the cattle do not form herd. They are trained in such a way that come back home during evening where they are fed with concentrate before the milking. Coconut and arecanut orchards occupying more than 14000 ha land are second major land use after forest in the Islands. The forage production in the grazing lands are not sufficient to meet out the demand of huge livestock population particularly during the lean period, form January to April. An attempt has been made to know the effect of four N fertilizer levels (0, 60 80 and 120 kg ha^{-1} N) on net soil N mineralization rate (NMR) and soil moisture (SM), and shoot biomass (SB), FP, shoot biomass/ root biomass ratio (SB/RB), N concentration in SB, N uptake and nitrogen use efficiency (NUE) of three grasses [Guinea (Panicum maximum Jacq.), Para (Brachiaria mutica (Forssk) Stapf) and Hybrid-napier (Pennisetum purpureum Schumach.)] under three canopy positions [under canopy (UC, representing high shade), between canopy (BC, representing low shade) and open] of coconut trees (*Cocos *nucifera L.) in a coconut based silvopastoral system in the Island conditions. The study has revealed that the tree reduces light 59% under UC and 32 % under BC positions, but the N fertilizer levels increase NMR by 11 to 51 % under UC and 3 to 44 % under BC positions compared to the open. SM does not differ across the canopy positions. Under all situations, FP of all grasses declines under UC (47 to*

78%) and BC (18 to 32%) positions compared to the open; the decline is greater in Hybrid-napier than Guinea and Para grasses. Forage production of all grasses increases with N fertilizer increments under all canopy positions reaching 32 t ha^{-1} dry matters for Hybrid-napier at 120 kg ha^{-1} N in the open. Both Guinea and Para grasses outyield Hybrid-napier grass under UC but not under BC or in the open. N concentration in the forage (SB) also increases as N fertilizer level increases. These observations suggest that forage production under coconut palms can be increased by the application of N fertilizer with both Guinea and Para grasses being more productive than Hybrid-napier grass under the high shade. Where light conditions are better Hybrid-napier produce higher forage than the other species.

Introduction

Andaman group of islands were once thickly covered with lush green tropical rain forests. Systematic modern human intrusion into these peaceful islands started form 1858 when virgin forests were felled of convict settlement by Britishers. Since then ruthless deforestation and mindless settlements have not looked back and now owing to modern transportation facilities the flux of immigrants is ceaseless. Existing human population of the influx of settler unchecked islands, there has been a regular unchecked inflow of livestock population according to 1992 census were 1.6 lakhs of which 33% were cattle, 9.1% buffaloes, 35.1% goats and 22.8% pigs (Table 1). Current green fodder requirement, estimated on daily average of 25 kg/cattle head and 1500 tones annually. Based on the available empirical data high yielding exotic fodder grasses, grown under rain fed conditions, indicates feasibility of production of more than 60-150 t ha^{-1} yr^{-1} fresh fodder. Based on this information, at least 800 hectares of lands exclusively required for perennial fodder crops to meat the year round demand of green fodder (Sharma *et al.*, 1990).

Table 1. Livestock population in Andaman and Nicobar Islands

Year	Cattle	Buffaloes	Goats	Pigs	Total
1996	7388	8357	8615	14217	38576
1966	10131	8078	10131	21314	49654
1972	18685	8013	11602	38648	76948
1979	27401	9719	17578	21243	75941
1982	36560	11869	11581	96429	178039

Source: Basic statistics (2001)

Free range grazing in nearby forests was an ages old practice of animal feeding in the Islands. But, now State Forest Department in the Islands has declared the forests as protected and has put ban on the free range grazing. Animals, therefore, are forced to graze on the verges of metal roads. This gave an insight to examine forage production potentials of different grasses under coconut trees and management of their forage production through N-fertilization. It is well known that light intensity, soil moisture and nutrients are three major growth resources that determine productivity of silvopastoral and savanna ecosystems (Belsky, 1994; Dodd *et al.*, 2005; Burner and MacKown, 2006; Koukoura *et al.*, 2009). Trees in the systems create a horizontal structure above grasses (Sarmiento, 1984; Tournebize and Sinoquet, 1994) and modify availability of the growth resources (Vetaas, 1992; Polley *et al.*, 1997; Benavides *et al.*, 2009) and thereby influence growths and yields of the grasses by facilitative or competitive mechanisms (Belsky, 1994; Callaway and Walker, 1997). Facilitative effects of trees on grass growth and yield are reported to occur when soil moisture and nutrient are growth limiting in the open (Wilson and Wild, 1991; Callaway and Walker, 1997; Anderson *et al.*, 2001), particularly when the soil moisture is a primary limiting factor (Ludwig *et al.*, 2004). Such conditions occur generally in arid, semi-arid (Belsky, 1994; Ludwig *et al.*, 2004) and mediterranean conditions, and also in sub-humid and humid climates particularly during dry season (Gea-Izquierdo *et al.*, 2009; Koukoura *et al.*, 2009) (Pandey *et al.*, 2007a; Pandey and Singh, 2009). Under such conditions, trees reduce evaporative losses by moderating temperature (both soil and air) (Belsky *et al.*, 1989; Belsky, 1994; Baumeister and Callaway, 2006; Guevara-Escobar *et al.*, 2007; Pandey *et al.*, 2010), and makes soil moisture relatively greater (Belsky *et al.*, 1989; Cruz *et al.*, 1993; Belsky, 1994), which stimulates net soil N mineralization rate and makes availability of nitrogen (N) relatively higher (Rhoades, 1997; Pandey *et al.*, 2007 a). The high soil moisture and N availability increase grass growth (Wilson and Wild, 1991; Cruz *et al.*, 1993; Gea-Izquierdo *et al.*, 2009). Once soil moisture becomes non growth limiting in the open, the facilitative effects of trees turn competitive and tree-shade causes reduction in grass yield (Somarriba, 1988; Mordelet and Menaut, 1995; Cruz, 1996; Ludwig *et al.*, 2004). Competitive effects of trees on grass growth and yield are common under irrigated conditions, and also in sub-humid and humid climates particularly during wet season (Somarriba, 1988; Cruz, 1996; Power *et al.*, 2001; Ludwig *et al.*, 2004). There is evidence that N fertilization compensates for the competitive effects of tree-shading when soil moisture is not growth limiting in the open (Cruz, 1996; Koukoura *et al.*, 2009). But, most of the N fertilization studies have been done *ex-situ* under simulated conditions of tree-shades and soil moisture (Cruz, 1996; Koukoura *et al.*, 2009). Data are lacking how different amounts of N fertilization moderate the competitive effects of tree-shade under field conditions when soil moisture is not growth limiting in the open.

The present chapter deals with grasses and fodder trees found naturally, integration of grasses with coconut trees as a silvopastoral system and ways to manage forage yield in the Andaman islands.

Major land uses

Land utilization pattern in the islands is given in Table 2. Maximum land (86%) of the total geographical area is under forest. Homegardens (including coconut and arecanut orchards) covering around 24000 ha land are second largest land use system in the islands. A limited land is used as pasture and grazing land.

Table 2. Land utilization pattern in Andaman and Nicobar Islands

Land use	Area (in 000 ha)
Total geographical area	829.40
Forest	717.10
Area under non- agricultural use	6.90
Barren and uncultivated area	6.41
Land under trees, hills and valley	36.13
Permanent pasture	5.15
Cultivable waste	5.43
Fallow land	3.15
*Net area sown	17.10

Source: Basic statistics, Andaman and Nicobar Islands, 1996

Herbaceous species of the silvo-pastoral system

At present around 24000 ha land are under coconut and arecanut plantations which include homegardens as well as monocultural plantation. Herbaceous species and their relative cover under coconut and arecanut trees are given in Table 3. In protected condition, *Themeda trioandra, Cyrtococcum* accrescens, and *Axonopus compressus* were dominant grasses contributing 49.4% to the total grass cover. Grasses contributed 74.6%, legumes 10.4% and forbs 15.0% to the total cover in protected condition. Growth in grasses starts with onset of rains in May and peaks in September. In temperate climate unimodel growth pattern is reported to occur where with the onset of snowfall herbaceous vegetation is buried completely (Ram *et al*., 1989). Howevere, in sub-humid and semi-arid climate bimodal growth pattern is report to found in grasses where maximum growth occurs during rainy season from South – West monsoon and a small but significant growth peak is supported by rains from North – East monsoon (Pandey and Singh, 1992). On the contrary in the present study new sprouts as well as flowering in grasses is visible even after September indicating that new sprouting continues to occur round the year. Grasses cease to grow during dry months (February-April). Legumes contribute only 7% to the total vegetation cover. Like other tropical grasslands, lower cover of legumes in the grassland Andaman may be attributed to low concentration of phosphorus in the soil. Grazing reduced the cover the *Themeda triandra, Cytococcum* species and *Axonopus compressus* due to their greater palatability. Grazing facilitated legumes, *Mimosa pudica, Gleissapsis cristata and Smithia* by reducing vegetable cover and increasing sunlight. Pandey (1990) reported that reduced light under tall grasses suppresses growth of legums.

Table 3. Relative cover of grasslands species in Andaman under protected, grazed and shaded conditions.

Species	Grazed	Ungrazed	Shade
Grasses			
Axonopus compressus	-	11.4	14.6
Bothriochloa pestusa	2.6	-	-
Cenhrus cilliaris	-	1.8	-
Centotheca lappacea	-	-	2.8
Chirysopogon aciculatis	31.2	-	-
Cynodon dactylon	1.6	2.7	-
Cyrtococcum accrescens	-	18.4	21.6
Dichanthium annulatum	1.2	-	-
Digitaria longiiflora	-	0.3	3.8
Digitaria sanguinalis	4.3	-	-
Digitaria spp.	-	-	3.8
Eragrotis tenella	0.3	-	-
Echinochloa colonum	-	0.2	0.8
Eleusine indica	-	0.2	-
Eragrostis atrovirens	-	0.8	-
Eragrostis unioloides	1.4	0.5	-
Heteropogon contortus	-	2.1	-
Ischaemum indicum	22.8	9.6	-
Oplismenus compoaitus	-	-	11.3
Oplismenus burmanii	1.2	-	-
Paspalum spp.	-	-	4.3
Paspalum commersoni	-	7.4	-
Pennisetum polystschion	-	4.8	-
Polytrias amara	-	0.6	-
Sacciolepis indica	0.4	0.2	-
Themada triandra	4.5	19.6	7.2
Other speces	1.2	1.4	4.8
Legumes			
Alysicarpus vaginalis	3.2	3.2	0.6
Clitoria ternatea	-	0.1	-
Desmodium spp.	-	-	0.3
Desmodium triflorum	1.4	0.9	0.6
Geissapsis cristata	2.8	0.8	-
Mimosa pudica	2.1	1.2	1.8
Smithia sensitive	1.6	0.8	-
Other species	0.2	3.4	-
Forms			
Ageratum conyzoides	0.7	0.5	0.6
Amaranthus gracillis	-	0.1	-
Blumea lacera	0.9	-	-
Borreria articularis	0.6	0.1	-
Borroria latifolia	-	0.8	-

(*Contd.*)

Species	Grazed	Ungrazed	Shade
Cassia tora	0.8	-	-
Centella asistica	0.8	00.3.3	-
Chromolaena odorata	3.4	1.8	2.4
Cyperus spp.	0.3	0.1	-
Daedacanrhus suddruticosus	-	-	0.6
Dentellaepens	0.2	-	-
Eclipta prostrate	-	0.2	-
Elephantopus scaber	0.4	-	-
Fimbristylis ferruginea	0.2	0.3	-
Hyptis capitata	2.0	1.2	2.4
Lindernia anagallis	0.2	-	-
Mikania cordata	1.2	2.2	1.4
Murdannia nudiflora	0.2	0.4	-
Ocimum sanctum	0.2	0.1	-
Oldenllandia diffusa	-	0.4	-
Seleria spp.	-	0.1	-
Sida rhomboiidea	-	-	0.1
Spilanthes paniculata	-	0.1	-
Stachytarphata jamaicensis	1.8	-	-
Striga asiatica	-	0.1	-
Urena labata	1.2	0.4	1.2
Vernomia cinerea	-	0.1	0.3
Other species	1.0	-	14.8

Source: Sharma *et al.* 1990.

Net primary production of natural grasses under different grazing conditions are given in Table 4. Maximum forage production occurs under ungrazed condition and minimum under grazed condition. Under coconut plantations partial shade occurs after 25 years, which is raise at the spacing of 7.5 × 7.5m. Fodder trees/ shrubs suitable for Andaman and Nicobar conditions are given in Table 5.

Table 4. Net primary production different grazing conditions

Grazing condition	Net primary production (t ha $^{-1}$yr $^{-1}$)
Grazed	5.27
Ungrazed	10.60
Ungrazed shaded	7.30

Source: Sharma *et al.*, 1990

Table 5. Fodder tree species of Andaman and their preferential use*

Species	Tree parts used			Animal's preference			
	Leaf	Seed	Fruits	Elephant	Goat	Cattle	Pig
A. gonezious	U	-	U	U	-	-	U
A. incise	-	-	U	-	-	-	U
Acacia andamanica	U	-	-	-	U	-	-
Acacia farnesiana	U	-	U	-	-	U	-
Artocarpus chaplasha	U	-	-	U	-	U	-
Azadirachta indica	U	-	-	-	U	-	-
Bauhinia variegate	U	-	-	-	U	U	-
Breidelia vitis-idaea	U	-	-	-	U	U	-
Barringtonia asiatica	U	-	-	-	U	U	-
Bauhinia acuminate	U	-	-	-	U	U	-
Bombax ceiba	-	U	-	-	-	U	-
Bridelia retusa	U	-		-	-	U	-
Cordia rothi	-	-	U	-	U	-	-
Desmodium umbellatum	U	-	-	-	-	U	-
Dracaena angustifolia	U	-	-	-	U	U	-
Erythrina variegate	U	-	-	-	-	-	-
Ficus religiosa	U	-	-	U	U	U	-
Ficus rumphii	U	-	-	U	-	U	-
Ficus hispida	U	-	-	U	-	-	-
Flacourtia indica	-	-	U	-	U	U	-
Garcinia cowa	U	-	-	-	U	-	-
Gliricidia sepium	U	-	-	-	-	-	-
Lxora arborea	U	-	-	-	-	-	-
L. monopetala	U	-	-	-	-	-	-
Lannea coromandelica	U	-	-	U	U	-	-
Litsea glutinosa	U	-	-	-	U	U	-
Murraya paniculata	U	-	-	-	U	U	-
Phoenix sylvestris	U	-	-	-	U	U	-
Plachchonella longipetiolata	U	-	-	U	U	-	-
Pongamia pinnata	U	-	-	-	U	-	-
Pterospermum acerifolium	U	-	-	-	U	-	-
Rhizophora apiculata	-	-	-	-	-	-	U
Sesbania grandiflora	U	-	U	-	-	U	-
Ziziphus mauritiana	U	-	-	-	U	-	-

*U: denotes tree part used

Source: Sharma *et al.* 1990

Strategies for increasing fodder productions

Topography, climate, stray cattle, non- availability of root slips of required during dry paucity of irrigation facilities during dry months are constraints which hinder required production of forage in the islands. In order to make the island self sufficient in forage production following steps are required to be taken.

Management of coconut plantation / homegardens as a silvopastoral system

Three grasses [Guinea (*Panicum maximum* Jacq.), Para (*Brachiaria mutica* (Forssk) Stapf) and Hybrid-napier (*Pennisetum purpureum* Schumach.)] have been tested under three canopy positions (under canopy (UC), between canopy (BC) and open) of coconut (*Cocos nucifera* L.) trees, for four N fertilizer levels (0, 60 80 and 120 kg ha^{-1}). The area under the trees between 0.5 to 2.5 m from the trunk is designated as 'under canopy' and the area in the centre of 4 coconut trees is designated as 'between canopy' position. Guinea, Para and Hybrid-napier grasses are transplanted in separate plots at a spacing of 50 x 50 cm. Nitrogen is applied in all lots in 4 equal doses after cutting of the grasses. Muriate of potash (K, 80 kg ha^{-1}) and single super phosphate (P, 100 kg ha^{-1}) are applied to all plots once with onset of growth in May in all grasses. Dimensions of coconut trees are: height = 11.9 ± 0.1 m, diameter at breast height = 22.8 ± 0.6 cm, canopy radius = 3.6 ± 0.1 m.

Light intensity is found to be reduced as much as 59 % under UC and 32 % under BC positions than in the open (Tables 6). The light condition under UC position is described as high shade, and under BC position it is also described as low shade in discussion section. SM levels do not differ significantly across the canopy positions (Tables 7). NMR is higher under UC (11 to 51 %) and BC (3 to 44 %) positions than in the open (Table 8). The NMR is correlated to the N fertilizer levels under all canopy positions (Table 9).

Table 6. Light (foot-candle) availability to different grasses in different canopy positions in a Coconut based silvopastoral system at South Andaman Island of India

*Grass	Under canopy				Between canopy				Open			
	Nitrogen (kg ha^{-1}yr^{-1})				Nitrogen (kg ha^{-1}yr^{-1})				Nitrogen (kg ha^{-1}yr^{-1})			
	0	60	80	120	0	60	80	120	0	60	80	120
Guinea	$^{a}183^{x}$ ± 5.42	$^{a}185^{x}$ ± 6.29	$^{a}184^{x}$ ± 8.34	$^{a}184^{x}$ ± 4.28	$^{b}304^{x}$ ± 3.56	$^{b}305^{x}$ ± 2.38	$^{b}306^{x}$ ± 7.13	$^{b}305^{x}$ ± 8.43	$^{c}450^{x}$ ± 7.95	$^{c}450^{x}$± ± 6.92	$^{c}449^{x}$ ± 7.15	$^{c}456^{x}$ ± 6.02
Para	$^{a}183^{x}$ ± 4.42	$^{a}185^{x}$ ± 6.29	$^{a}184^{x}$ ± 5.34	$^{a}184^{x}$ ± 3.28	$^{b}305^{x}$ ± 4.61	$^{b}305^{x}$ ± 5.76	$^{b}305^{x}$ ± 8.13	$^{b}305^{x}$ ± 6.45	$^{c}452^{x}$ ± 7.95	$^{c}450^{x}$ ± 8.43	$^{c}448^{x}$ ± 7.54	$^{c}456^{x}$ ± 7.47
Hybrid-napier	$^{a}183^{x}$ ± 5.32	$^{a}185^{x}$ ± 4.76	$^{a}184^{x}$ ± 6.54	$^{a}184^{x}$ ± 3.52	$^{b}305^{x}$ ± 4.23	$^{b}306^{x}$ ± 5.43	$^{b}306^{x}$ ± ±4.37	$^{b}305^{x}$ ± 7.64	$^{c}452^{x}$ ± 6.97	$^{c}459^{x}$ ± 7.12	$^{c}449^{x}$ ± 6.76	$^{c}456^{x}$ ± 6.93

*Guinea (*Panicum maximum*), Para (*Brachiaria mutica*), Hybrid-napier (*Pennisetum purpureum*)
Values prefixed with different superscript letters in a row are significantly different at P<0.05
Values suffixed with different superscript letters in a column are significantly different at P<0.05
Source: Pandey *et al.* 2011.

Table 7. Soil moisture (%) under different grasses in different canopy positions in a Coconut based silvopastoral system at South Andaman Island of India

*Grass	Under canopy				Between canopy				Open			
	Nitrogen (kg ha^{-1}yr^{-1})				Nitrogen (kg ha^{-1}yr^{-1})				Nitrogen (kg ha^{-1}yr^{-1})			
	0	60	80	120	0	60	80	120	0	60	80	120
Guinea	$^{a}25.2^{x}$ ± 1.01	$^{a}24.75^{x}$ ± 0.95	$^{a}25.5^{x}$ ± 1.02	$^{a}25.5^{x}$ ± 1.06	$^{a}24.4^{x}$ ± 0.98	$^{a}25.25^{x}$ ± 1.02	$^{a}26.0^{x}$ ± 1.05	$^{a}26.0^{x}$ ± 1.04	$^{a}24.4^{x}$ ± 0.94	$^{a}26.0^{x}$ ± 1.05	$^{a}25.75^{x}$ ± 1.02	$^{a}25.50^{x}$ ± 1.06
Para	$^{a}25.3^{x}$ ± 1.02	$^{a}24.75^{x}$ ± 0.94	$^{a}25.5^{x}$ ± 1.05	$^{a}25.5^{x}$ ± 1.06	$^{a}24.8^{x}$ ± 1.02	$^{a}25.25^{x}$ ± 1.02	$^{a}26.0^{x}$ ± 1.04	$^{a}26.0^{x}$ ± 1.04	$^{a}24.9^{x}$ ± 0.98	$^{a}26.0^{x}$ ± 1.04	$^{a}25.75^{x}$ ± 1.02	$^{a}25.50^{x}$ ± 1.05
Hybrid-napier	$^{a}25.8^{x}$ ± 1.06	$^{a}24.75^{x}$ ± 1.04	$^{a}25.5^{x}$ ± 1.05	$^{a}25.5^{x}$ ± 1.05	$^{a}24.6^{x}$ ± 1.03	$^{a}25.25^{x}$ ± 1.02	$^{a}26.0^{x}$ ± 1.03	$^{a}26.0^{x}$ ± 1.04	$^{a}25.0^{x}$ ± 0.93	$^{a}26.0^{x}$ ± 1.06	$^{a}25.75^{x}$ ± 1.05	$^{a}25.5^{x}$ ± 1.06

*Guinea (*Panicum maximum*), Para (*Brachiaria mutica*), Hybrid-napier (*Pennisetum purpureum*)
Values prefixed with different superscript letters in a row are significantly different at $P<0.05$
Values suffixed with different superscript letters in a column are significantly different at $P<0.05$
Source: Pandey *et al.* 2011.

Table 8. Net N mineralization rate (µg g^{-1} day^{-1}) in soils under different grasses in different canopy positions in a Coconut based silvopastoral system at South Andaman Island of India

*Grass	Under canopy				Between canopy				Open			
	Nitrogen (kg $ha^{-1}yr^{-1}$)				Nitrogen (kg $ha^{-1}yr^{-1}$)				Nitrogen (kg $ha^{-1}yr^{-1}$)			
	0	60	80	120	0	60	80	120	0	60	80	120
Guinea	$^{a}0.86^{x}$ ± 0.007	$^{b}1.26^{x}$ ± 0.003	$^{c}1.57^{x}$ ± 0.009	$^{d}2.25^{x}$ ± 0.006	$^{e}0.75^{x}$ ± 0.02	$^{f}1.23^{x}$ ± 0.006	$^{g}1.48^{x}$ ± 0.005	$^{h}2.01^{x}$ ± 0.005	$^{i}0.60^{x}$ ± 0.02	$^{j}1.12^{x}$ ± 0.003	$^{k}1.42^{x}$ ± 0.02	$^{l}1.96^{x}$ ± 0.003
Para	$^{a}0.80^{y}$ ± 0.02	$^{b}1.27^{y}$ ± 0.003	$^{c}1.55^{y}$ ± 0.009	$^{d}2.22^{y}$ ± 0.005	$^{e}0.78^{y}$ ± 0.03	$^{f}1.24^{y}$ ± 0.006	$^{g}1.46^{y}$ ± 0.006	$^{h}2.0^{x}$ ± 0.006	$^{i}0.54^{y}$ ± 0.03	$^{j}1.11^{x}$ ± 0.003	$^{k}1.40^{x}$ ± 0.02	$^{l}1.95^{x}$ ± 0.003
Hybrid Napier	$^{a}0.89^{z}$ ± 0.02	$^{b}1.26^{z}$ ± 0.003	$^{c}1.56^{z}$ ± 0.008	$^{d}2.26^{z}$ ± 0.005	$^{e}0.82^{z}$ ± 0.01	$^{f}1.25^{z}$ ± 0.005	$^{g}1.47^{z}$ ± 0.005	$^{h}2.02^{z}$ ± 0.005	$^{i}0.59^{x}$ ± 0.04	$^{j}1.10^{x}$ ± 0.005	$^{k}1.41^{x}$ ± 0.01	$^{l}1.94^{x}$ ± 0.003

*Gunea (*Panicum maximum*), Para (*Brachiaria mutica*), Hybrid-napier (*Pennisetum purpureum*)
Values prefixed with different superscript letters in a row are significantly different at $P<0.05$
Values suffixed with different superscript letters in a column are significantly different at $P<0.05$
Source: Pandey *et al.* 2011

Table 9. Pearson correlation coefficients and their level of significance between independent [N-fertilization levels (N-levels), nitrogen concentration in shoot biomass (N-conc.), nitrogen uptake by grasses (N-uptake), nitrogen use efficiency of the grasses, (NUE), soil moisture (SM), net soil N mineralization rate (NMR), light intensity (LI)] and dependent variables [forage yield (FY), mean shoot biomass (SB), shoot biomass/root biomass ratio (SB/RB), N-levels)] in a coconut based silvopastoral system at South Andaman Island of India

Dependent variables	Independent variables						
	N-levels	NConc.	Nuptake	NMR	NUE	SM	LI
			Data pooled across canopy positions				
			Guinea grass				
FY	0.753**	0.945**	0.985**	0.612*	-0.908**	0.697*	0.624*
SB	0.823**	0.880**	0.927**	0.736**	-0.860**	0.595*	0.501^{NS}
SB/RB	-o.432^{NS}	0.200^{NS}	0.244^{NS}	-0.577*	-0.103^{NS}	-0.197^{NS}	0.860**
N-levels	-	0.757**	0.726**	0.956**	-0.813**	0.730**	0.008^{NS}
			Para grass				
FY	0.738**	0.974**	0.989**	0.598*	-0.943**	0.687*	0.635*
SB	0.820**	0.900**	0.937**	0.724**	-0.869**	0.591*	0.498^{NS}
SB/RB	-0.385^{NS}	0.258NS	0.317^{NS}	-0.541^{NS}	-0.179^{NS}	-0.038^{NS}	0.890**
N-levels	-	0.749**	0.711**	0.958**	-0.797**	0.638*	-0.003^{NS}
			H. napier				
FY	0.552^{NS}	0.991**	0.980**	0.376^{NS}	-0.963**	0.508^{NS}	0.751**
SB	0.637*	0.974**	0.980**	0.492^{NS}	-0.907**	0.473^{NS}	0.658*
SB/RB	-0.605*	-0.489^{NS}	-0.568^{NS}	-0.632*	0.281^{NS}	-0.249^{NS}	0.140^{NS}
N-levels	-	0.556^{NS}	0.574^{NS}	0.945**	-0.453^{NS}	0.491^{NS}	0.003^{NS}
			Data pooled across grass species				
			Under canopy				
FY	0.721**	9.952**	0.982**	0.679*	-0.940**	0.087^{NS}	0.334^{NS}
SB	0.825**	0.899**	0.954**	0.762**	-0.867**	0.106^{NS}	0.412^{NS}
SB/RB	-0.429^{NS}	-0.950**	-0.935**	-0.409	0.935**	-0.123^{NS}	-0.126^{NS}
N-levels	-	0.560^{NS}	0.664*	0.953**	-0.497^{NS}	0.167^{NS}	0.478^{NS}
			Between canopy				
FY	0.918**	0.840**	0.980**	0.916**	-0.814**	0.878**	0.353^{NS}
SB	0.923**	0.857**	0.969**	0.953**	-0.822**	0.826**	0.200^{NS}
SB/RB	-0.886**	-0.833**	-0.931**	-0.945**	0.743**	-0.819**	-0.143^{NS}
N-doses	-	0.966**	0.949**	0.974**	-0.948**	0.956**	0.357^{NS}
			Open				
FY	0.867**	0.888**	0.982**	0.845**	-0.855**	0.546^{NS}	0.332^{NS}
SB	0.910**	0.909**	0.963**	0.944**	-0.820**	0.365^{NS}	0.412^{NS}
SB/RB	-0.634*	-0.667*	-0.882**	-0.642*	0.584*	-0.225^{NS}	-0.161^{NS}
N-doses	-	0.992**	0.902**	0.984**	-0.966**	0.604*	0.282^{NS}

**P<0.001; *P<0.01

Shoot biomass (SB), forage production (FP), Nitrogen (N) concentration in forage (SB), N uptake, and N use efficiency of the grasses under three canopy positions

Source: Pandey *et al.* 2011.

SB and FP of all grasses under all conditions (N fertilizer and no N fertilizer) are lower under UC and BC positions than in the open (Tables 10 and 11). The decline, however, under UC and BC positions is greater in hybrid-napier than Guinea and Para grasses. The decline in SB and FP of all grasses under UC (high shade) and BC (low shade) positions (0 kg ha^{-1} N fertilizer) of the tree than in the open is due to the effects of the shades. Occurrence of nearly equal amount of SM under all canopy positions suggests that it is not a limiting factor to the growths and yields of the grasses in the open. The equal amount of SM among the canopy positions is due to the high amount of rainfall during the growing season of the grasses. Rate of NMR is higher under UC and BC positions than in the open probably due to greater amount of soil organic matter contributed by leaf and root litters of the tree (Pandey and Singh, 2009). Reduction in SB and FP under canopies of trees is known to occur due to competitive interaction between trees and grasses for light when SM is not limiting in open in irrigated conditions, and in sub-humid and humid climates particularly during wet season (Grossman *et al.*, 1980; Somarriba, 1988; Scanlan *et al.*, 1991; Mordelet and Menaut, 1995; Cruz, 1996). Ludwig *et al.* (2004), Ludwig *et al.* (2001) have argued that tree shade limits productivity of understorey grasses during wet season when enough water and nutrients are available in open. However, during dry season when soil-water limits plant growth in the open, the shade increases growth and productivity of the grasses. Wilson and Wild (1991) report that understorey grasses respond N fertilizer positively when their growths and yields are limited by nitrogen in open. Therefore, we expect that the shades do not reduce SB and FP of the grasses in the N fertilized conditions because the sylvopastoral system is N limiting as evident from increase in SB and FP of all grasses in the open in response to the N fertilizer levels. The decline in SB and FP of all grasses with the N fertilizer levels under both the shades support our first hypothesis that SB and FP decline under the tree shades, both in N fertilized as well as no N fertilized conditions, when SM is not growth limiting in the open. However, amount of decline in the SB and FP depend on grass species and intensity of shades i.e. higher is the shade greater is decline in the SB and FP. Cruz (1996) observes that artificial shades reduce dry matter production under both N fertilized and no N fertilized conditions in a micro-plot experiment at Guadeloupe, France. Lesser reduction in SB and FP of all grasses under BC than UC position in our study seems due to the low shade. Koukoura and Nastis (1989) find lesser reduction in forage production under moderate shade (50%) than high shade (90%) at Thessaloniki farm of Aristotle University in Greece. Under all N fertilizer levels, lower reduction in SB and FP of Guinea and Para than Hybrid-napier under UC position in this study is attributed to their intrinsic characters, probably acquired through evolution, which enable them to produce more forage under the high shade.

Table 10. Shoot biomass (g m^{-2}) of different grasses under different canopy positions in a Coconut based silvopastoral system at South Andaman Island of India

*Grass	Under canopy				Between canopy				Open			
	Nitrogen (kg $ha^{-1}yr^{-1}$)				Nitrogen (kg $ha^{-1}yr^{-1}$)				Nitrogen (kg $ha^{-1}yr^{-1}$)			
	0	60	80	120	0	60	80	120	0	60	80	120
Guinea	^{a}180.7^{x} ± 5.89	^{b}315.5^{x} ± 2.97	^{c}399.5^{x} ± 6.13	^{d}474.0^{x} ± 7.96	^{e}224.7^{x} ± 3.80	^{f}406.8^{x} ± 8.51	^{g}412.3^{x} ± 10.28	^{h}606.0^{x} ± 9.09	^{b}316.2^{x} ± 5.99	^{d}473.3^{x} ± 12.14	^{h}563.5^{x} ± 11.09	^{i}808.3^{x} ± 10.0
Para	^{a}176.2^{y} ± 6.32	^{b}301.3^{y} ± 8.73	^{c}389.3^{y} ± 14.22	^{d}413.5^{y} ± 7.23	^{e}204.5^{y} ± 3.39	^{f}382.0^{y} ± 7.15	^{g}401.3^{y} ± 7.96	^{h}593.8^{y} ± 8.09	^{i}308.1^{y} ± 2.81	^{j}447.0^{y} ± 8.62	^{k}511.0^{y} ± 8.57	^{l}742.0^{y} ± 16.24
Hybrid-napier	^{a}151.4^{z} ± 3.63	^{b}239.5^{z} ± 4.07	^{c}262.0^{z} ± ±7.04	^{d}270.8^{z} ± ±9.46	^{e}196.8^{z} ± ±3.70	^{f}373.8^{z} ± ±8.95	^{g}510.8^{z} ± ±10.21	^{h}778.5^{z} ± ±6.75	^{i}362.0^{z} ± ±12.86	^{j}479.3^{z} ± ±8.30	^{k}708.8^{z} ± ±8.35	^{l}930.0^{z} ± ±11.18

*Guinea (*Panicum maximum*), Para (*Brachiaria mutica*), Hybrid-napier (*Pennisetum purpureum*)
Values prefixed with different superscript letters in a row are significantly different at $P<0.05$
Values suffixed with different superscript letters in a column are significantly different at $P<0.05$
Source: Pandey *et al.* 2011.

Table 11. Forage production (t ha^{-1}) in different canopy positions by different grasses in a Coconut based silvopastiral system at South Andaman Island of India

*Grass	Under canopy				Between canopy				Open			
	Nitrogen (kg ha^{-1}yr^{-1})				Nitrogen (kg ha^{-1}yr^{-1})				Nitrogen (kg ha^{-1}yr^{-1})			
	0	60	80	120	0	60	80	120	0	60	80	120
*Guinea	$^{a}4.64^{x}$ ± 0.20	$^{b}8.20^{x}$ ± 0.13	$^{c}10.20^{x}$ ± 0.94	$^{d}11.90^{x}$ ± 0.87	$^{e}7.46^{x}$ ± 0.23	$^{c}10.20^{x}$ ± 0.99	$^{f}15.40^{x}$ ± 1.02	$^{f}18.10^{x}$ ± 1.56	$^{g}8.81^{x}$ ± 0.18	$^{h}16.30^{x}$ ± 0.86	$^{i}19.20^{x}$ ± 1.30	$^{j}22.00^{x}$ ± 0.77
Para	$^{a}4.21^{y}$ ± 0.11	$^{b}7.40^{y}$ ± 0.27	$^{c}9.20^{y}$ ± 0.56	$^{d}10.85^{y}$ ± ±0.62	$^{b}7.18^{x}$ ± ±0.36	$^{e}12.10^{y}$ ± ±0.56	$^{f}14.40^{y}$ ± ±1.04	$^{g}17.30^{y}$ ± ±0.88	$^{h}8.10^{y}$ ± ±0.15	$^{i}15.30^{y}$ ± ±0.72	$^{j}18.20^{y}$ ± ±0.74	$^{k}21.10^{y}$ ± ±0.93
Hybrid- napier	$^{a}2.97^{z}$ ± 0.11	$^{b}4.30^{z}$ ± 0.55	$^{c}5.20^{z}$ ± 0.76	$^{d}5.60^{z}$ ± 0.55	$^{e}6.24^{y}$ ± 0.14	$^{f}13.30^{z}$ ± 1.46	$^{g}18.50^{z}$ ± 0.85	$^{h}24.60^{z}$ ± 2.33	$^{i}9.93^{z}$ ± 0.46	$^{j}20.40^{z}$ ± 0.95	$^{k}27.50^{z}$ ± 1.70	$^{l}32.10^{z}$ ± 1.32

*Guinea (*Panicum maximum*), Para (*Brachiaria mutica*), Hybrid-napier (*Pennisetum purpureum*)
Values prefixed with different superscript letters in a row are significantly different at $P<0.05$
Values suffixed with different superscript letters in a column are significantly different at $P<0.05$

Source: Pandey *et al.* 2011.

N fertilization and forage production coconut based silvopasture

SB and FP of all grasses increase with each increment of N fertilizer levels under all canopy positions. Under UC position the order of increase is: Guinea > Para > Hybrid-napier for all N levels. Under BC position, at 0 and 60 kg ha^{-1} N, the order of increase is: Guinea > Para > Hybrid-napier, but at 80 and 120 kg ha^{-1} N, it was: Hybrid-napier > Guinea > Para. The concentration of N in the forage (SB) and N uptake are affected by the tree canopy positions and levels of N fertilizer in all grasses. N concentration for all grasses increases progressively with each increment of the N fertilizer levels (Table 12). While differences in N concentrations between Guinea and Para grasses are generally minimal; N concentration in Hybrid-napier is lower than the other grasses under UC but almost similar under BC and the open positions. Concentrations reaches 2.5 % N at the highest N fertilizer level in the open. N uptake (uptake or nitrogen yield by the grasses is calculated as forage production × concentration of N in the shoot biomass) increases dramatically as N fertilizer level increases for all grasses under all canopy positions (Table 13). While uptake in Guinea and Para exceedes than in Hybrid-napier under UC position, uptake in Hybrid-napier greatly exceeds than in the other grasses under BC and in the open. Nitrogen use efficiency declines as levels of applied N increases; it is higher under UC than BC and in the open (Table 14). NUE is higher for Hybrid-napier than for other grasses under UC, but differences are less pronounced under BC position and in the open. SB/RB ratio declines under UC and BC positions than in the open in all levels of N fertilizer (Table 15). But, the decline is the highest in Guinea and the lowest in Hybrid-napier under UC. But, differences in SB / RB ratios are not much pronounced under BC position.

Table 12. Nitrogen concentration (%) in shoot biomass of different grasses under different canopy positions in a Coconut based silvopastoral system at South Andaman Island of India

*Grass	Under canopy				Between canopy				Open			
	Nitrogen (kg $ha^{-1}yr^{-1}$)				Nitrogen (kg $ha^{-1}yr^{-1}$)				Nitrogen (kg $ha^{-1}yr^{-1}$)			
	0	60	80	120	0	60	80	120	0	60	80	120
*Guinea	$^{a}0.88^{x}$ ± 0.04	$^{b}1.01^{x}$ ± 0.03	$^{c}1.20^{x}$ ± 0.02	$^{d}1.38^{x}$ ± 0.04	$^{a}0.87^{x}$ ± 0.03	$^{e}1.68^{x}$ ± 0.04	$^{f}2.01^{x}$ ± 0.02	$^{g}2.10^{x}$ ± 0.04	$^{b}1.00^{x}$ ± 0.04	$^{h}1.72^{x}$ ± 0.05	$^{g}2.11^{x}$ ± 0.16	$^{i}2.29^{x}$ ± 0.02
Para	$^{a}0.86^{x}$ ± 0.02	$^{b}1.00^{x}$ ± 0.06	$^{c}1.21^{x}$ ± 0.02	$^{d}1.36^{x}$ ± 0.01	$^{e}0.90^{y}$ ± 0.02	$^{f}1.71^{y}$ ± 0.02	$^{g}1.95^{y}$ ± 0.06	$^{h}2.08^{x}$ ± 0.05	$^{b}0.99^{x}$ ± 0.03	$^{i}1.81^{y}$ ± 0.07	$^{j}2.01^{y}$ ± 0.03	$^{k}2.32^{y}$ ± 0.08
Hybridnapier	$^{a}0.82^{y}$ ± 0.01	$^{b}0.76^{y}$ ± 0.06	$^{a}0.80^{y}$ ± 0.01	$^{a}0.80^{y}$ ± 0.007	$^{b}0.87^{x}$ ± 0.02	$^{c}1.30^{z}$ ± 0.01	$^{d}1.70^{z}$ ± 0.03	$^{e}2.10^{x}$ ± 0.06	$^{f}1.02^{x}$ ± 0.01	$^{d}1.70^{x}$ ± 0.02	$^{g}2.02^{y}$ ± 0.02	$^{h}2.53^{z}$ ± 0.07

*Guinea (*Panicum maximum*), Para (*Brachiaria mutica*), Hybrid-napier (*Pennisetum purpureum*)
Values prefixed with different superscript letters in a row are significantly different at $P<0.05$
Values suffixed with different superscript letters in a column are significantly different at $P<0.05$
Source: Pandey *et al.* 2011.

Table 13. Nitrogen uptake (kg ha^{-1}) by different grasses under different canopy positions in a Coconut based silvopastoral system at South Andaman Island of India

*Grass	Under canopy				Between canopy				Open			
	Nitrogen (kg $ha^{-1}yr^{-1}$)				Nitrogen (kg $ha^{-1}yr^{-1}$)				Nitrogen (kg $ha^{-1}yr^{-1}$)			
	0	60	80	120	0	60	80	120	0	60	80	120
*Guinea	$^{a}41.07^{x}$ ± 3.42	$^{b}82.98^{x}$ ± 3.82	$^{c}123.10^{x}$ ± 13.05	$^{d}165.01^{x}$ ± 15.14	$^{e}64.66^{x}$ ± 1.88	$^{d}171.62^{x}$ ± 17.69	$^{f}310.06^{x}$ ± 22.71	$^{g}379.83^{x}$ ± 32.62	$^{b}88.21^{x}$ ± 4.23	$^{h}279.08^{x}$ ± 10.71	$^{i}406.45^{x}$ ± 41.20	$^{j}504.25^{x}$ ± 20.95
Para	$^{a}36.28^{x}$ ± 1.65	$^{b}73.71^{y}$ ± 3.59	$^{c}110.78^{y}$ ± 5.11	$^{d}147.55^{y}$ ± 8.43	$^{e}65.52^{x}$ ± 3.46	$^{f}207.16^{y}$ ± 11.13	$^{g}278.76^{y}$ ± 14.11	$^{h}360.03^{y}$ ± 20.58	$^{b}80.17^{y}$ ± 2.21	$^{i}277.08^{x}$ ± 16.38	$^{j}365.14^{y}$ ± 11.48	$^{k}487.36^{y}$ ± 12.75
Hybrid- napier	$^{a}24.39^{y}$ ± 1.34	$^{b}32.65^{z}$ ± 4.87	$^{c}49.59^{z}$ ± 8.69	$^{b}37.45^{z}$ ± 9.44	$^{c}54.34^{y}$ ± 2.29	$^{d}173.01^{x}$ ± 19.25	$^{e}312.97^{x}$ ± 11.26	$^{f}515.83^{z}$ ± 48.55	$^{g}101.07^{z}$ ± 3.73	$^{h}346.57^{y}$ ± 15.38	$^{i}555.60^{z}$ ± 35.16	$^{j}815.52^{z}$ ± 54.78

*Guinea (*Panicum maximum*), Para (*Brachiaria mutica*), Hybrid-napier (*Pennisetum purpureum*)
Values prefixed with different superscript letters in a row are significantly different at $P<0.05$
Values suffixed with different superscript letters in a column are significantly different at $P<0.05$
Source: Pandey *et al.* 2011.

Table 14. Nitrogen use efficiency (kg ha^{-1} yr^{-1} kg^{-1} N) of different grasses under different canopy positions in a Coconut based silvopastoral system at South Andaman Island of India

*Grass	Under canopy				Between canopy				Open			
	Nitrogen (kg ha^{-1}yr^{-1})				Nitrogen (kg ha^{-1}yr^{-1})				Nitrogen (kg ha^{-1}yr^{-1})			
	0	60	80	120	0	60	80	120	0	60	80	120
Guinea	$^{a}114.37^{x}$ ± 4.41	$^{b}99.32^{x}$ ± 2.80	$^{c}83.43^{x}$ ± 1.42	$^{d}72.69^{x}$ ± 1.99	$^{e}115.45^{x}$ ± 3.91	$^{f}59.63^{x}$ ± 1.28	$^{g}49.76^{x}$ ± 0.36	$^{h}47.67^{x}$ ± 0.80	$^{b}100.57^{x}$ ± 3.86	$^{i}58.35^{x}$ ± 1.75	$^{j}48.38^{x}$ ± 3.27	$^{k}43.69^{x}$ ± 0.47
Para	$^{a}116.50^{y}$ ± 2.51	$^{b}101.26^{y}$ ± 5.48	$^{c}82.82^{x}$ ± 1.91	$^{d}73.56^{x}$ ± 0.65	$^{e}109.64^{y}$ ± 1.25	$^{f}58.50^{x}$ ± 0.49	$^{g}51.44^{y}$ ± 1.38	$^{h}48.18^{x}$ ± 1.08	$^{b}101.25^{x}$ ± 2.45	$^{i}55.54^{y}$ ± 1.99	$^{j}49.81^{x}$ ± 0.83	$^{k}43.29^{x}$ ± 1.44
Hybrid- napier	$^{a}122.09^{z}$ ± 2.02	$^{b}135.50^{z}$ ± 21.19	$^{c}125.15^{y}$ ± 2.17	$^{c}125.05^{y}$ ± 1.24	$^{d}115.29^{x}$ ± 3.15	$^{e}76.96^{y}$ ± 0.80	$^{f}58.89^{z}$ ± 1.01	$^{g}47.76^{x}$ ± 1.32	$^{h}98.74^{y}$ ± 5.70	$^{f}58.85^{x}$ ± 0.64	$^{i}49.52^{x}$ ± 0.49	$^{j}39.65^{y}$ ± 1.14

*Guinea (*Panicum maximum*), Para (*Brachiaria mutica*), Hybrid-napier (*Pennisetum purpureum*)
Values prefixed with different superscript letters in a row are significantly different at $P<0.05$
Values suffixed with different superscript letters in a column are significantly different at $P<0.05$
Source: Pandey *et al.* 2011.

Table 15. Shoot biomass /root biomass ratio of different grasses under different canopy positions in a Coconut based silvopastoral system at South Andaman Island of India

*Grass	Under canopy				Between canopy				Open			
	Nitrogen ($kg\ ha^{-1}yr^{-1}$)				Nitrogen ($kg\ ha^{-1}yr^{-1}$)				Nitrogen ($kg\ ha^{-1}yr^{-1}$)			
	0	60	80	120	0	60	80	120	0	60	80	120
*Guinea	$^{a}2.09^{x} \pm 0.08$	$^{b}2.01^{x} \pm 0.02$	$^{c}1.84^{x} \pm 0.02$	$^{d}1.80^{x} \pm 0.009$	$^{e}2.37^{x} \pm 0.08$	$^{f}2.30^{x} \pm 0.02$	$^{g}2.19^{x} \pm 0.02$	$^{h}2.08^{x} \pm 0.03$	$^{i}2.48^{x} \pm 0.08$	$^{j}2.40^{x} \pm 0.03$	$^{j}2.39^{x} \pm 0.03$	$^{f}2.31^{x} \pm 0.04$
Para	$^{a}2.14^{y} \pm 0.07$	$^{b}2.10^{y} \pm 0.02$	$^{c}1.92^{y} \pm 0.02$	$^{d}1.85^{y} \pm 0.02$	$^{e}2.36^{x} \pm 0.08$	$^{e}2.36^{y} \pm 0.02$	$^{f}2.24^{y} \pm 0.02$	$^{a}2.13^{y} \pm 0.03$	$^{g}2.55^{y} \pm 0.08$	$^{h}2.50^{y} \pm 0.01$	$^{i}2.44^{y} \pm 0.01$	$^{j}2.41^{y} \pm 0.02$
Hybrid-napier	$^{a}2.17^{z} \pm 0.17$	$^{b}2.13^{z} \pm 0.01$	$^{c}2.23^{z} \pm 0.02$	$^{c}2.25^{z} \pm 0.03$	$^{d}2.37^{x} \pm 0.11$	$^{e}2.31^{x} \pm 0.02$	$^{f}2.27^{y} \pm 0.02$	$^{g}2.05^{z} \pm 0.04$	$^{h}2.51^{z} \pm 0.09$	$^{i}2.46^{z} \pm 0.02$	$^{j}2.08^{z} \pm 0.009$	$^{k}1.95^{z} \pm 0.02$

*Guinea (*Panicum maximum*), Para (*Brachiaria mutica*), Hybrid-napier (*Pennisetum purpureum*)
Values prefixed with different superscript letters in a row are significantly different at $P<0.05$
Values suffixed with different superscript letters in a column are significantly different at $P<0.05$
Source: Pandey *et al.* 2011.

The increase in SB and FP of all grasses in response to the N fertilizer levels, compared to 0 kg ha^{-1} N, under the high and low shades suggest that the N fertilization compensates the shades effects. Koukoura et al. (2009) found that N fertilizer (225 kg ha^{-1} N) under two artificial shades (90% and 50% reduction in sunlight) increase herbage production of two grasses, i.e. *Dactylis glomerata* and *Festuca ovina* at Thessaloniki farm of Aristotle University in Greece. Eriksen and Whitney (1981) observe that N fertilizer increases dry matter production of six tropical grasses (*Brachiaria brizantha, B. millifirmis, Digitaria decumbens, Panicum maximum, Pennisetum clandestimum* and *P. purpureum*) under three artificial shades (30%, 55% and 73% reduction in full day light) at Paia, Hawii. But, the increase is the highest under the lowest shade. The increase in SB and FP of all grasses under both the shades with the N fertilizer levels in our study seems to have occurred due to the high N concentrations in their forage (SB) resulted probably due to the high rate of NMR. Mosquera-Losada *et al.* (2001), Soder and Stout (2003), Cruz *et al.* (1995) observed that N concentrations increase in the leaves of grasses under shade due to fertilization with nitrogen. High N concentrations in the leaves of grasses are reported to improve net CO_2 assimilation (Wong and Wilson, 1980; Cruz *et al.*, 1995), which compensates for low level of radiation under trees. Durr and Rangel (2000) report that high soil fertility compensates for reduction in photosynthesis in *Panicum maximum* due to cloth's shading at Townsville, north Queensland. But, they find a limit of the compensatory mechanism; under dense shades grass yields decline. In the present study N concentration in the forage of all grasses increases with N fertilizer levels peaking at 120 kg ha^{-1} N under both the shades, but the increase is greater in the low than high shade. Our these observations support the Field's (1983) theory that nitrogen use efficiency for photosynthesis is maximized when leaves have higher N content in greater illuminated than leaves having lower N in lesser illuminated region of tussocks. Under the high shade, in our study, increase in SB and FP due to the N fertilizer is greater in Gunea and Para than Hybrid-napier, but vice-versa under the low shade. These observations support our the second hypothesis that N fertilization increases SB and FP under the shades, but the increase depend on grass species, intensity of shades and amount of N fertilizer. In this study, though all levels of N fertilizer increases SB and FP of all grasses under both the shades, it does not increase them either equal to or over to that found in the open. These findings suggest that N fertilizer albeit increases SB and FP of the grasses under the shades, it can not annul the shades effects when soil-water is not growth limiting in the open. According to Callaway *et al.* (2002) a shift from competition to facilitation occurs only with increase in abiotic stresses.

N fertilization and nitrogen use efficiency in coconut based silvopasture

The SB and FP, under each canopy position of all grasses, and across the canopy positions and the grass species, are positively correlated to the N fertilizer levels, N concentration in forage, N uptake and NMR, but inversely correlated to NUE (forage production / nitrogen uptake) (Table 9). SB/RB ratio is inversely correlated to the NMR and N fertilizer levels under BC and the open (Table 9). The higher concentration of N in the forage of Guinea and Para than Hybrid-napier under both the shades for all N fertilizer levels is attributed to their greater N uptake. The greater N uptake by the grasses one can explain due to their lower SB / RB ratios. Decline in SB/RB ratios in 0 kg ha^{-1} N plot of all grasses compared to the open support the widely held view that under reduced light conditions plants divert more photosynthate belowground (Nambiar, 1990). Many studies find that lower shoot/root ratios in grasses under low light conditions (Margolis and Vezina, 1988). Greater uptake of N by Guinea and Para grasses than hybrid-napier for all N fertilizer levels under the high shade, and just reversed under the low shade particularly at high N fertilizer level indicate that Guinea and Para grasses are superior competitors for N in the high shade, however, Hybrid-napier is a superior competitor for N in the low shade conditions. Tilman (1988), and Tilman and Wedin (1991) are of the view that species which exploit growth limiting resources relatively greater is treated as a superior competitor. We observe lower SB/RB ratio in Guinea and Para than Hybrid-napier with higher N fertilizer under the high shade suggesting that superior competitors for N under shade divert more photosynthate to belowground plant part (root) for greater exploitation of available N. According to Tilman (1988) plant species modify their morphology to utilize nutrients and light under reduced conditions. Koukoura *et al.* (2009) report that N fertilization reduced shoot/root ratio in some grasses (*D. glomerata*), but does not affect others (*F.ovina*). Decline in NUE with the N fertilizer levels in all grasses under all canopy positions is not unexpected; others also report similar results (Delogu *et al.*, 1998). Inverse relation between NUE and N fertilizer levels is well documented (Delogu *et al.*, 1998; Hiremath *et al.*, 2002). Therefore, greater decline in NUE in Guinea and Para grasses than Hybrid-napier under the high shade suggest that Guinea and Para are more tolerant to the high shade; hence they may be exploited for silvopastoral systems.

Cut – and – carry system

Stall feeding has not been accepted socially in the Islands. Therefore, cut – and –carry system of fodder management is not practiced in the island. But for being self – sufficient in fodder system need to be popularized. Surplus fodder produced during rainy months may be stored as “Hay” that may be used along with green leaves of fodder trees during dry months.

Restricted grazing

Since free ranging is an ages old practice of cattle rearing in the islands, restricted grazing should be popularized on homestead level and Panchayat level. Silvo- pastural system should be divided into different compartments according to carrying capacity of the system and restricted grazing should be allowed. Bandyopadhyay (1998) reported carrying capacity Andaman pasture 5 head/ha. Restricted grazing is known to sustain the productivity of grasses for a longer time.

Social fencing

Free range grazing is one of the bottlenecks for maintaining the silvo-pastural system in the islands. People let their cattle to go on free range grazing in forests, orchards and paddy fields after the milking is made. Excessive grazing facilitates invasion of coarse alien species, which reduced the production potential of the forage on the one hand and make the quality poor on the one hand and make the quality poor on the other hand. Grazing is reported also to lead soil erosion, which ultimately makes the soil poor in nutrients. Free range grazing is one of the reasons, which ultimately makes the soil poor in nutrients. Free range grazing is one of the reasons for not raising grasses and fodder trees in thei9r orchards. Social fencing at homegarden level may help in adopting silvo- pastural system and making the islands self sufficient in fodder production.

Conclusions

The study concludes that coconut trees reduce SB and FP of Guinea, Para and Hybrid-napier grasses under UC (high shade) and BC (low shade) positions compared to that in the open in the humid tropical climate of South Andaman Island of India where SM is not growth limiting in the open. However, the reduction is higher in Hybrid-napier than Guinea and Para grasses. The grasses respond to the N fertilizer levels under the shades by lowering their SB/RB ratios, which probably facilitate N uptake and result in high N concentration in their forage (SB). Increase in N uptake and N concentration in the forage is greater in Guinea and Para than Hybrid-napier under the UC, but vice-versa under the BC (at high N level) and in the open. The higher N uptake and N concentration in the forage result in greater SB and FP in Guinea and Para under the high shade, but greater SB and FP in Hybrid-napier under the low shade. These results indicate that Guinea and Para are superior competitors for the high shade but Hybrid-napier is a superior competitor for the low shade. Though, under both the shades, SB and FP of all grasses increase due to the N fertilizer levels, the increase is neither at par nor above to that in the open. These observations suggest that Guinea and Para grasses can be exploited under the high shade and Hybrid-napier under the low shade, and high forage production can be managed with these grasses by N fertilization in the silvopastoral system of the humid tropics and other similar ecosystems.

References

Anderson, L.J., Brumbaugh, M. S., Jackson, R. B. 2001. Water and tree-understorey interactions: a natural experiment in a savanna with oak wilt. Ecology 82:33-49

Bandhyapadhyay, A. K. 1998. Agroforestry scientists in Andaman. Homecare. Agro. Climatic zonal planning unit for one Islands zone- XZ. CARI, Port Blair, Andaman & Nicobar Islands.

Basic Statistics. 2001. Andaman and Nicobar Islands: Basic statistics. Directorate of Economics and statistics, Andaman and Nicobar Administration, Port Blair, India

Baumeister, D., Callaway, R. M. 2006. Facilitation by *Pinus flexilis* during succession :a hierarchy of mechanisms benefits other plant species. Ecology 87:1816-1830

Belsky, A.J. 1994. Influence of trees on savanna productivity: test of shade, nutrients, and tree-grass competition. Ecology 75:922-932

Belsky, A.J., Amundson, R. G., Duxbury, J. M., Riha, S. J., Ali, A. R., Mwonga, S. M. 1989. The effects of trees on their physical, chemical and biological environment in a semi-arid savanna in Kenya. J. Appl. Ecol. 26:1005-1024.

Benavides, R., Douglas, G. B., Osoro, K. 2009. Silvopastoralism in New Zealand: review of effects of evergreen and deciduous trees on pasture dynamics. Agrofor. Syst. 76:327-350.

Burner, D.M., MacKown, C. T. 2006. Nitrogen effects on herbage nitrogen use and nutritive value in a meadow and Loblolly pine alley. Crop Sci. 46:1149-1155.

Callaway, R.M., Brooker, R. W., Choler, P., Kikvidze, Z., Lortie, C. J., Michalet, R., Paolini, L., Pugnaire, F.I., Newingham, B., Aschehoug, E.T., Armas, C., Kikodze, D., Cook, B. J. 2002. Positive interactions among alpine plants increase with stress. Nature 417:844-848.

Callaway, R.M., Walker, L. R. 1997. Competition and facilitation: a synthetic approach to interactions in plant communities. Ecology 78 (7): 1958-1965.

Cruz, P.1996. Growth and nitrogen nutrition of a *Dichanthium aristatum* pasture under shading. Trop. Grassl. 30:407-413.

Cruz, P., Munier-Jolain, N. M., Tournebize, R., Sinoquet, H. 1993. Growth and mineral nutrition in a Dichanthium aristatum sward shaded by tree. Proceedings of the XVII International Grassland Congress, Rockhampton, pp.2056-2057.

Cruz, P., Tournebize, R., Gaudichau, C., Haegelin, A., Munier-Jolain, N.M. 1995. Effect of shade on growth, nitrogen content and CO_2 leaf assimilation in a tropical perennial grass. In: Sinoquet H. and Cruz P. (eds.), Ecophysiology of Tropical Intercropping. Pp. 285-293. (INRA: Paris).

Delogu, G., Cattivelli, L., Pecchioni, N., DeFalcis, D., Maggiore, T., Stanca, A.M. 1998. Uptake and agronomic efficiency of nitrogen in winter barley and winter wheat. European J. Agron. 9:11-20.

Dodd, M.B., McGowan, A. W., Power, I. L., Thorrold, B.S. 2005. Effects of variation in shade level, shade duration and light quality on perennial pasture. NZJ Agirc. Res. 48: 531-543.

Durr, P.A., Rangel, J. 2000. The response of Panicum maximum to a simulated subcanopy environment: soil x shade interaction. Trop. Grassl. 34:110-117.

Eriksen, F.I., Whitney, A. S. 1981. Effects of light intensity on growth of some tropical forage species. I. Interaction of light intensity and nitrogen fertilization on six forage grasses. Agron. J. 73:427-433.

Field, C. 1983. Allocating leaf nitrogen for maximization of carbon gain: leaf age as a control on the allocation program. Oecologia 56:341-347.

Gea-Izquierdo, G., Montero, G., Canellas, I. 2009. Changes in limiting resources determine spatio-temporal variability in tree-grass interactions. Agrofor. Syst. 76:375-382.

Grossman, D., Grunow, J.O., Theron, G.K. 1980. Biomass cycling, accumulation rates and nutritional characteristics of grass layer plants in canopied and uncanopied subhabitats of Burkea savanna. Proc. Grassl. Soc. South Africa 15:157-161.

Guevara-Escobar, A., Kemp, P.D., Mackay, A.D., Mackay, A.D., Hodgson, J. 2007. Pasture production and composition under poplar in a hill environment in New Zealand. Agrofor. Syst. 69:199-213.

Hiremath, A. J., Ewel, J. J., Cole, T. G. 2002. Nutrient use efficiency in three fast growing tropical trees. Forest. Sci. 48: 662-672.

Koukoura, Z., Kyriazopoulos, A.P., Parissi, Z.M. 2009. Growth characteristics and nutrient content of some herbaceous species under shade and fertilization. Spanish J. Agric. Res. 7(2):431-438.

Koukoura, Z, Nastis, A.S. 1989. Effect of shade on production and forage quality of herbaceous species. Geotehnika Epistimonika Themata 1:17-25 [In Greek, English summary].

Ludwig, F., de Kroon, H., Berendse, F., Prins, H.H.T. 2004. The influence of savanna trees on nutrient, water and light availability and the understory vegetation. Plant Ecol. 170:93-105.

Ludwig, F., de Kroon, H., Berendse, F., Prins, H.H.T., Berendse, F. 2001. The effects of nutrients and shade on tree-grass interactions on an East African savanna. J. Veg. Sci. 12: 579-588.

Margolis, H.A., Vezina, L.P. 1988. Nitrate content, amino acid composition and growth of yellow birch seedlings in response to light and nitrogen source. Tree Physiol. 4:245-253.

Mordelet, P., Menaut, J.C. 1995. Influence of trees on aboveground production dynamics of grasses in a humid savanna. J. Veg. Science 6:223-228.

Mosquera-Losada, M. R., Lopez-Diaz, L., Rigueiro-Rodriguez, A. 2001. Sewage sludge fertilization of a silvopastoral system with pines in northwestern Spain. Agrofor. Syst. 53:1-10.

Nambiar, E.K.S. 1990. Interplay between nutrients, water, root growth and productivity in young plantations. For. Mgmt. 30:213-232.

Pandey, C. B., Singh. L. 2009. Soil fertility under homegarden trees and native moist evergreen forest in South Andaman, India. J. Sust. Agric. 33(30): 303-318.

Pandey, C. B., Rai, R.B., Singh, L. 2007. Seasonal dynamics of mineral N pools and N-mineralization in soils under homegarden trees in South Andaman, India. Agrofor. Syst. 71:57-66.

Pandey, C.B., Verma, S.K., Dagar, J.C., Srivastava, R.C. 2011. Forage production and nitrogen nutrition in three grasses under coconut tree shades in the humid tropics. Agrofor. Syst. 83:1-12.

Pandey, C.B., Singh, G.B., Singh, S.K., Singh, R.K. 2010. Soil nitrogen and microbial biomass carbon dynamics in native forests and derived agricultural land uses in a humid tropical climate of India. Plant Soil 333:453-467.

Pandey, C.B., Singh, J.S. 1992. Rainfall and grazing effects on net primary productivity in a tropical savanna, India. Ecology 73:2007-2021

Pandey, C. B. 1990. Spices composition, plant biomass and net productivity of savanna on topical dry forest area. *PhD. Thesis.* Banaras Hindu University, Varanasi. 210p.

Polley, H.W., Johnson, H.B., Mayeux, H.S. 1997. Leaf physiology, production, water use, and nitrogen dynamics of the grassland invader Acacia smallii at elevated CO_2 concentrations. Tree Physi. 17:89-96.

Power, I.L., Dodd, M.B., Thorrold, B.S. 2001. Deciduous or evergreen does it make a difference to understorey pasture yield and riparian zone management? Proc. N Z Grassl. Assoc. 63:121-125.

Ram, J., Singh, J. S., Singh, S. P. 1989. Plant biomass, species diversity and net primary production in central Himalayan High Altitude Grassland. J. Ecol. 77:456-468.

Rhoades, C.C. 1997. Single-tree influences on soil properties in agroforestry: lessons from natural forests and savanna ecosystems. Agrofor. Syst. 35:71-94.

Sarmiento, G. 1984. The ecology of neotropical savannas. Harvard University Press, Cambridge, Massachusets.

Scanlan, J.C., Wilson, B.J., Anderson, E.R. 1991. Sustaining productive pasture in the tropics.2. managing woody vegetation in grazing lands. Trop. Grassl. 25:85-90.

Sharma, A.K., Dagar, J.C., Bandyopadhyay, A. K. 1990. Fodder Resource of Bay Islands. Central Agriculture Research Institute, Port Balir.72.p.

Soder, K.J., Stout, W.L. 2003. Effect of soil type and fertilization level on mineral concentration of pasture: potential relationships to ruminant performance and health. J. Anim. Sci. 81:1603-1610.

Somarriba. E. 1988. Pasture growth and floristic composition under the shade of guava (*Psidium guajava* L.) trees in Costa Rica. Agrofor. Syst. 6:153-162.

Tilman, D. 1988. Plant strategies and the dynamics and structure of plant communities. Princeton University Press, Princeton, New Jersey, USA.

Tilman, D., Wedin, D. 1991. Plant traits and resource reduction for five grasses growing on a nitrogen gradient. Ecology 72:685-700.

Tournebize, R., Sinoquet, H. 1994. Light interaction and partitioning in a shrub/grass mixture. Agril. For. Meterol. 72:277-294.

Vetaas, O.R. 1992. Micro-site effects of trees and shrubs in dry savannas. J. Veg. Sci. 3:337-344.

Wilson, J. R., Wild, D.W.M. 1991. Improvement of nitrogen nutrition and grass growth under shading. In: Shelton HM, Stur WW (eds) Forage from plantation crops, (ACIAR Proceedings No.32: Canberra), PP 77-82.

Wong, C.C., Wilson, J.R. 1980. Effects of shading on the growth and nitrogen content of green panic and siratro in pure and mixed swards defoliated at two frequencies. Aust. J. Agric. Res. 31:269-285.

❑❑❑

13

Alley Cropping System: Traditional and Modified Version for the Humid Tropics

C.B. Pandey

Abstract: *Alley cropping system is known for several services like fodder and fuel production, nutrient management, and soil erosion control from hilly terrain in humid tropics, but it could not make in farmers field. This study explores the role of* Gliricidia *leaf manure on nitrogen management in soils. Simultaneously, the system is modified to make it highly profitable, so that it may become farmer's friendly in humid tropics. For nitrogen management in soil, two methods (leaves being incorporated into soil vs. surface) and three times of application, i.e. zero week after sowing (0 WAS), two week after sowing (2 WAS) and four week after sowing (4 WAS), of* Gliricidia *leaf in soil was considered. Two additional treatments, urea (120 kg ha^{-1}, equal to the leaves) and control (no urea + no leaves) are also maintained to compute recovery of nitrogen from the leaves by the crop, and to know if the leaf manure could produce grain yield equal to that of equivalent urea. The study reveals there is no effect of the method of application on the decomposition and release of nitrogen from the leaves, soil N mineralization, nitrogen uptake, shoot biomass and yield of the crop, but time of application affects these parameters significantly. Maximum 50 % nitrogen is released quickly from the leaves within 15 days and remaining 48-49 % gradually in 60 days. Rate of soil N mineralization; nitrogen uptake, shoot biomass and grain yield in maize are the highest in 2 WAS and lowest in 4 WAS treatment. Recovery of nitrogen from the leaves is very low ranging from 4.5 to 9.3 kg ha^{-1}. The leaves do not produce yield in maize equal to that of equivalent urea. However, for synchronization of maximum release of N from* Gliricidia *leaves and it's* *uptake by maize crop, the leaf manure need to be applied after two week of sowing. The system is modified by growing highly remunerative perennial crop like black pepper (*Piper nigrum *L.) on double hedgerows and maize and okra in alleys. Dimensions of the double hedgerows in the modified alley cropping system are: height 2.5 m and canopy width 3.0 m. Yields in maize and okra are maximum (5.04 t / ha and 7.47 t / ha, respectively for the first cropping year, which declines to 2.14 t / ha and 5.89 t / ha, respectively during the sixth cropping year. On the contrary, production in black pepper is the lowest during the second (0.21 t / ha) and the highest (2.38 t /ha) during the seventh cropping year. Investment in maize (Rs. 19,581 / ha) and okra (Rs. 30,179 / ha) did not differ among the cropping years. But, in black pepper, investment is maximum (Rs. 44 to 32 thousand / ha) for two cropping years in the beginning and declines to Rs. 11,880 to 15,675 / ha from the third year onwards. Economic analysis reveals that net profit from the black pepper is negative for the first and second cropping year (Rs. 44,399 and Rs. 6,630 / ha, respectively) in the beginning, but okra alone compensates it. From the third cropping year black pepper alone not only compensates its establishment cost, but also earns a reasonably good income for Rs. 97082 / ha. Moreover, net return in black pepper over the seven cropping years of the experiments is Rs. 12,97,292 that not only compensates the negative return from the system, but also make the alley- cropping system 4.46 times more profitable than without the black pepper.*

Introduction

Study of alley cropping system was started in late twentieth century to improve /remove crop- rotation with an idea to replenish nutrients and cultivate cereals simultaneously on the same piece of land (Kang *et al.*, 1981). The alley cropping involves hedgerows intercropping of food crops with fast growing shrub / tree species. The latter are preferably leguminous species because they contribute better to soil nitrogen and associated food crops due to biological N fixation (Amara *et al.*, 1996). The hedgerows are periodically pruned to prevent shading and reduce competition with the companion crops (Kang *et al.*, 1981). Prunings / leaves from the hedgerows are used for a number of purposes, an important one being as a source of nitrogen to the food crops. Out of the total amount of N present in the leaves / prunings, the fraction that is taken up by the crop is known as the nitrogen recovery value (Mafongoya and Nair, 1997). Management of alley cropping system aims mainly to increase the nitrogen recovery rate from the prunings/ leaves by the crops. Soil erosion control is another prominent use of alley cropping system in humid tropics (Pandey and Chaudhari, 2010).

Beneficial effects of trees in nutrient build up (Pandey *et al.*, 2007a), microclimate changes (Chen *et al.* 2002) and soil erosion control (Pandey and Chaudhari, 2010) have been well documented. But, these tree benefits are beyond the perception of farmers because they are intangible in nature. This is the reason it could not make dent in farmers fields. Nevertheless farmers follow either ages old traditional agroforestry like *Acacia nilotica* L. based system in central India, *Prosopis cineraria* (Sw.) DC. and *Acacia nilotica* systems in North and western India (Tejwani, 1994), coconut-arecanut based homegarden agroforestry in southern part of India (Pandey *et al.*, 2007b) or modern industry–linked highly remunerative poplar (*Populus deltoids* Bartr.) based agri-silviculture model in north-western India which feeds to match box industry (Gill *et al.*, 2009). The traditional systems fulfill primarily the basic needs of farmers whereas the modern industry-linked system provides high income to household. These clearly indicate that to make agroforestry farmers friendly, it should be made highly remunerative either by developing industry-linked models or by adding value on the existing agroforestry systems. Value added agroforestry systems might be designed either incorporating highly remunerative trees along with annual crops like popular + wheat or introducing highly remunerative crops on the existing agroforestry systems itself. In the present study I followed the second approach and developed a modified version of alley cropping system where the hedgerows of *Gliricidia sepium* (Jacq.) Walp. were used as standards (support) for growing black pepper, a highly remunerative crop, and the alleys were grown with annual crops like miaze (*Zea mays* L.) and okra (*Abelmoschus esculentus* L.) during rainy season and post-rainy season, respectively. Generally hedgerows in alley cropping is cut to 50 cm height from ground so that light travels to intercrops and pruning biomass is used as a supplement for nitrogen. But, in the modified alley cropping hedgerows are allowed to grow up to 2.5 m and canopy expansion up to 3 m so that they provide greater space to the pepper vines to spread with and also make harvesting of pepper berries easy, and ultimately to make the alley cropping system highly remunerative. Traditionally, the black pepper is grown on arecanut palm which is a tall tree (12 m height) having thin cylindrical bole (15 cm dia.). The pepper vines trails on the palm tree as high as 8-9 m and makes harvesting of the pepper berries difficult and labour intensive and ultimately the pepper cultivation uneconomical.

The present study describes role of alley cropping for N cycling and soil erosion control. It also describes how to make it highly remunerative so that farmers adopt it.

Experimental design and treatments

The experiment was carried out in a well established *Gliricidia* based double hedgerows alley cropping system. The system included 5 double hedgerows and 4 alleys (space between alleys). The alleys were 4 m in width and 21 m in length. The hedgerows were pruned to 1m to facilitate light in the alleys. The alleys were divided into twenty- four 3x3m plots. The plots were 50 cm away from the hedgerows, separated by 1 m deep root barrier.

The treatment included two methods (incorporated into the 0-15cm soil vs. surface application) and three times of application (at the time of sowing, called zero week after sowing (0 WAS), two week after sowing (2 WAS) and four week after sowing (4 WAS) of *Gliricidia* leaves as organic manure (4 t ha^{-1}). Concentration of nitrogen in the leaves was 3 % and C / N ratio 12.8. To compare the effect of the organic manure on the growth and grain yield of maize a treatment "Urea" (120 kg ha^{-1}) was also included. Urea was applied in two slit doses as a common practice of fertilization. However, to measure the recovery of N by the maize crop from the leaves, one more treatment control (No urea+ No leaves) was also maintained. The treatment was arranged in a completely randomized block design, replicated 3 times. *Gliricidia* leaves collected from the hedgerows of the alley cropping system during post monsoon were sun dried (d" 5% moisture) and stored in gunny bags. These sun-dried leaves were used for the experiment. Maize was sown in the plots in June 2002 at 50 cm row-to-row distance and 50 cm within a row. Single super phosphate (100 kg ha^{-1}) and muriat of potash (80kg ha^{-1}) were applied in all the treatments. Three random maize plants from each plot were harvested to ground level at 15 day, 30 day, 45 day and 60 day after the sowing to measure the shoot biomass. The sample was brought to laboratory, dried to constant weight at 80^0C and weighed. Dried samples were powdered in Wiley mill and sieved with 1mm sieve for chemical analysis. Grain yield was measured at the crop maturity.

To measure the amount and pattern of nitrogen release 20 g of the sun-dried leaves of *Gliricidia* were incorporated into a litterbag. Size of the litterbag was 20 x 20 cm. A set of 45 litterbags was placed into the soil (0-15cm) and another set of 45 litterbags was placed on the soil surface for decomposition. Five litterbags were retrieved at 8 days interval. Adhered soil on the leaves were removed thoroughly with brush, dried at 80^0C to constant weight and weighed. The samples were also powered like the maize sample. Maize as well as leaf samples were analyzed for total N by the microkjedhal digestion method using auto-N analyzer.

Soil samples (0-15cm depth) were collected from 4 random places and composited for each treatment. Soil sampling was done on the sampling dates of the maize. After removing the surface organic materials and fine roots carefully, each composited field-moist soil sample was divided into two parts. One part was transported to the laboratory for determination of NO_3^--N and NH_4^+-N. The other part was incubated *in situ* for estimation of N- minerlization rate. NO_3^- - N was measured by a phenol disulphonic acid method and NH_4^+-N by a phenate method (Wetzel and Likens, 1979).

Mineralization rate was measured by buried bag technique (Eno, 1960). Three soil samples (each about 150g) enclosed in polythene bags were buried at the 0-15cm soil depth for a period of 15 days. NO_3^- -N and NH_4^+-N were determined initially (at time zero) and after the recovery of buried bags. The increase in NH_4^+-N was considered as ammonification and the increase in NO_3^--N as the nitrification. The increases in the concentrations NH_4^+-N + NO_3^--N during the field incubation were estimated as the net N- mineralization.

Maize N uptake was calculated by multiplying the maize N concentration by corresponding shoot biomass at each sampling date. Nitrogen releases from the leaves were calculated by subtracting residual N in the leaves at each sampling date from total N applied at the beginning of the experiment. Nitrogen recovery rate was calculated following Mafongoya and Nair (1997) as:

$$\text{Nitrogen recovery rate} = \frac{\text{N uptake of treatment} - \text{N uptake of control}}{\text{N initially applied}} \text{X}100\%$$

Decomposition and N release from leaves

Weight loss and N concentration did not differ due to the method of application of the leaves Fig.1a & b. Around 50% leaves were decomposed quickly in 15 days in both the conditions. Remaining leaves were decomposed at a little faster rate in the incorporated than that in surface applied condition. At 30 days, decomposition in incorporated condition was 7% greater than that in surface condition. But, within two months maximum 98-99% leaves were decomposed in both the conditions. Quick decomposition of the leaves may be attributed to high concentration of nitrogen (high quality) (Pandey *et al.*, 2006). However, no difference in decomposition time owing to the method of application could be due to high humidity and rainfall, and optimum temperature. During the maize cropping season, high rainfall as well as humidity seem to have made the leaves wet. Whereas high intensity of rainfall could have mixed the leaves with soils by splashes and thereby ensured contact of the surface applied leaves with soil. Though data is lacking but it is most likely that the surface applied leaves could have provided stable temperature, and soil- moisture at the leaves - soil interface and thereby provided suitable breeding ground to microorganism similar to that found inside

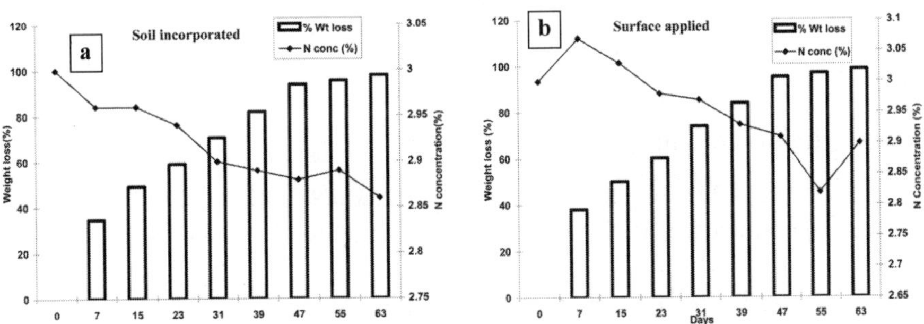

Figure 1. Per cent weight loss and concentration of nitrogen of *Gliricidia* leaves **(a)** in the soil incorporated and **(b)** on surface applied condition in an alley cropping system in the humid tropical climate of South Andaman.
Source: Pandey and Rai 2007

the soil. In sharp contrast to the humid climate of our study, decomposition and nitrogen release rates differed due to the method of application in semi-arid (Mafongoya and Nair, 1997) and temperate climate (Bross *et al.*, 1995). Direct release of nitrogen from the leaves without immobilization in our study, seem to have occurred due to high concentration of nitrogen. Mafongoya and Nair (1997) argued that the method of placement might have a less significant effect on decomposition and N release rates when organic matters are of better quality (higher N). Lehmann *et al.* (1995) also observed no net N immobilization during the decomposition of *Gliricidia* leaves in a sub- humid climate of West Africa.

Release of nitrogen was maximum (52 %) within 15 days in 0 WAS treatment that declined with time and was lowest (7 %) at 60 days (Table 1). However, in 2 WAS treatment the release was slow in the beginning, but it was 2.8 times greater at 60 days compared to that in 0 WAS treatment. In both 0 WAS and 2 WAS treatments, the amount of nitrogen released from the leaves in incorporated condition was similar to that in surface applied condition. In 4 WAS treatment no release of nitrogen occurred in the beginning for the first 15 days as the leaves were applied quite late.

Table 1. N release from the leaves under two methods of leaves application (incorporated into soil denoted with I and surface application, denoted with S) at different times at south Andaman.

Time of leaves application	N release (kg ha^{-1})									
	Surface					Incorporated				
	15 days	30 days	45 days	60 days	Total	15 days	30 days	45 days	60 days	Total
0 WAS	60.44	21.22	27.66	7.97	117.29	60.3	27.73	24.11	6.64	118.78
2 WAS	18.44	46.50	26.72	22.22	113.88	19.0	46.27	27.84	22.62	115.73
4 WAS	NA	36.88	32.55	58.81	91.36	NA	40.12	23.49	34.01	97.62

Values in a column with different superscript are significantly different at P<0.05
WAS denotes "week after sowing" LSD (P<0.05) compares means of a parameter in rows
Source: Pandey and Rai (2007)

Mineralization and inorganic nitrogen pool in soil

The NO_3^--N, NH^+_4-N and mineral N differed due to the method (P<0.0001) and time of application (P<0.0001) and their interaction (P<0.0001) (Fig. 2). In 0 WAS treatment NO_3^--N was high in the beginning until 45days but declined thereafter and was lowest at 60 days. However, in 2 WAS treatment, NO_3^--N pool size was lower at 15 days in the beginning, but it was 11 to 24 % higher across the rest of the sampling dates compared to that in 0 WAS. In 4 WAS treatment, $N0_3^-$-N pool increased from 30 days and was quite close to that in 2 WAS treatment at the later dates. In the control plots NO_3^--N was lowest of all the treatments across the sampling dates. The NH_4^+-N in 0 WAS treatment was high in the beginning until 30 days that declined thereafter and was lowest at 60 days. Pattern of variation in the pool sizes of NH_4^+-N across the sampling dates in 2 WAS was similar to that in 0 WAS, but it increased unlike 0 WAS up to 45days and declined thereafter. In 4 WAS treatment NH_4^+-N increased quite late from 30 days, however, it persisted for a longer period of time compared to that in 0 WAS and 2 WAS treatment. Like the leaf treated plots, in urea treated plot also NH_4^+-N was greater than that of NO_3^- - N. Urea is converted first to ammonium N that is oxidized to NO_3^--N (Campbell, 1995). The NH_4^+-N formation was faster than that of its oxidation due perhaps to continuous excessive rainfall during the experiment. John *et al.* (1992)

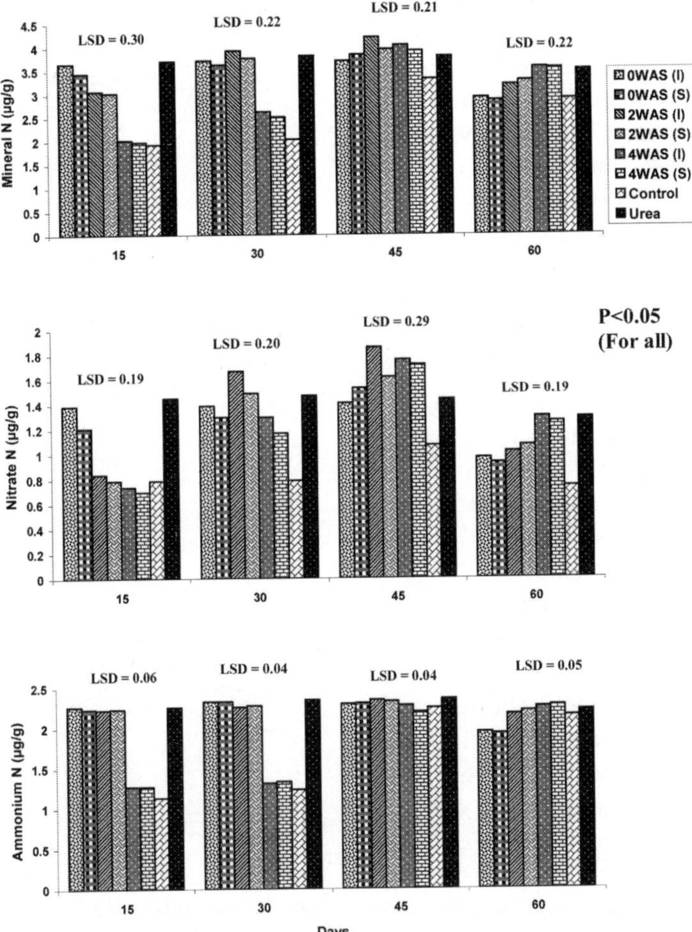

Fig. 2. Ammonium N, nitrate N and mineral N pool sizes in soil under 0 WAS, 2 WAS, 4 WAS, urea and control treatments in soil incorporated (I) and surface applied (S) condition in an alley cropping system in the humid tropical climate of South Andaman.
Source: Pandey and Rai 2007

also reported similar pattern in soil nitrate and mineral N in an organic manuring with cowpea to rice experiment at Laguna, Philippines.

Nitrification and ammonification rates in the soils varied due to the time of application ($P<0.0001$) (Fig.3). But, their variations did not differ due to the methods of application. It indicated that the method of application did not affect the mineralization in the soil. Nitrification increased quickly after application of the *Gliricidia* leaves in all the treatments. It was maximum in the beginning in the 0 WAS treatment, but declined after 30 days and was lowest at 60 days after application of the leaves. On the contrary in 2 WAS treatment, nitrification was lowest in the beginning and increased with time and was 24 to 77% higher at 45 days and 60 days, respectively compared to that in 0 WAS treatment. In 4 WAS treatment, nitrification was 17 to 35% lower compared to that in 2 WAS treatment. Lowest nitrification rate (0.05 to 0.08 μg g^{-1} day^{-1}) was observed in the control and highest (0.01 to 0.17 μg g^{-1} day^{-1}) in urea treatment (Fig. 3). Like nitrification, ammonification was higher in the beginning and declined with the time in the 0 WAS treatment. However, in 2 WAS treatment ammonification was lower in the beginning at 15 days, but at 30, 45 and 60 days it was 22, 15 and 60 % higher than that in 0 WAS treatment. Contrary to that of nitrification, ammonification was lower in urea treatment compared to that in the leaf manuring treatments at almost all the sampling dates. Pattern of soil N mineralization was similar to that of the nitrification and ammonification in the soil. Soil N mineralization was 9 % to 64% higher in 2 WAS compared to that in 0 WAS treatment across the observation dates. In 4 WAS treatment nitrogen mineralization

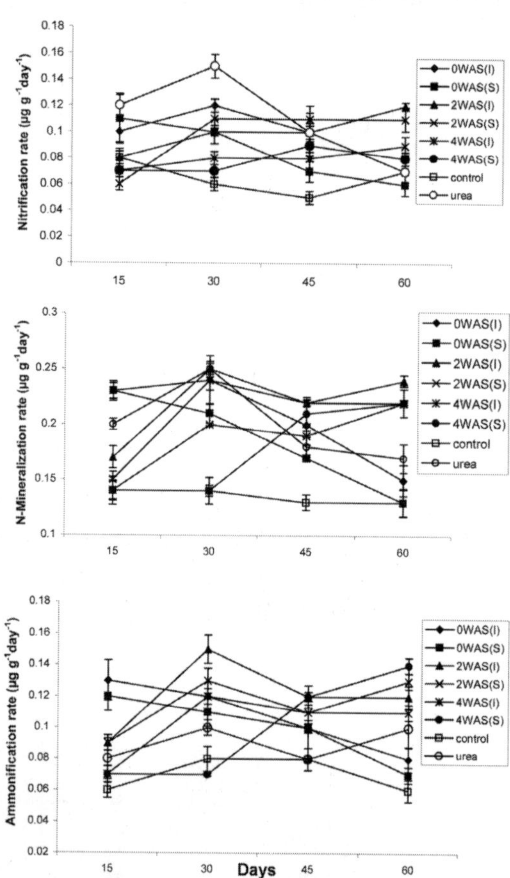

Fig.3. Ammonification, nitrification and N mineralization rates in soils under 0 WAS, 2 WAS, 4 WAS, urea and control treatments in soil incorporated (I) and surface applied (S) condition in an alley cropping system in the humid tropical climate of South Andaman. Bars represent ±ISE.
Source: Pandey and Rai 2007

until 15 days in the beginning was at par to that in control, but on the later dates it increased from 21 to 69 %. In urea treated plots soil N mineralization was though nearly equal to that in the leaf manuring treatments, but it was relatively more homogeneous.

Quick soil N mineralization in the leaf manuring treatments may be attributed to high concentration of nitrogen in the leaves, low C/N ratio, high soil-moisture, high temperature and high humidity (De Datta, 1995). Clement *et al.* (1995) and Haslam (1981) reported that a high tissue N concentration in green manures increased early N mineralization. Wong and Nortcliff (1995) are of the view that net mineralization occurs when nitrogen content of the plant material is more than 1.7%. However, Browaldh (1995) argued that C/N ratios < 20-25 are generally required for net minaralization of the organic material. Frankenberger and Abdelmajid (1985) also found C/N ratio or the N concentration important in determining the rate of mineralization. Most of the decomposition models have been described by single exponential decay functions indicating that decomposition proceeds at maximum rates immediately after incorporation (Wieder and Lang, 1982). According to Clement *et al.* (1995) plant residue N mineralization rate was related to a number of factors, the most important being the positive effect of N concentration. Lipid + polyphenol / N ratio, tannin / nitrogen, lignin /nitrogen and carbon / nitrogen are organic matter quality factors which affect residue N mineralization. Though in our experiment most of the quality parameters are not estimated, but Clement *et al.* (1995) reported lignin 103 g kg^{-1}, cellulose 133 g kg^{-1}, polyphenol 30 g kg^{-1} and tannin 3.3 g kg^{-1} in *Gliricidia* leaves. They are of the view that early N mineralization occurs due to higher tissue N concentration. However, high concentration of lignin (Fox *et al.*,1990) or polyphenols (Palm and Sanchez, 1991) have a little inhibitory effect on immediate net mineralization of plant N. Tannin is reported to precipitates proteins (Haslam, 1981) including bacterial exoenzymes which help organic matter decomposition.

Decline in N mineralization after 45 days in our leaf manuring experiment could be due to immobilization (Fig.1 a & b). Karen and Fownes (1992) also observed immobilization between 8 to 12 weeks in a green manuring with seven legume tree leaves. Based on visual observation they reported that late immobilization in organic manuring occurred because of the colonization of organic materials by lignin degrading fungi. High tannin content may cause N immobilization and could limit the N mineralization percentage (Clement *et al.*, 1995).

Nitrogen uptake and biomass production

ANOVA results indicated that method of application of the leaves did not affect the shoot biomass as well as yield of maize at all the observation dates (Table 2). But, time of application affected the parameters (P<0.001). Peak shoot biomass of the maize at 60 days was maximum in the plot applied with leaves at 2 WAS.

Interestingly the peak shoot biomass in the plots applied with the leaves at 2

Table 2. Shoot biomass (kg ha^{-1}) of maize in a *Gliricidia* based alley- cropping system under two methods of leaves application (incorporated into soil denoted with I and surface application, denoted with S) at different times at South Andaman.

Time of leaves application	Shoot biomass (kg ha^{-1})								Grain yield (t ha^{-1})	
	15 days		30 days		45 days		60 days			
	I	S	I	S	I	S	I	S	I	S
0 WAS	147^a	104^a	242^a	225^a	484^a	465^a	1348^a	1295^a	5.01^a	4.98^a
2 WAS	76^b	87^b	235^a	217^a	678^b	626^b	1588^b	1579^b	5.52^b	5.38^a
4 WAS	42^c	44^c	133^b	113^b	518^c	438^c	1183^c	1088^c	2.04^c	2.12^b
Control	43^c	43^c	56^c	56^c	276^d	276^d	680^d	680^d	1.94^d	1.94^c
Urea	150^a	150^a	271^d	271^d	669^b	669^b	1561^b	1561^b	6.28^e	6.28^d
LSD ($P<0.05$)	16.98		33.46		85.42		58.39		0.79	

Values in a column with different superscript are significantly different at $P<0.05$
WAS denotes "week after sowing"
LSD ($P<0.05$) compares means of a parameter in rows
Source: Pandey and Rai (2007)

WAS were equal to that applied with urea. In the beginning for 15 days, shoot biomass in the treatment 0WAS and urea was nearly equal, but with the passage of time biomass accumulation in the crop in the 0WAS treatment declined and finely at 60 days it was 15% lower than that of urea. Leaves application at 4 WAS was not much effective to the biomass accumulation as it produced lowest amount of shoot biomass among the treatments. Like the shoot biomass, grain yield of maize was highest in 2WAS and lowest in 4 WAS treatment.

In 0 WAS treatment nitrogen release from the leaves (>50 %) and nitrification rate in soil were high within 15 days which triggered the growth in the crop, but failed to sustain it at par to urea because the amount as well as the rate declined on the later dates. Maximum N demand and rapid growth phase in maize is reported to occur between 6 to 9 week after sowing (Mafongoya and Nair, 1997). When the leaves are applied at 2 WAS, maximum N is released after 15 days that continued up to the grain filling; and high nitrification rate persisted for later dates that translated comparatively greater grain production. It indicates that if *Gliricidia* leaves are applied to maize at 2 WAS, N release synchronizes greater with the nitrogen demand of the crop compared to that applied at 0WAS. However, when it is applied at 4 WAS the released N did not synchronize with the crop's demand and therefore did not register its impact on it's growth and yield.

The leaves applied at both 0WAS and 2WAS supported shoot biomass accumulation in maize to nearly equal amount up to 30 days (Table 3). However, the leaves applied at 2 WAS supported comparatively greater accumulation thereafter. Compared to 0WAS, shoot biomass in 2WAS treatment was 36 % greater at 45 days and 19% greater at 60 days of the crop. However, biomass accumulation in maize in urea-applied plots was always greater than that of the leaves applied plots. Like the shoot biomass, grain yield in 2WAS treatment was 8 to 10% greater than that in 0WAS treatment but 12 to 14% lower from that in urea treated plots.

Equal amount of shoot biomass in 0WAS and 2WAS treatments for 30 days could be due to similar amount of nitrogen uptake (Table 3). Up to 30 days there was no difference in the uptake of N in 0WAS and 2WAS treatments probably due to similar amount of it's release. However, at peak growth stage (60 days) N release in 0WAS treatment declined substantially which failed to support the crop to carry out the growth. The leaves applied contained nitrogen equal to that of urea and most of the N was released within the cropping season, but N uptake in the leaves applied plots were not equal to that applied with urea most likely because the urea was applied in split doses. It may also be attributed to lack of synchrony between N demand by the crop and N release from the leaves (Swift, 1987; Myers *et al.*, 1994). The differences in N uptake explain why grain yield in 2 WAS treatment was lower than that in urea treated plots though the peak shoot biomass was equal. Kang *et al.* (1981) reported that efficiency of maize crop in the utilization of nitrogen yield of *Leucaenea* prunings was low compared to nitrogen fertilizer in humid climate at Ibadan, South Nigeria. They further observed that the prunings were, however, when removed from the plots, a significant yield reduction in maize crop occurred. In our study concentration of nitrogen in maize plant in the urea treated plot was 16-23% greater than that in the leaf manuring treatments.

Table 3. Nitrogen uptake (kg ha^{-1}) by maize crop in a *Gliricidia* based alley- cropping system under two methods of leaves Application (incorporated into soil; denoted with I and surface application; denoted with S) at different times at South Andaman.

Time of leaves application	N uptake (kg ha^{-1})							
	15 days		30 days		45 days		60 days	
	I	S	I	S	I	S	I	S
0 WAS	1.56^{a}	1.06^{a}	2.57^{a}	2.30^{a}	5.13^{a}	4.74^{a}	14.29^{a}	13.21^{a}
2 WAS	0.80^{b}	0.88^{a}	2.46^{a}	2.19^{a}	7.11^{b}	6.33^{a}	16.75^{b}	15.95^{b}
4 WAS	0.42^{c}	0.44^{b}	1.34^{b}	1.13^{b}	5.23^{c}	4.42^{c}	11.95^{c}	10.91^{c}
Control	0.34^{c}	0.34^{b}	0.45^{c}	0.45^{c}	2.21^{d}	2.21^{d}	5.44^{d}	5.44^{d}
Urea	1.84^{d}	1.84^{c}	3.33^{d}	3.33^{d}	7.00^{b}	7.00^{e}	19.20^{e}	19.20^{e}
LSD (P < 0.05)	0.22		0.34		0.89		1.14	

Values in a column with different superscript are significantly different at P<0.05
WAS denotes "week after sowing"
LSD (P<0.05) compares means of a parameter in rows
Source: Pandey and Rai (2007)

N recovery by the maize crop differed due to the time of application of the leaves for all the observation dates (P<0.001). The nitrogen recovery was highest in the plot applied at 2 WAS (Table 4). N recovery from the leaves was nearly equal to that in the plots applied at 0 WAS and 2 WAS until 30 days, but increased later on and at 45 days the recovery was 63 to 68 % higher in the 2 WAS treatment compared to that in 0 WAS. The recovery of nitrogen in 2 WAS treatment reached almost equal to that in urea treated plot. However, finally maximum recovery occurred in the urea followed by 2 WAS treatment. Nitrogen recovery in 4 WAS treatment was lowest among all the treatments.

Low nitrogen recovery (4 to 8 %) from the applied leaves by the maize crop in our study could be due to its high immobilization in microbial biomass and, high run-off and leaching losses. Palm (1995), Giller and Cadisch (1995) reported that maximum 20 % of the N released from tree prunings or litter is taken up by the current crops. Much of the remaining part (40 % to 80 %) of the applied organic N is incorporated into soil organic matter (Haggar *et al.*, 1993). Gravelly- sandy-loamy soils of our study site seem to have caused heavy leaching loss. Runoff loss from the site is found 12 to 22 kg ha^{-1} under different landuses (Pandey and Venkatesh, 2003), which is expected to export a reasonably good amount of nitrogen from the soil.

Table 4. Nitrogen recovery (kg ha^{-1}) by maize in a *Gliricidia* based alley- cropping system under two Methods of leaves application (incorporated into soil; denoted with I and surface application; denoted with S) at different times at south Andaman

Time of leaves application	N recovery (kg ha^{-1})							
	15 days		30 days		45 days		60 days	
	I	S	I	S	I	S	I	S
0 WAS	1.00^{a}	0.59^{a}	1.75^{a}	1.53^{a}	2.41^{a}	2.09^{a}	7.29^{a}	6.39^{a}
2 WAS	0.38^{b}	0.44^{a}	1.66^{a}	1.44^{a}	4.04^{b}	3.40^{b}	9.31^{b}	8.66^{b}
4 WAS	0.06^{c}	0.08^{b}	0.74^{b}	0.57^{b}	2.49^{c}	1.82^{c}	5.36^{c}	4.50^{c}
Urea	1.23^{d}	1.23^{c}	2.37^{c}	2.37^{c}	3.95^{b}	3.95^{b}	11.33^{d}	11.33^{d}
LSD ($P < 0.05$)	0.20		0.40		0.90		0.13	

Values in a column with different superscript are significantly different at $P<0.05$
WAS denotes "week after sowing"
LSD ($P<0.05$) compares means of a parameter in rows
Source: Pandey and Rai (2007)

Experimental design of modified alley cropping and management

Lay out of the modified alley-cropping is given in Photo 1.a,b,c,d. Four years old alley-cropping system with six double hedgerows of *Gliricidia sepium*, each 20 m in length and planted 4 m apart from one another, were selected for the study. Hedgerow to hedgerow distance in a double hedgerow was 1m, and plant to plant distance within a hedgerow was 50 cm. Thus, there were 6 double hedgerows and 5 alleys. At one extreme end of the each alley, a plot without hedgerows, equal in width and 5 m in length, was present. A galvanized iron sheet was put 1 m deep at the interface between the hedgerows and the plots (alley) to keep them free from root competition. The plots served as controls for the crops grown in the alleys. Pits, 30 cm x 30 cm in size were dug at 50 cm interval in the center (1 m space) of the double hedgerows. Soils and FYM in 2:1 ratio were filled in the pits. Four rooted cuttings of black pepper (*Piper nigrum*) var. Panniyur1 were planted in the each pit. Two pepper vines from each pit were trailed on one hedge plant in one hedgerow and remaining two on a hedge plant in the opposite hedgerow. Pepper vines were thinned to best one later on when it was well established. Bordeaux mixture (1%) was spread fortnightly to the vine during the rainy season and at one-month interval during rest of the months in the beginning for two years and thereafter only during the rainy season. A 10 cm thick mulch of FYM: soil and *Gliricidia* leaves mixed in 4:2:1 proportion was filled in the space (1m) in the double hedgerows in January 2001 for soil - moisture conservation during the dry months. The double hedgerows were pruned four times in a year i.e. April, July, September and November for three consecutive years, and in March from the fourth year onwards. Pruning was done at 1 m during the first cropping year, at 2 m during the second cropping year and at 2.5 m from third year onwards. The pruning biomass production (oven dry) was estimated on per hectare basis by multiplying the number of the hedge plants (D) in a hectare with the weight of pruning biomass (leaf + stem) produced by a hedge plant. The number of hedge plants was calculated following Pandey *et al.* (2001) as follows:

Photo a. Double hedgerows of *Gliricidia*

Photo b. Black pepper grown in space between hedgerows

Photo c. Maize sown as an intercrop in the alley

Photo d. Okra sown as an intercrop in the alley

Photo 1, a, b, c, d. Lay out of the modified alley cropping system in South Andaman.
Source: Pandey (2011)

D = Area (100 x 100 m) / alley width (m) x plant to plant distance (m) in a hedgerow

All the alleys including their corresponding control plot were grown for maize during rainy months (early June to mid August) and okra during post-rainy months (Sept. to November) for seven consecutive cropping years i.e.2001 to 2007. Hybrid maize (Ganga 5) was grown at 50 x 50 cm distance. Fertilizer N,P,K was applied at the rate of 120:100:80 kg / ha. Okra (hybrid Arkanamica) was sown at 50 x 50 cm distance and fertilizer N,P,K was applied at the rate of 100:80:60 kg / ha. Urea was applied in two split doses to both the crops. Half at the time of sowing and remaining half after weeding. Weeding was performed after 15 days in maize and 20 days in okra. Okra was irrigated to field capacity generally 1 to 2 times as and when required during dry spell in November. Three random alleys and their corresponding control plots were sampled for the estimation of yields in maize as well as okra.

Hedgerows dimensions

Height of the double hedgerows of the modified alley-cropping system was maintained to 1m during the first cropping year, 2 m the second cropping year and 2.5 m from the third cropping year onwards by pruning as described in the materials and methods section (Table 5). Dimensions of the double hedgerows were manipulated according to the growth and the spreading of the black pepper vines. First year, the hedgerows were pruned to 1m to facilitate light to the vines. Shading is known to increase humidity (Pandey *et al.* 2007b), which causes foot-rot, a dreaded fungal disease to the vines. As the vines trailed high, the hedgerows height was increased to provide them cover to spread with on the one hand and avoid shading to the vines on the other hand. The pruning caused tillering in the *Gliricidia* hedge that resulted into widening of the hedgerows canopy. The pepper vines, however, grew quickly and covered the hedgerows up to 2.5 m within 3 years that subdued the lateral expansion of the hedgerows. Ultimately, the hedgerows canopy was stabilized to 3m within three years. The hedgerows produced maximum pruning biomass in the beginning for two consecutive cropping years, and then declined from the fourth cropping year onwards. However, the top of the hedgerows when pruned over the vines after third year, it grew further, which was evident from the low amount of the pruning biomass after the fourth year. It indicates that the hedgerows can be manipulated to any desired height. Decline in the pruning biomass with passage of time may be a disadvantage of growing black pepper on the hedgerows in the alley-cropping system. Pruning biomass production is regarded one of the major objectives of the alley cropping system. In our study, some amount of pruning biomass was produced every year. Pruning of the hedgerows is essentially required to maintain the vigor of the hedgerows. Pandey and Venkatesh (2007) have found that seven years old *Gliricidia* standards, when not pruned shed their leaves and commenced flowering in dry months and became stunted in growth within two years at the study site.

Table 5. Height and canopy width of *Gliricidia sepium* double hedgerowsa and pruning biomass production in the modified alley-cropping system in South Andaman

Cropping year	Double hedgerows of Gliricidia		Biomass Production
	Hight (m)	Canopy width (m)	(t/ha/yr)
1	1.01^{a}	1.11^{a}	12.02^{a}
2	2.02^{b}	2.53^{b}	10.01^{b}
3	2.51^{c}	3.02^{c}	5.04^{c}
4	2.52^{c}	3.01^{c}	2.11^{d}
5	2.53^{c}	3.02^{c}	2.12^{d}
6	2.52^{c}	3.02^{c}	2.11^{d}
7	2.51^{c}	3.03^{c}	2.10^{d}

Values of parameter in a column with different superscripts Are significant at $P<0.05$.
Source: Pandey (2001)

Crop production

Yields of the maize (5.04 t/ha for first year, 2.82 t/ha the second year, 2.41 t/ ha the third year, 2.24 t/ha the fourth year and 1.99 t /ha from fifth year onwards up to the seventh cropping year) and okra (7.47 t/ha for first year, 6.53 t/ha the second year, 6.22 t/ha the third year, 5.87 t/ha the fourth year and onwards up to the seventh cropping year) declined significantly in the alleys compared to controls all the cropping years. Decline in the yields of both the crops were the lowest during the first cropping year, increased thereafter up to third cropping year, and stabilized from the fourth cropping years onwards. Total decrease in the maize yield over the 7 cropping years was 54%. But, total decline in the okra yield, over the 7 cropping years, was 18 % lower than maize. Decline in the yield in maize during the first cropping year could primarily be due to competition for nutrients because the hedgerows were pruned to 1 m and the soil- water was in plenty during the rainy season. However, subsequent cropping years when the hedgerows grew taller, reduction in the yields of the maize was due to competition for both light and the nutrients. Delayed pruning in alley- cropping is reported to reduce crop yields, because of increased competition (Pandey *et al*., 2001). Maize yield is reported to decline in alley- cropping system in humid climate of Andaman (Pandey and Venkatesh, 2007). However, in okra yield reduction was mainly due to competition for soil- water and partly due to competition for light. During late rainy season, dry spells occur frequently. High inter-annual coefficient of variation (58%) in rainfall occurred in November across the cropping years. Eleven years data (1996 to 2007) revealed that evaporation in November was, an average, 103 mm (Pandey and Venkatesh, 2007) whereas average rainfall across the experiment years was 118 mm in November. Pandey and Venkatesh (2007) reported that whenever, rainfall is less than 150 mm per month in the growing season and evapotranspiration 4 mm / day, soil-water becomes limiting to crops. Black pepper started production from the second cropping year (0.21 t/ha), increased linearly and recorded maximum yield (2.38 t / ha) (Y, t / ha) from the sixth cropping year onwards (x, year) as: $Y = -1.124 + 0.76\ x$, ($r^2 = 0.998$, $P<0.01$). Early bearing in the black pepper in our study may be attributed to mulching with high amount of organic matter. Black pepper is reported to start bearing generally from third cropping year (Sivaraman *et al*., 2002).

Economic analysis

Cost of cultivation was highest for the black pepper followed by the okra and the maize for two cropping year in the beginning (Tables 6, 7 and 8). Thereafter the pattern was reversed and cost of cultivation was the highest in the Okra followed by maize and black pepper. Net profit from the black pepper was negative for the two cropping years in the beginning (Table 9). But, benefit started from the third cropping year and was maximum from the sixth cropping year. The intercrop okra earned profit invariably all the cropping years, with maximum in the beginning for three cropping years. But, the net return from the maize became negative from the second cropping year and persisted until the last year. For the two cropping years in the beginning, net return from the improved alley cropping system (across all the crops) was 40% lower than the traditional system (Table 5). However, from the third year net return was many folds (3.89 to 10.24) greater than traditional system.

Table 6. Cost of cultivation and benefit in maize in the modified alley cropping system in South Andaman.

Cropping years	Cost (Rs. / ha)					Total return (Rs.)	Benefit (Rs.)	Benefit cost ratio
	Labour	Fertilizer	Seed	Fungicides	Total			
1	6,930[a]	8,244[a]	850[a]	NAP	16,024[a]	25,200[a]	+9,176[a]	+0.57[a]
2	6,720[a]	8,244[a]	875[a]	NAP	15,839[a]	14,100[b]	-1,739[b]	-0.11[b]
3	9,700[b]	8,244[a]	915[ab]	NAP	18,859[b]	12,050[c]	-6,809[c]	-0·36[bc]
4	10,400[b]	8,244[a]	927[ab]	NAP	19,571[b]	11,200[c]	-8,371[c]	-0.43[c]
5	10,300[b]	9,635[b]	945[ab]	NAP	20,880[b]	17,600[d]	-3,280[d]	-0.16[d]
6	12,840[c]	9,635[b]	1,012[b]	NAP	23,487[c]	15,920[d]	-7,567[e]	-0.32[d]
7	11,760[c]	9,635[b]	1,014[b]	NAP	22,409[c]	17,120[d]	-5,289[e]	-0.24[d]
Average	[A]9,807	[B]8,840	[C]934	NPA	19,581	16,170	-3,411	-0.15

NAP= not applied; NA= not applicable
Values of a parameter in a column with different superscripts are significant at P<0.05
Values of parameters in a row with different superscripts are significant at P<0.05
Source: Pandey (2011)

Table 7. Cost of cultivation and benefit in Okra in the modified alley cropping system in South Andaman.

Cropping years	Cost (Rs. / ha)					Total return (Rs.)	Benefit (Rs.)	Benefit cost ratio
	Labour	Fertilizer	Seed	Fungicides	Total			
1	11,410[a]	7,245[a]	4,800[a]	875[a]	24,330[a]	89,640[a]	+65,310[a]	+2.68[a]
2	11,690[a]	7,245[a]	4,800[a]	875[a]	24,610[a]	78,360[b]	+53,750[b]	+2.18[b]
3	16,000[b]	7,870[a]	5,000[a]	900[a]	29,770[b]	74,640[b]	+44,870[c]	+1.51[c]
4	15,900[b]	7,870[a]	6,300[b]	930[a]	31,000[b]	70,920[c]	+39,920[d]	+1.29[d]
5	15,600[b]	8,630[b]	6,300[b]	1051[b]	31,581[b]	70,440[c]	+38,859[c]	+1.23[d]
6	18,600[c]	8,630[b]	6,700[c]	1051[b]	34,981[c]	71,160[c]	+36,179[c]	+1.03[d]
7	18,600[c]	8,630[b]	6,700[c]	1051[b]	34,981[c]	70,680[c]	+35,699[c]	+1.02[d]
Average	[A]15,400	[B]8,017	[C]5,800	962	30,179	75,120	44,941	1.56

Values of a parameter in a column with different superscripts are significant at P<0.05
Values of parameters in a row with different superscripts are significant at P<0.05
Source: Pandey (2011)

Table 8. Cost of cultivation and benefit in black pepper in the modified alley cropping system in South Andaman.

Cropping years	Cost (Rs. / ha)					Total return (Rs.)	Benefit (Rs.)	Benefit cost ratio
	Labour	Fertilizer	Seed	Fungicides	Total			
1	39,830[a]	NAP	3819	750[a]	44,399[a]	0	-44,399[a]	-1.0[a]
2	31,080[b]	NAP	NA	750[a]	31,830[b]	25,200[a]	-6,630[b]	-0.21[b]
3	12,800[c]	NAP	NA	750[a]	13,850[c]	1,23,600[b]	+1,10,050[c]	+8.12[c]
4	10,900[c]	NAP	NA	980[a]	11,880[c]	2,08,800[c]	+1,96,920[d]	+16.58[d]
5	11,500[c]	NAP	NA	1,270[b]	12,770[c]	2,29,200[d]	+2,16,430[e]	+16.95[d]
6	14,040[d]	NAP	NA	1,272[b]	15,312[d]	2,79,600[e]	+2,64,288[f]	+17.26[d]
7	14,400[d]	NAP	NA	1,275[b]	15,675[d]	2,85,600[e]	+2,69,925[f]	+17.22[d]
Average	[A]19,221	NAP	[B]3,819	[C]1,007	20,817	1,92,000	1,43,798	+10.7

Values of a parameter in a column with different superscripts are significant at P<0.05
Values of parameters in a row with different superscripts are significant at P<0.05
Source: Pandey (2011)

Table 9. Benefit from the modified and traditional alley cropping system in South Andaman.

Year	Modified System				Traditional		
	Maize (Rs.)	Okra (Rs.)	Black pepper (Rs.)	Total (Rs.)	Maize (Rs.)	Okra (Rs.)	Total (Rs.)
1	+9,176[a]	+65,310[a]	-44,399[a]	+30,087[a]	+9,176[a]	+65,310[a]	+74,486
2	-1,739[b]	+53,750[b]	-6,630[b]	+45,381[b]	-1,739[b]	+53,750[b]	+52,011
3	-6,809[c]	+44,870[c]	+1,10,050[c]	+1,48,111[c]	-6,809[c]	+44,870[c]	+38,061
4	-8,371[c]	+39,920[d]	+1,96,920[d]	+2,28,469[d]	-8,371[c]	+39,920[d]	+31,549
5	-3,280[d]	+38,859[c]	+2,16,430[e]	+2,52,009[e]	-3,280[d]	+38,859[c]	+35,579
6	-7,567[e]	+36,179[c]	+2,64,288[f]	+2,92,900[f]	-7,567[e]	+36,179[c]	+28,610
7	-5,289[e]	+35,699[c]	+2,69,925[f]	+3,00,335[f]	-5,289[e]	+35,699[c]	+30,410
Total	-23,879	+3,14,587	+10,06,584	[A]+12,97,292	-23,879	+3,14,587	[B]+2,90,706

Values of a parameter in a column with different superscripts are significant at P<0.05
Values of parameters in a row with different superscripts are significant at P<0.05
Source: Pandey (2011)

Negative net return from black pepper for the first and second cropping years was probably due to its long gestation period, a characteristic of most tree / perennial crop production systems and requirement of high number of labourers (569 per hectare) for the first cropping year. Labour requirement declined from third year onwards to 108 per hectare. Maize and okra together compensated the cost incurred in the establishment of the black pepper during the first and second cropping year. Maize did not earn income from the second cropping year, but okra always earned income mainly because of its high price. Labour requirement in these crops was relatively low 97 and 159 per hectare, respectively every year. The Island is always in deficit in vegetable, hence it is costly. Third year, black pepper alone not only compensated its cumulative establishment cost, but also earned a reasonably good income for Rs. 97,082 / ha. Comparing the cumulative net benefit with and without black pepper from the system across the seven cropping years of the experiments, it was estimated that black pepper introduction on the hedgerows increased the tangible profit 4.46 times greater from the modified alley- cropping than the traditional system, which made the alley cropping system highly remunerative. Pruning biomass, though an important product of the alley cropping system, is not considered for the economic analysis of the modified system because it is not a marketable item. Shi *et al.* (2005) are of the view that establishment costs are a major disincentive to adopt the tree / perennial based production system like hedgerow intercropping in the short term. Black pepper is known to yield for 20 years if managed properly (Sivaraman *et al.*, 2002). Hence, economic life of black pepper for 20 years seems sufficient to earn farmers a handsome income from the improved system.

The improved alley – cropping system appears quite attractive as it increases the yield of black pepper about three times compared to that on arecanut standard in the traditional black pepper cultivation system in the Island, and makes the harvesting of pepper berries easy due to shorter height of the hedgerows that saves labour input and ultimately reduces the cost of cultivation. Pepper production is reported an average 0.3 kg / standard on the arecanut in the Island (Pandey and Venkatesh, 2007), 0.6 kg / vine in Kerala and 303 kg / ha across the pepper growing states of India (Sivaraman *et al.*, 2002). Black pepper growers plant arecanut at 2.7 m x 2.7 m spacing in their homegardens in the Islands. Therefore, the modified alley cropping system seems comparatively more profitable and socially desirable as it requires no additional land and crop management inputs, but provides high return relative to small investment, once it is established.

Conclusions

Gliricidia leaves decompose quickly and release maximum nitrogen within 15 days and remaining part in two months in both surface applied as well as soil incorporated conditions. It increases mineralization and mineral N pools in soil, but the rate of mineralization as well as the amount of nitrogen pools are maximum in 2WAS and lowest in 4WAS treatment. It seems the rate of mineralization and nitrogen uptake synchronize maximum in 2WAS treatment that facilitate higher grain yield compared to that in 0WAS and 4 WAS treatments. Recovery of nitrogen from the *Gliricidia* leaves by the maize crop was in general low, but it also occurred highest in 2WAS treatment. These suggest that *Gliricidia* leaves should be applied as an organic manure after 2WAS with either of the two methods for maximum shoot biomass and grain production.

Black pepper vines trail on the double hedgerows of *Gliricidia* of the modified alley-cropping system quickly and starts bearing, though learning stage, from the second cropping year. Maximum production occurs from the sixth cropping year. Net return from the black pepper is negative in the beginning for two cropping years, but maize- okra crop rotation in the alleys offsets the negative return from the system. Net return in black pepper, averaged across the seven cropping years of the experiments, not only compensates the total investment incurred for its establishment, and reduction in the yields of maize but make the system tangibly profitable by 4.46 times greater than without the black pepper. It indicates that introduction of black pepper on the hedgerows in the alley – cropping on the long term will make the system highly remunerative and most attractive to farmers.

References

Amara, D. S., Sanginga, N. S., Danso, K. A., Suale D. S. 1996. Nitrogen contribution by multipurpose trees to rice and cowpea in an alley-cropping system in Sierra Leone. Agrofor. Syst. 34:119-128.

Bross, E. L., Gold, M. A., Nguyen, P. V.1995.Quality and decomposition of black locust (*Robinia pseudoacacia*) and alfalfa (*Medicago sativa*) mulch for temperate alley cropping systems. Agrofor. Syst. 29: 255 – 264.

Browaldh, M. 1995. The influence of trees on nitrogen dynamics in an agrisilvicultural system in Sweden. Agrofor. Syst. 30:301-313.

Campbell, C. A., Jame, Y. W., Akinremi, O. O., Cabera, M. L. 1995. Adapting the potentially minerlization N concept for the prediction of fertilizer N requirements. Fert. Res. 42: 61-75.

Casey, J. 2004. Agroforestry adoption in Mexico: using Keynes to better understand farmers decision making. J. Post Keynes 26 (3):505-521.

Chen, Y. B., Lin, W. C., Zhu, Z. L., He, Y. G. 2002. Studies on alley cropping system and its ecological and economical benefits. J. Soil Water Cons. 16(2):80-83.

Clement, A., Ladha, J. K., Chalifour, F. P. 1995. Crop residue effects on nitrogen minerlization, microbial biomass and rice yield in submerged soils. Soil Sci. Soc. Am. J. 59: 1595-1603.

De Datta, S.K. 1995. Nitrogen transformations in wetland rice ecosystems. Fert. Res. 42: 193-203.

Eno, C.F. 1960. Nitrate production in the field by incubating the soil in polyethylene bags. Soil Sci. Soc. Am. Procee. 24: 277-279.

Fox, R.H., Myers, R.J.K., Vallis, I. 1990. The nitrogen minerlization rate of legume residues in soils as influenced by their polyphenol, ligning and nitrogen contenets. Plant Soil 129: 251-259.

Frankenberger, W.T., Abdelmajid, H.M. 1985. Kinctic parameters of nitrogen minerlization rates of leguminous crops incorporated into soils. Plant Soil 87: 257-271.

Gill, R. I. S., Singh, B, Kaur, N. 2009. Productivity and nutrient uptake of newly released wheat varieties at different sowing times under poplar plantation in north-western India. Agrofor. Syst. 76:579-590.

Giller, K.E., Cadisch, G.1995. Future benefits from biological nitrogen fixation. An ecological approach to agriculture. Plant Soil 174: 255 –277.

Haggar, J. P., Tanner, E. U., Beer, J. W., Kass, D.C.L.1993.Nitrogen dynamics of tropical agroforestry and annual cropping systems. Soils Biol. Biochem. 25:1363 –1378.

Haslam, E. 1981. Vegetable tannins. Pp 527-556. *In*: E.E.Conn (ed.) *The Biochemistry of Plant*. Vol 7. Academic Press, New York.

John, P. S., Pandey, R. K, Buresh, R. J., Prasad, R. 1992. Nitrogen contribution of cowpea green manure and residue to upland rice. Plant Soil 142: 53-61.

Kang, B. T., Wilson, G. F., Sipkens, L. 1981. Alley cropping maize (*Zea mays* L.) and Leucaena (*Leucaena leucocephala* Lam) in southern Nigeria. Plant Soil 63:165-179.

Karen, A. O., Fownes, J. H. 1992. Effects of chemical composition on nitrogen minerlization from green manures of seven tropical leguminous trees. Plant Soil 143: 127-132.

Lehmann, J., Schroth, G., Zech, W. 1995. Decomposition and nutrient release from leaves, twigs and roots of three alley- cropped tree legumes in central Togo. Agrofor. Syst. 29: 21-36.

Mafongoya, P. L., Nair, P.K.R. 1997. Multipurpose tree prunings as a source of nitrogen to maize under semiarid conditions in Zimbabwe.1. Nitrogen- recovery rates in relation to prunings quliaty and method of application. Agrofor. Syst. 35:31- 46.

Myers, R.J.K., Palm, C. A., Cuervas, E., Gunatilleke, I.U.M., Brossard, M. 1994. The synchronization of nutrient mineralization and plant nutrient demand.pp.243 *In*: P.L.Woomer & M.J.Swift (eds.) The Biological Management of Tropical Soil Fertility. John Wiley and Sons Ltd, Chichester UK.

NBSSLUP. 1991. Soil Resource Atlas: Andaman and Nicobar islands. National Bureau of Soil Survey and Land Use Planning (ICAR) and Directorate of Agriculture, Andaman and Nicobar Administration, Port Blair, India.

Palm, C. A., Sanchez, P. A. 1991. Nitrogen release from the leaves of some tropical legumes as affected by their legning and polyphenolic contents. Soil Biol. Biochem. 23: 83- 88.

Palm, C. A.,1995. Contribution of agroforestry trees to nutrients requirements of inter cropped plants. Agrofor. Syst. 30: 105-124.

Pandey, C.B., 2011. A modified alley-cropping system of agroforestry: an analysis of production potential and economics. Indian J. Agric. Sci. 81 (7) : 616-621.

Pandey, C.B., Rai, R.B. 2007. Nitrogen cycling in gliricidia *(Gliricidia sepium)* alley cropping in humid tropics. Trop. Ecol. 48 (1) : 87-97.

Pandey, C. B., Chaudhari, S. K. 2010. Soil and nutrient losses from different land uses and vegetative methods for their control on hilly terrain of South Andaman. Indian J. Agric. Sci. 80 (5):50-55.

Pandey, C. B., Rai, R. B., Singh, L., Singh, A. 2007b. Homegardens of Andaman and Nicobar, India. Agric. Syst. 92:1-22.

Pandey, C. B., Rai, R. B., Singh, L. 2007a. Seasonal dynamics of mineral N pools and N-mineralization in soils under homegarden trees in South Andaman, India. Agrofor. Syst. 71:57-66.

Pandey, C. B., Sharma, D. K., Singh, A. K. 2001. *Leucaena*-Linseed compitition in an alley-cropping system in central India. Trop. Ecol. 42(2): 187-198.

Pandey, C. B., Venkatesh, A. 2007. Agroforestry for sustainable biomass production in the Andaman and Nicobar Islands. pp 147. Technical Report, Central Agricultural Research Institute, Port Blair, India.

Pandey, C. B., Venkatesh, A. 2003. Tree-soil-crop interactions in agroforestry practices in Andaman and Nicobar islands. pp 15-16. *In*: R.B. Rai, T.V.R.S. Sharma, R. Soundararajan, N. Sarangi, R.P. Medhi, G.S. Chaudhuri & S.C.Pramanik (eds.) Annual Report, Central Agricultural Research Institute, Port Blair, India.

Pandey, C.B., Sharma, D.K., Bargali, S.S. 2006. Decomposition and nitrogen release from Leucaena leucocephala in central India. Trop. Ecol. (in press).

Pandey, C.B., Lata, K., Venkatesh, A., Medhi, R.P. 2002. Homegarden: its strucuture and economic viability in south Andaman. Indian J. Agrofor. 4: 17–23.

Sanchez, I. A., Lassaletta, L., McCollin, D., Bunce, R.G. H. 2010. The effect of hedgerow loss on microclimate in the Mediterranean region: an investigation in Central Spain. Agrofor. Syst. 78:13-25.

Shi, D. M., Lu, X.P., Liu, L. Z. 2005. Study on functions of soil and water conservation by mulberry hedgerow intercropping of purple soil slopping farmland in Three Gorges Reservoir Region. J. Soil Water Cons. 19(3):75-79.

Sivaraman, K., Madan, M. S., Selvan, M. T. 2002. *Black Pepper Guide*, pp 79.. Directorate of Arecanut and Spices Development, Ministry of Agriculture and Department of Agriculture and Co-operation, Govt of India, Calicut, Kerala, India.

Swift, M. J. (ed.) 1987.Tropical Soil Biology and Fertility (TSBF). Inter- regional research planning workshop report. Biology International Special Issue 13, IUBS, Paris.

Tejwani, K. G. 1994. Agoroforestry in India. Oxford and IBH Publishing Co Pvt Ltd, New Delhi, India.

Wetzel, R. G., Likens, G. E. 1979. Limnological Analysis. Saunders, Philadephia.

Wieder, R., Lang, G. 1982. A critique of the analytical methods used in examining decomposition data obtained from litterbags. Ecology 63:1636-1642.

Wong, M. T. F., Nortcliff, S. 1995. Seasonal fluctuations of native available N and soil management implications. Fert. Res. 42: 13-26.

14

Mechanism of Competitive Interaction in Agroforestry: A Case Study from An Alley-Cropping System

C.B. Pandey

Abstract: *Trees are known to reduce growth and yield of crops under their canopy in agroforestry systems through competitive interaction. This chapter describes the mechanism of competitive interaction between Linseed (Linum usitatissimum L.) and Leucaena leucocephala (Lam) De Wit in an alley cropping system. To know the mechanism of competitive interaction in agroforestry, three competition situations are created: crop + Leucaena shrub neighbour, crop + L. hedge neighbour and sole crop. In this study competitive interaction is studied in an alley cropping system having two alley width sizes i.e. 4m and 8m. Sunlight at ground, soil organic C and total N are measured in all the competition treatments at different distances from shrub and hedge neighbors. Sunlight is not reduced in crop + hedge treatment like that in sole crop treatment. Sunlight is reduced in crop + shrub treatment (36 to 82%), being the highest in 4m alley and the lowest in 8m alley size. Soil organic C is high but total N is low in crop + hedge and crop + shrub neighbour than in sole crop treatment. Comparing the two competitive treatments (crop + hedge and crop + shrub neighbour) it is observed that growth rate (42 to 55%) and crop yield (71 to 72%) are reduced greater due to aboveground than belowground competition. Belowground competition is 3.6 times greater in 4m than in 8m alley, suggesting that spacing is an important factor, which regulates tree/crop competition so does the crop yield reduction in an agroforestry system.*

Introduction

In agroforestry, crop yields decline under trees due to competition (-ve interaction) between tree and crop components for aboveground (light) and belowground (soil water, nutrients) growth resources (Huxley *et al.*, 1989; Ong *et al.*, 1991; Pandey *et al.*, 1999). Trees generally utilize the growth resources proportionately greater and make them growth limiting. Competition starts between the tree and crop component for the growth limiting resources (Wedin and Tilman, 1993). Differences in establishment timing of competing species lead to asymmetries in their size and in resource capture (Miller, 1987; Weiner, 1990). Once difference in resource acquisition rate between tree and crop components are established, it is maintained or magnified during competition because of a positive feed back between growth and resource capture (Wedin and Tilman, 1993). Competition for light seems more asymmetrical than competition for nutrients (Weiner, 1990). Abilities to compete for light and nutrients are positively correlated because greater size and growth rate of tree allows it to preempt both above-and below-ground resources (Grime, 1973). This study describes (i) relative impact of above-and below-ground competition on the growth and productivity of linseed (*Linum utitatissimum* L.) in *Leucaena leucocephala* (Lam) De Wit shrub/*L. utitatissimum* intercropping, (ii) processes through which shrub neighbour modify growth resources and affect the crop.

Measurement of above-and below-ground competition in agroforestry

To separate above-and belowground competition in agroforestry, hedge rows of *Leucaena leucocephala* are raised in an agricultural field in east-west direction. Hedgerows are generally planted with *Leucaena leucocephala* seedlings, raised in a nursery. In this study hedgerows with two alley widths (8m distance 4m) are taken for the study. Within a row, distance of the seedlings is 50 cm. All the hedgerows are prunned to 1m from ground thrice in a year i.e. August, November and May as a standard practice of alley-cropping.

To measure the aboveground and belowground competition, half portion of two successive hedge are left unprunned randomly either in east or west side. Unprunned hedges grow to shrub. A plot without hedgerows is arranged in either side of each alley in the same east-west direction as is the hedgerows. *Linum usitatissimum* or other annual crop is sown in each alley including plot without hedgerows at an inter-row distance of 25 cm. Plant to plant distance is 5 to 6 cm. Thus there are three competition treatment i.e. sole crop (having no neighbour hence no competition), crop with *L.* hedge (neighbour having only root system hence only belowground competition) and crop with *L.* shrub (neighbour having both root and shoot systems hence both above-and belowground competition). Watering is done in the alleys including adjacent plots, as and when required. Fertilizer (N, P_2O_5, K_2O) is applied as basal dose at the rate of 60:30:30 kg ha^{-1} in each alley and adjacent plot at the time of the crop sowing. The hedge (crop +

hedge treatment) is pruned and fine branchlets are removed frequently from rainy season to stabilize death and decomposition of hedge's root, if any, resulted from pruning and branch removal and to ensure full availability of light to crop in the alley.

Three alleys each having shrub + neighbour crop, hedge neighbour + crop and adjacent sole crop treatment plots are selected randomly for sampling. Line transects are laid horizontally, across the competition treatments, in each alley at 50 cm, 1m, and 2m distance from the hedge and shrub neighbour from both alley sides in 4m alley. Similarly, line transects are laid in each alley, across the competition treatments from both the sides at 50cm, 1m, 2m, 3m, and 4m distance from the hedge and shrub neighbour in 8m alley. A quadrat 50 × 50 cm is placed at three random places along the transects in each treatment in each alley size, and number of tillers is counted. Shoot biomass (aboveground biomass) is determined by harvest method and belowground biomass by monolith method. Aboveground biomass of shrub is estimated on per m^2 basis as density (D) of the shrub x weight of a shrub. Density of the shrub is calculated as:

$$D = \frac{\text{area (100 x 100 cm)}}{\text{row to row (cm) x shrub to shrub (cm) distance}}$$

All shrubs in each row are cut at the time of crop harvesting at 1m from ground and biomass of their leaves and stems are estimated separately. Growth rate (r = biomass accumulation rate) of the crop is calculated as:

$r = \frac{W_2 - W_1}{t_2 - t_1}$ where, W_2 is final biomass at time t_2 and W_1 at time t_1

Harveat index (HI) is calculated as: grain yield divided by total biomass (grain yield + straw yield).

Intensity of competition (CI) is calculated following Wilson and Tilman (1991) as:

$CI = \frac{rNN\text{-}rSN}{rNN}$ where, CI is competition intensity, rNN is growth rate in crop without shrub neighbour and rSN is growth rate in crop in presence of shrub neighbour

Modification of above-and below ground growth resources under *Leucaena* shrub

Sunlight is reduced by 82% in 4m and 36% in 8m alley under the shrub in crop + shrub neighbour than in sole crop treatment (Table1). Sunlight under the hedge, both in 4m and 8m alleys does not differ from that in sole crop treatment. Sunlight in crop + shrub neighbour treatment in 8m alley is correlated to distance (r = 0.937, P<0.01). Soil organic C in crop + shrub neighbour treatment, being at par to that in crop + hedge neighbour, is 19% higher in 4m and 11% higher in 8m alley than in sole crop treatment. Soil organic C is inversely related to the distance in crop+ hedge neighbour and crop + shrub neighbour treatment in 4 m (P<0.05) and 8m (P<0.01) alley-size. Total N follows the pattern of organic C in both the competition treatments and alley sizes. C/N ratio in both the alley sizes varies due to the competitive treatment (P<0.001) and distance (P<0.01) and their interaction. Contrary to organic C and total N in the soil, C/N ratio declines in crop + hedge and crop + shrub neighbour treatments compared to sole crop treatment.

Crop growth and yield as a measure of competition

Harvest index is reduced in crop + shrub neighbour treatment, but greater in 4m alley (76%) than 8m alley (68%) (Table 2). Grain yield is correlated to the harvest index (P<0.01) in both the competition treatments (crop + hedge and crop + shrub neighbour) in both the alley-sizes. Aboveground biomass is reduced maximum (80%) in crop + shrub neighbour and minimum (73%) in crop + hedge neighbour treatment compared to sole crop treatment in 4m alley. Similarly, aboveground biomass in 8m alley-size is also reduced maximum (64%) in crop + shrub neighbour and minimum (9%) in crop + hedge neighbour treatment compared to that in sole crop treatment. Density of the crop is reduced maximum in crop + shrub neighbor (82% in 4m and 57% in 8m alley) and minimum in crop + hedge neighbour (36% in 4m and 16% in 8m alley) than in sole crop treatment. Similarly, grain yield is reduced maximum (96% in 4m and 89% in 8m alley size) in crop + shrub neighbour treatment and minimum (25 % in 4m and 17% in 8m alley sizes) in crop + hedge neighbour treatment compared to that in sole crop treatment. Estimate of sunlight intensity reveals that light to ground in crop + hedge neighbour treatment is equal to that in sole crop treatment. This indicates that reduction in growth of the crop due to hedge neighbour is primarily owing to belowground competition. However, decline in growth rate and yield of the crop in crop + *Leucaena* shrub treatment compared to that of sole crop suggests that both above- and belowground competition occurs under canopy of the shrub.

Table1. Light intensity, organic C and N of soil under different competition treatments in 4m and 8m alley-size in central India

Distance	Light intensity (foot-candle)			Organic C (%)			Total N (%)			C/N ratio		
from neighbour (m)	Crop+ Hedge neighbour	Crop+ Shrub neighbour	Sole crop	Crop+ Hedge neighbour	Crop+ Shrub neighbour	Sole crop	Crop+ Hedge neighbour	Crop+ Shrub neighbour	Sole crop	Crop+ Hedge neighbour	Crop+ Shrub neighbour	Sole crop
4 m alley												
0.5	$^{a}2113^{a}$	$^{b}349^{b}$	$^{a}2136^{a}$	$^{a}0.81^{a}$	$^{a}0.83^{a}$	$^{b}0.68^{a}$	$^{a}0.080^{a}$	$^{b}0.088^{a}$	$^{c}0.053^{a}$	$^{a}10.16^{a}$	$^{a}9.56^{a}$	$^{b}12.95^{a}$
1	$^{a}2114^{a}$	$^{b}390^{b}$	$^{a}2147^{a}$	$^{a}0.80^{a}$	$^{a}0.81^{a}$	$^{b}0.67^{a}$	$^{a}0.071^{a}$	$^{b}0.075^{b}$	$^{c}0.053^{a}$	$^{a}11.32^{b}$	$^{a}10.80^{b}$	$^{b}12.66^{a}$
2	$^{a}2121^{a}$	$^{b}403^{b}$	$^{a}2162^{a}$	$^{a}0.75^{b}$	$^{a}0.75^{b}$	$^{b}0.67^{a}$	$^{a}0.063^{c}$	$^{b}0.068^{c}$	$^{c}0.052^{a}$	$^{a}11.93^{b}$	$^{a}11.03^{b}$	$^{b}13.08^{a}$
8 m alley												
0.5	$^{a}2153^{a}$	$^{b}397^{b}$	$^{a}2148^{a}$	$^{a}0.82^{a}$	$^{b}0.83^{a}$	$^{b}0.67^{a}$	$^{a}0.082^{a}$	$^{b}0.086^{a}$	$^{c}0.053^{a}$	$^{a}10.05^{a}$	$^{a}9.71^{a}$	$^{b}12.67^{a}$
1	$^{a}2155^{a}$	$^{b}435^{b}$	$^{a}2150^{a}$	$^{a}0.80^{a}$	$^{a}0.78^{ab}$	$^{b}0.68^{a}$	$^{a}0.074^{b}$	$^{b}0.078^{b}$	$^{c}0.055^{a}$	$^{a}10.79^{a}$	$^{a}10.00^{a}$	$^{b}12.60^{a}$
2	$^{a}2162^{a}$	$^{b}1705^{b}$	$^{a}2165^{a}$	$^{a}0.77^{a}$	$^{b}0.73^{bc}$	$^{c}0.68^{a}$	$^{a}0.062^{c}$	$^{a}0.065^{c}$	$^{c}0.052^{a}$	$^{a}12.46^{b}$	$^{b}11.25^{b}$	$^{c}13.23^{a}$
3	$^{a}2154^{a}$	$^{a}2171^{c}$	$^{a}2175^{a}$	$^{a}0.65^{b}$	$^{b}0.70^{c}$	$^{c}0.67^{a}$	$^{a}0.054^{d}$	$^{a}0.056^{c}$	$^{b}0.053^{a}$	$^{a}12.39^{b}$	$^{a}12.50^{c}$	$^{a}12.73^{a}$
4	$^{a}2160^{a}$	$^{a}2193^{c}$	$^{a}2180^{a}$	$^{a}0.64^{b}$	$^{a}0.65^{c}$	$^{a}0.67^{a}$	$^{a}0.052^{d}$	$^{a}0.053^{d}$	$^{a}0.053^{a}$	$^{a}12.37^{b}$	$^{a}12.26^{c}$	$^{a}12.65^{a}$

Values in a column for a parameter suffixed with different superscripts are significantly different at $P<0.05$
Values in a row for a parameter suffixed with different superscripts are significantly different at $P<0.05$
Source: Pandey *et al.* 2001

Table 2. Grain yield, density, aboveground biomass and harvest index of *Linum usitatissimum* under different competition treatments in 4m and 8m alley-size in central India

Distance	Grain yield (g m^{-2})			Density (tiller m^{-2})			Aboveground biomass (g m^{-2})			Harvest index		
from neighbour (m)	Crop+ Hedge neighbour	Crop+ Shrub neighbour	Sole crop	Crop+ Hedge neighbour	Crop+ Shrub neighbour	Sole crop	Crop+ Hedge neighbour	Crop+ Shrub neighbour	Sole crop	Crop+ Hedge neighbour	Crop+ Shrub neighbour	Sole crop
4 m alley												
0.5	$^{a}17.35^{a}$	$^{b}0.83^{a}$	$^{c}42.0^{a}$	$^{a}172^{a}$	$^{b}48^{a}$	$^{c}400^{a}$	$^{a}77^{a}$	$^{b}15^{a}$	$^{c}148^{a}$	$^{a}0.19^{a}$	$^{b}0.05^{a}$	$^{a}0.22^{a}$
1	$^{a}32.53^{b}$	$^{b}1.53^{b}$	$^{c}40.5^{a}$	$^{a}270^{b}$	$^{b}74^{b}$	$^{c}398^{a}$	$^{a}105^{b}$	$^{b}35^{b}$	$^{c}163^{a}$	$^{a}0.23^{b}$	$^{b}0.05^{a}$	$^{a}0.20^{a}$
2	$^{a}41.46^{a}$	$^{b}2.73^{c}$	$^{a}39.5^{a}$	$^{a}321^{c}$	$^{a}101^{c}$	$^{c}402^{a}$	$^{a}112^{b}$	$^{b}47^{c}$	$^{c}156^{a}$	$^{a}0.27^{c}$	$^{a}0.06^{a}$	$^{a}0.20^{a}$
8 m alley												
0.5	$^{a}24.5^{a}$	$^{b}1.7^{a}$	$^{c}47.2^{a}$	$^{a}258^{a}$	$^{b}74^{a}$	$^{c}456^{a}$	$^{a}105^{a}$	$^{b}28^{a}$	$^{c}145^{a}$	$^{a}0.19^{a}$	$^{b}0.06^{a}$	$^{c}0.24^{a}$
1	$^{a}35.6^{b}$	$^{b}3.3^{b}$	$^{c}48.8^{a}$	$^{a}330^{b}$	$^{b}130^{b}$	$^{c}455^{a}$	$^{a}127^{b}$	$^{b}44^{b}$	$^{c}145^{a}$	$^{a}0.22^{b}$	$^{b}0.07^{c}$	$^{c}0.25^{a}$
2	$^{a}44.5^{c}$	$^{b}5.8^{c}$	$^{a}48.7^{a}$	$^{a}424^{c}$	$^{b}198^{c}$	$^{a}456^{a}$	$^{a}137^{c}$	$^{b}54^{c}$	$^{a}143^{a}$	$^{a}0.24^{c}$	$^{b}0.09^{c}$	$^{a}0.25^{a}$
3	$^{a}47.8^{d}$	$^{b}7.3^{d}$	$^{a}49.0^{a}$	$^{a}462^{c}$	$^{b}284^{d}$	$^{a}458^{a}$	$^{a}144^{d}$	$^{b}66^{d}$	$^{a}145^{a}$	$^{a}0.25^{d}$	$^{b}0.09^{c}$	$^{a}0.25^{a}$
4	$^{a}49.6^{d}$	$^{b}8.7^{c}$	$^{a}49.3^{a}$	$^{a}455^{c}$	$^{b}295^{d}$	$^{a}458^{a}$	$^{a}147^{d}$	$^{b}70^{d}$	$^{a}148^{a}$	$^{a}0.25^{d}$	$^{b}0.11^{d}$	$^{a}0.25^{a}$

Values in a column for a parameter suffixed with different superscripts are significantly different at $P<0.05$

Values in a row for a parameter suffixed with different superscripts are significantly different at $P<0.05$

Source: Pandey et al. 2001

Growth rate of *Leucaena leucocephala* is greater than that of its intercrop hence it utilizes greater growth resources than does its intercrop. This leads to competition between the L. leucocephala and Linseed (Table 3). When growth rate and yield of the crop in crop + shrub treatment are compared with growth rate and yield of the crop in crop + hedge treatment, reduction in the growth and yield was the greatest (42 to 55% and 71 to 72%, respectively) due to aboveground competition and least (9 to 37% and 17 to 25%, respectively) due to below ground competition (Table 4). Greater effect of above ground competition is evident also from results of step-wise multiple regression, which indicates that light explains maximum 84 to 86 % of the variability in growth rate and 74 to 83 variability in grain yield across the alley sizes. However, total nitrogen and organic carbon C in soil explains only 0.9 to 1.7 % and 0.7 to 14% of the variability, respectively in growth rate, and 6 to 10% and 6 to 20%, respectively of the variability in grain yield of the crop. Ong *et al.* (1992) in a similar experiment report that reduction in maize yield in *Leucaena* - maize intercropping is 46 to 55% on plots with over storey of *Leucaena* tree and 18 to 20% on plots with prunned hedgerows. Tilander *et al.* (1955) find that when trees are coppiced in the alley farming system, competition from tree is generally not strong enough to result in significant row differences in sorghum yield in both 5m and 8m alleys. Karim *et al.* (1991) and Kang *et al.* (1981) report aboveground competition for light greater that belowground competition for soil-water and nutrients in *Leucaena*-maize intercropping. Comparison of crop + hedge and crop + shrub treatment is done because effect of aboveground and belowground competition is additive. Greater reduction in the density and yield of the crop in alley with *Leucaena* shrub compared to that in alley with *Leucaena* hedge neighbour indicates that effect of above and belowground competition is additive. The assumption that effect of aboveground and belowground competition may be additive is also based on the fact that difference in transpiration rates of hedge row prunned at 50cm and under storey tree of *Leucaena* is negligible (Ong *et al.*, 1992). They report that this negligible difference is due to young, actively transpiring leaves of pruned hedge compared to proportionately greater older leaves and reproductive structures in *Leucaena* tree. Positive interactions between above-and belowground competition, such that their combined effects are greater than the sum of their separate effects are reported in replacement series type pot experiments by Donald (1958). Lower effect of belowground competition on the crop may be due to low amount of root of *Leucaena* in upper layer (0-15cm) of soil which is apparent from low neighbour root / crop root ratio.

Table 3. Growth and biomass production of Leucaena shrub (data are mean across the shrub rows and years)

Alley size(m)	November Height (m)	Biomass production Canopy diameter (m)	Leaf biomass (g m^{-2})	Stem biomass (g m^{-2})	Total aboveground biomass (g m^{-2})	Growth rate (g m^{-2} day^{-1})
4	3.45 ± 0.18	2.40 ± 0.09	288 ± 19	1130 ± 54	1419 ± 72	11.83 ± 0.60
8	3.83 ± 0.17	2.53 ± 0.03	130 ± 6	542 ± 26	672 ± 0	5.60 ± 0.27
Mean±SE	3.70 ± 0.18	2.47 ± 0.07	209 ± 13	836 ± 41	1046 ± 48	8.70 ± 0.44

Source: Pandey *et al.*, 2001

Intensity of competition in crop + shrub neighbour treatment is greater (2 times in 4m and 6.4 times in 8m alley size) than in crop + hedge neighbour treatment. Intensity of competition in crop + shrub neighbor treatment is 20% greater in 4m alley than in 8m alley (Table 4). Greater reduction in growth rate and yield in 4m alley may be attributed to its smaller width size (lesser space). Canopy of shrub neighbour extends 2.4 m from both sides in the alley and as a result intercept greater sunlight (Table 3). Belowground competition in 4m alley increases by 3.6 times compared to that in 8m alley most likely due to extension of roots of shrub from both sides in smaller volume of soil. But, it can not dominate over the effect of aboveground competition. Extension of root of shrub neighbour up to 2m in the plot is observed both in 4m and 8m alley sizes. Greater reduction in density and yield of the linseed crop due to the aboveground competition even in small alley size further substantiate the conclusion that light is major limiting factor under the canopy of *Leucaena* shrub and other agroforestry trees.

Biomass allocation pattern as a measure of competition

Aboveground / belowground biomass ratio {AB/BB) ratio declines by 28% in crop + hedge neighbour and by 48% in crop+ shrub neighbour compared to sole crop treatment in 4m alley (Table 4). AB/BB ratio is correlated to the distance in crop hedge and crop shrub neighbour treatments in both 4 and 8m alley sizes. Belowground biomass of the crop is correlated to its aboveground biomass in both crop + hedge and crop + shrub neighbour treatments in 4m and 8m alley sizes. Neighbour root/crop root ratio is also affected by the competition treatments. Low AB/BB ratio and harvest index of the crop in crop + shrub neighbour compared to that in crop + hedge neighbour treatment indicates greater impact of aboveground competition on the crop. Greater importance of aboveground competition is further evident from inverse relation of AB/BB ratio and harvest index with distance, and positive relation between sunlight and distance. Reduction in AB/BB ratio (Grunow *et al.*, 1980) and harvesting index (Ong, 1993) due to shade is well known. Well developed canopy at the time of crop sowing and 6 to 10 times higher growth rate (net accumulation rate of biomass) in *Leucaena* shrub over the crop seem to have allowed it to utilize proportionately greater mineral nitrogen. This proportionately greater utilization of available nitrogen seems to have caused belowground competition for it between the shrub neighbour and the crop. High amount of organic C and total N in the soil and low C/N ratio under the shrub neighbour indicates greater availability of mineral nitrogen. Greater mineralization rate is reported to occur under the tree shade due to relatively higher soil moisture in the soil (Belski, 1994). Soil moisture is found 110% greater under *Acacia nilotica* tree shade during the winter season in water limiting condition. Low population of the crop and higher neighbour shrub root/crop root ratio nearer to the shrub neighbour probably facilitate greater utilization of nitrogen (mineral N) by the shrub neighbour. One can expect to decline amount of total nitrogen in the soil under the shrub neighbour. But, it increases perhaps due to long term build up of organic carbon via dead roots and root nodules of the shrub which does not allow the system to reflect reduction in total nitrogen. Chulan and Waid (1981) report 94% transfer of net nodule nitrogen to soil in a tropical legume crop. Well developed canopy of the shrub intercepts asymmetrically greater light, which causes aboveground competition for light. Uptake of water and nutrient is known to be related to above grounddemand i.e. size of the leaf canopy and the aboveground sink strength for nutrients (Van Noordwijk *et al.*, 1996). One can expect that *Leucaena* shrub being a nitrogen fixer may not be competing for nitrogen. Contrary to this, belowground competition occurs perhaps for nitrogen due to substantially higher shoot growth rate (greater above ground demand) in it. Cruz (1997) finds low level of nodulation but strong growth in *Leucaena leucocephala* shrub under high N and soil-water condition (N uptake by the aboveground biomass more than 1000 kg ha^{-1} yr^{-1}) in a *L. leucocephala* shrub-*Dichanthum aristatum* stand in French Antilles. He argues that *Leucaena* legume shrub is more competitive for soil N rather than a N supplier. Karim *et al.* (1991) also observes below ground competition for nitrogen in a *Leucaena*-maize intercropping. Plants of nutrient-rich habitats are reported capable of rapidly consuming any surplus in available nutrients in pot experiments (Crick and Grime, 1987, Granato and Raper, 1989).

Table 4. Growth rate, AB/BB ratio, neighbour root/crop root ratio and intensity of *Linum usitatissimum* under different competition treatments in 4m and 8m alley-size in central India

Distance	Growth rate (g m^{-2} day^{-1})			AB/BB ratio			Neighbour root / crop root ratio			Intensity of competition		
from neighbour (m)	Crop+ Hedge neighbour	Crop+ Shrub neighbour	Sole crop	Crop+ Hedge neighbour	Crop+ Shrub neighbour	Sole crop	Crop+ Hedge neighbour	Crop+ Shrub neighbour	Sole crop	Crop+ Hedge neighbour	Crop+ Shrub neighbour	Sole crop
4m alley												
0.5	^{a}2.07^{a}	^{b}0.42^{a}	^{c}4.00^{a}	^{a}3.85^{a}	^{b}2.22^{a}	^{c}5.60^{a}	^{a}2.21^{a}	^{b}5.12^{a}	0	^{a}0.48^{a}	^{b}0.90^{a}	0
1	^{a}2.84^{bc}	^{b}0.94^{ab}	^{c}4.40^{a}	^{a}4.48^{b}	^{b}3.25^{b}	^{c}6.53^{a}	^{a}0.23^{b}	^{a}0.50^{b}	0	^{a}0.35^{b}	^{b}0.70^{b}	0
2	^{a}3.03^{c}	^{b}1.26^{bc}	c4.22^{a}	^{a}4.72^{b}	^{a}3.88^{c}	^{c}5.97^{a}	^{a}0.05^{b}	^{a}0.11^{b}	0	^{a}0.28^{b}	^{b}0.70^{b}	0
8m alley												
0.5	^{a}2.84^{a}	^{b}0.75^{a}	^{c}3.92^{a}	^{a}4.12^{a}	^{b}2.17^{a}	^{c}4.96^{a}	^{a}1.27^{a}	^{b}3.09^{a}	0	^{a}0.25^{a}	^{b}0.81^{a}	0
1	^{a}3.44^{b}	^{b}1.20^{b}	^{c}3.90^{a}	^{a}4.59^{bc}	^{b}3.37^{b}	^{a}4.92^{a}	^{a}0.24^{b}	^{a}0.48^{b}	0	^{a}0.12^{b}	^{b}0.69^{b}	0
2	^{a}3.70^{c}	^{b}1.47^{c}	^{a}3.87^{a}	^{a}4.69^{c}	^{b}3.81^{c}	^{a}4.93^{a}	^{a}0.05^{c}	^{a}0.10^{c}	0	^{a}0.06^{c}	^{b}0.62^{c}	0
3	^{a}3.89^{d}	^{b}1.79^{d}	^{a}3.95^{a}	^{a}4.65^{c}	^{b}3.93^{c}	^{a}4.60^{a}	^{a}0.02^{c}	^{a}0.01^{c}	0	^{a}0.03^{d}	^{b}0.55^{d}	0
4	^{a}3.96^{d}	^{b}1.89^{e}	^{a}4.00^{a}	^{a}4.30^{ac}	^{b}4.03^{c}	^{a}4.58^{a}	^{a}0.005^{c}	^{a}0.004^{c}	0	^{a}0.04^{d}	^{b}0.52^{e}	0

Values in a column for a parameter suffixed with different superscripts are significantly different at $P<0.05$
Values in a row for a parameter suffixed with different superscripts are significantly different at $P<0.05$
Source: Pandey *et al.* 2001

Conclusions

Leucaena shrub grows quickly and develops its canopy and as a result builds a big sink of nutrients aboveground which allows proportionately greater consumption of nitrogen. This disproportionate utilization of N leads to belowground competition between the shrub and the crop for the nitrogen. Canopy of the shrub neighbour intercepts light asymmetrically greater and as a result develops aboveground competition for light. Effect of aboveground competition on the growth and productivity of the crop under the shrub is greater than the belowground competition.

References

Belsky, A. J. 1994. Influences of trees on savanna productivity: test of shade, nutrients, and tree-grass competition. Ecology 75: 922-932.

Chulan, A., Waid, J. S. 1981. Loss of nitrogen from decomposing nodules and roots of the tropical legume Centrosema pubescens to soil. pp. 150-153, In: Westselaar, R., Simpson, J. R., Rosswall, T. (eds.) Nitrogen Cycling in South-East Asian Wet Monsoonal Ecosystems. Australian Academy of Science, Canberra.

Crick, J. C., Grime, J. P. 1987. Morphological plasticity and mineral nutrition capture in two herbaceous species of contrasted ecology. New Phytol. 10:403-414.

Cruz, P. 1997. Effect of shade on the growth and mineral nutrition of a C_4 perennial grass under field condition. Plant Soil 188:227-237.

Donald, C. M. 1958. The interaction of competition for light and for nutrients. Australian J. Agric. Res. 9: 421-435.

Granato, T. C., Raper, C. D. 1989. Proliferation of maize (Zea mays L.) roots in response to localized supply of nitrate. J. Exp. Bot. 40: 263-275.

Grime, J.P. 1973. Competitive exclusion in herbaceous vegetation. Nature 242:344-347.

Grunow, J. O., Groeneveld, H. T., DuToix, S. H. C 1980. Aboveground dry matter dynamics of the grass layer of a South African tree savanna. J. Ecol. 68:877-889.

Huxley, P. A., Akunda, E., Pinney, E., Darnhofer, T., Gatama, D. 1989. Tree-crop interface investigations: preliminary results with Cassia siamea and maize. pp. 361-370. In: Reifsynder W.S., Darnhofer, T. O. (eds.) Meteorology and Agroforestry. ICRAF, Nairobi, Kenya.

Kang, B. T., Wilson, G. F., Sipkens, L. 1981. Alley cropping maize (*Zea mays*) and Leucaena in Southern Nigeria. Plant Soil 63: 165-179.

Karim, A. B., Savill, S. P., Rhodes, E. R. 1991. The effect of young *Leucaena* leucocephala (Lam) De Wit hedge on the growth and yield of maize, sweet potato and cowpea in an agroforestry system in Sierra Leone. Agrofor. Syst. 16: 203-211.

Miller, T. E. 1987. Effects of emergence time on survival and growth in an early old-field comminity. Oecologia (Berlin) 72: 272-278.

Ong, C. K., Odono, C. W., Marshall, F., Black, C. R. 1991. Water use of agroforestry systems in semi-arid India. pp. 347-358. In: Calder, I.R., Hall, R.L., Adlard, P. G. (eds.) Growth and Water Use of Forest Plantations. Wiley, Chichester, UK.

Ong, C. K., Rao, M. R., Mathuva, M. 1992. Trees and crops: competition for resources above and below the ground. Agrofor. Today 4:4-5.

Ong, C. K. 1993. Cropping systems approaches to agroforestry research. pp. 51-69. In: Bentley, W.R., Khosla, P. K., Seckler, K. (eds.) Agroforestry in South Asia: Problems and Applied Research Perspectives. Oxford and IBH Publishing Co. Pvt. Ltd., New Delhi.

Pandey, C. B., Pandya, K. S., Pandey, D., Sharma, R. B. 1999. Growth and productivity of rice (Oryza sativa) as affected by Acacia nilotica in a traditional agroforestry system. Trop. Ecol. 40: 109-117.

Pandey, C.B., Sharma, D.K., Sngh, A.K. 2001. *Leucaena*-Linseed Competition in an alley-cropping system in central India. Trop. Ecol. 42 (2) : 187-198.

Singh, C. 1984. Modern Techniques of Raising Field Crops. Oxford and IBH Publishing Co. New Delhi, India.

Singh, R. P., Ong, C. K., Saharan, N. 1989. Microclimate and growth of sorghum and cowpea in alley cropping in semi-arid India. Agrofor. Syst. 9:259-274.

Statistical Package for Social Sciences. 1986. SPSS/PC for the IBM PC/XT/AT. SPSS, Chicago, III.

Tilander, Y., Ouedraogo, G., Yougma, F. 1995. Impact of tree cropping on tree-crop competition in park-land and alley farming systems in semiarid Burkina Faso. Agrofor. Syst. 30: 363-378.

Van Noordwijk, M., Lawson, G., Soumare, A., Groot, J. J., Hairiah, H. 1996. Root distribution of trees and crops: competition and or complementarity . pp. 319-364. In: Ong, C.K., Huxley, P. (eds.) Tree-crop Interaction: A Physiological Approach. CAB International Wallingford, U.K.

Wedin, D., Tilman, D. 1993. Competition among grasses along a nitrogen gradient: initial conditions and mechanisms of competition. Ecol. Monog. 63: 199-229.

Weiner, J. 1990. Asymmetric competition in plant populations. Trends Ecol. Evol. 5: 360-364.

Willey, R. W., Reddy, M. S. 1981. A field technique for separating above-and below ground interactions in intercropping: An experiment with pearl millet / groundnut. Exp. Agric. 17: 257-264.

Wilson, S. D., Tilman, D. 1991. Components of plant competition along an experimental gradient of nitrogen availability. Ecology 72: 1050-1065.

15

Agroforestry Systems for Resource Conservation and Sustainable Agriculture in Arid and Semi-arid Conditions

R.K. Singh, M. Osman and A.K. Parandiyal

Agroforestry systems essentially involve a perennial component and provide cover to the soil through litter fall and tree canopy, which contribute to the conservation of soil and water resources in arid and semi-arid areas. Moreover, arid and semi-arid areas are exposed to extensive soil erosion by wind and water, respectively. Planting of trees on erosion control structures as windbreaks, boundary plantations and tree-based systems are some of the potential ways for conserving natural resources (Singh, 2003). Agroforestry and soil conservation have major role to play in the context of global climate change likely to occur from extremes of weather events and erratic rainfall behaviour.

- Provide permanent vegetative cover to the soil or land surface and thus substantially control erosion caused by both runoff and wind.
- Improve soil quality through increased soil organic matter addition from litter fall and root turnover; and improved environment for soil microbe and earthworm activity.
- Reduce surface evaporation and weed growth and improve water use efficiency. Ensure rational utilization of soil moisture stored in various soil layers through differential rooting pattern of trees and crops.

- Provide good quality green fodder, which is in short supply to support fodder needs of livestock during lean period.
- Provide fuel, timber and minor forest products and lessen the farmers' dependence on forest reserves.
- Generate employment and provide livelihood support throughout the year, increase the level of income level and cash flow.
- Improve carbon sequestration potential through carbon storage in MPTs and SOM.

Agroforestry systems were classified by Nair (1985) into three sub-systems on a structural basis *viz.*, agrisilvicultural, silvipastoral and agrisilvipastoral. Growing of trees together with arable crops is termed as agrisilviculture and alley cropping is one of the forms of agrisilviculture systems (Table 1). Growing of trees together with forage crops is termed as silvipastoral system while agrisilvipastoral system is adopted for growing crop + animals/pasture + trees. Agrisilvipastoral is not a prevalent system in India. Katyal *et al.* (1993) have suggested land use alternatives involving agroforestry systems based on land capability class (LCC) and rainfall situation (Fig. 1).

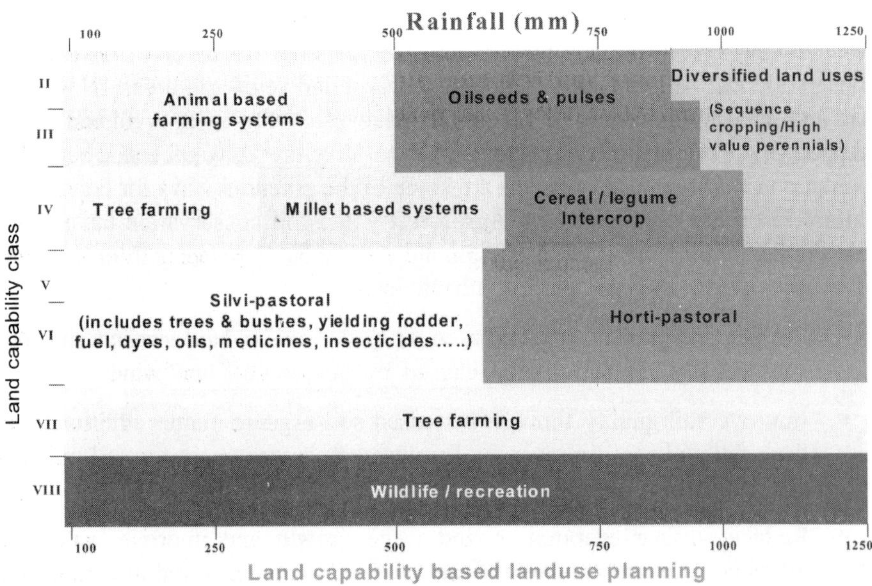

Fig. 1. Recommended Land uses for different land capability classes and different rainfall situations

Table 1. Agroforestry practices for resource conservation and sustainable agriculture

Agroforestry practice	Agro-climate	Description
Agrisilviculture	Arid and semi-arid areas Arable lands	Intercropping of *Prosopis cineraria*, *Tecomella undulata* and *Acacia* spp trees with arable crops in arid areas Alley cropping in semi-arid areas receiving >700 mm rainfall to control soil erosion by aligning alleys on contour. The distance between alleys depend on the slope and to be maintained as hedge by repeated cutting close to the ground. Suitable species for alley cropping are: *Leucaena leucocephala*, *Gliricidia maculata, Cassia siamea*.
Silvipastoral practices	Arid and semi-arid areas Non-arable lands	Trees and grasses can effectively control the soil erosion on mild to steep sloppy areas. Afforestation with suitable trees like *Acacia*, *Azadirachta indica, Prosopis cineraria, Dalbergia sissoo, Bambusa, Dendrocalamus* with grasses like *Cenchrus*, *Panicum*, *Dicanthium, Borthricloa, Cynodon, Sehima,* etc. will help in improving the productivity of degraded lands. Hardy fruit trees + *Stylos nthes hamata* (legume), Timber trees + grasses/ legume
Trees and shrubs as bio-engineering measures	All	Species like *Gliricidia maculata* (mulch), *Pongamia pinnata* (mulch cum biodiesel), *Lawsonia inermis* (Dye), *Jatropha curcas* (biodiesel), *Agave Americana* (fiber) *etc.* can be taken up on bunds for stabilizing and strengthening them and also on field boundaries. Livestock does not browse these species.
Wind breaks and shelter belts	Semi arid and arid zone	Proven potential to reduce wind erosion, stabilize sand dunes, provide supplementary products. *Acacia* spp., *Prosopis juliflora, Crotalaria burhia*, *Leptadenia pyrotechnica* and *Arva psuedotomentosa* are some of the trees suitable for arid regions.
Agrihorticulture	Arid and semi-arid areas Arable lands	Intercropping of hardy fruit spp. in cropped area such as *Phylanthus emblica*, *Achras zapota* (Chiku), *Citrus* spp., *Psidium guajava, Punica granatum,* etc. to ensure minimal livelihood support during drought years.

Agrisilvicultural Systems

Agrisilviculture systems are suitable to arable lands classified under land capability class II and III. Acceptance by farming communities for these systems depends largely upon severity of yield reduction of the main annual crop due to tree-crop competition for soil moisture in arid and semi-arid areas. Presence of a well developed tree canopy restricts light availability to under story crops which is also a limiting factor determining the production potential of the crop. Therefore, tree management is essential component to minimize competition and maintaining satisfactory crop yield levels. An acceptable harvest of an annual crop is possible through three different ways: pollarding of trees, maintaining of trees as hedgerows (alley cropping) and third by bush farming.

Pollarding of trees

Canopy management by pollarding of trees as observed by Hocking and Rao (1990) was found useful and gave an acceptable harvest of sorghum in semi-arid India. In a poor rainfall year when both leucaena and sorghum were under stress pollarding improved LER from 0.89 to 1.35 although yield of sorghum in agroforestry system was only 46 percent of sole crop yield (Table 2). A higher yield of leucaena under pollarded system resulted in yield advantage of 35 percent over sole system of tree or crop. Pollarding was found useful in not only maximizing crop yield but also yield of green fodder. Biomass production of stylo and cenchrus was 3.5 and 3.1 t/ha under pollarded trees compared to 2.0 and 1.7 t/ha under non-pollarded, respectively. The results reveal that efficient management of tree canopy is essential for agrisilviculture as well as silvipastoral system (CRIDA, 2002).

Table 2. Yield of sorghum and leucaena under different canopy management situations

Treatments of mixed crops	Yields (t/ha/yr)					Total biomass	LER
	Sorghum		Leucaena				
	Grain	Stover	Fodder	Fuel	Increment		
Leucaena un-lopped	0.01	0.79	-	-	3.55	4.34	0.88
Lightly lopped	0.13	1.62	1.45	0.30	2.60	6.10	1.08
Intensively lopped	0.25	1.71	1.40	0.29	2.36	6.01	1.10
Pollarded	1.31	3.92	1.15	3.84	1.79	12.11	1.35
Pure crops							
Leucaena un-lopped	-	-	-	-	4.52	4.52	-
Lightly lopped	-	-	1.21	0.25	3.53	4.99	-
Intensively lopped	-	-	1.35	0.28	3.05	4.68	-
Pollarded	-	-	2.27	5.16	1.89	9.32	-
Pure sorghum	2.85	5.67	-	-	-	8.52	-

Note: Data for sorghum are from 36 m^2 net plots; for Leucaena from 9-tree net plots, trees spaced 6m x 2m. 'Incr' column is increment in dry weight of gross standing biomass, calculated from dbh increments using a whole-tree function and dry matter factor except for pollarded treatments, where a double Samalian function was applied.

Alley cropping or hedgerow intercropping

In alley cropping, perennials are pruned close to ground and maintained as hedgerows to minimize competition with crops being grown in the alleys between two hedgerows (Duguma *et al.*, 1988; Kang *et al.*, 1990). In an experiment in semi-arid India, pruning of leucaena close to the ground at 15 cm height and wider alleys (7.8 m) were found beneficial and gave higher yield than sole crops when prunings were recycled as mulch (Singh *et al.*, 1987) (Table 3).

Table 3. Effect of cutting height on crop and leucaena yield

System	Grain yield (kg/ha) 60 cm	15 cm
	Pearlmillet	
Sole	1800	2010
7.8 m mulch	1190 (66)	2370 (118)
7.8 m fodder	1170 (65)	1730 (86)
	Soghum	
Sole	1520	1850
7.8 m mulch	840 (55)	1910 (103)
7.8 fodder	730 (48)	1670 (90)

Note: Figures in parentheses indicate crop yield as percent of sole

In the semi-arid tropics, competition for soil moisture is a major limiting factor causing the crop yield reduction in alley cropping studies (Singh *et al.*, 1989; Ong *et al.*, 1991; Rao *et al.*, 1991). However during normal rainfall year suppressed growth and yield of maize in the vicinity of hedgerows was attributed to shading, while the competition for moisture is more prominent factor during low rainfall years (Lal, 1989b). Experiments in semi-arid India indicated that growth of *Leucaena leucocephala* hedgerows continues even during the post-rainy period and yields 2 to 5 t ha^{-1} of green fodder essential for maintaining livestock when annual cropping is difficult (Ong, 1991).

Sorghum and pearl millet (C_4) are reported to be the most shade tolerant crops among various cereals with yield averaging about 75% of pure crop in wide alleys across various sites and years (Singh, *et al.*, 1989). Rao *et al.* (1991) reported that oilseed crops such as: sunflower, groundnut and castor bean are relatively less shade tolerant than sorghum. In a dry year, the adverse impact of *leucaena* was found to be more severe on sunflower than on sorghum. Lal (1989 b) observed a 10% decline in grain yield of maize (C_4) and up to 50% decrease in the yield of cowpea (C_3).in alley cropping on gross area basis compared to pure crop. Sorghum has also been found to be more compatible crop with Lucaena and *Eucalyptus tereticornis* compared to pigeon pea and black gram when trees are grown as boundary plantation (Prasad *et al.*, 1987; Prasad and Prasad, 1997). Prasad (1994) reported similar results with shelterbelt plantation along with improvement in soil fertility through addition of 371 kg/ha N through lucaena leaf litter.

Bush farming

Annual crops are more compatible with short growing shrubs than with trees in drylands. Perennial bushes are less competitive for light and moisture compared to trees because of smaller canopy. Plantation of economic shrubs and intercropping with arable crops would improve the production and income from rainfed areas, apart from bringing the required stability.

Some of the shrubs that have shown potential under rainfed conditions are henna (*Lawsonia inermis*), curry leaf (*Murraya koenigii*), annato (*Bixa orellana*) and jatropha (*Jatropha carcus*). Henna has a potential of yielding 2500 kg per ha per year dry leaf from second year onwards. Curry leaf yields about 10000 kg fresh leaf per ha per year from second year onwards. It can yield another additional 2500 to 3000 kg if the rainwater is harvested and recycled to irrigate the plants during the off-season. In places where there is demand for fresh curry leaf, this system is profitable. Jatropha is also a promising species for bush farming which is being planted extensively for producing bio-diesel. The survival of this plant is excellent under rainfed conditions however yield is low (1000 kg per ha per yr). Annato yields food grade dye. Under rainfed conditions it yields 600-700 kg seed per ha from third year onwards. A stable market and price would make this species popular. Further, low growing short duration legumes like black gram and green gram can be intercropped with curry leaf and annato; which would yield about 70% of the sole crop. Unlike trees, bushes have short gestation period, start yielding 1-2 years after planting and are readily accepted by the farmers.

Silvipastoral systems

Marginal drylands, are usually shallow in soil depth and poor in nutrients. Crops yields from these lands are generally low, uncertain and often not remunerative. The returns from these lands may improve if they are put to silvipastoral systems and integrated with small ruminants. These systems apart from yielding fuel wood and fodder, improve soil fertility. After one rotation of about 6-8 years with silvipastoral system, arable crops can be grown on the built up soil fertility without application of any fertilizer.

Soil erosion is often severe on pastures exposed to unregulated biotic pressures. Planting trees in the pasture lands alone usually does not effectively reduce soil erosion. Other basic requirements of pasture management like restriction of livestock numbers and rotational grazing need to be followed (Singh, 2003). Trees in silvipastoral systems supply protein-rich fodder during times when grass is absent or indigestible.

Fodder based silvipastoral system

The silvipastoral plot of *Leucaena leucocephala* planted with *Cenchrus ciliaris* or *Stylosanthes hamata* were established in 1981 at Hayatnagar Research Farm of

Central Research Institute for Dryland Agriculture (CRIDA) at Hyderabad. *Leucaena leucocephala* seedlings were planted at 2 m distance in contour furrows spaced 7.5 m apart. Half of the area under *Stylosanthes hamata* was planted through seeds and the rest under *Cenchrus ciliaris* through root slips. On an average, stylo yielded 4 t/ha/yr and cenchrus 2.5 t/ha/yr of dry matter. The tree growth in stylo system was marginally better than that under the cenchrus system. Leucaena dry matter accumulated at the rate of 4 t/ha/yr and yielded 32 t/ha after 8 years, mainly as stem and branch wood. The soil organic carbon increased by 0.04 per cent per annum in this system. The benefit cost ratio (BCR) with this system was 2.45 as against 1.6 to 1.8 with sorghum and castor (Neelam Saharan, 1989).

Fruit based hortipastoral system

In arid and semi-arid areas, inter-planting of grass strips with fruit trees will provide fodder as well as reduce wind erosion. Care should be taken that trees are planted one year ahead of the establishment of grasses. Hortipastoral systems of guava and custard apple with *Stylosanthes hamata* or *Cenchrus ciliaris* were studied at CRIDA, Hyderabad. The initial growth of fruit trees up to 18 months after pasture establishment was found to be poor with grass than legumes association. In terms of herbage yield, *Cenchrus* grass out yielded the stylo legume but fruit plants suffered severely due to moisture stress (Singh and Osman, 1995).

Timber based silvipastoral system

There is good potential of involving timber trees like *Tectona grandis* and timber cum fodder trees like *Hardwickia binata*, *Dalbergia sissoo* and *Gmelina arobrea* in a silvopasture system of semi-arid areas. The prunings of most timber trees excepting *Tectona* serve as good fodder. *Acacia nilotica* var. *cupressiformis* and *Azadirachta indica* are promising and hardy species which provide good fodder but low quality timber.

Short rotation forestry for livelihood enhancement from degraded lands

Short rotation forestry systems involving fast growing tree species ensure production of biomass for energy, besides reducing soil erosion and conserving rainwater. This can potentially generate round the year employment as well as meet the growing industrial need. There are two successful models of short rotation forestry in peninsular India involving *Prosopis juliflora* and *Leucaena leucocephala or Eucalyptus or Casuarina* grown in non-arable and arable land, respectively. Block plantations of leucaena as an alternative to tobacco and cotton is an important land use diversification ensuring not only higher income but also improvement in soil quality. A farmer is able to harvest about 100 t leucaena biomass from a hectare area in four years and is able to get a price of Rs. 1100-1300 per tonne, amounting approximately Rs.25,000 per annum with minimal investment. Similarly, the wood of *P. juliflora* is in great demand for charcoal making and also from wood based power-generating plants. A wasteland is able to produce a minimum of 5 t of biomass in a year and the net income works out to Rs. 10,000 per ha. Both leucaena and juliflora have high coppicing ability and need little investment on maintenance for second or third coppice growth.

The potential of agroforestry in the years to come is enormous considering climate change scenarios and clean development mechanism (CDM). The problem lies in successful establishment of tree component, which needs to be integrated with soil conservation measures, microsite improvement (pit modification), supplemental watering during summer and protection from stray animals. The success or failure of the system lies in right choice of appropriate perennial species, selection of annual crops/grasses/legume and timely canopy management. Preference to bushes over trees may be given to overcome the time lag in trees to yield the economic product. Takers of agrihorticulture system are more in number and need promotion on large scale to distribute risk between perennial sylvan and annual arable components. Farm income stabilization, nutritional security and employment generation are possible by harmonizing fruit / fodder / timber plants with arable crops.

References

CRIDA, 2002. Annual Report 2001-2002, Central Research Institute for Dryland Agriculture, Santoshnagar, Hyderabad, pp.111.

Duguma, B., Kang, B.T., Okali, D.U.U. 1988. Effect of pruning intensities of three woody leguminous species grown in alley cropping with maize and cowpea on an alfisol. Agrofor. Syst. 6: 19-35.

Hocking, D., Rao, G. 1990. Canopy management possibilities for araboreal leucaena in mixed sorghum and livestock small farm production systems in semi-arid India. Agrofor. Syst. 10: 135-152.

Huxley, P.A. 1983. Some characteristics of trees to be considered in agroforestry. In: Huxley, P.A. (ed.), Plant Research and Agroforestry, pp: 3-12, ICRAF, Nairobi, Kenya.

Kang, B.T., Reynold, L., Atta-Krah, A.N.. 1990. Alley farming. Adv. Agron. 43: 315-355.

Katyal, J.C., Das S.K., Korwar, G. R., Osman M. 1993. Technology for Mitigating Stresses: Alternative Land Uses. In: Virmani, S. M., Katyal, J. C., Eswaran, H., Abrol, I. P. (eds.), pp:291-305, Stressed Ecosystems and Sustainable Agriculture, Proceedings of International Symposium on Agroclimatology and Sustainable Agriculture in Stressed Environments, ICRISAT, India. Oxford and IBH Pub. Co., New Delhi, India.

Lal, R. 1989a. Agroforestry systems and soil surface management of a tropical alfisol. II: water runoff, soil erosion and nutrient loss. Agrofor. Syst. 8: 97-111.

Lal, R. 1989b. Agroforestry systems and soil surface management of a tropical alfisol, I: soil moisture and crop yields. Agrofor. Syst. 8: 7-29.

Makumba, W., Akinnifesi, F.K., Janssen, B., Oenema, O. 2007. Long-term impact of Gliricidia-Maize intercropping system on Carbon sequestration in Southern Malawi. Agric. Eco. Env. 118: 237-243.

Nair, P.K.R. 1985. Classification of agroforestry systems. Agrofor. Syst. 3: 97-128.

Neelam, S., Korwar, G.R., Das, S.K., Osman, M., Singh, R.P. 1989. Silvipastoral System in Marginal Alfisols for Sustainable Agriculture. Indian j. Dryland Agric. Res. Dev. 4:41-47.

Ong, C.K. 1991. Interactions of light, water and nutrients in agroforestry systems. In: Avery, M.E., Cannell, M.G.R., Ong, C.K. (eds.), Biophysical Research for Asian Agroforestry, pp: 107-124, Winrock International, New Delhi.

Ong, C.K., Corlett, J. E., Singh, R. P., Black, C. R. 1991. Above- and below-ground interactions in agroforestry systems. For. Ecol. Manage. 45: 45-47

Osman, M. 2003. Alternate Land Use Systems for Sustainable Production in Rainfed Areas. In: Pathak, P.S., Newaj, Ram (eds.), Agroforestry Potentials and Opportunities. pp-171-182, Agrobios Publisher, India.

Palm, C. A. 1995. Contribution of agroforestry trees to nutrient requirements of intercropped plants. Agrofor. Syst. 30: 105-124.

Prasad, A. 1994. Shelter belt of leucaena and it's effect on crop yield in rainfed Agroforestry system. In: Punjab Singh, Pathak, P.S., Roy, M. M. (Eds.), Agroforestry Systems for Degraded Lands. Vol. I., , pp. 499-506, Oxford and I.B.H., New Delhi, India.

Prasad, A., Prasad, S.N. 1997. Effect of white papinoc (*Leucaena leucocephala*) on field crops and nutrient addition in soil under agroforestry system. Indian J. Agric. Sci. 67 (11): 523-528.

Prasad, S. N., Srivastava, A.K., Narain, P., Bhola, S.N. 1985. Studies on the Compatibility of field crops with *Eucalyptus tereticornis* under agroforestry. Indian J. Soil Cons. 13(1) : 41-46.

Prasad, S.N., Verma, B., Prasad, A., Srivastava, A.K. 1989. Alley cropping in *Leucaena* for higher productivity and profitability in South-East Rajasthan. Indian J. Soil Cons. 17(1): 30-34.

Rao, K. V., Osman, M. 1994. Silvopastoral Systems for Drylands. In: Panjab Singh, Pathak, P.S., Roy, M. M. (eds.), Agroforestry systems for Degraded lands, pp: 755-760, Oxford and IBH Pub. Co., New Delhi.

Rao, M.R., Ong, C.K., Pathak, P., Sharma, M. M. 1991. Productivity of annual cropping and agroforestry systems on a shallow Alfisol in semi-arid India. Agrofor. Syst.15: 51-63.

Singh, R.P. 2003. Agroforestry for Conservation of Water Resources. In: Pathak, P.S., Newaj, Ram (ed.), Agroforestry: Potentials and Opportunities, pp. 1-6, Agrobios, India.

Singh, R.P., Osman, M. 1995. Alternate land use systems for Drylands. In: Singh, R.P. (ed.) Sustainable Development of Dryland Agriculture in India, Scientific Publishers, Jodhpur, pp. 375-389.

Singh, R.P., Vandenbeldt, R.J., Hocking, D., Korwar, G.R. 1989. Alley farming in the semi-arid regions of India. In: Kang, B.T., Reynold, L. (eds.), Alley farming in the humid and sub-humid tropics, pp. 100-122.

Singh, R.P., Vijayalakshmi, K., Korwar, G. R., Osman, M. 1987. Alternate Land Use Systems for Drylands of India. Research Bulletin No: 6, CRIDA, Hyderabad, p.37.

Young, A. 1986. Effects of trees on soils. ICRAF Reprint No:31 Nairobi, Kenya, p. 41.

❑❑❑

16

Role of Agroforestry Systems in Biodiversity Conservation: A Case Study from Coffee Based Agroforestry Systems, Karnataka, South India

Sathish, B.N., Kushalappa, C.G., Syam Viswanatha, and S. Raghavendra

Abstract: *Agroforestry, is a sustainable land use system being practiced since long time through out the country. A diverse climatic, edaphic, and environmental conditions prevailing in the country has made it possible to practice different kinds of agroforestry systems. Karnataka state has four major agro-ecological regions and coffee agroforests are prevalent in the region of Western Ghats of Karnataka, which has tropical climate. With in Karnataka, Kodagu district has highest area under coffee agroforests with a total area of 1,00,079 hectares and is the largest producer of coffee in the country. The area under coffee agroforests is increasing at the cost of natural forested landscapes, which is considered as a looming threat to biological diversity. But, at the same time there are arguments that these coffee agroforests also supports high level of biological diversity. There are very few studies to justify this argument. With this background the present study is an attempt to enumerate the biological diversity of coffee based agroforests of Kodagu.*

Introduction

Agroforestry is a farming system that integrates crops and/or livestock with trees and shrubs. The resulting biological interactions provide multiple benefits, including diversified income sources, increased biological production, better water quality, and improved habitat for both humans and wildlife. It has traditionally been practiced in several parts of India for a very long time.

Karnataka state is located in south-west India with an area of 1,91,791 sq km (19.179 million ha). Most part of the state lies on a plateau bordered in the west by costal strip along the Arabian sea, about (30 km wide; average annual rain fall 3000 to 3700 mm) and Western Ghats; a chain of hill mountains with an altitude of 1500 m and average rainfall 2500 mm. There are considerable variations in both quantity and distribution of monsoon rains. The south-eastern part of the state also receives rain fall from north-eastern monsoons. Nearly two third of the state in the north eastern part form the dry zone is classified as drought prone (Chalawadi and Jayakumar, 1987).

Kodagu, the second smallest district in Karnataka after Bangalore urban is one of the densely forested districts not only in the state but also in the country. Seventy three percent of the 4106 square kilometers of Kodagu is forested (FSI, 2005). Thirty percent of the total forest area of the district is under coffee based agroforestry systems and the area under coffee agroforestry systems is increasing at the cost of natural forested ecosystems. Many conservation biologists feel that the increase in area under coffee agroforestry systems is considered be the looking threat to biological diversity, at the same time people also say that coffee based agroforestry systems are also a refuge for biological diversity. The present chapter is an attempt to through light on role of coffee based agroforestry systems in conserving biodiversity through the following three hypotheses proposed by Schroth *et al.* (2004):

- Providing suitable habitat for forest dependent plant and animal species
- Reducing the pressure to deforest remaining forest land and degrade forest through the unsustainable extraction of its resources (agroforestry-deforestation hypothesis)
- Creating a biodiversity friendly matrix to facilitate movements among existing patches of natural habitat and buffer them against more hostile land uses.

Study area and methods

The study was conducted in Kodagu district which is located on the eastern slopes of the mountain range, extends between 11° 56' – 12° 52' N and 75° 22' - 76° 11' E. The lowest altitude in the district is about 300 meter and the highest peak, Thadiandamaol is at 1734 meter above sea level. Coffee agroforests were chosen in two vegetation types viz., semi-evergreen and moist deciduous vegetations, in each vegetation type, coffee agroforests were grouped into redeemed (Both the trees and land is owned by the farmer) and unredeemed tenure (farmer has rights over only the land and not the trees in his farm) depending on the kind of rights over tree and land.

The sampling technique adopted was randomized stratified sampling. The study area was grouped into different strata like vegetation types and size classes of plantations. To decide the ideal sampling size and shape, preliminary sampling was done using different shape and size of sampling units. Based on species area curve method it was decided that a rectangular quadrat of 25 X 50 meter dimension is the most appropriate size. Two plots of 25 X 50 meters were laid in each estate. Since, with an area of 2500 square meters 90 per cent of the tree species could be captured in the sampling units.

a. How coffee based agroforestry systems help in biodiversity conservation

b. Providing suitable habitat for forest dependent plant and animal species

Coffee based agroforestry systems contains forest like appearance. How ever structure and diversity are poorly known. The present study conducted in coffee agroforestry systems of Kodagu, Karnataka state indicates that the species richness of total, native and exotic tree species richness was higher in plantations in redeemed tenure in moist deciduous (93, 64 and 23 respectively), while it was higher in plantations in unredeemed tenure in semi evergreen vegetation type (86 and 19 respectively). This could be attributed to the age of the plantation (Elouard *et al*., 2000) and tree rights under different land tenure systems (Uthappa, 1998). The diversity was higher in coffee plantations unredeemed tenure in both moist deciduous (3.02) and semi evergreen vegetation (3.45). which could be again due to high restriction in tree felling in unredeemed tenure (Uttappa, 1998)

Table 1. Species richness and diversity in coffee plantations

Vegetation type		Species richness		Shannon's diversity index	
	Origin of species	Redeemed	Un-redeemed	Redeemed	Un-redeemed
Moist deciduous	Total	93.00	87.00	2.72	3.02
	Native	69.00	68.00	1.21	1.78
	Exotic	24.00	18.00	1.51	1.24
Semi evergreen	Total	27.00	105.00	2.75	3.45
	Native	23.00	86.00	2.14	2.32
	Exotic	4.00	19.00	0.60	1.13

Earlier studies by Shonil Bhagwat (2002) indicates that, the proportion of endemic trees were higher in reserve forests (49), followed by sacred forests (47) and coffee plantations (39). While the proportion of threatened tree species were higher in Sacred forests (32) followed by coffee plantations and reserve forests (25).

Since these coffee plantations are very rich in floristic composition, they are also providing habitat for higher bird diversity which is almost equal to moist and dry deciduous forests as indicated by Prakash (2003).

From the studies by Shonil Bhagwat (2002) indicates that the percentage of forest dwellers (birds) is decreasing with the increasing disturbance and the percentage of non forest dwellers are increasing with the disturbance and are high in coffee plantations than that of forest reserves and the sacred forests. From this it is evident that, coffee plantations too provide habitat for very rich flora and fauna (Fig. 1).

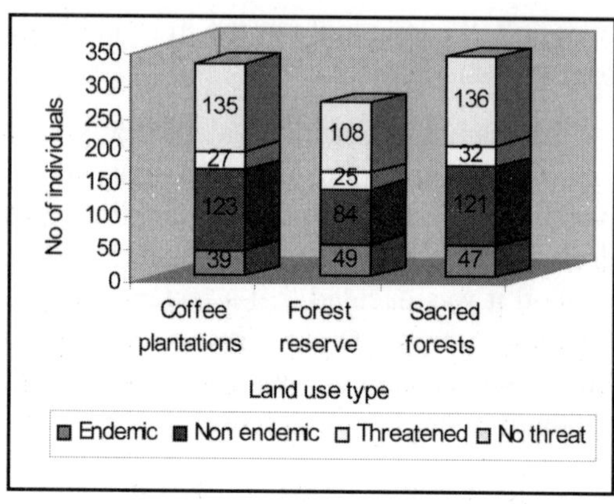

Fig: 1. Distribution of Endemic and threatened tree species in different landscapes

Reducing the pressure to deforest remaining forest land and degrade forest through the unsustainable extraction of its resources:

India is likely to face severe shortage of supply of timber to meet the supply of timber to meet both domestic and international front. It is estimated that the annual requirement of timber is 64 million cu m. (excluding fuel wood), the supply is of the order of 43 million cu. m. from all sources (www.tofac.org) and the projected demands by 2020 will be 153 million cubic meters and estimated supply will be 60 million cubic meters by 2020. The productivity of timber is only 0.7 cu. m/ha/ year. (www.wwfindia.org). The gap between the demand and supply of timber some times caused even official felling of timber beyond the silvicultural limits from the forests.

In coffee agroforestry systems of Kodagu the quantity of timber ranged between 60-80 cu. m/ha as per the studies conducted by Sathish (2004). In addition to the timber, fuel wood and other non-timber forest produce requirements are obtained from the trees in coffee agroforestry systems. A study from coffee agroforests of Muthappa (2000) the farmers opinion on the role of tree in the coffee agroforests indicates that, trees are mainly retained in the farm for shade and fuel wood (100 %), support for pepper and timber (98 %), religious value (96 %), food (76%), other (69 %). The results clearly indicates that the there is a reduction of pressure on the natural forests.

Of the total tree species in coffee plantations nearly 40 per cent were of timber value and 45 per cent trees were of fire wood value as per the information collected from the planters as shown in the Fig 2. In addition to which most of the trees were retained as a standard for pepper and also shade for the coffee (Sathish, 2006).

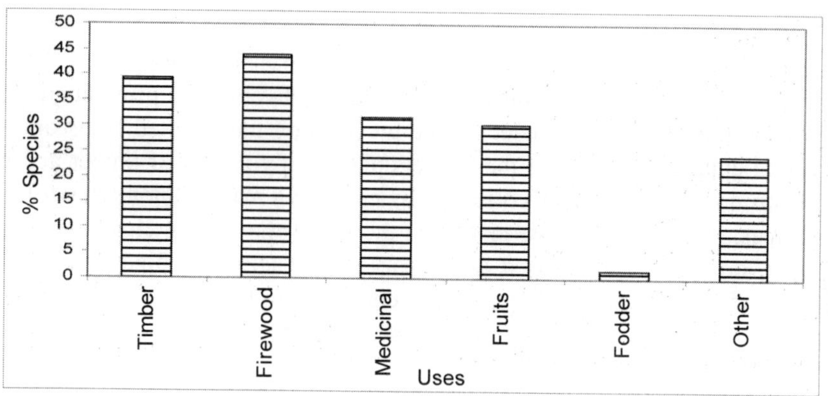

Fig. 2. Timber and non timber tree species in coffee plantations

Creating a biodiversity friendly matrix to facilitate movements among existing patches of natural habitat and buffer them against more hostile land uses.

The idea that landscape management should strive to increase spatial heterogeneity in order to conserve biodiversity is appealing. This prescription is often not practical in the tropics. Human society and biodiversity, in many ways, is antagonistic by nature. When tropical forests are impacted due to commercial logging, illegal felling, shifting cultivation and conversion to pasture and agriculture, ecosystem structure is simplified. The continuous expanses of natural habitats are interrupted with barriers. Simplification of ecosystem structure and habitat fragmentation results in reduction of biodiversity. To prevent loss of biodiversity in the tropics, the human activities that result in simplification and fragmentation should stop. However, these are not likely to happen. Human demands for forest products or conversion of forests into agriculture or human settlement usually take priority over forest preservation. Agroforestry may represent a way to help maintain the structural diversity required for high biodiversity, while at the same time providing for human needs.

Recent research studies have shown a surprising degree of connectivity on a landscape scale for trees in forest fragments that were previously assumed to be living dead for lack of near by mating partners. Coffee based agroforestry trees in the matrix may cross pollinate with trees of the same species in the forest fragments or may facilitate movements of pollinators and seed dispersal across the landscape.

Shade grown coffee plantations can be considered as corridors for fauna and flora, since, they are part of the scarcely vegetated areas with in the fragmented zones. (Moguel and Toledo, 1991)

Conclusion

The present study clearly indicates that, coffee based agroforestry systems of Karnataka play great role in conserving biological diversity by providing host to variety of flora and fauna, reducing the pressure on the forested landscapes by providing addition benefits to the farmers/planters by means of timber and non timber products and provides corridors by creating a matrix among the different landscapes. It is very fortunate that, even the forests have been cleared and have been converted into coffee based agroforests and are acting a refuge for biological diversity, if these forest would have been converted to other land use pattern like tea plantation, or other monocultures, the entire biodiversity would have been lost.

References

Chalawadi, S.M., Jayakumar, M.N. 1987. Land use system and agroforestry in Karnataka (India), paper presented at ICRAF, Nairobi, Kenya.

Elourd C., Chaumette, M., Pommery, H. 2000, The role of coffee plantations in biodiversity conservation (Coffee and Biodiversity Conservation – Chapter 6). *In:* Ramakrishnan, P.S., Chandrashekara, U.M., Elouard, C., Guilmoto, C.Z., Maikhuri, R.K., Rao, K.S., Sankar, S., Saxena, K.G (eds.), Mountain Biodiversity, Land Use Dynamics, and Traditional knowledge, pp. 120-149. Man and the biosphere Programme, Oxford and IBH Publishing Co. Pvt. Ltd., New Delhi.

Forest Survey of India, 2003, State of forest report. URL Forest Survey of India www.envfor.nic.in.

Moguel, P., Toledo, V.M. 1999. Biodiversity conservation in traditional coffee systems of Mexico. Cons. Biol. 13(1): 1-11.

Muthappa, P. P. 2000, A Resource economic study on tree diversity in coffee based plantations in the Western Ghats region of Karnataka. M.Sc. Thesis, University of Agricultural Sciences, Bangalore.

Prakash, C.B., Avifaunal diversity study under different habitats in Virajpet Taluk of Kodagu district (Western Ghats). M.Sc. Thesis, University of Agricultural Sciences, Bangalore.

Sathish, B.N., Kushalappa, C. G 2006. An insight the timber and non timber tree species in coffee based agroforestry systems of Kodagu, Central western Ghats. Int. J. Usuf. Mangt. 7(2):79-86.

Sathish, B.N. 2004. Assessment of tree diversity in coffee plantations under different land tenure systems in Virajpet taluk, Kodagu.

Schroth, G, Da Fonseca, G.A.B., Harvey, C.A., Gascon, C., Vasconcelos, H.L., Izac, A.M.N. 2004. Agroforestry and Biodiversity Conservation in Tropical Landscapes Washington, DC.

Shonil, B. 2002, Biodiversity and conservation of cultural landscapes in the Western Ghats of India, Ph.D. Thesis, University of Oxford, United Kingdom.

Uthappa, K. G 1998, Centenary celeberation of Virajpet Courts, Virajpet, Kodagu. pp 217-230

www.wwfindia.org, last date of visit 21st June 2008.

❑❑❑

17

Soil Erosion Control using Different Vegetative Methods on Hilly Terrain of South Andaman

C.B. Pandey and S.K. Chaudhari

Abstract: *This chapter reports soil and nutrient loss due to water erosion from five major land uses,* i.e. *vegetable fields, coconut plantation, arecanut plantation, home garden and moist evergreen forest on the undulating topography of the island. Soil loss under the land uses was quantified using replicated runoff plots. Vegetative methods, i e* Gliricidia *hedgerow + crop, hedgerow + mulch + crop, hedgerow + mulch + crop + grass barrier, for vegetable field under till conditions, and* Pueraria *cover crop (coconut +* Pueraria phaseoloides *cover crop) for coconut plantation were tested to know their potential to arrest soil erosion in the respective system. Soil loss from the vegetable field under No till conditions, coconut plantation, arecanut plantation, home garden and forest was 3.8, 12.4, 10.6, 8.4 and 2.3 tonnes/ha, respectively. Soil working in vegetable fields (till + crop treatment) provoked soil loss to124 tonnes / ha. Erodibility was the lowest (0.06) in the forest and highest (0.26) in the vegetable field under till condition. Among the nutrients, the highest loss, across the treatments, occurred for nitrogen and lowest for phosphorus. The hedgerow alone in vegetable fields reduced the soil loss substantially (66 %). However, it together with mulch and grass barrier reduced the soil loss nearly equal to that found in the forest. Like wise,* Pueraria *cover crop (19 years old) brought down the soil loss under the coconut plantation equal to that in the forest. These observations suggest that* Gliricidia *hedgerows for vegetable cultivation and cover crop for coconut plantation may be good vegetative methods for soil erosion control in the island.*

Introduction

Topographically the Andamans are undulating, characterized with hills, hillocks and flat bottomed vallies (Pandey *et al.*, 2007). The islands were once thickly covered with tropical rainforests. They were settled in mid 20th century. The settlers were provided with 2 ha forested lands on the hillocks and 2 ha lands in the valley. Those settlers removed the forest, constructed houses and planted trees round the houses that met their maximum basic needs because markets as well as roads were not available at that time. This led to the development of home garden in the islands. At present, coconut and arecanut plantations and home garden (29000 ha) are second major land uses after forest (86 % of total geographical area) in the islands (Pandey *et al.*, 2007). Some portion of the cleared forest lands are in use for vegetable cultivation by farmers. Vegetable fields cover about 3000 ha in the islands. Generally farmers perform deep soil working in the vegetable fields and remove weed for the vegetable cultivation. They follow vegetable-vegetable and maize-vegetable rotations in the vegetable fields. Before tsunami 12000 ha land in the valley was under bunded paddy cultivation (Pandey *et al.*, 2007).

The islands experience torrential and high rainfall during wet season from south-west monsoon (Pandey *et al.*, 2009). Agricultural activities particularly vegetable cultivation exposes surface soils to the rainfall which carry away a huge amount of fine soil particles to sea through low lying streams and makes soil poor in nutrients (Pandey and Singh, 2009). Loss of soils is known to increase many folds in hilly area due to accelerated rate of runoff (Narain *et al.*, 1998). The landmass of the islands is precious not only from the soil-fertility view point, but also for the existence of the islands as well. Tsunami in 2004 has already engrossed about 4000 ha low lying rice fields in the islands and several thousand hectare lands (rice fields) are still being inundated once in a day by high tides making them unfit for rice cultivation. After tsunami height of tides has been increased and frequency has been doubled.

Land-use wise information on soil erosion, estimated using runoff plot, so far is not available for the islands because it is expensive and time consuming (Velmurugan *et al.*, 2008). Recently Velmurugan *et al.* (2008) carried a soil erosion study in Dhanikhari watershed at South Andaman using revised Morgan, Morgan Finney model (RMMF) with the aid of remote sensing and GIS, but no ground truthing was carried out for a long term (even for one annual cycle) to validate the model's results. Objective of the present study was (i) to quantify soil loss under different land uses being practiced in the islands using runoff plots on long term basis (3yrs); (ii) to understand how soil erosion occurs in the land uses; (iii) and to find out suitable vegetative methods for arresting soils loss from the land uses in the islands.

Location of site and experiment design

The study sites were located at Sipighat and Garacharma within a radius of 5 km in South Andaman Island of India (10°31'-13°42'N lat. and 92°14' –94°14' E long.). The soils of the sites were Entisols. It was gravelly-sandy-loam in texture, low to moderate in nutrients and slightly acidic in reaction (Pandey *et al.*, 2007). Climate is an equatorial hot humid tropical having high temperature (23 to 30 ^{0}C) and relative humidity (71 to 85 %). About 12 years data (1994-2005) indicate that an average 3,000 mm rainfall occurs in the study region with mean monthly variation of 300-500 mm / month during wet season (May to October), 100-200 mm / month during post-wet season (November to January) and <100 mm / month during dry season (February to April).Generally no runoff is observed during the dry season.

The study included 5 land use systems, i e (1) vegetable field; (2) coconut plantation, (3) areanut plantation, (4) home garden and (5) native moist evergreen forest at similar slope. Detailed treatments are given in Table 1. To know the effect of soil working in the vegetable field, two treatments namely till and no till were included. To assesss how Gliricidia hedgerows and its mulch help arrest soil erosion in the vegetable fields, 3 treatments under till (till + hedgerows + crop, till + hedgerows + mulch + crop, till + hedgerows + mulch + crop + grass barrier) and two treatments under no till (no till + crop, no till + hedgerow + no crop) condition were arranged. One hedgerows was 4m away from its neighbour hedgerows. Within a hedgerow plant to plant distance was 50 cm. The hedgerows were planted against the slope, and were 4 yrs old. To know if grass cover help arrest soil erosion, two rows of napier grass 30 cm x 30 cm were planted between two plants of the hedgerows. At the time of sampling the grass tussocks overlapped each other. In the vegetable field, crop rotation was maize (*Zea mays*) (wet season) and okra (post-wet season). In the coconut plantation, coconut was planted at 7.5m x 7.5 m. To assess the potential of a cover crop for arresting soils from erosion, coconut plantation having *Pueraria* as a cover crop (19- year- old) was selected. Arecanut was planted at 2.7m x 2.7m distance. Age of both coconut and arecanut plantations was 25-year. Weeds (3-5 tonnes / ha) in both the plantations as well as the home gardens were removed once in a year in October by local made swards. Dominent trees in the forest were: *Dipterocarpus grandiflora* (Blanco) Blanco, *Calophyllum* spp., *Artocarpus chaplasha* Roxb, *Hopea odorata* Roxb. The forest (>100 yrs) was multistory similar to that of the homegarden. Thus, there were total 11 treatments under five land uses for the soil erosion study.

Three runoff plots, 2m x 5m in size, were installed in the each treatment at 22-25 % slope. Generally coconut plantations and vegetable fields are located at the above slope. Big size runoff plots were not possible due to hilly terrain. Dikes of the runoff plots were made by galvanized iron sheet. The each runoff plot was connected with a 500 L tank by 4" PVC pipe to collect runoff water from the plot.

Runoff volume in each tank was measured, total runoff water was computed and five liter runoff water was sampled after thorough stirring. One liter water sample was coagulated by alum, decanted off and dried on a water-bath to estimate soil loss. The other portion of the sample was preserved by addition of 1ml toluene and analyzed for total N, P, exchangeable K, Ca and Mg (Jackson, 1962). Each study year during dry season (when rainfall was over) soils were sampled from six random places in each runoff plot and composited as one replicate. The composited soil samples were analyzed for soil texture, soil organic carbon and pH. Soil erodibility was calculated following Wischmeier and Mannering (1969) as:

$$\text{Erodibility} = \frac{0.043R + \left(\dfrac{0.62}{O.M.}\right) + 0.0082S - 0.0062C}{Si}$$

where R = pH, O.M = organic matter (%), S = fine sand (%), C = clay (%), Si = silt (%). Soil texture analysis was done by hydrometer. Soil pH was measured with glass electrode (1:2, soil: water ratio). Organic C was estimated by Walkley and Black rapid titration method and total P was estimated after $HClO_4$ digestion. Total N was measured by microkjeldhal digestion method using a Kjel plus auto N analyzer. For exchangeable cations the soils were extracted with 1 M ammonium acetate at pH 7 and K was measured by a flame photometer, and Ca and Mg by Shimadzu (AA-6200) atomic absorption spectrophotometer.

Cover of trees in coconut and arecanut plantations, home garden and moist evergreen forest was measured by line intercept method. Cover of trees was calculated as: $3.14 \times r^2$ (r = radius of tree canopy). Cover of herbaceous vegetation (crop, grass and weed) was measured by point intercept method using ten 1 x 1m^2 random quadrat for pin contacts with 100 pins placed in a grid pattern over quadrat. Each grid consisted of 10 transects placed 10 cm apart. The points of contact were noted and cover was computed as per cent of total contacts made (Pandey *et al.*, 2007).

Runoff loss of fine soils

Proportion of clay and silt particles declined but proportion of gravels increased in the runoff plots in the almost all treatments across the land uses (Figs. 1a, b, c, d). Similar pattern in the change of proportions of clay and silt in all the treatments seems to have occurred due to same soil type. No significant change in fine sand particles was observed across the treatments (P<0.05). These changes in the proportions of the soil particles caused sheet erosion in all the treatments. The studied soils were poor in organic carbon and fine soil particles (clay and silt), but richest in gravels which made the soils suitable for the sheet erosion. Among the treatments in the vegetable fields, the highest loss of clay (72 %) and silt (61%) occurred in till + crop treatment and lowest (11 % and 9 % clay and silt, respectively) in no till + crop treatment. In no till treatment weed cover most likely provided greater protection to the soils from the rain beating and the weed roots acted as a binder to the soil particles. Introduction of hedgerows under till conditions in the vegetable field reduced the loss of fine soil particles (clay 49 %, silt 29 %) (till + hedgerow + crop vs till + crop). When mulch of the *Gliricidia* hedgerows was applied, the loss of fine soil particles (silt + clay) was reduced further by 6 %. In addition to the hedgerows and mulch when grass was also incorporated as a treatment, loss of silt and clay was reduced to the highest i e 99 % and 58 %, respectively. These indicated that loss of fine soil particles from the vegetable field can be managed to the lowest when mulch and grass barrier are also applied with the *Gliricidia* hedgerows (Senapati and Sharma, 2007).

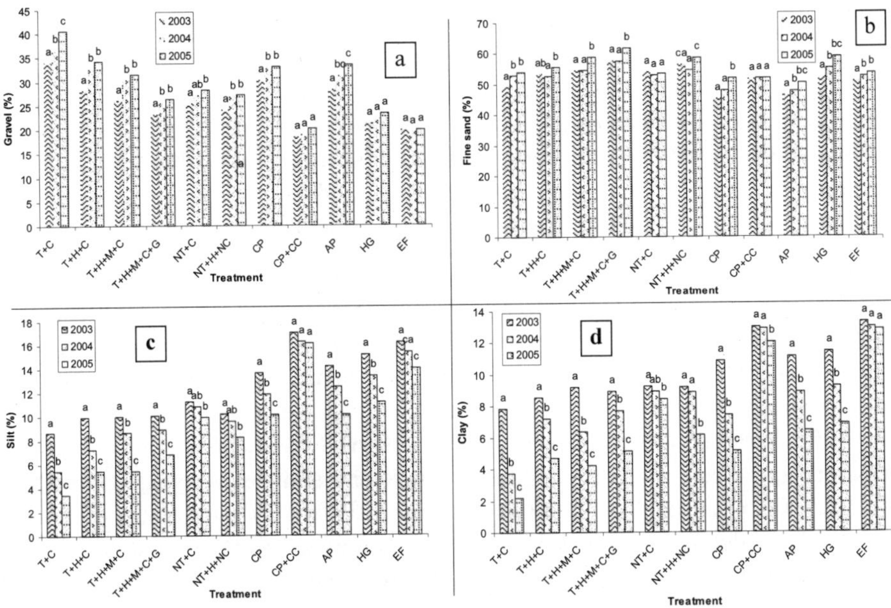

Fig. 1. Changes in the proportion of **(a)** gravel **(b)** fine sand **(c)** silt and **(d)** clay soil particles in the runoff plots under different treatments at South Andaman island. Data (bars) of a treatment with different letters are significantly different at P<0.05. Abbreviations for treatments are: T+ C = Till + crop; T + H + C = Till + hedgerow + crop; T + H + M+ C = Till + hedgerow +mulch + crop; T + H + M+ C + G = Till + hedgerow +mulch + crop + grass barrier; NT + C = No Till + crop; NT + H + NC = No till + hedgerow + No crop; CC = cover crop of *Pueraria phaseoloides*.

Source: Pandey and Chaudhari 2010

Erodibility of the soils in the vegetable field was the highest under till condition (till + crop) and lowest under no till condition (no till + crop treatment) (Table 1). This indicated that soil working exposed the fine soil particles and increased soil erodibility in the absence of vegetal cover as the weed was removed from the field after the soil working. Introduction of hedgerows in the till condition reduced the erodibility in the vegetable field. It occurred probably because the canopy of the hedgerows reduced the kinetic energy of the rains (Velmurugan *et al.*, 2008). Reduction in the soil erodibility in the hedgerows + mulch + crop + grass treatment in the vegetable field was the highest due to additional cover of mulch and grass besides the hedgerows. In the present study we found that the erodibility was inversely correlated with the vegetation cover ($r = -0.625$, $P<0.001$). Mulch and grass most likely provided additional protection to the soils from the high intensity rains (Lal, 1995; Senapati and Sharma, 2007). Among the tree based systems, the lowest erodibility was found in the forest and highest in the coconut plantation. The lowest erodibility in the forest was due to the highest vegetal cover. The cover crop reduced the erodibility in the coconut plantation equal to that in the forest.

Soil loss from the vegetable field under No till condition, coconut plantation, arecanut plantation, home garden and forest was 3.8, 12.4, 10.6, 8.4 and 2.3 tonnes / ha, respectively (Table 1). Soil working in the vegetable field increased the soil loss to 124 tonnes / ha. Soil loss, among the tree based systems, was the highest in coconut plantation and lowest in the evergreen forest. Velmurugan *et al.* (2008) found an average 25 tonnes / ha soil loss across agriculture field (rice / vegetable), plantations (coconut and arecanut) and evergreen forest by remote sensing and GIS methods from the Dhanikhari watershed at South Andaman.

Introduction of hedgerows in the vegetable field under till conditions reduced the soil loss by 66 % (till + hedgerow + crop vs till + crop treatment) (Table 1). Role of hedgerows intercropping to arrest soil loss on sloppy land is well documented (Kang *et al.*, 1990). In our study, mulch of *Gliricidia* prunings together with the hedgerows (till + hedgerows + mulch + crop) reduced the soil loss further by 91 %. Pacaredo and Montecillo (1983) observed that *Leucaena leucocephala* planted on the contour greatly reduced runoff and soil loss, especially so when stubbles and prunings were retained on the soil surface. Greater reduction in soil loss under hedgerows + mulch in our study seemed to have occurred due to two advantages of the hedgerows: first it decreased runoff velocity and reduced its sediment carrying capacity; and the second the mulch reduced splash and soil detachment (Lal, 1989 a, b). When napier grass barrier was included with the hedgerow and mulch (till + hedgerow + mulch +crop + grass barrier), the lowest loss (2.0 - 2.9 tonnes / ha) of soil was observed from the vegetable field. Our observations proved the hypothesis of Lal (1989 a, b) that tree + crop cover combinations may be more conservation effective than an arable land use system. Compared to the till treatment, no till treatment also reduced the soil loss substantially (97 %) from the vegetable field (no till + crop vs till + crop) (Table1).

Table 1. Vegetation cover, erodibility and soil loss under different treatments under prominent land uses at South Andaman island.

Treatment	Crop		Vegetation cover (%)	Erodibility			Average	Soil loss(t / ha)			Average (t/ha)
	Season			2003	2004	2005		2003	2004	2005	
	Wet	Post-wet									
Vegetable field											
till + crop	Maize	Okra	112^{a}	${}^{a}0.15^{a}$	${}^{b}0.24^{a}$	${}^{c}0.40^{a}$	0.26^{a}	${}^{a}122.3^{a}$	${}^{b}120.18^{a}$	${}^{c}130.5^{a}$	124.3^{a}
till + hedgerow + crop	Maize	Okra	123^{b}	${}^{a}0.13^{b}$	${}^{b}0.18^{b}$	${}^{c}0.25^{b}$	0.19^{b}	${}^{a}45.97^{b}$	${}^{b}40.27^{b}$	${}^{b}41.87^{b}$	42.7^{b}
till + hedgerow +mulch + crop	Maize	Okra	135^{c}	${}^{a}0.13^{b}$	${}^{b}0.15^{c}$	${}^{c}0.25^{b}$	0.18^{b}	${}^{a}12.29^{c}$	${}^{a}10.95^{b}$	${}^{a}11.87^{c}$	11.7^{c}
till + hedgerow +mulch + crop + grass barrier	Maize	Okra	140^{d}	${}^{a}0.14^{b}$	${}^{b}0.15^{c}$	${}^{c}0.18^{c}$	0.16^{c}	${}^{a}2.04^{d}$	${}^{a}2.57^{c}$	${}^{a}2.89^{d}$	2.5^{d}
no till + crop	Maize	Okra	140^{d}	${}^{a}0.10^{c}$	${}^{a}0.10^{d}$	${}^{a}0.11^{d}$	0.10^{d}	${}^{a}4.32^{d}$	${}^{bc}3.87^{d}$	${}^{c}3.21^{e}$	3.8^{e}
no till + hedgerow + no crop	Maize	Okra	130^{e}	${}^{a}0.12^{d}$	${}^{a}0.13^{e}$	${}^{b}0.16^{e}$	0.14^{e}	${}^{a}7.18^{e}$	${}^{a}5.18^{e}$	${}^{a}6.25^{f}$	6.20^{f}
Coconut plantation	-	-	104^{f}	${}^{a}0.07^{e}$	${}^{a}0.08^{f}$	${}^{c}0.11^{d}$	0.09^{f}	${}^{a}12.58^{e}$	${}^{a}13.87^{e}$	${}^{a}12.65^{f}$	13.0^{g}
Coconut plantation +cover crop	*Pueraria phaseo loides*	*Pueraria phaseo loides*	160^{g}	${}^{a}0.05^{f}$	${}^{a}0.05^{g}$	${}^{a}0.05^{e}$	0.05^{g}	${}^{a}2.80^{f}$	${}^{a}2.64^{f}$	${}^{a}2.25^{d}$	2.6^{d}
Arecanut plantation	-	-	110^{a}	${}^{a}0.06^{f}$	${}^{b}0.08^{f}$	${}^{c}0.10^{d}$	0.08^{h}	${}^{a}10.54^{g}$	${}^{a}10.01^{g}$	${}^{A}11.14^{f}$	10.6^{c}
Home garden	-	-	124^{b}	${}^{a}0.06^{f}$	${}^{a}0.07^{fg}$	${}^{a}0.09^{f}$	0.07^{h}	${}^{a}8.04^{g}$	${}^{a}9.0^{g}$	${}^{a}8.14^{g}$	8.4^{h}
Evergreen forest	-	-	180^{h}	${}^{a}0.05^{f}$	${}^{a}0.06^{g}$	${}^{a}0.06^{e}$	0.06^{g}	${}^{a}2.45^{f}$	${}^{a}2.25^{f}$	${}^{a}2.30^{d}$	2.3^{d}

Values in a column suffixed with different superscript letter are significant at $P<0.05$
Values in a row prefixed with different superscript letter are significant at $P<0.05$
Vegetation cover included ground weed cover + crop cover + tree canopy cover according to the treatment.
Source: Pandey and Chaudhari, 2010

Greater soil loss from the coconut compared to arecanut plantation was mainly due to its wider spacing. As a result, leaves (fronds) of neighbour coconut trees did not overlap, hence failed to provide that much cover as much areanut plantation did in closer spacing. Greater loss of soil under coconut could also be due to the high erodibility under the tall tree (height = 13 m). Soil loss, across the land uses, in our study was positively correlated to the erodibility ($r = 0.771$, $P<0.001$). Wiersum (1991) found relatively greater soil erodibility under tall trees (>12.5m). Narain *et al.* (1998) argued that tall tree (>6.5 m) increased terminal velocity of raindrop from tree canopy (leaf) and increased loss of soils. Greater reduction in soil loss under the evergreen forest than coconut and arecanut plantations, and home garden was due to multistory structure of the forest where height of the first storey (shrub and new growths) was less than 4 m, which reduced the erodibility of the soils to the lowest. We found that soil loss under coconut + cover crop (coconut plantation + cover crop) was 82 % lower than coconut plantation alone. This indicated that the cover crop together with tree cover was more effective in reducing the erodibility of the soils in the coconut plantation. Lower loss of soils from the home garden compared to the coconut and arecanut plantations could be due to its multistory structure (Pandey *et al.* 2007).

Amount of nutrients lost due to the soil erosion and concentrations of nutrients in the runoff soils are given in Tables 2 and 3. Across the treatments, loss of nutrients occurred in the following chronosequence: Nitrogen > K>Ca> Mg> P. The loss of nutrients followed broadly the pattern of soil loss. Among the land uses, maximum loss of the nutrients occurred from vegetable fields and lowest from the forest. Like the soil loss, the vegetative methods reduced the loss of nutrients also in the vegetable fields and under the coconut plantation (Table 1). Narain *et al.* (1994) also found almost similar pattern of nutrient loss in the western Himalayan valley.

Table 2. Loss of nutrients from different treatments under prominent land uses in South Andaman island. Data are mean for three yrs (2003-05).

Treatment	Nutrient loss (kg / ha)				
	N	P	K	Ca	Mg
Vegetable field					
till + crop	$^{a}50.97^{a}$	$^{b}1.24^{a}$	$^{c}45.25^{a}$	$^{d}14.17^{a}$	$^{e}8.21^{a}$
till + hedgerow + crop	$^{a}16.23^{b}$	$^{b}1.28^{b}$	$^{c}15.20^{b}$	$^{d}4.78^{b}$	$^{e}2.86^{b}$
till + hedgerow +mulch + crop	$^{a}6.79^{c}$	$^{b}0.47^{c}$	$^{c}5.39^{c}$	$^{d}1.42^{c}$	$^{d}0.78^{c}$
till + hedgerow +mulch + crop + grass barrier	$^{a}1.18^{d}$	$^{b}0.08^{d}$	$^{c}0.94^{d}$	$^{d}0.30^{df}$	$^{d}0.16^{d}$
no till + crop	$^{a}2.55^{e}$	$^{b}0.08^{d}$	$^{c}1.47^{e}$	$^{d}0.42^{d}$	$^{e}0.24^{e}$
no till + hedgerow + crop	$^{a}2.11^{e}$	$^{b}0.12^{e}$	$^{c}2.26^{e}$	$^{d}0.73^{eg}$	$^{d}0.42^{c}$
Coconut plantation	$^{a}11.31^{c}$	$^{b}0.39^{f}$	$^{c}6.92^{f}$	$^{d}1.56^{e}$	$^{d}0.86^{c}$
Coconut plantation + *Pueraria* cover crop	$^{a}3.98^{e}$	$^{b}1.07^{g}$	$^{c}1.52^{d}$	$^{d}0.39^{f}$	$^{d}0.18^{d}$
Arecanut plantation	$^{a}10.18^{f}$	$^{b}0.53^{h}$	$^{c}4.63^{e}$	$^{d}1.25^{gh}$	$^{b}0.67^{c}$
Homegarden	$^{a}9.24^{g}$	$^{b}5.12^{d}$	$^{b}5.16^{f}$	$^{c}1.02^{h}$	$^{c}0.56^{c}$
Evergreen forest	$^{a}3.20^{e}$	$^{b}0.05^{i}$	$^{c}0.55^{g}$	$^{d}0.24^{i}$	$^{d}0.15^{d}$

Values in a column suffixed with different superscript letter are significant at P<0.05
Values in a row prefixed with different superscript letter are significant at P<0.05
Source: Pandey and Chaudhari, 2010

Table 3. Nutrient concentration in runoff soils under different treatments under prominent land uses at South Andaman. Data are mean for three yrs (2003-05).

Treatment	Nutrient concentration						
	pH	Organic carbon (%)	Nitrogen (%)	Phosphorus (%)	Potassium (mg/kg)	Calcium (mg/kg)	Magnesium (mg/kg)
Vegetable field							
till + crop	6.33^{a}	^{a}0.58^{a}	^{b}0.041^{a}	^{c}0.001^{a}	^{d}364^{a}	^{e}114^{a}	^{f}66^{ad}
till + hedgerow + crop	6.30^{a}	^{a}0.56^{b}	^{b}0.038^{b}	^{c}0.003^{b}	^{d}356^{b}	^{e}112^{a}	^{f}67^{b}d
till + hedgerow + mulch + crop	6.30^{a}	^{a}0.62^{c}	^{b}0.058^{c}	^{c}0.004^{b}	^{d}461^{c}	^{e}121^{b}	^{f}67^{bd}
till + hedgerow + mulch + crop + grass barrier	6.20^{a}	^{a}0.54^{d}	^{b}0.047^{d}	^{c}0.003^{b}	^{d}375^{d}	^{e}118^{b}	^{f}63^{c}
no till + crop	6.19^{a}	^{a}0.85^{e}	^{b}0.067^{e}	^{c}0.002^{c}	^{d}387^{e}	^{e}111^{a}	^{f}64^{c}
no till + hedgerow + no crop	6.25^{a}	^{a}0.63^{c}	^{b}0.034^{b}	^{c}0.002^{c}	^{d}364^{c}	^{e}118^{b}	^{f}68^{d}
Coconut plantation	6.48^{a}	^{a}1.14^{f}	^{b}0.087^{f}	^{c}0.003^{b}	^{d}532^{f}	^{e}120^{b}	^{f}66^{ac}
Coconut plantation + *Pueraria phaseoloides* cover crop	6.72^{a}	^{a}1.81^{g}	^{b}0.153^{g}	^{c}0.041^{d}	^{d}583^{g}	^{e}148^{c}	^{f}68^{dc}
Arecanut plantation	6.60^{a}	^{a}1.18^{f}	^{b}0.096^{h}	^{c}0.005^{e}	^{d}437^{h}	^{e}118^{b}	^{f}63^{e}
Homegarden	6.40^{a}	^{a}1.31^{h}	^{b}0.110^{i}	^{c}0.061^{f}	^{d}614^{i}	^{e}121^{b}	^{f}67^{d}
Evergreen forest	5.94^{b}	^{a}1.62^{i}	^{b}0.139^{g}	^{c}0.002^{c}	^{d}237^{j}	^{e}104^{d}	^{f}66^{d}

Values in a column suffixed with different superscript letter are significant at $P<0.05$
Values in a row prefixed with different superscript letter are significant at $P<0.05$
Source: Pandey and Chaudhari, 2010

Conclusions

The study concludes that high rainfall increases erodibility particularly under till conditions in vegetable field and thereby removes fine soil particles, whereas vegetation cover of the trees, crops and grass in different combinations reduces the erodibility substantially. Loss of soils is more in coconut than arecanut plantation due to greater spacing and taller height of the tree, which together increases erodibility. The lowest soil loss occurs from the native forest due to highest vegetal cover (multistory structure) that reduced the soil erodibility. Cover crop of *Pueraria phaseoloides* reduces the erodibility in the coconut plantation equal to that in the native forest. Introduction of *Gliricidia* hedgerows in the vegetable field, reduces erodibility and soil loss. But, the hedgerows together with mulch and grass barrier reduce the soil loss equal to that of the native forest. Therefore, vegetable cultivation need to be advocated to be done under hedgerows of *Gliricidia*, and coconut plantation together with the cover crop for the lowest loss of soils from the land uses in the island.

References

Jackson, M.L. 1962. *Soil Chemical Analysis*. Prentice Hall, Enlewood Cliffs, NJ.

Kang, B.T., Reynolds, L., Atta-Krah, A. N. 1990. Alley-farming. Adv. Agro. 43: 315-359.

Lal, R. 1995. Erosion-crop productivity relationship for soils of Africa. Soil Sci. Soc. Am. J. 59(3):661-667.

Lal, R. 1989a. Agroforestry systems and soil surface management of a tropical alfisol. III. Changes in soil chmical properties. Agrofor. Syst. 8:113-132.

Lal, R. 1989b. Agroforestry systems and soil surface management of a tropical alfisol. IV. Effects on soil physical and chemical properties. Agrofor. Syst. 8:197-215.

Narain, P., Khybri, M. L., Tomar, H. P. S., Sindhwal, N. S. 1994. Estimation of runoff, soil loss and USLE parameters for Doon valley. Indian J. Soil Conser. 22:1-9.

Narain, P., Singh, R. K., Sindhwal, N. S., Joshie, P. 1998. Agroforestry for soil and water conservation in the western Himalaya valley region of India: Runoff, soil and nutrient losses. Agrofor. Syst. 39: 175-89.

Pacaredo, E.P., Montecillo, L. 1983. Effect of corn/ ipil-ipil cropping system on productivity and stability of upland agro-ecosystem. Annual Report. UPLLB-PCARRD Res. Project.

Pandey, C.B., Chaudhari, S.K. 2010. Soil and nutritent losses from different land uses and vegetative methods for their control on hilly terrain of South Andaman. Indian J. Agric. Sci. 80 (5) : 50-55.

Pandey, C.B., Rai, R. B., Singh, L., Singh, A. K. 2007. Homegardens of Andaman and Nicobar, India. Agricu. Syst.92: 1-22.

Pandey, C.B., Singh, L. 2009. Soil fertility under homegarden trees and native moist evergreen forest in South Andaman, India. J. Sus. Agri. 33(30): 303-318.

Pandey, C.B., Srivastava, R. C., Singh, R. K. 2009. Soil nitrogen mineralization and microbial biomass relation and nitrogen conservation in humid tropics. Soil Sc. Soc. Am. J. 73: 1142-1149.

Senapati, S.C., Sharma, S. D. 2007. Effect of buffer strips on soil erosion and yield of maize (*Zea mays* L.) in rainfed sloppy uplands in north-western plateau zone of Orissa. Indian J. Soil Cons. 35(1): 47-49.

Velmurugan, A., Swarnam, T. P., Kumar, P., Ravishankar, N. 2008. Soil erosion assessment using revised Morgan, Morgan Finney model for prioritization of Dhanikhani watershed in South Andama. Indian J. Soil Cons. 36(3): 173-179.

Wiersum, K F. 1991. Soil erosion and conservation in agroforestry systems. *In: Biophysical Research for Asian Agroforestry,* Avery M E, Cannell M.G.R. and Ong C (Eds.). Winrock International, USA. pp. 210-29.

Wischmeier, W.H., Mannering, J. V. 1969. Relation of soil properties to its erodibility. Proc. Soil Sci. Soc. Am. J. 33: 131-7.

18

Agroforestry and its Impact on Soil Fertility Improvement in North-east India

M. Datta

***Abstract**: North East India has a fragile inaccessible terrain with a sloping topography wherein agroforestry being diversified in nature, has the capability to eliminate the ill effects of land degradation through intensification in land use. Shifting cultivation widely practiced by the aboriginals is mainly instrumental in carrying out deforestation as well as land degradation but many indigenous land use systems present in practice since time immemorial are proven considering an approach to achieve eco-sustainability. A number of agroforestry systems such as agrihortisilviculture, multistoried homegradens, silvipastoral and agroaquaculture* etc *are dominantly present in the region. The information on soil fertility improvement by agroforestry systems is documented herein. Organic matter build-up was appreciably high in soils of agroforestry. Similar to soils under shifting cultivation, a low humification of soil humus was noted in soils of uplands without any forest cover but a reverse trend leading to high rate of humification could be initiated in soils under agroforestry systems. Moreover, decline in soil acidity, rise in water holding capacity and augmentation in nutrient availability were noted in soils under agroforestry systems.*

Introduction

The North Eastern Region of India, lies between 21° 50' and 29° 34' North latitudes and 85° 34' and 97° 50' East longitudes. The region (Basic Statistics,2002) has population of 39 million and geographical area of 262.2 thousand km^2, which is 3.85% and 7.97% of the population and area of the country, respectively. The region has inaccessible terrain, fragility, marginality, excessive sloping with rolling topography, rich biodiversity, unique ethnicity and socio-ecological set up. Morphologically, the region is marked by the development of series of ridges and valleys, terraces, scraps, several geomorphologic surfaces at different elevations (50 to 5000m and above). Out of a total geographical area of 262.2 thousand Km^2, 28.3 % has an elevation of more than 1200m, 17.9% has between 600m and 1200m and about 10.8%, between 300m and 600m above mean sea level.

Area under shifting cultivation

For over a million of years during the most of the period of the human's existence on this planet, their food source was by hunting and gathering. They did not know how to grow the plants and nurture the animals. The first major result in terms of food supply was the agricultural revolution, which occurred approximately 10,000 years ago as man entered the Neolithic Age. During that period, a system of slash and burn agriculture was developed and gradually spread over the most of the hilly and undulated arable lands of the world. This slash and burn agriculture regarded as the first step in food gathering and hunting to food production, is believed to have originated in Neolithic period around 7000 BC (Sharma,1976).This system is still prevalent in Africa and Southeast Asian countries as subsistence agriculture, where it is known as *Chena, Conuco, Jhum, Kaingin, Lading*, and *Milpa* (Chapman, 1975; Andriesse, 1989). In India, shifting cultivation or slash and burn agriculture is widely practised in the hills of the North-East, Orissa,Andhra Pradesh,and in some pockets of Chattisgarh. It is known as 'Jhum' in the NE region, 'Podu' in Andhra Pradesh, 'Bewar or Daihya' in Chattisgarh and 'Podu or Dunger Chasa' in Orissa. Indeed, it may be regarded as a successful form of agro-ecology in the hills because the system is also one of low input practices in terms of energy and technology.

In north eastern region, the mountain with its restricting influence on settlement and movement did not provide cultural advancement and, therefore, it has the considerable effect on the economic life of the inhabitants. They accepted the nature as it is and fought it valiantly, working on the mountain slopes for centuries in the way as their ancestor did. The prevalent primitive form of shifting agriculture (*jhum*) is the result of their urge for survival in the most difficult hilly terrains of the region. In 1983, the Task Force on shifting cultivation (Ministry of Agriculture, Govt.of India) estimated that the total area under this system was 14,66,000 hectares (3,86,900 hectares annually) and 4,43,336 tribal families were engaged (Table 1).The area (1.77 % of TGA) under shifting cultivation is minimum in Assam.

Table 1. Area under shifting cultivation in NEH region*

State	Area under shifting	Area sown sown at one cultivation	Col. 1 as % of point time TGA	Total families reported	Area (ha/family) involved	Fallow period (years)
Arunachal Pradesh	210.0	70.0	3.8	54.00	1.30	3-10
Assam	139.2	69.6	1.77	58.00	1.20	2-10
Manipur	360.0	90.0	16.3	70.00	1.28	4-7
Mizoram	189.0	63.0	8.99	50.00	1.26	3-4
Meghalaya	265.0	53.0	11.8	52.29	1.01	5-7
Nagaland	191.3	38.26	12.5	111.046	0.17	5-8
Tripura	111.5	22.3	10.6	43.00	0.46	5-9
Total	1466.0	406.2	6.5	443.336	0.87	

*Area in '000 ha; Families in 1000.

On the other hand, Forest Survey of India (1999) estimated the cumulative area of 1.73 m ha affected by shifting cultivation during the 1987-1997 (Table 2). On this basis, the annual area under shifting cultivation comprises 0.173 m ha/ annually, and maximum area affected by this practice exists in Nagaland, Mizoram, and Manipur. It is striking to note that Sikkim does not have any area under shifting cultivation and terraced cultivation with permanent agriculture is prevalent in the state.

Table 2. Cumulative area affected by shifting cultivation (m ha) in north east India

State	Area affected
Arunachal Pradesh	0.23
Assam	0.13
Manipur	0.36
Meghalaya	0.18
Mizoram	0.38
Nagaland	0.39
Tripura	0.06
North east Total	1.73

The practice of shifting agriculture (*jhum*) involves the manual cleaning of small area of a forest (commonly less than one hectare), drying and burning of the cut materials on sloppy lands. During the winter month (December to January), the undergrowth is slashed and small trees and bamboos are felled, leaving short tree stumps and large tree boles intact. The men as essential ingredients of a well-knit social organization do this laborious job jointly. The dried debris is burnt *in situ* before the onset of monsoon (end of March or the beginning of April). Farmers make holes with a pointed stick and a few grains of a single or mixed crops are dropped in each whole, which is closed by planter's foot after first few showers of monsoon. Upland rice is the main crop grown in mixture with maize, foxtail, finger millet, beans, cassava, yam, banana, sweet potato, ginger, cotton, chillies, tobacco, sesame, vegetables, *etc.* A single crop of paddy is usually preferred in second year of *jhuming* (Borthakur, 1992). Some 8-35 crop species are grown together.

Loss of forest cover and biodiversity

The deterioration in the productivity of the mountain environment has now been defined as a function of vegetative cover of uncultivated land. The assessment made by Forest Survey of India between 1995-97, observed that 1,875 km^2 forest areas were lost due to shifting cultivation. However, an area (1,700 km^2) under shifting cultivation came under forest cover as a regeneration (Table 3). It is noted that in most of the states, the forest cover tended to decline where *jhuming* is practised. In Tripura ,forest cover has a tendency to increase after 1997as a result of the settlement of *jhumias*.

Table 3. Loss or gain of forest cover (sq. km) in the NE states

State	Loss			Gain			Net change
	SC	Others	Total	NRSC	Others	Total	
Arunachal Pradesh	75	-	75	56	-	56	-19
Assam	257	159	416	163	16	179	-237
Manipur	603	-	603	463	-	463	-140
Meghalaya	75	2	77	20	-	20	-57
Mizoram	292	-	292	491	-	491	+199
Nagaland	573	-	573	503	-	503	-70
Tripura	-	3	3	4	7	11	+8
Total	1,875	164	2,039	1,700	23	1,723	-316

SC and NRSC = Shifting cultivation and natural regeneration in shifting cultivation

The forest vegetation undergoes a change from the continuous *jhum* cycle into semi-evergreen, moist deciduous, deciduous forest, grasses, bamboos and a number of aquatic and semi-aquatic plant species. In the course of time, a dominant stand of bamboos with the exclusion of tree growth, gradually succeeds these. It is particularly due to the extensive growth of bamboo rhizomes particularly in muli (*Melocanna baccifera*). The successive short *jhum* cycle (2-5 years) and lengthening of cropping periods at the same site have even eliminated the growth of bamboos. In some tracts, natural vegetation has been successively replaced by bamboo and thereafter by unproductive thatch grass (*Imperata cylindrica*), thus, ultimately leading to land abandonment.

The shifting cultivation has caused the destruction of forest and species habitat, which accounts for the most profound losses in biodiversity. During the recovery phase (abandoned periods), the vegetation evolves towards the original climax condition. If full recovery is achieved before the area is again cultivated, the system can be sustainable. The dynamics of vegetation in the abandoned *jhum* field was also studied. During the first 5 years, species diversity remained low and plots were dominated by herbaceous species. Between 5 and 15 years of abandonment, diversity increased rapidly as the vegetation passed into bamboo (*Dendrocalamus hamiltonii*)-dominated forest. Still later, it gradually passed into a mixed broad-leaved forest approaching the climax type, as compared with the species composition that of sacred groves in the locality. However, they could not follow the process beyond 20 years.

Indigenous landuse systems

Many indigenous farming systems are existing in North Eastern Hill region and performing well from the point of view of resource conservation and maintenance of ecological balance. These farming systems have been discussed in details by many workers from time to time (Sonowal *et al.*, 1989; Prasad and Sharma,1994; Sharma and Sharma, 2003).

Zabo system

Zabo is an indigenous farming system practised in Phek district of Nagaland and has a combination of forest, agriculture, livestock and fisheries(Sonowal *et al.,*1989). Zabo means impounding of water. In this land-use system, there is a forestland on the top of the hill followed by water harvesting system in the middle and livestock yard and paddy fields at the foothills. The slope of the catchment area is kept under natural vegetation. Water harvesting ponds are dug out with formation of earthen embankment below the catchment area. The usual capacity of the pond varies from 300 to 600 cubic meters with depth varying between 1.5 to 2.5 m. The ponds are having overhead silt retention tanks and compacted from the inner surface in order to arrest water loss. Below the pond, it is a common practice to make a cattle enclosure with wood and bamboo. Buffaloes, cows, pigs are the common animals with the farmers. Cattle yards are washed with run-off water that is then diverted to the paddy fields for manuring.

Rice is grown in the bench terraces constructed at the lower elevations. The area of paddy fields varies between 0.2 to 0.8 ha. Use of leaves and branches of *Alnus nepalensis, Albizzia lebbeck* and *Mekhonu* for green manuring is common. Rice fields are thoroughly rammed at the puddling time to minimize percolation losses. Seepage loss through shoulder bunds are checked with the use of paddy husk on the upstream side. Farmers also practise paddy cum-fish culture.

Rice-based farming system of Apatani Plateau

The Apatani plateau occupying a stretch of 26 sq. km area in Subansiri district of Arunachal Pradesh is inhabited by "Apatani" tribe. The valley is called "The Rice Bowl" of the Apatanis who practise wet rice cultivation and have a good knowledge of land and water management(Sharma and Sharma,2005). Apatani farmers grow a number of tree and bamboo species in the hills surrounding the valley. Rice is cultivated on terraces prepared on mild slope in the valley. They have developed a remarkable system of water management to maintain a 10 cm water table in the plots. The plot to plot outlets in rice fields are made with wood or bamboo. Nutrient and fertility management of the terraces is done mainly through the recycling of agricultural wastes. All rice varieties grown are local. '*Emo*' variety gives the highest yield (5.2 t ha^{-1}) followed by '*Pyaping*' (4 t ha^{-1}). Paddy-cum-fish culture is also attempted. Figermillet is generally grown on the risers. This system is highly sustainable and efficient and helps in conserving of natural resources, maintenance of ecological balance and flow of streams.

Bamboo drip irrigation system

This system (Singh,1989) is followed mainly in Jaintia and Khasi Hills of Meghalaya. This is a very useful irrigation system in the area where water is scarce, soils have poor water holding capacity, the topography is rocky and undulating and irrigation is required for such crops where water need is relatively low. The most convenient technique would be diversion of water from upper reaches of the hill top to lower reaches through gravitational flow. Different forms of bamboo pipes are used and the rate of water flow varies from 800 to 1200 litres hr^{-1}. Bamboo drip is an excellent system to provide irrigation to specific crops such as beetle-vines, arecanuts and black pepper *etc.*

Agriculture with alder trees

Alder (*Alnus nepalensis*) is grown in Nagaland on altitudes varying between 1000 –2000 meters. It is a leguminous tree having the ability to fix 50 to 100 kg N ha^{-1}. Alder trees have multipurpose uses. It is used for firewood, furniture and house construction. Alder trees become ready for pollarding after 6 to 8 years of planting. The alder is pollarded horizontally at a height of 2 to 2.5 m above ground level. Tree trunks are then stripped off sub-branches and leaves which are left in the field and burnt in heaps. Ash is spread uniformly in the field. From February to April, various crops, viz. Maize, job's tears, millet, potato, chillies, pumpkin and barley are sown. Harvesting is done in October or November. By this time, alder tree sprouts new coppices. Stubbles of the previous crops and the new leaves are collected, burnt and added to the soil to augment its fertility. Besides annual crops, alder tree is also useful for providing shade to coffee at lower altitudes and cardamom at higher altitudes.

Rice cultivation on terraces (Panikheti)

This system of rice cultivation has originated in Nagaland but is now prevalent in Manipur and Sikkim. Terrace construction on hill slopes becomes necessary to keep rain water standing in the terraces and thus it is locally known as 'Panikheti". The terraces are constructed and supported with strong bunds of width varying from 0.5 m to 1.2m and height from 0.2 to 1.4m depending on the slope of the land and size of the terrace. The bunds are rammed properly with sticks and local tools to make them firm. The terraces are levelled properly and clayey soils are spread to check water percolation appreciably. The water coming down from the hill surrounding the terraced lands is diverted to the rice fields through channels and sub-channels. Local rice varieties are grown on the terraces and colocasia and some other crops are sown on risers.

The perusal of the data as presented in Table 4 indicates that the indigenous farming systems have significantly large area under effective vegetative cover. Zabo system contains the maximum area under forest (46.5 %) followed by Apatani (41.4 %), Narrow valley cultivation (39.1 %), Agriculture with Alder (33.6 %), Panikheti (26.0 %) and Shifting Cultivation (19.8 %).

Table 4. Effective land cover(ha) under different land use systems.

Land use system	Forest (Open, dense and scrubs)	Grazing lands	Trees and groves	Cropped area	Total
Zabo	0.465	0.029	0.022	0.054	0.570
Apatani	0.414	0.019	0.017	0.090	0.540
Agri-with alder	0.336	0.038	0.034	0.096	0.504
Panikhati	0.260	0.014	0.023	0.043	0.440
Narrow valley cultivation	0.391	0.026	0.037	0.059	0.513
Shifting cultivation	0.198	0.034	0.022	0.050	0.304

Taungya system

This system first introduced by Brandis in Burma (now Myanmar) in 1856, is a method of establishing forest species in temporary combination with field crops. This system of agroforestry was adopted in Tripura by the State Forest Department since early fifties to cover up the areas left by the tribals after jhuming. After sowing of food crops, seedlings of trees, viz., *Gmelina arborea*, *Albizzia procera* and *Tectona grandis* are planted in clean and weed free site. Food crops are continued to be grown until overhead shade prevents satisfactory growth.

Homestead agroforestry system

In every homestead, farmers of Assam and Tripura grow number of tree species along with livestock, poultry and fish mainly for the purpose of meeting their own needs. Jackfruit, coconut mango, arecanut are grown and in their interspaces pineapple, vegetables and other crops are also grown.

These traditional agricultural land uses practised over centuries are well adapted to the local agro-climatic conditions thus providing food and employment to the rural masses as well as maintaining ecological balance. There is a need to bring about an improvement to the land uses for getting higher crop yields. This could be done through adoption of viable technology such as effective cropping sequence, new crop varieties and integrated nutrient management.

Scenario of agroforestry systems

The region is repository of floral and faunal diversity. Local people have screened many tree/shrub species having multiple uses for farming based on their indigenous knowledge system, and accordingly based on the needs of the people; the agroforestry models have been developed. Different tribal communities have developed different unique farming systems and the choice of the species also differs accordingly. State wise some of the important agroforestry systems (AFS) including promising agroforestry (AF) species have been mentioned below:

Arunachal Pradesh

The people of Arunachal Pradesh and its foothill areas bordering Assam have a tradition of cultivating a number of trees, fruit trees, cereals, vegetables, and rear livestock on the same land. Such practices are known to increase the ecological diversity within a landscape unit and optimize the sued of limited resources through the integration of complementary components. Some of the dominant systems present in Arunachal Pradesh are as follows.

- Agrihortisilvipisciculture AFS
- Agrihortisilviculture AFS
- Multistoried AFS/Home garden
- Agrisilviculture AFS
- Hortipastoral AFS

Major multipurpose tree species including Bamboo sp.

Livistonia jenkinsiana, *Anthocephalus chinensis*, *Aquilaria malaccensis*, *Duabanga grandiflora*, *Hippophae rhamnoides*, *Illicium griffithii*, *Mesua ferrea*, *Adenanthera pavonina*, *Altingia excelsa*, *Chukrasia tabularis*, *Eleaocarpus aristatus*, *E. floribundus*, *Michelia champaca*, *M. kisopa*, *Phoebe* spp., *Alnus nepalensis*, *Terminalia myriocarpa*, *Bambusa* sp., *Gmelina arborea*, *Dendrocalamus* sp., *Phyllostachys bambusoides*, *Morus laevigata*, *Thysanolaena maxima*.

Major fruit trees

Citrus reticulata, *Artocarpus heterophyllus*, *Musa* sp., *Ananas sqamosa*, *Areca catechu*, *Citrus limon*, *Pyrus malus*, *P. communis*, *P. domestica*, *Vitis vinefera*, *Juglans regia*.

Agri-horti-silvipiscicultural system

This is an age-old agrorforestry practice adopted by the Kalita community of Assam. This is one of the productive systems, where different agroforestry components are cultivated on the same land management unit. The farmers choose the crops and crop combinations based on their own wisdom and perceptions acquired over generations of experiences, the criterion being their day–to day requirements of food, fuel, fodder and timber (Table 5).

Table 5. Economic status of different traditional agroforestry systems

Agroforestry types	Agrihorit-silvipisci-culture system	Agrihorti-silviculture system	Agripisci-culture system	Bamboo-based silviculture system	Horti-silvipastoral system
Inhabitants	Kalita Group	Nyishi	Apatani	Apatani	Nyishi
Village name	Harmutty	Nirjuli & Doimukh	Ziro	Ziro	Nirjuli &Doimukh
Altitude (m asl)	120	118 & 126	1650	1700	118 & 126
Total yield (q/ha)	109.51	69.43	60.19	79.31	99.24
Gross return (Rupees)	9,200.00	4,500.00	7,570.00	3,185.00	5,983.75
Net return (Rupees)	6,781.00	3,188.00	5,858.00	2,410.00	5,119.00
Cost-benefit ratio	1 : 2 .8	1 : 1.4	1 : 3.4	1:3.1	1 : 5.9
Profit (%)	73.70	70.83	77.11	75.66	85.55

Recently, horticulture has become very profitable and popular because of the improvement in the production potential of edible fruits and timber species like *Muse* sp., *Citrus* sp., *Ananas comosus, Bauhinia* sp., *Erythrian indica, Gmelina arborea, etc.* From the socio-economic and cultural viewpoint, some species were maintained and utilized as cash crops. For example, arecanut tree plays a vital role in the economy of the local society. The labour input for managing the crops is less than that for many other crops, which makes it an ideal crops for the people engaged in other occupations. The economic advantage of the system is derived from the cash-sale of the agricultural products. The family income is greatly improved, as the farmers save the cash that otherwise would have been spent on food. The system also provides a more or less full-time employment to most participants who have no other source of income. The Kalitas also maintain fish ponds in the farmyard. Trees are planted surrounding the fish pond, and crops inter-planted forming an integrated biological production system. Common carps, silver carps and grass carps are generally preferred by the traditional society. The litter of many leguminous plants like *Leucaena* sp., *Moringa oliefera, etc.* has been found to serve as a good fish-feed when offered as pellets and improved the fish production. Further, the trees and shrubs in the traditional systems play an important role in regulating the microenvironment of the system. They are the principal source of rural energy and provide countless medicinal products used in the households.

Agri-horti-silvicultural system

This is a common agricultural system practiced in the foothills by the Nyishis (Table 6) (one of the dominant tribes in Papum Para district of Arunachal Pradesh).

Table 6. List of traditional agroforestry systems with dominant functional components

Species name	Component crop
Kalitas Fruit tree based *Artocarpus heterophyllus Lam. Mangifera indica Linn., Citrus reticulate Blance., Musa sp., Ananus comosus Merr, Zizyphus jujube Lamk., Cocos nucifera* Linn, *Areca catechu* Linn., *Dillenia indica* Linn., *Elaeocarpus floribundua* Blume., *Citrus limon* (L) Burm.	Winter vegetables *Brassica oleracea* Linn. Var. *capitata, Brassica oleracea* Linn. Var. *botrytis, Raphanus sativus* Linn. *Daucus carrato* Linn. *Coriandrum sativum* Linn. Summer vegetables *Solanum melongena* Linn., *Capsicum annum* Linn., *Cucurbita moschata* Poir., *Momordica charantia* Linn. *Lycopersicum esculentum* Linn., *Luffa cylindrical* Roxb., *Cucurmis sativus* Cereals *Oryza sativa* Linn., Zea mays Linn.,
Bamboo based *Bambusa tulda Roxb., Bambusa nutans* Wall., *Bambusa pallida* Munro., *Bambusa balcooa* Roxb.	**Cereals** *Oryza sativa* Linn.
Timber based *Duabanga sonnerioitedes* Buch., *Terminalia myriocarpa* Heurek. And Muell., *Michelia champaca., Gmelina arborea* Linn.,	**Cereals** *Oryza sativa* Linn., *Zea mays* Linn., Vegetables *Beta vulgaris* Linn. Var. *Benghaliensis, Raphanus sativus* Linn., *Solanum tuberosum* Linn., *Brassica campestris* Linn. Oil yielding *Sesamm indicum* Linn., *Glyaine max*(L) Merr.
Apatanis Fruit trees based *Pyrus malus* Linn., *Pyrus communis* Linn., *Prunus domestica* Linn., *Vitis Vinifera* Linn, *Citrus reticulate* Blanco.	**Cereals** *Oryza sativa* Linn., Eleusine *coracana* Linn., *Zem mays.* Linn.
Bamboo based *Phyllostachys bambusoides* Sieb and Zucc. Timber species based *Pinus wallichiana* A. B. Jack., *Morus laevigata* Wall., *Altingia excelsa* Noron.	

It is one of the most common and age-old farming practices in the northeast India. In this system, the local people plough (2-3 times) their field with bullocks and grow traditional crop species such as paddy, *Eleusin coracana, Zea mays, etc*. Among the tuber crops, *Manihot esculenta, Dioscorea* sp., *Colocasia* sp., *etc*. are the most widespread and chief subsidiary food crops. Vegetables include *Capsicum, Colocasia, Solanum tuberosum* and *S. melongena, etc*. Pineapple is a common floor crops grown along with the vegetables in the homesteads. A number of cultivars of banana are also cultivated. The farmers cultivate tree species on the boundary and banana are also cultivated. The farmers cultivated tree species on the boundary and agricultural crops in the middle. So, this farming practice is predominantly subsistence- oriented. The farmers have poor resource-base, their landholdings are small and fragmented, and they have diverse requirements of food, fodder, fuel and timber.

The Nyishis have a rich tradition of cultivating medicinal plants in their traditional agroforestry systems. In addition, other species that provide leaves spices and condiments (*Piper betel, Piper longum, Zingiber* sp., *Oscimum sanctum, Azadirachta indica, etc*.) are also cultivated. In practical terms, the main expectation from an intercropping system in a perennial plantation cropping system is that the overall return from a unit piece of land is increased without adversely affecting either the current or the long-term productivity of the main perennial crop. At the same time, the returns from the additional crops should justify the adoption of the intercropping practice and should contribute to the long-term productivity of the system.

Agripisciculture system

A stylish land, water and agricultural resource management system is seen among the tribes of 'Apatanis' in Arunachal Pradesh. The conventional societies adopt an exclusive paddy-cum-pisciculture in their field (locally called '*Aji*' system), which is extremely a dynamic system, based on a complex network of irrigation channel and firm water management practices involving the people. The water essential for the rice field are tapped from the nearby streams rising from adjacent catchment areas and steered into the fields through the means of small channels. Paddy is the chief crop cultivated by the Apatanis. They use two local varieties *viz.,* the early and the late growing varietes. The early variety (locally called Mipya, comprised four sub-varieties like Pyara, Pyapin, Pyani and Pyat) and the late variety (Emo, also has four sub-varieties: Lalang Enkhe, Elang, Empu).in general, the agricultural system of the Apatanis is well through-out to be extremely efficient in this area. The paddy-cum-pisciculture or the *Aji* system of this particular tribe is supposedly recent in origin, being and interaction promoted by the State Government in the mix-sixties. In this system, the local farmers introduce fishes like common carps, silver carps, *etc*. in the paddy fields. The effectiveness of the fish production in this system, nonetheless seems to be tremendously high. It is worth mentioning here that this outstanding efficiency is despite greater mortality of fingerlings. In addition to paddy-cum pisciculture system, the Apatanis also grow millets on the bunds sorting out the rice plots as well raise plots somewhere else.

Bamboo based silvicultural system

This system is also prevalently adopted by the Apatains. Bamboo is a commercially valuable non-timber species or proularly called as the "poor man's timber" and is common in the homesteads of the Apatains. For housing they maintain bamboo farms ('Bije') and individual forests ('Sansung'), in totaling to their clan forests. The Apatain farmers grow a sole species of their traditional bamboo locally called 'Apatain bamboo' (*Phyllostachys bambusoides)* in the agroforestry and also in the periphery of the field, which is a native of China and has been introduced in India (Tewari, 1992). Depsite its socio-economic importance and fast growing nature, information is lacking about its site characteristics, which otherwise can be very useful in introducing this bamboo to similar sites in other parts of the region or elsewhere. In the clan forests, the people mainly grow fruit species like apple, grapes, pears, peach, *etc. Pinus wallichiana* is also a component of agroforestry system in the Apatain plateau. These pines are mainly cultivated as timber species for house construction. The Apatains are more aware of the imporatance of medicinal plants and they call a meeting for time to time at local level to raise and preserve medicinal plants for future generation. The medicinal plants like *Taxus baccata, Illicium grifflhi, Geranium* sp., *Cymbopogan winterianus etc.* are found in their home-gardens. The Apatanis contrasting to their neighbours (the Nyishis) are confining to a narrow territory (geographical area, 1058 km^2) and therefore, the shortage of land has led to the growth of a commendable, efficient and well- managed land use system and water resource management worthy of replication in other areas.

Hortisilvipastoral system

This system is practiced by the Nyishis. Timber species such as *Terminalia myriocarap, Gmelina arborea, Mesua ferra* are grown in combination with the non-timber yielding species like bamboo. *Livistonia jenkinsiana* (locally called 'Toko') Cultivation is the unique feature of this tribe. This bamboo species is having a greater potential in agroforestry due to its shorter gestation period and recurring economic returns, and the Toka leaves are used as roofing material. The fruit trees like *Mangifera indica, Psidium guajava, Artocarpus heterophyllus, etc.* are common in this system. Crop-animal systems, in which livestock plan a multipurpose role yielded various products and performed different functions, such as recycling of nutrients and energy and helped to achieve self-sufficiency, stability and sustainability of the system, are common among the Nyishis. Most farm families rear cows, bullocks, mithun, chicken and ducks in their homesteads. Some families also rear pigs. The cow forms an important part of the household, not only to provide milk, but also generates organic manure. Increased productivity from livestock will be necessary in these systems to meet the increased demand for animal products, to alleviate poverty and to improve the livelihoods of resource-poor farmers (Davendra and Thomas, 2002). 'Mithun' (*Bos frotalis)* is common among Nyishis, which are used for every social traditional occasion. The biodegradable wastes from houses and the crops are also used as animal feed.

The horti-silvi-pastoral land used brings stability in total biomass production through fruit, fodder and fuelwood. During initial period, pasture would contribute most of the income and later the fruit trees. The cost incured during the preliminary stages would be substantially higher due to warranted performances of all operations like rubbing, ploughing, sowing, planting and use of materials like manure, seeds *etc.* however, the cost reduces substantially in the next year. The expenditure incured was mainly on collection of grass, seeds and harvesting of grasses and maintenance of pasture. Horti-silvi-pastoral system has proved to be quite beneficial in generating more income for the farmer especially during off-season when crops are not cultivated. The success of the system largely depends upon the proper selection of fruit tree species under different agroclimatic conditions to meet various objectives. More income could be generated if it is integrated with improved breed of milch animals. This system would be more stable if right type of tree and pasture species are grown together. So management strategies have to be improved for this system to attain the best return.

Assam

Assam has a geographical area (TGA) of 78,520 km^2 and out of which an area of 30,710 km^2 is legally notified forest area covering 39.1 % of TGA. Some of the dominant agroforestry systems are illustrated below.

- Agrisilviculture AFS (agroforestry system)
- Silvipastoral AFS
- Multistoried AFS/Home garden
- Agrihorticulture AFS
- Agroaquaculture AFS

Major multipurpose tree species including *Bamboo* sp.

Gmelina arborea, Acacia auriculiformis, Populus deltoides, Leucaena leucocephala, Cymbopogon martinnii, Bambusa sp., *Dendrocalamus* sp., *Albizia lebbek, A. procera, Anthocephalus chinensis, Castonopsis indica, Duabanga grandiflora, Ficus* sp., *Litsaea polyantha, Aquilaria agallocha, Michelia champaca, Tectona grandis, Shorea assamica, Bombax ceiba, Musua ferrea, Ailanthus excelsa, A. grandis, Dipterocarpus macrocarpus, Terminalis myriocarpa, Morus* sp.

Major fruit trees

Areca catechu, Cocus nucifera, Moringa oleifera, Musa paradisica, Ananas squamosa, Citrus sp.

Manipur

- Multistoried AFS/Home garden
- Agrihorticulture AFS
- Sericulture based AFS
- Agrisilviculture AFS
- Agroaquaculture AFS

Major Multipurpose Tree Species including *Bamboo* sp.

Parkia roxburghii, Bauhunia variegata, Albizia chinensis, Michelia champaca, Ficus sp., *Morus indica, Toona ciliata, Lagerstroemia indica, Duabanga grandiflora, Terminalia myriocarpa, Gmelina arborea, Bambusa* sp., Dendrocalamus sp., Alnus nepalensis.

Major fruit species

Parkia roxburghii, Ananas sqamosa, Artocarpus hetrophyllus, Magnifera indica, Phyllanthus emblica, Psidium guajava, Punica granatum, Tamarindus indica.

A diverse variety of forest tree species are derived from the forests which have been of use by the local people in Manipur for various purposes (Table 7).

Tree fodder and grasses are mainly collected by the villagers from the forests and homestead while paddy straw, sugar cane and crops residues are collected from agricultural lands. The per capita consumption of fuelwood is 4.37 kg/day, of which 2.07 kg is consumed in cooking, 0.4 kg in preparing food for piggery and cattle, 0.1 kg for festival use, 0.6 kg for warming of rooms and 0.4 kg for preparing rice beer. The quantity consumption was proportional to the family size (Table 8). For the stall-fed animals, nearly 15% of the total fodder demand is collected from the forests which is mostly drawn from 11 species while the remaining 85% fodder demand is met from the species grown in the farm land and in the agroforestry and farm lands including the bamboo stead. Although timber harvest is strictly prohibited from the forests, occasional felling is noticed. Out of the total fuelwood extraction, 66.0% comes from homestead, 17 % from bamboo stead and 17% from the forests. The requirement of fuelwood, fodder, tree foliage, small timber and bamboo for the whole village and their per cent extraction from various land uses is given in (Table 8).

Table 7. Forest tree/ shrub species used by the local populace in Manipur

Tree species	Local/ common name	Fuel	Fodder	Fruit	Fibre	Timber or other industrial use	Green Manure, medicine, dyeing and other use
Adhatoda vasica	Nomang-kha						+
Artocarpus integrifolia	Thei-bong		+	+		+	+
Azadirachta indica	Neem	+	+	+	+		+
Bamboosa tulda	Wa	+	+		+	+	+
Bauhinia variegate	-	+	+				+
Bombax malbaricum	Tera		+	+	+	+	
Cestrum nocturnum	Thabal-lei	+					+
Dendrocalamus strictus	Wa sanneibi	+	+		+	+	
Ficus bengalensis	Khongnang-boat		+	+			+
F. glomerata	Hei-bong		+	+			+
Ficus religiosa	Sana-khong-nang		+				+
Lagerstroemia indica	Saheb-lei	+			+		
Mangifera indica	Hei-nou	+	+	+			+
Michelia champaca	Lei-hao					+	
Morus indica	Kabrang-chak			+		+	+
Parkia roxburghii	Yong-chak			+			+
Phyllanthus emblica	Hei-kru	+		+			+
Psidium guajava	Pung-ton			+			
Punica granatum	Ka-foi			+			+
Tamarindus indica	Mange			+			
Toona ciliate	Tai-ren	+				+	+

Table 8. Consumption of fuelwood (kg/day) for various purposes

Purposes	Per capita consumption irrespective of family size	Family size		
		Small	Medium	Large
Piggery and feeding cattles	1.20	1.30	1.10	1.10
Festivals and rituals	0.10	0.08	0.12	0.15
Warming up rooms	0.60	0.30	0.50	0.08
Preparation of rice beer	0.40	0.20	0.40	0.60
Cooking	2.07	1.50	2.70	3.00
Total	4.37	3.38	4.82	5.65

Meghalaya

- Agrihortisilvipastoral AFS
- Multistoried AFS/homegarden
- Hortipastoral AFS
- Agroaquaculture AFS

Major multipurpose tree species including *Bamboo* sp.

Michelia champaca, M. oblonga, Schima wallichii, Prunus cerasoides, P. nepalensis, Exbucklandia populnea, Pinus kesiya, Lagerstromia speciosa, Bambusa pallida, Thysanolaena maxima.

Major fruit species

Citrus reticulata, Citrus limon, Prunus persica, Pyrus communis, Artocarpus heterophyllus, Musa sp., *Ananas, Psidium guajava.*

In all, four major agroforestry systems (AFS) viz., agrisilviculture (AS), agrihorticulture (AH), Silvihorticulture (SH) and silvipastoral (SP) have been identified prevalent at different sites (Table 9). At lower elevational sites (upto 700 m asl), agrisilviculture system is quite common. Agricultural crop such as paddy, ginger, *Colocasia,* Chilli, *Dioscoria,* pumpkin, sweet potato are grown in monoculture or polycuture on the permanent terraces and *bun* prepared along the hill slopes. While fodder, fuel/timber trees such as *Lagerstroemia speciosa, Schima wallichii, Acacia auriculiformis, Bambusa pallida* and horticultural trees like *Artocarpus heterophyllus, Litchi chinensis, Areca catechu, Psidium guajava, Citrus, Musa paradisica,* pineapple and betelvine are deliberately left or grown on the bunds of terraces and between the tree spacing.

Table 9. Important agroforestry systems in Meghalaya.

Elevational range	Agroforestry systems	Components
Up to 700 m asl	Agri-silviculture	• *Lagerstroemia speciosa* + Paddy • *Schima wallichii* + Paddy • *Schima wallichii + Ginger + Colocasia + Chilli+ Dioscoria + Pumpkin + Sweet potato.* • *Bambusa pallida (*Boundary plantation) + Paddy
	Agri-horticulture	• *Psidium guajava* + Maize + Turmeric • *Citrus grandis* + Maize • *Artocarpus heterophyllus* + *Litchi chinensis* + Ginger + *Colocasia* + Maize + Bottle gourd. • *Artocarpus heterophyllus* + *Litchi chinensis* + *Areca catechu* +Betel vine
	Silvi-horticulture Pure horticulture	• *Acacia auriculiformis* + Pineapple • *Acacia auriculiformis* + *Schima wallichii* + *Musa paradisica* + Pineapple • *Musa paradisica* + Pineapple • *Areca catechu* + Pineapple
700-1400 m asl	Agri-silviculture	• *Michelia oblonga* + Paddy • *Michelia oblonga* + Ginger • *Michelia champaca* + Paddy • *Pinus kesiya* + Paddy • *Pinus kesiya* + *Ginger* • *Pinus kesiya* + Turmeric + Maize • *Erythrina indica* (boundary plantation) + Ginger + *Colocasia* + Ladyfinger + Sweet potato + Chilli + *Perilla* • *Bambusa pallida (*boundary plantation) + Paddy + Ginger + Sweet poato+ Chilli+ Tapioca + Ladyfinger + *Colocasia* + *Perilla.* • *Michelia oblonga* + *Pinus kesiya* + Ginger + Chilli + Colocasia + Perilla + Maize + Turmeric. • *Bambusa pallida* + *Erythrian indica* +Maize + Sweet potato. • *Schima wallichii* + Paddy • *Schima wallichii* + Ginger + *Colocasia* + Chilli + *Dioscoria* + Pumpkin + Sweet potato.
	Silvipastoral system	• *Schima wallichii* + Broom grass • *Michelia oblonga* + Broom grass • *Michelia champaca* + *Schima wallichii* + *Pinus kesiya* + Broom grass + *Setaria.*

(*Contd.*)

Elevational range	Agroforestry systems	Components
	Horti-pastoral system	• *Musa paradisica* + Broom grass • *Musa paradisica* + *Citrus reticulate* + *Setaria* + Broom grass
1400 m asl and above	Agri-silviculture	• *Pinus kesya* + Paddy • *Pinus kesiya* + Ginger • *Pinus kesiya* + Turmeric + Maize
	Agri-horticulture	• *Pyrus communis* + Maize + Cabbage + Cauliflower • *Citrus reticulate* +Turmeric + Ginger + Mustard leaf • *Citrus grandis* + Maize + Turmeric + Cauliflowr + Mustard leaf + Potato.

Source: Bhatt *et al.*, 2006

In mid altitudinal range (700-1400 m asl), agricultural crops *viz.* paddy, ginger, turmeric, maize, ladyfinger, sweet potato, chilli, perilla, tapioca, *Colocasia,* pumkin are grown in large area in the interspaces of *Michelia oblonga, M. champaca, Pinus kesiya, Erythrina indicant, Bambusa pallida* and Schima wallichii. Broom grass and Setaria is also distributed in association of *Schima wallichii, Michella oblonga, P. kesiya* and *Musa paradisica.* Fruit trees are planted at regular spacing within the field and fodder or timber trees are left on the field bunds while the annuals are grown in the interspaces of the trees. In higher altitudinal range (> 1400 m asl), *P. kesiya* is the most dominant tree species and agricultural crops like paddy, ginger and maize are commonly cultivated. Agriculture crops mainly comprising leafy and rhizome types are grown in the interspaces of horticultural trees such as *Pyrus communic, Citrus reticulate* and *C. grandis.* Trees are uniformly distributed in the field (Tables 10, 11 and 12).

Table 10. Crop composition with forest tree species, irrespective of AFS

Tree species	Crop composition
Lagerstroemia speciosa, Michelia oblonga	Paddy Paddy Ginger Ginger + Chilli + Colocasia + *Perilla* + Maize + Turmeric Broom grass
Schima wallichii	Paddy Ginger + *Colocasia* + Chilli + Dioscoria + Pumpkin + Sweet potato Sweet potato Broom grass Broom grass + *Setaria* Pineapple
Michelia champaca	Paddy Broom grass + *Setaria*
Pinus kesiya	Paddy Turmeric Ginger Giner + Chilli + *Colocasia* + *Perilla* + Maize + Turmeric Turmeric + Maize Broom grass + *Setaria*
Bambusa pallida	Paddy Ginger + Sweet potato + Ladyfinger + Chilli + *Colocasia* + *Perilla* + Maize Sweet potato + Tapioca
Acacia auriculiformis	Pineapple
Erythrina indica	Ginger + *Colocasia* + Ladyfinger + Sweet potato + Chilli + *Perilla* Sweet potato.

Table 11. Crop composition with different fruit tree species, irrespective of AFS

Fruit tree species	Crop composition
Citrus grandis	Maize, Turmeric, Cauliflower, Cabbage
Pyrus communis	Maize, Cauliflower, Cabbage
Sidium guojova	Maize, Turmeric, Radish, Rice bean
Citrus reticulate	Turmeric, Ginger, Mustard leaf
Musa paradisica	Pineapple, Broom grass
Areca catechu	Pineapple, Betelvine

Table 12. Crop composition with different tree species, irrespective of AFS

Mixed plantation crop	Crop composition
Michelia oblonge + *Pinus kesiya*	Ginger + Chilli + *Colocasia* + *Perilla* + Maize + Turmeric
Bambusa pallida + *Erythrina indica*	Sweet potato
Artocarpus heterophyllus + *Litchi chinensis*	Ginger + *Colocasia* + Maize + Bottle gourd
Artocarpus heterophyllus + *Litchi chinensis* + *Areca catechu*	Betel vine
Acacia auriculiformsi + *Schima wallichii* + *Musa paradisica*	Pineapple
Michelia champaca + *Schima wallichii*	Broom grass + *Setaria* + *Pinus kesiya*
Musa paradisica + *Citrus reticulate*	Broom grass + *Setaria*

Species density of trees was higher in the agrihorticultural system compared to agrisilviculture and silvihorticultural system. As many as 16 MPTs and fruit species were recorded and among these, Khasi pine (*Pinus kesiya), Citrus reticulate* and *Citrus grandis* were the predominant ones. They have uniform distribution in the entire field. Pear (*Pyrus communis)* and guava (*Psidium guajava)* among fruit trees, and *Michelia oblonga, M. champaca* and *Lagerstroemia speciosa* among multipurpose tree species were the important components of this system.

In the agrisilviculture, silvihorticulture and silvipastoral system, *Pinus kesiya* contributed maximum fuelwood yield (117.47 q/ha), followed by *Schima wallichii* (111.49 q/ha) and *Michelia oblonga* (111.03 q/ha), while *Erythrina indica (*1645 q/ha) showed the minimum firewood yield. Tree foliage was produced maximum by *M. oblonga* (86.97 q/ha),followed by *L. speciosa* (70.08 q/ha) and *S. wallichii* (59.66 q/ha) (Table 5). Among horticultural tree species, guava showed the highest fruit (28.76 q/ha), followed by *Citrus grandis* (24.27 q/ha) and *Areca catechu* (17.66 q/ha).

Mizoram

- Agrisilviculture AFS
- Hortipastoral AFS
- Multistoried AFS/homegarden

Major multipurpose tree species

Tectona grandis, Gmelina arborea, Leucaena leucocephala, Meloccana baccifera, Dendrocalamus spp., Aleurities sp., *Schima wallichii, Mesua ferrea, Michelia champaca, Terminalia* sp., *Thysanolaena maxima, Elaegnus latifolia, Erythrina indica*

Major fruit species

Phyllanthus emblica, Citrus reticulata, Musa sp., *Parkia roxburghii, Ananas sqamosa.*

Some of the dominant agroforestry systems of Mizoram are illustrated below.

Tectona grandis based systems

Teak(*Tectona grandis)* is gaining popularity in the recent past due its higher market rates. It has comparatively less deleterious effect on associated field crops. Hence the teak is planted along field boundaries or as zonal planting. Teak based agroforestry model practice in the state is consisted of teak as tree crop, different field crops, hedges such as subabul, various grasses, fruit trees like papaya *etc.* Teak is planted at 10 m or 20 m apart with 2 m between plants and field crops in between two teak plants. Fodder are planted on either side of teak row as soil conservation measure and to obtain green fodder. The other models consisting of subabul + teak + lemon + paddy and *Cajanas cajan* + lemon + teak + paddy has been successfully developed at Sakawrtuichhum village of Aizawl. The intercropping of teak and rice (*Oryza sativa)* has also been successfully practiced by the farmers in the state. Paddy is sown during April-May an stumps of teak seedlings are planted on the boundaries in the month of July.

Leucaena leucocephala based systems

This model is developed for promoting sustainable agroforestry farming system. Subabul being a multipurpose nitrogen fixing tree species is of significance to agroforestry and farm forestry. The plants are often incorporated with agricultural crops in different special pattern in a view to enrich nutrient content in the soil through nitrogen fixation.

Fruits trees mixed with *leucaena leucocephala*

Common fruit trees are planted at 4m x 4m and inter-planted with subabul at 1 m x 1 m. The yield of fruit tree commence from 5th year onwards while fodder and fuelwood yield from subabul is available from 4th year onwards.

Medicinal plants mixed with *Leucaena leucocephala*

In the interspace of subabul (1m x 1m), various medicinal plants are raised to meet the traditional health care need of the people and for sale purposes.

Bamboo based systems

Bamboo area in Mizoram is highest in the country with 49.1 percent to the forest cover. A total of 21 bamboo species have been recorded in Mizoram. Out of 21 species, *Meloccana baccifera* (Mautak), *Dendrocalamus longisphathus* (Rawnal), *Dendrocalamus hamiltonii* (Phulrua), *Bambusa tulda* (Rawthing) *etc.*

are edible ones. The bamboo based agroforestry system is recently introduced by the farmers in the state. The models that have been successfully practiced are bamboo-ginger, bamboo-turmeric, bamboo-soya bean, bamboo-mustard *etc*. Bamboo conserves soil against erosion, creates a large shade under its foliage reducing evapo-transpiration loss. As more soil moisture exists under and around bamboo groves, a congenial atmosphere is created for microbial activities resulting better nutrient turn over and crop production.

Hedgerow intercropping

This model was introduced in the farmers field on experimental basis and is showing promising results. The nitrogen fixing trees such as subabul (*Leucaena leucocephala),* pigeon pea (*Cajanas cajan)* and non-nitrogen fixing species are planted at closer to spacing to establish hedges. The spacing between two contour hedges depends upon the slope percent. The spacing of fruit tree species and forestry species are fixed as per the silvicultural requirements of the species. The field crops are grown in between hedge rows. This farming system protects the land from degradation and keeps the land protective on sustained basis.

Coffee based systems

Coffee based agroforestry system had successfully practiced in Bualpui, Kolasib District of Mizoram. In this system, the existing jungle growths are used as shade and coffee is grown in understorey. The shade trees include *Gmelina arborea, Schima wallichii, Teminalia* sp., *and Michelia champaca etc.*

Sole banana cropping

The farmers plant banana in the farm with the belief that they are very stable and can withstand wall sorts of natural calamities, besides being providing high economic returns. Banana plantation provide cash, hold big chunks of soil, besides big leaves of banana dissipate a lot of monsoon rain drop kinetic energy resulting reduction in splash erosion. The practices is common with the chakmas.

Tung based systems

Tung (*Aleurities* spp) plantation was introduced little over a decade now and provide quick economic return. The plants are grown at a spacing between 5m x 5m, 8 m x 8m and 10m x 10m and in between the spaces, field crops such as ginger, turmeric, tapioca are grown. This practice is adopted in Aizawl and Lunglei districts of the state.

Nagaland

- Agrisilviculture AFS
- Agrihorticulture AFS
- Multistoried AFS/homegarden

Major multipurpose tree species including *Bamboo* sp.

Alnus nepalensis, Parkia roxburghii, Terminalia myriocarpa, Michelia champaca, Ficus sp., *Gmelina arborea, Tectona grandis, Mesua ferrea*

Major fruit species

Moringa oleifera, Musa sp., *Ananas sqamosa, Citrus reticulata, Artocarpus hetrophyllus, Parkia roxburghii*

The improved agroforestry practices existing in Nagaland are as follows.

Agri-horti-slvipastoral system

The system has been found ecologically and economically most viable for the region. *Dalbergia sissoo, Leucaena leucocephala, Albizia* spp., *Alnus nepalensis, Ficus* spp. *Terminalia myriocarpa, Mesua ferrae, Michelia champaca, Gmelina arborea, Moringa oleifera, Tectona grandis,* and fruit trees like Assam lemon, pineapple banana, papaya, litchi, pear, peach, plum, and Khasi mandarin are grown with a density of 400 plants/ ha. variety of crops like maize, paddy, ginger, turmeric, and seasonal vegetables like pea, cabbage, cauliflower, knoll-khol, radish, french bean, rice bean are cultivated in different combinations. Guinea, congo and broom are the important pastoral components of the system. While MPTs are recommended for cultivation on the top of the hillock, fruit trees are raised in the middle portion of the hillock. Lower area, however, can be used exclusively for light demanding food crops, particularly for rice cultivation.

Agri-horti system

In this system, maize, rice, rice bean, pea, radish, turnip, french bean, rajmah, cabbage, cauliflower, knoll khol, colocasia is grown in association of peach, plum, apricot, almond, apple and pear on half moon terrace along with intercrop perennial fodder grasses like red clover, white clover, setaria, *etc.*

Horti-silvi-pastoral system

Green pea, cabbage, cauliflower, radish, French bean, knoll khol, carrot, turnip, coriander, beans, ginger, turmeric, onion and garlic are cultivated in association of horticultural species like orange, areca not, guava, litchi, mango, banana, pineapple, jackfruit, jamun, citrus sp., apple, pear, peach, plum, apricot, and multipurpose tree species like *Acacai* sp., *Alnus nepalensis, Albizia* spp., *Michelia champaca, terminalia myriocarpa.* Fodder grasses like *Stylosanthes* sp., Napier and broom are also cultivated in the system.

Silvipastroal system

Tectona grandis, Gmelina arborea, Michelia champaca, Terminalia myriocarpa, Albizia spp., *Parkia roxburghii, Eucalyptus* spp., *Artocarpus chaplasa, A. heterophyllus etc* are cultivated. In the understorey, grasses like red clover, white clover, *Setaria* spp., and *Pennisetum* spp., are cultivated.

There is a strong traditional of homestead system throughout the Nagaland. As many as 40 species of various MPTs and fruit trees are cultivated along with fishery and livestock management. Varieties of crops are also cultivated in the underneath of multistoried systems.

Sikkim

- Agrihorticulture AFS
- Agrisilviculture AFS
- Silvipastoral AFS

Major multipurpose tree species including *Bamboo* sp.

Prunus cerasoides, P. nepalensis, Saurauia nepalensis, Schima wallichii, Symplocos theifolia, Toona ciliata, Alnus nepalensis, Ficus clavata, F. nemoralis, F. glomerata, F. hookerii, Castonopsis tribuloides, Litsaea polyantha, Dendrocalamus hamiltonii, Chimonobambusa hookeriana

Major fruit species

Citrus reticulata, Juglans regia, Pyrus communis, Psidium guajava

Traditional agroforestry systems of sikkim under different climatic zones are as follows.

The important agroforestry systems include agri-horticultural, agri-horti-pastoral, agri-silvi-pastoral, horti-silviculture, agro-horti-silvi-pastoral, livestock-based mixed farming, sericulture-based farming, bamboo-based farming, homesteads and tea plantation (Table 13).

Table 13. Traditional agroforestry systems of Sikkim

Agro-ecological zone and altitude	Agro-forestry systems	Dominant tree/ grass species	Crops/ animals
Sub-tropical (300 to 900 m amsl) to Mid hill temperate (900 to 1800 m amsl)	Agri-horticultural	Sikkim mandarin, avocado	Maize, ginger, turmeric, Buckwheat, mustard, pulses, beans, finger millet, potato, vegetables, mushroom, passion fruit, gladiolus, tuberose, marigold, orchids.
	Agri-horti-pastoral	Sikkim mandarin, guava, banana, avocado, lemon, gooseberry, *Ficus* sp. Broom grass, *Dalpizium* sp.	Maize, ginger, turmeric, tapioca, Buckwheat, wheat, pulses, oilseeds, beans, oat millets, rice vegetables.
	Agri-silvi- pastoral	*Alnus nepalensis, Schima wallichii, Prunus cerasoides, Terminalis myriocarpa, Castonopsis trubuloides, Litsea polyantha, Macranga denticulate, Ficus* sp., Broom grass.	Maize, wheat, pulses, Buckwheat, oilseeds, beans, finger millet.
	Horti-silviculture	*Alnus nepalensis, Schima wallichii, Macrange pustulata, Albizzia sp., Machilus edulis, Saurauia nepalensis, Terminalia myriocarpa, Jaglans regia*	Large cardamom (*Amomum subulatum* Roxb.)
	Agri-horti-silvipastoral	Sikkim mandarin, *Ficus* sp., avocado, peras, amls, guava, *Juglans regin, Schima wallichii, Prunus cerasoides,* Broom grass	Maize, tapioca, ginger, vegetables, beans, pulses, oilseeds.
	Livestock-based mixed farming	*Arundinaria* sp., *Setaria* sp., *Cynodon* sp., *Cyperus* sp., *Saccharum* sp., *Ficus* sp., *Erythrina arborescens, Agave* sp., *Prunus cerasoides, Urtica* sp., *Artemisia* sp., millets.	Cattle, pig, poultry, goat.
	Sericulture-based farming	*Morus laevigata,*	Pulses, oilseeds, Broom
	Bamboo-based farming	*Terminalia* sp. *Chimonobambusa* sp., *Bambusa* sp., *Drepanostachyum*	Grass, millets, oat Tender bamboo shoots collected, ginger, turmeric, large cardamom, rice bean (up to

(Contd.)

Agro-ecological zone and altitude	Agro-forestry systems	Dominant tree/ grass species	Crops/ animals
		intermedium, Himalayacalamus falconeri, Phyllostachys bambusoides.	11 to 15 m from bamboo rows).
	Homesteads	Sikkim mandarin, lime, Ficus sp., tree tomato (*Cyphomandra betaceae),* guava, pear, pomelo, payaya, pomegranate, avocado, banana, *Urtica* sp., *Artemisia* sp., *Moringa* sp., *Mangifera* sp.	Vegetables, Passion fruit, gladiolus,tuberose, marigold, Orchids, sugarcane, pig, poultry, cattle, goate, ducks, Wild edibles-ferns, nettles, fishery, mushroom, apiary.
Temperate zone (1880 to 2700m amsl)	Agri-horticultural	Sikkim mandarin, apple	Potato (table and seed), maize, barley, rice, Buckwheat, radish, cabbage, cauliflower, *Brassica juncea* var. *rugosa.*
	Horti-silvi-pastoral	Sikkim mandarin, apple, *Juglans regia, Alnus nepalensis, Prunus nepalensis, Quercus* sp., *Betula alnoides, Acer* sp., *Hippophae salicifolia.*	Maize, millets, large cardamom, potato (table and seed), peas, cabbage, cauliflower, beans, radish.
	Livestock-based mixed farming	*Betule uitlis, Acer* sp., *Imperate cylindrical, Arundinella* sp., *Avena* sp., *Eleusine* sp., *Setaria* sp., *Rubus* sp., *Viburnum erubescens, Berberis* sp., *Utica* sp., *Artemisia* sp.	Goats, pig, sheep, poultry, nomadic herds of yak (dzo's).
Sub-alpine (2700 to 4000m amsl)	Horti-pastoral-transhumance	*Quercus* sp., *Acer* sp., *Betula utilis, Sorbus* sp., *Carex* sp., *Trisetum* sp., *Egagrotis* sp., *Aralia* sp., *Allium* sp., *Iris* sp.,	Radish, peas, potato, beans, maize, cabbage, cauliflower *Brassica juncea* var. *rugosa,* yaks (dzo's), sheep, goats, mules.
Alpine (> 4000 m amsl)	Livestock- based mixed farming (beyond timberline) –transhumance	*Poa* sp., *Agrostis* sp., *Carex* sp., *Gentiana* sp.,*Brassica Rumex* sp. *Phlomis rotate, Urtica dioca*	Potato, cabbage, peas, *juncea* var. *rugosa,* yaks (dzo's), sheep and mules.

Tripura

Tripura has got 6,292 km^2 forest area comprising of nearly 60% of the total geographical area and the flora of Tripura (Deb,1981) is reported to have 1573 taxa having 1545 species,862 genera and 193 families of vascular plants.

- Multistoried AFS/homegarden
- Agrisilviculture AFS
- Hortisilvipastoral AFS
- Agroaquaculture

Major multipurpose tree species including *Bamboo* sp.

Albizia procera, Dalbergia sissoo, Melia azadirachta, Acacia auriculiformis, Vitex peduncularis, Gmelina arborea, Dipterocarpus turbinates, Meloccana baccifera, Bambusa tulda, Schima wallichii, Artocarpus chaplasa, Bombax ceiba, Terminalia arjuna, Teinostachyum dulloua, Dendrocalamus longisphathus and *D. hamiltonii.*

Major fruit species

Ananas sqamosa, Citrus limon, Artocarpus heterophyllus, Areca catechu, Cocus nucifera and *Musa* sp.

General characters of the area

The study pertaining to the impact of agroforestry systems on soil fertility was conducted in the research farm of Indian Council of Agricultural Research (ICAR) Complex for North Eastern Hill Region, Tripura, India. The area is characterized by dry winters and hot summers followed by heavy rains. The mean maximum temperature ranges from 29.2°C to 36.8°C and the mean minimum temperature from 4.1 °C to 21.4 °C with average maximum and minimum temperature being 35 °C and 10.5 °C, respectively. It receives 2255 mm of rainfall.In soils (Fine, Kaolinitic, Typic Kandiudults) in uplands (40-65 m above msl) toposequence, 12 multipurpose tree species (MPTs) suitable for humid subtropical region were planted in 1989 in randomized block design with 8 replications.. The tree species are *Acacia auriculiformis,* A. Cunn , *Morus alba,* Linn , *Leucaena leucocephala,* Lank, *Dalbergia sissoo,* Roxb, *Gliricidia maculata,* H.B. & K, *Azadirachta indica,* A. Jus, *Michelia champaca,* Linn, *Eucalyptus hybrid,* L. Herit, *Tectona grandis,* Linn, *Gmelina arborea,* Linn, *Samania saman,* Jacq, *Albizzia procera,* Linn ,Soils are acidic in nature with pH (4.1 to 4.4), low organic carbon (0.2% to 0.8%) and high aluminum saturation (53% to 63 %) as exchangeable cation.

Soil physico-chemical properties

The average values of soil physico–chemical properties of surface soil (0-15 cm) under MPTs in Tripura during 4-16 years are presented in Table 14.Soil pH was found to vary from 4.1 to 6.0. *A.indica* showed the maximum rise in soil pH (1.6 unit) while *G. maculata* showed a decline in soil pH (0.3 unit) as compared with open space (No tree). Exchangeable acidity (H^{+}+Al^{+3}) estimated in the surface soil (0-15 cm) was found to vary from 3.30 to 6.60 c mol (p^+) kg^{-1} .The lowest exchangeable acidity was recorded in soils under *A. indica* and the highest in soil under *L. leucocephala*. Water holding capacity showed an increase from 28.94 to 34.27 per cent in *A.indica* .Similar rise in water holding capacity was also recorded in soils under *A. auriculiformis* and *D.sissoo* but other MPTs showed a declining trend as compared to soils of open space. Mechanical analysis showed that sand, silt and clay in soils under MPTs varied from 45.5 to 62.8 %, 5.40 to 19.40 % and 17.8 to 42.8 %, respectively and soils of open space had 59.9 % sand, 3.93 % silt and 36.2 % clay.

Table 14. Effect of MPTs on soil physico-chemical properties[a] in Tripura, India

MPTS	Soil pH	Exchangeable acidity	Water holding capacity (%) [c mol (p^+) kg^{-1}]	Mechanical analysis (%)		
				Sand	Silt	Clay
A.auriculiformis	5.3	4.29	31.17	55.8	10.0	34.2
M. alba	4.7	4.29	26.50	59.9	15.20	24.3
L. leucocephala	4.6	6.60	25.90	60.8	20.3	18.9
D. sissoo	5.4	4.29	29.22	57.8	9.80	32.4
G. maculate	4.1	5.61	25.62	51.0	9.70	39.3
A. indica	6.0	3.30	34.27	60.6	9.61	29.8
M. champaca	4.1	4.62	40.82	60.8	17.40	21.9
E.hybrid	4.7	4.95	25.98	58.3	11.90	29.8
T. grandis	5.2	5.94	27.14	57.2	5.40	38.9
G. arborea	4.8	5.94	25.57	63.6	12.62	23.2
S. saman	5.0	4.29	27.82	62.8	19.40	17.8
A. procera	5.0	3.96	25.57	45.5	11.71	42.8
Openspace	4.4	6.27	28.94	59.9	3.93	36.2
Mean	4.8	4.95	28.93	53.2	12.11	29.9
CV (%)	10.4	18.4	14.5	12.7	39.22	26.2
LSD (P=0.05)	0.21	0.27	2.54	2.19	5.27	4.47

Note : [a]Average values from 4-16 years

In 3 sites in Arunachal Pradesh, India, changes in physico- chemical characteristics of soils were studied (Deb *et al.*, 2005) and the data are presented in Table 15. Sites I, II and III are inhabited by tribes of Kalita ,Nyishi and Apatani, respectively. The nitrogen content of soil in Site-I was recorded more than the other two sites. Moreover, the concentration of plant available nutrients in soil (ammonium, nitrate and available P) was maximum in the AFS of Site I dominated by Kalitas. This could be attributed to greater litter availability, as confounded by plant species of high richness and diversity (Table 15) in the systems. However, the carbon content in the AFS of the Nyishis was found higher. The Apatani people are practising bamboo-cum-pine plantation in their AFS, which reduce the decomposition and microbial population of soil as the leaves from pine are producing some resins which inhibit microbial growth thereby affecting the mineralization rates. The litter depth and biomass was found greater in Ziro AFS (Table 15) followed by Harmutty site.

Table 15. Physico-chemical properties of soils in three traditional agroforestry systems in Arunachal Pradesh

Properties	Soil depth (cm)	Site-I	Site-II	Site-III
Soil Textural Class	0.30	Sandy loam	Loamy sand	Loamy sand
SMC (%)	0-10	26.71±2.11	17.84 ± 1.34	15.80 ±1.06
	10-20	24.41 ± 2.13	16.54 ± 1.02	18.05 ± 1.36
Bulk density (g cm^{-3})	0-10	0.60 ± 0.08	0.36 ± 0.07	1.09 ± 0.11
	10-20	1.09 ± 0.09	0.43 ± 0.08	1.20 ± 0.13
pH	0-10	5.27 ± 0.21	5.95 ± 0.24	5.55 ± 0.19
	10-20	5.20 ± 0.20	6.03 ± 0.27	5.67 ± 0.14
	20-30	5.08 ± 0.24	5.89 ± 0.24	5.72 ±0.23
Organic C (%)	0-10	1.03 ± 0.08	3.95 ± 0.12	0.91 ± 0.08
	10-20	1.59 ± 0.01	3.79 ± 0.14	0.86 ±0.05
	20-30	0.89 ± 0.06	3.18 ± 0.25	0.89 ± 0.9
TKN (%)	0-10	0.78 ± 0.06	0.31 ± 0.03	0.40 ± 0.02
	10-20	0.69 ± 0.02	0.33 ± 0.01	0.42 ± 0.01
	20-30	0.52 ± 0.02	0.25 ± 0.01	0.22 ± 0.01
C/N	0-10	1.32	12.74	2.28
	10-20	2.30	11.48	2.05
	20-30	1.71	12.72	4.05
$NH_{4-}N$ (μg g^{-1})	0-10	13.43 ± 0.65	10.73 ± 0.89	10.04 ± 1.12
	10-20	11.50 ± 1.21	11.20 ± 1.51	10.61 ± 0.98
	20-30	7.54 ± 0.89	7.18 ± 0.97	11.46 ± 1.02
$NO_{3-}N$ (μg g^{-1})	0-10	2.18 ± 0.08	1.63 ± 0.09	0.89 ± 0.14
	10-20	2.96 ± 0.13	1.18 ± 0.17	1.40 ± 0.62
	20-30	1.87 ± 0.54	1.49 ± 0.64	1.64 ± 0.54
Available P (μg g^{-1})	0-10	32-40 ± 4.32	29.59 ± 3.25	20.76 ± 3.01
	10-20	29.38 ± 3.62	25.33 ± 3.05	22.20 ± 2.54
	20-30	25.10 ± 3.87	15.33 ± 2.14	21.02 ± 2.63

SMC-Soil moisture content, TKN-Total Kjeldahl nitrogen, NH_4-N- Ammonium nitrogen NO_{3-} N-Nitrate nitrogen
Source: Deb *et al.*, 2005

The available N, P and S were higher in agri-horti-silvipastoral and silvihorticulture compared to natural fallow and silvipastoral AFS (Table16).

Table 16. Effect of agroforestry systems on soil properties

Soil properties	Agroforestry systems					
	Agrisilvi culture	Agrihorti culture (khasi mandarin + crops)	Agrihorti culture (Assam lemon + crops)	Silvihorti pastoral (Alder + pine apple + fodder grass)	Multistoried AFS (Alder + tea+ black pepper + crops)	Natural forest
pH	4.65	4.62	4.80	4.25	4.61	4.62
Organic C (%)	1.62	1.55	2.02	2.19	1.91	1.92
Exchangeable Ca [cmol (p^+) kg^{-1}]	0.40	0.86	0.74	0.31	0.65	0.26
Exchangeable Mg [cmol (p^+) kg^{-1}]	0.75	0.51	0.33	0.48	0.71	0.16
Exchangeable K [cmol (p^+) kg^{-1}]	0.232	0.244	0.249	0.238	0.201	0.169
Exchangeable Na [cmol (p^+) kg^{-1}]	0.201	0.220	0.194	0.195	0.197	0.196
Exchangeable Al [cmol (p^+) kg^{-1}]	2.65	2.70	2.20	3.15	2.05	2.20
Available N (ppm)	190.1	180.8	203.6	199.4	216.9	167.2
Available P (ppm)	2.75	4.10	5.36	0.94	3.36	0.63
Available Fe (ppm)	8.9	10.4	12.8	10.9	13.9	7.3
Available Mn (ppm)	0.58	0.92	0.79	0.83	1.04	0.04
Available Zn (ppm)	0.08	0.05	0.07	0.006	0.08	0.025
Available Cu (ppm)	0.21	0.23	0.37	0.30	0.27	0.10

Moreover, the organic carbon, available P, exchangeable cations and available B contents in surface soil ranged between 2.0-2.5 %, 10.4-13.2 ppm, 5.9-8.4 cmol (p^+) kg^{-1} and 1.4- 2.4 ppm under jackfruit based AFS, while 1.5-1.8% organic carbon, 3.8-6.7 ppm available P, 3.9-5.9 cmol (p^+) kg^{-1} total cations and 0.8-1.4 ppm available B were found under arecanut/ khasi mandarin based AFS. In an another study, effect of agrihorticulture (comprising of khasi mandarin + agricultural crops, and Assam lemon + agricultural crops), agrisilviculture (multipurpose tree species + annual agricultural crops), silvi-horti-pastoral (alder + pineapple + fodder grasses) and multistoried AFS (alder + tea + black pepper + annual agricultural crops between the tree rows) on soil properties and fertility status was evaluated in acid Alfisol of Meghalaya after 15 years of plantation by comparing with natural forest of the same age group (Tables 17 and 18).

Table17. Effect of different land use systems on soil properties

Land use	Component	pH (1:2.5)	Organic C (%)	Bray's P-2P (ppm)	CEC [cmol (p^+) kg^{-1}]	Exchangeable Cations [cmol (p^+) kg^{-1}]		
						K	Ca	Mg
Agroforestry	Coffee, and black pepper	4.80	1.60	2.90	9.10	0.30	0.90	0.80
Natural fallow	Grasses	5.70	1.47	1.40	5.20	0.20	0.90	0.60

Table 18. Effect of different land use systems on soil properties in Meghalaya,India

Parameters			Agroforestry systems		
	Natural fallow	Abandoned jhum land	Silvipastoral	Agri-horti-silvipastoral	Silvihorti-culture
Soil texture					
pH	4.99 (4.90)	4.76 (5.20)	4.52 (5.10)	4.92 (4.90)	4.91 (4.90)
Organic C (%)	2.94 (1.85)	3.42 (1.90)	2.61 (1.80)	2.97 (1.82)	2.97 (1.80)
Exchangeable Ca [cmol (p^+) kg^{-1}]	1.96 (1.15)	1.57 (1.16)	1.25 (1.10)	2.11 (1.20)	2.00 (1.20)
Exchangeable Mg [cmol (p^+) kg^{-1}]	0.55 (1.15)	0.38 (1.16)	0.43 (1.20)	1.45 (1.20)	0.85 (0.60)
Exchangeable Al [cmol (p^+) kg^{-1}]	0.88	1.30	1.56	0.90	0.90
Available N (ppm)	179.2	251.1	214.5	220.3	210.9
Available P (ppm)	1.9	2.0	2.1	16.6	12.9
Available K (ppm)	175.6	130.8	98.0	162.7	265.0
Available S (ppm)	14.5	14.8	10.4	19.9	12.9

Figures in parentheses indicate the initial values at the start of the project

Soil organic carbon pool

Soil organic carbon pool (Table 19) was estimated in surface soils (0-15 cm) under MPTs in soils of Tripura, India. This showed a concomitant rise in SOC in soils under MPTs and a subsequent decline in soils of open space over 4-16 years. Maximum rise in SOC was noticed in soils of *A. indica* (28.6 Mg ha^{-1}) followed by *Acacia aurculiformisi* (21.9 Mg ha^{-1}), *Gmelina arborea* (21.8 Mg ha^{-1}), *Michelia champaca* (16.7 Mg ha^{-1}) etc. The minimum rise in SOC was noted in soils under *T.grandis*. So, an increase of SOC from 3.8 to 19.5 Mg ha^{-1} over soils of open space was noted in soils under MPTs after 16 years.

Table 19. Changes in SOC (Mg ha^{-1}) over the years in Tripura,India

MPTS	Years			
	4	8	12	16
A.auriculiformis	11.1	11.9	17.9	21.9
M. alba	9.9	9.9	9.9	15.9
L. leucocephala	11.5	11.5	12.8	16.7
D. sissoo	13.1	12.5	13.1	13.9
G. maculate	13.1	13.1	13.9	14.9
A. indica	10.9	10.9	14.7	28.6
M. champaca	13.9	13.7	13.9	16.9
E.hybrid	9.9	9.9	14.9	16.1
T. grandis	11.5	11.3	11.5	12.9
G. arborea	12.2	12.2	12.8	21.8
S. saman	10.6	11.3	11.3	13.9
A. procera	13.5	13.1	13.5	14.7
Open space	11.9	11.9	11.1	9.1
Mean	11.8	11.8	13.2	16.7
CV (%)	10.8	14.3	14.9	30.7
LSD (P=0.05)	2.27	2.54	1.58	3.45

Dilute alkali solouble humus as well as humin fractions were also estimated and the data are presented in Table 20.The amount of humin organic carbon content was of maximum (5.35 g kg^{-1}) in soils under *Acacia auriculiformis,* followed by *Gmelina arborea* (4.47g kg^{-1}). Soils under *Azadirachta indica,* though showed a high value of organic carbon, contained low humin carbon (2.13 g kg^{-1}). This low humin carbon in soils under *Azadirachta* indica showed only 7.6 per cent of organic matter humification. On the other hand, dilute alkali soluble humic carbon underwent a variation from 7.4 to 26.5g kg^{-1}. This indicated accumulation of alkali soluble humus in high amount in soils under *Azadirachta indica* (26.5g kg^{-1}) followed by *Gmelina arborea* (17.3g kg^{-1}), *Michelia champaca* (15.9g kg^{-1}), *Eucalyptus hybrid* (15.1g kg^{-1}) etc. This showed an increase in humus accumulation in soils under MPTs compared to open space (7.4 g kg^{-1}). The ratio of alkali soluble organic carbon and humin organic carbon (R) showed a variation from 3.1 to 13.2. The lower the value of the ratio, the higher would be the humification. So soils under *Acacia auriculiformis,* showing the least value (3.1) of the ratio indicated the high humification rate. On the other hand, the lower the value of inverse ratio, the lower would be the rate of humification and vise-versa. Comparatively, high value of inverse ratio in soils under *Acacia auriculiformis, Leucaena leucocephala, Gmelina arborea* indicated high humification rate. E_4/E_6 ratio of the absorbance of humic acid solution at 465nm and 665nm is also an index of humification. The low value of E_4/E_6 ratio indicated a high degree of humification of soil humic substances and high value indicated low degree of humification. Here, E_4/E_6 ratio was found to vary from 1.64 to 5.38, thus showing low degree of humification in soils of open space evidenced by high E_4/E_6 ratio (5.38). On the basis of E_4/E_6 ratio , it can be said that soils under the canopy of *Acacia auriculiformis, Michelia champaca, Tectona grandis, Dalbergia sissoo* showed low humification of the organic matter. But soils under the canopy of other MPTs showed high humification due to low value of E_4/E_6 ratio of soil humic substance. Soils of open space were of low humification due to high E_4/E_6 ratio and all other soils under MPTs indicated humification higher than soils of open space

Table 20. Humus[a] in soils under MPTS

MPTS	Humin O. carbon (g kg^{-1})	Alkali soluble O. carbon (g kg^{-1})	Humin O.carbon (%)	Ratio of alkali soluble & humin O. carbon (R.)	E_4/E_6 ratio	Inverse ratio of R
A. auriculiformis	5.35	16.6	24.4	3.1	4.47	0.32
M. alba	2.25	13.7	14.2	6.1	3.15	0.16
L.leucocephala	2.58	14.1	15.4	5.5	3.56	0.18
D. sissoo	1.21	12.7	8.7	10.5	4.73	0.09
G. maculate	1.14	13.8	7.7	12.1	1.91	0.08
A. indica	2.13	26.5	7.6	12.4	2.37	0.08
M. champaca	1.05	15.9	6.2	15.1	5.01	0.06
E.hybrid	1.04	15.1	6.5	14.5	1.64	0.07
T. grandis	1.13	11.8	8.8	10.4	4.00	0.09
G. arborea	4.47	17.3	20.5	3.9	3.35	0.26
S. saman	1.04	12.9	7.5	12.4	1.83	0.08
A. procera	1.04	13.7	7.1	13.2	2.30	0.08
Open space	1.75	7.4	24.7	4.2	5.38	0.24
Mean	2.05	14.7	12.3	9.5	3.70	0.14
CV (%)	14.3	28.2	53.8	43.6	36.8	59.3
LSD (P=0.05)	0.21	2.54	0.51	1.14	0.26	0.05

Soil nutrient availability

Available NPK and micronutrients (Fe Mn and Zn) were estimated in soils under canopy of MPTs and the data are presented in Table 21. Available nitrogen was found to vary from 370 to 497 kg ha^{-1}, thus showing the medium status of nutrient availability. Soil under *Albizzia procera, Tectona grandis, Eucalyptus hybrid, Morus alba* showed 7 to 14 per cent increase in nutrient availability over open space. Available phosphorus (Bray P_2), on the other hand was found to vary from 12.4 to 29.4 kg ha^{-1}, thus showing medium to high nutrient status. High P status was noted in soils under *Morus alba, Samania saman, Tectona grandis, Gmelina arborea* and *Eucalyptus hybrid.* Available potassium showed a variation from 85 to 143 kg ha^{-1}, thus showing low to medium nutrient status. Low potassium availability was noted in soils under *Gliricidia maculata, Michelia champaca, Tectona grandis* and Samania saman. But *Albizzia procera* showed the highest potassium availability, thus indicating 32 per cent increase over open space. Available Fe and Mn showed a variation from 59.2 to 118.8 mg kg^{-1} and 12.5 to 33.2 mg kg^{-1}, respectively thus indicating high micronutrient availability in soils under MPTs. On the other hand, available Zn showed a variation from 0.34 to 0.93 mg kg^{-1} thus indicating Zn stress in soils under MPTs planted in upland toposequence.

Table 21. Effect of MPTS on nutrient availability[a] in soils under MPTs in Tripura,India

MPTS	Macro nutrients ($kg\ ha^{-1}$)			Micro nutrients ($mg\ kg^{-1}$)		
	N	P	K	Fe	Mn	Zn
A.auriculiformis	372	18.6	128	68.0	33.2	0.72
M. alba	466	29.4	116	65.5	10.0	0.39
L. leucocephala	404	17.0	97	62.1	16.8	0.93
D. sissoo	466	12.4	128	118.8	10.3	0.68
G. maculate	466	12.4	124	56.7	13.5	0.59
A. indica	404	15.5	100	70.1	20.0	0.36
M. champaca	372	27.9	85	84.4	13.9	0.59
E.hybrid	466	24.8	124	59.2	24.6	0.68
T. grandis	466	27.9	97	23.9	14.3	0.36
G. arborea	372	24.8	112	86.5	18.5	0.37
S. saman	372	29.4	102	112.5	12.5	0.47
A. procera	497	21.7	143	68.8	24.6	0.45
Open space	435	26.3	108	107.9	18.2	0.34
Mean	427	22.1	113	75.7	17.7	0.53
CV (%)	10.4	27.3	19.8	33.1	27.6	32.2
LSD (P=0.05)	58	2.23	27	32	1.59	0.04

Interpretation of findings

The finding pertaining to rise in soil pH and water holding capacity followed by a decline in exchangeable acidity in agroforestry systems(Tripura,India) after 16 years of tree establishment was corroborated by the data (Datta *et al.* 1995) recorded during the initial periods of tree growth. Similar amelioration of soil acidity in various agro-ecosystems was reported by various workers (Nye and Greenland 1964 ; Ramakrishna and Toky 1981). Nutrient cycling from roots and foliage components might have augmented the nutrient turn over thus decreasing the soil acidity (Datta and Singh,2007). Similar rise in nutrient availability in soils under MPTs was also reported (Datta and Dhiman 2001 ; Datta *et al.*2004) probably due to substantial rise in organic carbon over the years from 4 to 16 years. Topsoil organic carbon content was found to increase significantly in Oxisols (Silva 1983) during ten years of growth of various timber species. One of the emerging areas of interest related to environments is carbon sequestration potential of agroforestry. Swamy *et al* (2003) estimated that a six year old *Gmelina arborea* based agrisivicultural systems in India sequestered 31.4 Mg hm^{-2} carbon. Similarly soils under *Azadirachta indica* showed the maximum SOC (28.6 Mg ha^{-1}) followed by *Acacia auriculiformis* (21.9 Mg ha^{-1}) and *Gmelina arborea* (21.8 ha^{-1}).On an average, 9.7 Pg of organic carbon was present in soils of India (Bhattacharjee *et al.*2000). In soils under AFS in Arunachal Pradesh, India, the excess amount of litter fall from bamboo and pine plant might have helped in preventing soil erosion in the hills. Leaf litter had higher N concentration (1.39%) than non-leaf (0.74%) litter in different AFS. Lignin content was, however, greater in non-leaf materials than in the leaves. The C/N ratio of different litter components reveal that the leaf litter of dicots (21.03) can be considered as good residues in these AFS. In Meghalaya, India, on an average, soil pH decreased in different AFS (except that of agrihorticulture AFS). In all the AFS, there was 1.17 to 1.65-fold increase in organic carbon, compared to initial status of organic carbon in surface soil. Silvi-horti-pastoral AFS contributed the maximum organic carbon in the soil, irrespective of soil depth. The same system also registered 43.2 per cent higher exchangeable Al compared to natural forest and consequently a maximum decrease of 0.50 units in pH. The exchangeable Ca, Mg, Na, K and Al and available N, P and K content were higher in all the systems compared to natural forest and the content of these nutrients decreased with increasing soil depth.

Organic carbon may be present in the soil either in the unaltered non-mineralised form or in altered mineralized form. The part of mineralized soil organic carbon which is soluble in 0.1 N NaOH or 0.5 N Na_2CO_3 solution is called alkali soluble organic matter or humic fraction of organic matter and this fraction includes both humic acid (HA) and fulvic acid (FA). But the part of soil organic carbon which is insoluble in 0.1 N NaOH and 0.5 N Na_2CO_3 solution is placed under the humin group (Kononova 1966).The loss of the capacity of humin in dissolving in alkali is attributed not to an alteration in the nature of the components, but to the firmness of the combination with the mineral part of the soil. So humin is the portion of humic/fulvic acids strongly bound to inorganic constituents of the soils. Humin carbon is the stored organic carbon of soil organic matter and it may be mineralised to supply the energy to soil microbes so that dilute alkali soluble humic carbon will be diminished in the soil. The storing of humic carbon depends on the minerals, metal compounds and the favourable condition of humification. The comparatively high humin carbon present in soils under *Acacia auriculiformis*, *Leucaena leucocephala* and *Gmelina arborea* indicated the enhanced storage of organic carbon pool in agroforestry systems. The low humification of soil humus was recorded in soils of open space as evidenced by high E_4/E_6 ratio. The magnitude of E_4/E_6 ratio (Chen *et al*., 1977) is inversely related to the particle size or particle or molecular weight of the humic material. A low ratio indicates a large particle size, where as a high ratio is indicative of the reverse. Datta *et al*. (2001) reported low humification in soils under shifting cultivation continuously over a period of 1-3 years.

Conclusions

Due to dwindling forest resource in the region, farming activities in the hills particularly on the slopes are the major areas of concern as it has resulted into soil erosion, declining fertility and loss of flora and fauna of the region. Scientific management of land, water and other resources in this Himalayan region is necessary for long-term sustainable production and environmental conservation. Sustainable development of resources is possible through adoption of technological innovations, organizational and institutional changes and building up of effective leadership through community participation.

Degraded lands of the region are typically characterized by impoverished or eroded soils, hydrological instability, reduced primary productivity and diminished biological productivity. Shifting cultivation is the major contributing factor to resource degradation. This practice of cultivation could not be replaced due to socio-cultural ethos and prevailing land tenure system in the region. However, it may be supplemented through various technological inputs so that productivity could increase. Hence, a comprehensive forest policy is required in the region for sustainability and augmentation of food, fuel, fodder and timber requirements. In this direction, agroforestry needs to be strengthened for long-term sustainable production and environmental conservation.

It is well understood that agroforestry in NE region of India is the mainstay of subsistence of rural folk Besides augmentation of biomass, tree/shrub species helps in soil fertility build up. It will be worth to mention here that tree alone in farming system may not be that much helpful to bring sustainability, however, development of integrated farming system with animal-crop-tree-fish as core components of the farming system will be more appropriate for the region. Unfortunately, so far nothing much has been done in this direction. Hence, the need of the hour is to develop such land use models in different agroclimatic zones with suitable combination of crops-livestock-fish-tree/shrub species of economic importance.

References

Andriesse, J. P.1989. Nutrient management through shifting cultivation: A comparative study on cycling of nutrients in traditional farming systems of Malaysia and Sri Lanka. In: Van der Heide, J. (ed.), Nutrient Management for Food Crop Production in Tropical Farming System, pp. 29-62. Institute of Soil fertility, Haren & University of Brawijaya, Malang, Indonesia.

Avasthe, R. K. 2006. Non timber forest products: Options for broad based agroforestry systems in Sikkim Himalaya. In: Bhatt, B. P., Bujarbaruah, K.M. (Eds.), Agroforestry in North East India: Opportunities and Challenges,ICAR Research Complex for NEH Region,Umiam, Meghalaya, India, 213-235,pp.

Basic Statistics. 2002. Basic Statistics of NE States. North Eastern Council, GOI, Shillong, Meghalaya, India, pp. 430.

Bhatt, B.P., Sachan, M.S., Singh K. 2006 Production potential of traditional agroforestry systems of Meghalaya: A case study. In: Bhatt, B.P., Bujarbaruah, K.M. (Eds.), Agroforestry in North East India: Opportunities and Challenges, pp. 337-349, ICAR Research Complex for NEH Region, Umiam, Meghalaya, India.

Bhattacharyya, T., Pal, D.K., Velayutham, M., Chandran, P. and Mandal, C. 2000. Total carbon stock in Indian soils: issues, priorities and management. In: Special publication for food, employment and environmental security, pp.1-46. Published by ICLRM, New Delhi, India.

Borthakur, D.N. 1992. Agriculture of the North Eastern region, with Special Reference to Hill Agriculture. BEE CEE Prakashan, Guwahati, India, pp. 245.

Chapman, E.C. 1975. Shifting agriculture in tropical forest of South East Asia. Proceedings of regional meeting on use of ecological guidelines for development in tropical forest areas of South East Asia. IUCN Publication, New Series 32:120-135.

Datta, M., Laskar, S. and Prasad, R.N. 1995. Effect of land use systems on physico-chemical characteristics and nutritional status of soils in Tripura. Indian J. Hill Farming 8: 22-27.

Datta, M., Bhattacharyya, B.K. and Saikh, H. 2001. Soil fertility-a case study of shifting cultivation sites in Tripura. J. Indian Soc. Soil Sci. 49:104-09.

Datta, M. and Dhiman, K. R. 2001. Effect of some multipurpose trees on soil properties and crop productivity in Tripura. Journal of Indian Society of Soil Science 49: 511-515.

Datta, M., Singh, N. P., Datt, C. and Dhiman, K. R. 2004. Agroforestry research in Tripura Publication 9, pp. 1-87, Published by ICAR research Complex for NEH Region, Lembucherra, Tripura, India.

Datta, M. and Singh, N.P. 2007. Growth characteristics of multipurpose tree species, crop productivity and soil properties in agroforestry systems under subtropical humid climate in India. J. For. Res. 18 (4): 261-270.

Deb, S., Tangjang, S., Arunachalam, A. and Arunachalam, K. 2005. Role of litter, fine roots and microbial biomass in soil C and N budget in traditionally managed agroforestry systems. In: Bhatt, B.P., Bujarbaruah, K.M. (eds.), Agroforestry in North East India: Opportunities and Challenges, pp. 507-522, ICAR Research Complex for NEH Region, Umiam, Meghalaya, India.

Deb, D. B. 1981.The Flora of Tripura State (Volume I),Vegetation and Ophioglosaceae–Staphyleaceae. Today and Tomorrow's Printers and Publishers, New Delhi, pp. 234.

Devendra, C. and Thomas, D. 2002. Crop-animal system in Asia: Importance of livestock and characterization of agroecological zones. Agric. Syst. 71:5-15.

Kononova, M. M. 1966. Soil organic matter, 2nd edition, Published by Pergamon Press, Inc., Elmsford, New York, USA, pp.544.

Nye, P. H. and Greenland, D.J. 1964. Changes in the soil after clearing tropical forest. Plant Soil 21:101-112.

Prasad, R. N. and Sharma, U. C. 1994. Potential Indigenous Farming Systems of North Eastern Hill Region, ICAR Research Complex for NEH Region, Barapani, Meghalaya, pp. 48.

Ramakrishnan, P. S. 1992. Shifting Agriculture and Sustainable Development: An Interdisciplinary Study from North-Eastern India. UNESCO, Paris and Oxford University Press, Delhi, India, pp. 421.

Ramakrishnan, P.S. and Toky, O. P. 1981. Soil nutrient status of hill ecosystems and recovery pattern after slash and burn agriculture (jhum) in north-eastern India. Plant Soil 60:41-64.

Silva, L. F.1983.Influencia de cultivose sistemas de manejo nas modificacoes eddficas dos Oxisols de tabuleiro (Haplorthox) do Sul da bahia. Balem (Brazil) Published by CEPLAC, Departamento Especial da Amazoniapp.1-236.

Sharma,T.C.1976.The pre-historic background of shifting cultivation. In: Proceedings of seminar on shifting cultivation in NE India. NE India Council for Social Sciences Research, Shillong, Meghalaya, India, pp. 125.

Sharma, U.C. and Sharma, V. 2003. The 'Zabo' soil and water management and conservation system in northeast India : tribal beliefs in the development of water resources and their impact on society- an historical account of a success story. In: *The Basis of Civilization-*

water Science (eds. Rodda, J.C., and Ubertini, L.). IAHS Publ. 286, pp 184-91, Rome, Italy.

Sonowal, D.K., Tripathi, A.K. and Krirekha, V. 1989. Zabo an indigenous system of Nagaland. Indian J. Hill Res. 2:1-8.

Tewari, D.N.1992. A Monograph on Bamboo, International Book Distribution, Dehradun, India.

19

Nutrient Cycling in Homegarden Agroforestry in the Andaman Islands

C.B. Pandey and Lalita Singh

Abstract: *This chapter reports variations in organic C, total N and P, mineral N and P, and exchangeable K, Ca and Mg, and microbial biomass C in soils in relation to leaf litter and root biomass (0 to 10, 10 to 20 and 20 to 30cm) at two canopy positions viz. under canopy and between canopy of homegarden trees* i.e. *coconut palm (Cocos nucifera L.), and clove (Eugenia cariophyllata Thunb) and nutmeg (*Myristica fragrans *Houtt. Nees) spice trees. It also deals with seasonal variations in pool sizes of mineral N (NH_4^+-N and NO_3^--N), and net N mineralization rate (0-15 cm) in relation to rainfall and temperature under the homegarden trees, and examines how high rainfall (incessant rainfall) influences water filled pore space (WFPS), net nitrification and net ammonification rates, pool sizes of NH_4^+-N and NO_3^--N and microbial biomass C in the soils and conserves nutrients in humid tropical climate. The chapter also describes how high rainfall makes phosphorus available in high regime of iron in the soils; and perennial cover crop like Pueraria separately and together with phosphorus increases soil N mineralization. Organic C, total N, total P, exchangeable NH_4^+, NO_3^-, exchangeable K, Ca and Mg are higher in the under canopy and between canopy positions of all the homegarden trees than that in open. Rate of mineralization is the highest during the post-wet season and the lowest during the dry season (February to April) under all homegarden trees. High rainfall during the wet season, however, reduces the rate of nitrification under all the trees. Anaerobic condition during incessant rainfall in wet season reduces net nitrification substantially (8 to 15 times) and convert NO_3^- to NH_4^+ and thus conserves nitrogen from loss; it* *converts ferric oxide into ferrous oxide and make fixed phosphorus available to plants. Pueraria cover crop protects fine soil particles from loss, add organic matter and thereby increases net soil N mineralization in humid tropical climate.*

Introduction

Homegardens are an example of a sustainable agroforestry system that provide low but sustainable yield for several decades, without use of fertilizers and pesticides, with no apparent loss in growth and yield (though low) and detrimental effects on the environment (Soemarwoto, 1987; Jensen, 1993; Pandey *et al.*, 2007). This sustainability in yield may be attributed to soil fertility accumulated under the trees as a consequence of continuous organic matter inputs through leaf litter and dead roots (Rhoades, 1997; Jensen, 1993; Pandey *et al.*, 2000). Moreover, these inputs depend on tree species (Schroth *et al.*, 1995; Mazzarino *et al.*, 1991; Pandey *et al.*, 2000). However, information on the innate characteristics of homegarden trees accumulating and releasing of nutrients is meager.

Nitrogen is one of the major nutrients, which is required the highest amount to plants. Organic matter input through leaf litter and roots are major sources of organic N in soils, which provide a readily hydrolysable source of C for microbes and thus promote microbial activity (Paul and Clark, 1996; Kemmitt *et al.*, 2008). The soil microbial biomass mineralizes humified soil organic matter at a remarkable rate over a long time (Gardner *et al.*, 1989; Joergensen *et al.*, 1990) and fresh litter inputs maintain the process. A positive relationship between microbial biomass C and net soil N mineralization is well documented for several ecosystems (Williams and Sparling, 1988; Pandey *et al.*, 2009). It is found that the largest proportion of nitrogen (>90%) in soils is found in organic forms that become available to plants upon mineralization to NH_4^+ (Wong and Nortcliff, 1995; Kelley and Stevenson, 1995). Under favourable conditions the NH_4^+ is oxidized further to NO_3^- (Wong and Nortcliff, 1995). Both, ammonification and nitrification are aerobic biological processes (Breuer *et al.*, 2002). However, they are influenced mainly by the quality and quantity of organic matter, rainfall amount and its distribution, temperature and allelochemicals (Powers, 1990; Montagnini *et al.*, 1989), as these factors influence soil microbial activity (Powers, 1990).

It is found that rainfall amount excess of plant demand does not act as a resource of biotic processes (Schuur *et al.*, 2001), rather it may suppress plant growth through variety of mechanisms (Schuur and Matson, 2001). For example, high precipitation removes mobile nutrients in soil solution via leaching (Radulovich and Sollins, 1991; Schuur and Matson, 2001), that leads to low pH and low nutrient condition (Schuur and Matson, 2001). It also restricts gas diffusion in soils and limits supply of oxygen from the atmosphere to the soil (Richards, 1987) and thereby creates anaerobic condition even in upland soils (Hobbie *et al.*, 2000), that causes reduction in soil N mineralization rate (Schuur and Matson, 2001).

Land use change in tropics is of great concern to ecologist as it alters biogeochemical cycle both regionally and globally (Scholes and Van Breemen, 1997), which is directly linked with productivity and stability of changed land use systems (Brown and Lugo, 1990). Forest clearing generally favours rise in temperature that increases biological activity, which causes to an increase in microbial mineralization of NH_4^+-N from soil organic matter (Piccolo *et al.*,1994; Gill *et al.*, 1995; Nunan *et al.*, 2000). The rate of production of NO_3^--N, which is extremely mobile in soil solution, influences N losses through runoff, leaching and denitrification (Neill *et al.*, 1997; Verchot *et al.*, 1999) that ultimately make the system fragile. Therefore, a better understanding of nitrogen transformation and its effect on soil fertility following forest conversion is critical for understanding the long-term implications of human induced land use changes (Piccolo *et al.*, 1994).

Tillage is another factor which influences soil N transformations. Farmers, the world over, till their fields deep several times before they grow a crop. They are of the perception that the soil working increases soil fertility, and removes weeds which makes their crops free from competition. Under such condition, the crops utilize maximum mineralized nitrogen (N) and give high productivity. But, as a side effect intensive tillage is known to break soil aggregates and exposes protected organic matters to microbial decomposition (Beare *et al.*, 1994) and high temperature causes excessive carbon loss through decomposition (Stockfisch *et al.*, 1999). Soils rank third after ocean and fossil fuel as a sink of carbon in the world (Duiker and Lal, 1999). Therefore, influence of tillage on green house gas emission, particularly CO_2, has been of great concern to environmentalists to understand its global balance. A great deal of studies, both short-term and long-term, has been conducted to know impact of tillage on soil organic carbon (SOC) (Duiker and Lal, 1999; Alvarez *et al.*, 2001). Now it is well established that no-tillage increases organic carbon in soils due to organic matter input and insulation of SOC from sun-heat (Hendrix *et al.*, 1988). But, conservation tillage has been found to decrease crop yields due to soil compaction (Ahl *et al.*, 1998) and increased weed and slug infestation (El Titi, 2003 a, b). Soil compaction is known to reduce aeration and oxygen exchange (Jensen *et al.*, 1996), which causes reduction in soil N mineralization (Pandey *et al.*, 2009). Several studies have been conducted, but mostly *ex-situ* to know mineralization potential of soils under different tillage conditions (Clay *et al.*, 1995). Few studies report on the impact of tillage on soil N mineralization *in-situ*, but their results are inconclusive. Some investigators have reported that no-tillage reduces soil N mineralization due to increased microbial immobilization (Doran and Power, 1983). Others have documented no difference in soil N mineralization after 16-yr of no-tillage Vs conventional tillage (Rice *et al.*, 1986). Although productive potential of agroecosystems has increased due to the crop varieties, farm mechanization and pest control, soil fertility limits the productive potential and reduces grain yields (Cole *et al.*, 1989). Current information suggests that sustainability in crop yields needs high SOC and steady soil N supply (N mineralization) (Pandey and Singh, 2009). However, there is a general lack of information on what level of tillage increases soil N mineralization and simultaneously maintains SOC and microbial biomass carbon (MB-C) pools to a reasonably high level?

Phosphorus is another important nutrient which is known as a master regulator of biological activity (Chadwick *et al.*, 1999; Cleveland *et al.*, 2002); it limits plant growth and productivity in many soils. Content of phosphorus in mineral soils, though compares favourably with that of nitrogen, its availability to plants is extremely low because of its marked fixation in inorganic form, and adsorption on clay mineral particles (Noemi *et al.*, 2006; Fankem *et al.*, 2006). The affinity of iron oxides and hydroxides for phosphorus is thought to make phosphorus unavailable to plants in acidic, highly weathered soils in humid tropics (Vitousek, 1984; Jordan, 1985; Herbert and Fownes, 1995; Noemi *et al.*, 2006). Despite high iron in soils, giant rainforests are found (Forest Working Plan, 1980), and vigorous growth in herbaceous vegetation occurs during wet season in humid tropics (Pandey and Venkatesh, 2007). This generates an interest to know how phosphorus becomes available to plants in iron rich soils in high rainfall regime of humid tropics.

Contrary to subsistence homegarden farming, productivity of commercial plantations and homegardens declines with time due to declines in soil fertility caused by heavy harvesting, loss of fine soil particles in runoff, and oxidation of soil organic carbon (Gajaseni and Gajaseni, 1999; Torquebiau, 1992; Szott *et al.*, 1999). In the commercial plantations and homegardens the most effective way to protect soil from erosion and to increase soil fertility is to establish a perennial cover crop such as *Pueraria phaseoloides* (popular name puero or tropical kadzu) and others (Schroth *et al.*, 2001a). The cover crop is reported to increase growth and yield in coconut palms (Zakra *et al.*, 1986; Pandey and Venkatesh, 2007). The species rapidly covers soil, fixes up to 150 kg ha^{-1} yr^{-1} of the atmospheric N_2 (Giller and Wilson, 1991), protects fine textured soils from erosion (Barrios *et al.*, 1997; Szott *et al.*, 1999; Pandey and Chaudhari, 2009) and adds N-rich leaf and root litter over long periods (Szott *et al.*, 1999). This provides a readily hydrolysable substrate for soil N mineralization (Gardner *et al.*, 1989; Joergensen *et al.*, 1990). The protected fine soil particles under the cover crop also provide adsorption sites for NH_4^+, the result from the hydrolysis of high molecular weight organic N into simpler compounds (Sinsabaugh *et al.*, 1991), followed by breakdown of dissolved amino acid compounds to NH_4^+(Gardner *et al.*, 1989), causing enhanced rates of net soil N mineralization. Growth in the cover crop declines with time, perhaps due to the low availability of P in soils (Pinkerton and Simpson, 1986; Szott *et al.*, 1994). Hence, P application to leguminous cover crops is found to increase their growth and biomass in P-limited soils. Phosphorus application to cover crops stimulates microbial activity and soil microbial biomass which enhance the rate of net soil N mineralization (White and Reddy, 2000).

This chapter describes seasonal changes in pool and fluxes of mineral N under coconut (*Cocos nucifera* L.), clove (*Eugenia caryophyllata* Thunb) and nutmeg (*Myristica fragrans* Houtt. Nees) trees in relation to rainfall and temperature; net soil N mineralization under high rainfall regime, land use change and tillage conditions; availability of fixed phosphorus due to anaerobiosis caused by high rainfall, and influence of a perennial leguminous cover crop on soil N mineralization and microbial biomass carbon (MB-C) under different land uses in humid tropical climate of the Andaman Islands.

Litter production and root biomass in homegardens

Tree species influence litter production in the Andamans homegardens with larger values for coconut followed by clove and nutmeg (Table 1). Total litter production, across the trees, is higher in the homegarden than that in the forest. But, root biomass is slightly (20 %) higher in the forest compared to that in the homegarden. Total root biomass (coarse + fine) is higher in the under canopy than that in the between canopy position of all the trees (Table 2). It is 24 to 26% lower under the spice trees than that under the palm. All the mineral contents, except P, are the highest in the litter and the roots of the spice trees and lowest in the palm. Except K, mineral contents in the forest litter and the roots are almost within the range found in the litter and roots of the trees. Total litter production (8.7 Mg ha^{-1} yr^{-1}) in the Andamans homegardens is comparable with that found in the Javanese homegarden (10 Mg ha^{-1} yr^{-1}), Indonesia (Jensen, 1993).

Table 1. Mineral content in leaf litter and roots of three homegarden trees and a native moist evergreen forest in the equatorial humid tropical climate of the Andaman islands.

Tree	Production (Mg $ha^{-1}yr^{-1}$)	(g/100g)				
		N	P	K	Ca	Mg
			Leaf litter			
Coconut	4.52^a	0.25^a	0.013^a	0.42^a	0.24^a	0.10^a
Clove	2.53^b	0.27^a	0.011^a	0.46^b	0.68^b	0.12^a
Nutmeg	1.69^c	0.39^b	0.004^b	0.77^c	1.16^c	0.13^a
Forest	5.04^d	0.49^c	0.008^b	0.11^d	0.89^d	0.11^a
			*Root			
Coconut	-	0.36^a	0.005^a	0.29^a	0.15^a	0.07^a
Clove	-	0.76^b	0.008^a	0.18^b	0.34^b	0.05^a
Nutmeg	-	0.14^c	0.021^b	0.53^c	0.45^c	0.12^b
Forest	-	0.35^d	0.006^a	0.08^d	0.52^d	0.05^a

Values in a column for a parameter suffixed with different superscript are significantly different at $p<0.05$.
- Not estimated
* Includes both live and dead
Source: Pandey and Singh 2009

Table 2. Root biomass in the under canopy and between canopy position of three homegarden trees in a coconut-tree spices plantation and a native moist evergreen forest in the equatorial humid tropical climate of the Andaman islands.

Crop	Soil depth (cm)	Root biomass (gm^{-2})	
		Under canopy	Between canopy
Coconut	0-10	591[a]	216[a]
	10-20	393[b]	172[b]
	20-30	321[c]	138[c]
	Average	[A]435	[B]175
Clove	0-10	351[d]	67[d]
	10-20	405[e]	46[e]
	20-30	213[f]	22[f]
	Average	[B]323	[A]45
Nutmeg	0-10	432[g]	112[g]
	10-20	329[c]	77[h]
	20-30	239[h]	57[i]
	Average	[C]333	[D]82
Forest	0-10	495[i]	
	10-20	476[j]	
	20-30	338[k]	
	Average	[A]436	

Values in a column for a parameter suffixed with different superscript are significantly different at $p<0.05$.
Values in a column for a parameter prefixed with different superscript are significantly different at $P<0.05$.
Source: Pandey and Singh 2009

Soil organic carbon and total N pool, C/N ratio and microbial biomass carbon (MB-C) under homegarden trees.

The soils are inherently slightly acidic in reaction in the Andaman's homegardens (Table 3). Soil organic C (0-30 cm) is as much as 83 to 119% and 18 to 75% higher in the under canopy and between canopy position, respectively of the palm and the spice trees than that in the open plot (Table 3). The difference is 26% higher under the spice trees than that under the palm. The pattern of total N and P variations among the treatments is broadly similar to that of organic C. As with organic C, N is 28 to 61 % higher under the spice trees, particularly greater under the nutmeg than that under the palm. P is also higher in the under canopy and between canopy positions of the trees than that in the open plot, and it is 15 to 30% higher under the spice trees than that under the palm (Table 3). Soil microbial biomass C, contributing 1 to 5% to the total organic C, is higher in the under canopy than that in the between canopy position for all the tree types, and it is 28% higher under the coconut than that under the spice trees (Table 4). The distribution of soil microbial biomass with soil depth is similar to that of the root biomass. It is correlated with the root biomass ($r = 0.71$, $P<0.001$) and the organic C ($r = 0.69$, $P<0.001$) across the trees. Microbial biomass C is slightly higher in the forest than that under the homegarden trees. Greater amounts of organic C and total N under the canopy positions of the palm and spice trees compared to that in the open plot are primarily due to accumulation of organic matter over long period through litter fall and dead roots similar to that of natural ecosystems (McClaugherty *et al.*, 1985; Boerner and Koslowsky, 1989) and different agroforestry systems (Kamara and Haque, 1992; Pandey *et al.*, 2000).

Table 3. Soil pH, organic C, total N and total P in the under canopy (UC) and between canopy (BC) position of three homegarden trees in a coconut-tree spices plantation, an adjacent open plot and a native moist evergreen forest in the equatorial humid tropical climate of the Andaman islands

Tree	Position	Depth(cm)	pH	Organic C(g/kg)	Total N(g/kg)	Total P(g/kg)
Coconut	UC	0-10	6.34^{a}	11.36^{a}	0.87^{a}	0.031^{a}
		10-20	5.95^{b}	9.30^{b}	0.68^{b}	0.030^{a}
		20-30	5.91^{b}	8.40^{c}	0.58^{c}	0.025^{b}
		Average	$^{A}6.07$	$^{A}9.78$	$^{A}0.71$	$^{A}0.029$
	BC	0-10	5.54^{c}	6.80^{d}	0.56^{c}	0.025^{b}
		10-20	6.09^{d}	6.33^{e}	0.44^{d}	0.022^{cd}
		20-30	6.08^{d}	5.80^{f}	0.36^{e}	0.021^{c}
		Average	$^{B}5.90$	$^{B}6.31$	$^{B}0.45$	$^{B}0.023$
Clove	UC	0-10	5.70^{d}	11.90^{g}	0.93^{f}	0.023^{d}
		10-20	6.36^{af}	10.30^{h}	0.79^{g}	0.021^{c}
		20-30	6.54^{e}	9.60^{b}	0.70^{h}	0.019^{c}
		Average	$^{C}6.20$	$^{C}10.60$	$^{C}0.81$	$^{BC}0.021$
	BC	0-10	6.39^{f}	9.90^{i}	0.64^{t}	0.022^{fc}
		10-20	6.50^{g}	8.00^{j}	0.56^{i}	0.019^{ij}
		20-30	6.54^{g}	6.80^{k}	0.36^{e}	0.018^{gei}
		Average	$^{D}6.48$	$^{D}8.23$	$^{D}0.52$	$^{CDE}0.020$
Nutmeg	UC	0-10	5.84^{i}	13.40^{l}	1.09^{j}	0.021^{c}
		10-20	6.04^{h}	11.80^{g}	0.94^{f}	0.019^{ij}
		20-30	6.04^{h}	9.80^{b}	0.75^{b}	0.019^{ij}
		Average	$^{B}5.97$	$^{E}11.67$	$^{E}0.93$	$^{CDE}0.020$
	BC	0-10	5.63^{l}	10.10^{h}	0.82^{a}	0.020^{i}
		10-20	5.69^{m}	9.63^{b}	0.78^{g}	0.019^{ij}
		20-30	5.60^{l}	8.30^{c}	0.69^{h}	0.018^{gei}
		Average	$^{E}5.64$	$^{A}9.34$	$^{F}0.76$	$^{E}0.019$
Forest		0-10	5.60^{l}	16.20^{m}	1.39^{l}	0.022^{cd}
		10-20	5.60^{l}	15.0^{n}	1.19^{m}	0.021^{c}
		20-30	5.55^{c}	13.40^{l}	1.02^{n}	0.021^{c}
		Average	$^{F}5.58$	$^{F}14.87$	$^{G}1.20$	$^{BC}0.021$
Open Plot		0-10	5.89^{i}	5.79^{o}	0.41^{o}	0.016^{k}
		10-20	5.82^{i}	5.80^{o}	0.36^{p}	0.013^{l}
		20-30	6.08^{d}	4.40^{p}	0.30^{p}	0.013^{l}
		Average	$^{B}5.93$	$^{G}5.33$	$^{H}0.36$	$^{F}0.014$

Values in a column for a parameter suffixed with different superscript are significantly different at $p<0.05$.
Values in a column for a parameter prefixed with different superscript are significantly different at $P<0.05$.
Source: Pandey and Singh 2009

Table 4. Microbial biomass C in the soil in the under canopy and between canopy position of the three homegarden trees, an adjacent open plot and a native moist evergreen forest in the equatorial humid tropical climate of the Andaman Islands

Crop	Soil depth (cm)	Microbial biomass C($\mu g\ g^{-1}$)	
		Under canopy	Between canopy
Coconut	0-10	529[a]	242[a]
	10-20	517[b]	225[b]
	20-30	315[c]	101[c]
	Average	[A]454	[B]189
Clove	0-10	358[d]	112[d]
	10-20	337[e]	105[e]
	20-30	201[f]	85[f]
	Average	[B]306	[A]101
Nutmeg	0-10	457[g]	161[f]
	10-20	456[g]	140[g]
	20-30	291[h]	88[e]
	Average	[C]401	[D]130
Forest	0-10	545[i]	
	10-20	540[i]	
	20-30	327[j]	
	Average	[D]471	
Open plot	0-10	142[k]	
	10-20	166[l]	
	20-30	153[m]	
	Average	[E]154	

Values in a column for a parameter suffixed with different superscript are significantly different at $p<0.05$.
Values in a column for a parameter prefixed with different superscript are significantly different at $P<0.05$.
Source: Pandey and Singh 2009

However, despite higher litter production and root biomass lower concentration of nutrients (organic C and total N) under the palm than that under the spice trees can be owing to lower concentration of nutrients in the leaves of coconut (Table 1). The greater amount of organic C, N and microbial biomass C (soil microflora) are due to larger litter inputs under the trees. Soil microflora immobilizes available N that ultimately lends to nutrients build up in soils (Singh *et al.*, 1989; Smith and Paul, 1990). Fenn *et al.* (1993) are also of the view that increases in total C with stand age is due to accumulation of organic matter and microbial products. Relatively low amount of the organic C and total N in the between canopy position among the trees may be attributed to low amount of root biomass and less leaf litter inputs. Roots are reported to play an important role in the build up of organic matter under the trees (Browaldh, 1995). Nadelhoffer *et al.* (1985) have reported that fine root biomass pool is replaced 0.5 to 2.2 times a year. The C: N ratio of the soil varies narrowly across the positions under the tree canopies except under clove trees (Table 5). It is slightly lower under the tree canopy positions than that in the open plot, but does not differ between the canopy positions. The C / N ratio is slightly lower in the surface soil and tends to increase with the depth in all the treatments. C: N ratio in the forest is quite close to that under the nutmeg, but lower compared to that found under the other trees. The C/P ratio is 52 to 73% higher under the spice trees than that under the palm. Unlike C/N ratio, the C/P ratio is the highest on the surface soil and declines with the depth.

Soil C and N and depth relation

Soil organic C is the highest in the surface soil (0 to 10 cm) invariably in the under canopy and between canopy positions of all the trees, including the open plot and declines with the depth (Table 3). However, the decline in the between canopy position is comparatively gradual. Organic C and total N in the forest soil follow the pattern found in the soil under the tree canopy positions of the homegarden trees. But, they are 27 to 52 % and 29 to 69 %, respectively higher in the forest than that under the trees. Total P in the forest is nearly equal to that found under the spice trees, but lower compared to that under the coconut trees. Greater nutrients in the surface soil (0-10 cm) compared to that in the deeper layers indicated that phenomenon like incorporation of nitrogen via through fall and stem flow (Jensen, 1993) and the subsequent incorporation of some mineral N into the organic phase, scavenging of nutrients from deeper soil depths and concentrating these below the trees through litter fall lend the nutrient accumulation under the trees in our study. However, greater nutrients under the spice trees (dicot) compared to that under the coconut (monocot) may be attributed to tap root system of the former ones.

Mineral N (NH_4^+-N, NO_3^--N) pool and NO_3^- / NH_4^+ ratio under homegarden trees

Exchangeable NH_4^+-N and NO_3^-- N is as much as 37 to 100 % higher in the under canopy and 14 to 73% higher in the between canopy position of the trees than that in the open plot, and they are 12 to 46% and 13 to 32 %, respectively higher under the spice trees compared to that under the palm (Table 6). Higher amounts of NO_3^--N and NH_4^+-N under the palm and the spice trees compared to that in the open plot may be attributed to greater microflora (microbial biomass C) which is known to provide readily available substrate for mineralization (Singh and Singh, 1996). Microbial biomass is reported to function both as a source (mineralization) and sink (immobilization) of available N (Smith and Paul, 1990). It may also be attributed to slightly lower C / N ratio. Both, exchangeable NH_4^+-N and NO_3^--N are the highest on the surface soil and declines with the depth in all the homegarden tree canopy positions. The pattern of variations in NH_4^+-N and NO_3^- - N in the forest soil are similar to that found under the trees, but these are 10 to 60 % and 14 to 50 %, respectively higher in the former. NO_3^- / NH_4^+ ratio is invariably higher in the under canopy position than that in the corresponding between canopy position of all the trees (Table 5). This difference is greater under the coconut

Table 5. C/N, C/P and NO_3^-/NH_4^+ ratios in the under canopy (UC) and between canopy (BC) position of three homegarden trees in a coconut-tree spices plantation, an adjacent open plot and a native moist evergreen forest in the equatorial humid tropical climate of the Andaman islands.

Tree	Position	Depth(cm)	C/N ratio	C/P ratio	NO_3/NH_4
Coconut	UC	0-10	13.39[a]	379[a]	2.87[a]
		10-20	13.73[a]	310[b]	2.73[b]
		20-30	14.57[b]	337[c]	2.80[b]
		Average	[BA]13.90	[A]342	[A]2.80
	BC	0-10	12.45[c]	274[dm]	2.34[c]
		10-20	13.30[d]	290[d]	2.14[d]
		20-30	16.13[e]	247[e]	2.31[e]
		Average	[A]14.13	[B]281	[B]2.26
Clove	UC	0-10	12.80[fc]	530[f]	2.80[b]
		10-20	13.12[fj]	504[g]	2.79[b]
		20-30	14.19[g]	520[g]	2.68[f]
		Average	[BE]13.37	[C]518	[C]2.75
	BC	0-10	15.93[e]	463[hb]	2.68[g]
		10-20	15.63[e]	425[i]	2.41[h]
		20-30	20.25[h]	380[a]	2.61[c]
		Average	[C]17.27	[D]423	[B]2.57
Nutmeg	UC	0-10	12.29[j]	638[l]	2.50[i]
		10-20	12.55[b]	627[m]	2.41[j]
		20-30	13.08[l]	515[n]	2.61[c]
		Average	[D]12.64	[E]591	[B]2.51
	BC	0-10	12.32[m]	505[o]	2.15[k]
		10-20	12.40[n]	507[o]	2.34[l]
		20-30	12.01[o]	461[hb]	2.11[m]
		Average	[D]12.24	[F]491	[D]2.20
Forest		0-10	11.66[p]	736[p]	2.39[j]
		10-20	12.62[q]	714[q]	2.59[i]
		20-30	13.15[r]	637[p]	2.77[o]
		Average	[D]12.48	[G]696	[B]2.58
Open plot		0-10	14.11[l]	362[r]	2.43[g]
		10-20	16.10[m]	444[s]	2.75[c]
		20-30	14.69[c]	337[s]	2.88[a]
		Average	[F]14.97	[H]381	[D]2.69

Values in a column for a parameter suffixed with different superscript are significantly different at $p<0.05$.

Values in a column for a parameter prefixed with different superscript are significantly different at $P<0.05$.

Source: Pandey and Singh 2009

palm. NO_3^- / NH_4^+ ratio in the forest is almost equal to that under the trees. Higher NO_3^--N /NH_4^+-N ratio at the under canopy compared to that at the between canopy position of the homegarden trees is due to two reasons. First, rapid ammonification followed by nitrification due to optimum moisture and optimum temperature (Roberston, 1984) like the sub-tropical rain forest of Australia where NO_3^- is found to predominate over the NH_4^+ (Chandler and Goosem, 1982). The second, preferential uptake of N as NH_4^+ by trees, inducing an equilibrium of mineral N forms in the soil that favors larger ratios. However, the higher ratio under the palm compared to that under the spice tees indicates greater demand of the former for the ammonium N. Coconut is an evergreen tree that flowers abundantly and bears fruits almost every month. On the other hand, clove flowers once in a year and nutmeg slowly round the year with a major flush in April.

Relatively greater accumulation of NO_3^-–N during the post- wet and dry seasons under all the tree species is probably be due to lack of leaching and run off losses owing to low amount of rainfall, and low vegetative uptake. The lowest accumulation of NO_3^--N under the clove trees at the end of post-wet season is most likely due to comparatively greater vegetative uptake. Maximum leaf fall in the clove trees occurs during February to March, and leaf formation from March to April that triggers root growth at rapid rate. Neill *et al.* (1995) also reports accumulation of NO_3^--N during dry season, the prolonged period of lower soil- moisture. Roy and Singh (1995) are of the view that temporal variations in the mineral N in soils are controlled by rainfall amount and its distribution, plant uptake dynamics and losses.

Exchangeable K, Ca and Mg pool in soils under homegarden trees

Exchangeable, K is 4.5 to 5.7 times higher in the under canopy and 2.7 to 2.8 times higher in the between canopy position of the trees than that in the open plot (Table 6). It is 25 % higher under the spice trees, particularly the clove, than that under the palm. The exchangeable K shows inverse relation with the depth. Both, the Ca and Mg are higher under the trees than that in the open plot; they do not show a clear pattern of variation between the tree canopy positions and among the soil depths. Except the exchangeable K, other cations like Ca and Mg in the forest are within the range found under the homegarden trees. Higher amount of exchangeable K, Ca and Mg under the palm and spice trees compared to that in the open plot indicate that the trees have an influence on the cations also. But, the lack of variations, particularly Ca and Mg, between the canopy positions and among the soil depths clearly reveal that the influence is not as much pronounced as much it is on the nitrogen. Kamara and Haque (1992) observe that *Faidherbia albida* does not influence cations (Ca and Mg) in the soil. The discrepancies in the concentration of K and Ca in the litter and roots of the palm tree and the tree spices, and that in the soil under the corresponding trees may be due to selective uptake system in their roots (Grubb, 1977) and variations in their accumulation in the leaves (Van den Driessche, 1974). Moreover, low concentrations of minerals in the tree litters, in general, may be attributed to their retranslocation (except Ca and Mg) during the senescence.

Table 6. Pool size of exchangeable NH_4^+, NO_3^-, available P and exchanageble K, Ca and Mg in the under canopy (UC) and between canopy (BC) position of three homegarden trees in a coconut-tree spices plantation, an adjacent open plot and a native moist evergreen forest in the equatorial humid tropical climate of the Andaman islands.

Tree	Position	Depth (cm)	NH_4^+ (mg kg^{-1})	NO_3^- (mg kg^{-1})	Av.P (mg kg^{-1})	Exchangeable cations(mg kg^{-1})		
						K	Ca	Mg
Coconut	UC	0-10	11.28^{a}	32.30^{a}	0.36^{a}	532^{a}	120^{a}	66^{a}
		10-20	10.52^{b}	27.75^{b}	0.28^{b}	439^{b}	150^{b}	67^{b}
		20-30	9.17^{c}	25.46^{c}	0.17^{c}	420^{c}	150^{b}	67^{b}
		Average	$^{A}10.32$	$^{A}28.50$	$^{A}0.27$	$^{A}464$	$^{A}130$	$^{A}66$
	BC	0-10	9.55^{d}	22.26^{d}	0.28^{b}	332^{d}	160^{c}	69^{d}
		10-20	8.63^{e}	18.46^{e}	0.21^{d}	287^{e}	150^{b}	65^{c}
		20-30	7.98^{f}	18.46^{e}	0.15^{e}	231^{f}	120^{a}	66^{a}
		Average	$^{B}8.72$	$^{B}19.73$	$^{B}0.21$	$^{B}283$	$^{B}143$	$^{A}67$
Clove	UC	0-10	12.94^{g}	36.39^{f}	0.45^{g}	665^{m}	140^{f}	66^{a}
		10-20	11.57^{a}	32.55^{g}	0.35^{a}	614^{n}	110^{g}	59^{g}
		20-30	10.21^{b}	27.43^{h}	0.25^{i}	456^{o}	90^{h}	59^{g}
		Average	$^{C}11.57$	$^{C}32.12$	$^{D}0.35$	$^{E}578$	$^{GE}113$	$^{C}61$
	BC	0-10	9.06^{c}	24.02^{i}	0.26^{i}	347^{p}	130^{i}	64^{h}
		10-20	8.69^{d}	20.76^{jt}	0.20^{d}	284^{q}	130^{i}	64^{h}
		20-30	8.04^{h}	20.99^{j}	0.16^{c}	223^{c}	100^{j}	64^{h}
		Average	$^{B}8.60$	$^{D}21.92$	$^{B}0.21$	$^{B}285$	$^{E}120$	$^{A}64$
Nutmeg	UC	0-10	16.15^{i}	40.18^{k}	0.41^{j}	574^{s}	170^{d}	66^{a}
		10-20	15.70^{j}	37.85^{l}	0.31^{k}	511^{t}	180^{k}	67^{b}
		20-30	13.37^{k}	34.83^{m}	0.21^{d}	316^{u}	160^{c}	65^{c}
		Average	$^{E}15.07$	$^{E}37.62$	$^{E}0.31$	$^{A}467$	$^{D}170$	$^{A}66$
	BC	0-10	14.60^{l}	31.12^{n}	0.24^{i}	318^{u}	170^{d}	65^{c}
		10-20	12.62^{g}	28.72^{o}	0.21^{d}	287^{c}	170^{d}	65^{c}
		20-30	12.00^{m}	25.24^{p}	0.17^{c}	215^{v}	160^{c}	67^{b}
		Average	$^{F}13.07$	$^{A}28.36$	$^{B}0.21$	$^{F}273$	$^{D}167$	$^{A}66$
Forest		0-10	18.82^{n}	45.67^{q}	0.71^{l}	237^{w}	104^{j}	66^{a}
		10-20	16.62^{o}	43.86^{r}	0.61^{f}	186^{k}	100^{j}	62^{i}
		20-30	14.05^{l}	38.90^{s}	0.50^{m}	181^{i}	120^{a}	60^{j}
		Average	$^{G}16.50$	$^{F}42.81$	$^{F}0.61$	$^{G}201$	$^{G}108$	$^{A}63$
Open Plot		0-10	8.46^{b}	20.51^{j}	0.20^{d}	130^{x}	120^{a}	54^{k}
		10-20	7.30^{c}	19.99^{t}	0.16^{c}	97^{h}	80^{n}	50^{l}
		20-30	6.89^{e}	19.90^{t}	0.09^{n}	78^{i}	80^{n}	49^{m}
		Average	$^{H}7.55$	$^{G}20.13$	$^{G}0.15$	$^{H}102$	$^{H}93$	$^{D}51$

Values in a column for a parameter suffixed with different superscript are significantly different at $p<0.05$.
Values in a column for a parameter prefixed with different superscript are significantly different at $P<0.05$.
Source: Pandey and Singh 2009

Soil N mineralization and nitrification under homegarden trees

The soil N mineralization occurs every month both the years under all homegarden tree species in humid tropics (Figure 1a,b,c). It increases quickly following the monsoon rains either in May or June, but declines quickly and remain low throughout the wet season. It increases further during the post-wet season when rainfall is moderately low and declines thereafter to the lowest during the dry season. The rate of mineralization (y), across the seasons and species, is correlated with the amount of rainfall (x), both the years, but the relation is quadratic: $y = a + bx - c\,x^2$, ($r^2 = 0.19$ to 0.34, $P<0.03$ to 0.001). The same quadratic relation is found between the mineralization rate and the mean monthly temperature, both the years, but it is not significant. High rate of mineralization is observed in two conditions, first just after dry month (March; April) and the second when high rainfall occurred in preceding month (November). The mineralization rate, averaged across the months and the species, during the wet season is observed 7 to 71 % lower compared to that during the post-wet season. Increase in the nitrification rate under all homegarden tree species with the onset of post- wet season can be due to moderate amount of rainfall (soil-moisture) and high amount of NH_4^+-N accumulated during the preceding wet season (Fig. 2 a, b). This observation supports the hypothesis that nitrification is an ammonia limited process (Robertson, 1984). However, persistent high rates in the nitrification throughout the post- wet season occurs mainly due to wetting and drying effects caused by the low amount of rainfall, but high coefficient of variation (258 to 537 %) in its distribution (Pandey *et al.*, 2007b). Davidson *et al.* (1993) also repots a very high nitrification rates after artificially rewetting soils from drought-stressed deciduous forests in Mexico at the end of dry season. The lowest rate of nitrification during the dry season in humid tropics indicate that high temperature perhaps overrides the effect of soil-moisture (14-16%). Barraclough (1995) observe through ^{15}N isotope dilution technique that nitrogen mineralization occurred during dry season when soil-moisture contents are in the range of 8-10°C, but rate of minerlization is very low. The rate of mineralization is the highest under the nutmeg and the lowest under the coconut trees almost all the months. This may be attributed to live and dead root dynamics (Fig. 3 a, b,c), nature of species (deciduous vs. evergreen) and amount of soil organic carbon in the growing habitat. Van der Krift and Berendse (2001), Roy and Singh (1995) found that plant species growing on high fertility microhabitats increased N mineralization greater than species growing on low fertility microhabitats. Maximum leaf fall in the clove trees occurs during February to March, and leaf formation from March to April that triggers root growth at rapid rate. On the contrary, leaf formation and leaf fall in the coconut and nutmeg trees occur every month (L. Singh, unpublished). The nitrification rate contribute maximum 96 to 98 % to the mineralization rate under all the tree species, therefore, its pattern of variations is closely similar to that of the latter. The nitrification rate

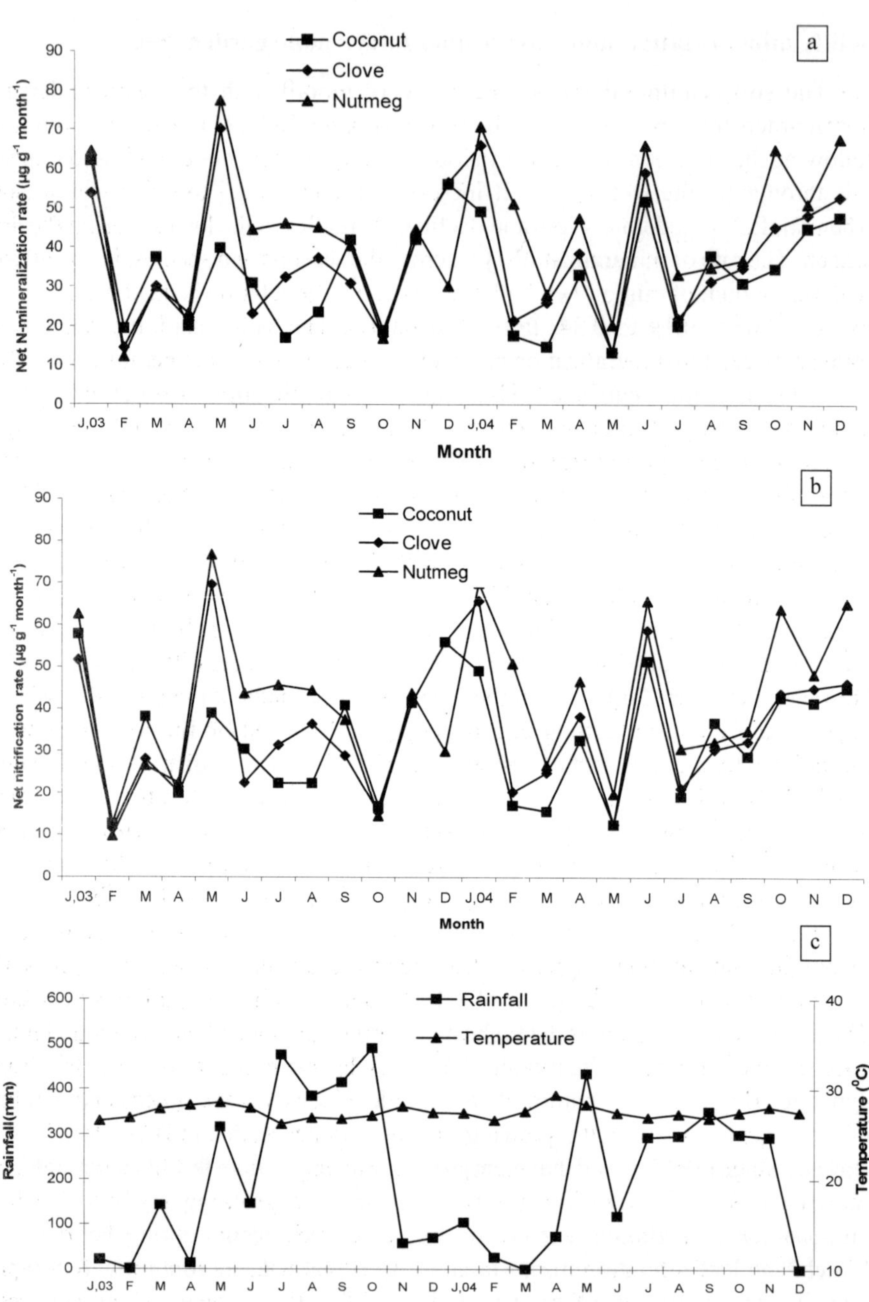

Fig.1. Seasonal pattern of variation in net nitrification, net soil N mineralization rate under coconut, clove and nutmeg trees and rainfall and temperature in the equatorial humid tropical climate of the Andaman islands.

Source: Pandey *et al*. 2007 b

shows the same quadratic relation with the rainfall amount (r^2= 0.20 to 0.36, P< 0.02 to 0.001) and temperature (r^2 = 0.13 to 0.05, P< 0.09 to 0.40) as the mineralization rate does.

Generally high rainfall with low coefficient of variation occurs during rainy season, leading to incessant rainfall for several days in a stretch. This necessitated to understand how incessant rainfall influences net soil N mineralization rate in humid tropics.

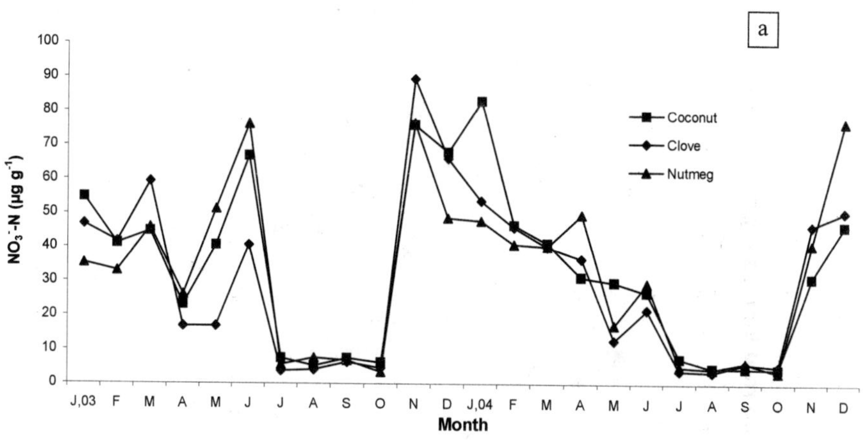

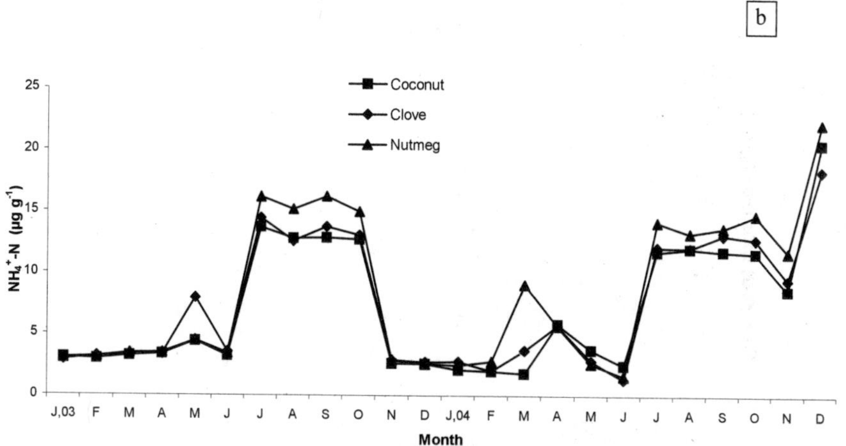

Fig. 2. Seasonal pattern of variation in **(a)** nitrate nitrogen, **(b)** ammonium nitrogen in soils under coconut, clove and nutmeg trees in the equatorial humid tropical climate of the Andaman islands.

Source: Pandey *et al.* 2007 b

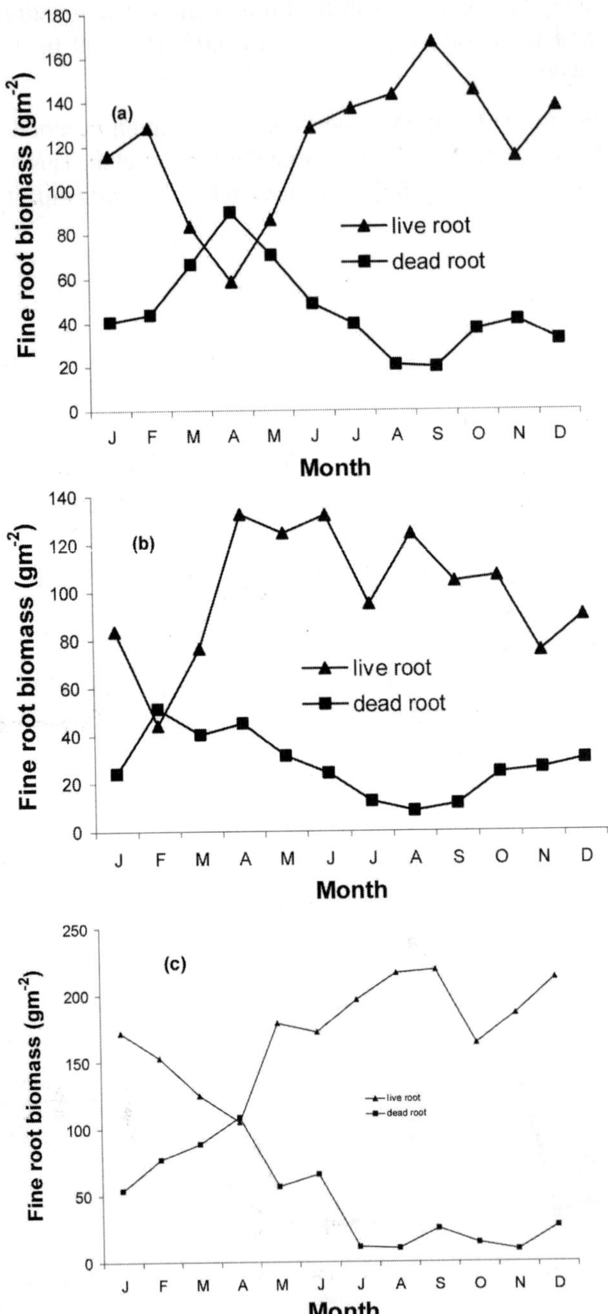

Fig. 3. Seasonal pattern of variation in fine live root and dead fine root in **(a)** coconut, **(b)** clove and **(c)** nutmeg trees in the equatorial humid tropical climate of the Andaman islands.
Source: Pandey and Venkatesh 2007

To know this, soils were sampled under two rainfall conditions *i.e.* 12 incessant rainfall (Sept. 9, 12, 13, 22; Oct. 3, 6, 13, 21, 24, 28; Nov. 3, 17) and 12 dry spell events (48 h to 72 h after rainfall) (Sept. 19, 26, 30; Nov. 7, 10, 14; Dec. 28; Jan 12; Feb. 15, 18, 21 and 24). Water filled pore space, synonymous with relative saturation, was calculated following Weier *et al.* (1993) as: WFPS = [(gravimetric water content x soil bulk density) / total soil porosity], where soil porosity = [(1-(soil bulk density / 2.65)] and 2.65 equals the assumed particle density of soil (Mg m^{-3}). Water filled pore space (WFPS) in the soils is affected significantly due to the rainfall conditions, *i.e.* incessant rainfall and dry spell. It is found from 68 to 84 % during the incessant rainfall and 25 to 47 % during the dry spells (Fig. 4a). The WFPS is always more than 1.5 times higher during the incessant rainfall compared to that during the dry spell in different land use systems. Net nitrification rate ranges from -0.8 to 8.2 $\mu g\ g^{-1} d^{-1}$ in the moist evergreen forest, -0.9 to 8.2 $\mu g\ g^{-1} d^{-1}$ in semi-evergreen forest and -0.6 to 4.1 $\mu g\ g^{-1}\ d^{-1}$ in homegarden (Fig. 4b). The net nitrification rate is 8-15 times lower during the incessant rainfall compared to that during in the dry spell. However, during the incessant rainfall it declines to the level that does not differ among the land use systems ($P < 0.05$). During the dry spell it is the highest in the semi-evergreen forest and the lowest in homegarden. The net nitrification rate is inversely correlated with the WFPS in the moist evergreen forest ($r = -0.668$, $P < 0.001$), semi-evergreen forest ($r = -0.717$, $P < 0.0001$) and homegarden ($r = -0.641$, $P < 0.002$). Net nitrification rate is inversely related with the WFPS ($r = -0.654$, $P < 0.0001$). However, unlike the latter, it tends to increase with the WFPS in the semi-evergreen forest and homegarden, but the increases is insignificant. Across the land use systems also the ammonification rate is positively correlated with the WFPS, but the relation is weak ($r = 0.11$, $P<0.30$) (Fig.4c). Ammonium N pool in the soils ranges from 2 to 14 $\mu g\ g^{-1}$ in the moist evergreen forest, 2 to 17 $\mu g\ g^{-1}$ in semi-evergreen forest and 4 to 14 $\mu g\ g^{-1}$ in homegarden (Fig.4 e). Ammonium N pool is nearly equal between the forest types, but it is 13 % higher in the homegarden compared to that in the forests. It was 2.4 to 3.8 times higher during the incessant rainfall compared to that during the dry spell in different land use systems. The ammonium N pool is positively correlated with the WFPS in the moist evergreen forest ($r = 0.453$, $P< 0.05$), semi- evergreen forest ($r = 0.498$, $P<0.02$) and homegarden ($r = 0.486$, $P<0.02$). The NH_4^+-N is also positively correlated with the WFPS (Fig. 5). Nitrate N pool, ranging from 2 to 86 $\mu g\ g^{-1}$ in the moist evergreen forest,2 to 86 $\mu g\ g^{-1}$ in semi-evergreen forest and 3 to 74 $\mu g\ g^{-1}$ in homegarden and is 2 to 3 times higher during the dry spell compared to that during the incessant rainfall (Fig.4 f).

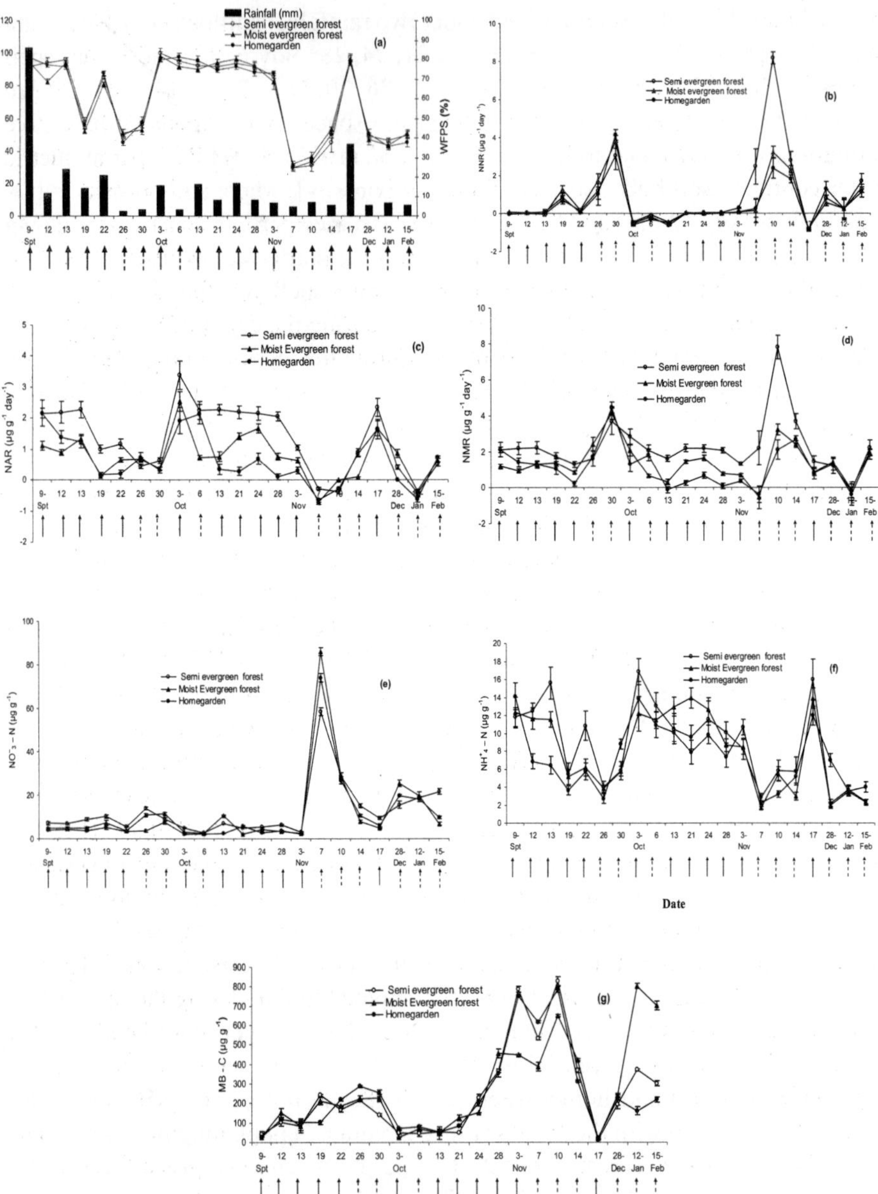

Fig. 4. Variation in **(a)** water filled pore space (WFPS) in relation to rainfall amount, **(b)** net-nitrification rate (NNR), **(c)** net-ammonification rate (NAR), **(d)** net N-mineralization rate (NMR), **(e)** nitrate N pool (NO_3^--N), **(f)** ammonium N pool (NH_4^+-N), and **(g)** microbial biomass C (MB-C) in the soils during the incessant rainfall (Sept. 9, 12, 13, 19, 22; Oct. 3, 13, 21, 24, 28; Nov. 3, 17) and dry spell events (Sept. 26, 30; Oct. 6; Nov. 7, 10, 14; Dec. 28; Jan 12; Feb. 15) in different land use systems in the equatorial humid tropical climate of the Andaman islands. Incessant rainfall dates are indicated with solid line arrows and dry spell dates are indicated with broken line arrows. Bars represent ±1SE. *Source:* Pandey *et al.*, 2009

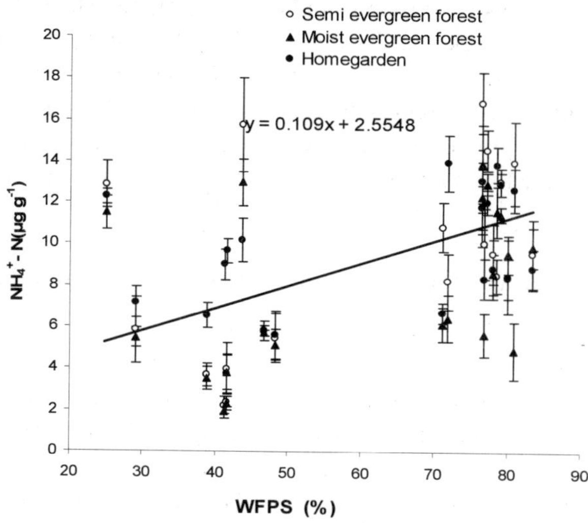

Fig. 5. Relationship between NH_4^+-N (Y, µg g^{-1}) and water filled pore space (WFPS) (X, %): Y = 2.5548 + 0.109 X, (r^2 = 0.7448, P<0.0001), across homegarden, moist-evergreen and semi-evergreen forests in the equatorial humid tropical climate of the Andaman islands. Bars represent +ISE.

Source: Pandey *et al.*, 2009.

Reduction in the net nitrification rate with concurrent accumulation of NH_4^+-N during the incessant rainfall is mainly due to the anaerobic condition caused by the high amount of WFPS. The high WFPS perhaps restricted gas diffusion in the soils and limited re-supply of oxygen from the atmosphere to the soils (Smith and Tiedje, 1979), and thus make the soils anaerobic (Hobbie *et al*., 2000). Linn and Doran (1984) finds a linear increase in relative microbial respiration with increasing water content to a maximum of 60 % WFPS, however, after it the microbial respiration declined in response to increasing anaerobic conditions. Reduction in the net nitrification rate partly may also be due to denitrification in the saturated zone (Jacobs and Gilliam, 1985), vegetation uptake from the unsaturated soil and capillary fringe (Fail *et al*., 1986) and leaching losses of NO_3^--N from the active site. Runoff loss of NO_3^--N at the study sites is found to vary from 3 to 5 kg ha^{-1}yr^{-1} under different land use systems, being maximum under grassland (Pandey and Venkatesh, 2007). The high WFPS (> 70 %) during the incessant rainfall in our study probably causes low redox potential quickly in the soils that perhaps facilitates rapid loss of the NO_3^--N as evident from its lower pool size during the incessant rainfall than that during the dry spell. Schuur and Matson (2001) observe that anaerobic condition caused due to high rainfall, reduces redox potential of soils in Hawaiian montane forest. Low redox potential is known to affect nutrient cycling adversely, which in turn reduces forest growth. On the contrary, at the mesic end of precipitation gradient, high resin N and P availability and high foliar nutrient concentrations were found similar in magnitude to that of fertile soils in a Hawaiian forest (Vitousek and Farrington, 1997). Nitrogen production during the incessant rainfall is 14 times lower compared to that during the dry spell (182 ± 20 kg $ha^{-1}yr^{-1}$). This study supports the observation of Schuur and Matson (2001) that decrease in net primary production with increased mean annual precipitation (MAP) is associated strongly with a decrease in N availability in the Hawaiian forest.

Microbial biomass carbon (MB-C) net nitrification relation in humid tropics

The microbial biomass C in all the land use systems declines simultaneously with decline in the rate of net nitrification during the incessant rainfall, and increases simultaneous with the increase in the latter during the dry spells (Fig.4 g). Positive relation between the microbial biomass C and net N mineralization rate in humid tropics contrasts with the general assumption that reciprocal relation existed between net nitrification and microbial biomass C, which act as a mechanism of nutrient conservation in the seasonally dry tropics (Fig.6) (Bonde *et al.*, 1988; Singh *et al.*, 1989; Grace *et al.*, 1993). This supports the findings of those who report that no change or gradual increase in microbial biomass occur during a growing season (Bauhus and Barthel, 1995; Wardle, 1998). Raubuch and Joergensen (2002) find that temporal changes in microbial biomass N are not linked to N mineralization events, neither in the C rich forest floor layer nor in the C poor mineral N horizon. Decomposition experiments with ^{14}C labeled material reveals that synthesis and decay of microbial products occur concomitantly (Wu *et al.*, 1993). It is also known that all soil micro-organisms simultaneously develop in one direction *i.e.* grow simultaneously under convenient or die under inconvenient conditions (Raubuch and Joergensen, 2002). Positive relation between net-nitrification rate and microbial biomass C, in humid tropics, therefore, suggests that soil N mineralization rates are a function of steady state microbial activity rather than fluctuations in microbial biomass.

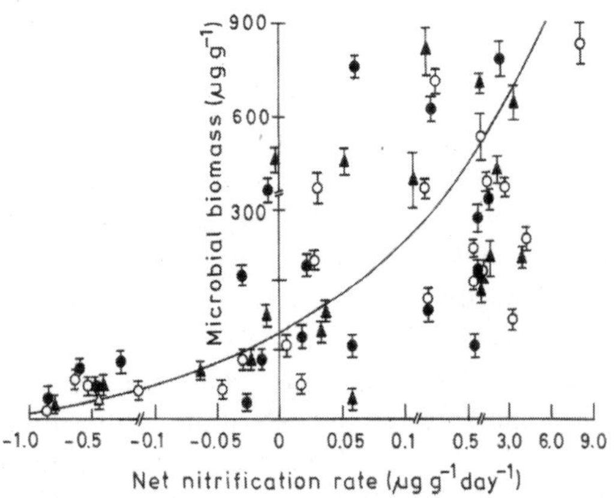

Fig. 6. Relationship between microbial biomass C (Y, $\mu g\ g^{-1}$) and net-nitrification rate (X, $\mu g\ g^{-1}day^{-1}$): $Y = 192.797 + 101.109X - 4.5822X^2$, ($r^2 = 0.2916$, $P < 0.0001$, $N = 63$), across homegarden, moist-evergreen and semi-evergreen forests in the equatorial humid tropical climate of the Andaman islands. Bars represent + ISE.

Source: Pandey *et al.*, 2009.

Hence, microbial biomass does not seem to help conserve nutrients in high rainfall regime of humid tropics. However, the high amount of accumulated NH_4^+ probably resists leaching losses during the incessant rainfall. The accumulated NH_4^+, however, is mineralized during the dry spell when O_2 is available and microbial biomass C is high.

Effect of land use change on net nitrification and net mineralization rates

Tropical land use change is of great concern to ecologist as it alters biogeochemical cycle both regionally and globally (Scholes and Van Breemen, 1997), which is directly, linked with productivity and stability of changed land use systems (Brown and Lugo, 1990). The net nitrification and net soil N mineralization rates are found to increase (39% to 3.8 times and 39% to 3.6 times, respectively) in derived land use systems (homegarden, silvopasture and grassland) than in native evergreen forests of the Andamans. However, the increase is the highest in grassland and the lowest in homegarden (Figs. 7 a, b). Soil temperature is found to increase (0.4 to 9.8°C) in the derived land use systems and moist deciduous forest (0.8 to 8.2 °C) compared to the evergreen forest (Fig. 8 a). The increase is the highest in the grassland and lowest in homegarden, and also the increase is the highest during the dry season and lowest during the early rainy season. Soil-moisture contrasts with the soil temperature and declines to the lowest during the dry season (Fig. 8 b). Net nitrification rate in the reported evergreen forest soils (0.27 $\mu g\ g^{-1}\ day^{-1}$ to 1.81 $\mu g\ g^{-1}\ day^{-1}$) is lower than in the tropical rain forest of Amazon Basin, Brazil (1.32 $\mu g\ g^{-1}\ day^{-1}$ to 3.51 $\mu g\ g^{-1}\ day^{-1}$) (Neill *et al.*, 1997), but higher than in the Malaysian rain forest (0.04 $\mu g\ g^{-1}\ day^{-1}$ to 0.75 $\mu g\ g^{-1}\ day^{-1}$) (Chandler, 1985). Increase in the rate of net nitrification is found to be accompanied with decline in the soil organic C, but it is always greater in the derived land use systems than in the evergreen forest. It is mainly due to oxidation of the soil organic carbon (Devevre and Horwath, 2000), resulted from the rise in temperature owing to the canopy opening of the forest (Cunningham, 1963). In sharp contrast to exponential relation under laboratory incubations (Wang *et al.*, 2006), quadratic relation between the net nitrification rate and soil temperature, and net N mineralization rate and soil temperature is found in the field conditions. Net nitrification rate in the reported study is found to increase linearly (0.8 $\mu g\ g^{-1}\ day^{-1}$ to 3.9 $\mu g\ g^{-1}\ day^{-1}$) with increase in temperature (26 ^{0}C to 32 ^{0}C) when soil- moisture is the highest (25 % to 29 %), drops a little and attains plateau (2.4 $\mu g\ g^{-1}\ day^{-1}$) with fluctuations at 32 to 35°C when soil moisture is 14 % to 15 % and declines to 1.5 $\mu g\ g^{-1}\ day^{-1}$ when temperature rises > 35°C and soil-moisture declines to the lowest (9.6 %). These indicate that increase in soil temperature enhances microbial activity till the soil-moisture is adequate. The highest rate of net N mineralization and net nitrification in the grassland in the present study may be attributed to the highest rise in temperature due to frequent grass cutting that exposes the soil to sun. Lower rate of net nitrification as well as net mineralization, however, in the silvopasture and homegarden compared to that in the grassland is probably due to the tree canopy cover that could maintain the lower temperature. This suggests that tree based land use systems in warm humid climate may be better options for soil fertility management.

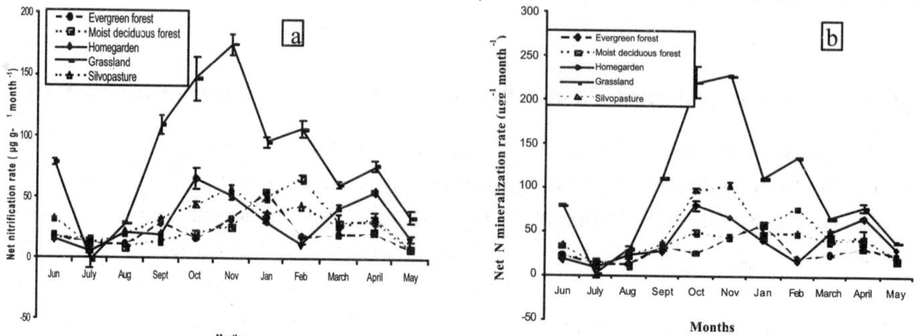

Fig. 7. Seasonal variation in **(a)** net nitrification rate and **(b)** net soil N mineralization native evergreen and moist deciduous forests; and derived land use systems i.e., homegarden, silvopasture and grassland in the equatorial humid tropical climate of the Andaman islands. Bars represent + ISE.

Source: Pandey *et al.* 2010a

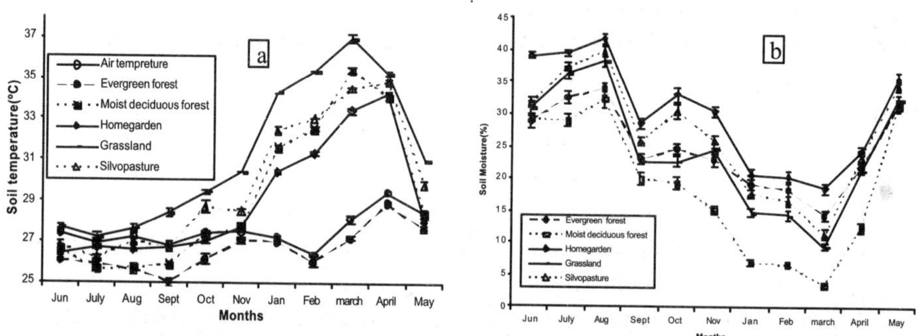

Fig. 8. Seasonal variation in **(a)** soil temperature and **(b)** soil moisture in the native evergreen and moist deciduous forests; and derived land use systems i.e., homegarden, silvopasture and grassland in the equatorial humid tropical climate of the Andaman islands. Bars represent + ISE.
Source: Pandey *et al.* 2010 a

Effect of season on net nitrification and net mineralization rates, and NO_3^--N and NH_4^+-N pools in relation to land use change

The net nitrification rate across the land use systems studied is the lowest (12.4 µg g^{-1} $month^{-1}$ to 34.6 µg g^{-1} $month^{-1}$) during the early rainy- season (May to August; e"350 mm $month^{-1}$ rainfall) and the highest (20.6 µg g^{-1} $month^{-1}$ to 143.6 µg g^{-1} $month^{-1}$) during the late rainy season (September–November; <350 mm $month^{-1}$) (Fig. 7). It has been observed to be 59 % to 2.4 times higher during the dry season than during early rainy season. Pattern of seasonal variations in the net mineralization rates is closely similar to that of the net nitrification rates as the latter contributed maximum (69 % to 83%) to the former. Net nitrification rates (Y, µg g^{-1} $month^{-1}$) show quadratic relation with temperature (X, °C): Y = -0.027 X^2 + 2.3148 X + 1.3515, r^2 = 0.1664, P<0.002). It tends to decline with increasing soil moisture and rainfall, but the relations are not significant. Like the net nitrification rate, net mineralization rate (Y, µg g^{-1} $month^{-1}$) also shows quadratic relation with temperature (X, °C): Y = -0.0274 X2 + 2.5547 X + 6.2419, r^2 = 0.1514, P< 0.004). Like the net nitrification rate, it also tends to decline with soil moisture and rainfall amount, but the relations are not significant. Pattern of net nitrification and net mineralization rates in the moist deciduous forest are broadly similar to that found in the evergreen forest. But, rate of the net nitrification as well as net mineralization during the late rainy (October) and dry season (<100 mm $month^{-1}$ from February to April) is 61% and 130 %, respectively higher in the moist deciduous forest than in the evergreen forest. Compared to the evergreen forest, higher rate of net nitrification in the moist deciduous forest on similar soil conditions may be attributed to the deciduous tree species that provides relatively greater amount of litter biomass. However, similar seasonal pattern of variations in the net nitrification and net mineralization rates in the forest types suggests that the climate over-rides the effect of the deciduous species. The lowest rate in the net nitrification and net N mineralization during the early rainy season in all the studied land use systems are probably due to high amount of rainfall and low variability in its distribution that perhaps creates anaerobic condition, which perhaps reduces aerobic nitrifier population as evident from low microbial biomass C. In another study WFPS (water filled pore space) is we found 70% to 80 % during incessant rainfall that declines to 38 % to 41 % after 4-hr at the study sites. According to Sexstone *et al.* (1985) anaerobic condition in sandy loamy soil develops quickly after addition of water and reaches maximum within 3 to 5-hr, but returns to pre-irrigated level within 12-hr. However, quick increase in the rate of nitrification during the late rainy season is probably due to plenty of leaf litters added to the soils in the tree based systems, and leaf cuttings and dead roots in the grassland and silvopasture systems due to the grass cutting; and their quick decomposition by termites. Higher amount of organic carbon (13 %) and net nitrification rate (20%) are found in termite soils than in native soils at the study sites. However, occurrence of relatively high rate of mineralization during the dry season in all the land use systems may be attributed to upward movement of water, condensation of water vapor and sporadic rains, which are well known phenomena in the equatorial humid climate of the island. Several studies have reported that soil-moisture does not have to be optimum to obtain high rates of N mineralization (Gill *et al.*, 1995).

Pool sizes of NO_3^--N, NH_4^+-N and mineral N are found to vary significantly among the land use systems, months and their interactions (P<0.0001). Mineral N, averaged across the months, declines (14 % to 24 %) in the derived land use systems compared to that in the evergreen forest (Fig. 9 a). However, the decline is the highest in the grassland and lowest in homegarden. Mineral N in the moist deciduous forest also is lower (21%) compared to that in the evergreen forest. The NO_3^-- N contributes maximum 65 % to 70 % to the mineral N; therefore, pattern of variation in the NO_3^--N in all the land use systems is similar to that of the mineral N (Fig. 9 b). The NH_4^+-N pool increases ($P<0.05$) whereas the NO_3^--N pool declines ($P<0.05$) during the rainy season in almost all the land use systems (Table 7). Low concentration of nitrate in the rainy season (early + late) in all the land uses studied is probably due to three reasons: (1) low rate of mineralization, (2) uptake by vegetation, and (3) losses due to leaching and denitrification. Greater concentration of nitrate N during the dry season, however, indicates its accumulation perhaps due to reduced uptake by the vegetation. Field observation indicates that the grass is almost dried whereas growth in the evergreen trees in the evergreen forest and the homegarden and silvopasture is higher during the rainy season compared to that during the dry season in the humid tropics (Pandey and Venkatesh, 2007). However, increase in the concentration of NH_4^+-N during the rainy season in the land use systems is obvious due to low rate of nitrification owing to high amount of rainfall (anaerobic condition). These indicate that the evergreen tree species in the equatorial humid tropical climate have acquired trade-on and trade-off mechanisms for the mineral N (NH_4^+-N and NO_3^--N) uptake.

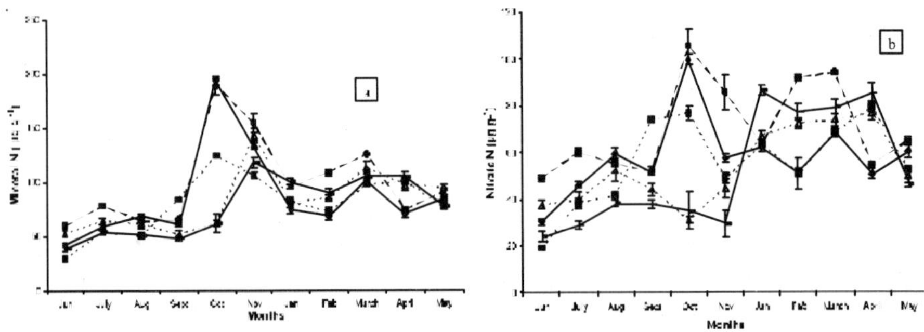

Fig. 9. Seasonal variation in **(a)** ammonium nitrogen and **(b)** nitrate nitrogen in the soil in the native evergreen and moist deciduous forests; and derived land use systems i.e., homegarden, silvopasture and grassland in the equatorial humid tropical climate of the Andaman islands. Symbols are same as in Fig. 7. Bars represent ± ISE.

Source: Pandey *et al.* 2010 a

Table 7. Seasonal concentrations (mean ± 1SE) of available nitrogen in the native evergreen and moist deciduous forests; and derived land use systems i.e. homegarden, grassland and silvopasture in the equatorial humid tropical climate of the Andaman islands.

Nitrogen	Evergreen forest	Moist deciduous forest	Homegarden	Silvopasture	Grassland
NO_3^--N (μg g^{1})					
Rainy season	67.44±7.24	50.23±7.09	57.82±7.42	42.80± 6.52	34.20±7.60
Dry season	80.15±10.45	66.07±6.77	56.73±4.69	74.19±11.03	0.41±12.01
NH_4^+-N (μg g^{1})					
Rainy season	32.43±8.56	24.45±4.88	33.25±9.51	32.49±7.13	29.54±9.45
Dry season	22.53±3.67	21.62±3.22	22.85±4.15	24.23±6.37	18.60±3.09

Source: Pandey *et al.* (2010 a)

Microbial biomass C variation and its relation with net nitrification rate in relation to land use change

Microbial biomass C varying significantly among the land use systems, months and their interactions (P<0.0001), declines (38 % to 47 %) in the derived land use systems and moist deciduous forest (24 %) than in the evergreen forest (Fig. 10). Among the derived land use systems, the decline is the lowest in the silvopasture and the highest in the grassland. Microbial biomass C follows broadly the pattern of net nitrification rate and is the lowest during the early rainy season like the latter one, but increases from the late rainy season and peaks in the dry season in all the land uses studied. High amount of microbial biomass C and comparatively low rates of net nitrification and net N mineralization in the evergreen forest indicate that nutrient cycling in the climax forest is stabilized and soil fertility is agile. However, reduction in the microbial biomass C pool in the derived land uses compared to the evergreen forest may be attributed to the soil disturbance (high bulk density). The lowest microbial biomass in the grassland is probably due to the grass cutting. Grazing and mowing are known to reduce root growth and rhizome carbohydrate reserves (Turner *et al.* 1993), which result in

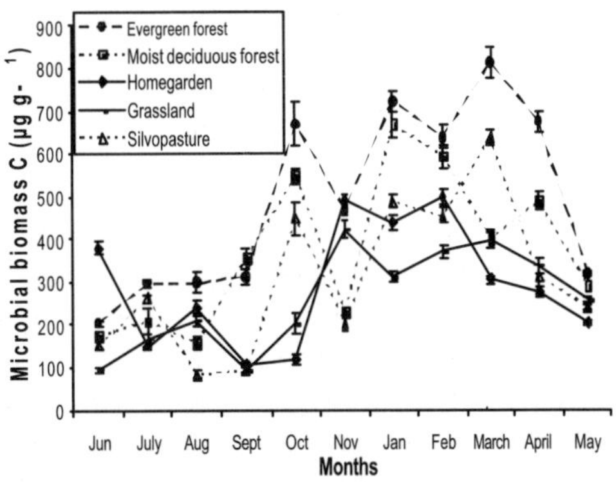

Fig. 10. Seasonal variation in microbial biomass C in the soils under native evergreen and moist deciduous forests; and derived land use systems i.e., homegarden, silvopasture and grasslandn in the equatorial humid tropical climate of the Andaman islands. Bars represent ± ISE.
Source: Pandey *et al.* 2010 a

reduction in microbial growth (Holland and Detling, 1990). Higher microbial biomass in the silvopastures and homegardens compared to that in the grassland is expected to be due to an additional tree roots, which are known to provide carbon to microorganisms through rhizodeposition (Leinweber *et al*., 1995). These clearly suggest that conversion of evergreen forest may upset the nutrient cycling, which will make the soil fertility fragile in the derived land use systems. Several studies have found inverse relation between net N mineralization and microbial biomass C in seasonally dry tropical soils (Singh *et al*., 1989; Nunan *et al*., 2000), that has been argued to conserve nutrients (mineral N) in the dry tropics (Shi *et al*., 2006). Positive relation between the microbial biomass C and net nitrification rate in the present study, however, do not support the hypothesis of nutrient conservation for the humid tropics. Our study, however, agrees with the view of Raubuch and Joergensen (2002) that net N mineralization and net nitrification rates are a function of steady- state microbial activity, rather than fluctuations in microbial biomass.

Land use change and its implications in equatorial humid tropics

Organic C is the highest (23.6 mg g^{-1}) in the evergreen forest and declines to the lowest (13.7 mg g^{-1}) in the grassland, however, on the contrary total N is the highest (1.33 mg g^{-1}) in the grassland and lowest in the silvopasture (0.93 mg g^{-1}); bulk density increases in the derived land uses, particularly grassland and silvopasture, than in the evergreen forest soil. Decline in the amount of soil organic carbon (SOC) with concurrent increase in the rate of net N mineralization, resulted from the increase in temperature, indicates that conversion of the evergreen forest in the humid tropics has regional as well as global implications. Comparing the amount of SOC (SOC concentration x bulk density) in the derived land use systems with that found in the evergreen forest, the highest amount (1430 kg ha^{-1} yr^{-1}) of SOC was lost from the grassland and lowest amount (127 kg ha^{-1} yr^{-1}) from the homegarden. Assuming that the loss of SOC occurred only through the soil N mineralization, oxidation of the SOC was estimated to had emitted 466 to 5248 kg ha^{-1} yr^{-1} CO_2 (amount of SOC x 44/ 12) in the atmosphere. However, it was the highest from the grassland and lowest from the homegarden. The CO_2 emission in the tree based land use systems was well within the range, however, in the grassland it was 8 times higher compared to that reported from white area of Alberta (140 to 669 kg CO_2 ha^{-1} yr^{-1}) (Sauve *et al*., 2000). In addition, the loss of SOC probably may reduce soil fertility, in long term, in the derived land use systems, particularly more in the grassland. The increased rate of nitrification was found to produce 85 to 728 kg ha^{-1} yr^{-1} higher nitrate N in the derived land use systems compared to that in the evergreen forest; and like the SOC loss, it was the highest in the grassland and lowest in the homegarden, which was expected to had accelerated N_2O and NO emissions significantly. Increased N availability has been found to favour activity of nitrifiers which increases N_2O and NO losses (Hall and Matson, 2003). The CO_2, being linked with N_2O and NO emissions, seems the most important amongst the green house gases. Increasing the net fixation of atmospheric CO_2

through C sequestration in soil is one of the options to lower the green house gas emissions. Therefore, tree based land use systems, as the present study suggests, may be better options to lower the green house gas emission in hot humid tropics.

Increase in the net nitrification and net mineralization rates; and subsequently greater loss of mineral N from the derived land use systems compared to that in the native forest indicate adverse effect of the land use change on the soil fertility in the island. Both human as well as livestock population has increased 13 times in the island within a-half century due to high birth rate and immigration from the mainland India (Basic Statistics, 2001). If this trend of population growth continues, it is extrapolated that in the next a-half century minimum fifty thousand hectare additional land will be required to meet their food, fuel and fodder requirements (5 head per family and 0.5 ha family^{-1}). It will pose an immense pressure on the forest in future. The highest interception and lowest runoff loss of rain water from the homegardens suggested that conversion of forest to agroecosystem like homegarden with multistory canopy cover would offer the possibility of supporting food needs while maintaining the water-cycle and restricted N losses (Salati and Vose, 1984). However, the forest clearance for cattle-grassland and its management through cut- and- carry system in the large exposed field would result in to the land use system which will lead to a major water runoff, soil erosion and soil degradation that will make the soil fertility more fragile than Amazon basin of Brazil as the runoff loss of dissolved NO_3^- in our study was 4 to 7 times higher compared to that found at Amazon Basin of Brazil (0.1 to 0.3 mg L^{-1}, Table 8) (Salati and Vose, 1984).

Table 8. Loss of NH_4^+-N, NO_3^--N and mineral N through runoff water from the native evergreen forest and its derived land use systems i.e. homegarden, silvopasture and grassland in the equatorial humid tropical climate of the Andaman islands.

Treatment	NH_4^+-N		NO_3^--N		Mineral N		Run off (%)	Canopy Interception of rainfall (%)
	(mg L^{-1})	(kg ha^{-1})	(mg L^{-1})	(kg ha^{-1})	(mg L^{-1})	(kg ha^{-1})		
Evergreen forest	0.62^{a}	2.81^{a}	0.72^{a}	3.26^{a}	1.34^{a}	6.07^{a}	22.3^{a}	63.4^{a}
Moist deciduous forest	-	-	-	-	-	-	-	-
Homegarden	0.78^{b}	5.64^{b}	0.83^{b}	6.000^{b}	1.61^{b}	11.64^{b}	35.6^{b}	41.6^{b}
Silvopasture	0.72^{ab}	5.86^{ab}	0.84^{b}	6.84^{b}	1.56^{b}	12.70^{b}	40.1^{d}	34.3^{d}
Grassland	0.64^{a}	6.96^{a}	1.23^{c}	13.36^{c}	1.87^{c}	20.32^{c}	53.5^{c}	12.2^{c}

- not estimated values of a parameter suffixed with different letters in a column are significantly different at $P<0.05$

Effect by tillage on net nitrification and net soil N mineralization rates

To find out a level of tillage that increases soil N mineralization and simultaneously maintains reasonably good amount of SOC and MB-C in the soils in a hot humid tropics, four tillage treatments, *i.e.* long term zero till, frequent till, low till and short term zero till, each replicated three times, are examined. In the long term zero till treatment crops are sown by dibbling, tillage is not done for past 20-yrs (Table 9). In the frequent till treatment, tillage is done three times before sowing of both the crops. Weeds are uprooted (three times, *i.e.* at the time of sowing, and 15- and 30-day after sowing) and removed; tillage is done frequently (6 times in a year) for past 20-yrs. In the low till treatment, the tillage is done once before sowing of both the crops. Weeds are uprooted three times like the frequent till treatment and after 6-hr (dead) buried (293 g m^{-2}) *in-situ*, tillage is not done for past 17-yrs in the beginning, thereafter it is done once before sowing of both the crops for two cropping seasons each. In the short term zero till treatment, the tillage is not done and crops are sown by dibbling. Weeds are cut three times close to the ground, like the long term zero till treatment, and mulched (287 g m^{-2}) *in-situ*. Tillage (15-cm deep) is done always by a local made plough, which mixed soils well.

The tillage is found to increase rates of the net soil N mineralization and net nitrification, but the highest in the frequent till and the lowest in the long term zero till treatment (Figs.11 a, b). The net N mineralization and net nitrification rates are the lowest during the wet season and the highest during the post-wet season in all tillage treatments. The net soil N mineralization as well as net nitrification rate is found positively correlated with the soil temperature ($r = 0.877$ and 0.935, $P< 0.01$ and 0.001, respectively), but inversely correlated with the water content of soils ($r = - 0.932$ and -0.942, $P<0.001$). The highest (760 kg ha^{-1} yr^{-1}) N is mineralized in the frequent till and the lowest (376 kg ha^{-1} yr^{-1}) N is mineralized in the long term zero till treatment (Table 10). In the tillage treatments maximum N is mineralized during the wet and the lowest during the dry season. The mineralized N is positively correlated with the soil temperature during the wet ($r = 0.927$, $P<0.05$) and post-wet seasons ($r = 0.904$, $P<0.05$). But, the mineralized N is inversely correlated with the water content of soil during the wet ($r = -0.911$, $P<0.05$) and post-wet seasons ($r = -0.905$, $P<0.05$).

Table 9. History of tillage treatments and management of crops in the experiment in the equatorial humid tropical climate of the Andaman islands.

Treatment	Tillage history	Crop management
Long term zero till	Tillage was not done from 1983 - 2002. From 2003-2006 a maize-okra crop rotation was practiced and the crops were sown by dibbling. The maize was cultivated during wet season (May to Oct.) and okra during the post-wet season (Nov. to Jan.).	Crops were not cultivated and fertilizers were not applied from 1983-2002. In the maize-okra crop rotation fertilizer was not applied and residues of the crops were removed. But, weeds were cut and mulched.
Frequent till	Tillage (15-cm deep) was done three times by a local made plough before each crop sowing in the maize-okra crop rotation from 1983-2002 and 2003-2006 as well.	From 1983-2002 weeds and crop residues were removed, and fertilizers N, P and K were applied at 60, 50 and 40 kg ha^{-1}, respectively to the maize, and 50, 40 and 30 kg ha^{-1}, respectively to the okra. From 2003-2006 weeds and crop residues were removed, but fertilizer was not applied.
Low till	Tillage was not done from 1983- 1999; then after the tillage was done, from 2000-2002 (two cropping seasons) and 2003-2006, by the local made plough, once before each maize and okra crop sowing in the maize-okra crop rotation.	From 1983-1999 crop management was done like the corresponding period in the long term zero till treatment. From 2000-2002 weeds (uprooted) and residues of the crops were removed, and fertilizers were applied to the crops as did in the frequent till treatment. From 2003- 2006 residues of the crops were removed, fertilizer was not applied, but weeds were uprooted and buried *in-situ*.
Short term zero till	Tillage history from 1983-2002 was the same as in the low till treatment. From 2003-2006 both the crops in the maize-okra crop rotation were sown by dibbling.	From 1983-2002 crop management was done like the low till treatment. From 2003 - 2006 crop residue and fertilizer management were like the low till treatment. But, weeds were cut and mulched *in-situ*.

Source: Pandey *et al.* 2010 b

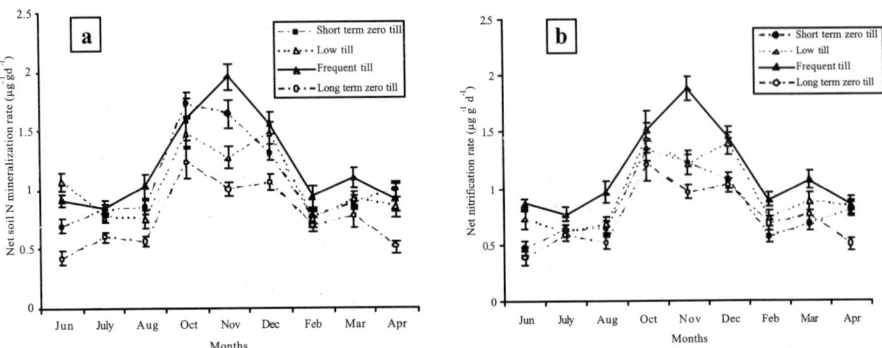

Fig.11. Seasonal variation in **(a)** net N mineralization rate and **(b)** net nitrification rate in the soils under different tillage conditions in the equatorial humid tropical climate of the Andaman Islands. Bars represent ± ISE.
Source: Pandey *et al.* 2010 b

Table 10. Amount of nitrogen (N) mineralized during different seasons under different tillage treatments in the equatorial humid tropical climate of the Andaman islands.

Treatment	N-mineralized ($kg\ ha^{-1}\ yr^{-1}$)			Total N mineralized ($kg\ ha^{-1}\ yr^{-1}$)
	Wet season	Post-wet season	Dry season	Short term zero tillage
Short term Zero tillage	$^{x}245^{a}$	$^{x}243^{a}$	$^{y}133^{a}$	621^{a}
Low tillage	$^{x}281^{b}$	$^{y}232^{b}$	$^{z}134^{a}$	647^{b}
Frequent tillage	$^{x}312^{c}$	$^{y}288^{c}$	$^{z}160^{b}$	760^{c}
Long term zero tillage	$^{x}143^{d}$	$^{x}147^{d}$	$^{y}86^{c}$	376^{d}

Values prefixed with different superscript letters in a row are significantly different at $P<0.05$
Values suffixed with different superscript letters in a column are significantly different at $P<0.05$
Source: Pandey *et al.*, 2010b

Soil organic carbon (SOC) and total N were the highest in the long term zero till and the lowest in the frequent till treatment. The highest rate of net soil N mineralization and the lowest amount of SOC in the frequent till treatment, and vice-versa in the long term zero till treatment supports the hypothesis that increase in rate of net soil N mineralization and decline in amount of SOC depends on frequency of tillage in the hot humid tropics. The lowest net soil N mineralization rate, but the highest SOC in the long term zero till treatment is probably due to: (1) reduction in soil aeration caused by compaction. The soil compaction in this study is found to occur due to high rainfall for 8-9 months followed by 3-4 months dry spell over a long time of 23-yrs. Alternate wetting and drying are known to develop crust / seals on soil surface that decreases air permeability and oxygen diffusion rates in soils (Jensen *et al.*, 1996); (2) comparatively greater fine sand particles probably forms more microaggregates in the long term zero till than other till treatments, which protects organic matter (OM) from decomposition. Roots of the weeds (346 ± 50 g m^{-2}) is expected to have provided root exudates (Campbell and Greaves, 1990), which favours microbial growth (Andrade *et al.*, 1998). The microbial growth together with the root exudates perhaps serve as cementing materials for the microaggregate formation (Dorioz *et al.*, 1993). The roots, upon death, is incorporated in to the microaggregates. According to Six *et al.*'s (1999, 2000) conceptual model, rate of macroaggregate formation and degradation, *i.e.* aggregate turnover, is reduced under no-till condition that leads to formation of stable microaggregates in which carbon is stabilized and sequestered in long term; (3) reduction in the soil temperature during the post-wet season is most likely due to vegetal cover of weeds, as evident from the positive correlation between the net N mineralization rate and the soil temperature. Net soil N mineralization is reported positively correlated with soil temperature (Wang *et al.*, 2006). Several soil N mineralization studies under laboratory conditions have reported that no-till soils have greater N mineralization potential (Kingery *et al.*, 1996). Similar result was expected in this study also, but despite the highest SOC, the lowest soil N mineralization rate in the long term zero till conditions suggests that in the field conditions the buried bag technique give an actual estimate of soil N mineralization as it covers variations in soil temperature and water content of soil.

One may argue that in the long term zero till and short term zero till treatments weeds can not be effectively controlled simply by cutting and mulching than unless they are completely killed. The weeds ultimately compete with the crops for nutrients and water content of soils, and causes reduction in yields. Competition of weeds with the crops for nutrients and water content of soil can not be overruled. But, when the weeds were cut, the crops grew taller and suppressed growth of the weeds due to shade. Pandey and Venkatesh (2007) observed that yields of the crops in the short term zero till did not differ significantly with the yields in the rests of the tillage treatments other than the long term zero till. This seems to have occurred due to greater N mineralization (Herman *et al.*, 2006), probably due to greater rhizodeposition (Vinton and Burke, 1995) and root exudation (Jaeger *et al.*, 1999) caused by the weed cutting, than N uptake by the weeds. Cutting / grazing of herbaceous vegetation is known to increase rhizodeposition and new roots formation as well (Pandey and Singh, 1992).

The highest rate of soil N mineralization in the frequent till treatment can be explained due to three reasons: (1) the frequent tillage probably broken soil structure and contributed to the N mineralization by exposing physically protected OM to microbial turnover (Beare *et al.*, 1994). Tillage simulated by sieving or grinding of soils has also been found to increase N mineralization (Kristensen *et al.*, 2000); (2) the frequent tillage might have formed macropores (*i.e.* relatively high gravels and low fine sand particles) and could have caused the highest aeration (Carter *et al.*, 1994). Aeration is known to favour soil N mineralization (Pandey *et al.*, 2007b);

(3) the frequent tillage removed weeds from the fields and exposed the soils to the sun and resulted in high soil temperature, which probably accelerated the rate of the net nitrification (Wang *et al.*, 2006).

Effect of tillage on seasonal variation in soil N mineralization and mineral N pool

The lowest rate of net nitrification during the wet season invariably in all tillage treatments is found (Fig. 12). This suggests that the tillage does not change the pattern of seasonal variation in the fluxes of nitrogen. The lowest rate of net nitrification during the wet season seems to have occurred primarily due to high amount of rainfall (water content of soil), which is found to make the soils anaerobic (Pandey *et al.*, 2009). Among the tillage treatments, the lowest rate of net nitrification in the long term zero till treatment during the wet season seems to have occurred probably due to prevalence of anaerobic conditions in the soils for a longer time owing to greater proportion of fine sand particles (micropores). Lindwall *et al.* (1995) have found that minimum tilled soils conserve moisture by more homogeneous distribution of micropores between aggregates and reduction of the amount of macropores. Contrary to this, higher rate of net nitrification in the frequent till soils is likely due to rapid drainage of rain-water through macropores (relatively low fine sand and high gravels) after the incessant rainfalls are ceased. Pandey and Srivastava (2009) have found that gravely, sandy-loamy soils in the study region become anaerobic during high rainfall, but upper layer (0-5 cm) become aerobic quickly within 2-hr after cessation of the rainfall. Among the seasons, NO_3^--N is the highest during the post-wet season and the lowest during the wet season (Fig. 12, Table 11). Unlike the NO_3^- -N, NH_4^+ -N is the highest during the wet season and the lowest during the post-wet season. NH_4^+ -N is positively correlated with the water content of soil ($r = 0.813$- 0.892, $P<0.01$-0.001) in all tillage treatments and pooled across the tillage treatments ($r = 0.671$, $P<0.001$). NO_3^- -N is inversely correlated with the water content of soil ($r =$ 0.391, $P<0.01$). High rate of net nitrification during the post-wet season in all tillage treatments is probably due to accumulation of NH_4^+ during the wet season. However, the highest rate of net nitrification in the frequent till treatment during the post-wet season corresponds to the highest accumulation of the NH_4^+. Accumulation of NH_4^+ during the wet season in all tillage treatments is most likely due to reduction of NO_3^- into NH_4^+ via a dissimilatory pathway (anaerobic conditions) (Silver *et al.*, 2001; Pandey *et al.*, 2009). The highest reduction of NO_3^- into NH_4^+ (highest NH_4^+/ NO_3^- ratio) in the low till conditions, however, seems to have occurred due to relatively greater fine soil particles (micropores), which maintains high water content of soil for longer time.

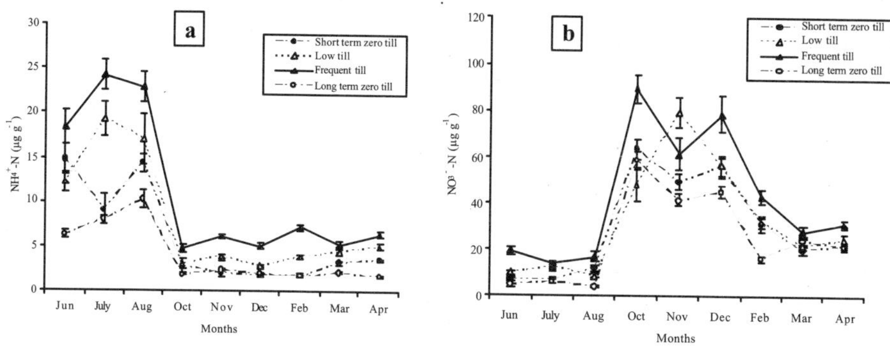

Fig. 12. Seasonal variation in **(a)** ammonium nitrogen and **(b)** nitrate nitrogen in the soils under different tillage conditions in the equatorial humid tropical climate of the Andaman Islands. Bars represent ± ISE.
Source: Pandey *et al.* 2010b

Microbial biomass -C and net N mineralization of soil under different tillage levels

The tillage treatments reduces soil MB-C significantly, but the reduction is the highest in the frequent till and the lowest in the long term zero till treatment (Fig.13). The MB-C, being a labile fraction, is positively correlated to the SOC (r = 0.961, P<0.01) across the tillage treatments. The highest MB-C in the long term zero till treatment is probably due to organic matter input (via leaf and root litters and root exudates) from weeds over a long period of 23-yrs (El Titi, 2003a). But, the highest decline in the MB-C in the frequent till treatment is due to removal of weeds from the fields; and the soil temperature induced mineralization of the labile carbon. The highest SOC and MB-C in the long term zero till treatment indicates the highest sequestration of carbon under the conservation tillage *i.e.* no till for the long period. Hernandez and Lopez-Hernandez (2002) have also found greater carbon sequestration in no till than conventional tillage in a tropical savanna in Venezuela. Across the seasons, MB-C is found to be the highest during the post-wet season and the lowest during the wet season in all tillage treatments (Table 11). High seasonal variations in the MB-C in all tillage treatments, other than long term zero till, indicates that the labile carbon plays an active role in the soil N transformation under the tillage conditions. The greater seasonal variation in the MB-C in the frequent till than long term zero till treatment

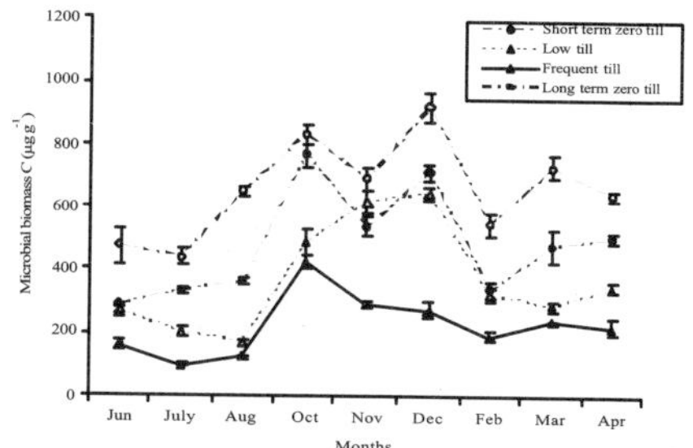

Figure 13. Seasonal variation in microbial biomass carbon in the soils under different tillage conditions in the equatorial humid tropical climate of the Andaman Islands. Bars represent ± ISE.

Source: Pandey *et al.*, 2010 b

Table 11. A sum up of seasonal variation in net nitrification rate (NNR), net mineralization rate (NMR), ammonium and nitrate N pool, microbial biomass (MB-C), water content of soil (WCS) and soil temperature (ST) in wet (WS), post wet (PWS) and dry (DS) seasons in different tillage treatments in the equatorial humid tropical climate of the Andaman islands.

Treatment	NNR(μg g $^{-1}$ d $^{-1}$)			NMR(μg g $^{-1}$ d $^{-1}$)			NH_4^+-N(μg g $^{-1}$)			NO_3^- N(μg g $^{-1}$)			MB-C(μg g $^{-1}$)			WCS(%)	ST(^{0}C)	
	WS	PWS	DS	WS	PWS	DS	WS	PWS	DS	WS	PWS	DS	WS	PWS	DS	WS	PWS	DS
Short term zero tillage	$^{a}0.58^{a}$	$^{b}1.21^{a}$	$^{c}0.68^{a}$	$^{a}0.79^{a}$	$^{b}1.56^{a}$	$^{c}0.88^{a}$	$^{a}12.3^{a}$	$^{b}2.1^{a}$	$^{b}2.8^{a}$	$^{a}8.8^{a}$	$^{b}56.1^{a}$	$^{c}24.3^{ab}$	$^{a}328^{a}$	$^{b}672^{a}$	$^{c}439^{a}$	$^{a}46^{a}$	$^{b}26^{a}$	$^{c}12^{a}$	$^{a}26.4^{a}$	$^{b}28.0^{a}$	$^{c}33.4^{a}$
Low tillage	$^{a}0.67^{b}$	$^{b}1.35^{b}$	$^{c}0.81^{b}$	$^{a}0.86^{b}$	$^{b}1.41^{b}$	$^{a}0.84^{a}$	$^{a}16.3^{b}$	$^{b}3.2^{b}$	$^{c}4.4^{b}$	$^{a}10.3^{a}$	$^{b}61.3^{b}$	$^{c}26.4^{b}$	$^{a}213^{b}$	$^{b}583^{b}$	$^{c}319^{b}$	$^{a}44^{b}$	$^{b}24^{b}$	$^{c}11^{a}$	$^{a}26.5^{a}$	$^{b}28.3^{a}$	$^{c}33.1^{a}$
Frequent tillage	$^{a}0.86^{c}$	$^{b}1.61^{c}$	$^{c}0.94^{c}$	$^{a}0.93^{b}$	$^{b}1.71^{c}$	$^{a}0.98^{b}$	$^{a}21.8^{c}$	$^{b}5.3^{c}$	$^{c}6.3^{c}$	$^{a}16.8^{b}$	$^{b}76.4^{c}$	$^{c}34.3^{c}$	$^{a}127^{c}$	$^{b}328^{c}$	$^{c}218^{c}$	$^{a}45^{ab}$	$^{b}24^{b}$	$^{c}10^{a}$	$^{a}26.3^{a}$	$^{b}28.2^{a}$	$^{c}33.2^{a}$
Long term zero tillage	$^{a}0.49^{d}$	$^{b}1.06^{d}$	$^{c}0.61^{a}$	$^{a}0.53^{c}$	$^{b}1.09^{d}$	$^{c}0.66^{c}$	$^{a}8.1^{d}$	$^{b}2.0^{a}$	$^{b}1.8^{d}$	$^{a}5.2^{c}$	$^{b}48.2^{d}$	$^{c}21.2^{a}$	$^{a}522^{d}$	$^{b}813^{d}$	$^{c}637^{d}$	$^{a}48^{c}$	$^{b}30^{c}$	$^{c}16^{b}$	$^{a}25.4^{a}$	$^{b}26.2^{b}$	$^{c}30.5^{b}$

Values prefixed with different superscript letters in a row for a parameter are significantly different at $P<0.05$

Values suffixed with different superscript letters in a column for a parameter are significantly different at $P<0.05$

Source: Pandey *et al.* 2010 b

may be due to the greater variations in the microenvironments of the soils (water content of soil and soil temperature). The micro-environmental changes are known to influence MB-C (Pandey *et al.*, 2009; Woomer *et al.*, 1994). The decline in the MB-C in all tillage treatments during the wet season can be owing to lyses of aerobic microbes (Rinklebe and Langer, 2006). But, relatively greater increase in the MB-C in the frequent and low till treatments during the post-wet season indicates that detritus of the microbes of the preceding wet season serves as readily available carbons for faster microbial growth. The lowest reduction in the MB-C during the wet season in the long term zero till treatment can either be due to greater proportion of anaerobes or adaptations of microbes to the rainfall induced compacted soil conditions (Jensen *et al.*, 1996); and presence of old micro-aggregates, which probably hold stable and recalcitrant organic carbon as evident from the highest SOC, but the lowest MB-C / SOC ratio (Hernandez and Lopez-Hernandez, 2002). MB-C / SOC ratio is found the highest (0.06) in the short term zero till and the lowest (0.03) in the long term zero till treatment. Kaiser *et al.* (1990) found that soil compaction results in a decrease in the activity and turnover of microbial biomass. MB-C is found positively correlated with the net soil N mineralization rate ($r = 0.746\text{-}0.859$, $P< 0.01\text{-}0.001$) and the net nitrification rate ($r = 0.750\text{-}0.906$, $P<0.01\text{-}0.001$) in all tillage treatments. This positive correlation between the net N mineralization and MB-C is, however, indicates that nitrogen mineralization is a steady state function of microbial carbon in the tillage treatments; and MB-C pool can be manipulated under tillage by incorporation of weed biomass in soils.

Implications of the soil tillage

Positive relation between the tillage frequency and net soil N mineralization in this study suggests that tillage may be used as a tool to manipulate soil fertility in the hot humid tropics. The highest SOC stock (34692 kg ha^{-1}), but the lowest mineralization rate in the long term zero till treatment, however, reflects that the long term zero tillage may be useful for soil carbon build up. Comparing the frequent till with the long term zero till, the highest (1190 kg ha^{-1} yr^{-1}) SOC is lost from the frequent till and the lowest (942 kg ha^{-1} yr^{-1}) from the long term zero till treatment. The nearly equal loss of SOC (981 kg ha $^{-1}$ yr $^{-1}$), but greater rate of net soil N mineralization in the low till than the short term zero till treatment in the present study supports the hypothesis that low tillage increases the rate of net soil N mineralization, but maintains SOC nearly equal to the short term zero tillage. Unexpectedly, we have found soil N mineralization during the dry season in all tillage treatments despite the lowest water content of soil. Gill *et al.* (1995) also have found that soil N mineralization occurs under low water content of soil. In our study, maximum amount of N mineralized during the dry season is likely to be lost during the following wet season. During the dry season plant's demand for the mineralized N is quite low like other parts of the tropics (Pandey *et al.*, 2007b). Runoff loss of mineral N is reported to be high during the wet season in the study region (Pandey and Venkatesh, 2007). But, the highest loss of the mineralized N is expected to occur from the frequent till and the lowest from the long term zero till treatment. However, despite nearly equal amount of SOC stock, greater rate of net N mineralization in the low till than short term zero till treatment indicates that the low tillage may be employed as a tool for management of soil fertility and built-up of organic carbon stock in the soils of the hot humid tropics.

Perennial cover crop and soil fertility management

Litter input from the cover crop

Productivity of commercial plantations and homegardens decrease with time due to declines in soil fertility caused by harvesting, loss of fine soil particles in runoff, and oxidation of soil organic carbon (Gajaseni and Gajaseni, 1999; Torquebiau, 1992; Szott *et al.*, 1999). In the commercial plantations and homegardens the most effective way to protect soil from erosion and to increase soil fertility is to establish a perennial cover crop such as *Pueraria phaseoloides* (popular name puero or tropical kadzu) and others (Schroth *et al.*, 2001). The cover crop adds nitrogen rich leaf and root litters to the soils over long periods (15-yr). Application of the phosphorus increased the vegetal cover of the crop, and inputs of the litters (Table 12)

Table 12. Organic matter (leaf and root litters), carbon and nitrogen inputs in the soils under different *Plueraria* cover crop treatments (No cover crop, NCC; No cover crop +P, NCC+P; Cover crop, CC; Cover crop +P, CC+P) in the equatorial humid tropical climate of the Andaman islands.

Treatment	Vegetal cover (%)	Production ($t\ ha^{-1}yr^{-1}$)			Input Carbon ($kg\ ha^{-1}$)	Nitrogen	C/N ratio ($kg\ ha^{-1}$)
		Leaf litter	Root litter	Total			
NCC	0	0	0	0	0	0	0
NCC+P	0	0	0	0	0	0	0
CC	108^{a}	1.24^{a}	0.86^{a}	2.10^{a}	1090^{a}	59.01^{a}	18.47^{a}
CC+P	162^{b}	1.63^{b}	1.01^{b}	2.64^{b}	1373^{b}	76.56^{b}	17.93^{b}

Data of a parameter suffixed with different superscript letters in a column is significant at $P<0.05$
Source: Pandey and Begum 2010

Soil-moisture and soil-temperature conservation in relation to cover crop

The soil-moisture across the seasons is 15% higher under the cover crop than the control (Fig.14). The difference in the soil-moisture between the cover crop and control treatments is the highest (29%) during the dry season and lowest (6%) during the wet season. Soil-moisture corresponds to amount of rainfall. The soil-moisture across the months and treatments is positively correlated to the rainfall amount ($r = 0.922$, $P<0.001$, $N= 36$).

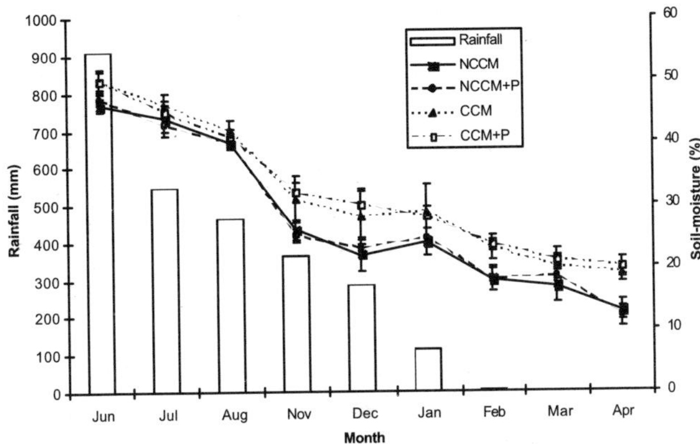

Fig.14. Seasonal rainfall amount and corresponding soil-moisture (M) in no cover crop (NCCM), no cover crop + P (NCCM+P), cover crop (CCM) and cover crop + P (CCM +P) treatments at the equatorial humid tropical climate of the Andaman Island. Bars represent ± ISE.

Source: Pandey and Begum 2010

Mineral N (NH_4^+-N and NO_3^- - N) pool in relation to the cover crop

The NH_4^+-N and NO_3^--N are significantly influenced by the perennial cover crop treatments and seasons. Compared to the control, average annual NH_4^+-N and NO_3^- -N increased by 40 % and 52 %, respectively in the soils under the cover crop (Fig.15). The phosphorus application increases the NH_4^+-N and NO_3^-- N in the no cover crop + phosphorus (NCC+P) treatment (19% and 13%, respectively) and cover crop + phosphorus (CC+P) treatment (27 % and 21%, respectively). The NH_4^+-N is the highest (14.2 to 26.8 µg g^{-1}) during the wet season and lowest (1.8 to 6.1 µg g^{-1}) during post-wet season in all the treatments, but the NO_3^- -N is the lowest (7.8 to 16.8 µg g^{-1}) during the wet and the highest (32.5 to 68.4 µg g $^{-1}$) during the post-wet season. The NH_4^+-N across the months and treatments is positively correlated to the rainfall amount (r = 0.786, P<0.001). But, the NO_3^- -N is inversely correlated to the rainfall amount and soil-moisture (r = -0.756 to 0.786, P<0.001).

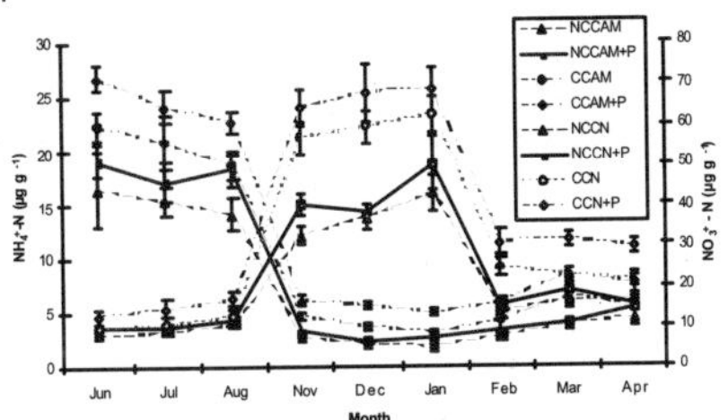

Fig. 15. Seasonal variation in NH_4^+-N (AM) and NO_3^--N (N) in no cover crop (NCCAM, NCCN), no cover crop + P (NCCAM+P, NCCN +P), cover crop (CCAM, CCN) and cover crop + P (CCAM +P, CCN +P) treatments at the equatorial humid tropical climate of the Andaman Island. Bars represent ± ISE.

Source: Pandey and Begum 2010

Net nitrification and net soil N mineralization in relation to the cover crop

The cover crop increases mean annual net nitrification and net soil N mineralization rates 40 % and 37 %, respectively compared to the control (Fig. 16 a, b). Application of the phosphorus to the cover crop (CC + P) increases the rate of net nitrification and net soil N mineralization further by 24% and 33%, respectively. Net nitrification as well as net soil N mineralization rate is the lowest (0.36 to 0.71 μg g^{-1} day^{-1} and 0.49 to 1.02 μg g^{-1} day^{-1}, respectively) during the wet season and the highest (1.12 to 1.8 μg g^{1} day^{-1} and 1.16 to 1.98 μg g^{1} day^{-1}, respectively) during the post-wet season. The phosphorus application to the soils (NCC + P treatment) does not increase the net nitrification and net soil N mineralization rates significantly during the wet season. But, the phosphorus application to the cover crop increases the net nitrification (16%) and net soil N mineralization (20%) rates during all the seasons. The net nitrification rate is positively correlated to the NH_4^+ pool (r = 0.380, $P< 0.05$). The net nitrification and net N mineralization rates are inversely correlated to the rainfall amount and soil-moisture (r = -0.698 to -0.795, $P<0.001$). The amount of N mineralized is higher under the cover crop and cover crop +P treatments by 39% and 73%, respectively than the control (Table 13). In the *ex-situ* experiment, net nitrification rate in the soils is the highest (1.24 μg g^{-1} day^{-1}) in 50 % field capacity (FC) and the lowest (0.03 μg g^{-1} day^{-1}) in 100% FC. On the contrary, net ammonification rate is the highest (0.40 μg g^{-1} day^{-1}) in 100% FC and the lowest (0.02 μg g^{-1} day^{-1}) in 25% FC. Addition of the phosphorus to the soils increases the net nitrification as well as net ammonification rate under all the soil water regimes, but their rates followed the pattern found in the soils without phosphorus (Table 14).

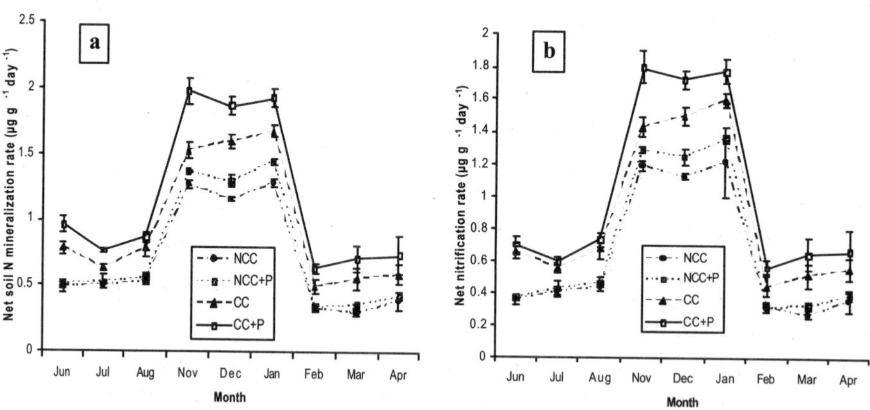

Fig. 16. Seasonal variation in **(a)** net soil N mineralization rate, and **(b)** net nitrification rate in no cover crop (NCC), no cover crop + P (NCC+P), cover crop (CC) and cover crop + P (CC+P) treatments at the equatorial humid tropical climate of the Andaman Island. Bars represent ± ISE. *Source:* Pandey and Begum 2010.

Table 13. Amount of nitrogen mineralized (kg ha^{-1} yr^{-1}) in soils under different cover crop treatments (No cover crop, NCC; No cover crop +P, NCC+P; Cover crop, CC; Cover crop +P, CC+P) in the equatorial humid tropical climate of the Andaman islands.

Season	Cover crop treatments			
	NCC	NCC+P	CC	CC+P
Wet	$^{a}93^{x}$	$^{a}97^{x}$	$^{b}135^{x}$	$^{c}172^{x}$
Post-wet	$^{a}114^{y}$	$^{b}126^{y}$	$^{c}147^{y}$	$^{d}178^{y}$
Dry	$^{a}31^{z}$	$^{a}33^{z}$	$^{b}48^{z}$	$^{c}61^{z}$
Annual	$^{a}238$	$^{b}256$	$^{c}330$	$^{d}411$

Data of a parameter prefixed with different superscript letters in a row is significant at $P<0.05$
Data of a parameter suffixed with different superscript letters in a column is significant at $P<0.05$
Source: Pandey and Begum 2010

Table 14. Net ammonification rate (NAR, µg g^{1} day $^{-1}$), net nitrification rate (NNR, µg g^{1} day $^{-1}$) and microbial biomass carbon (MB-C, µg g^{-1}) in soils with and without phosphorus under different soil-water regime in an *ex-situ* experiment in the equatorial humid tropical climate of the Andaman islands.

Field capacity	NAR		NNR		MB-C	
	Soil	Soil +P	Soil	Soil +P	Soil	Soil +P
100	$^{x}0.40^{a}$	$^{y}0.48^{a}$	$^{x}0.03^{a}$	$^{y}0.08^{a}$	$^{x}176^{a}$	$^{x}187^{a}$
50	$^{x}0.13^{b}$	$^{y}0.21^{b}$	$^{x}1.24^{b}$	$^{y}1.61^{b}$	$^{x}668^{b}$	$^{y}780^{b}$
25	$^{x}0.02^{c}$	$^{y}0.08^{c}$	$^{x}0.64^{c}$	$^{y}0.84^{c}$	$^{x}436^{c}$	$^{y}520^{c}$

Data of a parameter prefixed with different superscript letters in a row is significant at $P<0.05$
Data of a parameter suffixed with different superscript letters in a column is significant at $P<0.05$
Source: Pandey and Begum 2010

The higher rate of net soil N mineralization under the cover crop than the control occurs due mainly to the higher amount of SOC and lower C/N ratio of the soil. The leguminous leaf and root litters input with low C/N ratio for the long periods most likely increases the SOC and lowers the C/N ratio. The nitrogen rich litters (leaf and root) enhance carbon mineralization that lead to the lower C/N ratio (Persson *et al.*, 2000 a, b). The lower soil C/N ratio possibly reduces availability of carbon and thereby decreases microbial demand for nitrogen and result in the increase in the mineralization rate (Hart *et al.*, 1994). Schroth *et al.* (2001) report high mineralization rate in the soils under *Pueraria* cover, and under rubber trees (*Hevea brasiliensis*) with the cover crop, but they have observed low rate of mineralization under peach palm (*Bactris gasipaes*) and cupuacu (*Theobroma grandiflorum*) without the cover crop in central Amazonia. Steenwerth and Belina (2008) have found 2-4 times greater nitrification and N mineralization in cover crop soils than cultivation in a vineyard in California. Relatively greater (14.8%) proportion of the fine soil particles (clay and silt) also lends to the higher rate of soil N mineralization under the cover crop. Pandey and Chaudhari (2009) have found that the cover crop has potential to protect 10.4 t $ha^{-1}yr^{-1}$ fine soils from erosion loss at the study site. The greater proportion of fine soil particles under the cover crop probably increases adsorption sites as evident from the higher NH_4^+ pool in the soils. Fine soil particles are known to serve as adsorption sites to NH_4^+ (Pandey *et al.* 2000). Most likely the adsorbed NH_4^+ contribute to the increase in the rate of the net soil N mineralization. Several studies have shown that net soil N mineralization increases with decreasing particle size, possibly because fine soil particles have a narrower C:N ratio than coarse particles (Ladd *et al.*, 1977; Cameron and Posner, 1979).

Microbial biomass carbon (MB-C) in relation to the cover crop

The MB-C increases (41%) in the soils under the cover crop than the control (Fig. 17). Application of the phosphorus to the soils (NCC+P) and the cover crop (CC+P) increases (10 to 14 %) the MB-C, but the increase is 4 % higher in the CC+P than the NCC+P treatment. Like the net N mineralization rate, the MB-C is inversely correlated to the rainfall amount ($r = -0.662$, $P<0.001$). In the ex-situ experiment, MB-C is the lowest in the 100 % FC and the highest in 50 % FC. The MB-C is correlated to the net soil N mineralization rate ($r = 0.555$, $P<0.001$). These indicate that application of phosphorus increases growth in the cover crop indicating that the studied soils are phosphorus limiting. The increased growth result in greater input of readily hydrolysable carbon (Szott *et al.*, 1999), and thereby stimulate increase in the MB-C. Microbial biomass C is known to serve both as a source and sink (Singh *et al.*, 1989). The MB-C, being a labile fraction of SOC mineralizes quickly and increases the rate of mineralization as evident from the positive correlation between the MB-C and net soil N mineralization rate. Positive correlation between MB-C and net soil N meneralization is reported to occur in several ecosystems (Williams and Sparling, 1988; Pandey *et al.*, 2009). Increase in the MB-C simultaneously with the increase in the net soil N mineralization when the phosphorus is applied directly to the soils in the NCC + P treatment further indicated that microbial biomass mediated the net soil N mineralization in the soils. The *ex-situ* experiment also confirmed the results found in the field. We found 0.8 times increase in the mineralized N for a unit amount of P application to the soil, and 3.3 times increase in the mineralized N for a unit amount

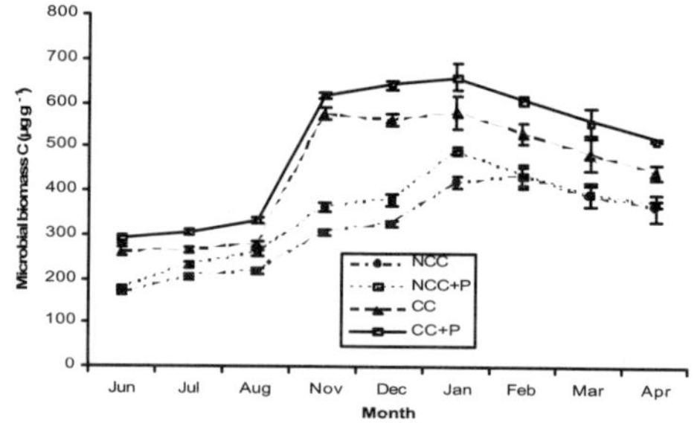

Fig. 17. Seasonal variation in microbial biomass carbon (MB-C) in no cover crop (NCC), no cover crop + P (NCC+P), cover crop (CC) and cover crop + P (CC+P) treatments at the equatorial humid tropical climate of the Andaman Island. Bars represent ± ISE.

Source: Pandey and Begum 2010

of P application to the cover crop in our P-limited soils. These observations indicate that nitrogen mineralization in the soils limited in P could be manipulated through phosphorus and the cover crop + P application. White and Reddy (2000) have found that P loading to oligotrophic soils of everglades of Florida increases soil N mineralization. They have observed that the P application to the oligotrophic soils increases heterotrophic activity, but total microbial biomass does not increase in short time. But, in humid tropics an increase in MB-C pool as well as net soil N mineralization rate due to the phosphorus application increases in 8 days. It indicates that both microbial population and microbial activity increases due to the phosphorus application in P-limited soils.

It is observed that during the wet season P application does not increase the net soil N mineralization in the NCC, but it does in CC treatment. No increase in the net soil N mineralization in the NCC treatment occur due to reduced size of microbial population and no response of the reduced population to the applied phosphorus in presence of plenty of native P (total P) during the anaerobic conditions. Pandey *et al.* (2009) have found that the Entisols soils of the study region contains a very high amount of iron which make the native P unavailable to plants and microbes during oxic conditions. But, during the anoxic conditions ferric-oxide is reduced to ferrouos-oxide complex of P and make the native P available to plants and microbes. The *ex-situ* experiment also conforms the results found in the field conditions as MB-C and net N mineralization are the highest in 50% FC and the lowest in 100% FC. In the CC and CC +P treatments, greater NH_4^+ accumulation during the wet season occurs mainly due to greater mineralization of organic N added via the nitrogen rich cover crop leaf and root litters (Sinsabaugh *et al.*,1991; Gardner *et al.*, 1989).

Available P in homegardens and native forests

Slightly acidic soils of the Andamans are highly rich in iron, which fixes phosphorus and makes it unavailable to plants and microbes. But, on the contrary *Dipterocarpus* tropical rainforests are found in the islands. This makes it interesting to know how phosphorus becomes available to plants and microbes. To know this, soils were sampled under two rainfall conditions *i.e.* 12 incessant rainfall (Sept. 9, 12, 13, 22; Oct. 3, 6, 13, 21, 24, 28; Nov. 3, 17) and 12 dry spell events (48 h to 72 h after rainfall) (Sept. 19, 26, 30; Nov. 7, 10, 14; Dec. 28; Jan 12; Feb. 15, 18, 21 and 24). Available P is found to range from 0.05 to 1.74 $\mu g\ g^{-1}$ in the moist evergreen forest, 0.05 to 2.02 $\mu g\ g^{-1}$ in semi-evergreen forest and 0.08 to 2.08 $\mu g\ g^{-1}$ in homegarden, and is affected due to the land use systems, rainfall conditions and their interactions. Averaged across the rainfall conditions, it is the highest in the homegarden and the lowest in the moist-evergreen forest. Availability of the P, across the land use systems, increases about 5 to 6 times during the incessant rainfall compared to that during the dry spell (Fig. 18 a). However, this increase is the highest in the homegarden and the lowest in the moist- evergreen forest. During the dry spell concentrations of P in the soils does not differ among the land use systems. Water filled pore space (WFPS) varies due to the rainfall conditions. It is more than two times higher during the incessant rainfall than that during the dry spell in all land use systems (Fig. 18 b). Available P is positively correlated with the WFPS in the moist-evergreen forest (r = 0.995, P<0.001), semi-evergreen forest (r = 0.908, P< 0.001) and homegarden (r = 0.948, P< 0.001). The available P is also positively correlated to the WFPS (Fig. 19).

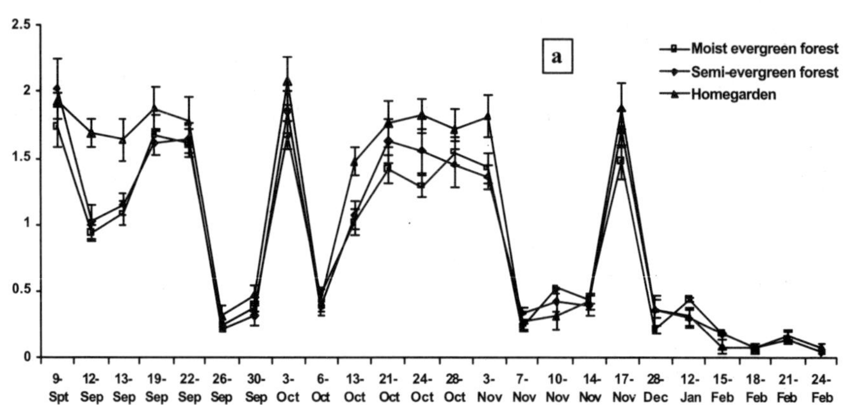

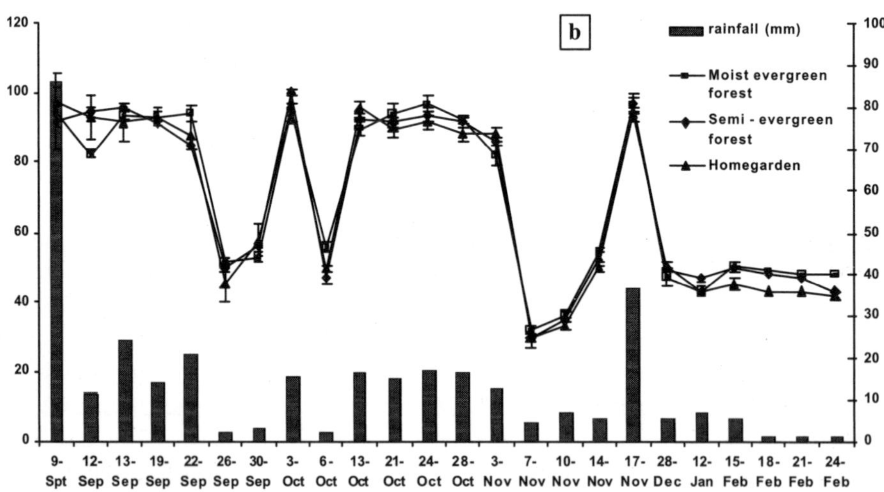

Fig. 18. (a) available P in the soils under different land use systems during incessant rainfall (Sept.9, 12, 13, 22; Oct.3, 6, 13, 21, 24, 28; Nov.3, 17) and dry spell events (Sept 19,26,30;Nov.7, 10, 14; Dec. 28; Jan. 12; Feb. 15,18,21,24); **(b)** Water filled pore space (WFPS) in the soils, relation to rainfall amount, in different land use systems in the equatorial humid tropical climate of the Andaman islands. Bars represent ± ISE. *Source:* Pandey and Srivastava 2009

Higher availability of phosphorus in the soils during the incessant rainfall compared to that during the dry spell invariably in all the land use systems may be attributed mainly to the rainfall. The incessant rainfall increases water filled pore space (WFPS) that probably make the soil anaerobic, which reduces redox potential of the soils. The reduction in the redox potential probably reduces ferrous oxide complexes that facilitate availability of phosphorus. High WFPS is known to restrict gas diffusion in soils and limit resupply of oxygen from the atmosphere to the soils (Smith and Tiedje, 1979) and thus makes the soils anaerobic (Hobbie *et al.*, 2000). Linn and Doran (1984) have found a linear increase in relative microbial respiration with increasing water content to a maximum of 60 % WFPS, however, after it the microbial respiration declines in response to increasing anaerobic conditions. Yavitt and Wright (2002) in a 5-yr experimental irrigation in lowland Panama have found that it increases permanent charge and cation retention that lead to less sorption of phosphorus and thus increases P mobility. Positive relation between WFPS and available P in humid tropics indicate that less sorption of the phosphorus can be another reason for the greater availability of phosphorus during the incessant rainfall. Wright *et al.* (2001) in a long term (6-month) artificial flooding experiment found that P availability increases significantly due to the flooding; however, it decreases following drainage in a Georgia floodplain forest.

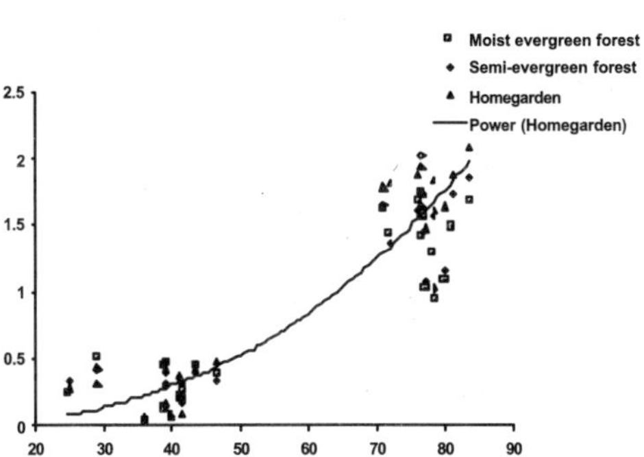

Fig. 19. Relationship between available P (y, μg g[1]) and water filled pore space (X, %): Y= 0.0002 X 2.1205; r^2 = 0.7477, P<o.oo1) across the different land use systems in the equatorial humid tropical climate of the Andaman islands. Bars represent ± ISE. *Source:* Pandey and Srivastava 2009

Following cessation of the incessant rainfall, the WFPS declines significantly within 2-hr in the upper layer (0-5 cm) in all the land use systems. However, in the deeper layer (5-15 cm) it does not decline significantly within 12-hr (Table 15). Quick decline in WFPS (within 2-hr) in the soils following the cessation of rainfall indicate that it probably facilitate uptake of the available P through aerobiosis to plants and microorganisms, which is evident from higher amount of microbial biomass P during the dry spell. Ability of microorganisms to effectively utilize P adsorbed on Fe and Al minerals has been recognized using culture media isolated from weathered soils (Shang *et al.*, 1996; He and Zhu, 1998). Studies of P cycling in microcosms using ^{32}P labeled bacterial cells have indicated that turn over of microbial P is greater than P mineralization from microbial detrital materials (Salas *et al.*, 2003). Noemi *et al.* (2006) have found that plant and microbial P demand serve as an important sink for released P, limiting the role of P re-adsorption in a humid tropical forest soil in Venezuela. They have observed that rates of P solubilization exceeded re- adsorption when labile carbon substrate is added. Uptake of nutrients from unsaturated soils and capillary fringes is also reported to occur during high rainfall (Fail *et al.*, 1986).

Table 15. Changes in water fillled pore space (WFPS) in the soils under different land use systems after cessation of incessant rainfall in the equatorial humid tropical climate of the Andaman Islands.

Land use system	Soil depth (cm)	WFPS (%)						
		0hr	2hr	4hr	6hr	8hr	10hr	12hr
Moist evergreen forest	0-5	$^{a}84.60^{x}$ ± 2.49	$^{b}81.30^{x}$ ± 1.02	$^{b}79.86^{x}$ ± 1.33	$^{bc}75.81^{x}$ ± ±1.63	$^{cd}73.84^{x}$ ± 1.83	$^{de}72.95^{x}$ + 1.67	$^{ef}71.53^{x}$ ± 1.06
	5-15	$^{a}85.20^{x}$ ± 2.35	$^{a}85.11^{y}$ ± 1.39	$^{a}83.34^{y}$ ± 1.46	$^{a}83.40^{y}$ ± 1.95	$^{a}82.17^{y}$ ± 1.46	$^{a}81.30^{y}$ ± 1.33	$^{a}80.86^{y}$ ± 1.42
Semi- evergreen forest	0-5	$^{a}80.80^{x}$ ± 2.08	$^{b}76.24^{x}$ ± 1.47	$^{b}74.81^{x}$ ± 1.01	$^{bc}67.97^{x}$ ± 1.76	$^{bc}66.41^{x}$ ± 2.19	$^{bc}66.27^{x}$ ± 5.33	$^{bc}65.54^{x}$ ± 3.39
	5-15	$^{a}82.70^{x}$ ± 3.39	$^{a}82.50^{y}$ ± 1.48	$^{a}81.50^{y}$ ± 1.24	$^{a}80.01^{y}$ ± 1.78	$^{a}77.30^{y}$ ± 4.03	$^{a}77.00^{y}$ ± 3.84	$^{a}76.80^{y}$ ± 4.77
Homegarden	0-5	$^{a}83.09^{x}$ ± 2.76	$^{b}80.19^{x}$ ± 1.28	$^{b}77.39^{x}$ ± 1.17	$^{bc}73.88^{x}$ ± 1.13	$^{bc}71.79^{x}$ ± 5.34	$^{bc}70.48^{x}$ ± 4.97	$^{bc}70.17^{x}$ ± 3.83
	5-15	$^{a}83.09^{x}$ ± 2.66	$^{a}83.68^{y}$ ± 1.05	$^{a}82.70^{y}$ ± 1.56	$^{a}80.49^{y}$ ± 1.81	$^{a}79.50^{y}$ ± 2.67	$^{a}78.40^{y}$ ± 2.33	$^{a}77.80^{y}$ ± 1.18

Values in a column for a land use system suffixed with different letters are significantly different at P<0.05

Values in a row for a land use system prefixed with different letters are significantly different at P<0.05

Source: Pandey *et al.* 2009

Conclusions

Homegarden trees like coconut palm and spice trees like clove and nutmeg increase organic C (including microbial biomass C), total N and P, mineral N and P, and exchangeable K, Ca and Mg under their canopies through continuous organic matter via leaf litter and dead roots in humid tropics. These increases most likely occur over a long period of time, but greater under the spice trees compared to that under the palm trees. Concentrations of the nutrients in the litters, and pattern of accumulation of the nutrients under the homegarden trees are similar to that of the native forest. However, seasonal variations in mineral N pools (NH_4^+-N and NO_3^--N) as well as net N mineralization and net nitrification rates are regulated greater by rainfall amount than temperature. Moderate rainfall and low temperature interactively together facilitate higher rates in the nitrification and soil N mineralization during the post-wet season, but the highest under the nutmeg trees the richest microhabitat and the lowest under the coconut trees the poorest microhabitat. Incessant high rainfall during the wet season makes soils anaerobic, which reduces redox potential of the soils and thereby causes reduction in net nitrification rate. Like the net nitrification rate, microbial biomass C also declines under the anaerobic condition during the incessant rainfall. But, simultaneously the incessant high rainfall promotes accumulation of NH_4^+ -N in the soils which resists leaching losses and thus conserve nitrogen. However, during dry spells when O_2 is available (low WFPS), the accumulated NH_4^+ -N mineralizes with simultaneous increase in microbial biomass C. Reduction in the nitrification rate with simultaneous increase in NH_4^+- N during the incessant rainfall may be regarded as an important mechanism of soil N conservation in the high rainfall regime of humid tropics. The reduces redox potential of the soils due to the incessant rainfall promotes precipitation of ferrous oxide complexes of phosphorus and releases fixed phosphorus in iron rich soils. The WFPS declines quickly within 2-hr in upper layer (0-5 cm) after cessation of the incessant rainfall and facilitates uptake of the released phosphorus. Perennial cover of *Pueraria* protects fine soil particles from erosion, and adds high quality leaf litter with a low C/N ratio and root litter to soils for long periods. The high quality litter builds up SOC, increases MB-C and lowers the C/N ratio of the soils. The fine soil particles adsorb the NH_4^+ resulting from the organic N (litter) mineralization. The decline in the soil C/N ratio increases net soil N mineralization together with increased MB-C and adsorbed NH_4^+ enhances the net soil N mineralization rate compared to the control. P application to the cover crop increases its cover, which increases inputs of N rich leaf and root litter, the readily mineralizable substrates. The application of P simultaneously stimulates microbial biomass pool and microbial activity, which further increases the rate of soil N mineralization. These results suggest that a perennial cover crop can be used to improve soil fertility in coconut plantations, home gardens and other similar plantations in humid tropics.

References

Ahl, C., Joergensen, R.G., Kandeler, E., Meyer, B., Woehler, V., 1998. Microbial biomass and activity in silt and sand loams after long-term shallow tillage in central Germany. Soil Till. Res. 49:93-104.

Alvarez, R., Alvarez, C.R., Lorenzo, G., 2001. Carbon dioxide fluxes following tillage from a mollisol in the Argentine rolling pampa. Eur. J. Soil Biol. 37:161-166.

Andrade, G., Linderman, R.G., Bethlenfalvay, G.J., 1998. Bacterial associations with the mycorrhizoshere and hyposphere of the arbuscular mycorrhizal fungus Glomus mosseae. Plant Soil 202:79-87.

Barraclough, D. 1995. ^{15}N isotope dilution techniques to study soil nitrogen transformations and plant uptake. Fert. Res. 42:185-192.

Barrios, E., Kwesiga, F., Buresh, R.J., Sprent, J.I., 1997. Light fraction soil organic matter and available nitrogen following trees and maize. Soil Sci. Soc. Am. J. 61: 826-831.

Basic Statistics, 2001. Andaman and Nicobar Islands: Basic statistics. Directorate of Economics and statistics, Andaman and Nicobar Administration, Port Blair, India.

Bauhus, J., Barthel, R., 1995. Mechanisms for carbon and nutrient retention in beech forest gaps. II the role of microbial biomass. Plant Soil 168-169: 585-592.

Beare, M.H., Hendrix, P. F., Coleman, D. C., 1994. Water-stable aggregates and organic fractions in conventional and no-tillage soils. Soil Sci. Soc. Am. J. 58:777-786.

Boerner, R.E.J., Koslowsky, S.D. 1989. Microsite variations in soil chemistry and nitrogen mineralization in a beech- maple forest. Soil Biol. Biochem. 21:795-801.

Bonde, T.A., Schnurer, J., Rosswall, T. 1988. Microbial biomass as a fraction of potentially mineralizable nitrogen in soils from long term field experiments. Soil Biol. Biochem. 20:447-452.

Bremner, J.M., 1965. Total nitrogen 8. In: Black, A., (Ed.), Methods of Soil Analysis, Agronomy 9. Pat2, Am. Soc. Agron, Madison, WI, pp.1149-1179.

Breuer, L.R., Kiese, Butterbach – Bahal K. 2002. Temperature and moisture effects on nitrification rates in tropical rain forest soils. Soil Sci. Soc. Am. J. 60:834- 844.

Browaldh, M. 1995. The influence of trees on nitrogen dynamics in an agrisilvicultural system in Sweden. Agrofor. Syst. 30:301-313.

Brown S. Lugo, A., 1982. The storage and production of organic matter in tropical forests and their role in the global carbon cycle. Biotropica 14: 161-187.

Brown, S. Lugo, A. E., 1990. Effects of forest clearing and succession on the carbon and nitrogen content of soils in Puerto Rico and Us Virgin Islands. Plant and Soil 124:53-64.

Cameron, R.S., Posner, A.M. 1979. Mineralizable organic nitrogen in soil fractionated according to particle size. J. Soil Sci. 30: 565-577.

Campbell, L.R., Greaves, M. P., 1990. Anatomy and community structure of the rhizosphere. In: Lynch, J.M. (Ed.) The rhizosphere. Chichester, John Wiley and Sons, pp.11-34.

Chadwick, O.A., Derry, L. A., Vitousek, P.M., Herbert, B.J., Hedin, L.O., 1999. Changing sources of nutrients during four million years of ecosystem development. Nature 397:491-497.

Chandler, G., 1985. Mineralization and nitrification in three Malaysian forest soils. Soil Biol. Biochem.17:347-353.

Chandler, G.E., Goosem, S. , 1982. Aspects of rainforest regeneration. III. the interaction of phenolics, light and nutrients. New Phytol. 92:369-380.

Clay, D.E., Schumacher, T. E., Brix-Davis, K. A., 1995. Carbon and nitrogen mineralization in row and interrow areas of chisel and ridge tillage systems. Soil Till. Res. 35:167-174.

Cleveland, C.C., Townsend, A.R., Schmidt, S.K., 2002. Phosphorus limitation of microbial processes in moist tropical forests: evidence from short term laboratory incubations and field studies. Ecosystems 5:680-691

Cole, C. V., Burke, I.C., Parton, W. J., Schimei, D.S., Stewart, J. W. B., 1989. Analysis of historical changes in soil fertility and organic matter levels of the North American Great Plains. pp. 436-438, *In:* Unger, P. W., Sneed, T. V., Jensen, R. W., (Eds.) Proceedings of the international conference on dryland farming, Bushland, Texas, August, 1988. Texas A and M University, College Station, Texas, USA.

Cunningham, R.K., 1963. The Effect of Clearing a Tropical Forest Soil. J. Soil Sci. 14:334-345.

Davidson, E.A., Matson, P. A., Vitousek, P.M., Riley, R., Dunkin, K., Garcia- Mendez G., Maass J. M.,1993. Processes Regulating Soil Emissions of NO and N_2O in a Seasonally Dry Tropical Forest. Ecology 74:130-139.

Devevre, O.C.; Horwath, W. R., 2000. Decomposition of rice straw and microbial carbon use efficiency under different soil temperatures and moistures. Soil Biol. Biochem. 32:1773-1785.

Doran, J.W., Power, J. F., 1983. The effects of tillage on the nitrogen cycle in corn and wheat production. pp. 441-445. *In:* Lowrance, R. R., Todd, R. L., Asmussen, L. E., Leonard, R. A. (Eds.). Nutrient cycling in agroecosystems. University of Georgia College of Agriculture Experiment Station Special Publication 23, Athens, Georgia, USA.

Dorioz, J.M., Robert, M., Chenu, C., 1993. The role of roots, fungi, and bacteria on clay particle organization. An experimental approach. Geoderma 56:179-194.

Duiker, S. W., Lal, R., 1999. Crop residue and tillage effects on carbon sequestration in a Luvisol in central Ohio. Soil Till.Res. 52:73-81.

El Titi, A., 2003 a. Implications of soil tillage for weed communities. In: El Titi, A. (Ed.), Soil Tillage in Agroecosystems. CRC Press. Boca Raton, USA, ISBN:0-8493-1228-0, pp. 147-185.

El Titi, A., 2003 b. Non-inversion tillage in integrated farming concepts: prospects and constraints of cropping systems in the southwest of Germany. In: Garcia-Torres, L., Benites, J., Martinez-Vilela, A., Holgado-Cabrera, A. (Eds.), Conservation Agriculture, Environment, Farmers Experiences, Innovations, Socio-ecinomy, Policy. Kluwer Academic Publishers, Dordrecht, The Netherlands, ISBN:1-4020-1106-7, pp. 211-219.

Fail, J.C., Hamzah Jr., M. N., Haines, B. L., Todd, R.L. 1986. Above and below ground biomass production, and element accumulated in riparian forests of an agricultural watershed. In: Correll, D. L. (Eds.), Watershed Research Perspectives. Smithsonian Institute Press., Washington. DC.

Fankem, H., Nwaga, D., Deubel, A., Dieng, L., Merbach, W., Etoa, F. X., 2006. Occurrence and functioning of phosphorus solubilizing microorganisms from oil palm tree (*Elaeis guineensis*) rhizosphere in Cameroon. African J Biotech. 5(24):2450-2460.

Fenn, M.E., M. A. Poth, P. H. Dunn, S. C. Barro. 1993. Microbial N and biomass, respiration and N mineralization in soils beneath two chaparral species along a fire-induced age gradient. Soil Biol. Biochem. 25(4):457-466.

Forest Working Plan, (1980) Working plan for the Middle Andaman forest division. Port Blair, India

Gajaseni, J., Gajaseni, N., 1999. Ecological rationalities of the traditional home garden system in the Chao Phrya Basin, Thailand. Agrofor. Syst. 46:3-23.

Gardner, W.S., Chandler, J. F., Laird, G. A., 1989. Organic nitrogen mineralization and substrate limitation of bacteria in Lake Michigan. Limnol. Oceanogr. 34:478-485.

Gill K.; Jarvis, S.C.; Hatch, D. J., 1995. Mineralization of nitrogen in long-term pasture soils: effects of management. Plant and Soil, 172:153-162.

Giller, K.E., Wilson, K.J., 1991. Nitrogen Fixation in Tropical Cropping Systems. CAB International, Wallingford, 313 pp.

Grace, P. R., Mac Rae, I.C., Myers, R. J. K., 1993. Temporal changes in microbial biomass and N mineralization under simulated field cultivation. Soil Biol. Biochem. 25:1745-1753.

Grubb, P.J., 1977. Control of forest growth and distribution on wet tropical mountains: with special reference to mineral nutrition. Ann. Rev. Ecol. Syst. 8:83-107.

Hall, S. J.; Matson, P. A., 2003. Nutrient status of tropical rain forests influences soil N dynamics after N additions. Ecolog. Monogra. 73:107-129.

Hart, S.C., Stark, J.M., Davidson, E.A., Firestone, M.K., 1994. Nitrogen mineralization, immobilization, and nitrification. In: Bigham, J. M. and Mickelson, S.H. (Eds.), Methods of Soil Analysis. Part 2. Microbiological and Biochemical properties. Soil Science Society of America, Madison, WI, USA, pp.1006-1014.

He Z, Zhu J., 1998. Microbial utilization and transformation of phosphate adsorbed by various charge minerals. Soil Biol Biochem 30:917-923.

Hendrix, P.F., Chun-Ru, H., Groffman, P. M., 1988. Soil respiration in conventional and no-tillage agroecosystems under different winter cover crop rotations. Soil Till. Res. 12:135-148.

Herbert, D.A., Fownes, J. H., 1995. Phosphorus limitation of forest leaf area and net primary production on a highly weathered soil. Biogeochem. 29:223-235.

Herman, D.J., Johnson, K.K., Gaeger C.D. III, Schwartz, E., Firestone, M.K., 2006. Root influence on nitrogen mineralization and nitrification in *Avena barbata* rhizosphere soil. Soil Sci. Soc. Am. J. 70:1504-1511.

Hernandez, R.M., Lopez-Hernandez, D., 2002. Microbial biomass, mineral nitrogen and carbon content in savanna soil aggregates under conventional and no-tillage. Soil Biol. Biochem. 34:1563-1570.

Hobbie, S.E., Schimel, J. P., Trumbore, S. E., Randerson, J. R., 2000. A mechanistic Understanding of carbon storage and turnover in high- latitude soils. Glob. Change Biol. 6 (supplimentary1) : 196-210.

Holland, E.A.; Detling, J. K., 1990. Plant response to herbivory and belowground N cycling. Ecology 71:1240-1049.

Jacobs, T.C., Gilliam, J. W., 1985. Riparian losses of nitrate from agricultural drainage waters. J. Environ. Qualif. 14: 472-478.

Jaeger, C. H. III, Lindow S. E., Miller, W. Clark, E., Firestone, M. K., 1999. Mapping of sugar and amino acid availability in soil around roots with bacterial sensors of sucrose and tryptophan. Appl. Environ. Microbiol. 65:2685-2690

Jensen, L.S., McQueen, D. J., Shepherd, T. G., 1996. Effects of soil compaction on N-mineralization and microbial-C and –N. I. Field Measurements. Soil Till. Res. 38:175-188.

Jensen, M., 1993. Productivity and nutrient cycling of a Javanese home gardens. Agrofor. Syst. 24: 187-201.

Joergensen, R. G., Brookes, P. C., Jenkinson, D. S., 1990. Survival of the soil microbial biomass at elevated temperatures. Soil Biol. Biochem. 22: 1129-1136.

Jordan, C. F., 1985. Nutrient cycling in tropical forest ecosystems, Wiley, New York

Kaiser, E.A., Walenzik, G., Heinemeyer, O., 1990. The influence of soil compaction in decomposition of plant residues and on microbial biomass. In: W. S. Wilson (Ed.), Advances in Soil Organic Matter Research: The Impact of Agriculture and the Environment. Royal Society of Chemistry, Cambridge, pp 207-216.

Kamara, C.S., Haque, I., 1992. *Faidherbia albida* and its effects on Ethiopian highland vertisols. Agrofor. Syst. 18:17-29.

Kelley, K. R., Stevenson, F. J., 1995. Farms and nature of organic N in soil. Fert. Res. 42:1-11.

Kemmitt, S.J., Lanyon, C.V., Waite, I. S., Wen, Q., Addiscott, T. M., Bird, N. R. A., O'Donnell, A.G., Brookes, P.C., 2008. Mineralization of native soil organic matter is not regulated by the size, activity or composition of the soil microbial biomass-a new perspective. Soil Biol. Biochem. 40: 61-73.

Kingery, W.L., Wood, C. W., Williams, J. C., 1996. Tillage and amendment effects on soil carbon and nitrogen mineralization and phosphorus release. Soil Till. Res. 37:239-250.

Kristensen, H.L., McCarty, G. W., Meisinger, J. J., 2000. Effects of soil structure distribution on mineralization of organic soil nitrogen. Soil Sci. Soc. Am. J. 64:371-378.

Ladd, J. N., Parsons, J.W., Amato, M., 1977. Studies of nitrogen immobilization and mineralization in calcareous soil – I. Distribition of immobilized nitrogen amongst soil fractions of different particle size and density. Soil Biol. Biochem. 9: 309-318.

Leinweber, P., Schulten, H. R. Korschens, M., 1995. Hot water extracted organic matter: chemical composition and temporal variations in a long- term field experiment. Biol. and Fert. Soils 20:17-23.

Lindwall, C.W., Larney, F. J., Carefoot, J. M., 1995. Rotation tillage and seeder effects on winter wheat performance and soil moisture. Can J. Soil Sci.,75:109-116.

Linn, D.M., Doran, J.W., 1984. Aerobic and anaerobic populations in no-till and plowed soils. Soil Sci. Soc. Am. J. 48:794-799.

McClaugherty, C.A., Pastor, J., Aber, J.D., Melillo, J.M. 1985. Forest litter decomposition in relation to soil N dynamics and litter quality. Ecology 66: 266-275.

Montagnini, F., Haines, B., Swank, W.,1989. Factors controlling nitrification in soils of early successional and Oak/Hickory forests in the Southern Appalachians. For. Ecol. and Manag. 26:77-94.

Nadelhoffer, K.J., Aber, J.D., Melillo, J.M., 1985. Fine roots, net primary production and soil nitrogen availability: a new hypothesis. Ecology, 66:1377-1390.

NBSS LUP, 1991. Soil Resource Atlas: Andaman and Nicobar Islands. National Bureau of Soil Survey and Land Use Planning (ICAR) and Directorate of Agriculture, Andaman and Nicobar Administration, Port Blair, India.

Neill C., Piccolo M.C., Steudler P. A., Melillo J. M., Feigl B. J., Cerri C. C., 1995. Nitrogen dynamics in soils of forests and active pastures in the western Brazilian Amazon Basin. Soil Biol. Biochem. 27:1167-1175.

Neill, C.; Piccolo, M.; Cerri, C.; Steudler, P.; Melillo, J.; Brito, M., 1997. Net nitrogen mineralization and net nitrification rates in soils following deforestation for pasture across the southwestern Brazilian Amazon Basin landscape. Oecologia 110:243-252.

Noemi, C., Whendee, S., Eric, D., Daniela, C. 2006. Iron reduction and soil phosphorus solubilization in humid tropical forests soils: the roles of labile carbon pools and an electron shuttle compound. Biogeochem. 78: 67-84.

Nunan, N.; Morgan, M.A.; Scott, J.; Herlihy, M., 2000. Temporal changes in nitrogen mineralization, microbial biomass, respiration and protease activity in a clay loam soil under ambient temperature Biology and Environment: Proceedings of the Royal Irish Acade. 100B:107-114.

Pandey C. B., Rai R.B., Singh L. and Singh A. K., 2007 a. Homegardens of Andaman and Nicobar, India. Agric. Syst. 92: 1-22.

Pandey, C.B., Singh, G.B., Singh, S.K., Singh, R.K., 2010a. Soil nitrogen and microbial biomass carbon dynamics in native forests and drived agricultural land uses in a humid tropical climate of India. Plant Soil 333:453-467.

Pandey, C.B., Chadhari, S.K., Dagar, J.C., Singh, G.B., Singh, R.K., 2010 b. Soil N mineralization and microbial biomass carbon affected by different tillage levels in a hot humid tropics. Soil Till. Res. 110:33-41.

Pandey, C.B., Begum, M., 2010. The effect of a perennial cover crop on net soil N mineralization and microbial biomass carbon in coconut plantations in the bhumid tropics. Soil use Manage. 26:158-166.

Pandey, C.B., Chaudhari, S.K. 2010. Soil and nutrient losses from different land uses; and vegetative methods for their control on hilly terrain of South Andaman. Indian J.Agri.Sci. 80(5):50-55.

Pandey, C.B., Rai, R.B., Singh, L., 2007b. Seasonal dynamics of mineral N pools and N-mineralization in soils under homegarden trees in South Andaman, India. Agrofor. Syst. 71:57-66.

Pandey, C.B., Singh, A. K., Sharma, D. K., 2000. Soil properties under *Acacia nilotica* trees in a traditional agroforestry system in Central India. Agrofor. Syst. 49:53-61.

Pandey, C.B., Singh, J. S., 1992. Influence of rainfall and grazing on belowground biomass dynamics in a dry tropical savanna. Can. J. Bot. 70:1885-1890.

Pandey, C.B., Singh, L., 2009. Soil fertility under homegarden trees and native moist evergreen forest in South Andaman, India. J. Sustaina. Agric. 33(3):303-318.

Pandey, C.B., Srivastava, R.C., 2009. Plant available phosphorus in homegarden and native forest soils under high rainfall in an equatorial humid tropics. Plant Soil, 316:71-80.

Pandey, C.B., Srivastava, R.C., Singh, R.K., 2009. Soii nitrogen mineralization and microbial biomass relation, and nitrogen conservation in humid-tropics. Soil Sci. Soc. Am. J. 73: 1142-1149.

Pandey, C.B., Venkatesh, A., 2007. Agroforestry for Sustainable Biomass Production in the Andaman and Nicobar Islands. Technical Report, Central Agricultural Research Institute, Port Blair, India, 147 pp.

Paul, E.A., Clark, F.E., 1996. Soil microbiology and biochemistry. Academic Press, San Diego.

Persson, T., Karlsson, P.S., Seyferth, U., Sjoberg, R.M., Rudebeck, A., 2000a. Carbon mineralization in European forest soils. *In:* Schulze E.D. (Ed.), Ecological studies- carbon and nitrogen cycling in European forest ecosystems. Springer, Berlin, pp.257-275.

Persson, T., Rudebeck, J.H., Jussy, J. H., Colin-Belgrand, M., Prieme, A., Dambrine, E., Karlsson, P.S., Sjoberg, R. M., 2000 b. Soil nitrogen turnover-mineralization, nitrification and denitrification in European forest soils. *In:* Schulze, E. D., (Ed.), Ecological studies-carbon and nitrogen cycling in European forest ecosystems. Springer, Berlin, pp.295-331.

Piccolo, M.C.; Neill, C.; Cerri, C. C., 1994. Net nitrogen mineralization and net nitrification along a tropical forest- to- pasture chronosequence. Plant and Soil 162:61-70.

Pinkerton, A., Simpson, J.R., 1986. Interactions of surface drying and subsurface nutrients affecting plant growth on acidic soil profiles from an old pasture. Australian J. Exp. Agric. 26:681-689.

Powers, R.F., 1990. Nitrogen mineralization along an altitudinal gradient: Interaction of soil temperature, moisture, and substrate quality. For. Ecol. and Manag. 30: 19-29.

Radulovich, R., Sollins P., 1991. Nitrogen and phosphorus leaching in zero-tension drainage from a humid tropical soil. Biotropica 23:84-87.

Ralte, V.; Pandey, H. N.; Barik, S. K.; Tripathi, R. S.; Prabhu, S. D., 2005. Changes in microbial biomass and activity in relation to shifting cultivation and horticultural practices in sub tropical evergreen forest ecosystem of north- east India. Acta Oecol, 28:163-172.

Raubuch, M., Joergensen, R.G. 2002. Net N mineralization in a coniferous forest soil: the contribution of the temporal variability of microbial biomass C and N. Soil Biol. Biochem. 4:841-849.

Raubuch, M., Joergensen, R.G., 2002. C and net N mineralization in a coniferous forest soil: the contribution of the temperal variability of microbial biomass C and N. Soil Biol. Biochem. 34:841-849.

Rhoades, C.C., 1997. Single-tree influence on soil properties in agroforestry: lessons from natural forest and savanna ecosystems. Agrofor. Syst. 35:71-94.

Rice, C. W., Smith, M. S., Blevins, R. L., 1986. Soil nitrogen availability after long-term continuous no-tillage and conventional tillage corn production. Soil Sci. Soc. Am. J. 50:1206-1210.

Richards, N. 1987. The microbiology of terrestrial ecosystems. Longman, Essex.

Rinklebe, J., Langer, U., 2006. Microbial diversity in three floodplain soils at the Elbe River (Germany). Soil Biol. Biochem. 38:2144-2151.

Roberston, G.P. 1984. Nitrification and nitrogen minerlization in a lowland rainforest succession in Costa Rice, Central America. Oecologia 61:99-104.

Roy, S., Singh, J.S., 1995. Seasonal and spatial dynamics of plant-available N and P pools and N-mineralization in relation to fine roots in a dry tropical forest habit. Soil Bio. Biochem. 27: 33-40.

Salas, A. M., Elliott, E.T., Westfall, D. G., Cole, C. V., Six, J., 2003. The role of particulate organic matter in phosphorus cycling. Soil Sci. Soc. Am. J. 67:181-189.

Salati, E.; Vose, P.B., 1984. Amazon basin: a system in equilibrium. Science 225:129-138.

Sauve, J. L.; Goddard, T.W.; Cannon, K. R., 2000. A preliminary assessment of carbon dioxide emissions from agricultural soils. Proceedings of 37th Annual Alberta Soil Science Workshop, February 22-24, Medicine Hat, Alberta.

Scholes, R.J.; Van Breemen, N., 1997. The effects of global change on tropical ecosystems. Geoderma 79:9-24.

Schroth, S., Salazare, E., Da Silva, J. P. 2001. Soil nitrogen mineralization under tree crops and a legume cover crop in multi-strata agroforestry in Central Amazonia: Spatial and temporal patterns. Exper. Agric. 37: 253-267.

Schuur, E.A.G., Chadwick O. A., Matson, P. A., 2001. Carbon cycling and soil carbon storage in mesic to wet Hawaiin montane forests. Ecology 82: 3182-3196.

Schuur, E.A.G., Matson, P. A., 2001. Net primary productivity and nutrient cyling across a mesic to wet precipitation gradient in Hawaiian montane forest. Oecol. 128: 431-442.

Sexstone, A. J.; Parkin, T. B.; Tiedje, J., 1985. Temporal response of soil denitrification rates to rainfall and irrigation. Soil Sci. Soc. Am. J. 49:99-103.

Shang, C., Caldwell, D. E., Stewart, J. W. B., Tiessen, H., Huang, P. M., 1996. Bioavailability of organic and inorganic phosphates adsorbed on short-range ordered aluminum precipitate. Microb. Ecol. 31:29-39.

Shi, W.; Yao, H.; Bowman, D., 2006. Soil microbial biomass, activity and nitrogen transformations in a turfgrass chronosequence. Soil Biol. Biochem. 38:311-319.

Silver, W.L., Herman, D. J., Firestone, M. K., 2001. Dissimilatory nitrate reduction to ammonium in upland tropical forest soils. Ecology 82(9):2410-2416.

Singh, J. S.; Raghubanshi, A. S.; Singh, R. S.; Srivastava, S.C., 1989. Microbial biomass acts as a source of plant nutrient in dry tropical forest and Savanna. Nature (London) 399:499-500.

Singh, S., Singh, J.S. 1996. Water –stable aggregates and associated organic matter in forest savanna and cropland soils of a seasonally dry tropical region, India. Biol. Fert. Soil. 22: 76-82.

Sinsabaugh, R.L., Antibus, R. K., Linkins, A. E., 1991. An enzyme approach to the analysis of microbial activity during plant litter decomposition. Agric. Eco. Env. 34:43-54.

Six, J. Elliott, E.T., Paustian, K., 1999. Aggregate and soil organic matter dynamics under conventional and no tillage systems. Soil Sci. Soc. Am. J. 63:1350-1358.

Six, J., Elliott, E.T., Paustian, K., 2000. Soil macroaggregate turnover and microaggregate formation: a mechanism for C sequestration under no tillage agriculture. Soil Biol. Biochem. 32:2099-2103.

Smith, J. L., Paul, E.A., 1990. The significance of soil microbil biomass estimations. pp. 357-396. In J.M. Bollag and G. Stotzky (eds.) Soil Biochemistry. Vol. 6, Dekker, New York.

Smith, M. S., Tiedje, J.M. 1979. Phases of denitrification following oxygen depletion in soil. Soil Biol. Biochem. 11: 261-267.

Soemarwoto, O., 1987. Homegardens: a traditional agroforestry system with a promising future. pp. 157-170. In H. A. Steppler and P. K. Nair (eds.) Agroforestry: A Decade of Development. ICRAF, Nairobi, Kenya.

Steenwerth, K., Belina, K. M., 2008. Cover crops and cultivation: Impacts on soil N dynamics and microbiological function in a Mediterranean vineyard agroecosystem. Appl. Soil Ecol. 40:370-380.

Stockfisch, N., Forstreuter, T., Ehlers, W., 1999. Ploughing effects on soil organic matter after twenty years of conservation tillage in Lower Saxony Germany. Soil Till. Res., 52:91-101.

Szott, L. T., Palm, C.A., Buresh, R. J., 1999. Ecosystem fertility and fallow function in the humid and subhumid tropics. Agrofor. Syst. 47:163-196.

Szott, L.T., Palm, C.A., Davey, C. B., 1994. Biomass and litter accumulation under managed and natural tropical fallows. For. Ecol. Manag., 67:177-190.

Torquebiau, E., 1992. Are tropical agroforestry homegardens sustainable? Agric. Eco. Env. 41:189-207.

Turner, C.L., Seastedt, T.R., Dyer, M. I., 1993. Maximization of aboveground grassland production: the role of defoliation frequency, intensity and history. Ecol. Appl. 3:175-186.

Van den Driessche, R., 1974. Prediction of mineral nutrient status of trees by foliar analysis. Bot. Rev. 40:347-394.

Van der Krift T.A.J.,Berendse F., 2001. The effect of plant species on soil nitrogen mineralization. J. Ecol. 89: 555-561.

Verchot, L.; Davidson, E. A.; Henriquq Cattanio, J.; Ackerman, I.; Erickson, H.; Keller, M., 1999. Land use change and biogeochemical controls of nitrogen oxide emissions from soils in eastern Amazonia. Glob. Geochem. Cycl. 13:31-46.

Vinton, M.A., Burke, I. C., 1995. Interactions between individual plant species and soil nutrient status in shortgrass steppe. Ecology 76:1116-1133.

Vitousek, P.M, Farrington, H., 1997. Nutrient limitation and soil development: experimental test of a biogeochemical theory. Biogeochem. 37: 63 - 82.

Vitousek, P.M., 1984. Litterfall, nutrient cycling and nutrient limitation in tropical forests. Ecology 65:285-298.

Wang C., Wan S., Xing X., Zhang L., Han X., 2006. Temperature and soil moisture interactively affected soil net N mineralization in temperate grassland in Northern China. Soil Biol. Biochem. 38:1101-1110.

Wardle, D.A., 1998. Controls of temporal variability of the soils microbial biomass: a global-scale synthesis. Soil Biol. Biochem. 30:1627-1637.

Weier K.L., Doran, J. W., Power, J. F., Walters, D.T., 1993. Denitrification and the dinitrogen / nitrous oxide ratio as affected by soil water, available carbon and nitrate. Soil Sci. Soc. Am. J. 57: 66-72.

White, J.R., Reddy, K. R., 2000. Influence of phosphorus loading on organic nitrogen mineralization of Everglades soils. Soil Sci. Soc. Am. J. 64:1525-1534.

Williams, B.L., Sparling, G. P., 1988. Microbial biomass carbon and readily mineralized nitrogen in peat and forest humus. Soil Biol. Biochem. 20:579-581.

Wong, M.T.F., Nortcliff, S., 1995. Seasonal fluctuations of native available N and soil management implications. Fert. Res. 42:13-26.

Woomer, P., Martin, A., Albrecht, D., Resck, D., Scharpenseel, H., 1994. The importance and management of soil organic matter in the tropics. In: P. Woomer and M. Swift (Editors), The Biological Management of Tropical Soil Fertility, Wiley/Sayce, UK, PP. 47-80.

Wright, R.B., Lockaby, B.G., Walbridge, M.R., 2001. Phosphorus availability in an artificially flooded southeastern floodplain forest soil. Soil Sci. Soc. Am. J. 65:1293-1302.

Wu, J., Brookes, P.C., Jenkinson, D. S., 1993. Formation and destruction of microbial biomass during the decomposition of glucose and ryegrass in soil. Soil Biol. Biochem. 25:1435- 1441.

Yavitt, J.B., Wright, S. J., 2002. Charge characteristics of soil in a lowland tropical moist forest in Panama in response to dry-season irrigation. Australian J. Soil Res. 40:269-281.

Zakra, N., Pomer, M., de Taffin, G., 1986. Premiers resultants d'une experience d'association cocotiers-cultures vivrieres en Moyenue Cote d'Ivoire. Oleagineaux 41:381-389.

20

Agroforestry: An Alternate Landuse for Wasteland Utilization in Central India

S.D. Upadhyaya and Aashutosh Sharma

Abstract: *The continouning decline in the availability of cultivable land, rising energy and land casts together with the mounting demand for fruits, fodder and timber have given thrust to the concept of agroforestry plantation in Madhya Pradesh. It is a very intensive form of cropping, which has high relevance to the food and nutritional security of our ever-increasing population. Modern cropping systems differ in many ways from the traditional systems. They require more capital to establish and must be more precocious and productive to be profitable. Agroforestry based land use development in Madhya Pradesh will claim sustainability, increased agricultural productivity besides fruit fuel-timber productivity and also helping small scale farmers by generating employment and income. Therefore, the utilization of wastelands has paramount importance and agroforestry is one of the landuse options.*

Wastelands

National Wasteland Development Board (NWDB) on 21st April 1986 unanimously adopted the definition *i.e.* "Wastelands are those lands, which are degraded and are presently lying unutilized except the current fallow due to different constraints". On degraded lands, it is becoming increasingly difficult to manage soil and water resources efficiently and economically. It is the need of the hour to check further degradation and requires re-vegetation of degraded lands, as we cannot take the soil for granted because it is not an inert substance but is a fragile entity of unrivalled complexity.

The term "Wasteland", although being used by the department of economics and statistics since the very inception of land use classes adopted by it, appears rather a misnomer. NWDB has classified wastelands into two broad categories *i.e.* firstly as culturable wastelands, which include salt affected land, degraded cultivation area, degraded forest land, strip lands, mining/ industrial wastelands, sand dunes *etc.* and secondly as unculturable wasteland *i.e.* barren rocky, steep sloping snow covered area, *etc.* The culturable wasteland, which has the potential for the development of vegetative cover, can easily be managed by appropriate agroforestry practices. It is likely that these practices may not be economically viable on the basis of actual return, but the investment is justified on the basis of social benefits.

The characterizations for culturable wasteland are low productivity of soil necessitating long periods of resting fallow, soil deterioration on account of erosion, land size beyond the requirement of owner, poverty, lack of irrigation facilities, scarcity of labour, damage by wild and stray cattle's, distant location and weed infestation like *Kans*

Causes of degradation

The land degradation is caused by excessive pressure on land to meet the competing demands of the growing population for food, fodder, fuelwood and fibre. Various human activities, such as the introduction of large-scale irrigation canals, industrialization, urbanization leads to deforestation which ultimately result in accelerated soil degradation through salinization, flooding, drought, erosion, leaching, soil compaction, waterlogging, diminished natural plant regeneration and disrupted hydrological cycle. These processes, in turn reduce agricultural productivity leading to social insecurity. Emmission of greenhouse gases to the atmosphere resulting into the global warming could be the major reason for soil degradation. Degradation of tropical land is a physical, chemical, and biological process set in motion by activities that reduce the land's inherent productivity. There is a strong relationship between inappropriate land-use practices and land degradation. In some places, degradation is manifest (*e.g.*, desertification), where in others it is inferred (*e.g.*, declining crop yields).

The improper cultivation practices and land management leads to degradation of lands. Land degradation is a major concern in different parts of the country. Many of the productive lands are degrading gradually and becoming as wastelands due to faulty land use practices coupled with alarming rate of deforestation.

Status of wastelands in the country

In India 57 per cent of land (187.7 million ha) is facing land degradation (148.9 million ha water erosion, 13.5 million ha wind erosion, 13.8 million ha chemical deterioration and 11 million ha physical deterioration). The present rate of deforestation is 1.5 million ha/year with soil loss at a rate of 16.35 tones/ha/year.

The information on the extent of soil degradation in the country has been assessed by various agencies. The estimates show wide variability, probably due to different approaches in defining degraded soils or variation in criteria of delineation (Table 1).

Table 1. Estimates of land degradation in India by various agencies

Agency	Estimated extent (Million ha)	Criteria for delineation
National Commission Agriculture (NCA 1976)	148.09	Based on secondary data only
Ministry of Agriculture (1978)	175.00	Based on NCA estimates. No systematic survey undertaken
Society for promotion of wastelands Development (SPWD, 1984)	129.58	Based on secondary collected data
NRSA (1985)	53.28	Mapping on 1:1 million scale based on remote sensing technique.
Ministry of Agriculture (MOA, 1985)	173.64	Land degradation statistics for the states
Ministry of Agriculture (MOA, 1994)	107.43	Elimination of duplication of area. Area reclaimed counted
NBSSLUP (1994)	187.70	Mapping on 1:4:4 million scale based on Global Assessment of Soil Degradation (GLASOD) guidelines

Source: Solanki *et al.* (1999).

Wastelands in Madhya Pradesh

After the formation of Chhattisgarh, the geographical area of Madhya Pradesh has reduced to 307.44 lac ha, out of which 49 per cent is under cultivation and 18.53 per cent under different types of wasteland. The majority of cultivated area is under rainfed (69%) with enormous unexploited agricultural production potential. Nearly one third of the net sown area is unutilized culturable wastelands (having less than 50% production). These situations warrant for an alternate land use and crop diversification for sustainable land use system.

The culturable wasteland, which has the potential for the development of vegetative cover, can easily be managed by adopting appropriate agro-forestry practices. There are nine major categories of wastelands in the State (Table 2) namely,

1. Gullied and/or ravinous land
2. Landwith/without scrub
3. Waterlogged/ Marshi land (Seasonal)
4. Underutilized/Degraded Notified forest land
5. Degraded Pastures/grazing land
6. Degraded Land under plantation crop
7. Mining wasteland
8. Barren rocky/Stone waste/sheet rock area
9. Steep sloping area

Table 2. Category-wise distribution of wastelands of Madhya Pradesh

Major category	Category Nos.	Wasteland categories	Wasteland area (in sq. km)
1.		Gullied and/or ravinous land	
	1	Gullied and/or ravinous land (shallow)	1732.11
	2	Gullied and/or ravinous land (Medium)	1427.55
	3	Gullied and/or ravinous land (Deep)	2114.66
2.		Land with/without scrub	
	4	Land with scrub	27080.57
	5	Land without scrub	1974.31
3.	6	Waterlogged/ Marshi land (Seasonal)	50.55
4.		Under utilized/Degraded Notified forest land	
	7	Under utilized/ Degraded Notified forest land	17549.17
	8	Under utilized/ Degraded Notified forest land (Agri.)	4188.35
5.	9	Degraded Pastures/ grazing land	29.64
6.	10	Degraded Land under plantation crop	6.94
7.	11	Mining wasteland	121.48
8.	12	Barren rocky/Stone waste/sheet rock area	779.19
9.	13	Steep sloping area	80.51
		Total	57134.03
		Total Geographical area	307440
		% to TGA	18.53

Source: 1:50,000 scale wasteland maps (Based on IRS satellite LISS- III imagery)

Wasteland in Madhya Pradesh has following characteristics

1) Sandy textured with very low organic matter content subject to fast erosion.

2) Low depth of fertile soils.

3) Undulating topography.

4) Soil erosion

5) Rainfed mono cropped.

6) Poor irrigation facilities.

7) Less green vegetation.

8) Farmers unawareness and negligence.

The major area belongs to gullied/ ravinous land, with & without scrub and under utilized degraded notified forest land. It is evident that in quiet a large number of districts of Madhya Pradesh; culturable wastelands constitute a fairly big proportion of net sown area. This 12% of net sown area can possibly become cultivable through alternate land use systems such as agroforestry which can provide higher total biomass per unit area in wastelands as compare to agricultural systems besides yielding different products such as fodder, fuel, fibre and fruits *etc*. Besides the systems also utilize off-season precipitation to the best advantage of perennials and also utilize the moisture below the root zone of crops. These systems help in soil and water conservation. The mechanical and vegetative soil conservation measures reinforce the developed eco-system and allow to sustain the system too.

The productivity of wastelands has been declined which is attributed to poor management and exploitative use. There is a great risk of growing food grains in extremely light textured soils and culturable wastelands of Madhya Pradesh. The culturable wastelands in different districts are scattered as small patches. A sizeable culturable wasteland comprised unoccupied holdings, not known to have ever been cultivated. This particular area could be brought under cultivation of suitable agroforestry tree species. Madhya Pradesh is blessed with climatic variability's and diverse soil types. This provides unique opportunities to the farmers to grow multipurpose trees in marginal and culturable wastelands under various agro-forestry systems. The technology of wasteland reclamation through multi-approach agroforestry systems will definitely frame a road map to convert waste into wealth.

Scope of wasteland reclamation

Large stretches of land, degraded and devoid of any vegetation are found in northern and central India. Madhya Pradesh has the largest area under wasteland in the country due to undulating topography, lack of irrigation and neglected natural resource management by the farmers. These sites have acute problems of soil and other natural resource management. However, some tree species are tolerant to salinity, sodicity and other site stresses. If plantations of such species with some soil amendments are to be raised, in due course of time leaf litter and humus added to the soil will make it biologically active and will improve its productivity.

Madhya Pradesh being a semi arid state having 1/3rd of Net Sown Area under culturable wasteland and 64 per cent rainfed is committed to the concept of sustainable land management through agroforestry system. In order to meet the challenges before the wasteland management, tree based farming is the only solution. Natural resources in the State are limited and declining every year, leading to major degradation. Such degraded areas, especially non-forest lands, require special treatment for their restoration to make them productive and profitable. The widely accepted "Agroforestry System", which is a known alternate land use system for its productive and protective role, can contribute significantly in developing the farmer's culturable wastelands, which has been significantly demonstrated by JNKVV, Jabalpur through its extension network (KVKs).

Agro-forestry systems provided a sustainable solution to overcome the problem of degradation of non-forestland. It is an alarming time to farmers for adoption of suitable agroforestry system for reclamation of wastelands and improvement in productivity and profitability. It is worth mentioning that agroforestry systems have capacity to generate additional income, provide fuel, fodder and ultimately additional job opportunities round the year.

Technology of agroforestry systems basically constitutes selection of region or even site-specific tree species to be incorporated in to the farming system. An agroforestry system allows flexibility for modification according to the local conditions and availability of resources.

The development of new technique for reclamation of wastelands through agroforestry is in its infant stage. Various experiments and trials are being conducted all over the country to evolve suitable methodologies for specific site condition of right species for specific locality, their socio-economic feasibility and adaptability of farmers is of prime importance.

Agroforestry systems

Agroforestry in Madhya Pradesh can be seen as phases in the development of productive agro-ecosystem and an alternate land use system, wherein, farmers can manipulate and manage their land by growing trees for the service or products. Numerous agroforestry systems have been evolved for different areas. One system differs from the other in respect of structure, function, even arrangement of components and composition. Nair (1993) classified the agroforestry systems based on nature of components, structure, function, socio-economic and ecological factors *etc*. One, which is very convenient, is based on the components, accordingly following are common agroforestry systems prevailing in India (Solanki *et al.* 1999).

- Agrisilviculture (trees + crops)
- Boundary Plantation (trees on boundary + crops)

- Block Plantation (tree + crops)
- Energy Plantation (tree + crops during initial stage)
- Alley cropping (hedges + crops)
- Agrihorticulture (fruit trees + crops)
- Agrisilvihorticulture (tree + fruit trees + crops)
- Agrisilvipasture (trees + crops + pasture/ animals)
- Hortipasture (fruit trees + pasture/ animals)
- Silvipasture (trees + pasture/ animals)
- Forage forestry (forage trees + pasture)
- Shelter belts (tree + crops)
- Wind breaks (tree + crops)
- Live fence (shrub under trees on boundary)
- Homestead (multiple combinations of timber/fruit trees, vegetables *etc.*)
- Entomoforestry (tree + sericulture)
- Aquaforestry (trees + fishes)

Potential agroforestry technologies have been identified for different agro-ecozones by AICRPs on Agroforestry and many ICAR institutes under the Natural Resource Management Divisions (Solanki *et al.* 1999), which can be extended at farmer's field on need based and location specific through Operational Research Projects (ORP). Agroforestry technology demonstration has been taken up by JNKVV and many other centers through the help of Ministry of Rural Development, Deptt. of Land Resources on large areas of degraded community lands and also the farmers field to motivate the farmers. This has promoted the farmers to make best use of class V and VI type of land.

Agroforestry helps to improve the soil ecosystem

The inclusion of leguminous trees on fragile soil ecosystem facilitates the soil productivity improvement through:

- Addition of soil organic matter through leaf litter and organic debris
- Nutrient cycling efficiency underneath the trees
- Efficiency in biological nitrogen fixation, nitrogen fixing trees can add the nitrogen ranging from 100-500 kg N/ ha/ yr
- Little loss of nutrient from the system, trees can withhold the nutrients that can otherwise be lossed by leaching.

- Little erosion by runoff, because the woody perennial component of agroforestry has the capacity to maintain soil cover.
- Trees take-up nutrients from deeper layers of soil and deposited it on surface layer via litter.
- Nutrients are also added in the soil by root decay
- Viable counts of beneficial microorganisms also added to the soil ecosystem favourably.
- Soil physical conditions like water holding capacity, aggregate stability, soil temperature and moisture regimes, permeability, drainage also improves favourably due to deliberate incorporation of trees with arable crops.
- Activation of mycorrhiza which can facilitate solubilization of unavailable nutrients, and
- Also help in mutual sharing of nutrients among agroforestry components.

Improvement in soil physical conditions such as water holding capacity, aggregate stability, soil temperature regimes, permeability, drainage *etc.* by

- Adding shoot/root biomass, about 3000 kg dry matter/hectare/year. Nitrogen supplement from Nitrogen Fixing Tree species (NFTs)

Approaches for development planning of non-productive culturable land

a) Vegetative measures

Agroforestry can be considered as a dynamic, ecologically based, natural resource management system that, through the integration of trees in farm and rangeland, diversifies and sustains small holder production for increased social, economic and environmental benefits. It conserves the ecosystem and at one and same time provides food, fodder and wood. Some form of agroforestry has long been practiced by many farmers, particularly those with low levels of technology and resource inputs, and in areas believed to be unsuitable for more profitable production systems. Hence, agroforestry can best be considered as a philosophy of integrated land-use particularly suited for marginal areas, low-input systems and also for problem soils.

Some areas are affected by salinity, alkalinity, acidity and waterlogging. Such degraded lands can often be reclaimed by agroforestry while providing poor farmers with some income. Many species of trees can grow well in these problem areas where most agricultural crops cannot.

(b) Agro ecological zonewise MPTs recommendation

Agro ecological zone wise management of wastelands should be more appropriately planned through watershed approach. Results of various trials conducted under AICRPs and watershed programmes under different agro climatic situations based on annual precipitation and soil type are summarized in wasteland management based on annual rainfall and soil types, which can be recommended for conducting on-farm trials to combat wastelands (Table 3).

Table 3. Wasteland management based on annual rainfall and soil types

Annual rainfall (mm)	Soil type	Land use systems	Suitable tree/ grass/ legume species
500-750	Shallow (0-0.30 m)	Silvipastoral system	*Acacia nilotica, Colophosphermum mopane, Dalbergia sissoo, Hardwickia binata, Cassia sturti, Albizia amara, Leucaena leucocephala, Cenchrus ciliaris*, C. *setigerus, Dicanthium annulatum, Panicum antidotale, Stylosanthes hamata*
	Medium (0-0.45m)	Horti pastoral system	*Annona squamosa, Zizyphus mauritiana, Syzigium cuminii, Emblica officinalis, Tamarindus indica, Feronia limonia, Aegle marmelos, Cenchrus ciliaris, Panicum anti-dotale, Urchloa mosambicensis, Stylosanthes hamata, Macroptilum atropurpureum, Clitoria tenatea.*
>750	Shallow (0-0.30 m)	Alley farming of silvi pastoral system	3 years *Stylosanthes hamata* and 4th year arable crop (sorghum on heavier soils, perl millet on lighter soils) silvipastoral system as above
	Medium (0-0.45m)	Alley farming or horti pastoral system	Ley farming as above. *Mangifera indica, Achras sapota, Psidium guajava, Emblica officinalis, Stylosanthes hamata/ Macroptilium atropurpureum*
	Deep (>0.45m)	Agrisilvi or agrihorti	*Prosopis cineraria, Acacia ferruginea, Tectona grandis, Hardwickia binata, Dalbergia sissoo*,+arable system crop *Mangifera indica, Achras sapota, Psidium guajava*+ arable crops

Madhya Pradesh with its tremendous diversity of ecological conditions from a dry region to the rainy places and from Patalkot, Chhindwara to the Satpura ranges, stores a very wide variety of biological communities. The role of afforestation in today's disturbed ecology is tremendous as we are rapidly loosing our forests, which are so essential for maintenance of biological life on earth.

The selection of appropriate fuel, fodder and fruit tree species for planting at a locality must be decided on the basis of the purpose for which tree is required and the physical and climatic characteristics of the area. It is also necessary to select an optimum combination of species, which will meet all the requirements in the light of varied physical and climatic conditions prevailing in peasants' district are summarized in Table 4.

Table 4. Different promising MPTs and fruit species for various agroforestry systems in farmer's field under agro-ecological zones of Madhya Pradesh.

Agro climatic zones	Systems	Promising tree species
i. Chhattisgarh Plain	Agri-silviculture	*Terminalia* spp., Shahtoot, bamboo, khamer
	Agri-horticulture	Mango, jackfruit, guava, drumstick, seedless lime, custard apple
ii. Northern Hills of	Agri-silviculture/ Agri-entomoforestry (boundary plantation is also included)	Traditional Palash based system, Chhattisgarh eucalyptus, Khamer, karanj
	Agri-horticulture	Jamun, bel, mango, guava, jackfruit, drumstick, lime, aonla
iii. Kymore Plateau & Satpura Hills	Agri-silviculture (boundary plantation is also included)	Traditional acacia-rice, palash, eucalyptus, teak, shisham
	Agri-horticulture	Guava, mango,aonla, lime, custard apple, jamun, shahtoot
iv. Central Narmada Valley	Agri-silviculture	Bamboo, khamer, shisham, arjun
	Agri-horticulture	Mango, seedless lime, ber, lime, aonla
v. Vindhya Plateau	Agri-silviculture	Khamer, neem, babul, eucalyptus, safed shirish
	Agri-horticulture	Mango, guava, lime
vi. Gird Region	Agri-silviculture	*Prosopis* spp., Babul, subabul, sissoo, Australian babul
	Agri-horticulture	Lime, ber, custard apple subabul, neem, karanj, bamboo, palash, mahua,
vii. Bundelkhand Region	Agri-silviculture (boundary plantation is also included)	
	Agri-horticulture	Aonla, drumstick, mango, ber, jamun
	Silvipasture (Although in this region no well-defined silvipasture is present, during monsoon the grasses appear, which are used as pasture)	sissoo, khejri
viii. Satpura Plateau	Agri-silviculture	Bamboo, teak, khamer, shisham
	Agri-horticulture	Santara, aonla, shahtoot, imli, karonda
ix Malwa Plateau	Agri-silviculture	Khamer, subabul, eucalyptus, kala shirish,
	Agri-horticulture	lime, drum stick, papaya, santara, karonda, aonla
x. Nimar Plains	Agri-silviculture	Neem, khamer
	Agri-horticulture	Banana, santara, papaya, lime, aonla, ber
xi. Jhabua Hills	Agri-silviculture	*Acacia* spp., guggal, neem, eucalyptus
	Agri-horticulture	Aonla, guava, mango

Source: Shrivastava (1995), Solanki *et.al.* (1999)

Some of the important grasses species identified for different habitats where they could be potentially used as vegetative barriers or in grasses waterways are summarized in Table 5.

Table 5. Grasses with potential use as live bunds (vegetative barriers) in different agro-ecological regions (AER).

S.No.	Name	Habit	AER	Soil preference
1.	*Andropogon gayanus*	Perennial	1,2	Alluvial
2.	*Bothriochloa pertusa*	Tufted perennial	1,2	Coarse textured
3.	*Cenchrus ciliaris*	Perennial Tussocks	1	Sandy loam, alluvium
4.	*C. setigerus*	Perennial Tussocks	1	Sandy loam, alluvium
5.	*Chrysopogon fulvus*	Tufted perennial	1,2	Clay, silty clay loamy
6.	*Dichanthium annulatum*	Perennial tufted	1,2,3	Silty loam
7.	*Imperata cylindrical*	Perennial	1,2	Light textured sandy loam
8.	*Panicum antidotale*	Tufted perennial	1,2	Clayey, sandy loam, sandy
9.	*Saccharum spontaneum* perennial	Rhizomatous	1,2	Sandy loam, clay
10.	*Sehima nervosum*	Tufted perennial	1,2	Sandy loam, alluvial clayey
11.	*Vetiveria zizanioides*	Tufted perennial	2,3	Sandy loam, swampy, ill drained soils

Modified from R.K. Gupta (1992)

Note: AER 1 = 400-1000 mm/year rainfall
AER 2 = 1000-1600 mm/year rainfall
AER 3 => 1600 mm / year rainfall

a) Agroforestry alternatives for different types of stress sites

Large areas in the tropic are affected by salinity, alkalinity, acidity and waterlogging. Unscientific land use practices have led to a further increase in the area affected by toxicities and deficiencies. Such degraded land can often be reclaimed by agroforestry while providing poor farmers with some income. Many species of trees can grow well in these problem areas where most agricultural crops cannot grow. Singh and Upadhyaya (1999), Upadhyaya *et al.* (2001) advocated alternate farming practices for degraded sites as under:

Desert Ravines lands and dunes

In India, there are about four million ha of ravines lands, a major portion of which are confined to Uttar Pradesh, MP, Rajasthan and Gujarat. These lands come under land capability classes VI and VII and are for below their economic utilization. Small to medium gullies can be put to silvipasture or hortipasture.

In Madhya Pradesh there are about 7924.40 ha of ravines lands and gullied area. A major portion of which are confined to Gwalior, Morena, Bhind, Shajapur, Guna districts. These lands come under land capability classes VI and VII and are for below their economic utilization. Small to medium gullies can be put to silvipasture or hortipasture.

Suitable species

Grasses : *Lasiurus sinduicus, Cenchurus cilliaris, C. setigerus*

Legumes : *Arylosia scarabaeoides, Lablab pupureum, Clitoria ternatea*

Trees : *Acacia tortilis, A. arabica, Prosopis cineraria, P. juliflora, Eucalyptus terminalis, E. camaldulensis, Azadirachta indica, Dichrostachys*

Acid Soils

In Madhya Pradesh 162.81 ha acid soils falls in the Bhind and Datia region. The soils affected by acidic condition can also be suitably utilized by growing acid tolerant species.

Suitable species

Grasses : *Ponnisolum polystachyon, P. pedicellatum, Paspalum nutatum, Pennisetum clandestinum* (Kikyo grass)

Legumes : *Centrocema pubescens, Desmodium intortum, Stylosanthes guynensis.*

Trees : *Ficus cumeralius, Albizia chinensis, Morus cerrata, Ulmus repelensis, Sehiama wallichi, Bucklandia pupulunea.*

Lands affected by shifting cultivation

Suitable species

Grasses : *Cenchurus cilliaris, C. setigerus, Pennisetum, Chrysopogon fulbus.*

Legumes : *Clitoria ternatea, Lablab pupureum,*

Trees : *Lucaena leucocephala, Hardiwicka binata, A. arabica, Acacia tortilis, Dalbergia sissoo, Albizzia* spp, *Anogeissus pendula.*

Cho and River bed affected area

Suitable species

Grasses	:	*Chrysopogon fulbus*, *Dichanthium annulatum*, *Pennisetum pedicellatus*, *Eucaliopsis binata* (babar grass).
Legumes	:	*Calaponium mucuoides*, *stylosinthes gracilis*.
Trees	:	*Dalbergia sissoo*, *Acacia catechu*, *A. nilotica*, *salix spp*, *Zizyphus spp*, *Psidium guajava*.

Through incorporating woody perennials under different agroforestry systems nutrient status can be improved in fragile ecosystems besides Fuelwood and nutritive value of animal feed and ultimately the environmental security can be achieved in totality. Different agroforestry interventions (Table 6) are also proposed for central India.

Table 6. Agroforestry interventions proposed for Central India

System	Agroforestry intervention proposed
Agri-Silviculture, Agri-Horticulture, Silvipasture	• *Acacia nilotica, Leucaena leucocephala, Dalbergia sissoo* and *Eucalyptus hybrid* with paddy, wheat, mustard, gram, sunflower *etc*. in irrigated areas. • *Azadirachta indica, Acacia nilotica, Hardwickia binata, Dalbergia sissoo etc*. with groundnut, sorghum, pigeonpea, gram, lentil, mustard, soybean, paddy, sesamum in rainfed areas. • *Mangifera indica, Psidium guajava* with paddy, wheat, mustard, gram, sunflower in irrigated areas. • *Zizyphus mauritianaa, Annona squamosa, Emblica officinalis, Citrus sp.*, Pomegranate with groundnut, sorghum, pigeonpea, gram, lentil, mustard, soybean, paddy, sesamum in rainfed areas. • Silvipasture with *L. leucocephala*, and *Hardwickia binata* and grasses viz, *Sehima nervosum, Cenchrus ciliaris, Stylosanthes hamata.* • Under rainfed conditions, legume crops are superior to other crops for higher green fodder yield of Leucaena hedge row. • *Acacia nilotica, Acacia tortilis, A. amara, D. sinerea, Prosopis juliflora* are superior for block plantation as energy plantation.

a) Agri-medicinal trees system

Agroforestry systems involve deliberately growing medicinal trees and shrubs along with arable crops and/or livestock seeking positive synergism. Traditionally several species of trees having medicinal values *e.g.* Neem, Babool, Harra, Bahera, Aonla were grown in wastelands. Farmers are growing these trees on farm land for different uses including fodder, fuel, timber, gum, medicine, oil, fruit, bee keeping, green manure *etc*. Further, these woody perennials also help in soil amelioration. Medicinal plants can also be grouped as suitable to different edaphic factors and for different soil problems like salinity and waterlogging. In this system, the medicinal trees are primary crops and healing herbs are secondary. Inter cropping can be taken for variable periods. Some of the medicinal herbs require shade, for which close planting trees with good coverage of canopy can be preferred; such systems are known as shaded perennial agroforestry systems, for instance; Aonla with turmeric, *Piper longum* or Khamer with Chitrak.

Suitable plants which have medicinal importance those do well on different problematic soils.

Shallow, rocky soil

Suitable species

Herbs : Punarnava (*Boerhavia diffusa*)Anantmool (*Hemidesmus indicus*), Ratti (*Abrus pricatorius*), Manjistha (*Rubia Cordifolia*), Gwar patha (*Aloe barbadensis, Aloe vera*).

Shrub: Gandh Babool (*Acacia farnisiana*), Jhand (*Prosopis cineraria*), Karonda (*Carissa carandus*) Gurmar (*Gymnema sylvestre*) Marorphali (*Helicteres isora*).

Trees: Neem (*Azadirachta indica*), Aonla (*Emblica officinalis*), Sitaphal (*Annona squamosa*), Bel (*Aegle marmelos*), *Cordia oblica*, *Holoptellia intigrifolia*, *Wrightia tinctoria* Malkagni (*Celastrus paniculata*).

Sandy soils

Suitable species

Herbs: Khus (*Vetiveria zizanioides*), *Cymbopogon* spp., Isabgol (*Plantago ovata*), Safed musli (*Chlorophytum borivilianum*, *C. tuberosum*), Kali Musli (*Curculigo orchioides*), Asgandh (*Withania somnifera*), Mulethi (*Glycyrrhiza glabra*), Senna (*Cassia angustifolia*), Akarkara (*Spilanthus acmela*), Sadasuhagan (*Catharanthus roseus*).

Shrub: Chitrak (*Plumbago zeylanica*), Sinduri (*Bixa orellana*)

Trees: Bel (*Aegle marmelos*), Katha (*Acacia catechu*), Jamun (*Syzygium cumini*), Arjun (*Terminalia arjuna*), Salai (*Boswellia serrata*), Khamer (*Gmelina arborea*), *Glyricidia sapium*.

Saline soils

Suitable Species

Herbs: Satawar (*Asparagus racemosus*), Sankh puspi (*Evolvulus alsinoides*), Khus (*Vetiveria zizanioides*), *Cymbopogon* spp., Chandrsur (*Lepidium sativum*), Kalongi (*Nigella sativa*)

Shrubs: Karonda (*Carissa carandus*), Adusa (*Adhatoda vasica*), Nirgundi (*Vitex negundo*), Ber (*Zizyphus* spp.)

Trees: Neem (*Azadirachta indica*), Aonla (*Emblica officinalis*), Katha (*Acacia catechu*), Arjun (*Terminalia arjuna*), Babool (*Acacia nilotica*), Bahera (*Terminalia bellerica*), Harra (*Terminalia chebula*), Sivnag (*Oroxylum indicum*).

Alkaline soils

Suitable species

Herbs: Sankh puspi (*Evolvulus alsinoides*), Khus (*Vetiveria zizanioides*), *Cymbopogon* spp., Jangali Pyaj (*Urginia indica*), Van lahsun (*Allium porrum*), Garlic (*Allium sativum*).

Shrubs: Karonda (*Carissa carandus*), Nirgundi (*Vitex negundo*), Guggal (*Commiphora* spp.)

Trees: Aonla (*Emblica officinalis*), Arjun (*Terminalia arjuna*), Babool (*Acacia nilotica*), Sissoo (*Dalbergia sissoo*), *Eucalyptus teritecornis*, Siris (*Albizzia procera*), Karanj (*Pongamia pinnata*)

Water logged soils

Suitable species

Herbs: Bach (*Acorus calamus*), Brahmi (*Bacopa monniera*), Mandook Parni (*Cintela asiatica*), Brangraj (*Eclipta alba*), *Cyperus* spp. Khus (*Vetiveria zizanioides*), Talmakhana (*Astercantha longifolia*).

Shrub: *Hedychium spicatum*

Trees: *Eucalyptus teritecornis*, Arjun (*Terminalia arjuna*), Salai (*Boswellia serrata*),

a) Water conservation interventions application

- Staggered trenches on hilly/sloppy terraces for water conservation.
- Construction of water harvesting ponds of size less than 1000cu m. capacity with seepage control arrangement.
- Water recycling by using micro irrigation with efficient pumping units, even small diesel engines shall be used.
- Tapping of adjoining nalas
- Construction of graded/ contour bunds

Techniques for ascertaining plant growth to establish fruit tree based agroforestry in wastelands

Protection against adverse weather conditions: There is heavy damage to fruit plants in hot as well as cold arid tracts because of frequent heavy storms not only due to transpiration losses but because of deposition of sand, mechanical damage to plants and soil erosion. Wind-breaks are narrow strips of trees planted around orchards to have protective barriers against hot or cold wind currents. Width of windbreaks depends upon the availability of land. Fast growing, deep rooted plants as per climatic requirements which can check or reduce the flow of air are preferred. Shelter belts are wide and long belts of several rows of trees and shrubs planted across the prevailing wind direction to defect wind currents and reduce wind movement of vast fields.

Pit preparation: Crescent shaped or semi circular bunds with a diameter of 6-13 meters (depending upon spacing requirements of individual fruit trees) are prepared and catch pits are also dug at the same time on the upper side of the slope. The trees are planted in the centre of the crescent. While the crescent bunds help in collection of rainwater, the catch pit conserves the same. New pits are opened as the old ones get filled with organic matter and pond sediments. Breaking of hard pan, leaching of harmful salts in saline soil and incorporation of amendments viz. gypsum/ pyrite (4-10 kg), press mud, pond sediments and rice bran are helpful in soil improvement for better establishment of plants.

Planting on terraces: In cold arid regions depending upon the space requirement and slope of land, broad terraces are prepared with slope towards inner sides and trees are planted in the centre. Such terraces help in conservation of moisture and support the plant growth.

Trench planting: Deep trenches (0.5 to 1.0 m) are dug across the contours, and fruit trees are planted in the centre. The trenches conserve the rainwater along with silt and organic matter and thus promote tree growth.

Popularization of *in situ* orchard establishment: It is a matter of common experience that seedlings plants have better and well-developed root system. Therefore, it is advisable to sow the seeds of rootstocks in the pits prepared as per layout or transplant the seedlings grown in polybags or polytubes. Budding/grafting with scion shoots obtained from the promising varieties may be perfomed in the appropriate season. This practice shall be helpful in better and uniform orchard establishment. In case, abiotic stress rootstocks are available, these should be preferred over common rootstocks.

Water harvesting in relation to fruit cultivation: Water harvesting is one of the old practices of collecting water in depressions for crop cultivation and drinking purposes. Excess rainwater of the area as per slope may be collected in ponds of appropriate sizes and reutilized through modern irrigation practices (drip or sprinkler irrigation). In hot arid regions where infiltration rate and run off losses are very high, the idea is to develop catchment area of each tree, size of which is determined by the slope of the land, water requirement, run off coefficient and canopy of the fruit tree (feeding zone).

Irrigation: Irrigation affects the soil environment making more water available for plant establishment making more water available for plant establishment and growth. Every care should be taken for efficient utilization of available water, reducing unproductive losses and thus improving soil environment.

Amongst the modern methods of irrigation, drip system is the most efficient and is gaining importance in arid/rainfed regions. It ensures uniform distribution of water, perfect control over water application, minimization of water losses during conveyance and seepage, reduces the weed population, keeping the harmful salts below the root zone and minimizes the labour cost.

Mulching: Covering of soil surface *i.e.* tree basin with organic waste materials, emulsions or black polythene is termed as mulching. Mulching reduces the evaporation by cutting of radiation falling on the soil surface, prevention of weed growth, maintenance of optimum temperature for normal root activity and moisture conservation. Continuous uses of organic mulches are also helpful in improving the organic matter content and thus water holding capacity and results in better plant growth and higher productivity.

Top working/ Frame working: Seedling plants of ber in hot arid regions and seedling plants of mango, aonla, bael and ber in semi arid regions may be converted to promising types by adaptation of top working/frame working with scion shoots of promising known varieties.

Use of hardy rootstocks: Few of the fruit plants are susceptible to adverse soil and climatic conditions, but with use of suitable rootstocks their cultivation is feasible under such situations. Rootstocks must have deep root system, capacity of reduced/ selective absorption of harmful salts, resistance to biotic and abiotic interferences. Similarly cold tolerant hardy rootstocks are available for temperate fruits, which may be utilized.

Closer plantation: It is matter of common experience that there is poor plant growth in problematic soils as compared with normal soils. Hence it is advisable to reduce the planting distance by 20-30 per cent advocated for different fruit crops under normal soil situations.

Soil coverage with inter/cover/filler crops: Most of the perennial fruit crops have long gestation period. In order to ensure early income, reduce evaporation, minimize weed growth, it has been observed that suitable cropping models should be encouraged instead of monoculture. The following models have been found ideal for rainfed conditions.

(i) Aonla + Ber

(ii) Aonla + Guava

(iii) Aonla + Ber + Karonda

(iv) Aonla + Ber + Phalsa

(v) Aonla + Guava + Karonda

(vii) Aonla + Guava + Phalsa

Besides combination of fruits, strip sowing/plantation of vegetables viz. pointed gourd, bottle gourd, tomato, brinjal, spinach or medicinal and aromatic crops viz. lemongrass, palmarosa, vetiver, oil bearing plants should also be tried. Growing of cover crops particularly legumes (organic farming) or green manuring with Dhaincha (*Sesbania* spp.) is helpful in improving the soil fertility.

Development of agroforestry models through Technology Development Extension and Training project on wasteland development: Some observations

One of the important wasteland plantation project entitled "Technology development, extension and training in agroforestry for wasteland development" was sanctioned in 2001 by Ministry of Rural development, Govt. of India for execution in 9 centers of JNKVV with its main coordinating centre at Department of Plant Physiology, JNKVV, Jabalpur.

A very remarkable change has been seen in the farmer's attitude regarding wasteland development technology adoption. They prefer aonla based agroforestry model. Aonla is very much promising species for Madhya Pradesh. Its performance in undulating terrain and muramy wasteland is very encouraging. In agroclimatic zone Kamore plateau and Satpura hills, one of the promising models, which have gained popularity among farmers of the region, is Khamer based agroforestry (Medicinal and timber tree) intercropped with Lemon Grass/ Rosa Grass/ *Tagetus* in agri-silvi-horti system. It is also proved very successful both in terms of biomass production and also in economic return and profitability.

At Sehore center under TDET project the plantation of various tree species has been done. The most of the species are showing good performance according to their suitability and some are failed to service, even under limited water resources condition the results in general are satisfactory. On the basis of three years observations it is found that some of species are in good condition having better survival. These are Khamer, Neem, Baheda, Bamboo, Ritha, Amaltas, Kapok, Deshi babool, Aonla, Deshi Ber, and Custard apple whereas; others are unsuitable for this region.

At KVK Jhabua, Khair, Guggal and Jatropha based agroforestry models are being developed at farmer's wasteland. At Jhabua, frequent severe drought conditions accompanied by very less topsoil warrants for suitable wild fruit/timber based agroforestry models as compared to traditional agriculture.

Agro-climatic situation of Chhindwara is very unique with respect to soil and climate, which is most suitable for horticultural crops. Hence, fruit based agroforestry system *i.e.* santara, aonla *etc.* is recommended for the region and which also gaining popularity among farmers.

Three year old plantations of neem, karanj, harra and aonla done at Shahdol attains good growth and height in the range of 1.5 m to 2.5 m and well established on very poor soils. These plants even under severe moisture stress survived very well. In general the performance of neem and karanj is best among others. Farmers are now much more interested for fruit tree plantation of improved variety seedlings. Survival percentage of fruit seedlings in farmers field are higher mainly because farmers of this area are showing interest in fruit based agroforestry models (Agri-horticultural system) followed by agri-silvicultural system to fulfill their requirements for fruits, fuel and timber. This region has major constraints of undulating terrain, which causes topsoil erosion and ultimately limits crop productivity. This can be mitigated by adopting suitable fruit tree particularly Aonla, Munga and Bel.

At Rajgarh center agroforestry models like Isabgol + Aonla, Isabgol + Khamer were developed. Plantations of saplings of different varieties of Mango, Aonla & Orange were done for improving the productivity of farmer's wasteland at different degraded sites of Rajgarh Block and these are performing very well.

The plantation of different fruit trees like Guava Var. Lucknow- 49, Aonla, Var. local, Kanchan, Chakaiya and NA-5, Mango Var. Lungra and Dashahari, Mahua, Emali local, Karonda local and forest tree like Eucalyptus, Khamer and Bamboo have been planted under different agroforestry models at farmers field at Sidhi. The plantations of fruit plants have been done under agri-horticulture system and forest trees are planted under agri-silviculture, silvi-pastoral and boundary plantation. Guava, Aonla, Mango, Eucalyptus and Khamer based agroforestry system has been developed at farmers field as well as KVK farm in which crop component included Arhar, Urd, Moong, Til and Niger at different locations. In silvi-pastoral system eucalyptus + jowar, Bamboo + natural grasses and karonda has been planted as boundary plantation.

Soil of the Sagar region varies from clay to sandy loam with predominantly medium black soils. The depth is variable that varies from shallow to deep. Medicinal and aromatic plants like Palmarosha, aloe, satawar, bel, arandi and chandrasur were planted in the institutional land for demonstration to the farmers under different agroforestry models. Mango, eucalyptus, subabool, gulmohar, Glyricidia, mango, aonla *etc* were performing well under agroforestry model at ZARS, Sagar with respect to tree height, diameter growth and branching pattern. Medicinal crop chandrasur was studied in detail under interaction with aonla var. chakaiya in agroforestry model and found very promising one.

Efforts made to motivate the farmers for adoption of agroforestry

Following determined attempt has been made to convince the farmers to adopt agroforestry at their wastelands.

- Need based targeted planting campaign under site-specific agroforestry models for environmental security and employment generation
- Organized training's at village level and for extension functionaries on nursery, technologies, planting techniques, wastelands reclamation, benefits of agroforestry, *etc.*
- Kisan Nursery in Watershed programmes.
- Bulletins, folders and Training manuals in Hindi published by different centers.
- Kisan Mela/Agroforestry Day organized.
- Radio/TV talk, popular articles in local newspapers.

Technological achievements

Farmers have developed their own wastelands through the technologies developed and extended, such as adoption of proper pit size digging, their solarization and use of Nadep and vermicompost organic base in the pits. Most of the farmers adopted plant and row distance 8 meter which is most suitable for long term production of associated inter crops in the interspace of Aonla trees. Farmers also planted some fruit trees in trenches along with the field bunds just to facilitate the limited irrigation water to the fruit plants. Farmers also adopted proper distance of horticultural plants, Eucalyptus and Khamer. Inter-cropping in the agri-horticulture model gave the additional source of income to the farmers. Farmers also aware about the limited use of irrigation water to the fruit plants using different drip methods and moisture conservation methods.

The implementation of agroforestry at farmers field has shown a major impact on elevating fertility status, productivity of land, biomass yield as well as water holding capacity of the soil from low to high level in terms of high NPK and organic content of the soil, fruit, fuel wood and fodder production *etc.* This has also pave way to generate various source of employment with regard to engagement in agroforestry models. The barren lands of farmers were brought under active cultivation as a source for livelihood to the peasant community of the project area. Farmers have been educated with these technological innovations.

Scopes have been focused as future thrust on D&D survey for growing MPTs under Agrisilviculture, Agri-horticulture and silvipastoral systems at different agroclimatic regions of different crop zones of the State. Further scientific investigations are needed to assess the suitability of these systems under different tree species of different crop zones for its sustainability under different agroclimatic conditions. The studies need to be focused on biological interaction including light, nutrient and water use efficiency.

Social achievements

Farmers certainly got the social recognition due to the plantation of fruit/ timber trees. The significance of the planting trees under agro-forestry system created awareness amongst the rural people as fruit & other important tree species not only provided the environmental safety but also it will be the major source for their subsistence. The other farmers are also taking interest to grow some fruit plants in their cultivated lands and near the open well.

Plantation on wastelands created awareness among farmers and they realize the importance of plantation for balance of nature. They understand the needs of nation and society as well as economic return they will get in future. Farmers have accepted that plantation is not only national wealth but also base and lifeline for their coming generations.

Future prospectus

Low and late cash returns due to long gestation period from agroforestry system are the foremost problem needs future attention. When farmers executed an agroforestry system in a particular area, the local people or individual farmers should be made aware about the programme. It has been noticed that farmers when do not get reasonable price for their produce in due course then they do not like to participate in tree farming. Providing moisture to tree in wasteland is often neglected rather cumbersome. Additional allotment/ attention must be done to ensure this before taking up plantation work.

Successful adoption of any agroforestry system needs a multidisciplinary approach *i.e.* it requires knowledge of soil science, agronomy, horticulture, forestry, meteorology, economics *etc*. Scientist personal contact with farmers is the best method to educate the cultivator about agroforestry technology and guiding them time to time but this method rather slow and could cover only limited number of farmers. PRA tool should be adopted to know their problems, basic needs, selection of crop and suitable multipurpose trees.

Technology broadcasted on radio, telecast on TV and newspaper, farmers fair, meeting repeatedly are the effective methods to develop wasteland. Financial help in the form of planting material, seed & seedling *etc* has supported otherwise it will become difficult to develop wasteland through agroforestry system.

Technology gaps

Farmers are still unaware about the disease and pest management in different agro-forestry systems. They are also unaware about use of fertilizers and other management practices. Cost effective water conserving technologies must be generated, failing which the growth, even its survival is difficult.

Farmers preferred agri-horticultural system and boundary plantation for development of own wastelands. There is quite a large scope for popularization of other systems to meet the requirements of fuel, fodder, medicinal crop, timber *etc*.

Methodological deficiencies

Most of the farmers preferred to develop their uncultivated lands/ wasteland by adopting proper methodology but after due course of time they could not look after the plantation. Most of the plantations done by the government / non-government bodies have been failure due to lack of aftercare in adverse conditions especially in summer. *In-situ* moisture conservation may work for some weeks hence adequate provision of watering through water harvesting measures should be practiced.

Development of wasteland through agro-forestry systems initially required investments and income would generate in coming months to years. This situation is troublesome for small and economically weak farmers. Even resourceful farmers have desire for quick return from their investment. Utilization of inter-row space of widely spaced fruit plants by cultivation of high market value crops is one of the best solutions and required technological as well as methodological expertise.

Service required

Farmer's expectation is more from the scientists for the development of their wastelands. They also requested to give more number of genuine planting materials to cover larger area of their wastelands. They also requested for the distribution of medicinal plants and some important tree species of multiple use. Some farmers need to establish their processing unit for the medicinal crops. Expectation of the farmers is more from the KVK for the development of their wastelands. They demanded more inputs and frequent supervision. They like fast growing plants *i.e.* clonal Eucalyptus, Khamar, Papaya, Guava *etc*. Extension programmes run by KVK creates not only awareness but also deep desire for wasteland development.

Advantages of agroforestry systems

- It yields extended range of products, viz. food, fodder, fuel, fibre, fruits, fertilizer, cosmetics, medicines, oils, resins and gums.
- It helps in utilization of off-season precipitation. Most of the drylands are cropped for a single season but if a perennial tree component is there in

the system, it would make use of the same efficiently. A dryland ecosystem without perennial component can never be a sustainable system.

- Sustainable option for reclamation and management of wastelands such as sand dunes, salty lands, ravines, mine spoils, *etc.*
- Helps in maintaining hydrological balance of ground water. Integration of trees having higher transpiration rate with annual crops helps in lowering water table in regions having rising water trends.
- Agroforestry in dryland helps soil and water conservation. The perennials component of the agroforestry system either on contours or in some other geometry helps in reducing soil and water losses.
- It imparts stability and results in risk reduction of the 2 or more components of the system, even if one fails, the other would give a harvest of some economic value.
- A viable option for building organic matter in the soil by moderating temperature under tropical conditions.
- Presence of trees in agroforestry system gives extended management options for the farmers, *i.e.*, the trees can be harvested either for food or fuel or pole or small timber or for nitrogen as per farmer's requirements and market demands.
- Disposal option for poor quality waters such as domestic and industrial wastes including heavy metal loaded waters.
- Agroforestry system checks the rise in soil temperature, especially during summer months. Thus, it protects the soil microorganism, which would be of great benefit in crop production. Trees also provide as shade for cattle.
- It helps in recycling of nutrients from lower soil horizons. The deep-rooted nature of most trees results in tapping of nutrients from deeper soil layers and returning them to plough layer through leaf drop and litter.
- Mitigation of green house effect, depletion of ozone layer and pollution check.
- It helps in off-season family labour utilization. During off-season there would be no fieldwork for family labour in dryland areas. Time can, thus, be effectively used for harvesting, pruning and marketing of the tree components in agroforestry systems.
- Tree component of agroforestry system results in improvement in the microclimate of the whole system.

- Agroforestry results in improving and stabilizing the income of farmers.

Constraints of agroforestry systems

- Farmers, in general, are reluctant to grow trees on their farm lands owing to long gestation period in getting returns from trees.
- Proven agroforestry technologies are available only for limited situations/ locations/ regions.
- Lack of quality seed/ planting material sources of promising agroforestry species.
- Rigid legal laws restricting harvesting, transporting and sale of trees.
- Lack of assured financial support for popularizing agroforestry; non-availability of assured market and support price for agroforestry produce; lack of post harvest processing facilities for agroforestry produce.
- As in case of other state departments such as agriculture, horticulture, forestry, *etc.* wherein a well-developed network exists for transfer of technology, no such system for agroforestry extension, which requires different complex approach, is available. This is a serious lacunae in the adoption of agroforestry technologies at district, block or taluka level.
- Small land holdings and defective land ownership.
- Lack of time series analysis of economics.
- Limited knowledge about choice of the species.
- Some reduction in yields of annual crops.
- Lack of trained manpower and infrastructure for agroforestry.

Suggestion for promoting agroforestry

- Laying out adaptive research trial on the farmer's fields.
- Establishment of pilot scale demonstrations in collaboration with developmental departments such as forest, agriculture, soil conservation, horticulture, animal husbandry, *etc.* on their research and demonstration farms.
- Media supppoert and publication/ distribution of literature in local language.
- Organization of Kisan Melas/ Kisan Diwas at the research centers and at demonstration sites.

Future thrust

- There is a need to develop appropriate location specific agroforestry systems for Central India region of the country for higher productivity, sustainability and better economic returns to the farmers.
- To upgrade and refine the already identified agroforestry models/ practices for value addition, quality improvement and quick returns coupled with higher productivity. Work in this direction is already in progress at some of the lead centers.
- Role of agroforestry in mitigating/ reversion of environmental degradation processes such as control of wind and water erosion, sand dune stabilization, reclamation of salt affected and acidic soils, rehabilitation of ravines and gully lands will continue to form an integral part of future research in agroforestry. Agroforestry systems as a source of sink for atmospheric carbon sequestration to balance the predicted climatic changes will be one of the key areas of research in the near future. Hydrologists are predicting major fluctuations in ground water hydrology of major food grain production zones in the country. Researcher are, therefore, needed to identify suitable forest and fruit species which can be incorporated in Central India region for balancing ground water level without jeopardizing productivity and economic returns to the farmers.
- Practices which will promote the build-up of organic matter (OM) in the soil will largely determine the future prospectus of increasing production of food, fibre, timber and fuel production in India. Our long-term field trials with most of the arable crops cultivation in different regions indicate no build up of OM in the soil. In most cases, depletion of OM with cultivation under tropical conditions has been reported. On the other hand, where trees were incorporated with arable farming, net build-up in organic matter status of soil has been noticed. Therefore, the role of trees in improving soil quality for sustainable agricultural production aiming to reduce dependence on artificial inputs such as fertilizers will be one of the main issues for future research in agroforestry.
- Development and standardization of agronomic/ management practices such as irrigation, nutrient management, spacing, pruning and plant protection, *etc.* for the already identified agroforestry systems for improving productivity will continue to form an integral part of future research in agroforestry. The emphasis should be on how total income of farmer can be increased by incorporating trees without affecting crop yields and degrading the land.
- Intensive research efforts will be required to improve the already identified promising tree species such as shisham, teak, poplar, babool, eucalyptus,

tamarindus, vilayati babul, neem, khejri, *etc.* using latest plant breeding and genetic tools for improved productivity, value addition and quick returns. There is also an urgent need to standardize techniques for their asexual propagation, establishment of seed / propagule orchards at representative sites for speedy multiplication of germplasm for distribution to agencies involved in afforestation/agroforestry programmes in the country.

- To develop viable agroforestry systems in the future and to refine the present ones, it will be highly desirable to strengthen basic research on agroforestry *viz.* (i) Isolation of tree-intercrop-soil-root relationships in terms of sharing space, moisture, nutrients, light and other growth factors amongst the agroforestry components, (ii) Nitrogen fixation, role of VAM, physical, chemical and biological transformations in soil as governed by the species mixtures *etc.*
- In the light of GATT agreement and globalization of agriculture, there is an urgent need to identify high value medicinal, aromatic, industrial and aesthetic trees, bushes, grasses and crops and to develop technology for their cultivation in a unified agroforestry system where ever possible for export purpose.
- At present, there is no mechanism/ machinery for transferring agroforestry technologies from the research centers to the user agencies *i.e.* farmers. Whatever efforts in TOT are being presently made are by the agroforestry scientists themselves. There is a need to develop a system through which farmers can be motivated to adopt agroforestry practices. Efforts will be required to publish agroforestry information generated at different centers in local languages for free distribution to the farmers.
- There is also an urgent need to develop infrastructure facilities for organizing national and international training courses on agroforestry research, teaching and extension.

Conclusion

Agroforestry systems hold promise to provide sustainable land management, by arresting degradation and loss of productivity due to excessive use and lower population supporting capacity. Through diversification of agriculture, it is possible to reduce the risk of crop failure due to uncertain weather and erosion hazards. Such interventions lead to the supply of food, fodder, fuel, and timber in a more sustainable way for the survival of man and animal. When the needs of farmers are satisfied in respect of forest products, the dependence on forests will come down, leading to ecological restoration and conservation of soils and forests.

In this context the implementation of agroforestry systems at farmers field and institutional land shown a major impact on elevating fertility status, productivity of land, biomass yield as well as improving water holding capacity of the soil from low to high level in terms of high NPK and organic content of the soil, fruit, fuel wood and fodder production *etc*. This has also pave way to generate various source of employment with regard to engagement in agroforestry models.

Several scopes have been arisen as future thrust on D&D survey for growing MPTs under Agrisilviculture, Agri-horticulture and silvipastoral systems at different agroclimatic regions of different crop zones of the State. Further scientific investigations are needed to assess the suitability of these systems under different tree species of different crop zones for its sustainability under different agro climatic conditions. The studies need to be conducted on biological interaction including light, nutrient and water use efficiency.

The traditional land use systems have become very fragile and unsustainable due to unbalanced utilization of natural resources, resulting in large-scale degradation and consequent occurrence of wastelands. The primary cause of degradation is the demographic pressure on land, leading to loss of vegetative cover through deforestation. Harvesting of timber, collecting firewood, overgrazing and encroachment of forest areas are some of the reasons for deforestation in State. Besides that, ignorance of proper soil conservation measures, non-judicious use of fertilizer and pesticides, faulty irrigation and water management, discharge of industrial effluent, sewage/sludge are also responsible for soil degradation to a considerable extent. The rationale of developing agroforestry in Madhya Pradesh is to obtain a higher total, a more diversified and/or a more sustainable production of the combined tree and other agricultural and/or livestock resources that is possible with other forms of land use.

Soils research in agroforestry leading to ecological sustainability and organic farming is now scientifically recognized. During the last two decades understanding of the ecological potential of agroforestry systems under ecozones has increased significantly. The greatest ecological potential of agroforestry systems, to lie in buffering and maintaining the production capacity of agro ecosystem, rather than in increasing crop production. Properly managed woody perennials on croplands under different agroforestry combinations/ systems contribute positively to present resource losses. On degraded land the reintroduction of woody perennials may assist in restoring rather improving the production potential of the land. Even if trees exert a negative effect on crops (e.g. through shade or root competition,) it may still make economic sense for a farmer to practice agroforestry as a sustainable form of land use and more over it is analogous to organic farming. It is now recognized, that highly productive organic based, ecologically sustainable agroforestry systems is the viable option to be considered for land use practices especially for culturable wastelands.

Agroforestry thus concerns the ways towards organic farming and sustainability, a mechanistic approach towards anticipatory research concentrating on tree-crop interface, nutrient cycling and C-sequestration is needed to strengthen soil productivity research in agroforestry.

References

Nair, P.K.R. 1993. An Introduction to Agrofoestry. Kluwer Academic Publishers. Dordrecht. The Netherlands. 450.

Shrivastava, S.S. 1995. Agroforestry. Central Book House, Raipur (CG) pp. 384.

Singh, P., Upadhyaya, S. D. 1999. Alternate land use system for problem soils. In: Proceedings of the Training programme on management of problematic soils at CAS, Jabalpur: 156-160.

Solanki, R. R., Bisariya, A. R., Honda, A. K. 1999. A Decade of Research (1988-1999). National Research Centre for Agroforestry, Jhansi 1- 44.

Upadhyaya, S.D., Tiwari, G, Sharma, A. 2001. "Scope of Medicinal Plants cultivation in wastelands of Madhya Pradesh" Published in the SOUVENIR of National Seminar on "*Commercial Cultivation Processing and Marketing of Medicinal and Aromatic Plants.*" Organized by Department of Plant Physiology, College of Agriculture, JNKVV, Jabalpur-482004 (MP) held during Nov. 27-29, pp.24-31.

Wasteland Atlas of India 2000. Ministry of Rural Development, DOLR GOI, New Delhi.

21

Rehabilitation of Degraded Lands through Silvipastoral Systems: A Balanced Approach for Alternative Land use in the North-Western Himalayas

Charan Singh and O.P. Chaturvedi

Introduction

India has about 18% of the human and 15% of the livestock population of the world that needs to be supported from only 2% of the earth's land area. The per capita land availability has declined from 0.89 ha in 1951 to 0.37 ha in 1991 and is likely to be 0.20 ha by 2035 (Singh, 2005). It is estimated that demand for green and dry fodder would be short by 64 and 24% by 2020, while it was short by 62 and 22% in 2005. The current sustainable production of firewood is 17 million tons from forest areas and 98 million tons from non-forest areas, leaving a net deficit of 86 million tons which is removed illicitly from forest and scrub formations all over the country. Nearly 32.67% of India's geographical area is affected by various forms of soil erosion and land degradation. The area reported to be degraded in India varies widely, ranging from 63.9 million ha (Anonymous, 2000) to 187.70 million ha (Sehgal and Abrol, 1994), due to variations in approaches used to define degraded lands. The area under wastelands is estimated to occupy 63.85 million ha which is 20% of the country's geographical area. In the foothills of the Western Himalayas, large areas are categorized as bouldery wastelands which have been formed due to deposition of huge amounts of coarse soil forming material that has been brought down by seasonal torrents originating in the higher reaches. Deforestation, faulty landuse practices and road construction in the upper reaches lead to soil displacement which moves down the slopes and is deposited in the foothills in the form of gravel bars, with coarser fractions spreading out over the original soil layers. Recent surveys indicate that the area affected by these deposits is about 2.73 million ha in the Himalayan foothills and 0.3 million ha occur in Doon Valley alone (Arora and Vishwanatham, 1995). These lands are unsuitable for cultivation of field crops and covered by uneconomic vegetation which needs immediate attention for rehabilitation to prevent further extension. Utilization of these degraded areas and wastelands in developing countries through silvipastoral systems is a widely recommended practice for meeting demands of firewood and fodder (Upadhyay and Pathak, 1994). Trees and grasses together simulate the multi-storeyed physiognomy of natural forests and help in the recovery of degraded habitats through nutrient cycling and enrichment (Nair, 1983).

The successful establishment of trees and grasses together depend on balanced access to solar energy, water and nutrients. Establishment and growth of herbaceous vegetation underneath trees is directly affected by tree density (Walker *et al.*, 1986), shade tolerance (Walters *et al.*, 1982), light interception (Wilson, 1996) and moisture and nutrient availability (Raison *et al.*, 1992). Growth and herbage yield from the understorey component begin to be affected only when tree crowns are fully developed (Mathew *et al.*, 1992). In several cases, like water stressed sites, it has been observed that herbaceous vegetation affects tree growth due to competition for water and nutrients (Balocchi and Philips, 1997). Tree canopy management in a silvipastoral system is thus necessary to maximize system productivity, by lopping, pollarding, coppicing or thinning. Crown pruning is known to modify rooting patterns and processes occurring below ground (Jones *et al.*, 1998) in trees and probably enrich soil by greater fine foot turnover (Jeschke and Hartung, 2000). Management of tree canopy by lopping and pollarding improves ground vegetation tree productivity by enhancing light penetration (Debroy, 1994). The impact of tree canopy management on grass productivity has been reported earlier from the north west Himalayas by Vishwanatham *et al.* (1999), from semi-arid central India by Pathak and Roy (1987) and from the arid zone of Rajasthan by Shankarnarayan and Singh (1990). Perennial vegetation such as multipurpose trees (*Grewia, Bauhinia, Albizia, Robinia,* etc.) and hardy grasses (Napier, Gorda, *Panicum, Eulaliopsis,* etc.) are not only the alternative for economic utilization of these degraded lands unsuitable for agricultural crops but also protects them by considerable reduction the soil loss and runoff.

Silvipasture is the most promising alternate land use system which integrates multipurpose trees, shrubs, legumes and grasses, mostly on non-arable, degraded and marginal lands for optimising land productivity. However, in arable land, boundary plantations, hedges and live fences, perennial crops with pasture, fodder banks, trees on cropland, hedgerow intercropping, shelter belts and windbreaks are the other form of silvopastoral practices. Silvipastoral practices help in conservation of natural resources and provide forage, timber and fuel wood on a sustainable basis. Small and marginal farmers who are engaged in raising livestock can benefit a lot by adopting this system.

Fodder production from marginal lands

Marginal and sub-marginal lands are usually uneconomical for crop cultivation and are often used for grazing by livestock. These lands have been stripped off the useful and adequate vegetative cover for livestock grazing and are continuously subjected to soil erosion. Silvipastoral systems attract forage production from these lands. The systems approach for these areas is based on the mixtures of shrubs/trees/grasses/legumes.

Table 1. Grassland types and herbage production.

Grassland types	Harvestable dry Forage yield (tha^{1})
Heteropogon contortus and Eremopogon foveolatus (natural)	2.4
Sehima nervosum, Cenchrus ciliaris, Albizia lebbek and *Clitoria ternatea*	7.3
Heteropogon contortus and *A. lebbek*	4.1

(*Source:* Patil, 1982)

The productivity from natural grasslands is generally low (Table 1). The research results of IGFRI, Jhansi and CSWCRTI, Dehra Dun indicate the possibility of obtaining forage productivity from lands under silvipastoral systems of the order of 8 to 11 tha^{-1} which includes forages from understorey grasses and also loppings from the fodder trees. The trees have also an additional advantage of providing green fodder during the extreme dry season. The system of integrating tree species along with arable crops/range grasses not only combine for their complimentarily but also serve to conserve and improve the productivity of the site and at the same time optimise the combined production. The woody species which form the upper layer is very important for its proper selection which will suit to specific ecosystem and the need based production. The choice of multi-species crop depends on their complementarity and not competitiveness.

Fodder production from problem soils

Nearly 50 per cent of land area in the country is in various stages of degradation. A large chunk of land under problem soils such as ravines, deserts, saline, alkaline and acidity, swampy and flood and cyclone prone conditions, waste areas along the canal and river banks, in watershed areas of the dams, areas along the railway line which all can form the reserves of forage production. These soils not only have the soil and water management problems, but many a time accentuated with the problems of water logging, soil salinity, alkalinity and acidity depending upon their agro-climatic locations. The biomass production strongly depends on plant population and their growth, which depends on availability of nutrients, moisture and proper aeration in soils. The problem soils have single or more limited factors which may come in the way of normal plant growth. The problem soils have to be tackled through specific soil and crop management techniques and selection of appropriate vegetation which can make use of such environment in best possible way.

A number of crops have been identified for their suitability to be grown under saline and sodic conditions but it has been observed that the crops like berseem, teosinte, sesbania, trigonella, Sainji can tolerate some degree in salinity sodicity but karnal grass is very suitable under saline-alkali conditions. Some crops like teosinte and hybrid Napier can tolerate to some degree of water logging at latter growth stages. Similarly, para grass could be grown under marshy habitat.

Viable silvipastoral systems have been designed for better utilisation of alkaline soils. Abandoned alkali lands can be improved by growing *Prosopis juliflora* with *Leptochloa fusca* (Karnal grass). After 52 months with this combination, soil pH decreased from 10.3 to 9.4 and electric conductivity from 2.20 to 0.42 dS, while organic carbon in the top soil increased from 0.18 to 0.43% (Singh *et al*., 1991). These two species improved the soil to such an extent that it was possible to plough under the Karnal grass after 4 years and grow less tolerant but more palatable fodder species like *Medicago* and *Trifolium alexandrium* under *Prosopis* trees.

Reclamation of alkali soils by addition of organic manure, gypsum followed by crop of Sesbania (Dhaincha) and molasses have been carried out (Abrol, 1982). The following filling mixtures for filling the planting holes in sodic soils have been suggested;

a) Shallow pit 3 kg gypsum + 3kg FYM + 3 kg rice husk

b) Deep pit 5 kg gypsum + 5kg FYM + 5 kg rice husk

c) Deep pit for fruit trees 10 kg gypsum + 10kg FYM + 5 kg rice husk

Unlike alkali soils, amendments of gypsum or pyrites are not required in saline soils. Instead leaching with good quality water is done to wash out the salts from the rooting zones. Sub-surface planting (30-45 cm below) and ridge-trench planting method helps to keep the salts away from the rooting zones.

Soil working techniques

To rehabilitate the degraded lands, innovation in the technological aspects of afforestation is needed, particularly to ensure in situ soil moisture conservation to enhance the productivity of land. Further, suitable plant species and proper management practices which may be site-specific, based on socio-economic needs are vital for achieving the success. Fodder and fuel wood production could be increased several-folds from the existing levels through appropriate silvipastoral practices and proper management of wastelands in the watersheds.

The soil working techniques for silvipasture systems depend upon the edaphic factors, such as the soil depth, texture, proportion of gravel and boulders, rockiness and rooting habit of the species etc., which governs the depth and size of the pits. Soil working conserves moisture in the rooting zone, improves aeration of soil and removes weeds from the site. Filling mixture should consist of the top soil and farm yard manure. In very poor sites, good soil is required to be brought from other areas for filling the pits. A brief description of the types of pits required for this area is given in Table 2.

Table 2. Types of pits for different sites

Type	Suitable for
1. Ordinary pits (30-50 cm^3)	Clayey, saline/alkaline in all rainfall areas.
2. Saucer pits (50-75 cm^3)	All types of loamy soil for the dry zone.
3. Ridge-ditch (large)	Deep soil, low rainfall zone.
4. Shallow filled shelfed trenching	Sandy, gravely soil, low rainfall areas.
5. Trench mound	In areas with low erratic rainfall.

Trenching is also an integral part of watershed development especially in non-arable lands and it is important that such activities be carried out prior to planting of seedlings. Trenches are made on contours at horizontal interval of 4-5 m and are always interrupted contour trenches. Trenches are 3-4 m long, 0.5 m wide and 0.5 m deep. In dry localities trench mound planting on down slope is usually taken up with the selected species.

It is also important to carry out some moisture conserving practices, like contour furrowing, trenching to improve grass yields and sowing of legumes like *Stylosanthes hamata* in a ratio of 4:1 in order to improve fodder quality.

Development of non arable land for productive use

Cultivable wastelands are actually those lands where the productivity is less than its optimum and can be reclaimed and put to use for producing multiple benefits that shall benefit rural communities. Technologies for the restoration of these lands exist and if properly implemented, wastelands can be restored to half their productive potential. Besides environmental stability, the restoration of wastelands provides employment opportunities and creates alternate sources of income for farmers.

While it is impossible to recreate the complex structure of a natural forest, properly planned mixed species plantations can serve as 'foster' ecosystems.

Plantations that are raised, protected and managed efficiently can mimic natural forests to a large extent and they perform the same functions. Thus trees and grasses can be managed to provide fuelwood, fodder, fruits and non-wood forest produce in wastelands of different types.

(a) If the objective is to improve degraded lands and thereby reduce runoff and erosion, then it is necessary to built and /or improve forest areas. This can be done by protection from grazing, fire; planting with suitable multi-purpose tree species; supplementing with contour staggered trenches, furrows etc.

(b) If the objective is to maintain good watershed conditions, while expecting timber from forest and grazing, then all activities have to be done in a controlled manner. Only 30-40% of the basal area (of trees) should be harvested and if soil losses are expected to be high than only 25-30% should be removed. Grazing by animals is to be controlled and pockets of natural regeneration are to be strictly protected. Trees have to be harvested by following the selection system.

The selection of species for plantation on degraded lands depends upon the choice of ultimate end user. For example, trees planted for fuel wood should be fast growing, capable of producing coppice shoots rapidly, have high calorific value wood and easy to harvest and transport. Trees planted for fodder should be fast growing, nitrogen fixing, rich in crude protein, produce nutritious non-toxic leaves and coppice easily. Species planted for timber production should be tall with straight stem and have strong fine grained wood with good machining ability.

Establishment of silvipasture systems

Silvipasture system establishment particularly in Class V, VI and VII lands, which are not suitable for crop production can be accomplished through afforestation; pasture establishment and range improvement measures.

A. Pasture establishment

Silvi-pastoral systems can be raised on marginal lands with suitable fodder grasses like *Dicanthium annulatum, Leptochloa fusca, Brachiaria mutica, Panicum antidotale, Sehima nervosum, Chloris gayana, Cenchrus ciliaris.* Trees are planted at wide spacing, 5x5 m or more depending up on the area available. In the initial years, grasses are harvested manually and later on when the trees have attained sufficient height, animals are permitted to graze. Trees which can be planted on marginal lands for silvi-pastoral system are: *Acacia nilotica, A. catechu, Terminalia alata, Madhuca latifolia, Anogeissus latifolia, Adina cordifolia, Hardwickia binata,* etc. To improve grass yields and fodder quality of pastures, sowing/planting of legumes like *Stylosanthes hamata, Medicago* and *Trifolium alexandrium* can be done. Silvipasture systems involving *Dalbergia sissoo* + *Chrysopogon fulvus; Acacia catechu* + *Eulaliopsis binata; Bauhinia purpurea* or *Albizia lebbek* with *Eulaliopsis binata* were found promising in the utilization of degraded lands for fuel wood and fodder production.

B. Range improvement

In this kind of silvipasture system, trees are grown at wide space (8x8 m; 10x10 m or 12x12 m) in the rangeland (grassland) and controlled grazing by the livestock is allowed. The optimum productivity of the range is maintained by appropriate grazing systems with the required livestock and periodic improvement measures.

C. Grassland improvement

1. Before taking up improvement, it is necessary that protection is provided to the area, which should not be a very large area in the first phase (4-5 ha), with the area increasing slowly in phases. Protection, with the active involvement of farmers is desired to first ensure recovery of the site and secondly to avoid social conflict. Areas close to a village, with shallow soils (< 30 – 45 cm) and moisture stress for a longer period which cannot be put to other use, can best be developed as grazing areas. Use of thorny bushes, *Agave*, staking with dried thorny branches or stone walls are the common and cheap methods of protecting such areas.

2. Moisture conservation: - To improve seed germination or establishment of slips, soil and water conservation in these areas becomes necessary. Contour furrowing, contour trenching and contour bunding (on 5-8% slopes) have been found to be effective in moisture conservation and the consequent increase in forage production. Furrows can be made by tractor drawn implements, with the contours being 5–7 m apart, depending on slope features and rainfall pattern. Yield increases, ranging from 100–165% have been reported from the dry regions.

3. Agro techniques: - Reseeding of grasses, although economical, does not guarantee seed germination if the soil condition is not conducive. The best time to sow seeds is after the first heavy shower. The seed is mixed with sandy soil 3-4 times the seed volume and sown by seed drills, 50-75 cm apart, 3-5 cm deep. Grass species suitable for planting or seeding in silvi-pastoral land use systems and their seed rate (kg ha^{-1}) given in the brackets are - *Brachiara mutica* (5-7), *Cenchrus ciliaris* (4-6), *Themeda triandra* (7-8), *Chrysopogon fulvus* (5-7), *Panicum maximum* (4-5), *Panicum antidotale* (4-5), *Apluda mutica* (5-7) and *Bromus* spp (5-6) *C. setigerus* (5-7), *Lasirus sindicus* (6-8), *Dicanthium annulatum* (2.5-3.5). A list of suitable grass and legumes for different agro-climatic situations is given in Table 3. Periodic weeding is carried out frequently in the first year and continued till the stand is well established.

Table 3. Suitable grass and legume species for different sites

Region	Grass species	Legumes
North eastern Humid region	*Arundinella bengalensis, Imperata cylindrica, Pennisetum purpureum, P. polystachyon, Imperata cylindrica, Setaria sphacealata.*	*Medicago sativa, Gycine javanica, Stylosanthes hamata,Stylosanthes gracilis, Dolichos axillaris*
Northern sub-humid region	*Dicanthium annulatum, Chrysopogon fulvus, Cenchrus ciliaris, Pennisetum pedicillatum, Setaria sphacealata, Panicum maximum, Hybrid napier*	*Stylosanthes hamata,Dolichos lablab, Phaseolus atropurpurem*
Se mi arid (<800 mm/year)	*Sehima nervosum, Cenchrus ciliarisDicanthium annulatum, Chrysopogon fulvus Eragrostis curvula,Chloris gayana, Panicum antidotale, Heteropogon contortus Brachiara mutica,Isaelema laxum*	*Stylosanthes hamata,Phaseolus Lasirus atropurpureum, Dolichos lablab,Dolichos axillaris*
Arid (<500 mm/year)	*Lasirus sindicus,Cenchrus setigerus, Panicum antidotale, Cenchrus ciliaris, Dicanthium annulatum*	
Saline – alkali (Salt affected)	*Leptochloa fusca* (karnal grass), *Chloris gayana* (rhodes grass), *Cynodon dactylon* (bermuda grass), *Brachiaria mutica* (para grass), *Panicum maximum* (Guinea grass), *Pennisetum purpureum* (Hybrid napier).	*Desmodium* sp.*Rhychosia minima*

i) Planting of rooted slips of suitable grass species is costly but a certain method of grassland improvement. Slips are planted 0.5–0.7 m apart in shallow holes and pressed firmly into the ground. The entire area is covered in this way, after he first heavy showers, when the soil layer is moist. Application of DAP @ 20kg/ha is beneficial for growth. In difficult areas, pelleted seeds are broad casted in the beginning of the rainy season. Pellets are prepared by mixing seeds with cow dung, clay, sand in a ratio of 1:1:3:1, using water to prepare round pellets of 0.5 cm diameter.

ii) Introduction of legumes: Legumes are known to fix nitrogen in the soil and being rich in crude protein, they improve the fodder value when mixed with poor quality or dry grass. A list of the important legumes for different agroclimatic regions as given in table 3. Legumes are usually mixed with grasses in a ratio of 1:4. Application of small doses of DAP are required to allow for nodulation to occur in poor soil, where much of the top soil has already been washed away.

iii) Management: Once planting/seeding has been completed, it is necessary to provide adequate protection (illegal grazing, fire, harvesting etc) to the area in the first year. In the 2nd year, depending on the establishment, a few manual harvests can be allowed. From the third year controlled grazing can be permitted based on the carrying capacity. Even then a few small patches need to be rested and allowed to seed, so that seed collection for further multiplication, of the desired species can be carried out. After 5 years, the area needs to be examined once again for any improvement if required, which is most easily seen from decline in yields.

Selective silvipastoral practices for arable land

A. Boundary plantations of fodder species

In this practice fodder species such as *Acacia* spp., *Adina cordifolia, Ailanthus excelsa, Albizia lebbek, A. procera, Anogeissus latifolia, Celtis australis, Dalbergia sissoo, Derris indica, Ficus* spp., *Grewia optiva, Leucaena leucocephala, Melia azadirach, Sesbania grandiflora, S. sesban* are planted along field boundaries or other borders e.g. along footpaths or irrigation channels as per the agroclimatic condition.

B. Trees on erosion control structures

Trees such as *Leucaena leucocephala, Melia azadirach, Sesbania grandiflora, S. sesban, Gliricidia, Calliandra, Grevillea robusta, etc.* are added to earth structures employed for soil and water conservation, such as terrace risers, bunds, ditches and grass strips. These trees help in stabilizing the structures and make productive use of the land which they occupy. Grasses such as napier, guinea, vetiver are also planted in between the trees.

C. Wind breaks and shelter belts

Planting of suitable MPTS wind breaks around the field boundaries is very important for crop production. MPTS are planted in lines across the direction of damaging winds, to reduce wind erosion and wind or frost damage to crops. They also provide shelter for animals. Some of the trees suitable for wind breaks/shelter belts are: *Acacia senegal, Aegle marmelos, Albizia lebbek, Bauhinia purpurea, Cordia myxa, Dalbergia sissoo, Holoptelia integrifolia, Kigelia pinnata, Lagerstroemia indica, Leucaena leucocephala, Madhuca indica, Pongamia glabra, Prosopis cineraria, Tamarindus indica, Syzygium cumini,* etc.

D. Hedgerow intercropping and contour hedgerows

This is also called alley cropping or alley farming. Hedges of *Leucaena leucocephala, Sesbania grandiflora, S. sesban, Gliricidia, Erythrina, Albizia or*

Cassia are planted as more or less parallel rows with the plants closely spaced and crops/grasses are frown in the alleys between them. The rows are generally kept 4-10 m apart. The hedges are regularly pruned. Prunings may either be removed as fodder and fuel wood or retained on the soil. A contour hedgerow is a variation of this system, practised on sloping lands and with the primary objective of soil and water conservation.

E. Fodder Banks

Trees are planted as blocks and managed for fodder production. This can make good productive use of areas of poorer soil on the farm.

Management of trees under silvipasture systems

The management considerations are influenced by site conditions, inputs, demand for forest products, market prices, etc. Trees have the potential to provide the maximum returns (through biomass), only when they are managed effectively and efficiently. Some of the popular techniques adopted to manage tree as are given below.

A. Tending

Tending is carried out for the benefit of a forest crop by creating best possible conditions of growth. Tending helps in producing high quality wood and maximizing returns per unit area, it is therefore, an important silvicultural operation. Tending operation include weeding, cleaning, thinning, pruning, climber cutting, etc.

Immediately after the planting is over, pits should be mended, and if climatic conditions permit, casualties are replaced before fourth week of July to enable the seedlings to get sufficient moisture from the rains that follow. Limited casualty replacement is desirable in second year and only if there is a big gap due to insufficient moisture or other climatic hazards. Simultaneously, first weeding and hoeing is carried out around the plants within 2-3 weeks after planting. *In situ* water harvesting using the tie ridge technique is desirable to combat the moisture stress. The second and third weeding should follow mulching with dry grass/cut weeds around the pits. This operation should be completed before withdrawal of monsoon.

It is necessary to carry out the two or three weeding and hoeing during first year and another two during second and third year. The sites with woody weeds and scrub growth may need additional one or two shrub cuttings or pruning to keep the plants free from their adverse effect. Use of herbicide and weedicide may be useful particularly for the woody plants.

B. Canopy management

In arable lands as the farm holdings are small and the farmers' first consideration is to grow food and he cannot afford to sacrifice crop yield for tree growth. Therefore, canopy management besides selection of compatible shade tolerant crops (millets, rhizomatous crop) and trees with minimum shade *(Albizia* etc.) is the important consideration. Canopy management options in relation to trees include pruning, lopping, pollarding and coppicing. Trees in wood lots may be coppiced to produce fuel wood, while trees in fodder lots or hedge rows intercropping systems are often lopped to produce fodder and mulch and where necessary to reduce shade on the adjacent crops. Trees in pastures or on boundaries are pollarded for the same purposes. In crop lands, on boundaries or in homegardens, trees are pruned to reduce shading, to increase crop productivity and provide fuel wood and fodder.

C. Pruning

Pruning is removal of live or dead branches or multiple leaders from standing trees for the improvement of the tree or its timber. Pruning is the principal method for managing the vegetative yield of trees. Pruning stimulates the outgrowth of dormant buds that are often present at old nodes over the whole shoot system. Pruning near ground level produces longer shoots than pruning higher up. Pruning reverses tree ageing and the new shoots are vigorous in growth. Pruning operations are commonly done in forestry practices for improvement of timber volume of trees.

D. Lopping

Lopping means cutting of branches of a tree. Incidentally the lopped trees produce new shoots which are annually or periodically lopped for various purposes. *Acacia, Anogeissus, Hardwickia binata, Melia azadirach, Moms spp., Terminalia tomentosa, Kydia calycina,* etc. are annually lopped for leaf fodder. The optimum cutting frequency to ensure survival and rapid recovery will depend on the species, environment and amount of foliage left.

E. Pollarding

Pollarding in forestry is an operation in which the stem of a tree is cut off at a height beyond the reach of browsing animals with the object of producing a crown of new shoots from buds below the cut. The flush of new shoots is cut down periodically so that the pollard may produce fresh shoots again.

F. Coppicing

Coppice is to cut broad-leaved trees close to the ground to produce sprouts and regrowth. In fact in forestry practices, coppice is a method of vegetative reproduction in which trees when cut from near the ground level, produce coppice shoots. However, its success depends on the factors such as species, age of tree, season of coppicing, height and method of cutting and rotation. Regular coppicing promotes the growth of leafy shoots which are accessible for harvesting or browsing, and by restricting the development of the boles and woody frames, coppicing increases the proportion of dry matter availability tone shoots. However, all species do not coppice and even in the species that coppice, the power varies with species. Some examples of good coppicer tree species are, *Dalbergia sissoo, Acacia catechu, Shorea robusta, Tectona grandis* and species of *Albizia, Salix, Morus, etc.*

Rehabilitation of degraded lands for multiple outputs

A number of MPTs viz. *Grewia optiva, Morus serrata, Celtis australis, Robinia pseudoacacia, Ulmus wallichiana, Quercus* spp., *Bauhinia variegata Melia azedarach, Parkinsonia aculeata. Eucalyptus tereticornis,* Cassia *siamea, Acacia auriculiformis, Dalbergia sissoo, Albizia procera, Acacia nilotica, Thespesia populnea, Ailanthus excelsa, Derris indica* etc. have been found suitable for establishment of silvipastoral systems for rehabilitation of degraded lands. Trees with forage are preferred so that they can be lopped during lean periods and roughages supplemented with palatable and nutritious leaves for livestock feeding. *Hippophae rhamnoides*, an important indigenous multipurpose shrub of J & K, is used as fuel and fodder by the local inhabitants. In Spiti Valley of Himachal Pradesh 1ha plantation of this shrub can meet the fuelwood needs of about 20 families (ICFRE, 1993).

In general, trees for fodder should be fast growing, nitrogen-fixing, a good coppicing ability and production of nutritious leaves. A list of the tree/shrubs/ grasses suitable for the given area which can be used for fodder is presented in Table 4. These species can be grown on a wide variety of sites and lopping can be carried out when the trees have attained sufficient growth.

Table 4. Promising trees, grass and legumes for silvipastoral systems in different regions of India

Agroclimatic zones	Fodder trees	Grass and legumes	
Temperate, (J&K, Himachal Pradesh and NW Himalaya)	Wet region *Robinia pseudoacacia, Celtus australis, Populus alba, Quercus semecarpifolia, Q. dilatala, Acer campbelli.* Dry region *Salixfragilis, P. ciliata, P. euphratica, Belula utilis, Acer candatum, Q. ilex*	Lucerne *(Medicago satira),* Red clover (*Trifolium pratense),* white clover *(T. repens),* Cocksfoot *(Dactylis glomerala),* Timothy (*Pheleum pratense*)	Singh, 1986
Kangra Valley, HP	*Erythrina suberosa Dalbergia sissoo, A. catechu, Albizia stipulata*	*Eulaliopsis binata, Chrysopogonfulvus, Bauhinia purpurea, Albizia lebbek, Napier bajra, Setaria anceps*	Vishwantham *et al*., 1998
North-eastern Hill region	*i) Albizia lebbek, Bauhinia purpurea, Leucaena leucocephala.ii)Salix* spp., *Populus ciliata, Castanopsis, Celtis australis, Sachima\ vallichii, Ficus* spp., *Bauhinia variegata*	*Panicum maximu, Setaria sphacealata, Pennicetum polystachyon, Stylosanthes guyanensis,* Bajra x Hybrid Napier, *Panicum maximum* (HamiU *Doctylis glorneruta, Lolium perenne)*	Verma, 1986, Dhyani and Chauhan, 1990 Singh *et. al*., 1996
High rainfall, Doon Valley	*Leucaena leucocephala, Grewia optiva, Acacia* sp., *Bauhinia retusa, Salix tetrasperma.*	*Eulaliopsis binata, Chrysopogon fulvus*	Dhyani *et al*., 1988
Saline-alkali soils	*Prosopis juliflora, Acacia nilotica, Terminalia qrjuna, Pongamia pinnata*	*Leptochloa fusca, Chloris gayane, Brachiana mutica, Sporobolus* sp.	Singh *et al*., 1994
Semiarid tropics and red lateritic soils of Chattisgarh	*Acacia leucophloea Leucaena leucocephala*	*Pennisetum purpuream Hybrid napier, Guinea, Stylosanthas*	Rai and Suresh 1988' Rao and Osman, 1994, Nangraiya and Puri, 1994.
Arid semi arid	*Acacia tortilis, A. Senegal, Hardusckia binata, Ailanthus excelsa*	*Cenchrus ciliaris, C. stetigenus, Laxirus sindicus Cenchrus ciliaris, C. stetigenus, Laxirus sindicus*	Gupta, 1994 Mann *et al*., 1994

Trees planted for fodder as top feed along with suitable grass species should be widely spaced, so that moderate shade is available. At scattered locations some closely planted, shady trees are required to be planted so as to serve as resting spots for animals, along with a watering hole (for drinking and wallowing). Species with top feed value (Table 5) are preferred, so that some fodder is also available during extended dry periods. A list of species suitable for planting as trees in wastelands is provided in Table 6. Many of these species have other uses as well. It is however advisable to mix these with other species so that farmers are involved in their protection. The list is not comprehensive and neither is it necessary to plant all of them. The final choice of species should normally be obtained from the end-users. Participatory approaches and joint forest management plans should come into play in the very beginning of all development projects.

Table 5. Promising top feed species for silvi pastoral use in the subtropical zone

Sl. No	Species	Common name	Rating	Crude protein %
1.	*Acacia nilotica*	Babool	Good	14
2.	*Acacia catechu*	Khair	Good	13 - 18
3.	*Anogeissus latifolia*	Bakli	Good	7.5 – 11.5
4.	*Ailanthus excelsa*	Maharukh	Excellent	20
5.	*Anogeissus pendula*	Kardhai	Good	8 – 10
6.	*Albizzia lebbek*	Siris	Good	17 – 26
7.	*Bauhinia variegata*	Kachnar	Good	11 – 16
8.	*Cordia dichotoma*	Lasora	Good	12.4 – 15
9.	*Ehretia laevis*	Chamror	Medium	13.5
10.	*Ficus glomerata*	Umar	Good	11.2 – 15.2
11.	*Ficus lacor*	Pakar	Good	16
12.	*Kydia calycina*	Safed dhaman	Excellent	11 – 13.6
13.	*Lannea coromandelica*	Jhingan	Good	11.4 – 16
14.	*Prosopis cineraria*	Khejri	Excellent	15.2
15.	*Syzygium cuminii*	Jamun	Good	10.2

Mixing 2-3 kg of gypsum and 7-8 kg FYM with alkali soil is better as filling mixture. Species like–*Prosopis juliflora*, *Acacia nilotica*, *Acacia catechu*, *Butea monosperma*, *Madhuca latifolia*, *Terminalia belerica*, *Salvadora persica* are better in low rainfall areas. In areas with a higher rainfall (1000–1500mm) *Syzygium cuminii*, *Pongamia pinnata*, *Trewia nudiflora*, *Cordia dichotoma* are better. These are intercropped with suitable grasses as described earlier.

Table 6. List of fast growing tree species suitable for planting as permanent vegetation

Sl. No.	Name	Common name	Habit	Coppicing	Uses ability	Min. Rainfall (mm)
1.	*Acacia nilotica*	Babool	SP	Medium	Fw/Fd/T	200
2.	*Acacia senegal*	Gum acacia	SE	Good	Fw/Fd/Gum	200
3.	*Acacia tortilis*	Israeli babool	SP	Good	Fw/Fd	100
4.	*Ailanthus excelsa*	Maharukh	SP	Good	Fw/Fd	600
5.	*Adina cordifolia*	Haldu	SP	Medium	Fw/Fd	500
6.	*Albizzia lebbek*	Siris	SP	Medium	Fw/Fd	400
7.	*Azadirachta indica*	Neem	SP	Medium	Med/Fw	400
8.	*Boswellia serrata*	Salai	SP	Medium	Fw/Fd	400
9.	*Cassia fistula*	Amaltas	SP	Good	Fw/Fd/M	500
10.	Cassia siamea	Kassod	SP	Excellent	Fw	600
11.	*Dalbergia sissoo*	Shisham	SP	Excellent	Fw/Tim	400
12.	*Dalbergia latifolia*	Rosewood	SP	Good	Tim/Fw	600
13.	*Diospyros melanoxylon*	Tendu	SP	Good	Bidi leaves/	500
14.	*Erythrina suberosa*	Madar	SP	Good	Fw/Fd/T	500
15.	*Gmelina arborea*	Gamhar	SP	Good	Fw/Fd/T	500
16.	*Leucaena leucocephala*	Subabool	SP	Excellent	Fd/ Ml/sT	400
17.	*Hardwickia binata*	Anjan	SP	Medium	Fw/Fd/T	400
18.	*Holoptellia integrifolia*	Kanju	SP	Medium	Fd/Fw/T	500
19.	*Madhuca indica*	Mahua	SP	Good	Fd/Fw/Fr	500
20.	*Terminalia belerica*	Bahera	SP	Medium	Fr/ Med.	500

In the Shiwalik foothills, agri-horticulture and horti-pastoral systems have been developed for the highly eroded lands. Intercropping of guar, cowpea or pearl millet with peach; turmeric with papaya; *Chrysopogon fulvus* or Napier grass with *aonla* or *ber* has been identified as an economically-viable and eco-friendly system for rehabilitation of these lands. Aonla gave the highest yield of 86 kg fruits/tree in association with pigeonpea but the yield was reduced by 17 and 23% respectively in association with *Chrysopogon* and Napier grass. Silvi-pasture systems involving *Eulaliopsis binata*, *Saccharum munja* or *Vetivaria zizanoides* with *Acacia nilotica* were also found promising. Further, *Eucalyptus tereticornis* + *Eulaliopsis binata* has been identified as an economically viable and eco-friendly system for rehabilitation of these lands.

For highly degraded lands in semi-arid regions of central India, silvipastoral systems involving multi-purpose species are most suitable. *Albizia amara* based silvipastoral system produced 7.95 t ha^{-1} yr^{-1} fodder followed by *Acacia tortilis* (7.60 t ha^{-1} yr^{-1}). At Jhansi in a silvipastoral system consisting of *A. amara* + *L. leucocephala* + *D. cinerea* + *C. fulvus* + *Stylos* and natural grassland the total biomass yield of 12.62 t ha^{-1} yr^{-1} (4.55 t from pasture + 0.51 t from tree leaves and pods through pruning + 1.66 t from fuelwood through pruning + 5.90 t from small timber, fuelwood, leaf fodder through harvesting of trees) was recorded (Table 7).

This was about 4 times higher than yield obtained from natural grassland (3.16 t ha^{-1} yr^{-1}). These results showed that it is possible to get more than 12 t ha^{-1} yr^{-1} biomass through established silvipasture on these degraded lands which otherwise was producing only 3 t ha^{-1} yr^{-1} through natural vegetation (Rai *et al.*, 1999).

Table 7. Total biomass production under silvipastoral system and natural grassland at 8 years rotation (Rai *et al.*, 1999).

Components	Production (DM tha^{-1} yr^{-1})
Grasses and legumes	4.55
Tree leaves and pods at 50% pruning	0.51
Fuel wood at 50% pruning	1.66
Timber at 8 years harvesting	2.39
Fuel wood at 8 years harvesting	3.18
Leaf fodder at 8 years harvesting	0.33
Natural grassland	3.16
Total production	12.62

In shallow (15 cm) red chalka soils of Andhra Pradesh having about 2% slope, a silvipastoral system comprising *Leucaena leucocephala* (K-8) and *Cenchrus ciliaris* and *Stylosanthes hamata* produced on an average, 6 t ha^{-1} yr^{-1} dry matter yield from *S. hamata* and 3 t ha^{-1} yr^{-1} from *C. ciliaris*, besides, 6 t ha^{-1} yr^{-1} (biomass) from *L. leucocephala*.

About one-third of the lands in the Himalayan region are bouldery wastelands formed due to excessive erosion and deforestation on the hill slopes. These lands are virtually devoid of any economic vegetation and can be rehabilitated by establishing silvipastoral systems. In the Shiwalik foothills, silvipasture systems involving *Eulaliopsis binata*, *Saccharum munja* or *Vetivaria zizanoides* with *Acacia nilotica* were promising. Further, *Eucalyptus tereticornis* + *Eulaliopsis binata* has been identified as an economically - viable and eco-friendly system for rehabilitation of these lands. In the southern hilly region, a silvipastoral combination of *Acacia melanoxylon* and *Digitaria decumbens* on steep slopes was promising for soil conservation and meeting the fodder requirement. Similarly mixed plantation of *Eucalyptus globulus* and *Acacia mearnsii* - a system of a legume and non-legume ideal for soil and moisture conservation and also resulted in higher wood yield.

On the basis of researches and potentials under agro-climatic situations following areas were identified (Table 8) for different regions of the country. All the areas indicated above are interspersed in all the zones in various proportions along with cultivated lands. Forage floristic composition for each of these systems and for each of these zones may be different which need to be kept in view and researched upon.

Table 8. Potentials of various regions for silvipasture system in India.

Region	Components	Results
I. Himalayan Region	Alpine and legumes	Only pastures with suitable grasses *Agrostis* spp., *Chrysopogon gryllus, Dactylis glomerata, Poa pratensis, Trifolium repens, T. pratense, Medicago falcata*
II. Temperate	Intensive pasture production, temperate horticultural plant combined with pasture legumes as ground floor, fodder forest legumes	5.7 tha^{-1}, on an av. 8 to 10 kg $tree^{-1}$ dry matter
III. North-eastern region	Silvipasture,*i)Alnus nepalensis+ Stylosanthesi guyensis + Panicum maximumii)Ficus hookerii + Thysanolaena maxima*	13.5 tha^{-1}(DM) fodder yr. 12.3tha^{-1} fodder yield
IV. Outer Himalaya	Silvipasture systems, *Dalbergia sissoo +Chrysopogon fulvus Acacia catechu + Eulaliopsis binata Leucaena leucocephala* (high density) + Hybrid Napier (NB 5)*Bauhinia purpurea, Albizia lebbek + Eulaliopsis binata*	Suitable system for utilization of degraded lands for fuel, fodder and fibre 64-71 t ha^{-1} fuelwood 5.2-5.5 t ha^{-1} yr^{-1} fodder grassfor short rotation of 4 -5 yearsMost suitable for gravelly riverbed lands of Doon Valley.
V. Shiwalik foothills	Silvipasture,Fodder production from wastelands and watershed management	Fodder production increased by 267%
VI. Indo-Gangetic plains	Tree-grass combination Forage production on dry lands	Annual production of 8-10 t ha^{-1} (4.5-5 t ha^{-1} fuelwood) 7-10t $ha^{-1}yr^{-1}$ biomass production*Leucaena* + sorghum-safflower- Forage yield of 12.3 t ha^{-1}
VII. Humid and sub-humid	Use of quick growing leguminous fodder plants in crop rotations, plantation crops with pastures, Farm forestry. Deenanath (*Pennisetum pedicellautm*) + *Stylosanthes* and Deenanath + *Leucaenai*) *Leucaena* (Hawaiian, Hybrid-28) +fodder grass. *Glyricidia/Sesbania/ ii)Ficus hookerii Litsea+Thysalaena /Andropogon nardus*	Best for pasture development in the eroded marginal and sub-marginal hilly area of Southern BiharSuitable combination for plain area in West BengalSuitable for hilly areas
VIII. Coastal area (Kerala etc.)	Fodder trees (Subabul/ Agathis /Desmenthes) + Congo signal + *Stylosanthes or Centrosema*	Total fodder yield of 22-29 t ha^{-1} could be obtained in coconut based fodder production systems.
IX. Arid and semi-arid region	Grazing land and silvipasture and Horti-pasture in arid and integration of agri-silvipasture,	Very good potential for silvipastoral enterprise

(*Contd.*)

Region	Components	Results
	Forage-cum-coppicing farming on the marginal and sub-marginal lands with intercropping of dryland cereals and legumes, *Acacia tortilis, Albizia lebbek, Prosopis cineraria* with *Chrysopogon fulvus, Cenchrus ciliaris, C. Setigeresm*	

In Doon valley, silvipasture systems, such as *Dalbergia sissoo* + *Chrysopogon fulvus*; *Acacia catechu* + *Eulaliopsis binata*; *Eucalyptus* hybrid + *Chrysopogon fulvus*; and *Leucaena leucocephala* + Napier / *Panicum maximum* and *Albizia lebbek*, *Grewia optiva*, *Bauhinia purpurea* and *Leucaena leucocephala* with *Chrysopogon fulvus* and *Eulaliopsis binata* were found suitable for utilizing the degraded lands for productive as well as protective functions (Vishwanatham *et al.*, 1999; Samra *et al.*, 1999, Table 9). Biomass production of *Eulaliopsis binata* (4.6 t ha^{-1}) was considerably higher than *Chrysopogon fulvus* (2.7 t ha^{-1}). There was a marked decrease in soil pH and increase in organic carbon, available P and available K in the surface layer due to the growing of the tree species of *Albizia lebbek* and *Leucaena leucocephala.*

Table 9. Biomass production (kg ha^{-1}) of a silvipasture system after 14 years of establishment in degraded bouldery lands of Doon valley

Treatments	1993				Pooled (1980-1993)			
	Tree biomass				Tree biomass			
	Leaf	Branch wood	Grass Bio-mass	Total	Leaf	Branch wood	Grass Bio mass	Total
Tree species								
Albizia lebbek	359	474	1078	1911	364	573	3592	4529
Grewia optiva	124	254	1994	2372	262	754	3461	4477
Bauhinia purpurea	395	459	2075	2929	362	629	3878	4869
Leucaena leucocephala	158	236	1338	1732	194	361	3655	4210
CD (0.05)	172	223	325	-	75	174	170	-
Lopping intensity								
50%	239	323	1572	2134	285	583	3567	4435
75%	279	389	1671	2339	306	575	3727	4608
CD (0.05)	NS	NS	NS	-	NS	NS	159	-
Grass species								
Chrysopogon fulvus	302	423	1068	1793	371	723	2687	3781
Eulaliopsis binata	216	289	2174	2679	219	436	4607	5262
CD (0.05)	65	63	320	-	25	53	166	-

Total biomass production

Green biomass production under two canopy management practices along with hybrid napier grass was compared with control in a silvipastoral system for different components and system as a whole (Table 10). The maximum green grass fodder (260 q/ha) from the Napier was obtained under pollarding management followed by grass alone without tree (control) while tree canopy of lopping has lower yield may be due to less penetration of light. Whereas, pollarding canopy showed compatible association with Napier grass. The total biomass yield from trees was also recorded higher with tree pollarding alone (89 q/ha) without grass association followed by pollarding + grass (36 q/ha), which about 53 per cent lower. Singh *et al.* 1998 has also recorded highest leaf biomass (28.4 t ha^{-1}) from pollarding. The decline in the tree biomass yield is because of the positive effect of grass association (Napier). However, the total biomass from the silvipastoral system as a whole was highest with pollarding+grass (296 q/ha), followed by lopping+grass (229 q/ha). It was concluded that the pollarding+grass (management) was the best technique to obtain grass biomass from degraded lands following by lopping+grass. Secondly tree pollarding alone was the best technique for biomass production as compared with the canopy management practice with grass combinations.

Table 10. Green biomass production (q/ha) under two canopy management practices from Hybrid Napier - *Grewia optiva* based silvipastoral system on degraded bouldery riverbed lands (Av. of 10 yrs.)

Treatments	Green grass fodder	Total tree biomass	Total biomass from SPS
Lopping + Napier Grass	201	28	229
Pollarding + Napier Grass	260	36	296
Lopping only	-	30	30
Pollarding only	-	89	89
Napier Grass only	221	-	221

Conservation of forages

In order to sustain animal production, it is essential that the optimum feeding should be maintained throughout the year. However, there are 2 to 3 months of lean period (October to December/January and April to July) when the fodder availability to the animal is at its low. In the summer months it becomes more difficult to meet the maintenance requirements of the animals. It is, therefore, essential to conserve surplus forages at appropriate stage of maturity to feed the animals adequately during lean period for optimum animal production. The conservation of forages could be done in various ways especially in the form of silage from cultivated fodders (legumes and cereals) and also pasture grasses. Forages could also be conserved in the form of hay when dried to about 80-85 per cent DM which will preserve more of its nutrients including carotene. This feed is quantitatively and qualitatively important from both maintenance and nutritional point of view.

Thrust areas in research and development on silvipastoral systems

The increased productivity of forages is to come through various concerted efforts by researchers and developmental agencies. The priority area is dependent on the soil, climate, cropping pattern, traditional practices and other socio-economic conditions. On the basis of researches and potentials under agro-climatic situations following areas were identified for different regions of the country (Table 2).

All the areas indicated above are intersparsed in all the zones in various proportions along with cultivated lands. Forage floristic composition for each of these systems and for each of these zones may be differential which need to be kept in view and researched upon.

Conclusion

A dynamic approach to tackle the problem of fodder and feed scarcity must consider how best to utilize the areas otherwise unproductive for crop production both within and outside the forests, for forage production. Obviously, such areas have poor soil fertility and often suffer from moisture deficit. In such situations when one would think of forage promotion, one has to keep in mind a composition of forage plant types complementary to each other and not competitive in growth pattern and yield potentials with an ability to fulfill the nutritional deficiencies of each other and also that they together yield balanced nutritional feed to livestock. Thus greater emphasis needs to be placed on the leguminous component through the use of rhizobial cultures.

References

Anonymous 2000. Agricultural Statistics at a Glance. Directorate of Economics and Statistics, Dept. of Agric. and Cooperation, Min. of Agriculture, Govt. of India, New Delhi.

Arora, Y.K., Vishwantham, M.K. 1995. Alternate land use system for marginal and bouldery riverbed lands. In: G. Sastry *et al.* (eds.), Torrent Menace: Challenges and Opportunities, pp. 156-163. CSWCRTI, Dehradun.

Balocchi, O.A., Phillips, C.J.C. 1997. Grazing and fertilizer management for establishment of *Lotus uliginosus* and *Trifolium subterraneum* under *Pinus radiata* in Southern Chile. Agrofor. Syst. 37:1-14.

Dhyani, S.K., Dadhwal, K.S., Katiyar, V.S. 1988. Biological measures for rehabilitation of abandoned mine area. Indian Farming 38(7):15-18.

Deb-Roy, R. 1994. Common agroforestry systems and their management for optimizing production. In: Singh, P., Pathak, P.S., Roy, M.M. (eds.), Agroforestry Systems for Degraded Lands, pp. 379-387, Vol. I. Oxford and IBH Publishing Co. Pvt. Ltd., New Delhi.

ICFRE 1993. Annual Research Report for 1991-92. Indian Council of Forestry Research and Education, Dehra Dun, India.

Jeschke, W.D., Hartung, W. 2000. Root-shoot interactions in mineral nutrition. Plant Soil 226: 57-69.

Jones, M., Sindair, F.L., Grime, V.L. 1998. Effects of tree species and crown pruning or root length and soil water content in semi-arid agroforestry. Plant Soil 201:197-207.

Mathew, T., Kumar, M., Suresh, B.V.K., Verma, M. K. (1992). Comparative performance of four multipurpose trees associated with four grass species in the humid regions of Southern India. Agrofor. Syst. 17:205-218.

Nair, P.K.R. 1983. Soil productivity aspects of Agroforestry, ICRAF, Nairobi, Kenya, 85 pp.

Pathak, P.S., Deb Roy, R. 1987. Silvipastoral system for forage production on wastelands. Indian Farming 36(8): 10 -11.

Raison, R.J., Myers, B.J., Benson, M.L. 1992. Dynamics of *Pinus radiata* foliage in relation to water and nitrogen stress: I. Needle production and properties. For. Eco. Manag. 52:139-158.

Rai, P., Solanki, K.R., Rao, G.R. 1999. Silvipasture research in India–a review. Indian J. Agrofor. 1(2): 107-120.

Sehgal, J.L., Abrol, I.P. 1994. Soil Degradation in India – Status and Impact. Oxford and IBH Publishing Co. Pvt. Ltd., New Delhi, India.

Shankarnarayan, K.A., Singh, K.C. 1990. Rehabilitation of degraded pasture land in the arid regions of Rajasthan. In: Abrol, I. P., Dhruvnarayana, V.V. (eds.), Technologies for Wasteland Development, pp. 21-31, ICAR, New Delhi.

Samra, J.S. 1997. Status of Research on Watershed Management. CSWCRTI, Dehra Dun.

Samra, J.S., Vishwantham, M. K., Sharma, A.R. 1999. Biomass production of trees and grasses in a silvipasture system on marginal lands of Doon Valley of north-west India. 2. Performance of grass species. Agrofor. Syst. 46:197-212.

Singh, S. 2005. Management of degraded lands for sustainable agricultural production – An overview. In: Watershed Management for Sustainable Production, Livelihood and Environmental Security – Issues and Options. Lead Papers presented at the *National conference on Watershed Management*, May 19-21, 2005, Soil Conservation Society of India and G.B. Pant University of Agriculture and Technology. p. 30-41.

Singh, K.A., Yadav, B.P.S., Goswani, S.N. 1996. Forage resource management for sustained livestock productivity in Himalayan Agro-ecosystems. Research Bulletin No.39, ICAR Research Complex for NEH Region, Barapani, pp.94.

Singh, P. 1986. Forage production in India, RMSI, IGFRI, Jhansi, pp.213.

Singh, B.P., Dhyani, S.K., Prasad, R.N. 1994. Traditional agroforestry systems and their soil productivity on degraded alfisols/ultisols in hilly terrain. In: Singh, P., Pathak, P.S., Roy, M. M. (eds), Agroforestry systems for Degraded Lands. pp.205-214. Oxford & IBH Pub. Co. Pvt. Ltd., New Delhi.

Singh, C., Agarwal, M.C., Kumar, N., Puri, D.N. 1998. Biomass production of *Morus alba* under different management practices on degraded bouldery riverbed lands of Doon Valley. The Indian For. 124 (3):252-260.

Upadhyay, V.S., Pathak, P. S. 1994. Silvipastoral system evaluation for secondary production in Central India. In: Singh, P., Pathak, P.S., Roy, M.M. (eds.), Agroforestry Systems for Degraded Lands, pp. 780-787, Vol. II. Oxford and IBH Publishing Co. Pvt. Ltd., New Delhi.

Vishwanatham, M.K., Samra, J.S., Sharma, A.R. 1998. Biomass production of silvi-pasture systems on gravelly lands of Doon Valley. Bulletin No.T-39/D-26, CSWCRTI, Dehra Dun, pp. 126.

Vishwanatham, M.K., Samra, J.S., Sharma, A.R. 1999. Biomass production of trees and grasses in a silvipasture system on marginal lands of Doon Valley of North West India. I. Performance of tree species. Agrofor. Syst. 46: 181-196.

Walker, J., Robertson, J.A., Penridge, L.K., Sharpe, P.J.H. 1986. The influence of an *Acacia karoo* tree on grass productivity in its vicinity. J. Grass. Soc. Southern Africa 4(3):83-48.

Wilson, J.R. 1996. Shade stimulated growth and nitrogen uptake by pasture grasses in a sub-tropical environment. Australian J. Agric. Res. 47 (7): 1075-1093.

Wolters, G.L., Maetris, A., Pearson, H.A. 1982. Forage response to over storey reduction on loblolly-shortleaf-pine hardwood forest range. J. Range Manage. 35(4): 443-446.

22

Role of Windbreaks/shelterbelts for Conservation of Natural Resources in Arid Conditions

Rajendra Prasad, R.S. Mertia, and S.K. Dhyani

Introduction

Dry lands cover nearly one-third of the earth's total surface and characterized by high climatic variability in rainfall and temperature along with higher potential of wind and water erosion (Mainguet, 1991; Okin, 2000). Wind erosion has been identified as one of the overriding factors of desertification in regions characterized by strong and persistent winds through out an annual dry season (Le Houerou, 1996). Clear sweep of strong winds across sandy desert is the great hindrance in sustenance and progress of agricultural activities. The strong blowing winds on one hands takes away top fertile soil thus causing irreparable loss to soil productivity, on the other hand deposition of airborne soil particles (sand, silt and clay) block the roads, railways, water bodies, open canals and burry agricultural fields (Gupta and Aggrawal,1978; Prasad *et al*, 2004). To minimize erosion hazards of speedy winds and optimize production of agricultural crops, various efforts have been made in the past by adopting different soil conservation measures among which the windbreak/ shelterbelt plantations rank high due to their expected effects on wind velocity, microclimate and productivity.

Windbreaks and shelterbelts are designed to reduce wind velocity thus soil erosion, provide protection to roads, railways, water bodies, canals *etc* and create favorable environment for cropping, horticulture and livestock enterprises. Beside shelterbelts are also viewed as means to fulfill objectives of social forestry in supplementing local demands of fuel, fodder, shelter and timber. The focus of current chapter is on concept of windbreaks and/ or shelterbelt technology and its contributions towards conservation of natural resources- soil and water, and augmentation of agricultural production in hot arid region of India.

Description of windbreak/shelterbelt

Windbreaks can be defined as a plantation having spaced row of shrubs or trees designed to protect a space by baffling the speed and direction of wind. It is different from the hedge in which trees / shrubs are closely spaced defining an edge or boundary. The shelterbelt is generic term for windbreak and often both are used as synonyms. For a natural resource conservationist the windbreak and/or shelterbelt means planting of vegetative barriers consisting of trees/shrubs/bushes in single or multiple rows across prevailing wind direction to minimize hazardous effect of speedy wind and provide sheltered area on leeward and windward sides. The windbreak and/or shelterbelt plantation is the model practice of protection forestry in arid regions where continued effect of various natural processes such as low and erratic rainfall, intense heat, high evaporation, low relative humidity, high wind speed etc yields fragile ecosystem. The main objectives with which these plantations are raised include reducing wind speed and drifting sand thus wind erosion; improving microclimate, irrigation efficiency and soil environment; supplementing supply of fuel and fodder and providing shelter and habitat to wildlife.

Design of windbreaks/shelterbelts

The effectiveness of windbreak depends on its plantation geometry, direction, shape, foliage density, uniformity and length. In India, Central Arid Zone Research Institute (CAZRI), Jodhpur (Rajasthan) has done commendable research on various aspects of windbreak/ shelterbelt. The early research work (in late fifties and early sixties) on windbreak was aimed to develop its suitable design having optimum planting geometry, porosity, thickness (No. of rows) and shape (Mertia, 1986). Simultaneously, the emphasis was also given on screening of most suitable trees and developing techniques for establishment of planted seedlings. For providing maximum shelter zone, the windbreaks having 40 –50 % porosity with evenly distributed gaps was assessed to be the best (Nageli, 1946). For arid condition in Thar Desert, Kaul (1969) suggested five row windbreaks with a pyramidal shape having one row of tall trees such as *Acacia tortlis, Tamarix articulata, Azadirachta indica* followed by two rows of smaller trees like *Acacia senegal, Prosopis juliflora, Parkisonia aculeata* etc. and then followed by two rows of shrubs such as *Aerua tomentosa, Zizyphus spinachristi* and *Calligonum polygonoides* at the edge in the flank rows. To increase the shelter efficiency of windbreaks following design parameters should be considered.

Direction: The orientation of windbreaks/ shelterbelt should be more or lesss perpendicular to the prevailing wind.

Shape: The ideal form is pyramidal with fast growing tall trees in the central row, flanked by fast growing and high branching trees of medium height. The outer rows should be planted with low and much branched shrubs.

Space: It should be 5-10 times height of trees and varies with availability of land.

Density: It should vary from 50 to 80%.

Length: Depends on land availability. Ideally it should not be less than 100m and 10 -13 times height of tree.

Growth: It should be uniform with out any gaps.

Selection of suitable tree species and their planting techniques are crucial factors in establishment of windbreak plantations. In selection of species, following facts should be given due consideration.

- Choose species as per aim.
- Give preference to native species.
- Take expert opinion.
- Consider landscape.
- Know silvi-cultural characteristics of the species.
- Avoid invasive, short-lived and disease prone plants/species.

Some time the protective forestry plantations such as windbreaks are also supposed to meet objectives of social forestry like producing fodder (top feed), fuel and small timber. Under such circumstances multi-purpose tree species (MPTs) should be screened and evaluated. For different purposes, suitable species and design of windbreak have been given in (Table 1).

Table 1. Design and species for windbreak plantations

Purpose	Design	Suitable species
Road side	3 to 5 staggered rows	*A tortilis, A. lebbeck, A. indica, D. Sissoo, P. aculeata, P. Juliflora and Tamarix articulata*
Railway lines	6 rows	*P. acculeata, P. juliflora, T. articulata*
Canal side	6 rows	*A. nilotica, Eucalyptus spp. T. undulata, A. tortilis, P. juliflora, D. sissoo, P. cineraria A. nubica*
Farm boundary (rainfed)	1 / 2 / 3 rows	*Accacia tortilis, A. Lebbeck, A. indica, D. sissoo, P. aculeata, P. juliflora* (most common)
Farm boundary (irrigated)	2 rows	*A tortilis, A. lebbeck, Dicrostachys cineraria, P. juliflora*

In arid region major thrust is on protection of roads and railways through side-belt plantations. The Rajasthan state forest department has undertaken large scale roadside planting in eleven districts of Western Rajasthan. To check engulfing of railway line from blown sand in part of Bikaner, Sikar, Churu, Jaisalmer, Barmer and Jodhpur districts of Western Rajasthan, a program on railway line plantations was initiated in 1980s. In beginning a section of 100 km along Sikar-Loharu, Sikar-Fatehpur and Palsana-Deshnoke lines was planted with a six row tree belt using *Parkinsonia aculeata, Prosopis juliflora* and *Tamarix articulata* (Mann and Muthana, 1984). With the advent of IGNP in Western Rajasthan, massive tree belt plantation work was undertaken to check silting of canal through wind blown sand. Six row tree belts consisting of *Acacia nilotica, Eucalyptus* spp, *Tecomella undulata, Acacia tortilis, Prosopis cineraria, Dalbergia sissoo, Tamarix aphylla* etc have been planted on either side of the open canal. In Jaisalmer district of western Rajasthan about 1305 Rkm roads-side and 27812 Rkm canal-side tree-belts have been raised. In addition, about 14245 ha area has been covered under sand dunes, range lands and farmlands.

Contribution of windbreaks/ shelterbelts

Reduction in wind erosion: Windbreaks have been found effective in reducing wind speed and thus soil erosion (Mertia, 1992). The wind speed reduction is higher during monsoon season than in summer due to thicker canopy growth (Table 2). Different windbreaks are associated with different degrees of wind reduction in different seasons and over the surrounding area.

Table 2. Percent reduction in wind speed due to different windbreaks in Jodhpur.

Species	Distance from windbreak					
	Summer season			Monsoon season		
	2 H	5 H	10 H	2 H	5 H	10 H
Prosopis juliflora	33	17	12	38	26	21
Cassia Siamea	36	17	13	46	36	24
Acacia tortlis	36	25	13	46	36	20

Due to reduction of wind speed the soil loss in windbreak area was reportedly reduced by 50 percent as compared to an unsheltered area (Gupta *et al.*, 1984). The windbreak based on *Cassia Siamea* proved most effective in conserving soil. While assessing impact of windbreak in arid ecosystem, Mertia *et al.* (2006) observed that different windbreaks had varying effects on reduction of wind speed in leeward side of the belt. The maximum extent of reduction in wind speed of leeward side was observed at distance of 2H (H is the average height of the windbreak in meter) which slowly got nullified up to 20H (Table 3). Rajendra Prasad *et al.* (2009[a]) observed that on an average, irrespective of the species, double row windbreak were more effective in reducing wind speed (Fig 1) as compared to single row belt. However, single row belt provided more effective area in comparison to double row belts.

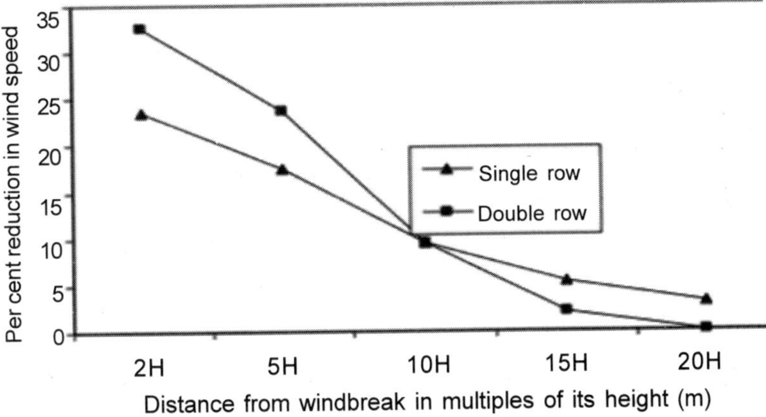

Fig. 1. Effect of type of windbreak on reduction in wind speed in leeward side of the windbreak

Table 3. Effect of windbreak on wind speed in leeward side

Windbreak	Direction and number of rows	Mean height (m)	Per cent reduction in mean wind speed				
			2 H	5H	10H	15 H	20 H
Tube well command, Lathi series							
Acacia tortilis	S-W 2	8.9	29.6	24.5	10.7	3.3	0
Eucalyptus camaldulensis	S-W 1	15.5	24.8	18.6	12.5	9.4	7
Dalbergia sissoo	S-W 2	7.8	32.4	17.9	5.5	1.5	0
IGNP command, Mohangarh							
Dalebergia sissoo	S-W 1	6.2	21.5	14.2	6.2	3.2	0.2
Dalbergia sissoo	S-W 2	8.8	36	28.7	12.2	1.4	0
Tecomella undulata	S-W 1	10.2	24	19.7	9.7	3.5	2

The structure of canopy, density (proportion of solid material such as foliage, branches etc), age, height and direction of the windbreak plantations are major factors which decide its effectiveness in controlling the wind flow. Mertia *et al.* (2006) reported that fairly tall and dense double row *D sissoo* reduced maximum wind speed but its effectiveness was only up to distance of 15H whereas, 20 years old single row *E. camaldulensis* provided sheltered area up to distance of 20H. For maximizing effectiveness of windbreak plantations, it is desirable to plant those species which can grow fast and attain maximum height in short period. Belt of multiple rows apparently provide more resistance due to dense canopy resulting in decrease of sheltered area in the leeward side of the belt. The turbulence causing tunneling effect at ends is more pronounced in shelterbelts having more height, density and short length .The wind speed at ends decreased with the increase in length: height (L/H) ratio (Fig 2). Minimum turbulence at the ends was observed in double-row *A. tortilis* shelterbelt having L/H ratio of 13 whereas, it was maximum in double-row D. sissoo with L/H ratio of 8.3.

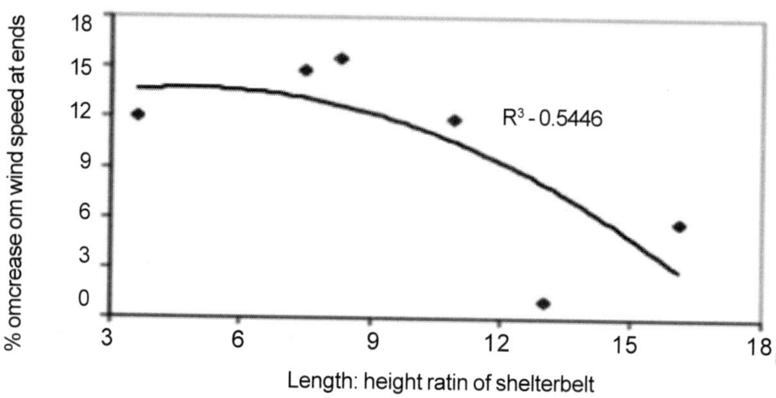

Fig. 2. Effect of L/H ratio of shelterbelts on wind speed at the ends in arid Western Rajasthan

Protection of Roads and Railways: The main objective of road side plantations or windbreaks along roads is to abate deposition of sand on the road and reduce velocity of moving wind. Study by Mertia *et al.* (2006) reported that road side plantations have been efficacious in reducing deposition of blown sand which, otherwise deposited and blocked roads (Table 4).

Table 4. Road blockages due to blown sand in absence of road-side plantation

Sr. No.	Name of Road	Location (Km)		Length (Mtrs)	Height (Mtrs)
1.	Ramgarh - Longewala	11.20	11.50	300	3
		14.00	14.35	350	3
		28.20	28.30	100	1
2.	Ramgarh – Ashutar -Ghotaru	37.10	37.50	400	3
		44.60	44.80	200	1

Protection of canals: Sand deposition in Indira Gandhi canals has reduced considerably after windbreak plantations along canal sides. Shifting dunes, in the absence of any check in the form of plantations, deposited huge quantities of blowing sand in the main canal and its distributaries thereby either completely choking or reducing the flow of the water. In order to de-silt the canals to restore flow of water, huge expenditures were incurred. The seriousness of the problem can be gauged from the fact that about 0.57 million cu m of blown sand had to be removed from the main canal (km 383.2 to km 444. 8) at an approximate cost of

Rs. 72 lakhs in December 1986 (Bithu, 1989). A study of Upadhaya (1991) in IGNP stage II area in Jaisalmer have revealed that quantum of sand deposited in canal was reduced by 513 m^3 and 1023 m^3 per Km at one and two years old canal side tree-belt plantations , respectively as compared to unplanted sites (control). The corresponding savings on de-silting of deposited sand was estimated to be in the tune of Rs. 6,156 and 12,276 per km. The study further revealed that quantum of sand movement varied significantly with respect to various months of a year (Table 5).

Table 5. Effect of canal side plantation on sand deposition (M^3 / RM of canal) in canal

Month	Control	One year old plantation (M^3)	Two years old plantation (M^3)
Nov. 89	0.089	0.071	0.041
Jan. 90	0.094	0.068	0.041
March 90	0.150	0.109	0.069
May 90	0.261	0.225	0.141
June 90	0.323	0.289	0.157
July 90	0.198	0.101	0.051
Sept. 90	0.068	0.041	0.021
Oct. 90	0.039	0.033	0.016

Available information on removal of blown sand from canal beds revealed that inverse relationship exists between the age of tree belt and the quantity of sand deposited. The older the plantations the lesser the quantity of sand deposited in canals (Table 6).

Table 6. Reduction in blown sand deposition in canal (M^3 per km) after canal side plantation

Branch/Distri/MInors	1994-95	1996-97	1997-98	1998-99
Sagarmalgopa Sakha	4125	2877	2079	1269
Mandau Distributry system	2296	777	373	718
Sankla minor	1949	1165	1009	388
Tibrewala Minor	1875	563	1250	1250

Microclimate moderation: Significant changes in microclimatic conditions have been assessed under windbreaks as compared to an unsheltered area. Favorable impact of canal side (Table 7) plantations on microclimate has also been noticed (Mertia *et al.*, 2006). Trees having good crown cover markedly affect the inside temperature and relative humidity levels irrespective of their age. Wherever, crown density is poor, very little or simply no difference in the inside and outside values is observed. Unlike canal side plantations, the roadside plantations did not show any significant variations in microclimatic conditions between inside and outside of the tree belt (Table 8). This could be due to the emission effects of running vehicles on road. It is inferred that canal side windbreak plantation and other afforestation activities have brought about significant change in microclimate at different places, which may have some impact on macroclimate of the district as a whole. Rise in decadal mean rainfall due to the effects of an increase in vegetation cover has been reported (Ramkrishna and Rao, 1991).

Table 7. Impact of canal side plantation on micro climate

Sr. No.	Site	Time	Temperature (°C)		RH %	
			Inside	Outside	Inside	Outside
1.	40-50 RDSBS	2.20 PM	33.0	36.0	47	42
2.	14-20 RD	5.00 PM	38	38	28	28
3.	30-40 RD SMGS	12.10 PM	36.5	40.0	42	27
4.	0-10 RD Sadha disty R/S	10.35 AM	32.5	24.0	52	51
5.	70-80 RD Sultana Disty	9.30 AM	37.0	39.0	31	29
6.	12-20 RD Sadha Disty R/S	5.00 PM	39.5	39.5	23	22
7.	165-175 RD Sadha B/S	6.30 PM	32.0	32.0	58	58

Table 8. Impact of road side plantation on microclimate

Sr No.	Site	Time	Temperature (°C)		Relative Humidity (%)	
			Inside	Outside	Inside	Outside
1	54RDBS to Digha	8.00 AM	28.0	28.0	81	70
2	Sultana Kabir Basti	5.00 PM	40.0	40.0	27	27
3	86-104 SBS R/S	1.30 PM	38.0	38.0	38	38
4	190-212 RD	10.30 AM	33.0	33.0	47	47

Mertia *et al.* (2006) revealed that windbreaks brought significant changes in microclimatic conditions on farmer's fields. Considerable reduction in air temperature in sheltered area of leeward side was noticed which varied with the type of windbreak plantations. Irrespective of type of windbreak plantation, maximum reduction in air temperature was observed beneath the windbreak. Rajendra Prasad *et al.* (2009[b]) reported that presence of shelterbelt had reduced daily air temperature on down wind side in comparison to reference up wind (-5H) air temperature with maximum reduction beneath the tree canopy, which nullified up to10H in decreasing order. The decrease in air temperature on downwind side of 15 year old double-row belt was more pronounced as compared to 10 year old single-row belt (Fig 3). As a general rule, under tree canopy the extreme temperatures are modified ie. the maxima are reduced and minima are raised as compared to those prevailing in open areas (Tournebize *et al.*, 1996). In savannah regions reduction in maximum soil temperature of 5 to 12°C under isolated trees has been reported (Blesky *et al.*, 1983). The reduction in temperature plays an important role in conserving surface moisture and affects soil fauna which are effective in litter breakdown and nutrient re-cycling.

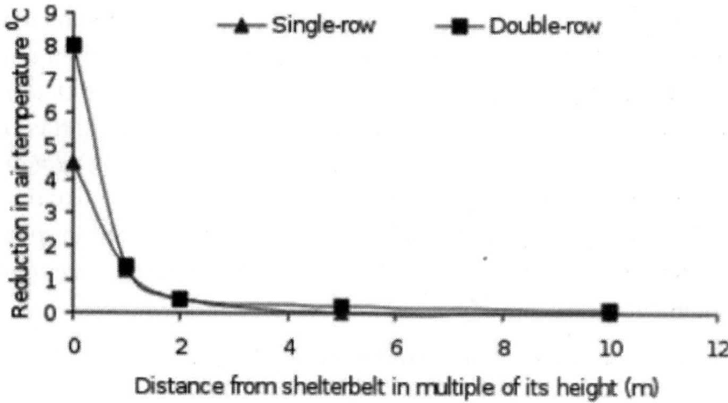

Fig. 3. Reduction in air temperature on leeward side of *Dalbergia sissoo* shelterbelt in IGNP command area in arid Western Rajasthan

The reduction in temperature decreases water loss from the soil through evaporation. Muthana *et al.* (1984) reported that the evapotranspiration (PE) is largely influenced by the wind in arid region and windbreak had reduced 5-4% pan evaporation values on either side of the belt (Table 9). The microclimatic variations within windbreaks and their modification with respect to seasons have significant impact on associated agricultural crops (Table 10). Positive effects of windbreak on conservation of soil moisture on its leeward side have been reported (Mertia, 1992).

Table 9. Effect of Cassia *siamea* windbreak on pan evaporation

Location	Pan evaporation (mm /day)			
	April	May	June	July
Windward side	11.0	14.5	13.4	6.0
Leeward side	9.5	13.4	12.5	5.7
Percent decrease	14	8	8	5

Table 10. Effect of windbreak on yield of pearl millet

Distance from windbreak	Pearl millet yield (q / ha)	Control	Consumptive use (mm)	Water use efficiencies (Kg / ha / mm)
5 H	3.7	-	174	1.5
10 H	6.1	-	223	1.51
15 H	6.3	-	216	3.09
20 H	8.8	-	216	3.79
25 H	9.8	-	227	5.62
30 H	5.9	-	194	2.65
-	-	4.8	-	-

Influence on soil quality: Presences of windbreak plantations on farmer's field have significant influence on soil characteristics of sheltered area in leeward side. Mertia *et al.* (2006) found that presence of windbreak increased soil organic carbon (Table 11). These effects were more pronounced up to the distance of 2H and nullified after Distance of 5H. On an average, maximum accumulation of OC was found in sheltered area of *A. tortilis* followed by *T. undulate* and *Dalbergia sissoo*. The accumulation of organic carbon decreased with increase in distance from the belt. In general, organic carbon (OC) decreased with depth of soil in sheltered area of all the windbreaks (Fig 4). The built up of OC in sheltered area appears to be due to decomposition of leaf litter.

Table 11. Effect of shelterbelts on surface soil properties on its leeward side

Shelterbelt	Location	Soil properties on leeward side at distance from shelterbelt in multiples of its height (m)				
		0H	1H	2H	5H	10 H
Organic carbon content (%)						
Dalbergia sissoo	IGNP Mohangarh	0.11	0.08	0.05	0.04	0.04
Tecomella undulata	IGNP Mohangarh	0.37	0.21	0.05	0.08	0.07
Acacia tortilis	Tube well command Lathi series	0.46	0.21	0.16	0.14	0.15
Soil reaction ($pH_{1:2}$.)						
Dalbergia sissoo	IGNP Mohangarh	7.84	7.88	7.99	7.98	8.04
Tecomella undulata	IGNP Mohangarh	7.75	7.8	7.96	7.96	8.05
Acacia tortilis	Tube well command Lathi series	8.23	8.55	8.93	8.95	8.84
Soil electrical conductivity (dSm^{-1})						
Dalbergia sissoo	IGNP Mohangarh	0.4	0.38	0.28	0.27	0.28
Tecomella undulata	IGNP Mohangarh	0.88	0.42	0.22	0.21	0.21
Acacia tortilis	Tube well command Lathi series	0.67	0.34	0.34	0.36	0.34

Influence on crop production: The irrigation facility created by IGNP has changed the cropping pattern of the region. In irrigated areas of desert region, orchards of pomegranate, guava, grapes, *Citrus* spp. and *Cordia* spp. are maintained. Today more than two mha lands in Thar Desert grow brinjal, chillies, tomatoes and potatoes. Ber fruit especially a fleshy grafted variety is an important source of income. The above advancements are directly related to advent of canal irrigation, however, indirectly the windbreak plantation carried out along field boundary for protection has helped in sustaining changed cropping pattern. In addition, tree-belt plantation along canals to protect the channels from sand deposition has immensely helped in maintaining the flow and volume of water in the channels. In the absence of windbreaks, the much needed water for irrigating agricultural fields would not have reached the farmers of Jaisalmer.

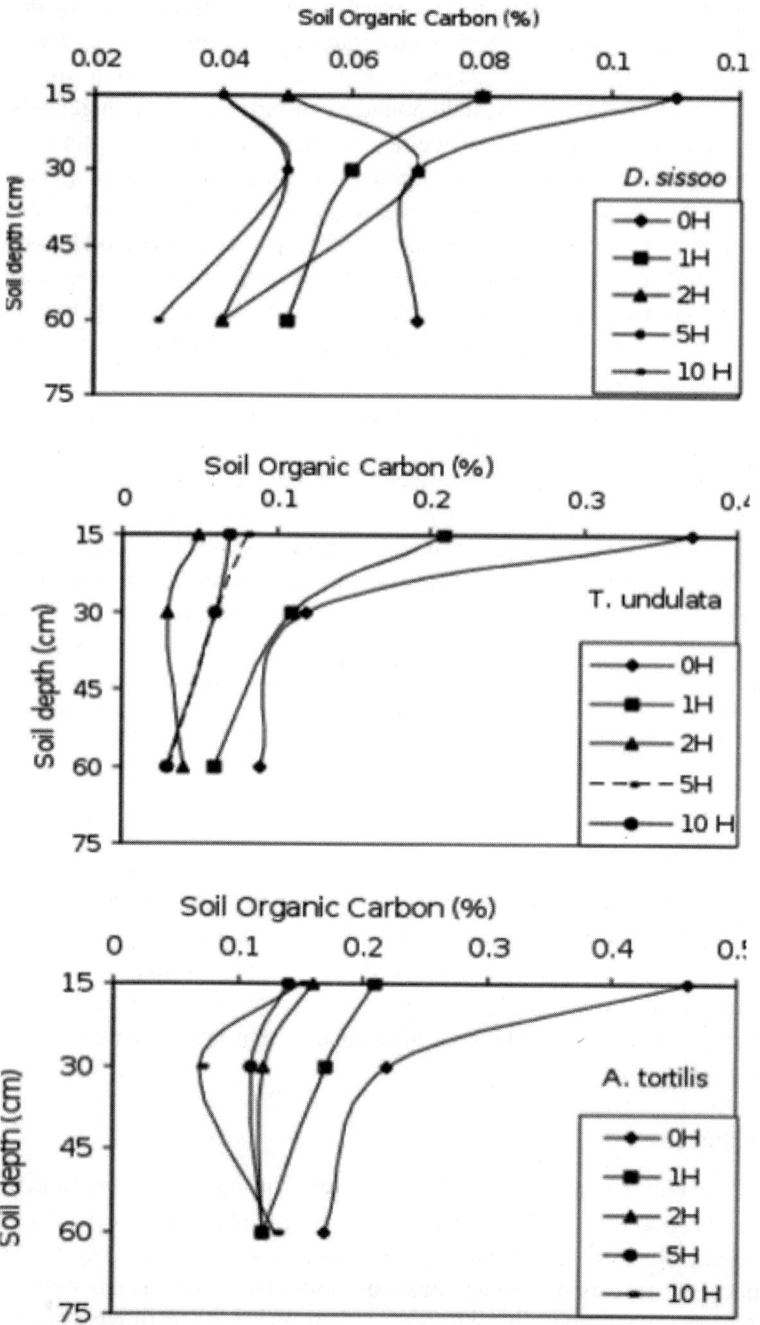

Fig. 4. Effect of shelterbelt on distribution of organic carbon in soil profile in sheltered area

Reduction in biotic pressure on natural forest: In arid region natural grasslands and forests are often subjected to excessive biotic interference in the form of overgrazing and removal of fuel woods. Introduction of multipurpose trees (MPTs) like *Accacia tortilis, Albiziz lebbeck, Azadiracta indica, Acacia nilotica* etc for tree-belt along roads, canals, field boundary of farm land and rangelands is aimed to supplement local demand of fuel and fodder along with its main objectives of reducing wind, drifting of sand and providing shelter. Trees on rangeland provide nutritious top feed in the form of leaves and pods to grazing animal (Ahuja, 1977). The top feed is an alternative source to supplement decreasing fodder productivity of grasslands in traditionally pastoral economy of the region. Some of the species like *Calligonum polygonoides,* which used to be the main vegetation in sandy desert of Jaisalmer, Barmer and Bikaner districts of Rajasthan, are now facing serious threat of their extinction due to excessive extraction and uprooting by the locals to meet demand for fuel and fodder. Supply of fuel wood and small timber from windbreak would certainly put a break on such devastating processes. It is assumed that seven to eight years old belt plantations can provide sufficient quantity of fuel wood and timber to meet some of the needs of local people. Due to improvement in agricultural production and availability of fodder for livestock, there has been a change in migration pattern of nomads. The migration of sheep to neighboring areas has reduced to some extent (Table 12).

Table 12: Nomadism in western Rajasthan

Years	No. of sheep migrated	Percentage of total population sheep in district
1988-89	65320	13.48
1990-91	53140	10.97
1992-93	24610	5.08

Employment generation: Massive afforestation work taken by State Forest Department along canals, road and railways has created employment opportunities for both men and women. For raising plantations and their maintenance, skilled as well as unskilled work force has been utilized. According to an estimate, employment of about 2.4 million man-days has been generated in Jaisalmer district of Rajasthan during 1990-2002.

Conclusions

Tree-belt plantations, in arid region, were conceived, designed and raised with the primary objectives of providing protection against harsh effects of high velocity of winds. Selection of suitable species and design holds the key for efficacious impacts of tree-belts on various components of arid ecosystem. Different windbreaks/ shelterbelts have varying effects on reduction of wind speed. Double-row windbreaks are more effective in reducing wind speed as compared to single-row belt. However, the single-row belt provides more effective area (up to 20H) in comparison to double row belts (up to 15 H). The structure of canopy (density), age height and direction of the shelterbelt are major factors which decide its effectiveness in controlling wind flow. For maximizing effectiveness of windbreaks, it is desirable to plant those species which can grow fast and attain maximum height in short period. Belt of multiple rows provide more resistance due to dense canopy resulting in decrease of sheltered area. The turbulence or tunneling effect is more pronounced at ends of shelterbelts having more height, density and short in length. The length of shelterbelt directly controls its turbulence effects on ends. Uninterrupted and uniform shelterbelt plantations appear to be more effective than those with interruption or gaps. Significant reduction in sand deposition on roads and canals, favorable changes in microclimate, improvement in soil quality and agricultural productivity signifies usefulness of windbreaks in arid zone.

Considering huge and irreparable losses due to speedy winds, it can be concluded that shelterbelt plantations provide relatively cheap and long-term option for reducing wind erosion and associated hazards. For harnessing maximum benefits of the shelterbelt plantations, careful consideration should be given to the location, species, desired height, length, density and composition of the tree belt. Over all, the basic objective of the shelterbelt should guide the whole farm planning and strategies. The judicious use of shelterbelt technology would not only substantiate potential productivity of natural resources but also help in their sustainable managements in arid ecosystem.

References

Ahuja, L.D. 1977. CAZRI monograph, 24: 161-166.

Bithu, B.D. 1989. Problem of soil erosion and its control in western Rajasthan. In: Proceedings of International Symposium on Managing Sandy Soils, Part II, pp. 551 – 555.

Blesky, A. J., Mwonga, S. M., Amundson, R.G., Duxbury, J.M., Ali, A.R.1993. Comparative effects of isolated trees on their undercanopy environment in high- and low- rainfall savannas. J. Appl. Ecol. 30:143-145

Chepil, W.S. 1949. Agron. J. 41: 127-129.

Gupta, J. P. and Aggrawal, R.K. 1978. Sand movement studies under different land use conditions of western Rajasthan. In: Proceedings of International Symposium of Arid Zone Research and Development., CAZRI Jodhpur India

Gupta, J. P., Rao, G.G. S.N., Ramakrishna, Y.S. and Ramana Rao, B.V. 1984. Role of shelterbelts in arid zone. Indian Farming October: 29-30.

Kaul, R.N. 1969. Windbreaks to stop creep of the desert. Science and culture, 24:406-409.

Le Houerou, H. 1996 Climatic Change, drought and desertification. Journal Arid Enviroments. 34:133-185.

Mann, H.S. and Muthana, K.D. 1984. Arid Zone Forestry (with special reference to Indian arid zone), CAZRI Monograph, 23:1-48.

Mainguet, M. 1991. Desertification : natural background and human mismanagement. Springer-verlag, Berlin

Mertia, R.S. 1986. Windbreak studies in India's Thar Desert. M.Sc. Thesis. University of Wales, U.K.

Mertia, R.S. 1992. Windbreak research in arid zone. J. Trop. Forestry, 8 (3):196-200.

Naegeli, W. 1946. Weitere untersuchungen iiber die windverhaltnisse im bereich von windschutzanlagen (Furthar investigations of wind conditions in the range of shelterbelts) Mitt Schweiz Anst Forstl Versuchswesen 24:660-737.

Okin, G.S. 2000. Wind-driven desertification: Process modeling, remotemonitoring and forecasting. Doctor of Philosophy Thesis, California Institute of Technology, Pasadena.

Rajendra Prasad, R.S. Mertia, R.N. Kumawat and J.S. Samra (2009[a]). Shelter efficiency of Dalbergia sissoo shelterbelt in IGNP command area in arid Western Rajasthan, India. Indian Journal of Agroforestry. Vol 11(1): (in press).

Rajendra Prasad and R.S. Mertia (2009[b]). Tree Windbreaks and their Shelter benefits on Farmland in Arid Region of Western Rajasthan. J. Soil & Water Cons. (in press).

Ramakrishna, Y. S., Rao, A. S., Singh, R. S., Kar, A. and Singh S. (1990). Moisture, Thermal and Wind measurements over selected stable and unstable sand dunes in the Indian desert. Journal of Arid Environments, 19: 25-38.

Said, M. 1954. Pakistan Journal of Forestry, 4: 175-179.

Tournebize, R., Sinoquet, H. and Bussiere, F. 1996. Modelling evapotranspiration partitioning in a shrub/grass crop. Agricultural and Forest Meteorology 81:255-272.

Upadhyaya, A.K. 1991. Windbreak plantations effectively check sand deposition in Indira Gandhi canal, Indian Forester 117 (7): 511-514.

23

Scope and Potential of Medicinal Plants in Agroforestry: An Appraisal

P.S. Thakur and C.L. Thakur

Abstract: *This article gives an overview of the possibility and scope of successful cultivation of medicinal and aromatic herbs as intercrops with multipurpose tree species under agroforestry conditions. The challenges are many. Sustainable production under any agroforestry system is the main goal to be achieved through short and long term strategies. Diversification of the existing conventional cropping sequences coupled with developing of suitable technology packages is the need of the hour to cope up with ever increasing demand for diversified products. The hypothesis is, if intercropping of annuals, biennials and perennial medicinal plants in association with fuel, fodder, timber, pulp, fruit trees including tree species with medicinal values on the farmland, an alternate land use system; can provide economically viable option for sustainable systems. This chapter describes the potential and benefits of medicinal plants based agroforestry. Our experience is that planting timber, fuel and fodder or fruit tree species in combination with medicinal crops paves the way for diversified products and better economic returns to the farmers. Multipurpose tree species outside the forest have played very important role in catering to the day to day requirements of the rural population. Different farming systems consisting of crops + trees have yielded encouraging results and have not only helped diversification but enhanced productivity of land per unit area. Intercropping medicinal and aromatic herb species further makes agroforestry more remunerative. Rough estimate is that 95% of medicinal and aromatic plants are collected and harvested from their natural habitats, which has not only* *reduced the availability of medicinal plants but severely affected their regeneration and even posed threat to the existence of some of the valuable species. The ever increasing demand by the pharmaceutical firms at national and international scenario has further led to over exploitation of these precious herbs. The commercial cultivation of medicinal herbs, which are in great demand or high conservation priority species (endangered species) on the farmland in association with woody perennials, seems to be a promising strategy with great conservation impacts.*

Introduction

Diversification of existing cropping patterns coupled with developing of suitable technology packages seem to be the need of the day to cope up with the ever increasing demand for diversified products. Insofar as the know-how about the conventional agriculture and forest productivity is concerned there has been a wealth of information regarding various aspects involved therein. This knowledge has undoubtedly, helped in maximizing the productivity over the years, but the rising population, land degradation, soil erosion and shrinking and over stretchibility of cultivable land are some of the major constraints for sustainable production. Water is another one of the most important critical resources, which will remain a decisive constraint for sustainable production in future. Major cultivable land in India is under rainfed conditions, which may increase further on account of continuously depleting water sources. Nevertheless the limits of agriculture production even using the most intensive high input agriculture have already been reached. Integration of multipurpose tree species on the farmland with agricultural and / or high value cash crops (agroforestry practice) is a sensible step to increase number of arboreal population out side the forest. This land use pattern will pave the way for diversification of farming systems, account for higher returns from the system as well as function as good sinks for carbon storage for longer duration. There exists a tremendous imbalance between demand – supply equilibrium for fulfilling the basic needs. The continual dependence on forests for timber, fuel fodder, fiber, other minor forest products and health needs will adversely affect environment; the possible repercussion seen is climate change.

Importance of medicinal herbs

The tremendous awareness for herbal medicines (phytomedicines) world over has led to the extensive use of these medicines for two reasons, i) because these are less expensive and within the reach of common people who can not afford costly health services and ii) that the side effects are minimum. This trend over the years has sharply increased the demand for raw materials by the multinational pharmaceutical and cosmetic industries in India and abroad. The domestic manufacturers of herbal based medicines have further widened the gap between demand and supply. To meet this demand several thousand tones of medicinal and aromatic herbs are unscientifically extracted from the forest ecosystem in the Himalayas resulting a trade, which runs into millions of dollars every year. In the recent years, there has been a shift from growing traditional food crops towards high value cash crops such as medicinal and aromatic plants, which are in great demands. Medicinal and aromatic plants offer multiple benefits including one of monetary benefits to unemployed people world over. Besides securing the healthcare needs of a large number of people, medicinal plants are the exclusive source of some drugs even for the modern medical treatment. The global importance of medicinal and aromatic plants is evident from a huge volume of trade at national and international levels. The demand for medicinal plant based raw material is increasing roughly by 15-20% annually. Conservative estimate put world trade of medicinal plants at 5 trillion US Dollar by 2050. The natural sources are in no position to keep pace and meet this skewed demand. India is considered treasure house of valuable medicinal and aromatic plant species. Out of approximately 15000 plant species found in India; 3000 are medicinal species. North-West Himalayas in Himachal Pradesh is the natural home for about 1100 species of medicinal and aromatic plants and 350 species of ethnobotanical importance. The government of Himachal Pradesh has initiated appreciable step at the right time to make this hilly state a 'Herbal State'. National Medicinal Plants Board, India has already prioritized medicinal and aromatic herb species for cultivation under different agroclimatic zones of the country.

Cultivation of medicinal herbs

Roughly 80% of the rural population in India and other developing countries of the world, exclusively exploit forest resources for meeting their health requirements. This *per se* is the area of great concern to every one may be environmentalists, planners or governments. Unscientific, unsystematic and uninterrupted extraction / removal (slaughter harvesting) of medicinal and aromatic plants from their natural habitats are collectively contributing to the demise of a number of species. The increasing demand for herbal medicines is threatening to extinct several precious medicinal herb species and can spark global healthcare crisis. This practice has resulted in shrinking bioresource availability year after year. In case the extraction of these herbal species continues unchecked for few more years, which seems to be a real possibility, several precious medicinal and aromatic species may acquire the status of endangered species. The scenario is, indeed, grim. Though India has rich diversity but the growing demand is exerting heavy pressure on the existing resources causing a number of species to be either threatened or endangered category. Approximately 230 medicinal and aromatic species in India are threatened. The present practice of use and over exploitation of medicinal species from their natural home to meet huge demand for raw material may push many more herbal species into threatened status. There are no strict policies or rules for judicious extraction of these herbs. Several natural sites (home) of these herbal species in Himalayas have completely been denuded. Sooner or later the commercial cultivation of medicinal and aromatic herbs, which are in greatest demand by the pharmaceutical and cosmetic industries at home and abroad, will have to be taken up in order to bridge the gap between demands and dwindling supplies. The shift from collection to cultivation approach will assure sustained supply of raw material and reduce *in situ* harvesting pressure. The truth is that most of the medicinal and aromatic plant species grow wild under the forest canopies, so are shades tolerant. Agroforestry, which is an intersectoral intervention spanning two important and at times competitive land use systems i.e. agriculture and forestry, seems to play crucial role, where cultivation of medicinal and aromatic herbs along with multipurpose tree species including trees of medicinal importance will ensure production as well as conservation of herbal biodiversity. Cultivation of medicinal and aromatic plants on the farmland will have several socio-economic impacts on rural unemployed population. Large scale cultivation of medicinal herbs is one option to generate income. For threatened medicinal and aromatic species cultivation *per se* is convenient conservation strategy.

Agroforestry with medicinal herbs

Several medicinal and aromatic plant species are shade loving so multipurpose tree species will not only protect herb species from erratic high heat, frost, wind velocity but also provide additional monetary returns in the form of commercial timber or other products. Moreover, the preference in all countries is towards greater proportion of cultivated material. Agroforestry, indeed, is a convenient and strategic option for cultivation of medicinal and aromatic plants beneath canopies of multipurpose tree species (fuel, fodder, timber, fruits, pulp *etc.*).The integration of trees on the farmland paves the way for the improvement and diversification of existing systems, but creates complex biological interactions (Lawson and Kang, 1990; Ong *et al.*, 1991; Khybri *et al.*, 1992; Rao *et al.*, 1998; Gillespie *et al.*, 2000; Thakur and Singh, 2002; Thakur and Dutt, 2004). Diversification and sustainability in production are the two main goals and there is an opportunity to explore the possibility of growing commercial tree species with high value cash crops (HVCP) such as medicinal plant species on the farmland without compromising quality of either product. The option of simultaneous growing of crops + trees under well managed agroforestry practice helps to reduce gaps between demand and supply (Thakur and Dutt, 2003; Rao *et al.*, 2004; Thakur *et al.*, 2007). The critical resources like solar radiation, water and soil nutrients can be put to optimum use through diversification methods like agroforestry and intercropping (Noguchi and Ichimura, 2004; Thakur and Kumar, 2006; Thakur and Dutt, 2007). Growing of trees on the farmland leads to better economic returns and diversification at the farm level (Yadava and Singh, 1996; Chauhan, 2000; Thakur *et al.*, 2005; 2007). Trees provide farmers with marketable products, such as timber, fuel wood, fodder, fiber, fruits and medicines, all of which earn extra income. Trees increase ecosystem biodiversity and can help mitigate global climate change by sequestering carbon. The appropriate management practices are useful for minimizing competition between the components of the systems (Cannel, 1983; Thakur and Dutt, 2003; Thakur and Kumar, 2006; Thakur *et al.*, 2007). Different agroforestry systems have been developed and numerous tree crop combinations have been tried by researchers all over the world, with an eye on increasing the economic situation of the farmers (Chauhan *et al.*, 1997; Gillespie *et al.*, 2000; Rao *et al.*, 2004; Thakur and Kumar, 2006; Thakur *et al.*, 2007).

In addition, agroforestry practice is essential for various reasons. For example, the existing traditional land use systems with separate allocation to agriculture and forest are inadequate to meet the demands for diversified products like food, fuel, fodder, timber, paper, pulp, fruits etc. The exponential growth in human and livestock population over the years coupled with the declining production, soil erosion and declining cultivable land have increased gaps between supply and demand for timber, fuel and fodder. The growing appetite for trees for fuel, fodder and construction in both rural and urban areas has largely been responsible for much of the deforestation. The scenario on fuel, fodder and timber production is not satisfactory. There is a chronic deficiency of timber, fuel wood and fodder in the Southeast Asian countries. For example in India, according to the available estimates the annual consumption and production of fuel wood, timber and forage is as shown in Table 1.

Table 1. Fuel wood, timber and fodder production and consumption status in India

	Consumption	Production	Deficiency
Fuel wood	235 million m^3	90 million m^3	62 %
Timber	28 million m^3	12 million m^3	57 %
Fodder	900 million tones (dry)	441 million tones	51 %
Fodder	1100 million tones (green)	250 million tones	77 %

Table 1 shows imbalances between the demand and supply for fuel, forage-fodder and timber. The gap is being met from the farmland and other lands outside the forests. The situation likewise is no better in the remaining developing countries. Therefore, agroforestry seems to be panacea for overcoming most of the problems related to the alleviation of poverty and socio-economic instability. The gains of agroforestry research till date are certainly impressive. There are good number of success stories where tree based cropping systems have yielded good results. However, in order to transform present day agroforestry into high tech scientific adventure in the 21st century, there is an urgent need to formulate hypothesis through inductive reasoning, test and validate the hypothesis through deductive experimentation and strive to attain predictable understanding that will have wide applicability. Integration of traditional knowledge with scientific research is required for the development of improved technologies. Further, agroforestry research in context of the present global scenario is bound to have changed vision and strategies in order to make this alternate land use more viable and profitable venture. Today we are equipped with fair amount of technologies to handle present day agroforestry, however, there is every probability that the present package may not suffice to meet the challenges and problems namely, alleviation of poverty and hunger, reversal of environmental degradation, reduction of deforestation, exponential demand for timber, fuel, fodder, pulp etc., which agroforestry, in new millennium will encompass from time to time. We further need to refine the technology package based on the past experiences and failures to make agroforestry more beneficial, viable, environmentally sound and widely acceptable option. Diversification and sustainability in production are the two main goals to be achieved through short and long term strategies. There is immediate need to explore the possibilities of growing multipurpose tree species with high value cash crops on the farmland to attain predictable understanding that will have wide applicability and the information so generated would go a long way in developing need based, acceptable and economically viable diversified agroforestry models. Agroforestry holds potential for increased availability of on-farm wood products, energy sources and improved carbon sequestration (Semwal *et al.*, 2002; Thakur and Sehgal, 2003; Shively *et al.*, 2004; Dutt and Thakur, 2004). Agroforestry is being promoted and popularized by the Government and wood based industries in India in order to address the crisis of food, timber, fuel wood and fodder. After eucalypt species poplar based agroforestry has become very popular and remunerative. *Acacia mangium* plantation in Southeast Asia, particularly in Indonesia and *Paulownia* and *Populus* based agroforestry systems in East Asia, particularly in China, can be reasonably mirrored in the remaining developing countries. Short rotation plantations consisting eucalyptus, poplars, acacias and pines will go a long way in improving the socio-economic status of smallholders.

Common agroforestry systems

In general, the Asian countries encompass two peculiar situations, i) hilly terrain and ii) plain areas with contrasting agroclimatic conditions. The main agroforestry systems existing on the farmer's field are as under:

- Agri-silviculture
- Agri-horticulture
- Agri-silvi-horticulture
- Agri-horti-silviculture
- Silvi-pastoral
- Horti-pastoral
- Horti-silvi-pastoral
- Home gardens

- Energy plantations (Fuel-fodder banks)
- Commercial plantations

The functional components of agroforestry systems, however, differ greatly in size, texture and nature depending on the necessity of rural population. The fuel, fodder and timber tree species dominate in the hilly regions, whereas commercial species like poplar, teak, eucalyptus, sandalwood, rubber *etc*., constitute the major component of agroforestry systems in the plains. The cultivation of medicinal and aromatic herb species as intercrops with arboreal component on the farmland will ensure diversification of farming systems and better economic returns. The need is to undertake concerted efforts to achieve a balance between consumption and conservation. The cultivation of medicinal plants on the farmland is certainly one strong and viable option to fill the gap between demand and supply of herbal raw material and nevertheless eases pressure on their natural habitats. The practice to bring wild species under cultivation programme shall have several benefits, i) assures sustained supply of raw material to pharmaceutical and cosmetic industries, ii) uplift of economic status of farming community, iii) diversification, iv) conservation etc. The need is to develop low- cost technologies to bring down cost of cultivation.

Management of functional components

It is true that the appropriate radiation environment under tree canopies is the deciding and driving force for triggering on and off of vital physiological and biochemical processes associated with the performance and production potential of under storey crops. The transmission of photosynthetically active radiation (PAR) through tree canopy can be regulated by adopting suitable canopy management options. Medicinal plants are shade loving and love to grow and satisfactorily complete their phenological phases even at shade intensity up to 75%. Canopy management of multipurpose tree species integrated on the farmland as agroforestry tree with crops, is essential to minimize competition for critical resources. However, some time the removal of aboveground portion of trees results in enormous decrease in photosynthesis and vigour of the trees, which is not desirable. There exists a positive correlation between the amount of intercepted incident solar radiation and total aboveground production of foliage and wood biomass. Thus, it is of utmost importance to have full understanding of the crown architecture and growth in order to design appropriate canopy management strategies. The main purpose of canopy management and tree spacing is to make multipurpose tree species suitable and compatible with the associated arable crops in any agro forestry system. Reports are available, where the positive and/or negative impacts of the presence of tree canopies on the performance of under storey vegetation have been indicated. The major management options for manipulating trees in agroforestry are based on the alternation of light (solar radiation) profile and moisture distribution.

Canopy management options

The options for managing tree canopy are numerous e.g. pruning, coppicing, pollarding, lopping, partial crown removal, appropriate plant spacing and orientation *etc*. However, each one of these options has equal number of advantages and disadvantages. The impacts of such practices can some times have drastic implications as the removal of aboveground portion results in enormous decrease in photosynthesis and the plant may die (Thakur and Sehgal, 2003). In addition, the management of tree canopy is bound to have direct bearing on the root characteristics as well as growth, vigour and biomass production of the tree itself (Kang *et al*., 1981; Basappa, 1986; Dutt and Jamwal, 1987; Heering, 1995; Puri and Gargya, 1995; El-Fadl, 1997; Singh *et al*., 1998; Thakur, 2000; Thakur and Sehgal, 2001, 2003, 2004; Thakur and Singh, 2003; Thakur *et al*., 2005). Author's experience with canopy management of some of the important tree species of temperate region, indicated enormous influence of management options on vigour and biomass production potential. In agroforestry, tree canopy reduces and modifies incident solar radiation intensity available to the under storey crops; this situation will be highly beneficial for supporting cultivation of medicinal plants under agroforestry intervention.

Intercropping of medicinal plants with poplar – a profitable option

The burgeoning demand and a steady increase in the per capita consumption of building materials, pulp, paper and other wood products is intensifying the loss of increasingly scarce natural forests. The improved hybrid clones of poplar are excellent multipurpose short rotation tree species. Our experience is that poplar planted at 8x3m, 6x4m and 4x4 m spacing proved suitable spacing for successful cultivation of medicinal and aromatic crops under agroforestry systems. The medicinal and aromatic species, namely *Digitalis lanata, Matricaria chamomilla, Salvia sclarea, Valeriena jatamansi, Andrographis paniculata, Withania somnifera, Tagetes minuta, Ocimum sanctum, Ocimum basilicum, Spilanthes acmella, Chlorophytum borivilianum, Stevia rebaudiana,* species providing active ingredients used in herbal medicine formulations, have been successfully cultivated as intercrops along with poplar plantation on the farmland. There were no or little adverse effects on the performance and production of medicinal crops. The total income from diversified systems (medicinal crops plus poplar) was substantially higher than monocropping. The presence of poplar trees were not observed to adversely affect herb yield in the plots. Cultivating medicinal and aromatic plants with commercial timber species like poplar in agroforestry systems, therefore, can be a viable and promising option for better economic returns to the farmers. The benefit cost ratio (BCR) ranging from 1:1.50 to 1:2.00 was achieved for the above mentioned systems. This can further be enhanced by developing suitable and low-cost agro techniques, which lower the cost of cultivation of herbal plants. There will be relatively higher initial investment because of the purchase of planting materials; however, this cost in the subsequent years will be nil as there is no need to buy the planting material since the portion of the already cultivated population can be maintained for the purpose of production of seeds, suckers, roots and other forms of planting materials. Use of fertilizers and organic manures to the medicinal herbs was found very useful for enhancing growth and production ability as well as photosynthetic efficiency of all the herb species. Nitrogen @100 kg/ha, FYM @ 4 t/ha, vermicompost @2t and enriched manure @2t/ha proved most effective. Growth with respect to height, leaf number, plant spread, flower number, branch number and leaf area index; all indices of better growth, was significantly higher in fertilizer and organic manure treated plants in comparison to control plants without fertilizer and manures. Similar was the trend for the yield of economic organs/parts, where production was greater with the use of fertilizers and manures even in the presence of tree species. The herb yield was more under the systems when applied with different doses of fertilizers and organic manures in comparison to sole crops. Nitrogen was found most effective. The use of fertilizers and organic manures was found to reduce competition between the functional components. This is desirable since better growth and production in herbal species means higher herbal yield, thus better economic returns. Our findings reveal that appropriate doses of fertilizers and / or organic manures may be used for successful cultivation of important medicinal herb species under poplar based agroforestry systems. The cultivation of high value cash crops (HVCC) like medicinal and aromatic herbs with multipurpose tree species, especially commercial timber is a profitable alternative for traditional cropping systems. We have observed that high value cash crops like, medicinal herbs can successfully be cultivated as intercrops with arboreal component without any adverse impact on the performance and production ability of these crops. The presence of tree species, particularly commercial timber species provide additional monetary benefit. The bio-economic appraisal of poplar based agroforestry systems incorporating medicinal and aromatic herb species has exhibited better cost-benefit ratio over the sole cropping.

Medicinal tree species as agroforestry component

Beside commercial timber species as one of the functional components of medicinal and aromatic plant based agroforestry intervention, the scientific plantation of tree species with immense medicinal values on the farmland with medicinal herb species will be highly profitable. The current emphasis on the integration of medicinal tree species on the field is far from sufficient. The potential tree species with medicinal values can be prioritized for cultivation with annual, biennial and perennial medicinal and aromatic plant species for different agroclimatic zones. There is long list of tree species with known medicinal values; some of them commonly found on the farmers field, which does not require any mention here. However, more systematic approach shall be fruitful.

Water: a critical constraint for large scale cultivation of medicinal herbs

Cultivation of medicinal and aromatic herbs, possessing abundant medicinal values, is fast becoming a convenient and attractive approach in India to partially substitute for conventional agricultural crops. Farmers are taking up cultivation of important medicinal and aromatic herbs mainly to increase economic returns. Cultivation of medicinal and aromatic herb species is considered a viable and lucrative option to enhance economy of farmers. Water, however, is one of the most important critical resources, which will remain a decisive constraint for large scale cultivation of medicinal and aromatic plants. Major cultivable land in India is under rainfed conditions, which may increase further on account of continuously depleting water sources. It is expected that in the future, climate variability will increase leading to more frequent droughts. The consequences would be overall decline in production level. There is need to critically examine responses of medicinal and aromatic plant species to water stress and to identify herb species capable of coping and sustaining growth and production even under water stress. Integrated agronomic approaches to enhance water productivity of medicinal and aromatic plants shall be useful.

Climate change and medicinal herbs

Implications of global warming and climate change on flora and fauna including production ability of food crops have been the hot topic of discussion and debate at national and international level during the last decade. Alarmed by the possible adverse impacts of global and regional climate change on flora and fauna, there has been serious concern world over in understanding vulnerability of agricultural, fruit and tree crops and finding out short and long term strategies to reverse the negative effects. The Intergovernmental Panel on Climate Change (IPCC) in its Fourth Assessment Report (AR 4) published in December 2007 has already indicated warming trend during the 20th century and its repercussions on biodiversity and existing flora and fauna. The report gives greater attention to the integration of climate change with sustainable development and the interrelationships between mitigation and adaptation with special mention of regional issues, uncertainty and risks, climate change and water. The climate change in the form of elevated carbon dioxide, drought and extreme temperature will not only result in drastic decline in production potential of annual and perennial crops but may also cause elimination of some of the important germplasms. The senior author in a recently concluded study (Thakur *et al.*, 2008) has for the first time reported that regional climate change exerts significant influence on the phenological phases of multipurpose tree species. The findings indicate that phenological phases namely, leaf emergence, flower initiation phase and growth period of eleven tree species advanced during a short period of eight years, which could be the result of climate change at the regional level. The increase in average maximum temperature by 3.7 °C within 8 years (1999-2006) and a sharp decline in annual rainfall from 1213.9 mm during 1999 to 969.4 mm in 2006 probably were deciding factors. There is every possibility that fast changing climate, especially temperature, level of CO_2, moisture unavailability, etc. will have adverse impact on survival and production ability of all categories of plants. Medicinal and aromatic plants will be no exception. Uncongenial microsite climate, result of global warming or environmental degradation, may threaten medicinal and aromatic plants at their natural habitats. Several overexploited medicinal and aromatic herb species may be put at 'risk' in the near future. The result can be the bulging list of endangered species. The short and long term strategies to bring wild medicinal and aromatic plant species under cultivation needs serious rethinking on scientific lines. Development of promising cultivation methodologies seems to be the need, both for better production and biodiversity conservation of medicinal and aromatic plants. Climate variability at the regional rather than global level needs to be looked into more closely in order to more precisely understand impacts on medicinal and aromatic herbs under Indian conditions.

Research agenda

Future agroforestry research is bound to have changed vision and strategies in order to make this alternate land use more lucrative and profitable venture. Today we are equipped with adequate technologies to handle present day agroforestry, however, there is every probability that this package may not suffice to meet the challenges and problems in future. We don't have the comprehensive information on the advantages and disadvantages of different canopy management practices on various aspects related to growth, vigour, root characteristics, biomass production, longevity, *etc*. We have not yet completely understood the intricate resource sharing between different components of the system; extent of adverse impact on the production potential of crops, which undergo drastic change at the advent of canopy management. A great deal of work has been done on various aspects in agroforestry although the underlying physiological processes are yet to be understood completely. There is no specific data to establish any kind of relationship between canopy management practices and carbon sequestration. Quantification of above and belowground biological interactions, especially at tree-crop interface still remains inconclusive. It is important to workout the proportion of woody and arable components, which can be accommodated per unit area, in order to enhance yield advantage and benefits from the agroforestry systems by minimizing competition. Suitable management of tree canopies, permitting adequate transmission of photosynthetically active radiation beneath canopies has been found more critical for improving productivity of associated arable crops. Appropriate tree management practices not only regulate biomass production ability and maintain vigour for extended period but also help associated crops to perform better under agroforestry intervention. Diversification of farming systems by growing short rotation timber, paper, pulp, fuel and fodder tree species with high value cash crops on the farmland is a viable option for increasing income of farmers. So we need to continuously refine strategies and update the technology package based on the past experiences and failures to make agroforestry more viable, beneficial and widely acceptable option. The task is not yet over. More and more wild medicinal and aromatic plant species need to be brought under cultivation. Upgradation of cultivation methodologies for individual medicinal and aromatic species is essential for production (profit) as well as conservation. In brief, priority areas like i) identification of suitable medicinal plants for agroforestry, ii) development of suitable and low-cost cultivation agrotechniques, iii) shift from collection to cultivation, iv) screening of water stress tolerant medicinal species, v) species able to cope with high temperature and CO_2 (tolerate climate change), vi) technology to improve production and utilization, vii) conservation, viii) post harvest handling and ix) strong and organized market linkages, need immediate attention before scope and potential of medicinal and aromatic plants under agroforestry intervention can be realized.

Conclusions

Diversification of farming systems by cultivating annuals, biennials or perennials medicinal and aromatic plant species as intercrops on the farmland, especially with commercial timber and medicinal tree species is a viable option for increasing income of farmers. There is tremendous scope of large scale cultivation of important medicinal and aromatic plant species on the farmland under agroforestry intervention. The approach will make the raw material available for pharmaceutical and cosmetic industries, increase the income of farming community and provide convenient conservation option.

References

Bassapa, B. 1986. Coppicing of silver oak (*Grevillea robusta).* My Forest. 22: 1-22.

Cannel, MGR. 1983. Plant management in agroforestry: manipulation of trees, population densities and mixtures of trees and herbaceous crops. In: Huxley P. A. (Ed.), pp. 455-487, Plant Research and Agroforestry Nairobi: ICRAF.

Chauhan, H. S. 2000. Performance of poplar (*Populus deltoides*) based agroforestry system using aromatic crops. Indian J. Agrofor. 2: 17-21.

Chauhan, H.S., Kamla, S., Patra, DD. 1997. Performance of aromatic crops in eucalyptus based agroforestry system. Journal Med. Arom. Plant Sci. 19: 724-728.

Dutt, A. K., Jamwal, U. 1987. Effects of coppicing at different heights on wood production in *Leucaena*. Luciana Res. Rep. 8:27-28.

Dutt, V., Thakur, P.S. 2004.Bio-economics of cropping systems combining medicinal and aromatic herbs with commercial timber tree species. Indian J. Agrofor. 6:1-7.

El-Fadl, M. A. 1997. Management of *Prosopis juliflora* for use in agroforestry systems in Sudan. Tropical For. Rep. Helsinki 16:107.

Gillespie, A. R., Jose, S., Mengel, D.B., Hoover, W.L., Pope , P.E., Street, J.R., Biehle, D.J., Stall, T., Benjamin, T.J. 2000. Defining competition vectors in a temperate alley cropping system in the Midwestern USA. A Production Physiology. Agrofor. Syst. 48:25-40.

Heering, J.H. 1995. The effect of cutting height and frequency on the forage weed and seed production of six *Sesbania sesban* accessions under irrigated conditions. Agrofor. Syst. 30:340-350.

Kang, B.T., Wilson, G.F., Sipkens, L. 1981. Alley cropping with maize and *Luciana* in Southern Nigeria. Plant Soil. 63:165-169.

Khybri, K.L., Gupta, R.K., Ram, S., Tomer, H.P.S. 1992. Crop yields of rice and wheat grown in rotation as intercrops with three tree species in the outer hills of western Himalayas. Agrofor. Syst. 17: 193-204.

Lawson, T.L., Kang, B.T. 1990. Yield of maize-cowpea in alley cropping system in relation to available light. Agric. For. Met. 55: 347-357.

Noguchi, A., Ichimura, M. 2004. Effects of environmental factors on the growth, flowering, and essential oil concentration and composition of sweet basil and spearmint. Hort. Res. Japan. 3: 67-70.

Ong, C.K., Corlett, J.E., Singh, R.P., Black, C.K. 1991. Above and belowground interactions in agroforestry systems. Forest Ecol. Manage. 45: 45-57

Puri, D.N., Gargya, G. R.1995. Management of *Morus alba* and *Grewia optiva* for degraded lands. Van Vigyan 33:109-113.

Rao, M. R., Nair, P.K.R., Ong, C. K. 1998. Biophysical interactions in tropical agroforestry systems. Agrofor. Syst. 38: 3-50.

Rao, M.R., Palada, M.C., Becker, B.N. 2004. Medicinal and aromatic plants in agroforestry. Agrofor. Syst. 60: 107-122.

Semwal, R. L., Maikhuri, R.K., Rao, K.S., Singh, K., Saxena, K.G. 2002. Crop productivity under different lopped canopies of multipurpose trees in central Himalaya, India. Agrofor. Syst. 56: 57-63.

Shively, G.E., Zelek, C.A., Midmore, D.J., Nissen, T.M. 2004. Carbon sequestration in a tropical landscape: an economic model to measure the incremental cost. Agrofor. Syst. 60:189-197.

Singh, C., Agarwal, M.C., Kumar, N., Puri, D.N. 1998. Biomass production of *Morus alba* under different management practices on degraded bouldry riverbed lands of Doon Valley. Indian Forester. 124:252-260.

Thakur, P.S. 2000. Comparative performance of multipurpose tree species growing at the degraded site. Indian Forester 126:895-900.

Thakur, P.S., Dutt, V. 2003. Performance of wheat as alley crop grown with *Morus alba* hedgerows under rainfed conditions. Indian J. Agrofor. 5: 36-44

Thakur, P.S., Dutt, V. 2004. Diversification of farming systems by combining commercial tree species with medicinal plants to boost economy of farmers under rainfed conditions. 4th International Crop Science Congress, Brisbane, Australia, pp. 1-6.

Thakur P.S., Dutt, V. 2007. Cultivation of medicinal and aromatic herbs in agroforestry6 for diversification under submontane conditions of western Himalayas. Indian J. Agrofor. 9:66-76.

Thakur, P.S., Dutt, V., Sehgal, S., Kumar, R. 2005. Diversification and improving productivity of mountain farming systems through agroforestry practices in northwestern India. North American Agroforestry Conference, University of Minnesota, Rochester, USA, Abstract No 25.

Thakur, P.S., Dutt, V., Thakur, A. 2008. Impact of inter-annual climate variability on the phenology of eleven multipurpose tree species. Current Sci. 94:1053-1058.

Thakur, P.S., Kumar, R. 2006. Growth and production behaviour of medicinal and aromatic herbs grown under hedgerows of *Leucaena* and *Morus*. Indian J. Agrof. 8:12-20.

Thakur, P.S., Sehgal, S. 2001. Effect of canopy management on root parameters in agroforestry tree species of temperate region. Indian J. Agrofor. 2:75-78

Thakur, P.S., Sehgal, S. 2003. Growth, leaf gas exchange and production of biomass in coppiced and pollarded agroforestry tree species. J. Trop. For. Sci. 15: 432-440.

Thakur, P.S., Sehgal, S. 2004. Influence of canopy management on the performance of multipurpose tree species. Indian For. 130:639-646.

Thakur, P.S., Sehgal, S., Dutt., V., Thakur, C. 2007. Strategies to improve production ability of medicinal and aromatic herbs under rainfed agroforestry systems in northwestern India. Proceedings of International Workshop on Medicinal and Aromatic Plants Held at Chiang Mai University, Thailand from January 15-18.

Thakur, P. S., Singh, S. 2003. Influence of canopy size modification of *Morus alba* on productivity of *Vigna mungo* and *Pisum sativum* in agroforestry system. Indian J. Plant Physio. (special vol.): 104-109.

Thakur, P.S., Singh, S. 2002. Effect of *Morus alba* canopy management on light transmission and performance of *Phaseolus mungo* and *Pisium satium* under rainfed agroforestry. Indian J. Agrofor. 4: 25-29.

Yadava, A. K., Singh, K. 1996. Performance of the lemongrass *Cymbopogon flexuosus* under poplar based agroforestry systems. Journal Med. Arom. Plant Sci. 18:290-294.

24

Tree Improvement in Tamil Nadu

M.P. Divya, R. Jayaramasoundari and N. Chandra Sekaran

Introduction

Historically, foresters did not view trees as typical plants having the systems of heredity similar to all other living organisms. Previously it was felt that the tree development depended only upon the environment in which it was grown. Recently, the "forest tree parentage" in the tree improvement programme has gained importance. Although there are several breeding methods for improvement of any crop, the most commonly used breeding method in tree improvement is "selection" of superior trees from existing or established population. The objective of selection programme is to obtain significant amount of genetic gain as quickly and inexpensively as possible, while at the same time maintaining a broad genetic base to ensure the future genetic general principle; that is choosing the most desirable individuals for the use as parent in breeding and production programmes.

The most successful tree improvement programmes are those in which proper vegetative propagules and provenances are used. Success in the establishment and productivity of forest plantations is determined largely by the species used and the source of seed or vegetative propagules used within the population.

The tree improvement work done in Tamil Nadu is summarized as below.

1. Eucalyptus

Krishnaswamy *et al.* (1981) recorded high heritability for height in *E. tereticornis*. In the same species heritability was consistent over different growth stages for girth at base, number of branches, leaf length, leaf length/breadth ratio. The leaf length/breadth ratio recorded highest heritability value (44.85) and leaf breadth recorded the least value (31.80) (Surendran, 1982). Rathinam *et al.* (1982)showed positive and significant correlations between total height, girth at breast height and weight of green bark in *E. tereticornis*. Heritability and genetic advance were high for number of branches so that this trait could be useful in a selection programme for higher biomass in *E. tereticornis* (Surendran and Chandrasekaran, 1984). Pugazhendi (1994) reported that stem volume and cellulose yield showed larger variability among the locations. At one location, cellulose yield exhibited high GA%. On the basis of cumulative scoring for their quantitative and qualitative characters of two trees *viz.*, tree number 9 at germplasm and tree number 11 at Ayyankolly among the candidate trees were selected as plus trees. These can be mass multiplied by micro propagation technique. The nature and extent of correlation differed between the two locations studied. Path analysis revealed that height to possess the largest positive direct effect on cellulose yield followed by dbh and cellulose content. These characters may be considered as selection indices for improvement of cellulose yield. The study indicated that for high cellulose yield and basic density, a rotation period of 18 years would be more remunerative. With respect to effect of rotation on growth and wood traits, the original plantation was found to be superior to coppiced stands. Coppice stands recorded cellulose yield less by one-fourth in the first coppice and by less than half of the original stand. Effect of number of coppice shoots per stump on growth and wood traits indicated that two shoots per stump resulted maximum total height, merchantable height and cellulose yield.

Gokul (1997) reported that heritability was greater in respect of characters like height, leaf breadth, leaf area, flower weight, diffusive resistance and volume. The characters like height, dbh and PAR were found to be positively associated with wood volume. Path analysis showed the same traits to have both positive direct and indirect effects on wood volume. The D^2 analysis resolved the five eucalyptus in to three clusters. The inter cluster between cluster I and III was higher and hence the species of these clusters can be exploited for hybridization. The canonical analysis and principal component analysis were consistent with the clustering pattern as that of D^2 analysis. Heritability was higher in respect of characters like basic density, fibre length, moisture content and double wall thickness. The characters *viz.*, double wall thickness, fibre length and fibre diameter were found to be positively associated with basic density and the same traits had both direct and indirect effects.

Higher heritability and genetic advance was reported for growth attributes like height and dbh, in four species of *Eucalyptus* viz., *E. alba E. camaldulensis*, *E. microtheca* and *E. tereticornis* and 12 hybrids of *Eucalyptus* (Paramathma *et al.*, 1997a) and high heritability coupled with higher genetic advance as percentage of mean for height, merchantable weight, bark, twig and leaf weight in *E. globulus*

(Paramathma *et al.*, 1997b).Venkataramanan (1996) in *E. grandis* found that traits like height, dbh were positively and significantly correlated with wood volume. Inter-correlation between them was also positive and highly significant. Balaji (1997) recorded high GCV, PCV, heritability and genetic advance for flower weight and capsule weight in five *Eucalyptus* species. Heritability was greater in almost all the morpho-physiological characters except fibre length.The characters *viz.*, height, dbh and transpiration rate were found to be positively associated with wood volume. Path analysis also showed that the same traits to have positive direct effect on wood volume. The characters *viz.*, double wall thickness and runkel ratio had positive association with wood basic density. Path analysis revealed the same traits to have both positive direct and indirect effect on wood basic density. The D^2 analysis resolved the five species and a hybrid into three clusters. Canonical analysis and principal component analysis confirmed the results. The inter-cluster distance between cluster I and III was higher in all the three clusters and hence the species of these clusters can be exploited for hybridization.

Jude Sudhagar (1999) recorded high GCV and PCV for volume, moderate heritability for characters such as diameter at breast height, basal area, crown height and total height in *E. grandis*. Balaji (2000) recorded highest GCV, PCV, heritability and genetic advance for volume in *E. tereticornis*. He reported that *E. tereticornis* expressed positive and highly significant correlation between volume and other traits except clear bole length. Inter-correlation among them was also positive.

Sasikumar (2003) identified sixteen plus trees of *E. tereticornis* from the existing eight progeny trials based on regression analysis. In the selected *Eucalyptus* population, volume exhibited positive and significant correlation with dbh, followed by height and clear bole length. In clonal evaluation trials of *E. tereticornis,* 4 clones *viz*., ET_{12}, ET_9, ET_1 and ET_6 recorded superiority in terms of growth characteristics and volume production respectively. Volume recorded the highest PCV and GCV, and the heritability values were high for all the traits investigated. Positive and significant phenotypic and genotypic correlation with volume was registered by plant height and collar diameter. Path analysis revealed that all the traits had positive direct effect on volume in *E. tereticornis*.Genetic diversity study revealed that clustering pattern obtained through hierarchial analysis exhibited a wide range of genetic divergence among the clones investigated.

Jude Sudhagar (2005) revealed that in *E. camaldulensis*, volume registered the maximum PCV and GCV while total height, number of branches and clear bole height registered moderate values. In *E. tereticornis,* branch angle recorded the minimum GCV while volume and dbh registered high GCV. In *E. urophylla,* volume registered the maximum GCV while basal diameter, dbh and total height recorded moderate values of GCV. In *E. camaldulensis,* heritability values were high for basal diameter, dbh, total height, number of branches, branch angle, clear bole height and volume. Volume registered the maximum genetic advance in this species while branch angle and clear bole height recorded considerably low genetic advance. Heritability values were high for volume and dbh in *E. tereticornis.* Genetic advance as percentage over mean was found to be maximum for volume while branch angle and clear bole height registered minimum genetic advance. In *E. urophylla*, volume contributed maximum to the heritability followed by dbh .Maximum genetic advance over mean was exhibited by volume while the minimum was registered by branch angle.

In *E. camaldulensis,* positive and highly significant phenotypic and genotypic correlation and inter-correlation was recorded for basal diameter, dbh, total height, number of branches and clear bole height with volume while branch angle recorded positive and non significant weak correlation with volume. In *E. tereticornis,* basal diameter, dbh, total height, number of branches and clear bole height showed strong, positive and highly significant phenotypic and genotypic correlation and inter- correlation with volume while branch angle registered positive and non significant weak correlation.

In *E. urophylla,* basal diameter, dbh, total height and clear bole height showed positive and highly significant correlation while number of branches had non significant weak correlation with volume. Branch angle showed negative and non significant correlation with volume. In *E. camaldulensis,* path analysis revealed that dbh exerted the highest and positive direct effect on volume followed by clear bole height and total height while branch angle exerted negative direct effect on volume. In *E tereticornis,* dbh exerted the highest and positive direct effect on volume followed by total height and basal diameter while clear bole height and number of branches exerted a negative direct effect on volume.

In *E. urophylla,* dbh exerted the highest and positive direct effect on volume followed by total height and branch angle. Clear bole height and basal diameter exerted a negative direct effect on volume.Phenotypic stability studies in *E. camaldulensis, E. tereticornis* and *E. urophylla* revealed that five out of sixty five seedlots viz. FCRI 6, 19615/16, 15825/2, 15825/5 and 19615/5088 in *E. camaldulensis,* three out of thirty two seedlots viz., 16558/JD1630, 19960/K1320 and 19960/K1326 in *E. tereticornis* and two out of ten seedlots viz. 17567/20 and 17567/22 in *E. urophylla* were phenotypically stable.

The genetic diversity study using morphometric approach revealed that sixty-five seedlots in *E. camaldulensis* were resolved into eleven clusters and thirty-two seedlots in *E. tereticornis* into six clusters. Molecular marker studies in *E. tereticornis* revealed that all the ten arbitrary random primers used exhibited polymorphic electromorphs across the 32 seedlots.The primers viz., OPF 09 and OPR 11 were highly polymorphic which exerted more than or equal to 90 per cent polymorphism followed by OPAB 15.

Ashok kumar (2006) reported that among the 27 CPTs tried for propagation, 17 were successfully rooted. The rooting study revealed that the clones exhibited high variation. The rooting percentage ranged from 7.19 (RMD-3) per cent in July to 99.33 (MTP-5) per cent during November. The best season of planting the cuttings for rooting was found to be November. The number of roots produced per cutting was maximum during January (4.55) and minimum during October (1.58). A similar trend was found for the root length also, while the shoot length has registered its maximum during December (12.01 cm) and a minimum in October (2.06 cm). The clone MTP-5 exhibited highest rooting per cent (99.33) and in contrast, the lowest rooting per cent was shown by RMD-3 (7.19). The clones MTP-5, TCR-2 and SG-2 have recorded the maximum value for number of roots per cutting, root length and shoot length respectively and the lowest value was exhibited by NGL-3, MTP-2 and MTP-4 respectively. The results showed that maximum rooting percentage was obtained for the cuttings from mother plants of age four years, while, lowest rooting per cent was obtained for cuttings collected from 27 year old mother plants. The mother plants of age 25 years exhibited a poor rooting ability in cuttings.

Among the different traits, the genotypic variation and phenotypic variation was high for height and stomatal conductance. The environmental variation was very low for all the characters viz., height, transpiration and stomatal conductance. Similarly, phenotypic coefficient of variation was very high for all the characters studied. Very high genotypic and phenotypic correlation was exhibited between transpiration and stomatal conductance, height and stomatal conductance and these traits have shown negative environmental correlation. Simple correlation was very high for transpiration vs stomatal conductance and transpiration vs height.

The D^2 analysis using transpiration, stomatal conductance and height resolved 17 clones into 9 genetically distinct clusters. Among the 9 clusters, the cluster 1 to cluster 8 have 2 clones each and the cluster 9 has 1 clone. The highest genetic distance was recorded between cluster 5 and cluster 7. The clones within the cluster 8 had shown very high intra cluster distance.

The ecophysiological estimations in clones revealed that the clonal members in cluster 7 and cluster 3 have very high productivity, transpiration and stomatal conductance. The clonal members in cluster 5 showed very low performance in both biometric and ecophysiological traits. The study also showed an array of decreasing performance of clonal members from cluster 7, 3, 4, 1, 6, 8, 9, 2 in terms of biometric and ecophysiological assessments.

The diurnal variation study of all 17 clones on a sunny day from morning 8.00 am to 5.00 pm by estimating the ecophysiological traits viz., transpiration, photosynthesis, stomatal conductance and intercellular CO_2 concentration at an hourly interval showed significant variations among the clones. The high productive clones in 7th cluster showed high performance of all ecophysiological parameters from 8.00 am to 5.00 pm with a peak performance at 11.00 am and 3.00 pm. It is interesting to note that these clones have not shown any mid-day depression at 12.00 noon indicating their suitability for high biodrainage efficiency. On the contrary clonal members in 5th cluster have shown very low performance with minimal mid-day depression. An array of difference has been observed from the cluster 7, 3, 4, 1, 6, 8, 9, 2 on ecophysiological performance in relation to diurnal variation. All the clonal members in these clusters have shown moderate performance to low performance without any mid-day depression.

The result revealed that the clones MTP-5, SG-1, SG-2, MTP-3 and MTP-4 are highly amenable for vegetative propagation. The clones in cluster 7 (MTP-3 and MTP-4) and cluster 3 (MTP-5 and TCR-3) are having very high biodrainage efficiency in terms of biometric and ecophysiological estimation. Hence these clones can be efficiently used for biodrainage to alleviate water logging problem.

Dhayalan (2007) revealed that among the 26 clones and one seed source of *E. saligna*, seven clones viz., MTP-1, MTP-2, MTP-4, MTP-5, MTP-6, MTP-7 and *E. saligna* recorded higher biometric value in terms of height, basal diameter and volume index. Another 11 clones viz., PD-6, PD-9, PD-10,ALR-1, ALR-2, NGL-2, SGc-2, TCR-2, ITC-286, EU-1 and EU-2 showed very poor performance and a moderate performance was observed in the remaining clones viz., MTP-3, MTP-8, SG-1, SGc-1, SGc-3, KK-1, KK-2, EC-1 and ALR-5. The clonal efficiency was also analysed in terms of four ecophysiological characters viz., transpiration rate, stomatal conductance, photosynthetic rate and inter cellular CO_2 concentration. The transpiration rate was ranged from 0.21 m mol m^{-2} s^{-1} (ALR-5) to 0.61 m mol m^{-2} s^{-1} (MTP-1).Among the 26 eucalypts clones and one seed source of *E. saligna*, highest transpiration rate, stomatal conductance, photosynthetic rate and inter cellular CO_2 concentration was observed among the seven clones viz., MTP-1, MTP-2, MTP-3, MTP-4, MTP-5, MTP-6 and *E. saligna*.The lowest performance was observed in ALR-5.

The analysis revealed that highest genotypic and phenotypic variations were observed for the traits viz., height and volume index respectively. The environmental variation was minimal when compared to genotypic and phenotypic variations. The estimation of genotypic, phenotypic and environment coefficient of variation revealed that volume index registered highest GCV (89.06), PCV (96.19) and ECV (36.36) followed by transpiration rate with the value of 30.73, 37.83 and 22.07 respectively for GCV, PCV and ECV. The heritability and genetic advance were very high for volume index (85.71 %) followed by height (87.08 %), transpiration rate (65.98 %) and stomatal conductance (51.67%).The volume index achieved maximal genetic gain of 169.84 per cent. The correlation studies revealed that very high genotypic correlation (0.993) exist between height vs transpiration rate. Whereas high phenotypic correlation (0.942) was observed between height *Vs*., volume index.The path analysis showed very high direct effect of stomatal conductance and transpiration rate on volume index with the path value of 0.525 and 0.709 respectively. Negative indirect effect on volume index was observed by transpiration rate through height (-0.239).

The data subjected for multivariate analysis using D^2 statistics helped to genetically grouped 26 clones and one Australian seed source into seven genetically distinct clusters. Among the seven clusters, the high productive clones viz.,MTP-1, MTP-5, MTP-6 and SG-1 were grouped in the cluster IV and the low productive clones viz., SGc-3, PD-9, ALR-2, ALR-5 and TCR-2 were grouped in the cluster VII. The high inter cluster distance of 9.420 was observed between the cluster I and IV and high intra cluster distance of 5.329 was found in the cluster VI. Among the six traits considered for genetic grouping, height contributed maximum (39.32 %) towards genetic diversity followed by volume index (32.48%) and transpiration rate (12.54%). He observed that the highest transpiration rate, photosynthetic rate, stomatal conductance and inter cellular CO_2 concentration were registered in the clonal members in cluster IV. The lowest transpiration rate, stomatal conductance and photosynthesis were observed in the cluster VII. The inter cellular CO_2 concentration was the lowest among the clonal members constituted in the cluster III.

The diurnal variation study of the ecophysiological traits among the clones revealed that highest transpiration rate throughout the day was observed among the clonal members present in the cluster IV by registering highest transpiration rate of 0.79 m mol m^{-2} s^{-1} at 12.00 noon. The lowest transpiration rate throughout the day was observed among the clonal members in the cluster VII. A moderate transpiration rate was observed among the clonal member in the cluster III and V. It is also interesting to note that all the clusters have maintained steady transpiration rate throughout the day without prominent midday depression. However, transpiration rate remarkably varied among the clusters. Similar to transpiration rate, stomatal conductance, photosynthetic rate and inter cellular CO_2 concentration were exhibited same trend among the clusters. The clonal members in the cluster IV maintained highest photosynthetic rate, stomatal conductance and inter cellular CO_2 concentration throughout the day exhibiting their suitability for high biodrainage efficiency. On the contrary, the clonal members in the cluster VII maintained very low stomatal conductance and photosynthetic rate throughout the day representing their inability to fit for biodrainage efficiency. The moderate biodrainage efficiency has observed among the clonal member in the cluster III and V by maintaining moderate level of photosynthetic rate, stomatal conductance and inter cellular CO_2 concentration. It is also important to note that all the cluster members varied in their ability to maintain different levels of photosynthetic rate, stomatal conductance and inter cellular CO_2 concentration but none of the cluster showed any prominent mid-day depression for all the traits under consideration.

2. Neem

Kumaran (1991) reported significant differences among 28 one-parent families of *Azadirachta indica* collected from various parts of Tamil Nadu in respect of seed parameters viz., seed length, seed breadth, seed length to seed breadth ratio and 100 seed weight. Seed length varied between 1.21 cm to 1.90 cm, seed breadth from 0.55 to 0.86 cm, seed length to breadth ratio from 1.81 to 2.76 and 100 seed weight between 17.02 to 29.09g. Significant variations for various biometric traits *viz.*, height, basal diameter, root length, number of branches, number of leaves, leaf size (length and breadth) and different components of dry weight (shoot, leaf and root dry weights) among twenty eight one parent families of neem was reported.

Surendran *et al.* (1993) evaluated one hundred and forty nine plus trees in neem and reported that, variation in oil content was narrow and found no correlation between seed parameters and oil content.In *Azadirachta indica,* GCV and PCV for seed length, seed breadth, seed length breadth ratio and 100 seed weight were moderate, whereas the heritability and genetic advance as per cent of mean were high (Kumaran,1997; Philomina, 2000). Kumaran (1997) observed that seed length to breadth ratio showed maximum positive direct effect on oil content and azadirachtin content. Sivagnanam *et al.* (1997) noticed a significant variation in all seed attributes except seed germination due to location. Among 26 traits studied in *Azadirachta indica,* root fresh weight exhibited maximum positive direct effect on oil content. Seed length showed positive indirect effects via seed breadth and seed length to breadth ratio on oil content (Philomina, 2000).

In neem, the seedling parameters viz., volume index, basal diameter, number of branches and number of leaves recorded fairly high values of heritability while sturdiness quotient showed moderate value (Kumaran, 1991). Among the seedling traits studied in neem, collar diameter, leaf dry weight and shoot length showed moderate GCV and PCV, whereas root length, number of leaves and root to shoot ratio showed low value (Kumaran, 1997). Philomina (2000) observed that oil content, shoot length, root length, root to shoot ratio, collar diameter, number of leaves and total dry weight recorded high PCV, GCV, heritability and genetic advance as percentage mean in neem and she reported significant variations among 73 one-parent families in respect of seed characters *viz.*, seed length, seed breadth, seed length to breadth ratio and 100 seed weight. The seed length ranged from 1.14 cm to 1.70 cm, seed breadth from 0.48 cm to 0.76 cm, seed length to breadth ratio varied between 1.86 to 3.38 and 100 seed weight between 14.53 g to 24.95g.

3. Pungam

Kumaran (1991) reported significant variations between 28 one-parent families for seedling traits in pungam. His study showed the genotypic coefficient of variation in basal diameter and volume index to be 21.82 and 49.74 per cent respectively. He observed a significant variation among the six agro-climatic zones of Tamil Nadu for oil content and concluded the influence of agro-climatic zones on the oil content of *Pongamia pinnata*.

Harini (2007) reported that the candidate plus trees exhibited significant variation in five seed physical attributes investigated, in which PC1, PC2, PC3, PC7, PC8 and PC23 had an edge over others. In the nursery experiment, the clones *viz.,* PC2, PC3 and PC7 performed better than other clones. The evaluation under field conditions showed that the clones PC7 expressed superiority for all the four traits *viz.,* plant height, basal diameter, number of branches and volume index while the clones PC1, PC2, PC3 and PC4 also performed well for a minimum of three morphometric traits.

Hundred seed weight recorded the highest PCV, GCV, heritability and genetic advance as percentage of mean. In morphometric traits, volume index had registered the highest PCV, GCV, heritability and genetic advance as percentage of mean followed by basal diameter. Correlation studies revealed that seed oil content exhibited highly positive and significant correlation with 100 seed weight at phenotypic level but with seed breadth at genotypic level. Volume index was highly and positively associated with basal diameter at phenotypic level and with plant height at genotypic level.

Path analysis indicated that 100 seed weight exerted maximum direct positive effect on seed oil content followed by seed breadth, seed length-breadth ratio and seed length. Basal diameter and plant height had direct positive effect on volume index. Hence, basal diameter and plant height could be used as selection indices in *Pongamia pinnata*. Genetic divergence studies using morphometric approach revealed that the clustering pattern obtained through hierarchical analysis expressed similarity with geographic distribution of clones.

4. Mahua

George Jenner (1995) observed in *Madhuca latifolia*, that among the seedling parameters investigated, shoot dry weight displayed high GCV, heritability and genetic advance as percentage mean. The total dry weight possessed moderate heritability and genetic advance as percentage of mean.He reported distinct variation among the 23 one-parent families of *Madhuca latifolia* in respect of seed characters viz., seed length, seed breadth, seed length: seed breadth ratio and 100 seed weight. Seed length varied between 3.92 cm and 2.16 cm, seed breadth from 0.94 cm to 1.35 cm, seed length: seed breadth ratio from 0.94 to 1.35 and 100 seed weight between 134.06 g and 262.04 g. Mahua exhibited significant variations for shoot length and collar diameter among 23 one-parent families investigated at all four growth phases. Significant variations were also found in root length, number of leaves and total dry weight among the progenies investigated. He also found that seed length to breadth ratio showed maximum direct effect on oil content in Mahua, but 100 seed weight exhibited positive indirect effects through seed breadth and seed length to breadth ratio. Similarly, total dry weight showed maximum direct effect on oil content. Umesh Kanna (2001) also recorded significant variations in seed length, seed breadth, seed length: seed breadth ratio and 100 seed weight due to progenies. He found significant and positive correlation in height and collar diameter with oil content in *Madhuca latifolia* whereas, number of leaves and leaf length: breadth ratio had exhibited negative and non-significant correlation with oil content.

5. Simaruba

Sekar (2003) reported that the 15 seed sources exhibited significant variations in terms of plant height, collar diameter, number of leaves and leaf area index in early stages of growth in simaruba which thereby help to improve the species. He also stated that among seed attributes,100 seed weight exhibited highest heritability followed by seed length, seed perimeter, seed germination and seed 2D surface area. Highest positive direct effect on oil content was exhibited by seed 2D surface area, aspect ratio, seed breadth, 100 seed weight, seed germination and seed protein content. Seed perimeter expressed indirect positive effect via seed 2D surface area, seed breadth, seed aspect ratio, 100 seed weight, seed protein and seed germination among 15 seed sources of *Simarouba glauca*. Thus, the morphological variation in seed characteristics among the population of a species is useful in selection programme for genetic improvement. He reported that among the biometric traits, plant height and collar diameter expressed high phenotypic and genotypic coefficient of variation, high heritability and genetic advance as percentage of mean.

Nesamani (2005) reported that the evaluation under field condition showed that seed sources from Akola at Mettupalayam, S.K.Nagar at Coimbatore and GKVK 2 and GKVK 1 at Tindivanam performed superior for four traits *viz.*, height, basal diameter, number of branches and volume index. The performance of these seed sources were consistent through out the growth period.

Volume index registered high phenotypic and genotypic coefficient of variation at all the three locations. Regarding heritability higher values were exhibited by volume index at Mettupalayam and Coimbatore and height at Tindivanam. The genetic advance as percentage of mean was high at Tindivanam, Coimbatore and moderate at Mettupalayam for volume index. Correlation studies revealed that volume index was highly and positively correlated with height and basal diameter at both phenotypic and genotypic levels in all the three locations. Path analysis indicated that height and basal diameter had positive direct effect on volume index at all the three locations. Mahalonobis D^2 statistic resolved the 15 seed sources into three clusters at Mettupalayam, two clusters at Coimbatore and five clusters at Tindivanam. Volume index contributed the maximum towards genetic divergence followed by height at all the three locations. Stability analysis restricted that the seed sources viz., S.K. Nagar and Bhubaneshwar from Gujarat and Orissa to be capable of giving stable performance.

6. Tamarind

Divakara (2002) carried out investigations to elicit information from thirty-five CPTs in *Tamarindus indica* on variability in pod characters, seed, seedling, pulp biochemical traits and estimation of genetic parameters, association and genetic diversity during 1999-2002. He found that KA BA NG performed well in terms of height and KA BA GR performed better with respect to collar diameter and volume index. KA BA NG expressed significant superiority for majority of pod characters. Candidate plus trees AP CW KU expressed superiority for seven seed parameters. Half-sibs variation revealed that, CPTs AP CW KU and KA BA NG recorded significant superiority for all the traits under study. Pulp biochemical traits indicated that, CPTs KA BA NG and TN TI JM were superior in terms of sugar content, TSS and anthocyanin pigment content.

Estimation of genetic parameters revealed that, among the biometric traits, volume index recorded moderate GCV and PCV and high heritability and GA as percentage over mean. Pulp weight and 2D surface area expressed high GCV and PCV with high heritability among all pod and seed parameters respectively. Among half-sib and pulp biochemical traits, shoot dry weight and anthocyanin pigment registered high GCV and PCV with moderate heritability and GA as percentage over mean respectively.

Association studies indicated that volume index expressed a positive genotypic and phenotypic correlation with plant height and collar diameter. 2D surface area and hundred seed weight registered significant and positive correlation with germination percentage. Significant and positive correlation was observed between vigour index and shoot length.

Path coefficient analysis indicated that, biometric traits viz., collar diameter and height exhibited direct positive effect on volume index. Pod weight, pulp percent, seed percent and shell percent exhibited positive direct effect on pulp weight. Mean daily germination exerted maximum direct positive effect on germination. Estimation of genetic diversity using biometric and isozyme approach revealed that, the clustering pattern attained through hierarchical analysis did not express similarity among the clones for their geographic distribution.

7. Casuarina

Half-sib progenies of *Casuarina equisetifolia* exhibited high variability for height, dbh and total biomass at half rotation age (Jambulingam and Surendran, 1989). The consistency of heritability estimates over three growth stages analysed indicated that genetic factors were more important than site factors in determining growth rates. Genetic advance as percentage of mean was also consistently high for all the traits investigated (Jambulingam *et al.*, 1990). Paramathma *et al.* (1994) reported high variability for dbh, total height and total biomass and significant genetic variation in 55 clones of *Casuarina equisetifolia* were also documented (Ashok Kumar and Gurumurthi, 1998). Significant genetic variation in height, diameter at ground level, diameter at breast height and mean bole volume was recorded (Ashok Kumar *et al.*, 1998; Kumaravelu, 1997).

Estimation of broad sense heritability for various characters *viz.* height, diameter at ground level, diameter at breast height, and mean bole volume in 42 clones of *Casuarina equisetifolia* at 12 months age, showed that heritability for height was highest and little less for mean bole volume. Nevertheless, the percentage of heritability was moderately high for all other traits studied (Ashok Kumar and Gurumurthi, 1998). They also observed genetic correlations of plant height with any of the characters *viz.*, diameter at ground level (DGL), diameter at breast height (DBH) and mean bole volume (MBV) studied was very high in 42 clones. DGL and main bole volume were found to have minimum genetic correlation.

Ashok Kumar and Paramathma (2005) reported that plant height, collar diameter, number of branches, survival per cent, suitability index and chlorophyll content were strongly associated with volume index except needle length. Ashok Kumar *et al.* (1998) observed that the relation between the traits viz., plant height, diameter at ground level, diameter at breast height and number of branches with volume showed that diameter at breast height was the most important trait for selection. Kumaravelu (1997) also observed similar association in the same species.

8. Jatropha

Interspecific hybrids of *Jatropha* are available in India through natural hybridization as well as artificial hybridization. A new species of *Jatropha* identified by the Botanical Survey of India *viz.*, *Jatropha tanjorensis* was proved to be an interspecific hybrid between *Jatropha curcas* and *J. gossypifolia.* Earlier hybridization was attempted between *J.curcas* and *J. integerrima* as a possible source of introgressive bridge between *Jatropha* and *Ricinus* (Sujata and Prabakaran, 1997).They collected different species of *Jatropha* from different states of India and studied their characters like chromosome number, fertility status, morphological features and desirable attributes. They also tried interspecific crosses among the species and found that crossability was high in cases involving *J. curcas* as female parent and also cases where *J. integerrimma* was used as male parent.Crossing between *J. curcas* and *J. integerrimma* showed high success and the F_1 showed high variability opening the way for further selection and production of suitable ornamentals and economic crops (Sujata and Prabakaran, 2004).

Interspecific hybridization has been successful between two economically important species of *Jatropha*, *viz.*, *J. curcas* and *J. integerrima*. The interspecific hybrids exhibited morphological intermediacy for various vegetative characteristics but produced flowers with three distinct colours. Backcrossing of the F_1 hybrids resulted in a number of flower colours varying from dark pink through green to white enhancing the ornamental value of the genus (Sujatha and Prabakaran, 2004).

Prabakaran and Sujatha (2004) reported a natural interspecific hybrid, *Jatropha tanjorensis* ,a species found abundantly in Tanjore, Pudukottai, Trichirapalli and Ramnad districts of Tamil Nadu state, India, showed intermediacy in phenotypic characters of *J. curcas* L. and *J. gossypifolia* L. A detailed survey at its place of occurrence supplemented with data employed from cytological and peroxidase isozyme studies revealed that *J. tanjorensis* is a natural interspecific hybrid between these two species.

Selvam (2006) reported that seed sources exhibited significant variation in six seed physical parameters investigated, in which three seed sources *viz.*, TNMC 2, TNMC 7 and TNMC 4 had expressed superiority. Hundred seed weight registered high heritability and genetic advance as percentage over mean. Correlation studies revealed that seed oil content expressed positive and significant correlation with hundred seed weight both at phenotypic and genotypic levels. Path analysis indicated that seed length followed by hundred seed weight exerted direct positive effect on seed oil content.

In the nursery experiment, three seed sources *viz.*, TNMC 7, TNMC 5 and TNMC 2 performed better than other sources. The evaluation under field conditions showed that seed sources viz., TNMC 7 and TNMC 26 expressed superiority for four biometric traits *viz.*, plant height, basal diameter, number of branches and sturdiness quotient.In biometric traits, plant height and number of branches had recorded high heritability, phenotypic and genotypic coefficient of variations and higher genetic advance as percentage over mean followed by sturdiness quotient and basal diameter. Sturdiness quotient was highly and positively associated with plant height both at phenotypic and genotypic levels followed by number of branches and basal diameter. Path analysis indicated that plant height had direct positive effect on sturdiness quotient.

Analysis of physico-chemical properties of Jatropha seed sources showed that the five seed sources were found to have higher acid value and free fatty acid (%) and three seed sources had higher iodine number, which may not be useful for the production of biodiesel, other seed sources are in good agreement with biodiesel standards which can be used to produce biodiesel. All 25 seed sources fulfill Cetane and saponification number requirement for biodiesel production.

The saturated fatty acids *viz.*, myristic acid (C14:0) was found to be absent in two seed sources and 14 seed sources recorded significantly higher percentage of myristic acid. Palmitic acid (C 16:0) was found to be highly significant in 13 seed sources. Stearic acid (C18:0) was significantly higher in 11 seed sources. The unsaturated fatty acids *viz.*, oleic acids (C18:1) and linoleic acid (C18:2) were found to be significant for twelve and fourteen seed sources respectively.

Reeja (2007) stated that seed yield, number of branches and basal diameter registered the maximum PCV and GCV while plant height registered low values. Heritability values were high for seed yield, plant height and basal diameter at 12, 24 and 36 MAP. Seed yield registered the maximum genetic advance while plant height recorded moderate genetic advance at 36 MAP. Positive and highly significant phenotypic and genotypic correlation and *inter* correlation was recorded for all the three traits *viz.*, plant height, basal diameter and number of branches with seed yield. Path analysis revealed that number of branches exerted the highest and positive direct effect on seed yield followed by basal diameter while plant height exerted negative direct effect on seed yield.

Phenotypic stability studies in *Jatropha curcas* over three growth periods *viz.*, 12, 24 and 36 MAP revealed that seven out of twenty seed sources *viz.*, TNMC 1, TNMC 2, TNMC 3, TNMC 4, TNMC 6, TNMC 7 and TNMC 19 were phenotypically stable over three growth phases. The performance of backcross progenies revealed that five out of 10 families in the BC_2F_1 generation *viz.*, 3-66, 3-97, 3-75, 3-82 and 3-60 were found to perform consistently well for seed yield, number of branches and basal diameter over the two growing periods of 15 and 30 MAP .

Variability studies for BC_3F_1 generation in the nursery at 3 MAS revealed that thirty four out of one hundred and fifty progenies of fifteen families exhibited significantly superior mean performance for both plant height and collar diameter. The assessment of BC_3F_1 progenies for plant height, basal diameter and number of branches revealed that the progeny 3-97-7 exhibited high mean performance at all three locations. The family *viz.*, 3-66-16 exhibited high mean performance for two traits *viz.*, height and basal diameter at two locations *viz.*, Mettupalayam and Coimbatore. The family *viz.*, 3-66-25 had better performance for the same traits at Coimbatore and Bhavanisagar while the family *viz.*, 3-97-5 had best mean performance at Mettupalayam and Bhavanisagar for plant height and basal diameter. Besides, the progeny 3-75-9 showed consistent mean performance for all three traits in Mettupalayam location, whereas progenies *viz.*, 3-66-16 and 3-66-23 exhibited significant mean performance for all three traits in Coimbatore and Bhavanisagar respectively.

Among the ten families evaluated for seed traits, the members of the following families *viz.*, 3-60-5, 3-66-16, 3-66-26 and 3-97-7 showed consistently superior mean values for three traits *viz.*, hundred seed weight, seed width and germination percentage. All the families recorded high phenotypic coefficient of variation but genotypic coefficient of variation, heritability and genetic advance as percentage of mean were low for all families for the trait number of seeds/capsule. Moderate to low phenotypic and genotypic coefficient of variation were observed for germination percentage in all the families except family 3-90. The families *viz.*, 3-56, 3-60, 3-66, 3-75, 3-82, 3-90, 3-91, 3-95 and 3-97 recorded highest heritability. Among the six seed traits studied, hundred seed weight showed the highest heritability and in the family 3-56 and 3-60 followed by germination percentage. Among the traits, hundred seed weight recorded the highest genetic advance as percentage of mean followed by germination percentage indicates that selection can relay upon these traits for improvement.

The traits *viz.*, germination percentage, number of seeds per capsule and seed width exhibited positive and significant correlation with hundred weight for both phenotypic and genotypic correlations in the families *viz.*, 3-93, 3-95 and 3-97.The traits *viz.*, seed width, seed aspect ratio and number of branches exhibited positive and highly significant *inter* phenotypic and genotypic correlation in the families *viz.*, 3-60, 3-66, 3-93, 3-66, 3-95 and 3-97.

9. Prosopis

In a study conducted by Masilamani (1992), pods of *Prosopis juliflora* collected from different locations *viz.*, Coimbatore, Mettupalayam, Palladam, Pollachi and Tirupur of Coimbatore district of Tamil Nadu showed significant variation in physical, physiological and biochemical characteristics of seed. The pod and seed characters vary with geographical locations and hence the best location must be selected for seed collection. Seeds of this species collected from the middle portion of the pod were found to be more vigorous compared to that from provincial and distal position.

Reeja (2004) stated that seed sources exhibited significant variation in ten seed physical attributes investigated, in which seed sources from Tuticorin, Rameshwaram, Ramnad and Keezhakarai had an edge over others. In the nursery experiment, the seed sources *viz.*, Ramnad and Rameshwaram performed better than other sources. The evaluation under field conditions showed that seed sources from Ramnad, Rameshwaram, Keezhakarai and Madurai expressed superiority for five traits *viz.*, plant height, collar diameter, number of branches, sturdiness quotient and volume index.

Hundred seed weight recorded high heritability coupled with moderate genetic advance as percentage over mean. In morphometric traits, volume index had registered high heritability, phenotypic and genotypic coefficient of variations and moderate genetic advance as percentage over mean followed by collar diameter and sturdiness quotient. Correlation studies revealed that seed germination percentage exhibited positive and significant correlation with hundred seed weight and viability percentage both at phenotypic and genotypic levels. Volume index was highly and positively associated with plant height both at phenotypic and genotypic levels followed by collar diameter.

Path analysis indicated that seed perimeter followed by hundred seed weight and aspect ratio exerted maximum direct positive effect on germination percentage. Collar diameter and plant height had direct positive effect on volume index. Hence, collar diameter and plant height could be used as selection index for biomass production in *Prosopis juliflora*. Genetic divergence studies using biometric approach revealed that the clustering pattern obtained through hierarchical analysis expressed similarity with geographic distribution of seed sources.

10. Bamboos

Antony Joseph Raj (1999) reported that the genotype variance and co-efficient of variation were higher for number of internodes per shoot, average internodal length, average culm height and number of culms in *Bambusa bambos* and *Dendrocalamus strictus*.High heritability coupled with genetic advance was observed for leading shoot length , average shoot length, number of culms and average culm height for both *Bambusa bambos* and *Dendrocalamus strictus*.The character like leading shoot, average shoot length, number of internodes per shoot, average internodal length, average culm diameter, leaf length and leaf width were found to be positively correlated with number of culms in *Bambusa bambos* and average culm diameter and leading shoot diameter in *Dendrocalamus strictus*.Path co-efficient analysis revealed that the leading shoot diameter and average shoot length in *Bambusa bambos* and leading shoot diameter, average culm height and diameter in *Dendrocalamus strictus* have both direct and indirect effect on number of culms. The principal component analysis resolved 34 and 25 clones of *Bambusa bambos* and *Dendrocalamus strictus* respectively into five clusters . Clones from cluster II and IV in *Bambusa bambos* and IV in *Dendrocalamus strictus* can be exploited for further improvement.

Paramathma *et al.* (2000) reported wide range of variability for number of culms, culm height, culm diameter and culm internode length in *Dendrocalamus strictus*. Average culm height recorded high heritability coupled with high genetic advance in *Bambusa bambos* and in *Dendrocalamus strictus* (Antony Joseph Raj, 1999). Antony Joseph Raj (1999) and Satheesh (2000) recorded that characters height and culm diameter had significant and positive correlation with number of culms in *Bambusa bamboos*. In the case of *Dendrocalamus strictus*, the trait average culms diameter showed highly positive significant correlation with number of culms. In case of *Bambusa bambos,* average internodal length recorded maximum positive direct effect on number of culms whereas in case of *Dendrocalamus strictus* leading shoot diameter had highest positive direct effect.

11. Acacias

Jayaprakash (2000) observed that in biometric triats analysis, variability estimates revealed that volume registered maximum phenotypic and genotypic coefficients of variation whereas basal diameter registered lower PCV and GCV. Heritability values were high for all the traits except number of branches. In terms of correlation studies, basal diameter recorded highest positive correlation with volume. Association studies observed that plant height recorded maximum positive direct effect on volume. Regression analysis revealed that two half sibs *viz.*, TNPKM1 and KRKL2 were plotted above the regression line for three characters and recommended for further tree improvement work. In case of biochemical traits approach, total tannin content registered maximum PCV and GCV . Heritability values were high for all the traits except total tannin content, whereas GCV were higher for total starch. All the traits had negative and nonsignificant correlation with total weight. Association studies between yield components revealed that total starch registered positive direct effect on total dry weight. In the biochemical experiment, eight half sibs *viz.*,TNRVK1, TNKMD1, TNKPT, KRGB1, KRGB2, KRMS2, KRKL1 and KRKL2 were plotted above the regression line for all the four traits and these half sibs are recommended for further improvement programme.

Genetic diversity studies based on biometric and biochemical approaches revealed that the half sibs viz., TNPKM1 and KRKL2 , and TNTR and TNSA respectively were highly divergent. The half sibs of TamilNadu origin were not clustered together in distinct sub cluster whereas halb sibs from Karnataka were grouped together in distinct small sub clusters as per their geographic origin. In the isozyme based genetic diversity investigation, three enzyme systems *viz.*, peroxidase, esterase and malate dehydrogenase were tested and a total of 19 markers among three enzyme systems were obtained. Esterase and malate dehydrogenase gave maximum polymorphic percentage of 100 and it was concluded that all the three markers used brought out different pattern of diversity.

12. Teak

Arun Prasad (1996) found that heritability was high (70-90%) for all the seed and seedling characters studied in teak (*Tectona grandis*). He observed that height recorded maximum positive direct effect with wood volume in *Tectona*. Parthiban (2001) conducted study with 30 seed sources *viz.*, 28 from India and one each from Lao PDR and Bangladesh. The results revealed that the seed sources exhibited significant differences in seed physical parameters. Seed sources from moist localities expressed superiority in terms of filling per cent , germination per cent and other seed physical parameters. The biometric observations in the nursery indicated a wide range of variabilities among seed sources for biometric traits. However, in the field, only four seed sources viz., Kerala VI, Kerala VII, Tripura and Tamil Nadu III expressed superiority for four traits. Volume index had registered higher PCV and GCV followed by number of leaves, plant height and collar diameter.Number of leaves expressed high heritability and GA as per cent over mean. Correlation studies revealed that volume index was highly and positively associated with collar diameter, plant height and number of leaves both at phenotypic and genotypic levels. Path analysis indicated that collar diameter exerted maximum positive direct effect on volume index followed by plant height and number of leaves.Genetic divergence studies using biometric and RAPD approaches revelaed that the clustering pattern obtained through hierarchical analysis expressed similarity among the seed sources collected within state barring few exceptions.

13. Albizia

Radhakrishnan (2001) revealed that the seed sources expressed significant variation in seed attributes for which the Mettupalayam seed source had an edge over others. The seed Gudag expressed superiority and consistency over others in both nursery and field in terms of volume index. In the biochemical traits, none of the seed sources exhibited superiority.However augustimuni, Gaucher Doiwala and Gudag seed sources registered significance for three traits viz., total chlorophyll , soluble protein and total carbohydrate contents. Germination per cent in seed attributes and volume index in nursery and field, chlorophyll 'b' expressed highest GCV, PCV, heritability and genetic advance as per cent of mean in the genetic parameters. In association studies, germination per cent with roundness, aspect ratio and 100 seed weight and shoot length and collar diameter under nursery condition and height and collar diameter under field condition and most of the biochemical traits with volume index expressed positive association for seed, biometric and biochemical characteristics. Path analysis revealed maximum direct positive effects of seed diameter and germination per cent and shoot length and collar diameter on volume index under nursery and field conditions and chlorophyll 'a' on volume index for seed, biometric and biochemical attributes. The genetic diversity studies using biometric, isozyme and RAPD approaches revealed that the cluster pattern attained through hierarchical analysis did not expressed similarity among the seed sources for their geographic distribution.

14. Kapok

Rajendran (2001) reported significant variation among twenty three candidate plus trees in respect of seed characters *viz.*, pod length, pod diameter, pod weight, husk weight, hundred seed weight, number of seeds per pod, seed breadth, seed length and floss weight per pod. Clonal variation for root characters *viz.*, number of sprout, sprout length, number of roots, root length and per cent rooting. Clonal traits *viz.*, sprout height, sprout diameter, number of leaves, sturdiness quotient, volume index in nursery and plant height, basal diameter, number of branches and volume index in field, biometrical attributes *viz.*, total phenol content, total carbohydrate content, total soluble protein chlorophyll and oil content. The candidate plus trees TN TI VN, TN TI KG, TN TI PK and TN CB MP proved significantly superior in all the pod and seed traits. The ramets TN CB MP and TN TI VN recorded the highest values for number of sprout, sprout length, number of root, root length and per cent rooting in the nursery at 60 days after planting. The ramets TN CB MP and TN TI PI was found to be consistently superior in sprout height, sprout diameter, number of leaves, sturdiness quotient and volume index in the clonal evaluation trial in nursery at twelve months after planting. The ramets TN CB MP and TN TI VN proved significantly superior in all the morphometric traits *viz.*, plant height, basal diameter, number of branches and volume index. Among the pod and seed parameters, pod length and seed weight per pod recorded moderate PCV and GCV, high heritability and genetic advance as percentage mean.

Per cent rooting, sprout length and number of sprout and clonal traits viz., volume index, sprout height, sturdiness quotient recorded high values for PCV and GCV, heritability and genetic advance as percentage mean at twelve month after planting. In clonal evaluation trial, volume index and plant height registered high values for PCV and GCV, heritability and genetic advance as percentage mean at nine month after planting. Among the biometrical attributes, chlorophyll recorded maximum PCV, GCV heritability and genetic advance as percentage mean.

The correlation studies revealed that pod length, pod weight and husk weight showed positive and significant association with floss weight per pod. Among clonal attributes for root characters, number of sprout, sprout length, number of roots and root length showed positive but non significant association with per cent rooting. Among the clonal attributes, sprout height, and sprout diameter recorded positive and significant correlation with volume index. Plant height, basal diameter and number of branches showed positive and significant correlation with volume index. The oil content showed positive and highly significant correlation with chlorophyll content.

Among the ten traits studied in pod and seed parameters, seed weight per pod, pod diameter, husk weight , hundred seed weight and pod length exhibited maximum positive direct effect on floss weight per pod. Number of sprout, sprout length and number of roots showed positive direct effect on per cent rooting. Sprout diameter, number of leaves and sturdiness quotient registered positive direct effect on volume index.

Genetic diversity studies based on biometric and biochemical approaches revealed that the ramets *viz*., TN KK NC and TN DG DG and TNSM VI and TN CB CB respectively were highly divergent and it revealed that the clustering pattern obtained through hierarchial analysis did not show similarity with geographic distribution of ramets. In the isozyme based genetic diversity investigation, three enzymes *viz*., peroxidase, esterase and malate dehydrogenase were tested and a total of 15 markers among 3 enzyme systems were obtained. Peroxidase gave maximum polymorphic percentage of 100 and it was concluded that all the markers used to brought out different pattern of diversity.

References

Antony Joseph Raj. 1999. Studies on biometrical genetics in bamboos. M.Sc. (For.) Thesis, Forest College and Research Institute, Tamil Nadu Agricultural University, Coimbatore.

Arun Prasad, K.C.A. 1996. Variability studies in teak (*Tectona grandis* L.F.). M.Sc. (For.) Thesis, Forest College and Research Institute, Tamil Nadu Agricultural University, Coimbatore.

Ashok Kumar. 2001. Provenance variation and clonal evaluation studies in *Casuarina equisetifolia*. L. Johnson and *Casuarina junghuniana* Miq. M.Sc. (For.), Thesis, Forest College and Research Institute, Tamil Nadu Agricultural University, Coimbatore.

Ashok Kumar, S. 2006. Assemblage and evaluation of Eucalypts clones for higher biodrainage efficiency. M.Sc. (For.) Thesis, Forest College and Research Institute, Tamil Nadu Agricultural University, Coimbatore.

Ashok Kumar, K. Gurumurthi. 1998. Path coefficient studies on morphological traits in *Casuarina equisetifolia.* Indian For. 122(8): 727-730.

Ashok Kumar, Paramathma, M. 2005. Correlation and path coefficient studies in *Casuarina equisetifolia* L. Johnson. Indian For. 131(1): 47-55.

Ashok Kumar, Gurumurthi K., Kumar A. 1998. Genetic assessment of clonal material of *Casuarina equisetifolia.* Indian For. 124(3): 237-242.

Balaji, B.1997.Multivariate analysis and interspecific hybridization in *Eucalyptus* species. M.Sc.(For.) Thesis, Forest College and Research Institute, Tamil Nadu Agricultural University, Coimbatore.

Balaji, B. 2000.Estimation of genetic parameters and genetic diversity using morphometric, isozyme and RAPD approaches in *Eucalyptus tereticornis* Sm. Ph.D. Thesis, Forest College and Research Institute, Tamil Nadu Agricultural University, Coimbatore.

Dhayalan, K. 2007. Evaluation of eucalyptus clones for biodrainage. M.Sc. (For.) Thesis, Forest College and Research Institute, Tamil Nadu Agricultural University, Coimbatore. Divakara, B. N. 2002 Clonal evaluation and genetic diversity studies using biometric and isozyme approaches in Tamarindus indica. Ph.D. Thesis submitted to the Tamil Nadu Agricultural University, Coimbatore

George Jenner, V., Dasthagir, M.G., Kumaran K., Divya M.P. 1999. Genetic analysis of one parent families of mahua (*Madhuca latifolia* Roxb. Macbride) in Tamil Nadu. *In:* National Symposium on Forestry towards 21st century, 27-28 September, Tamil Nadu Agricultural University, Coimbatore. 110 pp.

Gokul, R. 1997. Genetic analysis and hybridization in *Eucalyptus* species. M.Sc (For.). Thesis, Forest College and Research Institute, Tamil Nadu Agricultural University, Coimbatore.

Harini, V. 2007.Clonal evaluation of P*ongamia pinnata* (linn.) Pierre. M.Sc(For.) Thesis Forest College and Research Institute, Tamil Nadu Agricultural University, Coimbatore.

Jambulingam, R., Lankany, E. L., Turnbull, M. H., Brewbaker, J. L. 1990. Recent developments in research on Casuarina in Tamil Nadu. Variation in population, Advances in Casuarina Research and Utilization, 45-54. In: Proceedings of the Second International Casuarina Workshop, January, 15-20, Cairo, Egypt.

Jambulingam, R., Surendran C.1989.Breeding strategy for *Casuarina equisetifolia.* National seminar on Casuarina, December, 18-19, Neyveli, India.

Jayaprakash, J. 2000. Studies on genetic analysis among sources of *Acacia nilotica* (Linn) wild ex.del, using biometric and biochemical approaches. Ph.D. (For.) Thesis, Forest College and Research Institute, Tamil Nadu Agricultural University, Coimbatore.

Jude Sudhagar, R. 1999. Genetic analysis in *Eucalyptus grandis* Hill ex. Maiden and interspecific hybridization in Eucalyptus species. M.Sc. (For.) Thesis, Forest College and Research Institute, Tamil Nadu Agricultural University, Coimbatore.

Jude Sudhagar, R.2005 .Genetic analysis of morphometric traits in three species of eucalyptus and molecular marker studies in Eucalyptus *tereticornis* sm.Ph.D (For.)Thesis Forest College and Research Institute, Tamil Nadu Agricultural University, Coimbatore.

Krishnaswamy, S., Srinivasan, V.M., Vinaya Rai, R. S. 1981.Growth increment and genetic variability in relation to age in saplings of *E. tereticornis* Sm. National Seminar on Tree Improvement, Trichirapalli, January 8: 122.

Kumaran, K. 1991.Genetic analysis of seed and juvenile seedling attributes in neem (*Azadirachta indica* A. Juss.), pungam (*Pongamia pinnata* Linn. Pierre). M.Sc. (For.) Thesis, Forest College and Research Institute, Tamil Nadu Agricultural University, Coimbatore.

Kumaran, K. 1997.Selection of one-parent families for higher growth, oil and azadirachtin content in neem (*Azadirachta indica* A. Juss.). Ph.D. Thesis, Forest College and Research Institute, Tamil Nadu Agricultural University, Coimbatore. 131 pp.

Kumaravelu, G.1997.Tree improvement in *Casuarina equisetifolia*. Ph.D. Thesis, Bharathiyar University, Coimbatore. 243 pp.

Masilamani, P. 1992. Production, processing and storage technology of seeds of *Cassia siamea* Lab. *Hardwickia binata* Roxb. and *Prosopis juliflora*, Swartz. DC. M.Sc. (Ag.) Thesis, Tamil Nadu Agricultural University, Coimbatore p- 641 003.

Nesamani, K.2005. Studies on seed sources, micropropagation and molecular characterization of sex specificity in *Simarouba glauca* dc. M.Sc. (For.) Thesis, Forest College and Research Institute, Tamil Nadu Agricultural University, Coimbatore.

Paramathma, M., Surendran, C.1999. Genetic divergence in *Eucalyptus* species, 44: In: Abstract of National Symposium on Forestry towards 21st Century, 27-28 September, Tamil Nadu Agricultural University, Coimbatore.

Paramathma, M., Surendran, C., Rai, R.S.V. 1997a.Studies on heterosis in six *Eucalyptus* species. J. Trop. For. Sci., 9(3): 283-293.

Paramathma, M., Surendran, C., Rai, R.S.V. 1997b. Heritability in *Eucalyptus globulus*, 239. In: Proceeding of the IUFRO-Conference on Silviculture and Tree Improvement of *Eucalyptus,* 24-29 August, Salvador, Brazil.

Paramathma, M., Surendran, C., Natarajan, V., Annamalai, R.. 1994. Trait selection in *Causuarina equisetifolia,* 190-202: Wheeler, C.T., Narayanan, R., Parthiban, K.T., Kesavan, A. and Surendran, C. (eds.). In: Proceeding of National Seminar, 6-9, December 1994, Tamil Nadu Agricultural University, Coimbatore.

Paramathma, M., Antony Joseph Raj, M., Surendran, C., Viswanathan, P., Sundersingh Rajapandian, J., Balaji, S. 2000. Variability in *Dendrocalamus strictus*. In: Proceeding of the National Workshop on Bamboo for Wasteland Development, 20-21st January, Raipur, U.P.

Parthiban, K. T. 2001. Seed source variations, molecular characterization and clonal propagation in teak (*Tectona grandis* Linn. F.). Ph.D. Thesis, Tamil Nadu Agricultural University, Coimbatore.

Philomina, D. 2000. Genetic analysis of one-parent families for variability, diversity, stability and propagation techniques in neem (*Azadirachta indica* A. Juss). Ph.D. Thesis, Tamil Nadu Agricultural University, Coimbatore.

Pugazhendi, 1994. Studies on variability in pulp wood traits and its causes in *E. grandis* Hill. ex. Maid. M.Sc. (For.) Thesis, Forest College and Research Institute, Tamil Nadu Agricultural University, Coimbatore.

Radhakrishnan, S. 2001. Genetic divergence and DNA based molecular characterization in *Albizia lebbeck* (L.) Benth. Ph.D. (For.) Thesis, Forest College and Research Institute, Tamil Nadu Agricultural University, Coimbatore.

Rajendran, P. 2001. Clonal propagation, evaluation and genetic diversity in Kapok (*Ceiba pentandra* Linn. Gaertn.). Ph.D. Thesis, Tamil Nadu Agricultural University, Coimbatore.

Rathinam, M., Surendran, C., Kondas, S. 1982. Inter-relationship of wood yield components in *Eucalyptus tereticornis*. Indian For. 108: 465-470.

Reeja, S. 2007. Evaluation of seed sources in J*atropha curcas L*. and assessment of interspecific backcross progenies.Ph.D. (For.) Thesis. Forest College and Research Institute,Tamil Nadu Agricultural University. Coimbatore

Sasikumar, K. 2003. Clonal evaluation of *Eucalyptus tereticornis* sm. and A*cacia* hybrids, and interspecific hybridization between *A. uriculiformis* A. Cunn. ex. Benth. and *A. Mangium* willd. Ph.D. Thesis. Tamil Nadu Agricultural University. Coimbatore

Satheesh, N. 2000. Selection, propagation and genetic diversity in bamboos. M.Sc. (For.) Thesis, Forest College and Research Institute, Tamil Nadu Agricultural University, Coimbatore.

Sekar, I. 2003. Seed storage, seed source variations, molecular characterization and *in vitro* propagation in *Simarouba glauca* DC. Ph.D(For.) Thesis,Forest College and Research Institute, Tamil Nadu Agricultural University, Coimbatore. 272 pp.

Selvam, P.2006. Seed source evaluation of *Jatropha curcas* linn. For oil, biometric and other biochemical properties. M.Sc Thesis. Tamil Nadu Agricultural University. Coimbatore.

Sivagananam, K., Vanangamudi, K., Umarani, R. 1997. Effect of provenance on seed quality on neem (*Azadirachta indica*). In: IUFRO Symposium on "Innovations in forest tree seed science and nursery technology". November 22-25, Pt. Ravishankar Shukla University, Raipur, India. P. 144.

Sujatha, M., Prabakaran, A.J. 1997. Characterization and utilization of Indian Jatropha. Indian J. Pl. Genet. Resources, 10(1): 123-128.

Sujatha, M., Prabakaran, A.J. 2004. New ornamental Jatropha hybrids through interspecific hybridization. Genetic Resources and Crop Evolution. 50(1): 75-82.

Surendran, C. 1982. Evaluation of variability, phenotypic stability, genetic divergence and heterosis in *Eucalyptus tereticornis*. Ph.D. Thesis, Tamil Nadu Agricultural University, Coimbatore.

Surendran, C., Chandrasekharan, P. 1984. Heritable variation and genetic gain estimates in half sib progenies of *Eucalyptus tereticornis* Sm. J. Tree Sci., 3(1-2): 1-4.

Surendran,C., Chandrasekaran, P. 1988. Genetic divergence among half-sib progenies of *Eucalyptus tereticornis* In: Trends in Tree sciences (Eds. Khosla, P.K. and Sehagal, R.N), ISTS: 218-224.

Surendran, C., Paramathma, M., Rai, R.S.V. 1993. Use of biometrical techniques in tree improvement. In: Proc. Inter. 14th Commonwealth Forestry Conference. 13-18th September 1993. Kuala Lampur, Malaysia.

Umesh Khanna, S. 2001. Genetic analysis, biochemical and molecular characterization of *Madhuca latifolia* (Mach.) Ph.D. Thesis, Tamil Nadu Agricultural University, Coimbatore.

Venkataramanan, J. 1996. Studies on genetic parameters in *Eucalyptus* species. M.Sc. (Ag.) Thesis, Tamil Nadu Agricultural University, Coimbatore.

❑❑❑

25

Package and Practices of Mass Production Through Vegetative Propagation and Cultivation of Bamboos

R. Kaushal, S.K. Tewari, R.L. Banik, J.M.S. Tomar and O.P. Chaturvedi

Introduction

Bamboos are group of woody perennial evergreen plants in the true grass family Poaceae, subfamily Bambusoideae, tribe Bambuseae. Bamboos are the fastest growing woody plants (up to 60 centimeter/day) in the world. Bamboo cultivation, especially in intensively managed and productive plantation has attracted a great interest in recent years leading to huge demand of good quality planting material. The easiest and most common method for propagation is seeds. However, in bamboo due to a long flowering cycles, seeds are not available every year and being a cross-pollinating species seedling population show much variability in characters. Species like *Bambusa vulgaris* and *Bambusa balcooa* flowers rarely and do not produce seeds making propagation absolutely dependent on vegetative means. Vegetative propagation offers a tremendous scope for mass multiplication of many commercially important species. Bamboos are vegetatively propagated through clump divisions, rhizomes, offsets, layering, marcotting, culm cutting, branch cutting, macroproliferation and through in vitro techniques. Propagation through clump divisions, rhizomes and offsets yield limited number of plants. Layering and marcotting techniques are not fit for commercial production. *In vitro* methods offer great scope for mass multiplication, but require high initial cost and skilled manpower and thus are not viable for decentralized forest nurseries. Culm and branch cuttings along with macroproliferation offer simple and useful solution for mass multiplication of bamboos. True-to-type progeny with genetic qualities identical to the mother plant are obtained through vegetative propagation.

Large sized and thick walled bamboo species are usually multiplied through culm and branch cuttings which are cost effective and popular. In this chapter, an attempt has been made to discuss various steps involved in propagating bamboos by culm and branch cuttings (Appendix I) and its cultivation techniques in the field are discussed below

Selection of plus clump

Only a few bamboo species mainly considering their utility in housing and making agricultural implements have been domesticated and many bamboos are still in the wild. They possess enormous natural variability, which can be gainfully utilized to improve quality and productivity. Superior clones based on phenotypic characters (number of culms per clump, height, diameter and wall thickness of culms, length of internode, fibre length, and resistance to diseases and pests, etc.) can be selected from the adult wild populations, plantations and even homesteads. In selection of Candidate Plus clumps (CPC's) the following desirable characters should be considered:

- The clump should be healthy and not infected by a disease.
- Branching mostly at the top and none or less at the bottom.
- Somewhat open clump (not congested) for facilitating easy harvesting.
- Wider growing period of culm emergence.
- High number of culm production per clump.
- No or low mortality at juvenile stage of culm emergence.
- Succulent and palatable shoot.
- Comparatively capable of growing in waterlogged/flooded and/or drier areas.
- Easy to propagate vegetatively.
- None or only partial death after flowering (part-flowering in nature).
- Simultaneous sexual (seeding) and vegetative growth (shoot production).
- High capacity of viable seed production.

Candidate Plus clump (CPC) selection criteria quantification and assessment is done by Selection Index Method. The clumps having higher scores are designated as Candidate Plus Clump (CPC) and selected for conservation.

Preparation of propagation beds

Permanent nursery needs to be developed for large scale production. In permanent nursery, propagation beds are made on cemented platform with proper drainage to avoid water logging and check the infection. If cemented platforms are not available, base portion should be filled with stones and gravels to accelerate drainage. The size of a bed is kept 1.2 meter wide, 6.0 meter long and 21 cm deep. However, the length of the bed is variable with the availability of area. The beds are demarcated and three layers of bricks, one over the other is placed along the dimension of the bed. Bricks are not joined with mortar as it leads to water stagnation which is undesirable for plant growth. Besides, it becomes easy to collect the rooted cuttings from the bed just by removing the bricks from edges.

Rooting media and misting

Sand is used as rooting media for rooting because it is cheap, easily available, neutral and prous. Further, sand does not harbor any fungal infection and rooted cuttings can be easily taken out of the sand without damaging the roots. The propagation beds are filled with coarse sand up to half of the height of the bed as rooting media. On the top, fine sand is added. The beds are provided with sprinklers for intermittent misting to maintain the humidity. Water supply is provided through a main feeder pipeline along the centre of the bed with 90 cm long uprights having baffle-type nozzles spaced 1.0 m apart. A pressure of at least 30 pound per square inch (psi) is maintained for operating the sprinklers. The sprinkler system is fitted with monoblock pump of half or one horse power. The pump is connected to the water tank or directly to bore, if water table is high. The spraying of water through misting nozzles is mostly required in day time. The operation for misting should be intermittent and is controlled by electric timer which automatically starts and stops the pump. The frequency and duration of watering will depends on the season.

Preparation of cuttings

A. Culm cuttings

Culm cuttings are, usually 1 or up to 2-3 nodes segments with buds or branches. In this technique, branches are forced to grow in a manner which encourages the basal region to resemble a rhizome and produce roots. Culm cuttings offer many advantages over the traditional methods of offsets/rhizome propagation. On an average a well developed clump provides only 3-4 rhizomes/offsets each year without a severe reduction in clump vigour and productivity. However, about 75-100 culm segments are taken from a mature clump without much affecting clump productivity. Success and survival of culm cutting by this method is also higher as compared to traditional method.

After selecting the mother clump, culm is felled with saw. Culms are further, crosscut into segment with two nodes. The harvesting of cuttings is best before sunrise or after sunset, to maintain the turgidity of the cuttings and to avoid any wilting. The culm segments of young culms possess root primordial at their nodes which get activated after culms are cut into individual segments. Cuttings with damaged nodal bud are discarded. The culm segments are transported to nursery without any delay. The cut ends are waxed or wrapped with moist gunny bags with moist sawdust to minimise water loss from cut ends. Maximum care should be taken to prevent any wilting or mechanical damage to the shoots.

Rooting of cuttings is affected by a number of factors like age of plant part used, collecting season, source of planting material, auxin levels, variation in bud development, physiological intervals, food reserves in supporting tissues, maintenance of overall condition at propagation nursery, the seriousness and skill of the nursery men. The plant part taken for propagation should have distinct healthy growth buds that can grow to produce plants. The buds on the culm are located on alternate sides marginally above the nodes.

One-two year old culms give best results for rooting in most of the bamboo species. The culms emerged in the early part of the growing season show better response in root induction. In a study conducted with *B balcooa,* at Pantnagar, it was observed that culm segments collected from 2 year old culm performed better compared to one year old culms. Culm cuttings collected in April from middle portion of the culm showed 45 per cent success with the application of 200 ppm IBA. Culms with strong branching gave better success than those with small branches. The best time for propagation by culm cuttings is just before spring growth commences, when the buds are ready to burst. In bamboos of North India, the cessation of winter and onset of spring, results in resumption of active extension growth and upward mobilization of stored photosynthates and axillary substances from the underground rhizomes which become available for differentiation and growth of organs such as culms and adventitious roots, thus show better response in root induction.

Physiological state of the culm plays a crucial role for subsequent shoot production. The plants in the morning hours have no water deficit or water stress conditions. Therefore, the cuttings should be collected early in the morning. The culm cuttings should have enough carbohydrate for rooting. If the carbohydrate content is low it is advisable to apply sucrose or glucose at the ends of the cuttings. The culm cuttings should be taken from the middle and basal portion of the culm. Top portion should be discarded for taking cutting. In a study conducted, *B balcooa,* cuttings collected from base portion showed better rooting followed by middle and top portion In *B.vulgaris, D.giganteus, D.hamiltonii,* culm segments taken from the mid-culm nodal positions rooted well, where as mid to lower culm nodal positions (basal portion of the culms) were found suitable for *B.bambos, B.tulda.*

B. Branch cuttings

Species having stout branches with rhizomatous swelling at bases (*B balcooa, B. vulgaris, D asper* and *D hamiltonii*) are easily propagated through branch cuttings. These cuttings require about 40-60 days for rooting, and 9- 12 months for the rhizome development. However, regeneration time of branch cuttings can be reduced by selecting branch that has spontaneous *in situ* rooting and rhizome tips at their base. Such pre-rooted and pre-rhizomed cuttings perform better than the normal branch cuttings. Pre-rooting is obtained by chopping off the top of the culm or removal of newly emerging culms from the clump. The latter, if done continuously for 2 years, is more effective. The method is however unsuitable for thin walled bamboo species having comparatively thin branches (*Melocanna baccifera, Schizostachyum dullooa, Thyrsostachys oliveri* and *T. siamensis*). The best time for collecting branch cutting is after 1-2 showers in July-August in north India

Branch cuttings are collected from 1.5-2 year healthy and disease free old culms by cutting them at conjecture of the branch and the culm along with the rhizomatous swelling using a saw without causing any damage to the buds. Cuttings are made by trimming leaves, small branches and the branch tip with secateurs. The branch is trimmed to 2-6 nodes with healthy buds which usually is 50-80 cm long depending on the species. The cut ends are waxed or wrapped with moist gunny bags to minimize water loss from cut ends during transportation. The final trimming is done before placing the cuttings in the propagation bed.

Cutting treatments

A. Prophylactic treatments: The prepared culm cuttings are given prophylactic treatment against fungal infection during rooting in the propagation beds. For this 0.2 % solution of Bavistin is prepared by dissolving weighed quantity in water (20 gm/10 litre). The cuttings are than dipped in fungicidal solution for 5 minutes.

B. Application of rooting hormone: Rooting hormones like Indole Butyric Acid (IBA), and Naphthalene acetic acid (NAA) have been found to increase rooting. Difficult to root species are treated with hormones to get better rooting by drilling a hole into the internodal cavity in culm cuttings in which 200 ppm concentration IBA/NAA solution is poured. Generally 50-100 ml of hormonal solution is used for *B balcooa, B bambos, D hamiltonii, B nutans* etc. For large diameter bamboos like *D giganteus* about 300 ml of solution is required. The hole is than covered with cellophone tape or polythene strip. Single noded cuttings and branch cuttings are dipped in hormonal solution for 12-24 hours depending on the species. Hormonal solution (200 ppm) is prepared by dissolving 20 g of the pure chemical in 100 ml of 90 per cent ethyl alcohol and than making final volume to 100 litre while adding concentrated solution to water. The solution should be stirred while adding to water to avoid precipitation. Besides auxins, other chemicals like boric acid and coumarin have also been reported to enhance rooting and rhizome initiation.

Placing of culm cuttings in the nursery beds

Immediately after fungicidal and hormone application, the cuttings are inserted into sand in propagation beds at spacing of 15-30 cm. The beds are saturated with water before planting the cuttings. Care is taken so that the root zone remains wet. The culm segments are placed horizontally at 5-7 cm below the surface of the sand in the propagation beds for shoot and root initiation. The base of central branch on the node is kept in the culm segment as this has been seen to encourage root production. Water movement in the inner or outer walls of the cutting, branch base, or the new shoots cannot take place until development of root system therefore, hollow portion of the cuttings are filled with sand at both the ends so as to meet the water requirement. *Dendrocalamus* species are planted a little deeper than *Bambusa* species.

All dormant buds on the culm and on the central and lateral branch bases should be covered with sand so that shoots arising from them have their basal regions in the sand. The cuttings are kept in the propagation beds till the development of shoots and an independent root system. After rooting, the cuttings are shifted to polythene bags and the sand is taken out from the propagation beds, cleaned, spread over in sunlight and is refilled into the beds for the next set of propagation. In propagation bed of 1.2 m × 6.0 m, three 2-node culm segments can be placed horizontally in 1 row, so, 60-75 pieces of 2-nodes culm cuttings (120-150 potential cuttings) are accommodated in one bed. Considering 70% success, one bed may provide 84-105 rooted cuttings. If propagation is done 4 times in a year, 336-420 cuttings can be expected from single bed in one year.

Branch cuttings are placed vertically at 7-10 cm depth at a distance of 8-10 cm in sand beds in such a way that the rhizomatous swelling and one node remain below the soil surface. The sand should be pressed firmly around the cutting with fingers. In propagation bed of 1.2 m × 6.0 m, about 55 branch cutting can be placed in one row along the length (6.0 m) of a bed and 8 number in the width, so 55 × 8 = 440 branch cuttings can be accommodated. Considering 70% success, one bed can provide 308 rooted cuttings. If propagation is done 3 times in a year, 924 branch cuttings can be expected from single bed in one year

Management of culm and branch cuttings in propagation beds during rooting

The beds are watered immediately and thereafter regularly with sprinkler until rooting take place. The first thirty days of the planting of cuttings is very important. The rooting media in which the cuttings are placed must not dry out. During this period, light and temperature should be maintained at the optimum level. Temperature should never be allowed to go beyond 35 ^{0} C as excessive heat increases the respiration rate of the cutting resulting in depletion of the stored foods. Cuttings and the new sprouts should therefore be covered with partial shade to prevent drying due to intense heat. Shade should be removable so that it may be taken off during overcast days. For providing shade, strips of bamboo joined with ropes can be used. Alternatively agro shade nets can also be erected on propagation bed. These shade nets can be lifted depending on the requirement of light.

Care should be taken that cuttings don't wilt. It is also important to maintain humidity of the air surrounding the cuttings as high as possible to reduce water loss from the leaves to a minimum level. Beds can also be covered with polythene sheets to maintain the humidity. If watering or heavy rain exposes the culms they should be immediately covered with sand again. Sprouting is noticed within 15 days. The sprouts grow to 1 meter or more in height and produce leafy branches before beginning to root. Root development takes 5-10 weeks depending on the season and species.

In branch cuttings, sprouting is usually seen after 15 days, and good root development is seen by 30-60 days (for cuttings taken April-August) or 55-70 days (for cuttings taken October-November). New culms emerge from the base of the cuttings within 30-60 days.

The newly emerged sprouts are treated with 0.2 per cent bavistin to avoid any fungal attack. Emergence of roots can be observed by inserting the finger adjacent to the sprout and moving it along the cutting. Rhizome development and new shoot formation take 4-6 months.

Transplanting of rooted cuttings in polythene bags

The cuttings are kept in the propagation beds till the development of shoots and sufficient amount of roots. The rooted cuttings are than removed from sand bed by removing the bricks from the edges of the propagation beds. The removal of rooted cuttings from propagation bed is done in the evening or rainy /cloudy days when temperature is moderate. The cuttings are then washed carefully with clean water and are transplanted to polythene bags containing a mixture of soil, sand and FYM in ratio 2:1:1. Bag size of 15 x 23 cm for branch cuttings and 40 x 50cm for culm cutting are used. Drainage holes are kept in polythene bags at the lower portion to avoid any kind of water logging in root zone. Roots should be given light pruning to avoid coiling in the polythene bags. After filling the polythene bag with potting mixture, the soil is pressed gently near the base of the cutting. Cuttings are than watered and are immediately transferred to hardening unit to promote further growth.

Hardening of rooted cuttings

The 50 per cent net shades are most suitable for hardening the freshly shifted rooted cuttings. The hardening unit is provided with sprinklers which are used intermittently to provide sufficient moisture on the leaf surface to avoid shock and drying up. Watering is regulated and gradually reduced over a period of two weeks. Over watering is avoided as it rots the cuttings. Shading from the sun is not important after the strong new shoots have emerged but protection from frost is necessary. The rooted plants are kept in the net shade till new growth start coming in 20 -25 days depending upon the species. The cuttings in polythene bags are than gradually shifted under open sky until planting.

The duration of stay of plants in the open nursery is 3-4 months or till the time the plant system fully consolidates itself. Thus, care should be taken to avoid the coiling of roots in the polythene bags. For this regular shifting of polythene bags is necessary till the plants are planted out.

Multiplication of planting materials using macroproliferation technique

Bamboos possess an inherent proliferating capacity. The plants produced by culm and branch cuttings can be further multiplied using macroproliferation. For macroproliferation, the rooted plants with adequate rhizome and root formation in the polythene bags are first exposed by removing the soil. The individual shoots are separated at the rhizome neck region with the help of a secateur such that each shoot retains a portion of the rhizome systems and roots. The upper portion of the shoots are trimmed leaving two nodes, to restrict apical dominance and to enable production of more shoots. The cut surface are sealed with wax. The cuttings are dipped in bavistin (0.2 %) solution before planting them in multiplication bed or in polybags. The active growing season is the best time for macroproliferation operations. Comparative chart for the production of plants from the methods discussed is given in Table 1.

A bamboo plants can be multiplied two to three times with survival rate of 90-100%. A macroproliferated plants can be used as original plant to develop proliferating pieces after three months. The plants obtained are then hardened under shade before bringing them to the nursery bed under the sun. By raising the plants in nursery beds and fertilizing them with 25 kg N/ha and 75-150 kg/ha of P_2O_5, better rhizome development and shoot formation are achieved. By this method planting stock can be multiplied by 5-6 times. Combination of culm and branch cutting with macroproliferation technique in bamboo are economical. A road map for quality planting stock production in bamboo is given in Fig 1. The key steps envolved in propagation by different methods is given in Appendix I. A Comparative chart depicting the advantages and disadvantages of different methods is given in Table 1.

Table 1: Clonal Propagation of Bamboo - A Comparative Chart

SlNo.	Particulars	Clonal Propagation Technique			
		Culm Cutting	Branch Cutting	Macroproliferation	Conventional (Offset /Rhizome) method
1.	Size of propagule	Medium	Small to Medium	Small	Medium to Very large
2.	Time required	3 months	3 months	4 months	6 months
3.	Number of times can be done in a year	4 times	3times	4 times	Once
4.	Optimum time	March- Oct.	July- Oct.	March- Oct.	Feb-April
5.	Success rate	60-85 %	40-85 %	>90 %	30-50 %
6.	Unit cost of production (Ist year)	Rs. 10.0/-	Rs. 10.0/-	Rs. 5.0/-	Rs. 20-25/-
7.	Advantages	Simple and easy, good survival, source material easily available and extractable	Easy to handle, time saving	Cost effective, suitable for large scale plantation	Good survival
8.	Disadvantages	Not suitable for species with long internodes and thin wall	Not suitable for species with thin wall and thin branches	Plant vitality reduced overtime, need fresh stock is required	Difficult to handle, labour intensive, not available in mass numbers. Unsuitable for large scale plantation

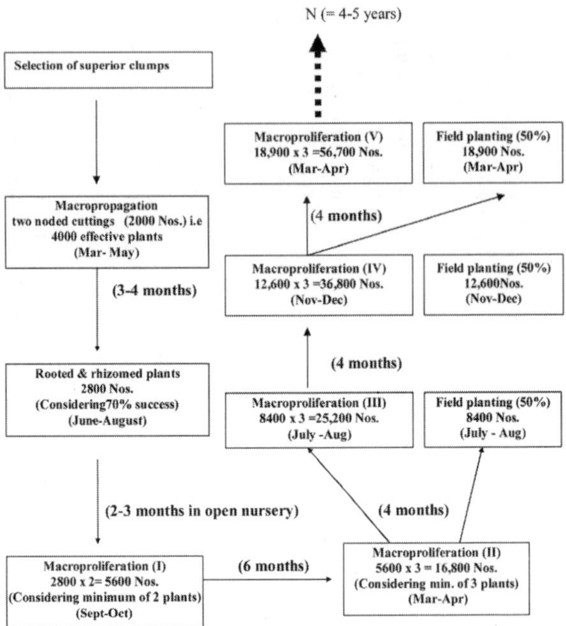

Fig 1: Road Map for Quality Planting Stock Production in Bamboo

Bamboo cultivation

Brief description of different steps involved in raising bamboo plantation is as under

A. **Site preparation:** Bamboo prefer sandy loam soil and the pH range ideally would be between 5.5-6.5. Places with water logging should be avoided. Bamboo can withstand partial shade thus even if the selected plot has few standing trees it can be used for plantation purpose. The selected site should be cleared of weeds and jungle growth in lines.

B. **Layout and planting:** The planting is usually done at a spacing of 5m x 5m. However for larger sized bamboos like *D giganteus*, planting can be done at 8m x 8m. Plantations of diverse flowering populations should be raised in *Mosaic* so that all clumps will not flower and die at same time. Best time for planting is during the pre-monsoon showers so that establishment during the monsoon is successful and requires less watering. The planting is normally done in 45 cm x 45 cm x 45 cm pits. The nursery raised plants are placed in the pit without disturbing the roots and rhizome of the plant. These pits should be filled with mixture of 5 kg FYM, 100 gm neem cake and soil. Soil around the plants should be consolidated tightly by pressing with feet making a small cavity, which is around 5cm Application of anti-termite (chloropyriphos/phorate) should be done in termite prone areas.

Immediately after planting one watering @ 20 litres per pit is given in case it is not raining. The process of planting should be finished as quickly as possible and must be carried out on a cloudy day and early morning or late evening. There should also be a provision for gap filling caused due to death of seedlings.

C. **Aftercare and protection:** After planting in the first year 3 to 4 weedings would be required for better survival of the plants. In subsequent years weeding could be reduced i.e. 3 weeding in second year, 2 weeding in 3rd and subsequent years. The soil around the plant should be loosened twice every year for better rhizome growth and shoot production. Mulching increase soil temperature to varying degrees (2 to 5°C), notably during winter (December-January) and make early shoots emergence and prolonged the shooting period and thus should be encouraged with bamboo leaves or with the weeded debris. Young plantation should be fenced to protect it from cattle.

D. **Irrigation:** During the first year of planting watering helps to give higher survival rate especially in areas where the dry period is longer than two months. Irrigation generally helps to increase the productivity at least by three times. Natural water conservation methods like ditches or crescent shaped trenches helps in moisture conservation. In marshy areas, mound plants should be practiced to ensure drainage and create a favourable condition of O_2/CO_2 exchange for underground rhizome system.

E. **Fertilizer application:** The bamboos in India are generally cultivated without manuring or fertilizing. On the contrary countries like China, where the productivity is much higher than our country use fertilizer profusely. Experiments have shown that adding farm yard manure and fertilizers helps in boosting the farm productivity. Fertilizer application is important to increase the yield. Nitrogen, phosphorus and potassium are most essential for proper growth. While planting nursery raised plants, single super phosphate (300 g), potash (150 g) and cow-dung (10 kg) are recommended for application in the pit as basal dose. After the seedling establishment, Urea @ 200 gm per pit can be applied when the soil is moist. In the second year to 3rd year of planting a total quantity of NPK per clump in the ratio of 4:1:2 is to be given as first dose. From 4th to 5th year, the quantity should be increased by 50% and 100% respectively and fertilizer should be applied all along the circumference of the clump thrice in the calendar year. The fertilizer can also be added in the trench made (about 15 cm deep) around the clump. The dose amount need to be increased with the increasing clump size.

F. **Clump management:** Pruning of side branches (thorns) are carried out annually for proper growth of the clump without congestion. Cleaning of the clump is done by removing the dead and dying culms from the clump and thinning the clump by removal of weaker culms and which facilitate proper growth of new shoots. Removing of the upper part of the culm is also recommended in areas prone to heavy wind. Every year new culms are added to the clump which can be marked by painting rings with the same or different coloured paints. This technique is useful in removing the correctly aged culms for different purpose.

G. **Harvesting:** If the clumps are scientifically managed by pruning and fertilizer applications, harvesting is initiated at fifth year and then annually. For construction purpose generally 4-5 year old culms are used. For mat, basket making 2 -3 year old culms are used. For edible purpose, newly emerging shoots are extracted. Harvesting should not be carried out during monsoon or when the growth of new shoots is observed except for shoot production. Culms should be cut to about 30- 45 cm above ground from the base leaving only one internode. The cut culm is then pulled down and sized to required length, dried and stacked. After harvesting the culm, the debris is removed and disposed elsewhere. Clear felling of the clumps lead to degeneration into a bushy form, resulting in a gap of 5-6 years to produce extractable culms. Hence, it should be avoided as far as possible. The area around the clump should be kept clean after removing the cut debris to minimize fire and attack by fungi and insects. The yearwise activity for cultivation of bamboo is given in Table 2.

Table 2: Yearwise activities for cultivation of bamboos are as under

Activity Item	Time
First year	
Survey, weeding/jungle cleaning in the area	March-April
Site cleaning, staking at required spacing	April-June
Pit digging	May-June
Fencing and closure	May-June
Fertilizer application, sowing & refilling of pits	June-July
Insecticides application	July-Sept
Weeding and Mulching	July-Sept
Second & Third Year Activities	
Ring (0.5m radius) weeding, soil mounding around the plant in 10 and 15 cm height	April-May
Fertilizer application	March, July and September
Maintenance of Fence	May-June
Weeding, vine cutting and mulching	July-September

Growth and productivity

From seedlings to formation of clumps that can yield mature culms it takes about 4 to 5 years. After harvestable age, about 7-9 culms per clump can be harvested annually. A well managed plantation can yield 20- 30 tons of culms/ha. Fresh weight of one culm varies from 30 to 90kg depending on species and site conditions. Price for one bamboo varies from Rs. 20 to 50. The cost and income incurred from 1 ha plantation at 6x6 m spacing is given in Tables 3 and 4.

Table 3: Cost norms for creatino and maintanance of 1ha plantation

Sl. No.	Particulars of Works	Amount (Rs) (1st Year) (Rs)	Amount (Rs) (2nd Year)	Amount (Rs) (3rd Year)	Amount (Rs) Total	MD
1.	Site clearing	1000			1000	10
2.	Alignment of pits	400			400	4
3.	Excavation of pits	4000			4000	40
4.	Cost of seedlings (Rs. 20/- per seedling) - 275 nos.	5500			5500	
5.	Transportation	500			500	
6.	Planting of seedlings	1000			1000	10
7.	FYM	5000			5000	
8.	Cost of fertilizer and their application	1000	1500	1500	4000	
9.	Watering	1500	1500	500	3500	35
10.	Weeding (three times) + hoeing (two times) per year	1500	1000	500	3000	0
11.	Trenching/ fencing	4000	1000	1000	6000	60
12.	Casuality replacement	1500	500	2000	20	
13.	Soil working/ mounding/ mulching	1500	1500	1500	4500	45
	Total	**26900**	**8000**	**5500**	**40400**	

Table 4: Cumulative expenditure, income and net profit

	Year wise statement									
	1	2	3	4	5	6	7	8	9	10
Expenditure	26900	8000	5500	2000	2000	2000	3000	3000	3000	3000
Cumulative expenditure	26900	34900	40400	42400	46400	48400	51400	54400	57400	60400
Income					16500	33000	49500	66000		
Cumulative income				16500	49500	99000	165000			
Net profit							-34900	-490041600		104600

In 7th year 4 years bamboo 2 nos. culm from each clump may be extracted = 2 X 275 X 30.00 = 16500.00

8TH 4 NOS. = 4 X 275 X 30.00 = 33000.00

9TH 6 NOS = 6 X 275 X 30.00 = 49500.00

10TH Year Onward 8 NOS. = 8 X 275 X 30.00 = 66000.00

Appendix - I

Propagation with cuttings: Key steps

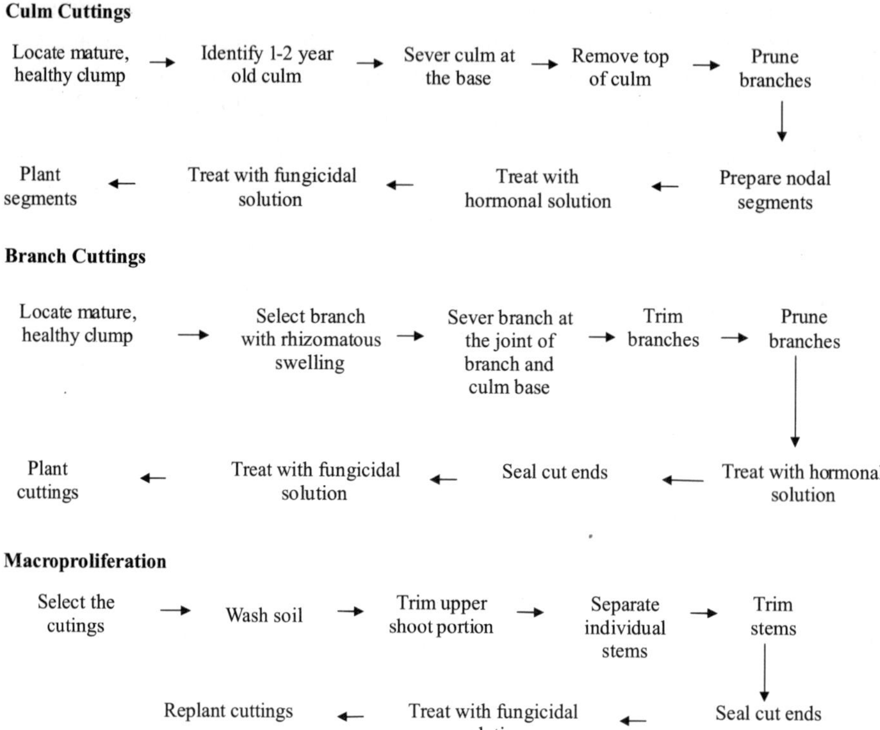

Acknowledgements

The publication is output from a research project Vegetative Propagation Centre funded by of National Mission on Bamboo Applications (NMBA), Department of Science & Technology, New Delhi whose financial support is duly acknowledged. Thanks are also due to guidance and unstinted support of Sh. Sudhir Pande, Advisor, NMBA during the entire project.

Further Reading

Anonymous. 2004. Propagating Bamboo. Training manual. TM 03 11/04. National Mission on Bamboo Applications. Technology Information, Forecasting and Assessment Council (TIFAC), New Delhi

Anonymous. 2007. Establishing a Vegetative Propagating Centre for Bamboo. Info-Sheet VPC establishment. IS 06 09/07. National Mission on Bamboo Applications. Technology Information, Forecasting and Assessment Council (TIFAC), New Delhi.

Banik, R.L. 1987. Techniques of bamboo propagation with special reference to prerooted and prerhizomed branch cuttings and tissue culture. In: Rao, A. N., Dhanaranjan, G., Sastry, C. B. (Eds.), pp. 160-169, Recent research on Bamboos, Proc. of the International Bamboo Workshop. IDRC, Hangzhou, China.

Banik, R.L. 1994. Review of conventional propagation in bamboos and future strategy. In: Constraints to production of bamboos and rattan. Report of a consultation held in Bangalore, India, 9-13 May 1994, pp.115-142. INBAR, Technical Report No. 5. International Network for Bamboo and Rattan, New Delhi, India.

Banik, R. L. 1995. A manual for vegetative propagation of bamboos. INBAR Technical Report 6. INBAR, New Delhi: 66p.

Banik, R.L. 1997. Domestication and improvement of bamboos. UNDP/FAO Forest and Tree Improvement Pro. IDRC/INBAR Working Paper No. 10, New Delhi, Guangzhou pp. 53.

Banik, R. L. 2000. Silviculture and field – guide to priority bamboos of Bangladesh and South Asia. BFRI, Chittagong. pp.187.

Banik, R.L. 2008.Issues on production of bamboo planting materials - Lessons and Strategies. Indian Forester, Bamboo Issue 134(3): 291-304.

Kaushal, R., Tewari, S. 2009. Technology for vegetative propagation of bamboos. Research Bulletin No. 166. Directorate of Experiment Station. G.B. Pant University of Agriculture & Technology, Pantnagar, Uttarakhand, India.

Pandalai, R.C., Seethalakshmi, K. K., Mohanan, C. 2002. Nursery and silvicultrual techniques for bamboos. KFRI Handbook No. 18, Kerala Forest Research Institute, Kerala.

Stapleton, C.M.S. 1986. Propagation of *Bambusa* and *Dendrocalamus* species by culm cuttings: Preliminary practical recommendations. Forest Survey and Research Office Publication No. 44.

26

Economics of Traditional Agroforestry Models in Orissa, India

P.K. Singh and S.M.S. Quli

Abstract: *The Santhals of Orissa raise Teak (*Tectona grandis*) and Cashew nut(*Anacardium occidentale*) on their marginal and wastelands randomly to meet their contingencies. Upland is utilized for short duration upland paddy cultivation for food. The sporadically grown Teak and Cashew is used as insurance crops for meeting the high risk of dry land paddy. This work is unique and innovative as it was carried out in participatory mode on the pooled land from 14 Santhal households. It is a maiden attempt in Orissa to develop Agroforestry models using forestry (teak) and horticulture (cashewnut) interventions into the paddy cultivation on uplands by Santhals without much alteration in their age old traditional agricultural practice. The basic objective of this research is to develop most profitable Agroforestry system(s) to provide best land use option for the risk prone dry land paddy cultivation to improve the farmers' economy with simultaneous natural resource conservation.*

The Agroforestry models Silvi-horticulture (Teak and Cashewnut), Agri-silviculture (Paddy and Teak), Agri-horticulture (Paddy and Cashewnut) and Agri-silvi-horticulture (Teak, Cashew nut and Paddy) were developed by the Santhal tribe in Mayurbhanj District of Orissa. The inputs and planting material were sponsored by a leading NGO- Gamin Vikas Trust, which is working on poverty alleviation through Natural Resource Management of the State. The uniform spacing for all the models was 4.5m x 4.5m based on minimum space required for movement of draught animal for paddy cultivation. The economic analysis was done through the standard economic indicators like Present Net Worth (PNW), Internal Rate of Return (IRR), Benefit Cost ratio (B:C) ratio, Equivalent Annual Income (EAI) and Rice Equivalent Value (REV) to illustrate the overall economics. Sensitivity analysis was done to project the impact of price fluctuations on the profits. The data trends of first 3 years extrapolated to project the economic profits revealed highest B :C ratio of 12.39 in case of Silvihorticulture (Teak + Cashewnut) and lowest B :C ratio of 3.67 in Agrisilviculture (Teak + Paddy) at 7% discounting at 15th year. At rotation (45 years) the highest B :C ratio (68.78) in Silvihorticulture (Teak + Cashewnut) and lowest in Agrihorticulture (Cashew nut+ Paddy) 24.72 at 7% discounting were recorded. Where as at 15% discounting the highest C: B ratio was in Silvihorticulture (Teak + Cashewnut) and lowest in Agrisilviculture (Teak + Paddy) at rotation.

Introduction

Agroforestry provides a landowner the opportunity to develop a portfolio of short- and long-term investments which mitigates financial risk on number of products through diversification. The diversification of investment provides financial advantages, although it requires additional management expertise to deal with the added complexity of the farm operation. The tree-crop combination of Agroforestry provides best options for unsuitable land use (paddy) over an extended period to woody perennials (trees), for optimizing both the intangible as well as tangible outputs fostering socio-economic development on sustainable basis. The contribution of Agroforestry directly to sustainable improvement in rural economy and environment, along with its outstanding long term positive impact on agricultural productivity and forest conservation in recent era has established its credentials as the best land use option. The report (World Bank, 2002) on developing countries shows that about 1.2 billion people rely on Agroforestry farming systems to sustain agricultural productivity and generate income. The data of this research were analysed to identify the best Agroforestry models in terms of profits.

Target Area

State profile

Orissa occupying tropical zone between latitude 17°47`and 22°34` N longitude 81°22` and 87°29`E, with geographical area of 15.57 million ha constituting 4.74% area of the country has land use pattern as recorded in Table 1 showing very unique feature of equal percentage of forest and net sown area (37.33 and 37.4%) with 11.83% area not available for cultivation. The very low availability of grazing land and permanent pasture to the tune of just 2.85% reflects very high pressure of grazing by 23.39 million livestock on the forest of the state causing severe damage to the regeneration which is expected to cause serious ecological consequences.

Table 1. Land use of Orissa state

Sl.No	Land use	Area in'000ha	Percentage
1.	Total geographical area	15571	
2.	Reported area for land utilization	15571	100.00
3.	Forests	5813	37.33
4.	Non available for cultivation	1842	11.83
5.	Permanent pastures and other grazing lands	443	2.85
6.	Land under misc. tree crops and groves	482	3.10
7.	Culturable wasteland	392	2.52
8.	Fallow lands other than current fallows	430	2.76
9.	Current fallows	340	2.18
10.	Net area sown	5829	37.43

Source: Land use statistics, Ministry of Agriculture, GOI, 2005

The forest cover of the state based on satellite imagery (Plate 1) of Oct-Dec, 2004 shows 31.37% of total geographic area with 0.35% very dense, 17.76% moderately dense, 12.96% open and 65.88% non forest area (Annon, 2005). Physiographically this state has four distinct regions viz. Northern plateau, Eastern Ghats, Central Table land and Coastal plains, having very rich mineral resources including coal, iron, bauxite, chromite and nickel. The annual rainfall varies between 1200-1600mm with distinct seasons. The state is drained by three major rivers, Mahanadi, Brahmani and Baitarni. It is a tribal dominated state of India located in eastern part of the country. Out of total 437 ethnic groups inhabiting eastern part 62 are found in this state, consisting of 22.21% of the 36.7 million total population (Census, 2001). The role of forest in rural economy of tribal area is quite high since most of the ethnic group follow risk prone dry land short duration paddy cultivation.

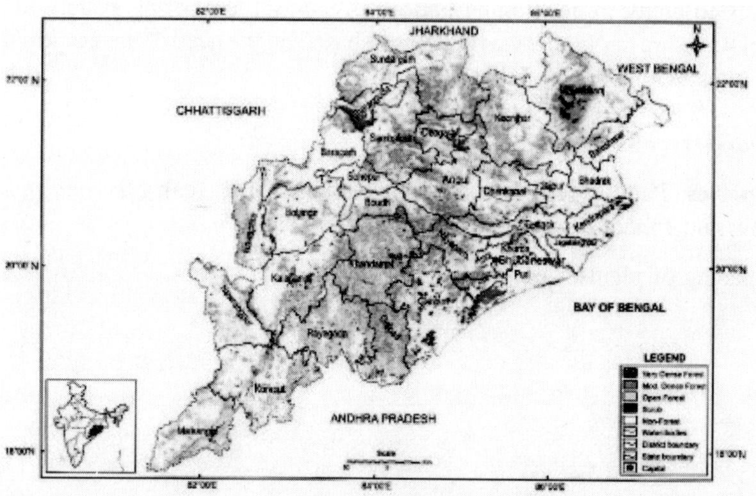

Plate 1: Map of Forest in Orissa (SFR-2005)

The target village

The study has been conducted at farmers' field in Sirasahi village of Bangriposi block in Mayurbhanj district of Orissa. It lies between 21.25°N - 22.6°N latitude and between 85.7° E - 87.2°E longitudes. The experimental area is located in Plateau and falls in sub-humid, sub-tropical climatic zone.

The target population

The Santhal tribes in Mayurbhanj district of Orissa constituting the major population, were the stakeholders of this research. Most of the stakeholders were small farmers having average holding of 2.4 acres consisting of barren uplands. The key problems are nutrient deficiency in soil, with almost no irrigation facility limiting the land use to single dry land paddy option with low yield, fodder, fuel wood shortages and low incomes from farming. Imperative cost of these problems include widespread poverty and severe food insecurity – most of the families have limited food security ranging from three to nine months only giving way to migration/ exodus and environmental degradation.

The main agriculture crop is Upland Paddy (*Oriza sativa*) cultivated by the Santhals on bunded uplands. The return from the upland paddy is not so remunerative; still the crop is being cultivated to meet the food demand of the household and more so because of age old tradition. The farmers on the barren wasteland/upland are raising Teak (*Tectona grandis*) and Cashew nut(*Anacardium occidentale*) on their marginal and wastelands randomly with an objective to acquire economic profit and utilize their land. The demands for both timbers from teak and Cashew nutfruits are rising. Teak is an undisputed global leader for high quality tropical timbers. Cashew is an important nut crop that provides food and hard currency to many in developing nations. Keeping the existing system intact the present Agroforestry model was developed based on the need of the farmers through facilitating NGO- Gramin Vikas Trust

Agroforestry models

Species: Paddy (*Oriyza sativa* var Kalinga-III) , Teak (*Tectona grandis)* , Cashew nut(*Anacardium occidentale*)

Spacing of plants= 4.5mx4.5m

Table 2. Different Agroforestry Systems

Sl. No	Components	Agroforestry System
1.	Teak + Cashew nut (alternate row)	Silvi-horticulture
2.	Teak + Paddy	Agri-silviculture
3.	Cashew nut + Paddy	Agri-horticulture
4.	Teak + Cashew nut + Paddy	Agri-silvi-horticulture

Note: The different Agroforestry systems (Table 2) were developed on participatory basis in farmer's field. The financial support was provided bwy NGO and the labour inputs was from the stakeholders. Observations were taken up to 30th month.

Economic analysis of the agroforestry model

The following indicators were adopted for the projection of economics of the various models under this study.

Present Net Worth (PNW)

The PNW reflects the relative probability of a model profit or surplus income from a project after the project has satisfied the rate of return on capital invested.

$$\text{PNW} = \sum_{t=1}^{n} \frac{(B_t - C_t)}{(1+i)^t}$$

B_t = benefits in each year t

C_t= Cost in each year t

n = number of year to the end of the project

i = Discount rate (7%)

Benefit Cost (B:C) ratio –

Net return (Rs/ha) = Gross return (Rs/ha) - Cost of cultivation (Rs/ha)

B: C ratio = Gross return/Total Cost

Discounting

The compounded/discounted values of the cost and returns were calculated by the formula

$Pv = Fv /(1+i)^n$

Fv = Future value; Pv = Present value; i = discount rate; n = number of years

Rate of interest was calculated at 7% and 15%

Internal Rate of Return (IRR)

IRR is an interest rate, much like the rate earned on a certificate of deposit or a saving account. It is the average interest rate earned on all cost accrued before the investment mature. The interest rate at which PNW=0 is here the present value of revenues equals the present value of the cost.

Equivalent Annual Income (EAI)

EAI is the annualized value of cash flow. This value is the PNW converted to an average annual income. This can be used to compare forestry with other land uses which could have provided annual income.

$$EAI = PNW \frac{i(1+i)^n}{(1+i)^n - 1}$$

PNW= Net Present Worth; n= number of years in the rotation;

i= interest rate

Rice Equivalent Values (REV)

The monetary gains of the Agroforestry was further projected in terms of Rice Equivalent Value to illustrate the output in terms of total rice produced/ increase in purchasing capacity for the rice (prevailing market price) which is the staple diet of target population. REV for such studies can be a wonderful motivational factor for replication of the most profitable Agroforestry model in rest of the state.

Sensitivity Analysis

In order to enhance the reliability of calculation against uncertainties in price fluctuation the sensitivity analysis was done to predict the fluctuation in the values of products. The fluctuation range of 80-120% was used.

PNW

i) **At 15th year**: Highest PNW (Table 3) was observed for Cashew nut+ Paddy (Rs 311,192.00/ha) at 15th year of plantation and lowest in Teak + paddy (Rs 77,560.00) at 7% rate of discounting. The lower PNW was observed in teak paddy combination due to the return obtained from paddy cultivation was lower compared to investment but in teak cashew and paddy cultivation the recurring yield of Cashew nuthas made the system profitable. Through these model assets have been created within the village supporting the finding of Jha and Ranjan (1993). At 15% discounting the PNW is highest in Cashew nut+ paddy followed by Teak+ Cashew nutand lowest in Teak + Cashew nut+ paddy model where as negative worth was reported in Teak + paddy (Rs -1149.00) this may be due to cost involved in paddy cultivation and the volume of teak at 15th year is not sufficient for positive return. The finding of Patil *et al.* (2000) shows that teak + sapota + field crop gave 46% higher income than field crop + sapota.

ii) **At Rotation:** The PNW (Table 3) calculated at rotation revealed that maximum worth is in Teak + Cashew nut & Teak+ Cashew nut+ paddy combination at 7% rate and lowest in Casehewnut + paddy. At 15% rate of discounting the highest PNW was observed in Cashewnut+paddy and lowest in Teak + paddy. PNW at rotation age the teak + cashew was found to be best in turns of return from Silvihorticulture combination at 7% rate of interest but if rate of interest have been increased to 15% the value obtained from Cashew + paddy was best for the combination of Teak + Cashew nut once again endorsing the depreciative impact of 15% rate of interest for longer duration on teak. The PNW for treatment on combination with paddy has always been negative making the whole issue very complicated on first sight but this negative trend is resultant of the fact that the labour input is considered free in dry land paddy farming practices while carrying out scientific calculation of this experiment the labour component was duly recognised that made upland paddy combination cultivation PNW negative.

Table 3. PNW and IRR of Agroforestry at 15th year and at rotation (45 years) @ 7% and 15% rate of discounting

Component	PNW at 15th year (Rs)		PNW at rotation age (Rs)		IRR at 15th Year (%)	IRR at rotation (%)
	7%	15%	7%	15%	7%	15%
Teak and Cashew nut	209,180	66,617	1,245,254	87,726	39.00	37.13
Teak and Paddy	77,560	-1,149	932,360	12,500	21.95	19.73
Cashew nutand Paddy	311,192	107,451	703,291	122,736	44.75	44.75
Teak, Cashew nut and Paddy	186,455	49,543	1,241,879	70,784	34.11	31.86

IRR

The IRR of the Agroforestry models (Table 3) reveal that the rate of return is highest in Cashew nut+ paddy (44.75%) and lowest in teak +paddy combination (21.95%) respectively at 15th year correspondingly the rate of return of the model at rotation is highest in Cashew nut+ paddy (44.75%) and lowest in Teak + paddy (19.73%) respectively. Reports of FAO-1985 from Gujrat reflects an IRR of 129% in case of Eucalyptus with cotton in first rotation.

The result of IRR calculated at 7% after 15th year designates Cashew nut+paddy as most profitable crop followed by Teak+ Cashew nutand further Teak+ Cashew nut+ paddy. Such a trend if viewed judiciously in light of risk involved in monoculture, makes Teak + Cashewnut+ paddy combination most safer in terms of profit an sustainable basis since in combination of three component with paddy the risk factor appears to reduce by 33% on one hand and the key horticulture component introduction into exiting upland farming system is more acceptable to the families that asking them to replace the old age practice of paddy cultivation in one stroke by switching over to combination of Teak and Cashewnut. Although there is reduction in terms of IRR by around 10% but the option of using tree and horticulture intervention into upland paddy cultivation was found to be more readily acceptable to the participants who are in habit of carrying over their traditional upland rice cultivation system even if it was not profitable to them. The value of IRR obtained at rotation age i.e. 45 years again reconfirm almost similar trend in case of 7% with little reduction in case of Teak +Cashewnut+ paddy at 15%.

B: C Ratio

The benefit cost ratio of Agroforestry model (Table 4, Fig. 1) at 15th year was highest in Teak+ Cashew nut (12.39) and lowest in Teak + paddy (3.67) the tendency show that Teak + Cashew nut, Cashewnut+ paddy and Teak +Cashew nut+ paddy gave higher return compared to teak + paddy. The same situation is at 15% rate of discounting the highest benefit cost ratio is in Cashew nut+ paddy (5.26) and lowest in Teak + paddy (0.95) the trend show that land use with Teak + Cashew nut, Cashew nut+ paddy and Teak + Cashew nut+ paddy showed higher return compared to teak + paddy.

Table 4. Benefit cost ratio of Agroforestry at 15th year and at rotation @ 7% and 15% discounting

Component	Benefit Cost ratio at 15th year		Benefit Cost ratio at rotation	
	7%	15%	7%	15%
Teak and Cashew nut (alternate row)	12.39	5.20	68.78	6.53
Teak and Paddy	3.67	0.95	33.00	1.50
Cashew nut and Paddy	11.49	5.26	24.72	5.87
Teak , Cashew nut and Paddy	7.34	2.98	43.25	3.83

The benefit cost ratio calculated at rotation (45 years) revealed that highest benefit cost ratio is observed in Teak + Cashew nut(68.78) and lowest in Cashewnut+ paddy (24.72) at 7% discounting. Likewise at 15% discounting the highest benefit cost ratio is in Teak+ Cashew nut (6.53) and lowest in Teak + paddy (1.50) respectively. Shukla *et al.* (2003) reported a benefit cost ratio of (3.98) for Teak + paddy model in comparison to paddy alone 2.11. Sohkhlet (2001) reported that economic returns were highest in case of agri-horticulture system where as minimum economic return accrued from agri-silviculture system supporting the present finding. Similarly the finding of Patel (1988) showed that teak plantation raised with cotton and other agriculture crop showed much greater profitability than of pure agriculture usually, of 1.5 to 2 times supporting Hazra *et al.* (2000) whereas in Gujrat 1:5 B: C ratio was reported by Saxena, (1989) , several finding indicate that growing trees and agricultural crops together is a better land use option in terms of productivity, maintenance of soil conditions and economics contrary finding has been reported by Pethiya (1999) that Agroforestry is not remunerative as agriculture.

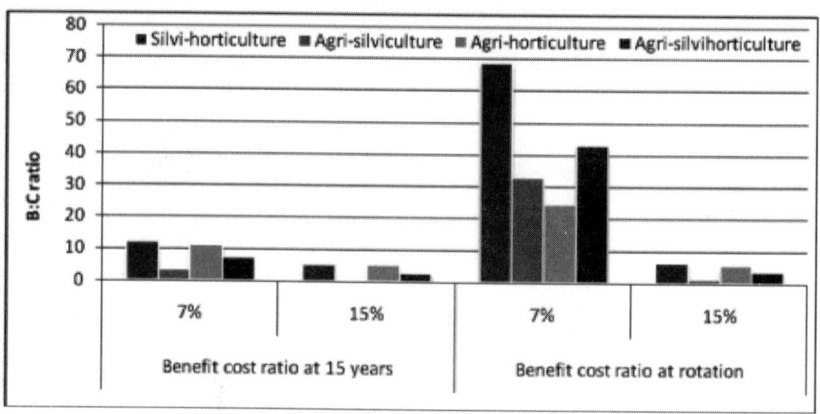

Fig. 1. B: C ratio of Agroforestry at 15th year and at rotation @ 7% and 15% discounting

Upland paddy cultivation gave negligible return when the labour involved in agriculture operation was calculated and the cost benefit ratio was negligible, the Santhals cultivate paddy for food but the family labour involved in its cultivation is never taken into account but when it was taken into the return was very low thus upland paddy cultivation should be discouraged on upland. Similar trend was reported by Gupta and Mohan (1982) too who reported that growing tree in marginal land in Rajasthan indicated much higher economic return from tree than of traditional agriculture.

The analysis of the B:C ratio data trend (Table 4) further substantiate that for encouraging farmers to switch over to the Agroforestry from their traditional upland paddy cultivation system the best option was Teak +Cashew nut+ paddy which had lowest risk factor ensuring the high profit on sustainable basis for longer period of time.

EAI

Equivalent Annual Income (EAI) of the Agroforestry model (Table 5, Fig. 2) calculated at 15th year reveal that at 7% rate of discounting the highest EAI have been reported in land use Cashew nut + paddy and lowest in teak + paddy. EAI calculated at 15th year was highest in Cashew nut + paddy combination (Rs 16148.00) and lowest/ negative in teak + paddy (Rs 173.00) at 15% discounting.

EAI calculated at rotation was highest in Teak+ Cashew nut combination (Rs 91526.00) and lowest/negative in Cashew nut+ paddy (Rs 51692.00) at 7% discounting likewise at 15% rate the highest EAI was observed in cashew nut + paddy (Rs 18445.00) and lowest in Teak + paddy (Rs 1879.00). The tend show that at 7% discounting Cashew + teak model is profitable where at higher discounting cashew nut paddy combination is profitable.

Table 5. EAI of Agroforestry at 15th year and at rotation @ 7% and 15% discounting

Component	Equivalent Annual Income at 15th year (Rs)		Equivalent Annual Income at rotation (Rs)	
	7%	15%	7%	15%
Teak and Cashew nut (alternate row)	15,375	10,011	91526	13183
Teak and Paddy	5,701	-173	68528	1879
Cashew nut and Paddy	22,872	16,148	51692	18445
Teak, Cashew nut and Paddy	13704	7445	91278	10637

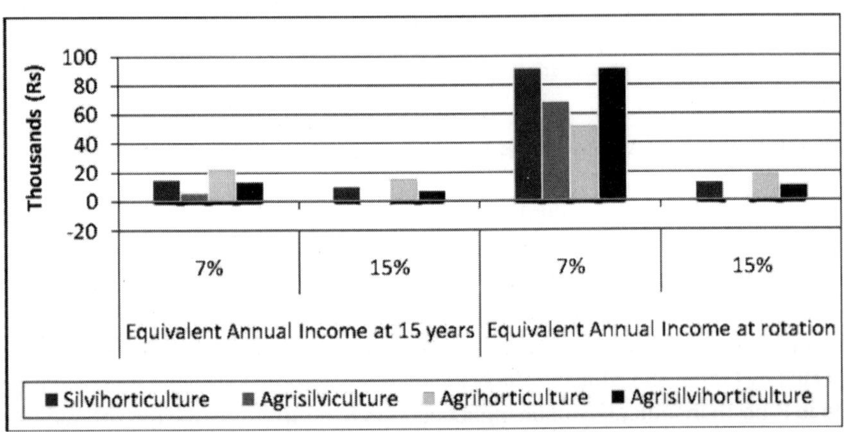

Fig. 2. EAI of Agroforestry 15th years and at rotation @ 7% and 15% discounting

REV

Rice Equivalent Value (REV) calculated by converting the monitory gains into the equivalent quantity of rice variety used by the Santhals at prevalent market rate @ Rs 5/kg at time of study of the Agroforestry model (Table 6). The REV at 15th year and 7% rate of discounting the highest (4574.4kg) in Cashew nut+paddy and lowest in teak+ paddy (1140.2 kg). REV at 15th year at 15% discounting was highest in Cashew nut+paddy (3229.6kg) and lowest/negative in teak+ paddy (-34.6kg).

REV at rotation was highest in both the models Teak+ Cashew nut & Teak+ cashew nut + paddy and lowest in Cashew nut + paddy (10338.4kg) at 7% discounting likewise at 15% rate the highest REV was observed in cashew nut +paddy (3689kg) and lowest in Teak+ paddy (375.8kg). The tend show that at 7% discounting Cashew + teak model is profitable where at higher discounting cashew nut paddy combination is profitable. The average production of the upland paddy variety being cultivated traditionally was around 2000 - 3000 Kg ha^{-1} yr^{-1}.

Table 6. Rice Equivalent Value at 15th year and at rotation @ 7% and 15% discounting

Component	Rice Equivalent Value at 15th year (kg)		Rice Equivalent Value at rotation (kg)	
	7%	15%	7%	15%
Teak and Cashew nut (alternate row)	3075	2002.2	18305.2	2636.6
Teak and Paddy	1140.2	-34.6	13705.6	375.8
Cashew nut and Paddy	4574.4	3229.6	10338.4	3689
Teak, Cashew nut and Paddy	2740.8	1489	18255.6	2127.4

Sensitivity analysis

The sensitivity analysis of the value of product at rotation show maximum changes in benefit when the cost is increased by 10% and benefit remains the same (Table 7). The IRR of the treatments were in order with maximum in Teak+ Cashew nut (82.96) >Teak + Cashew nut + paddy (77.43) >Teak + paddy (73.38) and lowest Cashenwut + paddy (64.15) where as lowest change is observed when the benefit reduces by 20% and cost increased by 20% then the IRR would be in order with highest in Teak+ cashew nut (79.92) > Teak + Cashew nut + paddy (74.31) >Teak + paddy (70.38) > Cashew nut + paddy (61.06) respectively.

The sensitivity analysis of the model at rotation showed maximum changes with cost increased by 10% and benefit remaining same the IRR of the land use where as lowest change have been observed when the benefit reduces by 20% and cost increased by 20.

As regards combination of paddy with tree component and horticulture the most profitable was Cashew nut+ paddy at 7% rate of interest calculated at 15th year and the same tend prevailed at 15% rate. The unique feature observed was the negative value of (-1149) PNW for Teak+ paddy with 15% rate at 15th year such a negative PNW is outcome of the fact that Sole paddy had always been negative in terms of PNW and the PNW of treatment at 15% could not surpass the negativity due to huge rate of interest at 15th year.

Here it will be pertinent to record that paddy component continue with the Silvihorticulture only up to initial few years after which it automatically would get eliminated because of excessive shade effect. The follow up of the system reveal that in case of present study also after three year period paddy component has been replaced by stakeholder with shade bearing agriculture crops like mesta a fibre crop to reinforce the profitability of the system for further years.

Plate 2. Teak+ Cashew nut + paddy

Plate 3. Teak + Cashew nut

Conclusion

Three general economic benefits are ascribed to agroforestry: (1) spreading (sharing) of fixed costs because of the joint-production relationship; (2) reducing the initial time period required to produce income from land devoted exclusively to tree production; and (3) diversifying income sources, in effect spreading the risk generally associated with a monoculture. The Santhal tribes have been successful in establishing these Agroforestry systems in Bangriposi village. There is general criticism from Agroforestry that the initial cost of establishment, in terms of capital and labor, may be prohibitive if no early income is possible but here the farmers are getting short term initial return from paddy during the first three years and the short income is then added from cashew nut production which continues there on. The other criticism is growing more than one crop at a time in the same field can complicate management as well as lack of knowledge but this self developed model is overcoming the problem as there is replacement of crops.

The result of this research reflects gradual reduction in risk factor with increasing number of interventions. The Teak +Cashew nut+ paddy gave the best sustainable return to the farmers. Judicious interpretations of the returns as corroborated by the sensitivity analysis clearly recommend the Teak + Cashew nut + paddy as best Agroforestry model.

References

Annon, 2005. State of Forest Report-2005. Forest Survey of India, MOEF,GOI,Dehradun,171

Gupta,T., Deepinder, M. 1982. Economics of Tree versus Annual crops on Marginal Agricultural Lands, Oxford IBH Publication Co. New Delhi, pp. 139.

Hazra,C.R., Dipankar, S., Saha, D. 2000. Agroforestry in watershed management: adoption and economic perspective from the central plateau and hills region of India. Agrofor. Today 12: 23-28

Jain, S. 1988. Case studies in Farm Forestry in Gujrat, F.A.O., Rome

Jha, L., Ranjan. 1993. Impact of Farm/Agroforestry on village socio-economic life (A case study). Agroforestry- Indian Perspective. pp. 227-300.

Patil, S.J., Nadagoudar, B.S., Mutanal, S.M., Madiwalar, S.L., Devaranavadgi, S.B. 2000. Suitability of tree species in an agroforestry system in hill zone of Karnataka. Indian For. 126: 1187-1190.

Patel, V.J. 1988. A new strategy to High Density Agroforestry: Jivrajbhai Patel Agroforestry Centre, Surendrabag, Gujrat, pp. 57.

Pethiya, B. P. 1999. Comparative profitability of agriculture, agroforestry and farm forestry in Maharashtra State, India. International-Forestry-Review. 1: 236-241.

Saxena, N.C. 1989. Wasteland development, environmental protection for the poor. Indian J. Public Adm. 35(3):487-497.

Shukla P.K., Sarkar, A., Dixit, S., Shrivastava, J.L. 2003. Cost Benefit Analysis of Agro-Forestry Model practiced in Madhya Pradesh. ENVIS Bulletin on Grassland Ecosystem and Agroforestry 1(2):11-32.

Sonkhlet, B., Singh, B., Rethy, P., Sood, K. K., Upadhyaya, K. 2001. Identification and evaluation of existing homestead Agroforestry in Cherapunjee, Meghalaya. Range Mgmt. Agrofor. 22 (1):106-112.

World Bank. 2002. A Revised Forest Strategy for the World Bank Group. World Bank, Washington DC.